AF776007

Unconventional Hydrocarbon Resources

Advanced Textbook Series

Unconventional Hydrocarbon Resources

Techniques for Reservoir Engineering Analysis

Editors

Reza Barati Ghahfarokhi

Mustafa M. Alhubail

This Work is a co-publication of the American Geophysical Union and John Wiley and Sons, Inc.

AGU
ADVANCING EARTH AND SPACE SCIENCE

WILEY

This Work is a co-publication between the American Geophysical Union and John Wiley & Sons, Inc.

This edition first published 2021 by John Wiley & Sons, Inc., 111 River Street, Hoboken, NJ 07030, USA and the American Geophysical Union, 2000 Florida Avenue, N.W., Washington, D.C. 20009

Published under the aegis of the AGU Publications Committee

Brooks Hanson, Executive Vice President, Science
Carol Frost, Chair, Publications Committee
For details about the American Geophysical Union visit us at www.agu.org.

Wiley Global Headquarters
111 River Street, Hoboken, NJ 07030, USA

For details of our global editorial offices, customer services, and more information about Wiley products visit us at www.wiley.com.

Library of Congress Cataloging-in-Publication data

Names: Barati Ghahfarokhi, Reza, author. | Alhubail, Mustafa M., author.
Title: Unconventional hydrocarbon resources : techniques for reservoir engineering analysis / Reza Barati Ghahfarokhi, Mustafa M. Alhubail.
Description: Hoboken, NJ : Wiley-American Geophysical Union, [2020] | Includes bibliographical references and index.
Identifiers: LCCN 2020029620 (print) | LCCN 2020029621 (ebook) | ISBN 9781119420323 (paperback) | ISBN 9781119420491 (adobe pdf) | ISBN 9781119420675 (epub)
Subjects: LCSH: Oil shale reserves–Textbooks. | Shale gas reservoirs–Textbooks. | Oil reservoir engineering–Textbooks. | Hydrocarbon reservoirs–Textbooks. | Hydraulic fracturing–Textbooks. | Petroleum–Geology–Textbooks.
Classification: LCC TN858 .B37 2020 (print) | LCC TN858 (ebook) | DDC 553.2/83–dc23
LC record available at https://lccn.loc.gov/2020029620
LC ebook record available at https://lccn.loc.gov/2020029621

Cover Design: Wiley
Cover Images: © Maximov Denis/Shutterstock

Set in 9.5/12.5pt STIXTwoText by SPi Global, Pondicherry, India
Printed and bound by CPI Group (UK) Ltd, Croydon, CR0 4YY

10 9 8 7 6 5 4 3 2 1

Contents

Contributors *xiii*
Preface *xv*

1 Introduction to Unconventional Hydrocarbon Resources *1*
Mustafa M. Alhubail and Reza Barati Ghahfarokhi
1.1 Background *1*
1.2 Overview of Shale Revolution *2*
1.2.1 What Led to the Shale Phenomenon? *3*
1.2.2 Importance of Recent Unconventional Resource Discoveries *4*
1.3 Basic Definitions and Classifications *5*
1.3.1 Conventional and Unconventional Resources *5*
1.3.2 Unconventional Oil-Bearing Sediments *6*
1.3.3 Unconventional Natural Gas Resources *6*
1.4 Global Description of Unconventional Plays *8*
1.4.1 North America Unconventional Plays *9*
1.4.2 South America Unconventional Plays *20*
1.4.3 Europe Unconventional Plays *24*
1.4.4 Middle East Unconventional Plays *30*
1.4.5 Africa Unconventional Plays *31*
1.4.6 Asia Unconventional Plays *33*
1.4.7 Australia Unconventional Plays *36*
1.5 Unconventional Resources Interpretation Workflow *37*
1.5.1 Workflow of Unconventional Reservoirs *37*
1.6 Future Projection and Challenges *38*
1.7 General Remarks *39*
1.8 Problems *39*
Additional Reading *40*
References *40*

2 Petrophysical Properties of Unconventional Reservoirs *45*
Negar Nazari, Mustafa M. Alhubail, Sherifa E. Cudjoe, and Reza Barati Ghahfarokhi
2.1 Background *45*
2.2 Petrophysics *45*
2.2.1 Evaluation of Rock Properties *46*

2.2.2 Shale Volume 46
2.2.3 Gamma Ray Spectroscopy 48
2.3 Lithology Evaluation 51
2.3.1 Lithology Measurements Using Cross-Plots 52
2.3.2 Lithology Measurements Using a Combination of Logs 53
2.3.3 Lithology Measurements Using the Diffuse Reflectance Infrared Fourier Transform Spectroscopy Techniques 57
2.4 Porosity 58
2.4.1 Porosity Measurement 60
2.4.2 NMR Core Porosity for Shales 71
2.5 Pore-Size Distribution 75
2.5.1 Pore-Size Distribution Using NMR Logging 75
2.5.2 Pore-Size Distribution Using Nitrogen Adsorption 78
2.6 Permeability 80
2.6.1 Unsteady-State Permeability Measurement Methods 81
2.6.2 Single Phase Permeability Measurements 89
2.6.3 NMR Permeability 91
2.6.4 Relative Permeability 95
2.6.5 NMR Capillary Pressure 99
2.6.6 Relative Permeability from NMR Pseudocapillary Pressure 99
2.7 Saturation 100
2.7.1 Techniques for Calculating Water Saturation 100
2.7.2 Resistivity Logs 100
2.7.3 NMR Saturation Estimation 104
2.8 Wettability 105
2.8.1 Wettability Measurement 105
2.9 Hydrocarbon Pore Volume and Reserve Estimation 110
2.9.1 Volumetric Analysis Theory 110
2.10 Problems 115
Additional Reading 117
References 117

3 Petroleum Geochemistry in Organic-Rich Shale Reservoirs 131
Sherifa E. Cudjoe, Mustafa M. Alhubail, and Reza Barati Ghahfarokhi
3.1 Background 131
3.2 Evolution of Organic Matter 131
3.2.1 Diagenesis 132
3.2.2 Catagenesis 132
3.2.3 Metagenesis 133
3.3 Total Organic Carbon (TOC) 133
3.4 Kerogen, Bitumen, and/or Pyrobitumen 134
3.4.1 Classification of Kerogen 136
3.5 Vitrinite Reflectance 138
3.6 Solid Bitumen Reflectance 139
3.7 Organic Porosity 140
3.8 Methods of Determining Source Rock Potential 142

3.8.1 Direct Combustion 143
3.8.2 Indirect Method 143
3.8.3 Rock-Eval Pyrolysis Method 143
3.8.4 In-Situ Measurements 151
3.9 Original TOC and Hydrocarbon Yield Determinations 160
3.9.1 Organic Porosity from Rock-Eval Parameters 163
3.10 Thermal Maturity and Source Rock Assessment 165
3.10.1 Biological Markers (Biomarkers) 165
3.10.2 Diamondoids 168
3.11 Raman Spectroscopy Analysis of Thermal Maturity in Kerogen 170
3.11.1 Thermal Maturity Controls of Organic Matter Types in LEF Samples 171
3.11.2 Maturity-Related Changes 176
3.12 Drifts Analysis of Kerogen Maturity 176
3.13 Problems 179
Additional Reading 182
References 182

4 Application of Imaging Techniques in the Characterization of Organic-Rich Shales *189*
Sherifa E. Cudjoe and Reza Barati Ghahfarokhi
4.1 Background 189
4.2 X-ray Microcomputed Tomography (X-ray Micro-CT) 190
4.2.1 Operation of X-Ray Micro-CT 192
4.2.2 Sample Preparation 194
4.2.3 X-ray Micro-CT Scanning Procedure 196
4.2.4 Image Reconstruction 198
4.2.5 Application of X-ray Micro-CT on Shale Samples 200
4.2.6 Image Visualization and Processing 203
4.2.7 Estimating Porosity from CT number (CTN) of CT Images 209
4.2.8 Permeability Estimation from CT scanner 212
4.2.9 Two-Phase Fluid Saturations 212
4.3 X-Ray Nano-CT 213
4.3.1 Sample Preparation for X-Ray Nano-CT 214
4.3.2 In-Situ Wettability and Spontaneous Imbibition at Nanoscale 214
4.4 Electron Microscopy 217
4.4.1 Scanning Electron Microscopy (SEM) 217
4.4.2 SEM/BSE Images of Various Ultratight, Organic-Rich Formations 220
4.4.3 Energy-Dispersive X-Ray Spectrometry (EDS/EDX) 228
4.4.4 Quantitative Evaluation of Minerals by Scanning Electron Microscopy (QEMSCAN) 235
4.4.5 Focused Ion Beam-Scanning Electron Microscopy (FIB-SEM) 241
4.4.6 Three-Dimensional (3D) Rock Model 243
4.4.7 Pore Network Model (PNM) and Pore Size Distribution (PSD) 245
4.4.8 Permeability Estimation 249
4.5 Broad Ion Beam-Scanning Electron Microscopy (BIB-SEM) 249
4.5.1 Sample Preparation, BIB-SEM Acquisition and Processing 250
4.6 Acknowledgment 251

4.7 Problems on Micro-CT and Nano-CT 252
4.8 Problems on Electron Microscopy 252
Additional Reading 257
References 257

5 Geomechanical Properties of Unconventional Reservoirs 265
Mustafa M. Alhubail, Anil Misra, and Reza Barati Ghahfarokhi
5.1 Background 265
5.2 Basic Concepts and Definitions 265
5.2.1 Stress 265
5.2.2 Strain 266
5.2.3 Elastic Constants 266
5.2.4 Poisson's Ratio 267
5.3 Stresses and Pressure Gradients 269
5.3.1 Vertical Stress and Overburden Pressure 269
5.3.2 Effective Vertical Stress 270
5.3.3 Effective Horizontal Stress 271
5.3.4 Biot's Poroelastic Constant 271
5.3.5 Horizontal Stresses and Fracturing Pressure 272
5.4 Well-Logging Measurements to Determine the Elastic Parameters 277
5.4.1 Calculating the Dynamic Moduli 277
5.4.2 Correlations for Static Moduli 281
5.5 Identifying the Geomechanical Sweet Spots 282
5.5.1 Brittleness Index 283
5.6 General Remarks 290
5.7 Problems 291
Additional Reading 294
References 294

6 Hydraulic Fracturing 299
Mustafa M. Alhubail and Reza Barati Ghahfarokhi
6.1 Background 299
6.2 Fundamentals of Hydraulic Fracturing 300
6.2.1 Fracture Geometry 301
6.2.2 Fracture Conductivity 301
6.2.3 Folds of Increase 305
6.2.4 Multistage Hydraulic Fracturing 307
6.2.5 Stress Shadow 310
6.2.6 Zipper Fracturing 311
6.2.7 Fracture Hits 312
6.2.8 Surface Pumps 315
6.2.9 Minifrac and DFIT Tests 316
6.2.10 Microseismic Monitoring 320
6.2.11 Stimulated Reservoir Volume 321

6.3 Fracturing Fluids 323
6.3.1 Purpose 323
6.3.2 Fracturing Fluid Types, Properties and Selection Process 324
6.3.3 Rheology of Fracturing Fluids 326
6.3.4 Damage of Fracturing Fluid and Fracture Cleanup 329
6.3.5 Fracturing Fluids Additives 329
6.4 Proppant 333
6.4.1 Purpose 334
6.4.2 Proppant Characteristics and Selection Process 334
6.4.3 Proppant Types 337
6.4.4 Proppant Flowback 339
6.4.5 Proppant Transport 340
6.4.6 Proppant Schedule 341
6.5 Modeling of Hydraulic Fractures 341
6.5.1 Importance of Modeling 342
6.5.2 Governing Processes of the Models 342
6.5.3 Modeling History 342
6.6 Problems 351
Additional Reading 352
References 352

7 Phase Behavior of Shale Oil and Gas 359
Xiaoli Li, Jyun-Syung Tsau, Qinwen Fu, and Reza Barati Ghahfarokhi
7.1 Background 359
7.2 Compositional Analyses of Shale Fluids 359
7.2.1 Subsurface Sampling 360
7.2.2 Surface Sampling 363
7.3 Phase Behavior and PVT Experiments 368
7.3.1 Phase Diagrams 368
7.3.2 PVT Experiments and Data Quality Check 372
7.4 Equation of State (EOS) 379
7.4.1 Cubic Equation of State 379
7.4.2 Stability Analysis 381
7.4.3 Confinement/Pore Proximity Effect on Phase Behavior 382
7.4.4 Phase Diagrams of Bakken, Eagle Ford, and Wolfcamp Fluids 390
7.4.5 Diffusion Coefficient 393
7.5 EOS Regression to Experimental Data 393
7.6 Minimum Miscibility Pressure 395
7.6.1 Experimental Methods 396
7.6.2 Analytical Methods 402
7.6.3 Numerical Methods 403
7.6.4 Correlation Methods 403
7.7 Problems 403
Additional Reading 408
References 408

8 Fluid Flow Through Nanosized Pores 413
Mohammad Kazemi, Ali Takbiri-Borujeni, Sherifa E. Cudjoe, and Reza Barati Ghahfarokhi
8.1 Background 413
8.2 Pore Size Distribution 414
8.3 Adsorption 414
8.4 Flow Regimes 418
8.5 Modeling Techniques 421
8.5.1 Fluid Transport in Confined Enclosures 421
8.5.2 Apparent Permeability of Shale 421
8.5.3 Transport in Organic Nanopores 423
8.5.4 Molecular Simulations 424
8.5.5 Molecular Structure of Kerogen 424
8.5.6 Multiscale Modeling Techniques 427
8.6 Lattice Boltzmann Model (LBM) 428
8.6.1 LBM Simulation 431
8.6.2 Implementation of LBM Simulation in Organic Nanopores 432
8.6.3 Apparent Permeability 433
8.7 Problems 434
Additional Reading 434
References 434
Appendix 439

9 Decline Curve and Rate Transient Analysis 445
Mustafa M. Alhubail, Mojdeh Rasoulzadeh, and Reza Barati Ghahfarokhi
9.1 Background 445
9.2 Purpose of Decline Curves 445
9.3 Decline Curve Assumptions and Limitations 446
9.4 Traditional Decline Curve Models 447
9.5 Arps's Models 447
9.5.1 Exponential Decline Model 448
9.5.2 Determination of Exponential Decline Graphically 448
9.5.3 Harmonic Decline Model 451
9.5.4 Determination of Harmonic Decline Graphically 451
9.5.5 Hyperbolic Decline Model 452
9.5.6 Determination of Hyperbolic Decline Parameters 454
9.6 Modern Decline Curve Models 456
9.6.1 Modified Hyperbolic Model 457
9.6.2 Power-Law Exponential Model 459
9.6.3 Stretched Exponential Model 460
9.6.4 Logistic Growth Model 462
9.6.5 Duong Model 462
9.7 Rate Transient Analysis 464
9.7.1 Purpose and Features of RTA 464
9.7.2 RTA Concept 465

9.7.3 Type Curves 466
9.7.4 Type Curve Methods 467
9.7.5 Square Root Time and Flowing Material Balance Plots 471
9.7.6 Flow Regimes 472
9.8 Problems 478
Additional Reading 494
References 494

10 Petroleum Economics of Unconventional Shale Reservoirs 499
Mustafa M. Alhubail, Hajar Aghababa, and Reza Barati Ghahfarokhi
10.1 Background 499
10.2 Effect of Shale Oil/Gas Developments on Economics and Energy Security 499
10.3 Fundamentals of Petroleum Economics 503
10.3.1 Business Expenditures 503
10.3.2 Basic Cash Flow 504
10.3.3 Interest Rate 508
10.3.4 Future Value 509
10.3.5 Present Value 509
10.3.6 Net Present Value (NPV) 510
10.3.7 Rate of Return (ROR) 511
10.3.8 Payout 512
10.4 Fiscal Regimes and Contracts 513
10.4.1 The Concessionary System (Royalty/Tax System) 513
10.4.2 The Production Sharing Contracts (PSCs) 513
10.4.3 Service Contracts 515
10.5 Decision, Uncertainty, and Risk Analysis 516
10.6 Chapter Project 517
10.6.1 Project Data 518
10.6.2 Project Calculations 518
10.7 Problems 525
Additional Reading 526
References 526

11 Environmental Aspects of Shale Hydrocarbon Reservoir Developments 527
Reza Barati Ghahfarokhi, Stephen Randtke, and Edward Peltier
11.1 Background 527
11.2 Water Management and Reuse 527
11.2.1 Basic Terminology Related to Water Management and Reuse 527
11.2.2 Water Cycle in Oil and Gas Production 529
11.2.3 Water Acquisition for Hydraulic Fracturing 531
11.2.4 Flowback and Produced Water Quantity and Quality 533
11.2.5 Flowback and Produced Water Reuse 535
11.3 Chemicals used in Fracturing Fluids 539
11.4 Potential Impacts on Drinking Water Resources 546

11.5 Induced Seismicity 549
11.6 Air Pollution 551
11.7 Problems 552
Additional Reading 552
References 552

Index 555

Contributors

- **Hajar Aghababa** PetroEconomic Solutions LLC, Lawrence, Kansas, USA
- **Mustafa M. Alhubail** The University of Kansas, Lawrence, Kansas, USA
- **Reza Barati Ghahfarokhi** The University of Kansas, Lawrence, Kansas, USA
- **Sherifa E. Cudjoe** The University of Kansas, Lawrence, Kansas, USA
- **Qinwen Fu** The University of Kansas, Lawrence, Kansas, USA
- **Mohammad Kazemi** Slippery Rock University of Pennsylvania, Slippery Rock, Pennsylvania, USA
- **Xiaoli Li** The University of Kansas, Lawrence, Kansas, USA
- **Anil Misra** The University of Kansas, Lawrence, Kansas, USA
- **Negar Nazari** Stanford University, Stanford, California, USA
- **Edward Peltier** The University of Kansas, Lawrence, Kansas, USA
- **Stephen Randtke** The University of Kansas, Lawrence, Kansas, USA
- **Mojdeh Rasoulzadeh** The University of Alabama, Tuscaloosa, Alabama, USA
- **Ali Takbiri-Borujeni** West Virginia University. Morgantown, West Virginia, USA
- **Jyun-Syung Tsau** The University of Kansas, Lawrence, Kansas, USA

Preface

The development of ultra-tight hydrocarbon unconventional resources, which started in the early 21st century and has disturbed the energy sector, is an expensive and risky business. However, if the unconventional resources are properly developed, the business becomes quite lucrative. Consequently, the shale boom, which primarily occurred in the United States, has become a phenomenon that is spreading all over the world. Advanced technology (i.e., horizontal drilling and multistage hydraulic fracturing) has assisted in the development of unconventional resources; however, advanced technology must be supported with technical understanding of the fundamentals by the industry to commercially develop the unconventional resources.

This book offers a comprehensive guide to the methods and available techniques that are utilized in the industry to commercially develop unconventional resources with practical examples and projects using real-world data sets of major unconventional plays such as the Bakken and Eagle-Ford.

This book aims to cover the entire scope of hydrocarbon unconventional resources from discovery to commercial development based on the current understanding of the industry. Chapter 1 is an overview of unconventional resources and their importance, and it highlights the differences between the conventional reservoirs and the unconventional reservoirs. It also points out the top unconventional plays in North America and around the world stating the geological age, the estimated recoverable resources, and a summary of the petrophysical properties of each unconventional play in addition to the average cost of drilling and completing the wells with multistage hydraulic fracturing.

Chapter 2 discusses the petrophysical properties of unconventional reservoirs and the methods that are utilized in the industry to evaluate the rock properties, such as lithology, porosity and pore-size distribution, permeability, relative permeability and capillary pressure, as well as water saturation and wettability. This chapter provides detailed explanation with examples of the techniques that assist in characterizing the unconventional reservoirs.

Chapter 3 presents the fundamentals of geochemical properties of unconventional reservoirs. The evolution steps of organic matter and the classifications of kerogen are also discussed. The methods to determine the source rock potential, such as the rock-eval pyrolysis method, are explained in detail with examples. The thermal maturity and source rock assessment techniques are also explained in this chapter. Furthermore, Raman spectroscopy and diffuse reflectance infrared Fourier transform spectroscopy (DRIFTS) are presented as valuable methods to analyze the thermal maturity of kerogen.

Chapter 4 deals with advanced imaging techniques and their applications to characterize organic-rich shales. The different imaging and visualizing methods as well as the processing procedures are explained in detail with real-world examples from different unconventional reservoirs, such as the Eagle-Ford play.

Chapter 5 presents concepts and definitions of rock mechanics such as stress, strain, Young's modulus, and Poisson's ratio. The geomechanical properties of unconventional reservoirs, such as the minimum principal stress and its importance, closure stress, brittleness index, and the mineral-based brittleness index, are explained in detail with examples. The methods to interpret well logs to obtain the geomechanical properties of unconventional reservoirs and to locate the optimum sweet spots for hydraulic fracturing are also discussed in this chapter.

Chapter 6 presents the fundamentals of hydraulic fracturing treatments, such as fracture geometry, fracture conductivity, multistage hydraulic fracturing, complex fracture networks, stimulated reservoir volume (SRV), and the fracture clean-up process with case studies from some well-known unconventional plays in North America. The microseismic and minifrac tests are also described. The purpose of the fracturing fluids, the different types of fracturing fluids, and their properties as well as the fracturing fluid design and selection process are discussed in detail. The purpose of proppant, the essential properties of the proppant, the different types of proppant, and the proppant selection process are also discussed in detail with examples. The modeling process of hydraulic fractures and the governing equations are described here.

Chapter 7 deals with the phase behavior of shale oil and gas. The compositional analysis of hydrocarbon unconventional fluids is explained in detail with PVT experiments. The equations of state (EOS) are also discussed with phase diagram examples. The laboratory experiments used to develop analytical and numerical methods required for EOS regression in addition to the minimum miscibility pressure estimation are explained with examples from different unconventional reservoirs, such as the Bakken, Eagle-Ford, and Wolfcamp plays.

Chapter 8 discusses fluid flow through nanosized pores. The pore-size distribution and its effects on fluid flow is explained. The complex flow regimes and the modeling techniques of fluid flow in nanopores are discussed in detail. The lattice Boltzmann model (LBM) as well as the molecular dynamic simulation methods and how they are implemented to conduct simulation in organic nanopores are also explained with examples.

Chapter 9 presents the history and purpose of the decline curve analysis (DCA) and rate transient analysis (RTA). The fundamental concepts of the DCA, such as the decline rate and the estimated ultimate recovery (EUR), are explained in detail with examples. The types of decline curves, such as the conventional Arps's models, as well as the modern decline curves, such as the modified hyperbolic model, the stretched exponential model, Doung's model, and the multisegment model, are discussed. In addition, methods to determine the models' parameters are explained in detail with examples. The concept of RTA and the type curve methods are also discussed with examples and a chapter project.

Chapter 10 discusses the effects of unconventional resources on economics and energy security. The fundamentals of petroleum economics, such as business expenditure systems, net present value (NVP), and rate of return, are explained in detail with examples. The fiscal regimes, types of contracts, decision-making processes, and risk analysis are also discussed with examples and a chapter project.

In Chapter 11, the environmental aspects of developing the hydrocarbon unconventional resources are explained. The concerns of water management, water reuse, and the impact on drinking water resources and air pollution are discussed. The types of chemicals that are used in the fracturing fluids, and the effects of induced seismicity are also covered in this chapter.

This book is intended to serve as a textbook for undergraduate and graduate students in petroleum engineering, geology, petrophysics, and for other readers interested in hydrocarbon shale resources. It is also intended to be a handbook for petroleum engineers, geologists, petrophysicists, petroleum economists, and employees with different backgrounds working in the oil and gas industry.

We would like to acknowledge different people and corporates who have helped in the preparation of this book. In particular, we thank all the contributing authors for their involvement in this book and for sharing their experience and knowledge. We would like to thank our publishing editor, Dr. Rituparna Bose, for her immense support. Our special thanks is extended to Slawson Exploration for donation of the Bakken data set, Kansas Geological Survey (KGS) for its open access database and for sharing valuable measurements on the Chattanooga shale, Chesapeake Energy for providing us with core and fluid samples and giving permission for publication to parts of the data.

Specifically, we would like to thank David Mohrbacher of Chesapeake Energy, Dr. Younes Jalali, Dr. Behruz Shakershiran from Uni Research CIPR, and Dr. Belladona Maulianda Hidayat from Universiti Teknologi Petronas for kindly reviewing chapters of this project and giving us feedback that improved the content. We also want to thank Schlumberger, IHS Markit, and NSI Technologies for providing us with data sets and core samples as well as licenses for advanced software packages.

Reza Barati Ghahfarokhi
Mustafa M. Alhubail
Lawrence, Kansas, USA
May 2019

1

Introduction to Unconventional Hydrocarbon Resources

Mustafa M. Alhubail and Reza Barati Ghahfarokhi

The University of Kansas, Lawrence, Kansas, USA

1.1 Background

To better define the term unconventional resources, and to understand why they have been called unconventional, the history of oil should be reviewed.

Petroleum seeps are one of the earliest sources of oil, along with wood and coal, used for heating and illumination. Another source of oil for illumination was whales. Whaling, which is the term that describes hunting of whales, became a competitive industry in the eighteenth century. Factories were built to make ships, and well-equipped fleets were sent for one purpose and one purpose only: hunting whales. Whale oil was mainly used in oil lamps (Jackson, 1998). The amount of oil from the above two sources was not sufficient enough for the high demand. With the industrial revolution, the need for an alternative source of energy, other than wood and coal, to maintain machinery and for transportation, had never been greater. Coal mining was a very dangerous occupation, and it resulted in a high number of casualties (Natgas, 2013). Hence, in the nineteenth century, oil exploration began, and it has changed the way the world operates in all aspects of life.

People started to dig wells using basic percussion tools, and the rubble was carried out by loading it into baskets and pulling it up to the surface. The drilling process improved a bit by using the cable-tool drilling method, which helped drill wells faster. In the early twentieth century, rotary drilling was employed with the ability of cleaning the wellbore while drilling by using fluids. Rotary drilling has changed the whole petroleum industry and, over time, rotary drilling technology has been improved drastically (Hogg, 2016). With the ability to drill wells more conveniently using rotary drilling technology, there still exists a more challenging mission, which is where to drill. To this day, with the advancement in imaging the Earth's strata by using seismic waves, there is still uncertainty about whether hydrocarbon exists in a new area or not, and drilling is the only way to find out for sure.

The global demand for oil surged, and thousands of wells were drilled both onshore and offshore. Keep in mind that this whole process is an investment, and profit is the main motivation. To recover oil faster and without the need of drilling new wells from the start, horizontal drilling and directional drilling were employed. In addition to that, horizontal drilling also helped decrease the drilling footprint, which greatly helped minimizing the environmental impact from the drilling process. With directional drilling technique and the secondary and enhanced oil recovery methods (conventional methods), more hydrocarbons became

Unconventional Hydrocarbon Resources: Techniques for Reservoir Engineering Analysis, Advanced Textbook Series, First Edition. Edited by Reza Barati Ghahfarokhi and Mustafa M. Alhubail.

accessible. However, even with directional drilling and the well-established recovery methods, there is way more hydrocarbons that are trapped in certain reservoirs, and those hydrocarbons cannot be recovered by the conventional methods because of complex petrophysical properties of the reservoirs. Hence, the question now is how to extract the hydrocarbons that are trapped in those reservoirs. Those reservoirs are called *unconventional reservoirs,* and the hydrocarbons that are trapped in them are called *unconventional resources*. Thus, in general, the unconventional reservoirs can be defined as the reservoirs that contain a huge quantity of hydrocarbons that cannot be extracted by the conventional methods due to their petrophysical properties, which will be discussed in detail in Chapter 2. The unconventional resources are classified into different categories, and they will be discussed later in this chapter.

1.2 Overview of Shale Revolution

The shale boom primarily occurred in the United States. The unconventional resources have changed the energy sector and, in 2012, the number of wells that were brought back online in the United States alone exceeded those that were put online in the rest of the world excluding Canada (Maugeri, 2013). Figure 1.1 shows the number of the issued drilling permits just for the Eagle Ford play. Furthermore, other countries (e.g., Argentina and Colombia) are currently working on attracting talented workers, petroleum engineering experts, and foreign investments in order to replicate the U.S. shale boom to become independent on energy imports from other countries, and to become energy exporters as well (Kuuskraa et al., 2013). Therefore, it is fair to say that the shale boom is a U.S.

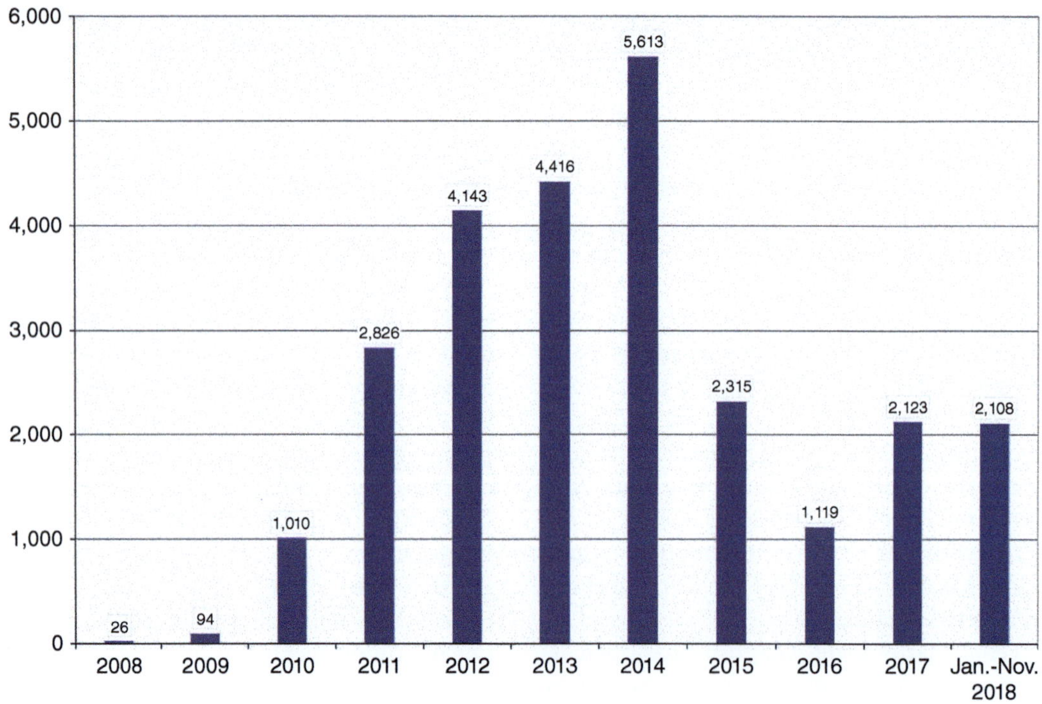

Figure 1.1 Drilling permits issued for Eagle Ford play, 2008 through November 2018 (Railroad Commission of Texas).

phenomenon that is spreading all over the world. However, that is easier said than done.

It is true that advanced technology (i.e., horizontal drilling and multistage fracturing) had existed for more than a decade. However, advanced technology is just a tool to unlock the unconventional resources and help develop them. Hence, advanced technology by itself is not practically sufficient for large-scale unconventional resources development. For instance, even though advanced technology exists today, the commercial development of the shale gas resources in China is more complex than one would think, and that is due to the lack of knowledge and experience in exploration and development of such resources in China in addition to the remote location of some of the resources (Dong, et al., 2016). Furthermore, conducting effective hydraulic fractures, in terms of hydrocarbon productivity and profitability, is a difficult task in China due to complex tectonic stress (Xie et al., 1993; Wang & Wang, 2016; Yong, et al., 2018).

The development of shale resources in the United States started in the early 1980s, and it has undergone intensive academic and industrial research as well as pilot testing before the breakthrough of a revolutionary technology of drilling and hydraulic fracturing in the late 1990s. After years of research and testing, this advanced technology was fully employed in large-scale unconventional resources field development in 2005, and since then, thousands of wells have been drilled. The amount of knowledge and information that has been gathered through the years (e.g., subsurface strata database), the experience in exploration and development, and the delicate implementation of such advanced technologies cannot be passed on to other nations. This is due not only to economic and national security reasons but also because those technologies might not be applicable there, or they might require further enhancements in order to become feasible in those nations (Ning et al., 2015; Dong, et al., 2016). To show how well the technology is developed in the United States, one can compare the number of the completed horizontal wells in the Eagle Ford play (3,560 wells), after the first 6 years of development, with the number of the completed horizontal wells in Vaca Muerta in Argentina (140 wells), after the first 6 years of development. Figure 1.2 shows the number of the completed horizontal wells in Eagle Ford play, Permian-Midland basin, and the Vaca Muerta play after the first 6 years of development.

1.2.1 What Led to the Shale Phenomenon?

It may be well known that hydrocarbons exist in unconventional reservoirs, but is it possible to extract those unconventional resources economically? This question is actually a very complicated one because of the word *economically*. Advancements in the petroleum industry have surely made extracting the most difficult unconventional resources possible. However, these developments have to be economically viable for producers. Improvements in horizontal drilling and hydraulic fracturing techniques made multistage fracturing a standard practice in the industry. Multistage fracturing is the process of dividing the horizontal section of the well into subsections that are fractured independently in stages, hence the name *multistage fracturing*. Multistage fracturing will be discussed further in Chapter 6. One major factor that led to the shale boom is the ability to extract hydrocarbons from source rocks (i.e., shale rocks) (Brown & Yücel, 2013). The other major factor that led to the shale phenomenon is the spike in oil prices due to high demands in Asia and other regions of the world (Brown & Yücel, 2013).

The high oil prices attracted investors to the oil and gas sector. Hundreds of millions of dollars were invested in research and development, and those investments resulted in an innovative technology that lowered the cost of extracting the unconventional resources. For instance, by integrating lab work and modeling, researchers have developed more

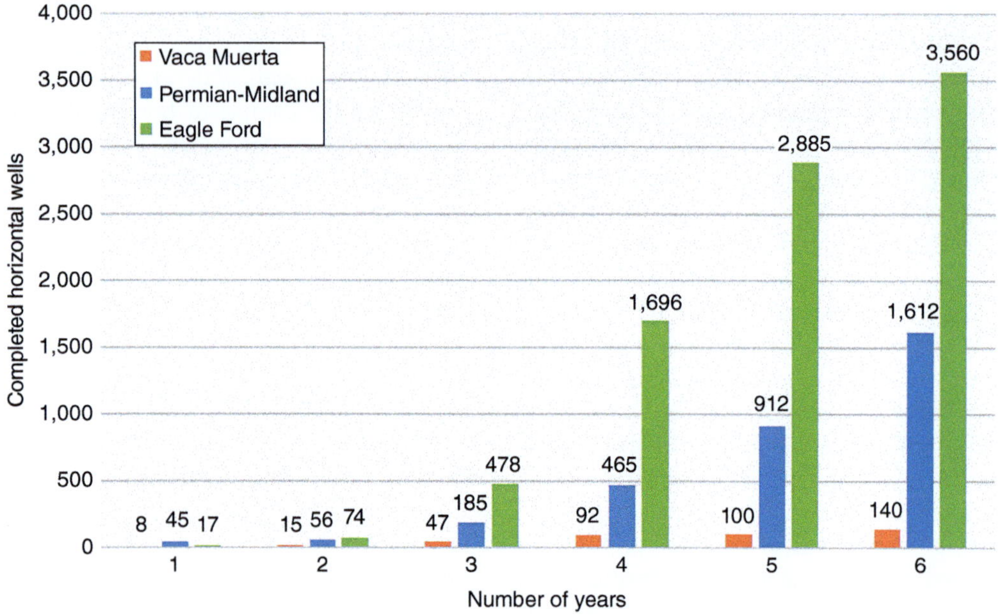

Figure 1.2 Number of completed horizontal wells in Eagle Ford, Permian-Midland, and Vaca Muerta after the first 6 years of development (modified from Zborowski, 2018).

efficient fracturing fluids and designed sophisticated pumping schedules that yield a significant increase in production from the unconventional reservoirs, minimizing the cost of hydraulic fracturing treatments and reducing environmental impact from the fracturing treatments. Those innovations have made production from the unconventional resources feasible and profitable even at low oil prices. For instance, in 2008, the average cost of a horizontal well with multistage fracturing in some plays was approximately 16 million U.S. dollars; however, the average cost in 2016 was 8 million U.S. dollars. This huge reduction in cost became possible because of the investments that were made in research and development for unconventional resources. Consequently, production from the unconventional resources has become profitable even at low oil prices.

1.2.2 Importance of Recent Unconventional Resource Discoveries

The benefits of the unconventional resources can be looked into from different points of view. First and foremost, no one can deny the fact that a discovery of an unconventional resource in a country will most likely change the economy of that nation for decades to come (e.g., discovery of Vaca Muerta in Argentina). It also helps nations to become energy independent. For instance, the total natural gas production in the United States between 2008 and 2012 has increased substantially and a significant part of that increase, almost 60%, was due to shale gas (U.S. Energy Information Administration, 2018). Consequently, the surge in gas production has made the United States the world's largest natural gas producer (U.S. Energy Information Administration, 2017). The recent discoveries of unconventional resources can also benefit the stock markets. Once a company unveils the discovery of a new unconventional play or uncovers more hydrocarbons in an existing play, its stock price surges rapidly (Egan, 2016). The recent discoveries of unconventional resources also help communities and impact the public education system around the newly discovered plays because when the exploration and development companies start working in that area, they will have to make some infrastructure that benefits people (Ratledge & Zachary, 2017). Moreover, the small businesses near that area will benefit and flourish. For example, in the United States, when oil prices dropped in 2014, petroleum companies, especially those in the shale sector, started laying off their employees. Consequently, a great number of workers were also laid off from other business because those businesses were solely dependent on serving the petroleum industry personnel in that area (Beattie, 2018).

There are benefits the petroleum operator itself gains from the recent unconventional resources discoveries, and those benefits without a doubt extend to other industries as well. For instance, consider the innovative technology in exploration and development. Service companies spend billions of dollars on research in order to invent new equipment that facilitates the discovery of new unconventional plays as well as increases production from the existing ones. As a result, the cost of operation in the shale sector drops significantly.

There are valid concerns that the recent discoveries of unconventional resources could affect water sources and harm the environment. However, the petroleum industry is serious about these concerns, and it is working in developing new technologies and techniques to minimize the usage of water in hydraulic fracturing treatments as well as to handle the large amounts of produced water. The environmental concerns that are related to the unconventional resources will be discussed in Chapter 11.

1.3 Basic Definitions and Classifications

1.3.1 Conventional and Unconventional Resources

By analyzing the resources triangle, Figure 1.3, conventional resources can be described as the small volume resources that can be produced with well-established production mechanisms (i.e., primary recovery, secondary recovery, and tertiary recovery processes). Specific methods, such as conventional enhanced oil recovery (EOR) methods as well as simple hydraulic fracturing and acidizing, are usually practiced to produce hydrocarbons from these reservoirs. Due to their simple and cheaper development, the conventional reservoirs are typically the first hydrocarbon targets in the industry (Hanania et al., 2013). The conventional resources are very profitable at low oil and gas prices; however, since they are very small in volume, they become depleted in a few years.

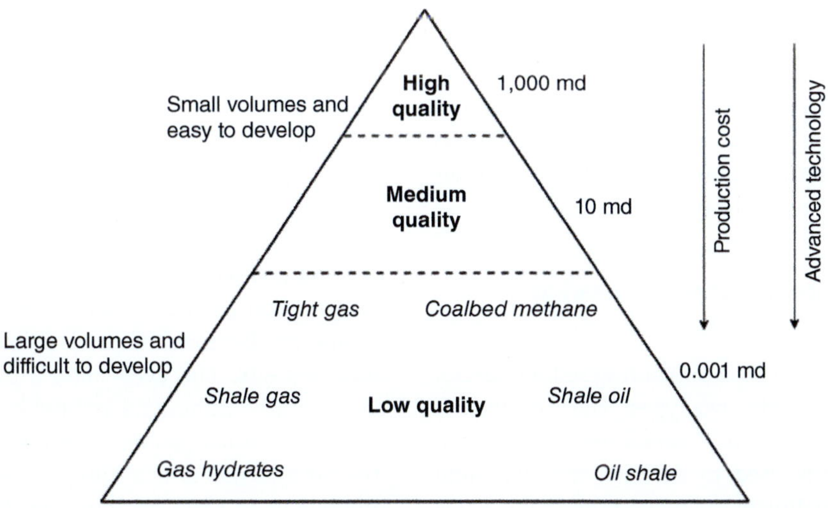

Figure 1.3 Resource triangle (modified from Holditch & Ayers, 2009; Masters, 1979; Gray, 1977).

On the other hand, the unconventional resources (e.g., shale gas, shale oil, oil shale, and methane hydrates) can be defined as the formations that have extremely low permeability and porosity as well as complex geology. The unconventional reservoirs show complex mineralogy, lateral and vertical heterogeneity, and anisotropy. Thus, detailed characterization and advanced exploration technologies are required for their development to improve their production mechanisms (Hameed et al., 2014). The unconventional reservoirs are very large in volume. The world energy resources report estimated that global deposits of shale oil resources are equivalent to 6.05 trillion barrels of oil in place (World Energy Resources, 2016). However, they require advanced technology (i.e., horizontal drilling and multistage fracturing) for commercial production. The innovative technology that is involved in extracting those unconventional resources is extremely expensive; hence, the unconventional resources are not profitable at low oil and gas prices. Furthermore, multistage fracturing requires a huge amount of water, which is a very valuable resource, and it is generally associated with the public health as well as the environment. The water source for the multibillion mega unconventional projects always disquiets the public because they are concerned about their health as well as the environment. Consequently, that could cause a negative perception to the whole unconventional industry. It is important to note that the quality of the extracted hydrocarbons (e.g., API gravity and composition) from the conventional and unconventional resources is not really different. See a comparison in Table 1.1.

1.3.2 Unconventional Oil-Bearing Sediments

The unconventional oil-bearing sediments can be classified into two types. The first type is called *tight oil rocks*. *Tight oil rocks* is a general term that is used to describe the tight and complex sedimentary rocks other than shale (e.g., middle Bakken formation.) The middle Bakken formation is not a shale formation but rather a fine-grained carbonate formation. The second type of the oil-bearing sediments is the oil that is found in shale. The term *oil shale* describes shale rocks that are rich in organic matter (i.e., kerogen). The oil shale contents include inorganic matrix, bitumen, and kerogen (Hameed et al., 2014) Hence, oil shale are solid rocks, and the oil that is trapped inside of them cannot be produced directly out of the ground. Therefore, to extract the hydrocarbons from oil shale rocks, first those rocks have to be mined and then processed. Different methods, such as pyrolysis, hydrogenation, and thermal dissolution, are used for this purpose (Bartis et al., 2005). The pyrolysis procedure will be discussed in Chapter 3. If it is found that less than 20% of the kerogen is converted to hydrocarbons, then those rocks are rather called low-kerogen rocks instead of oil shale rocks. The term *oil shale* should not be mixed with the term *shale oil*. Overall, oil shale has the capability of producing hydrocarbons after it reaches certain temperatures and pressure, which are required for the kerogen to become mature. On the other hand, shale oil, which can be produced from underground by the use of hydraulic fracturing, is the actual product that is produced geologically from the oil shale rocks, and it is the common unconventional oil resource.

1.3.3 Unconventional Natural Gas Resources

The main resources of unconventional natural gas are classified as follows.

Deep Natural Gas

Deep natural gas is considered unconventional natural gas because the gas is found in pretty deep deposits. The deep natural gas deposits are typically thousands of feet deeper than the conventional gas deposits (Natgas, 2013). The advancements in drilling have made the production of deep gas possible; however, it is still very costly to produce. Hence, deep

Table 1.1 Comparison of Conventional and Unconventional Hydrocarbon Resources

Classification	Conventional reservoirs	Unconventional reservoirs
Volume	Small	Large
Development	Relatively easy	Difficult
Resource estimation uncertainty and risk of development	Low	High
Subsurface structure and stratigraphic	Relatively distinct	Complex
Hydrocarbon migration	Hydrocarbon migrates from source rocks to reservoir rocks	Source rocks are reservoir rocks
Cost of hydrocarbon extraction	Relatively low	Extremely high
Porosity	High to moderate	Moderate to low
Permeability	High to moderate	Extremely low
Horizontal drilling	Not necessarily required to achieve commercial flow	A must to achieve commercial flow
Stimulation treatments (e.g., multistage fracturing)	Not necessarily required to achieve commercial flow	A must to achieve commercial flow

natural gas is not profitable to be produced at low gas prices.

Geopressurized Zones

Geopressurized zones are formations that are under extremely high pressure considering their depths. These formations are under high pressure due to the way the layers were deposited (Natgas, 2013). Even though these gas reservoirs might not be at very great depths, due to the high pressure, they are considered unconventional reservoirs because they must be handled differently compared with typical conventional gas reservoirs.

Tight Natural Gas

The term *tight gas* is not solely associated with unconventional resources, as conventional reservoirs with a very low permeability can also be considered a tight gas resource. However, since the boom of unconventional gas resources, the term *tight gas* or *tight natural gas* has been used to characterize one of the six main unconventional natural gas types. Tight natural gas is trapped in a very complex impermeable formation, and to extract the gas from these tight formations, hydraulic fracturing and acidizing must be performed (Natgas, 2013). Hydraulic fracturing and acidizing are very expensive procedures. Hence, economic conditions control the production of the tight gas reservoirs. Figure 1.4 shows the major tight gas plays in the United States.

Shale Gas

Shale gas is basically the natural gas that is found in shale rocks. It has been known for years that natural gas exists in shale deposits. However, it was not possible to economically extract the gas from the shale deposits at commercial levels because of the complex nature of the shale formations. The advanced technology (i.e., horizontal drilling and multistage fracturing) has made the production from shale gas formations commercially profitable. Shale gas has become one of the largest natural gas sources in the United States. Hence, the gas supply and gas prices are less sensitive to the disturbance in gas production that is caused by severe weather conditions (e.g., hurricanes) in the Gulf of Mexico.

Coalbed Methane

Coalbed methane (CBM), which is also known as coal seam gas, is basically coal deposits that contain natural gas. Once the coal mining

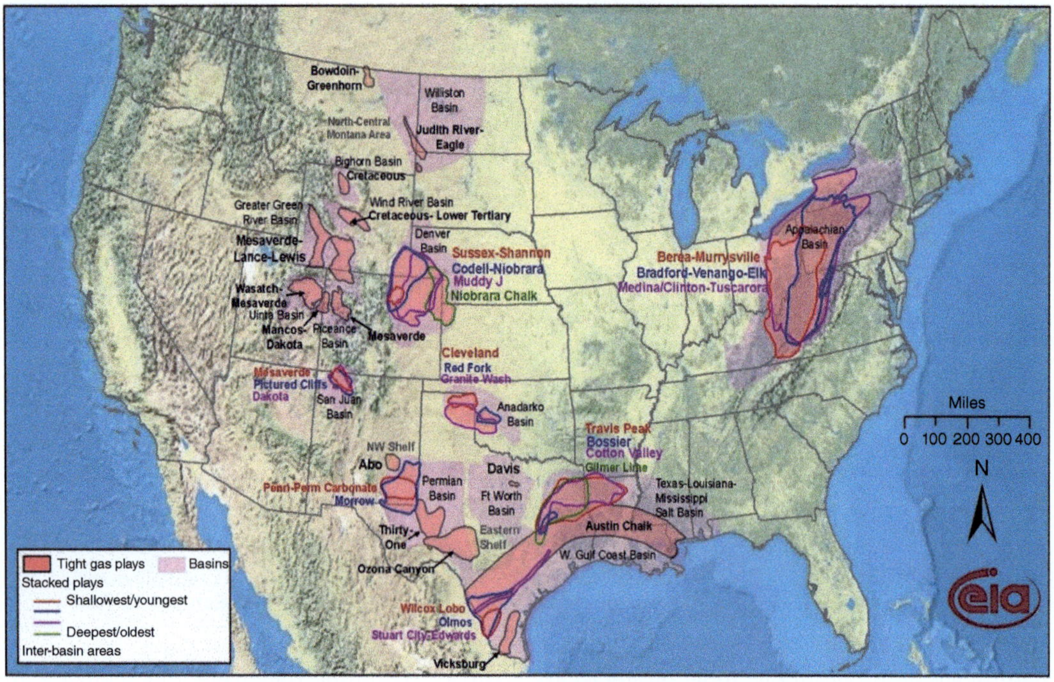

Figure 1.4 The major tight gas plays in the United States (www.eia.gov).

activity starts, the trapped natural gas (i.e., methane) is usually released out of the coal seams. In the past, the gas that was released out of coal seams was a serious concern because the escaped gas, especially at high concentrations, could cause safety issues to the coal miners (Natgas, 2013). In recent years, the industry has realized the potential and the abundant quantity of the gas that is coming out of the coal seams. Hence, the released gas is now collected and used for a purpose similar to natural gas. Thus, CBM has become a popular source of unconventional natural gas. Figure 1.5 shows the major coal basins and CBM fields in the United States

Methane Hydrates

Methane hydrate, which could be the most complex unconventional natural gas resource, is basically molecules of methane that are trapped inside frozen water, and it was first discovered in regions of the Arctic. Methane hydrate generates water and natural gas when it is warmed or depressurized. The quantity of this unconventional natural gas is enormous, and it is estimated to be more than 70,000 Tcf (Natgas, 2013). Recent studies by the U.S. Geological Survey (USGS) estimates that methane hydrates might contain more organic carbon than all conventional natural gas, oil, and CBM combined. However, the U.S. Department of Energy (DOE) and other organizations are still researching for ways to accurately locate this unconventional resource before drilling, and they are working on innovative technologies to better study its physical properties and to determine if production of methane hydrates could cause any environmental implications.

1.4 Global Description of Unconventional Plays

The unconventional reservoirs can be classified into three types: halo plays, geostratigraphic plays, and shale plays. The halo plays are the

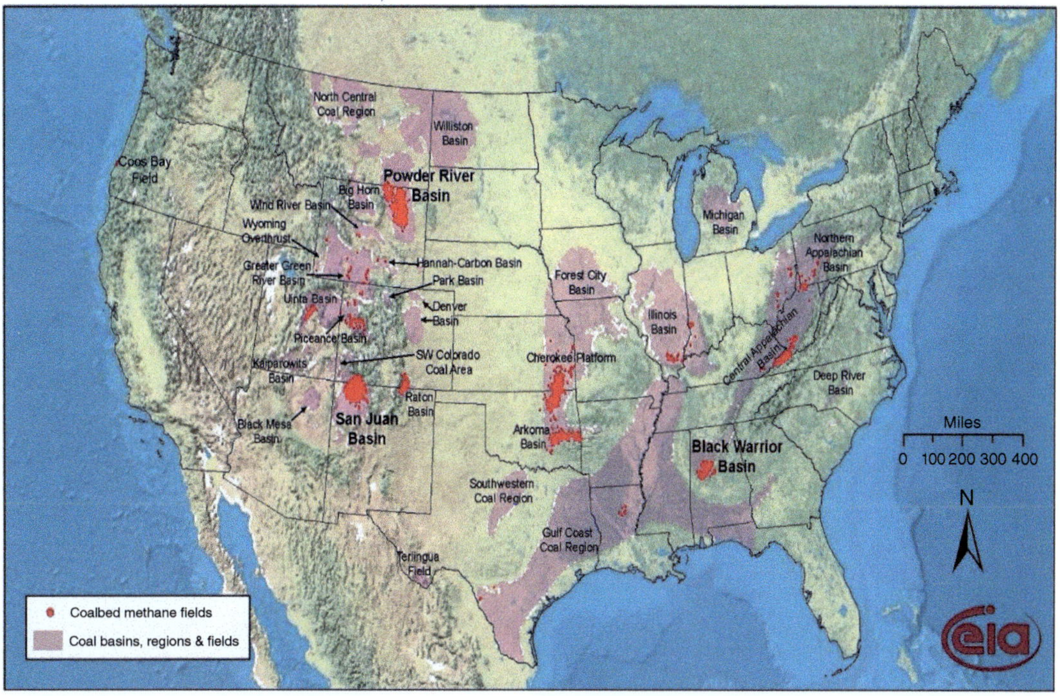

Figure 1.5 The major coalbed methane fields in the United States (www.eia.gov).

surrounding areas of an already developed formation, and those surrounding areas, fringe regions, contain hydrocarbons. However, the hydrocarbons could not be not produced by conventional methods because those formations (e.g., cardium formation in Canada) are severely complex and require advanced technology (i.e., multistage fracturing) to produce from them (CSUR, 2017c). Hence, the halo plays are considered unconventional formations.

The geostratigraphic plays are formations that contain an enormous amount of hydrocarbons over a wide geographic area, and those types of plays require multistage fracturing for commercial production. A famous geostratigraphic play is the middle member of the Bakken play. The middle Bakken formation is located in parts of North Dakota and Montana in the United States and the Saskatchewan province in Canada. It is a limestone formation that is trapped between two organic-rich shale formations (i.e., upper Bakken shale and lower Bakken shale) (CSUR, 2017c).

The shale plays, which are the most common unconventional plays, are organic-rich shale rock deposits. Those organic-rich shales are the source rocks as well as the reservoir (CSUR, 2017c). The permeability of the shale plays is extremely tight, and their petrophysical properties are very complex. Thus, these plays require multifracture networks for commercial production. Each part of a shale play that has a network of fractures is considered as a separate region of that main play, and that region is usually referred to as the stimulated reservoir volume (SRV.) The concept of SRV will be discussed further in Chapter 6.

1.4.1 North America Unconventional Plays

In this section, we discuss unconventional shale plays in North America (Fig. 1.6).

Barnett Shale

The Barnett shale, which is one of the largest shale gas plays in the United States, is located in the Fort Worth basin in north central Texas,

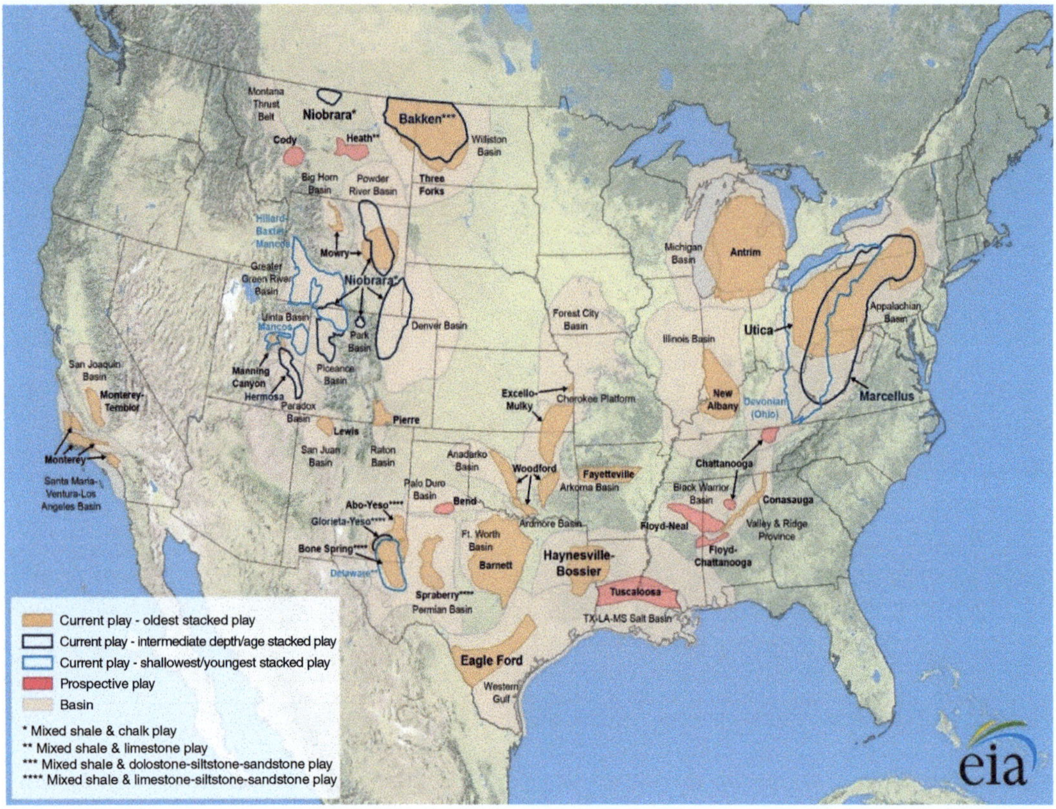

Figure 1.6 The major shale plays in the United States (www.eia.gov).

and its geologic age is the Late Mississippian (Fig. 1.7). The Barnett shale play was discovered in 1981. It is fair to say that the Barnett shale is what started the shale boom in the United States as most of the pilot testing of the cross-linked and slick water hydraulic fracturing treatments that took place in the 1980s and 1990s were attempts to better develop the Barnett shale play. Horizontal drilling was also practiced in the Barnett shale play in the early 1990s. However, the first two horizontal wells did not yield desired production rates compared with the production rate from the vertical wells (International Energy Agency, 2013).

In 2003, the horizontal wells were accompanied by multistage fracturing, and the production rates improved significantly in certain areas of the play. Shortly after that, the drilling activity increased rapidly in the Barnett shale play, and many companies got involved because of the high gas prices at that time. It was estimated by Baker Hughes that almost 5% of all the active rigs in the world were operating in the Barnett shale play. However, the advanced technology (i.e., horizontal drilling and multistage fracturing) did not always yield high recovery rates mainly because many of the areas in the Barnett shale play are not that productive (Ahmed & Meehan, 2016). As a result, companies started to experiment with different completion designs (e.g., number of fracturing stages and spacing) to further improve recovery rates in such low-productive areas. After outstanding success in the Barnett shale play, the advanced technology has become a standard practice in commercial unconventional plays development.

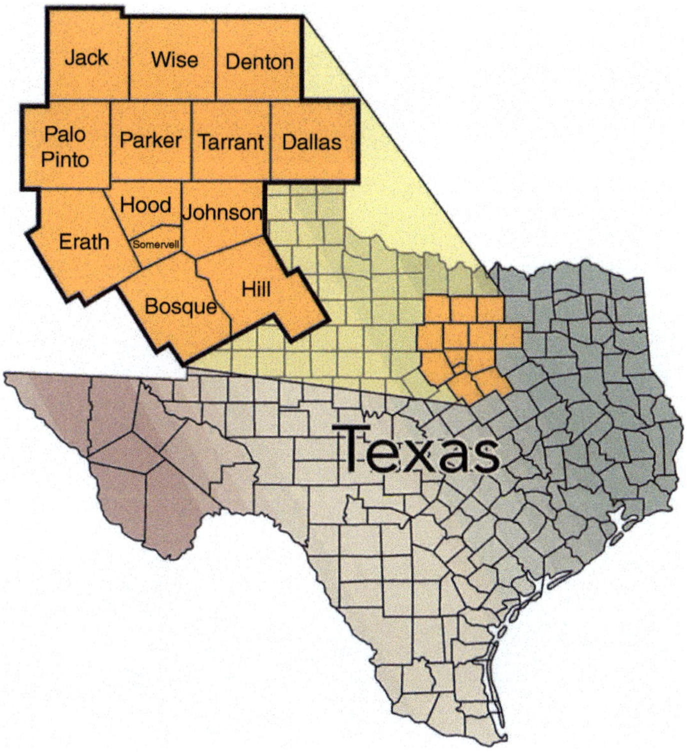

Figure 1.7 Barnett play (www.ogj.com).

Bakken Formation

The Bakken, which is from the Late Devonian and Early Mississippian geological ages, is located in the Williston basin in parts of North Dakota and Montana in the United States and also the Canadian provinces of Saskatchewan and Manitoba (Fig. 1.8). The word *Bakken* by itself usually includes the Bakken, Three Forks, and Sanish formations. However, the Bakken formation is subdivided into three layers by itself: upper Bakken, middle Bakken, and lower Bakken. The upper and lower Bakken formations are organic-rich shales, which are the source rocks, with high total organic carbon (TOC) values whereas the middle Bakken formation is mostly carbonate. In the early 1950s, oil production from the Bakken started with a very few producing wells with an average daily oil production that was less than 1,000 bbl/d in the entire decade. However, in 2008, advanced technology was fully deployed in the Bakken, and the drilling activity significantly increased between 2008 and 2014 due to high oil prices. There are more than 11,400 producing wells in the Bakken as of August 2017, and the oil production was more than 1 MMbbl/d. The Bakken holds between 3.6 to 7.4 billion bbl of technically recoverable oil in the United States and approximately 1.6 billion bbl of technically recoverable oil in Canada (U. S. Energy Information Administration, 2011).

Marcellus Shale

The Marcellus shale play, which is currently the largest shale gas play in the United States with approximately 410 Tcf technically recoverable resource (TRR) is from the Middle Devonian geological age, and it is located in the Appalachian basin (Fig. 1.9). The Appalachian basin is also one of the largest coal basins in the world, which makes it a major basin for coalbed methane as well. There are at least

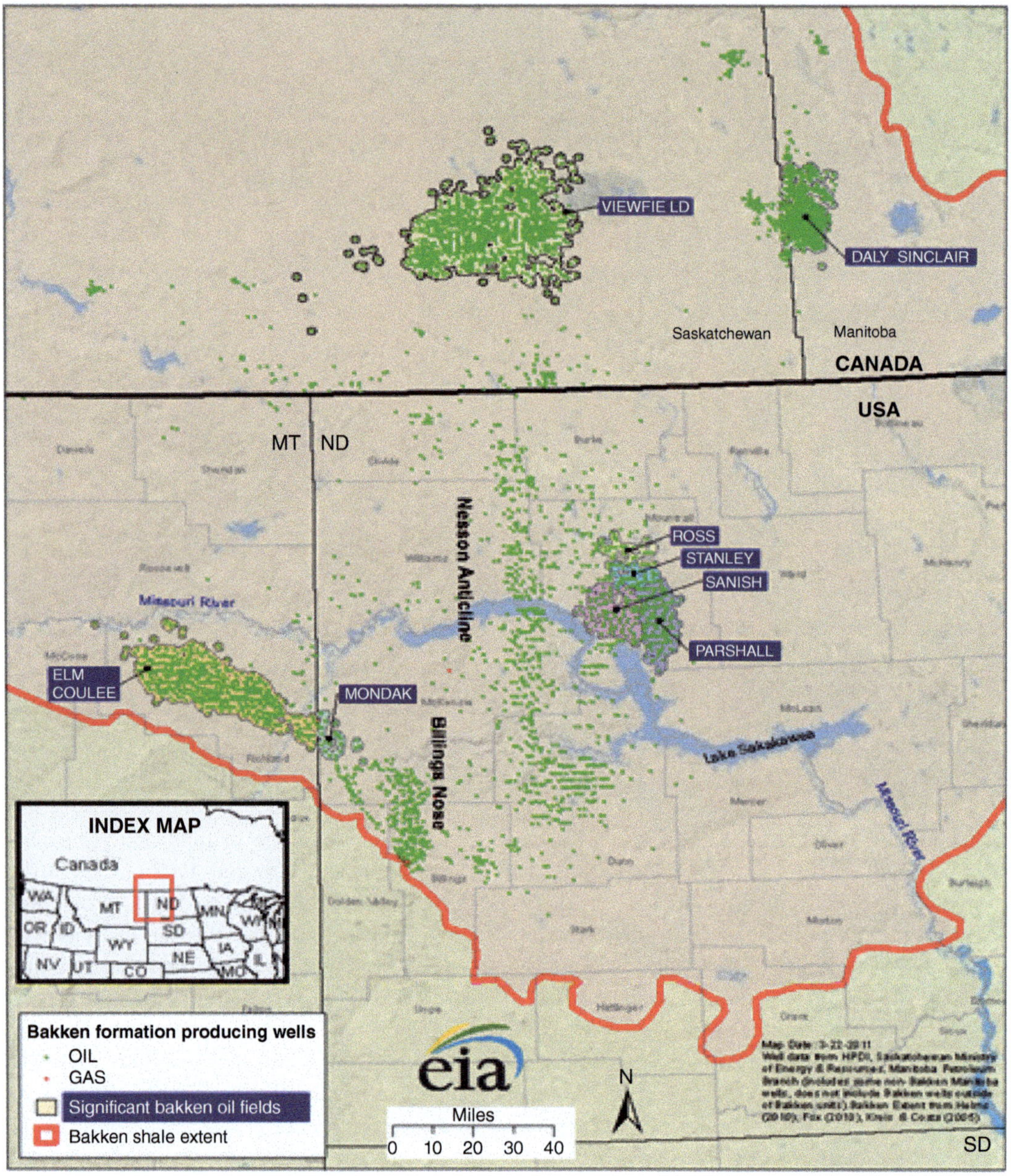

Figure 1.8 Bakken play (www.eia.gov).

six states that contain the Marcellus shale play according to the USGS. They are Pennsylvania, West Virginia, New York, Ohio, Virginia, and Maryland. Horizontal drilling and multistage fracturing were employed in the Marcellus shale play for commercial development in 2005 and, shortly after that, the Marcellus shale play surpassed the other U.S. natural gas resources to become the largest natural gas source in the United States. The average drilling cost of a horizontal well in the Marcellus shale play was estimated to be between \$3 million to \$4 million U.S. back in 2008 by Deutsche Bank (U.S. Energy Information

Figure 1.9 Marcellus and Utica play (www.eia.gov).

Administration, 2011). In 2018, a few reports stated that the average cost of one well is approximately $8 million U.S. However, that cost includes drilling, completion, and all the required facilities (Adams, 2018).

Utica Shale

The Utica shale play, which is from the Middle Ordovician geological age, is located in the Appalachian basin, which is the same basin as the giant Marcellus shale play. It is located in New York, Pennsylvania, Ohio, and West Virginia in the United States, and it extends to Quebec province in Canada (Fig. 1.9). The Utica shale play is deeper than the Marcellus shale play, and it is economically challenging to develop due to its great depth. The estimated technically recoverable resources of the Utica shale play in the United States are 38 Tcf of shale gas and 0.9 billion bbl of oil as well as more than 200 million bbl of natural gas condensate according to the USGS (Kirschbaum et al., 2012). In Canada, the Utica shale play holds approximately 31 Tcf of technically recoverable natural gas (Kuuskraa et al., 2013).

Chattanooga Shale

The Chattanooga shale play, which is of Devonian geological age, is contained in two basins, and is mainly a shale gas play. The Chattanooga shale play consists of two members. One member is located in the Appalachian basin in Tennessee, and is an extension to the giant Marcellus shale play. The second member is located in the Black Warrior basin in Alabama. There are very limited drilling and exploration activities in the Chattanooga shale play since limited fracturing activities, such as horizontal drilling and hydraulic fracturing, were not allowed in Tennessee until 2013. The low gas price is also a major factor in the limited activities in the Chattanooga shale play. The USGS estimated that the undiscovered resources in the Chattanooga shale play is approximately 610 Bcf of shale gas (Higley, et al., 2016). There is also a Devonian-Mississippian shale play called Chattanooga in southeastern Kansas that is correlative to the Woodford shale play in Oklahoma (Cudjoe et al., 2016). There are no reported estimates of the technically recoverable resources of the Devonian-Mississippian Chattanooga shale play in Kansas.

Eagle Ford Shale

The Eagle Ford shale, which is from the Early Cretaceous geological age, is considered as one of the most developed unconventional plays in the world (Figs. 1.11). It is located in the Maverick basin in south Texas, and it extends to the Burgos basin in Mexico (U.S. Energy Information Administration, 2011). However, PEMEX, the Mexican state-owned petroleum company, has not disclosed much information about the wells that have been drilled and tested in the Mexican region. The Eagle Ford shale is the second most active shale play in the United States in terms of running rigs and drilling activity (Fig. 1.10). It is only surpassed by the Wolfcamp shale play, which is located in the Permian basin (Fig. 1.12). There are more than 15,000 active oil and gas wells in the Eagle Ford shale play with an approximate daily oil production rate of more than 870 Mbbl as of 2018, and the average initial oil production rate per well was approximately 2,500 BOPD (U.S. Energy Information Administration, 2015). The daily production

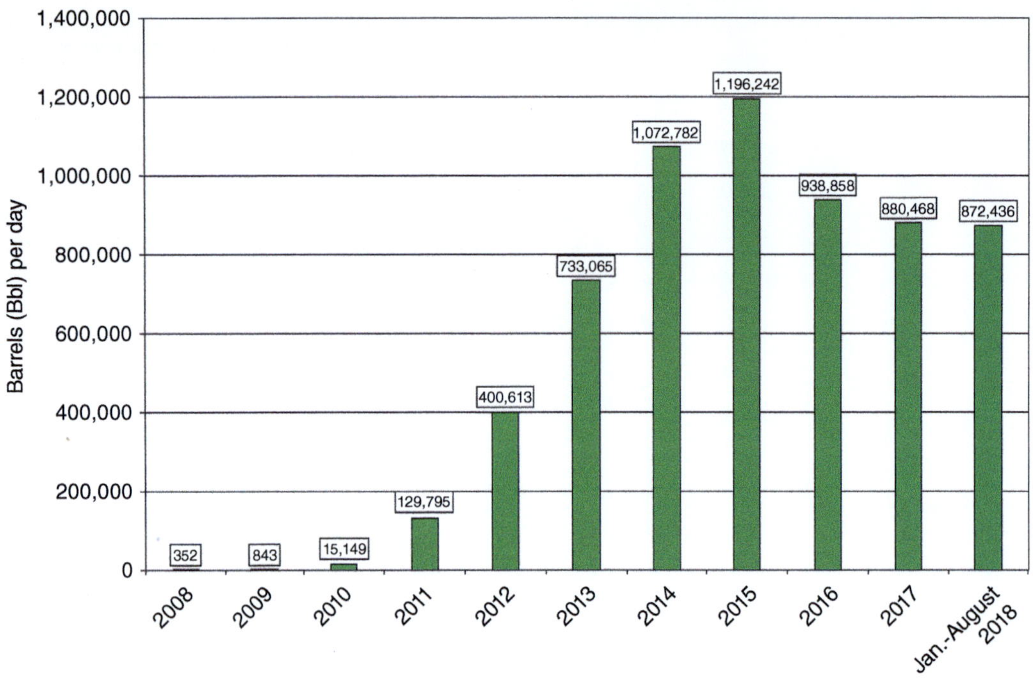

Figure 1.10 Oil production for Texas Eagle Ford shale play (Railroad Commission of Texas, 2017).

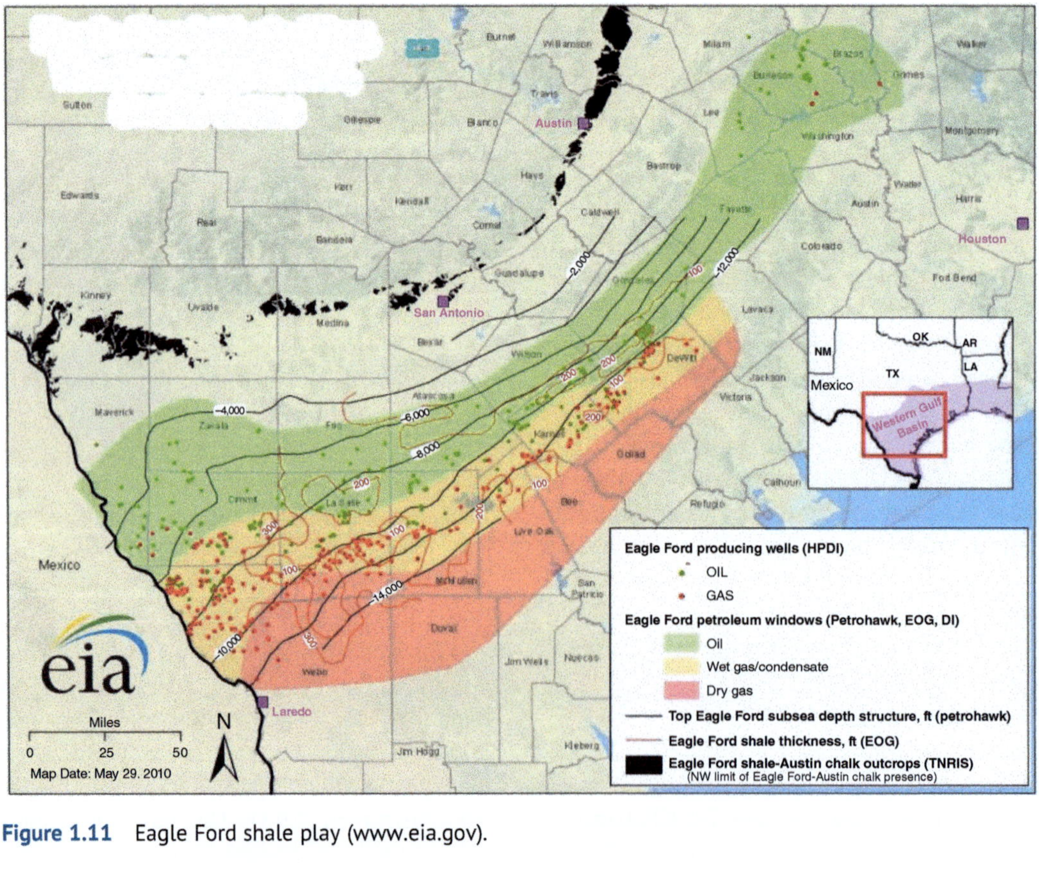

Figure 1.11 Eagle Ford shale play (www.eia.gov).

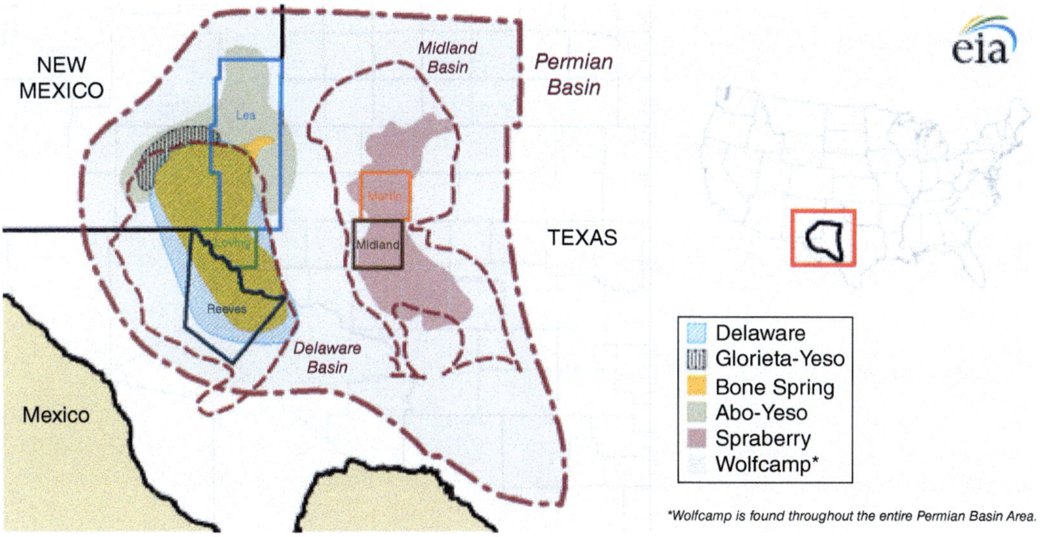

Figure 1.12 Permian basin (www.eia.gov).

figures in July 2017 for the Eagle Ford shale play were actually disappointing because they were below the expectations that were made by the EIA by more than 80 thousand BOPD. The average cost of a well in the Eagle Ford play is between $4 million and $9 million U.S. (U.S. Energy Information Administration, 2011).

Wolfcamp Shale

The Wolfcamp shale play, which is from the Permian geological age, is located in the giant Permian basin in west Texas and southeast New Mexico. The Permian basin contains three subbasins: the Delaware basin, the Midland basin, and the Central basin platform. The Permian basin is one of the top oil productive basins in the world, and it contains massive hydrocarbon resources, both conventional and unconventional resources. Exploration and development started in the Permian basin in the 1920s, and most of the conventional recovery methods (i.e., secondary recovery and enhanced oil recovery techniques) have been practiced in the Permian basin in order to maintain its high oil production. The Permian basin is the home of several giant oil fields in North America including the second largest oil field in the United States, which is the Sparberry field. The Sparberry field is only surpassed by the Eagleville field, which was discovered in 2009 in the Eagle ford shale play. The daily production from the Permian basin is expected to be more than 4 million bbl per day by 2021 (IHS Markit, 2017).

The Wolfcamp formation is divided into several target zones, which are located in all three subbasins: Midland basin, Delaware basin, and central basin. The target zones are called Wolfcamp A, Wolfcamp B, Wolfcamp C, Wolfcamp D, and Wolfcamp XY. Wolfcamp C and D are also subdivided into upper and lower zones due to slight lithology changes (Shale Experts, 2017). The Wolfcamp formation has been producing from the Midland basin since the 1950s with vertical wells. Nowadays, with advanced technology, greater depths have become accessible, and the Wolfcamp unconventional resources are now being recovered. There are more than 3,000 completed horizontal wells in the Wolfcamp shale play, most of which are in the Midland basin in Wolfcamp A and B zones, and more than 1,000 horizontal wells were drilled in 2014 alone (Blomquist, 2016). There are different estimates of the technically recoverable resources in the Wolfcamp shale play. Some companies state that the Wolfcamp shale play still holds at least 35 billion bbl of technically recoverable oil (Blomquist, 2016), and with time as the technology advances, the technically recoverable resources in the Permian basin could reach up to 100 billion bbl of technically recoverable oil. Nevertheless, the Wolfcamp shale play is one of the largest crude oil resources in the world. The Wolfcamp shale play in the Permian basin has made the United States a decisive player in the oil market and a competitive oil producer with Saudi Arabia and Russia.

Niobrara Shale

The Niobrara shale play, which is from the Late Cretaceous geological age, is located in the Denver-Julesburg basin (Sonnenberg, 2012). The Denver-Julesburg basin is usually referred to as the D-J basin. The Niobrara shale play is located in northwest Kansas, northeast Colorado, southwest Nebraska, and southeast Wyoming (Fig. 1.13). The Niobrara shale play has a great potential with more than 2 billion bbl of recoverable oil (Kelly, 2012). However, the Niobrara shale play is still in its early stages of development, and there are challenges in locating its sweet spots due to the complex geology of the formation.

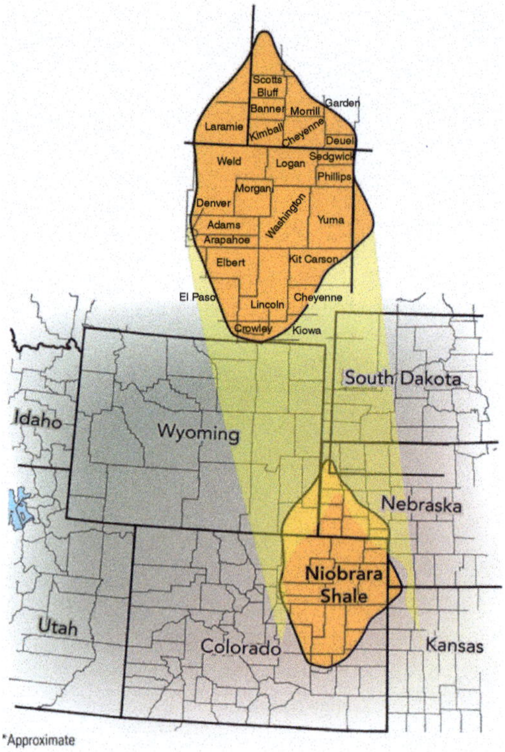

Figure 1.13 Niobrara play (www.ogj.com).

Woodford Shale

The Woodford shale play, which is from the Late Devonian and Early Mississippian geologic age, is one of the Magnificent Seven shale gas plays in North America along with Barnett, Marcellus, Fayetteville, and Haynesville in the United States and Montney and Horn River in Canada (Fig. 1.14). It is located in southeastern Oklahoma and western Arkansas in Ardmore, Anadarko, and Arkoma basins (U.S. Energy Information Administration, 2011). The production from the Arkoma basin has been mainly gas and condensate. However, there are oil wells in the Anadarko basin, hence, the Woodford play contains oil as well. In fact, when the oil prices were high, the developed oil wells in the Woodford shale play exceeded the number of developed shale gas wells (Ahmed & Meehan, 2016).

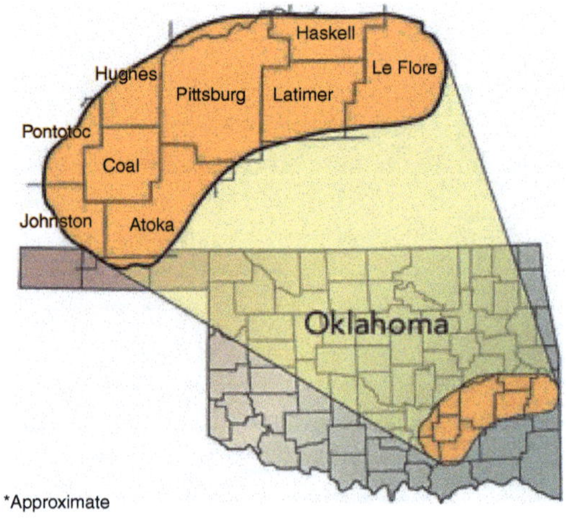

Figure 1.14 Woodford play (www.ogj.com).

Fayetteville Shale

The Fayetteville shale play, which is from the Mississippian geological age, is mainly a dry gas play. It is located in Arkansas in the Arkoma basin (Fig. 1.15). Operations in

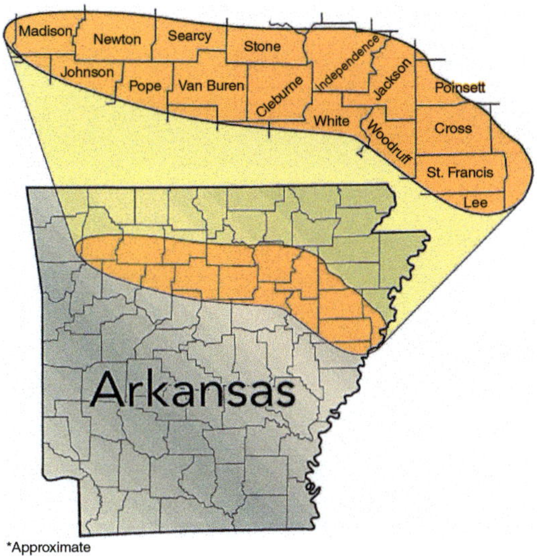

Figure 1.15 Fayetteville play (www.ogj.com).

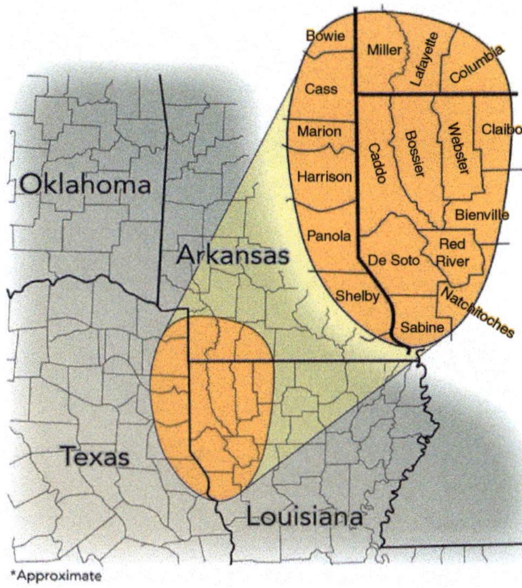

Figure 1.16 Haynesville play (www.ogj.com).

Fayetteville spiked in 2008 because of the high natural gas prices; however, today with low natural gas prices, many drilling companies have moved to oil-rich zones such as the Bakken and the Permian basins. In 2009, the cost of a well in the Fayetteville shale play was estimated to be between \$1.75 and \$3.05 million USD according to Deutsche Bank. The technically recoverable resource of the play is approximately 31.96 Tcf (U.S. Energy Information Administration, 2011).

Haynesville Shale

The Haynesville shale play, which is also known as the Haynesville-Bossier shale play, is from the Late Jurassic geological age, and it is a dry gas play. It is currently one of the top shale gas plays in the United States along with Marcellus and Barnett shale plays with approximately 74 Tcf technically recoverable resource (U.S. Energy Information Administration, 2011). It is located in western Louisiana, east Texas, and southwestern Arkansas (Fig. 1.16). It was known for many years that the Haynesville play contained abundant natural gas, however, it was not economical to extract it at low gas prices. The advanced technology has made extracting the natural gas feasible and profitable from the Haynesville play. Yet, hydraulically fracturing the Haynesville play is still a very complex procedure because of the lithology of the formation; which is 30% to 45% clay (Kennedy et al., 2016). Deutsche Bank reported in 2008 that the cost of a well in the Haynesville shale play was from \$6 to \$7 million USD, however, in 2010, active companies in the play reported the average cost of a well to be \$9 million USD (U.S. Energy Information Administration, 2011).

Monterey Shale

The Monterey shale play, which is from the Early Neogene geological period, is located in the San Joaquin basin in California (Fig. 1.17). The Monterey formation has various natural fractures, and it is the source rock of many conventional reservoirs in California. The Monterey shale play is very challenging to develop due to the San Andreas Fault, and horizontal drilling and hydraulic fracturing not being that effective in the play (Kennedy et al., 2016). In 2011, the EIA estimated that the Monterey shale play contained approximately 24 billion bbl of oil with 15 billion bbl

Figure 1.17 Monterey play (www.ogj.com).

of technically recoverable resource (U.S. Energy Information Administration, 2011). However, in 2014, the EIA reevaluated the play and dropped its previous estimate of the technically recoverable resource by 96% from 15 billion bbl of oil to only 600 million bbl of oil (NGI, 2017).

Muskwa Shale

The Muskwa shale play, which is from the Devonian geological age, is located in the Horn River basin and the Cordova embayment in British Columbia, and it extends to the NW Alberta area basin as well (Fig. 1.18). The Muskwa formation holds 94 Tcf technically recoverable shale gas in the Horn River side and 20 Tcf in the Cordova embayment side. It also holds 31 Tcf technically recoverable shale gas and 2.1 billion bbl of technically recoverable oil in the NW Alberta area basin (U.S. Energy Information Administration, 2015).

Besa River Shale

The Besa River shale play, which is from the Devonian geological age, is located in the Liard basin in Canada. The Besa River formation holds 158 Tcf of technically recoverable shale gas, which makes it one of the biggest shale

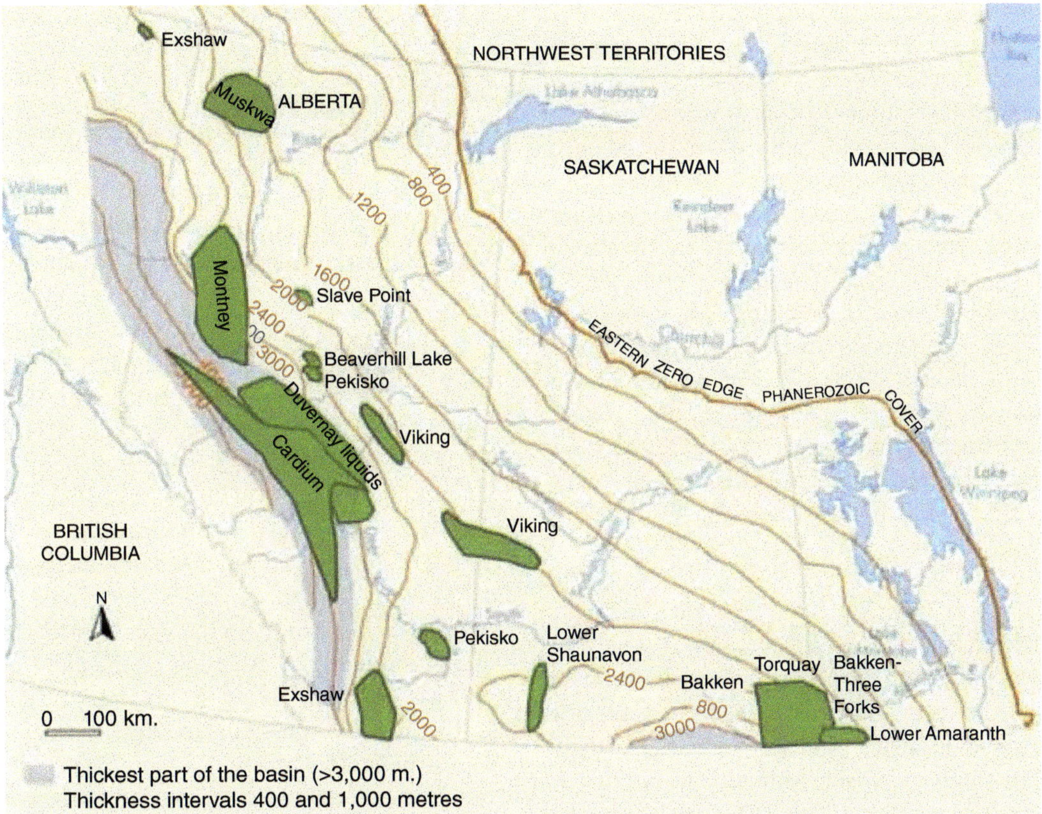

Figure 1.18 Muskwa play (www.neb-one.gc.ca).

gas plays in North America. However, the geology of the Besa River formation is very complex with a great number of faults (U.S. Energy Information Administration, 2015).

Colorado Shale

The Colorado group shale play, which is from the Cretaceous geological age, is located in the southern Alberta basin. The technically recoverable resources in the Colorado shale play varies between 14 Tcf to 43 Tcf of shale gas. Mainly the shallower area of the Colorado shale play is being developed with limited exploration wells (U.S. Energy Information Administration, 2015).

Eagle Ford Shale in Mexico

The Eagle Ford shale play extends to the Burgos and Sabinas basins in Mexico. The Burgos basin correlates to the Maverick basin, which is located in Texas, United States. The technically recoverable resources of the Eagle Ford shale play in the Burgos basin are estimated to be 343 Tcf of natural gas and 6.3 billion bbl of oil. In the Sabinas basin, the Eagle Ford shale play holds approximately 100 Tcf of technically recoverable shale gas (Kuuskraa et al., 2013). The Mexican unconventional plays are still in their initial stages of development, and PEMEX, which is the state-owned petroleum company in Mexico, is the sole operator of the unconventional plays in the country. The exploration wells in the unconventional plays in Mexico cost from $20 to $25 million USD. PEMEX has started producing from the unconventional plays in 2015. However, there are still challenges that are limiting a rapid expansion in the shale sector in the country (Slav, 2017). The public concern over fracturing activities and the high cost of operations, which is due to the limited number of local unconventional experienced service companies in Mexico, are two major challenges that PEMEX has to deal with (Kennedy et al., 2016). Furthermore, foreign investments in the upstream oil and gas industry are very restricted in Mexico; hence, developing unconventional plays will remain costly in the near future. Thus, at low oil and gas prices, unconventional plays development could remain unattractive in Mexico. Also, Mexico has plenty of conventional assets as well as offshore assets that are their top priority. See Fig. 1.19.

Tithonian and Tithonian La Casita

The Tithonian formation, which is from the Late Jurassic and Early Cretaceous geological ages, is located in the Burgos basin, and the Tithonian La Casita formation is located in the Sabinas basin. The Tithonian shale play is one of the most prospective shale plays in Mexico. The Tithonian formation in the Burgos basin holds 50 Tcf of technically recoverable shale gas, and the Tithonian La Casita formation in the Sabinas basin holds 24 Tcf of technically recoverable shale gas (Kuuskraa et al., 2013).

Pimienta Formation

The Pimienta formation, which is from the Late Jurassic geological age, is located in the Tampico basin in Mexico. The Pimienta formation correlates to the U.S. Haynesville shale play. The Pimienta formation is the second most oil bearing unconventional play in Mexico, after the Eagle Ford play, and it holds 23 Tcf technically recoverable shale gas and 5.5 billion bbl of technically recoverable oil (Kuuskraa et al., 2013).

Tables 1.2 and 1.3 summarize North American unconventional plays.

1.4.2 South America Unconventional Plays

Vaca Muerta Formation

The Vaca Muerta unconventional play, which is from the Late Jurassic and Early Cretaceous geological ages, is located in the Neuquen basin in Argentina. The Neuquen basin is the major basin in South America for conventional and unconventional reservoirs, and it contains the two major unconventional plays in Argentina (i.e., Vaca Muerta and Los Molles) (U.S. Energy Information Administration, 2015). The Vaca Muerta formation is the biggest shale play in South America in terms of the technically recoverable resources with 308 Tcf shale gas and 16.2 billion bbl of oil (Kuuskraa et al., 2013; U.S. Energy Information Administration, 2015). The announcement of the abundant hydrocarbon discovery in the Vaca Muerta formation brought the attention of the giant oil and gas operators to Argentina. YPF, which is the primary operator in Argentina, has signed billion-dollar joint venture deals with multiple experienced shale operators such as Chevron to develop the unconventional resources in the country (U.S. Energy Information Administration, 2015; Abogados, 2016).

Los Molles Formation

The Los Molles shale play, which is from the Jurassic geological age, is also located in the Neuquen basin in Argentina. The Los Molles formation is the second major shale play in Argentina, and it holds 275 Tcf of technically recoverable shale gas and 3.7 billion bbl of technically recoverable oil (U.S. Energy Information Administration, 2015; Kuuskraa et al., 2013).

Estratos con Favrella Formation

The Estratos con Favrella shale play, which is from the Cretaceous geological age, is located in the Austral-Magallanes basin (Fig. 1.20). The Austral-Magallanes basin is located in Argentina, and it extends to Chile as well. The Austral-Magallanes basin contains two

Figure 1.19 Mexico unconventional plays and basins (Kuuskraa et al., 2013).

shale plays. The Estratos con Favrella play is located in Chile with 48 Tcf of technically recoverable shale gas and 2.3 billion bbl of technically recoverable oil (Kuuskraa et al., 2013). The other play (i.e., L. Inoceramus-Magnas Verdes) is located in the Argentinean side with 129 Tcf of technically recoverable shale gas and 6.6 billion bbl of technically recoverable oil. The L. Inoceramus-Magnas Verdes has not been developed yet and the resources are just initial estimates (Kuuskraa et al., 2013).

Ponta Grossa Formation

The Ponta Grossa shale play, which is from the Devonian geological age, is located in the Parana basin (Fig. 1.21). The Parana basin is mainly located in Brazil, however, it extends to Argentina, Paraguay, and Uruguay. The Ponta Grossa formation covers three nations (Argentina, Brazil, and Paraguay). The majority of the unconventional resources are located in the Brazilian side with 80 Tcf of technically recoverable natural gas and 4.3 billion bbl of technically recoverable oil. The other two countries hold 11 Tcf of technically recoverable shale gas and 500 million bbl of technically recoverable oil combined (Kuuskraa et al., 2013).

Los Monos Formation

The Los Monos shale play, which is from the Devonian geological age, is located in Chaco basin in Paraguay and Bolivia (Fig. 1.22). The Los Monos formation has not been fully tested and evaluated. However, initial estimates show that the Los Monos holds 103 Tcf of technically recoverable shale gas and 3.8 billion bbl of technically recoverable oil most of which are located in the Paraguayan side of the play (Kuuskraa et al., 2013).

Table 1.2 Summary of North America Unconventional Plays, Part 1

Unconventional play	Barnett	Bakken	Marcellus	Utica	Eagle Ford – U.S. region	Wolfcamp	Niobrara	Woodford
Geological age	Late Mississippian	Late Devonian and Early Mississippian	Middle Devonian	Middle Ordovician	Early Cretaceous	Permian	Late Cretaceous	Late Devonian and Early Mississippian
Location	North central Texas	North Dakota and Montana U.S. /Canadian provinces Saskatchewan and Manitoba	Eastern United States	Eastern United States and Quebec province in Canada	South Texas	West Texas and Southeast New Mexico	Northeast Kansas, northeast Colorado, southwest Nebraska, and southeast Wyoming	Southeastern Oklahoma and Western Arkansas
Basin	Fort Worth	Williston basin	Appalachian	Appalachian	Maverick basin	Permian	Denver-Julesburg	Ardmore, Anadarko, and Arkoma
Average depth (ft)	7,500	6,000	6,750	5,000–11,000	7,000	10,000–12,000	3,000–14,000	5,000–9,500
Average thickness (ft)	300	22	125	100–400	200	1,200–2,000	450	150–250
Porosity (%)	5.0	8	8	2.0–8.0	9	5.0–9.0	6.0–9.0	6.0–7.0
TOC (%)	4.0–5.0	NA	12	0.5–8.0	4.25	2.0–5.0	3	4.0–6.5
TRR, Oil (BBO)	NA	3.6–7.4	0.14	0.95	10	35–100	2	NA
TTR, Gas (Tcf)	43	NA	410	38	34	16***	NA	22
Number of active rigs	2*	50*	54*	23*	70*	435**	26*	8*
Average well cost ($millions)	2.0–4.0	5.5–9.0	5.0–6.0	6.0–8.0	4.0–9.0	7.0–8.0	5.5–8	5.0–8.0

Table 1.3 Summary of North America Unconventional Plays, Part 2

Unconventional play	Fayetteville	Haynesville	Monterey	Muskwa	Besa River	Eagle Ford–Mexico region	Pimienta formation	Tithonian and Tithonian La Casita
Geological age	Mississippian	Late Jurassic	Early Neogene	Devonian	Devonian	Early Cretaceous to Late Cretaceous	Late Jurassic	Late Jurassic to Early Cretaceous
Location	Arkansas	East Texas and Western Louisiana	California	Canada	Canada	Mexico	Mexico	Mexico
Basin	Arkoma	Texas-Louisiana Salt	San Joaquin and Los Angeles	Horn River, Cordova Embayment and NW Alberta	Liard	Burgos and Sabinas	Tampico	Burgos and Sabinas
Average Depth (ft)	4,000	12,000	11,250	6,000–8,000	10,000–15,000	3,500–10,500	5,500–8,000	11,500
Average Thickness (ft)	110	250	1875	200–380	400–1000	160–210	490–655	200 -240
Porosity (%)	5	8.5	11	6.3	3.0–8.0	7.0–9.0	7.0	5.0
TOC (%)	6.9	2.25	6.5	2.5–3.5	3.0–6.0	5.0	3.0	2.0–3.0
TRR, Oil (BBO)	NA	NA	1.0–15	2.1	NA	6.3	5.5	NA
TTR, Gas (Tcf)	32	74	1.5–2.0	114	158	330	23	74
Number of active rigs	1*	52*	NA	NA	122****	NA	NA	NA
Average well cost ($millions)	3.0	9.0	5.0–7.0	9.4****	NA	NA	NA	NA

* Based on BHGE data published in February 2018.
** Based on BHGE data published in February 2018 for entire Permian basin.
*** Initial estimates based on companies operating in the play.
**** Estimates based on companies operating in the play.
NA= not available/not announced at this time.

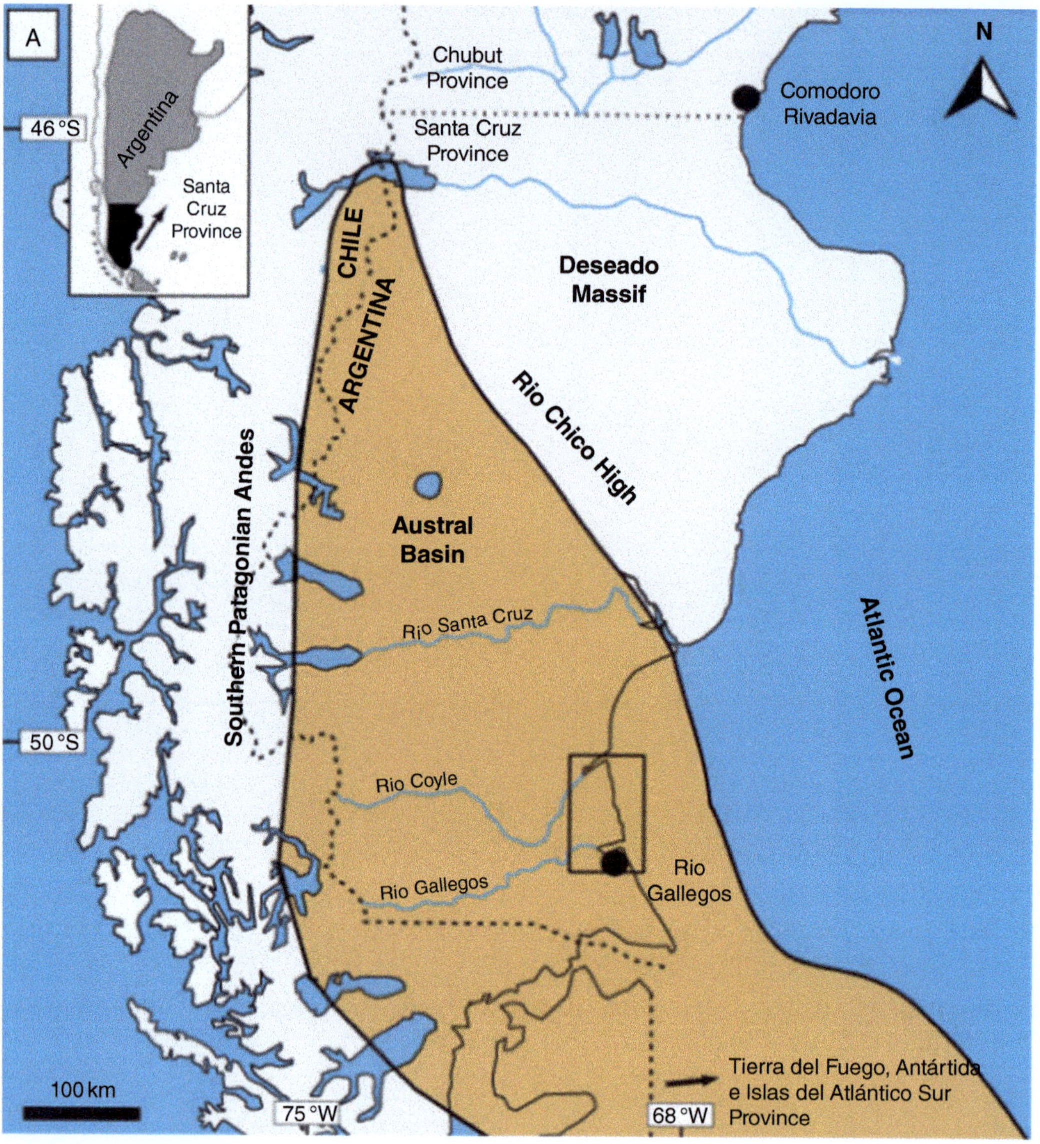

Figure 1.20 Estratos con Favrella play (Kuuskraa et al., 2013).

La Luna and Capacho Formations

The La Luna and Capacho shale plays, which are from the Cretaceous geological age, are located in the Maracaibo basin in Colombia and Venezuela (Fig. 1.23). In Colombia, the Maracaibo basin is also referred to as the Catatumbo basin. The Maracaibo-Catatumbo basin is still under development for shale resources especially in the Venezuelan side (Maracaibo basin). However, initial drilling results show that the Maracaibo-Catatumbo basin holds 202 Tcf of technically recoverable shale gas and 14.8 billion bbl of technically recoverable oil (Kuuskraa et al., 2013). The Maracaibo-Catatumbo basin is the home of an enormous conventional resources as well.

Table 1.4 summarizes South American unconventional plays.

1.4.3 Europe Unconventional Plays

Table 1.5 summarizes European unconventional plays.

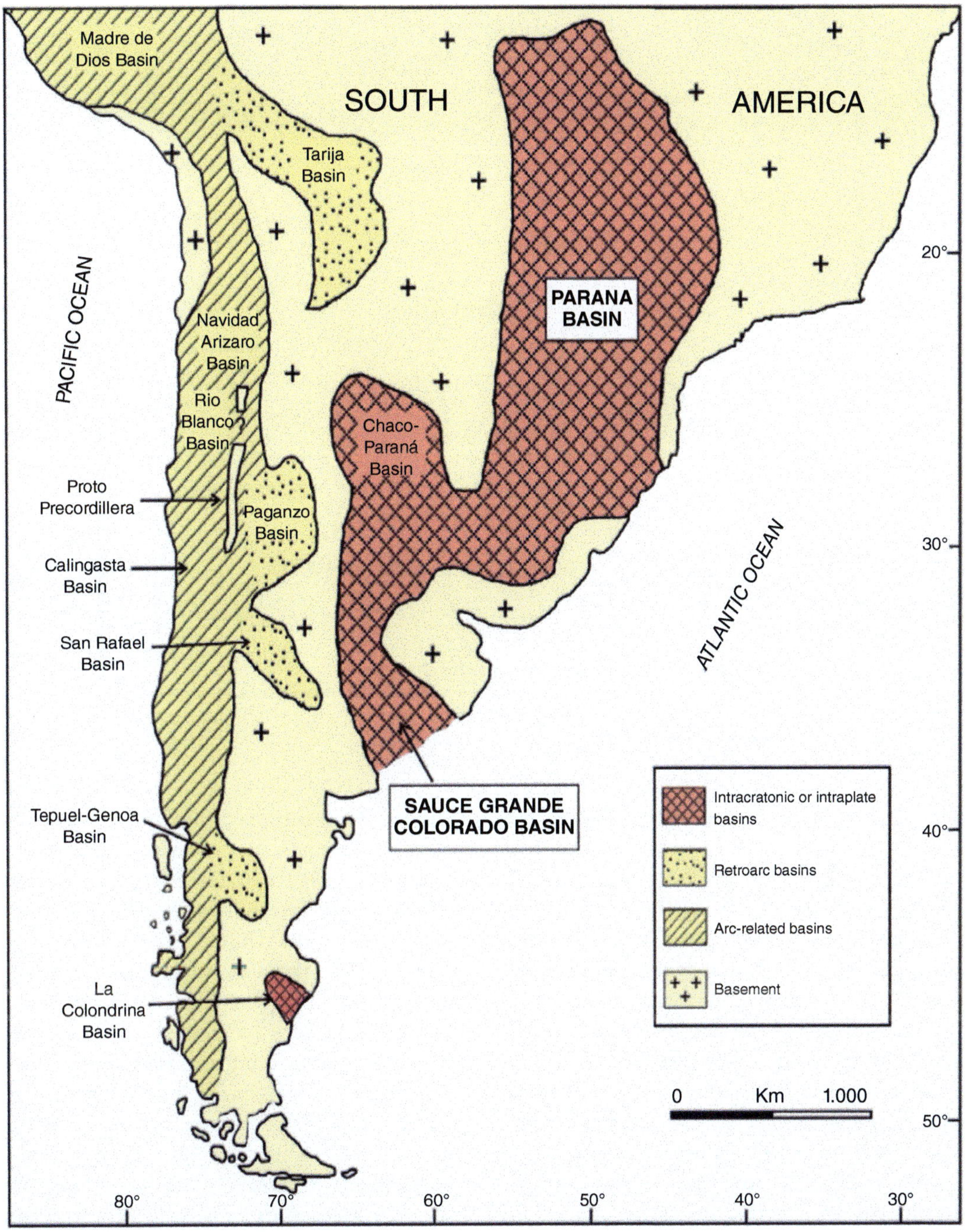

Figure 1.21 Ponta Grossa play (Kuuskraa et al., 2013).

Bazhenov Formation

The Bazhenov shale formation, which is from the Late Jurassic and Early Cretaceous geological ages, is located in the west Siberian basin in Russia (Fig. 1.24). The west Siberian basin is the largest oil and gas basin in the world, and it is the home of several megafields, such as the Samotlor field, which is the largest oil

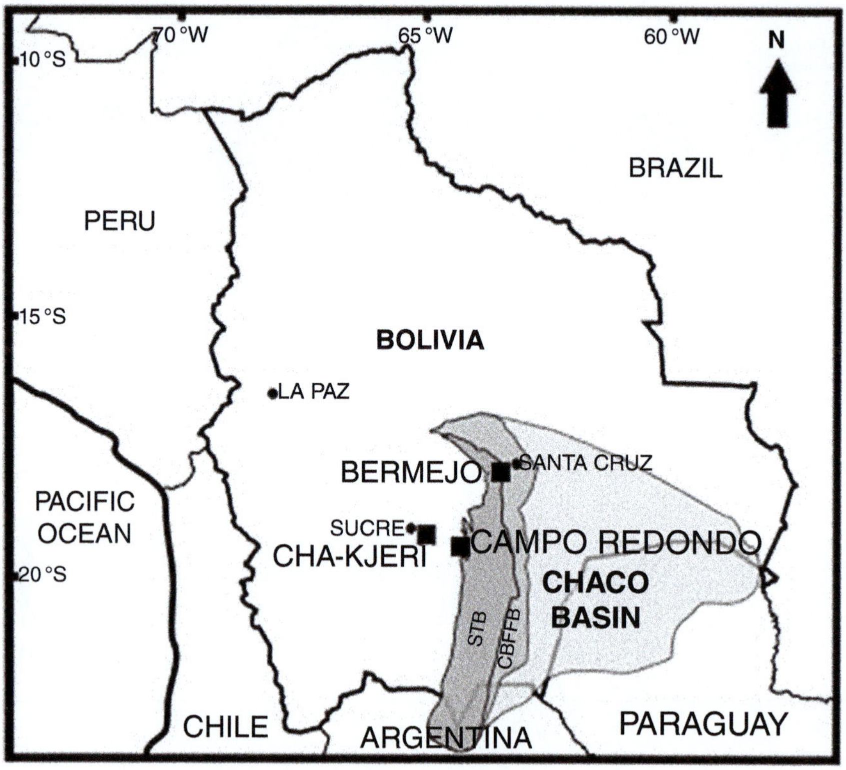

Figure 1.22 Los Monos play (Kuuskraa et al., 2013).

field in Russia. The oil production from the conventional reservoirs in Russia is declining, and exploring and developing new conventional reservoirs is very costly because they are located in remote areas. Consequently, Rosneft, which is the Russian national oil company, has planned on developing the massive unconventional resources in the country. Early estimates show that the Bazhenov play contains more hydrocarbons than the Bakken and the Eagle Ford plays combined. The Bazhenov play holds 285 Tcf of technically recoverable shale gas and 74.6 billion bbl of technically recoverable oil (Kuuskraa et al., 2013). Rosneft has signed several agreements with ExxonMobil and Statoil to employ advanced technology and unlock unconventional resources. However, there are two main issues that are restricting a shale boom in Russia. The first issue is the recent sanctions on Russia. The sanctions limited the partnership between Russian and American companies, hence, the advanced technology to extract unconventional resources, which is already developed in the United States, cannot be shared with Russian companies (Rapoza, 2017). Thus, the Russians might have to develop their own technology to extract the unconventional resources, which will take time and a lot of financial investment, which is the second issue. Foreign investment in the oil and gas industry in Russia are highly encouraged. However, the tax code for resource extraction in Russia is inefficient especially for unconventional reservoirs. The Russian government is working on more efficient mineral extraction tax (MET) reforms to create an attractive environment for investments in the oil and gas industry (Katona, 2017).

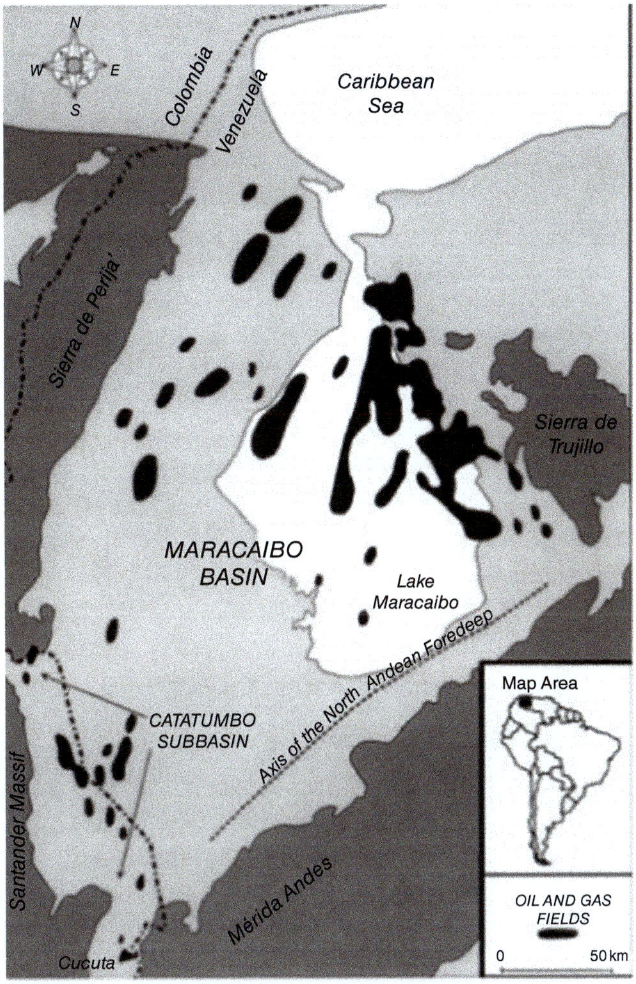

Figure 1.23 La Luna and Capacho plays (Kuuskraa et al., 2013).

Table 1.4 Summary of South America Unconventional Plays

Unconventional Play	Vaca Muerta	Los Molles	Estratos con Favrella	Ponta Grossa	Los Monos	La Luna and Capacho
Geological age	U. Jurassic/L. Cretaceous	M. Jurassic	L. Cretaceous	Devonian	Devonian	U. Cretaceous
Location	Argentina	Argentina	Chile	Brazil	Paraguay/ Bolivia	Colombia/ Venezuela
Basin	Neuquen	Neuquen	Austral-Magallanes	Parana	Chaco	Maracaibo
Average depth (ft)	6,500	11,500	12,000	12,000	10,000	11,000
Average thickness (ft)	325	300	400	300	450	500
Average TOC (%)	5	2	3.5	2	2.5	5
Estimates of total oil in place (BBO)	270	60	47	107	75	297
Estimates of total gas in place (Tcf)	1200	980	228	450	455	970
TRR, oil (BBO)	16	3.7	2.3	4.3	3.8	14.8
TTR, gas (Tcf)	308	275	48	80	103	202

Table 1.5 Summary of Europe Unconventional Plays

Unconventional play	Bazhenov	L. Silurian	L. Carboniferous	Llandovery	Permian-Carboniferous
Geological Age	U. Jurassic/ L. Cretaceous	L. Silurian	L. Carboniferous	L. Silurian/ Devonian/ U. Cambrian	Permian - Carboniferous
Location	Russia	Ukraine	Ukraine	Poland	France
Basin	West Siberian	Carpathian Foreland	Dniepr-Donets	Baltic	Paris
Average depth (ft)	8,200	10,000	12,000	10,000	10,000
Average thickness (ft)	85	400	350	451	83
TOC (%)	10	2	4.5	3.9	9
Estimates of total oil in place (BBO)	1,240	0	23	25	79
Estimates of total gas in place (Tcf)	1,920	365	312	532	666
TRR, oil (BBO)	74.6	0	1.1	1.2	3.2
TTR, gas (Tcf)	285	72	76	105	127

Silurian and L. Carboniferous Formations

The L. Silurian and L. Carboniferous shale plays are located in the Carpathian Foreland basin and the Dniepr-Donets basin, respectively. Both shale plays are located in Ukraine, and early estimates show that the L. Silurian play holds 72 Tcf of technically recoverable shale gas whereas the L. Carboniferous play holds 76 Tcf of technically recoverable shale gas and 1.1 billion bbl of technically recoverable oil (Kuuskraa et al., 2013). Ukraine has already signed a production sharing agreement for shale exploration and development with Shell for 50 years.

Llandovery Formations

The Llandovery shale play, which is from the Early Silurian, Devonian, and Late Cambrian geological ages, is mainly located in Poland, and it is contained in the Baltic basin as well as two other subbasins: the Lublin subbasin and the Podlasie subbasin. The Llandovery shale play in Poland holds approximately 124 Tcf of technically recoverable shale gas and 1.8 billion bbl of technically recoverable oil (Kuuskraa et al., 2013).

Permian-Carboniferous Formation

The Permian-Carboniferous shale play, which is from the Permian and Carboniferous geological ages, is located in the Paris basin in France. The Paris basin contains two formations. They are the Lias shale play, which is from the Jurassic geological age, and the the Permian-Carboniferous shale play. Initial estimates show that the Permian-Carboniferous shale play holds 127 Tcf of technically recoverable shale gas and 3.2 billion bbl of technically recoverable oil, whereas the Lias shale play holds 2 Tcf of technically recoverable shale gas and 1.5 billion bbl of technically recoverable oil (Kuuskraa et al., 2013). There is some exploratory activity in the Lias formation, which is the shallow play in the Paris basin. However, it might take a long time before any major activities occur in the Permian-Carboniferous shale play because of the strict laws in Europe, particularly in France, on hydraulic fracturing (Jolly, 2013).

Figure 1.24 Bazhenov play (Kuuskraa et al., 2013).

Table 1.6 Summary of Middle East Unconventional Plays

Unconventional play	Qusaiba	Shilaif	Diyab	Batra	Athel	Dadas
Geological age	Silurian	M. Cretaceous	U. Jurassic	Silurian	L. Cambrian	Silurian/ Devonian
Location	UAE	UAE	UAE	Jordan	Oman	Turkey
Basin	Rub' Al-Khali	Rub' Al-Khali	Rub' Al-Khali	Hamad	S. Oman Salt	SE Anatolian
Average depth (ft)	15,000	6,500	11,000	8,500	10,000	9,500
Average thickness (ft)	80	330	135	80	444	207
TOC (%)	4	4	3	2	3.2	3.6
Estimates of total oil in place (BBO)	0	367	8.7	0	41	91
Estimates of total gas in place (Tcf)	248 (UAE side)	157	409	33.4	126	130
TRR, oil (BBO)	0	22	0.5	0	2.9	4.6
TTR, gas (Tcf)	62 (UAE side)	18	122.8	6.7	17.7	17

1.4.4 Middle East Unconventional Plays

Table 1.6 summarizes Middle East unconventional plays.

Qusaiba Formation

The Qusaiba shale play, which is from the Early Silurian geological age, is a massive shale formation located in the Arabian peninsula. The majority of the Qusaiba shale is located in Saudi Arabia. However, the estimates of the technically recoverable resources of the Qusaiba shale in Saudi Arabia is not disclosed and not much information is published nor released publicly. In the United Arab Emirates (UAE), the Qusaiba shale is located in the Rub'Al-Khali basin. The Rub'Al-Khali basin is also the host of two other shale formations in UAE: Diyab and Shilaif. The Qusaiba shale in UAE holds approximately 62 Tcf of technically recoverable shale gas (U.S. Energy Information Administration, 2015). The Qusaiba shale is one of the hot shale formations in the Middle East. Those formations are called hot shales because they contain high uranium content, which leads to high gamma-ray readings.

Shilaif Formation

The Shilaif formation, which is from the Middle Cretaceous geological age, is located in the Rub'Al-Khali basin in the UAE. Early estimates show that the Shilaif formation holds approximately 19 Tcf of technically recoverable shale gas, 157 Tcf of associated gas, and 22 billion bbl of technically recoverable oil (U.S. Energy Information Administration, 2015).

Diyab Formation

The Diyab formation, which is from the Late Jurassic geological age, is located in the Rub'Al-Khali basin in the UAE. The Diyab formation is mostly a dry gas formation with approximately 123 Tcf of technically recoverable shale gas (U.S. Energy Information Administration, 2015).

Athel Formation

The Athel formation, which is from the Early Cambrian geological age, is located in the south Oman salt basin in Oman. The south Oman salt basin holds three shale formations: Thuleilat shale, Athel shale, and U shale. The Athel formation is the main shale play in the basin because it has the most technically recoverable resources. It holds approximately 18 Tcf of technically recoverable wet gas and 2.9 billion bbl of technically recoverable oil and condensate (U.S. Energy Information Administration, 2015).

Batra Formation

The Batra formation, which is from the Silurian geological age, is located in Hamad and Wadi Sirhan basins in Jordan. The total technically recoverable resources of the Batra shale play in both basins are approximately 6.7 Tcf of shale gas (U.S. Energy Information Administration, 2015). The Batra formation is also a hot shale formation. There are reported drilling activities in the Batra shale play, however, the production data are not disclosed.

Bahrain Shale

In April 2018, the official Bahrain News Agency reported a new discovery of a massive quantity of offshore tight oil in the Khalij Al-Bahrain basin. The initial estimates of the new find are approximately 80 billion bbl of oil and 14 Tcf of deep gas based on independent studies made by DeGolyer and MacNaughton consulting services and Halliburton (OGJ, 2018). However, the Bahrain Petroleum Co. (Bapco) is still looking into how much can be actually recovered out of the total resources. They have signed contracts with Halliburton to drill two additional appraisal wells in 2018 to further estimate the total resources in place. They have set a 5 year plan to start production from this new discovered unconventional resource.

Dadas Formation

The Dadas shale play, which is from the Late Silurian and Early Devonian geological ages, is located in the Anatolian basin in Turkey. Early estimates show that the Dadas formation holds 17 Tcf of technically recoverable shale gas and 4.6 billion bbl of technically recoverable oil (Kuuskraa et al., 2013). TPAO, which is the Turkish national energy company, is currently working to further explore the Dadas shale play by drilling more exploration wells, and seeking partnerships with giant shale expert companies such as ExxonMobil and Shell (Coskun, 2014).

1.4.5 Africa Unconventional Plays

Table 1.7 summarizes unconventional African plays.

Whitehill Formation

The Whitehill formation, which is from the Permian geological age, is located in the Karoo basin in South Africa. The Karoo basin holds approximately 390 Tcf of technically recoverable shale gas. The Karoo basin also contain two other major shale gas plays. The Whitehill shale play holds 211 Tcf of technically recoverable shale gas. The other two plays, Prince Albert and Collingham, hold 96 Tcf and 82 Tcf of technically recoverable shale gas, respectively (Kuuskraa et al., 2013). South Africa is working on developing the Karoo basin to become independent on natural gas and stop importing it from other countries. Several technical cooperation permits have been issued for shale exploration in the Karoo basin. Major companies, such as Shell, have already signed shale exploration contracts with the South African companies. Other major companies have formed joint ventures such as Chesapeake and Statoil to explore the Karoo basin (Kuuskraa et al., 2013).

Frasnian Formation

The Frasnian shale play, which is from the Late Devonian geological age, is located in Algeria, Libya, and Tunisia (Fig. 1.25). It is contained in four basins: Ghadames basin, Reggane basin, Timimoun basin, and Ahnet basin. The Frasnian formation holds 241 Tcf of technically recoverable shale gas and 7 billion bbl of technically recoverable oil (Kuuskraa et al., 2013). Most of the unconventional resources are located in the Algerian side with approximately 224 Tcf of technically recoverable shale gas and 4.3 billion bbl of technically

Table 1.7 Summary of Africa Unconventional Plays

Unconventional play	Whitehill	Frasnian	Tannezuft	Tannezuft	Sirte	Khatatba
Geological age	L. Permian	U. Devonian	Silurian	L. Silurian	U. Cretaceous	M. Jurassic
Location	South Africa	Algeria	Algeria	Libya	Libya	Egypt
Basin	Karoo	Ghadames/ Berkine	Ghadames/ Berkine	Ghadames	Sirte	Abu Gharadig
Average depth (ft)	8,000	9,500	10,000	11,000	11,000	12,000
Average thickness (ft)	100	248	90	104	200	300
TOC %	6	6	2.8	5.7	2.8	4
Estimates of total oil in place (BBO)	0	78	9	104	406	47
Estimates of total gas in place (Tcf)	845	496	731	240	350	326
TRR, oil (BBO)	0	3.9	0.5	5.2	16.2	1.9
TTR, gas (Tcf)	211	106	176	42	28	65

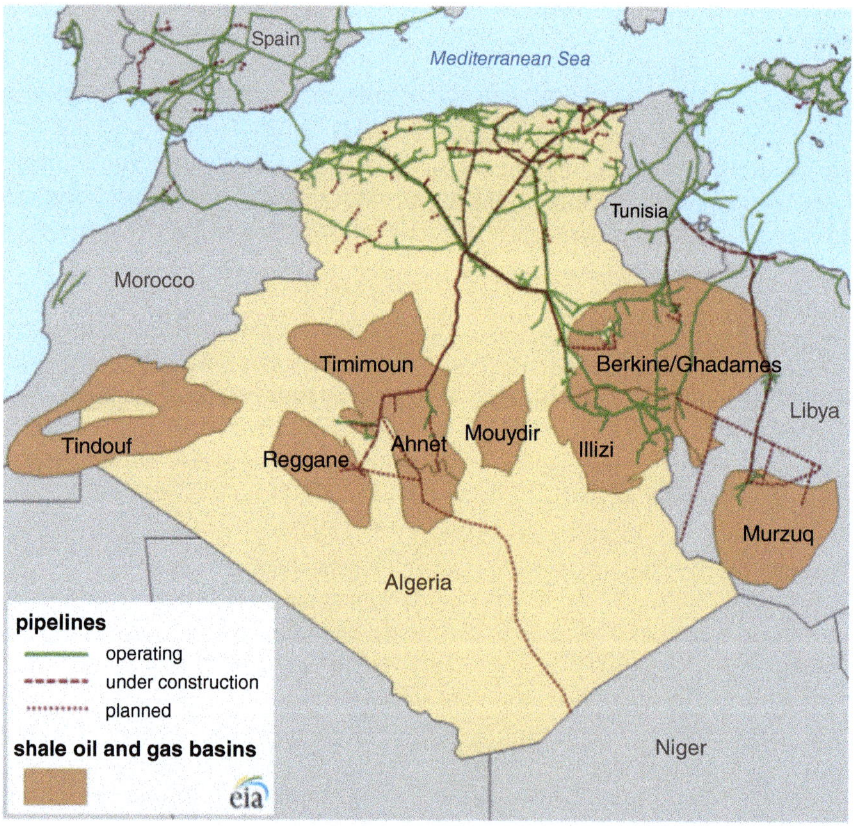

Figure 1.25 Frasnian play (Kuuskraa et al., 2013).

recoverable oil. Libya and Tunisia share the rest of the unconventional resources with 5 and 11 Tcf of technically recoverable shale gas and 1.3 and 1.4 billion bbl of technically recoverable oil, respectively. Very limited activities have occurred in the unconventional reservoirs in Algeria due to the constant protests by the public over the hydraulic fracturing treatments and the associated health issues they object to. The Algerian government is currently working on educating the public about the hydraulic fracturing treatments, and that the unconventional resources must be recovered to maintain the country's budget for the coming years.

Tannezuft Formation

The Tannezuft formation, which is from the Silurian geological age, is also located in Algeria, Libya, and Tunisia. It is contained in eight basins: the Ghadames basin, Reggane basin, Timimoun basin, Ahnet basin, Illizi basin, Mouydir basin, Tindouf basin, and Murzuq basin. The Tannezuft formation holds 538 Tcf of technically recoverable shale gas and 7.9 billion bbl of technically recoverable oil (Kuuskraa et al., 2013). In the Algerian side, there are approximately 483 Tcf of technically recoverable shale gas and 1.4 billion bbl of technically recoverable oil. Libya holds a share of 44 Tcf of technically recoverable shale gas and 6.5 billion bbl of technically recoverable oil, whereas Tunisia holds the rest of the unconventional resources with 11 Tcf of technically recoverable shale gas.

Sirte Formation

The Sirte shale play, which is from the Cretaceous geological age, is located in Sirte basin in Libya. The Sirte basin also contains another shale play, which is the Etel formation. The Sirte formation holds 28 Tcf of technically recoverable shale gas and 16.2 billion bbl of technically recoverable oil (Kuuskraa et al., 2013). The Etel formation holds 45 Tcf of technically recoverable shale gas and 2 billion bbl of technically recoverable oil. The Sirte, Etel, Frasnian, and Tannezuft formations make Lybia an attractive place for the giant shale oil and gas operators in the world as it holds approximately 21 billion bbl of technically recoverable oil and 122 Tcf of technically recoverable shale gas.

Khatatba Formation

The Khatatba shale play, which is from the Jurassic geological age, is located in four basins in Egypt: Abu Gharadig, Alamein, Natrun, and Shoushan-Matruh basins (Fig. 1.26). The Khatatba formation holds approximately 100 Tcf of technically recoverable shale gas and 4.6 billion bbl of technically recoverable oil (Kuuskraa et al., 2013). The major unconventional resource of the Khatatba formation are located in Abu Gharadig basin with 65 Tcf of technically recoverable shale gas and 1.9 billion bbl of technically recoverable oil.

1.4.6 Asia Unconventional Plays

Table 1.8 and Fig. 1.27 summarize Asian unconventional plays and show basins of unconventional resources in China.

Longmaxi Formation

The Longmaxi shale play, which is from the Early Silurian geological age, is located in the giant Sichuan basin in China. The Sichuan basin contains three major shale gas plays in China: the Longmaxi formation, the Qiongzhusi formation, and the Permian formation. The Sichuan basin holds approximately 630 Tcf of technically recoverable shale gas, 287 Tcf of which is located in the Longmaxi formation (Kuuskraa et al., 2013). The Qiongzhusi formation and the Permian formation hold 125 Tcf and 215 Tcf of technically recoverable shale gas, respectively. There are major activities in

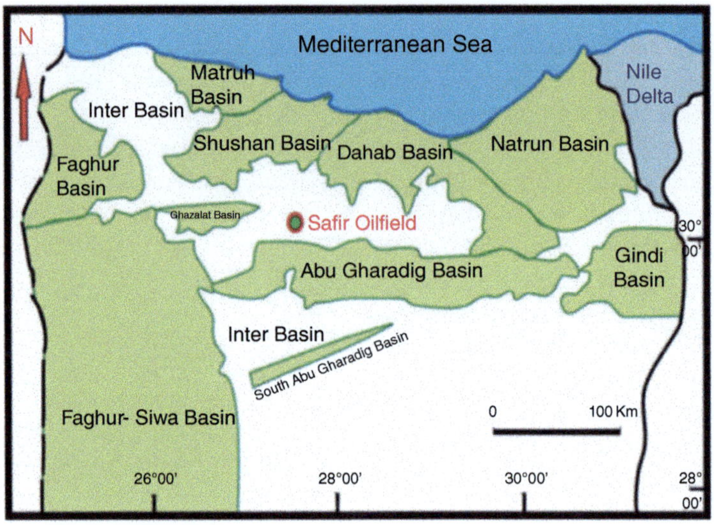

Figure 1.26 Khatatba play (Kuuskraa et al., 2013).

Table 1.8 Summary of Asia Unconventional Plays

Unconventional play	Longmaxi	Ketuer	Sembar
Geological age	L. Silurian	L. Triassic	L. Cretaceous
Location	China	China	Pakistan
Basin	Sichuan	Tarim	Lower Indus
Average depth (ft)	10,000	13,000	8,000
Average thickness (ft)	197	200	250
TOC %	2	3	2
Estimates of total oil in place (BBO)	0	129	145
Estimates of total gas in place (Tcf)	1,100	161	531
TRR, oil (BBO)	0	6.5	5.8
TTR, gas (Tcf)	280	16	101

fully developing the Sichuan basin, and the Chinese government has already set production goals for 2020. However, there are several challenges in commercial development of the shale plays in China. One challenge is the complex lithology in the shale plays might delay developing progress for some time as well as the limited shared data, as the Chinese government consider these data confidential and state secrets (Dong, et al., 2016). Another challenge is the high cost of drilling and completing a single unconventional well, which in some regions exceeds $12 million USD (U.S. Energy

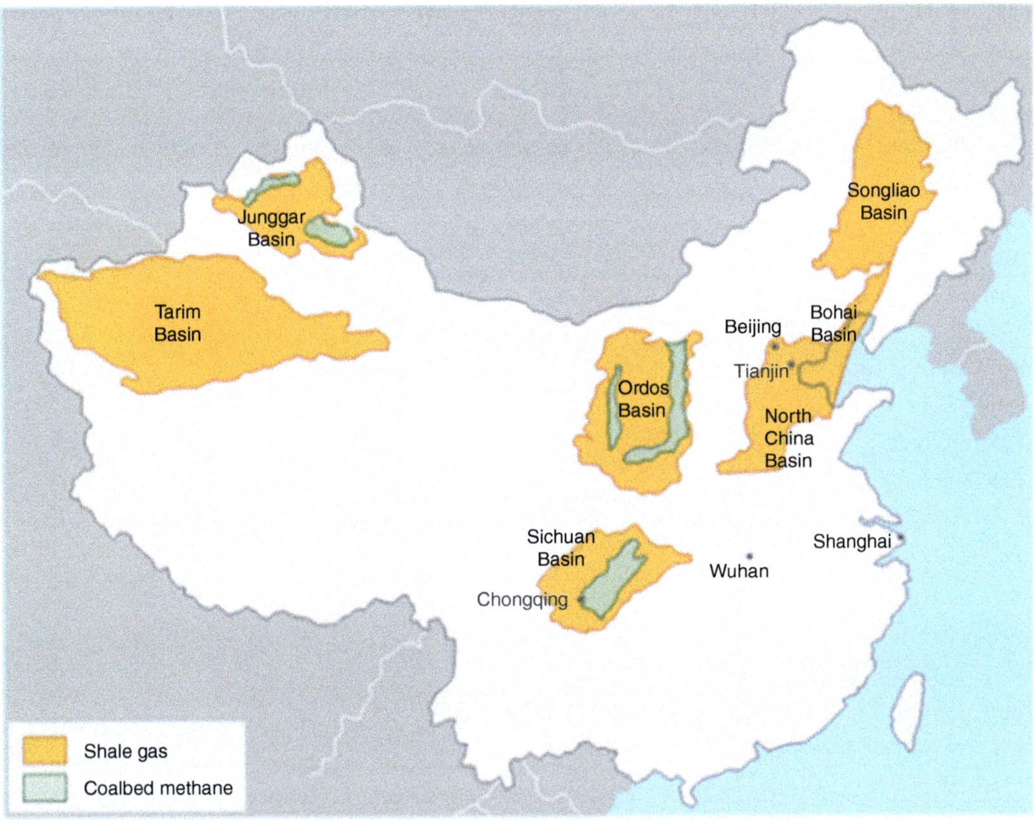

Figure 1.27 Basins of unconventional resources in China (Kuuskraa et al., 2013).

Information Administration, 2015). The Chinese government can overcome these challenges with time; however, there might be a risk in not developing the shale gas resources soon enough to reduce the coal usage in China.

Ketuer Formation

The Ketuer shale play, which is from the Triassic geological age, is located in the Tarim basin in China. The Tarim basin also holds four shale plays. They are the Ketuer shale play with 16 Tcf of technically recoverable shale gas and 6.5 billion bbl of technically recoverable oil, the L. Cambrian shale play with 44 Tcf of technically recoverable shale gas, the L. Ordovician shale play with 94 Tcf of technically recoverable shale gas, and the M-U Ordovician shale play with 61 Tcf of technically recoverable shale gas and 1.6 billion bbl of technically recoverable oil (Kuuskraa et al., 2013).

Sembar Formation

The Sembar shale play, which is from the Cretaceous geological age, is located in the lower Indus basin in Pakistan. Early estimates show that the Sembar shale play holds 101 Tcf of technically recoverable shale gas and 5.8 billion bbl of technically recoverable oil (Kuuskraa et al., 2013). However, no development has occurred yet.

Table 1.9 Summary of Australia Unconventional Plays

Unconventional play	Goldwyer	Nappamerri	L. Kyalla
Geological age	M. Ordovician	Permian	Precambrian
Location	Australia	Australia	Australia
Basin	Canning	Cooper	Beetaloo
Average depth (ft)	8,800	8,000	5,500
Average thickness (ft)	250	300	130
TOC %	3	2.6	2.5
Estimates of total oil in place (BBO)	244	17	65
Estimates of total gas in place (Tcf)	1,227	307	100
TRR, oil (BBO)	9.7	1	3.3
TTR, gas (Tcf)	235	89	22

1.4.7 Australia Unconventional Plays

Table 1.9 and Fig. 1.28 summarize Australian unconventional plays and show basins of unconventional resources in Australia.

Goldwyer Formation

The Goldwyer shale play, which is from the Ordovician geological age, is located in the Canning basin in Australia. The Goldwyer shale play is estimated to hold 235 Tcf of technically recoverable shale gas and 9.7 billion bbl of technically recoverable oil (Kuuskraa et al., 2013). The major operators in the Canning basin have signed agreements and formed joint ventures with shale expert operators such as ConocoPhillips to accurately examine the play and develop it (U.S. Energy Information Administration, 2015).

Nappamerri Formation

The Nappamerri shale play, which is from the Permian geological age, is located in the Cooper basin in Australia. The Cooper basin also contains two other shale plays: Patchawarra and Tenappera. Most of the unconventional resources are located in the Nappamerri shale play, which holds 89 Tcf of technically recoverable shale gas and 1 billion bbl of technically recoverable oil (U.S. Energy Information Administration, 2015). Several joint ventures have been signed with giant companies such as Chevron to further explore the Cooper basin and develop it to extract unconventional resources. However, recent reports indicate that Chevron has quit the project because it does not align with its worldwide strategic plans (Wilkinson, 2015).

L. Kyalla Formation

The L. Kyalla shale play, which is from the Precambrian geological age, is located in the Beetaloo basin in Australia. The Beetaloo basin contain another shale play, which is the M. Velkerri shale play. The Beetaloo basin holds approximately 44 Tcf of technically recoverable shale gas and 4.7 billion bbl of technically recoverable oil (U.S. Energy Information Administration, 2015). The L. Kyalla shale play holds 22 Tcf of technically recoverable shale gas and 3.3 billion bbl of technically recoverable oil. Falcon Oil and Gas Ltd., which is the principle operator in the Beetaloo basin, is seeking joint venture with companies that are expert in shale exploration to further explore and develop the basin (U.S. Energy Information Administration, 2015).

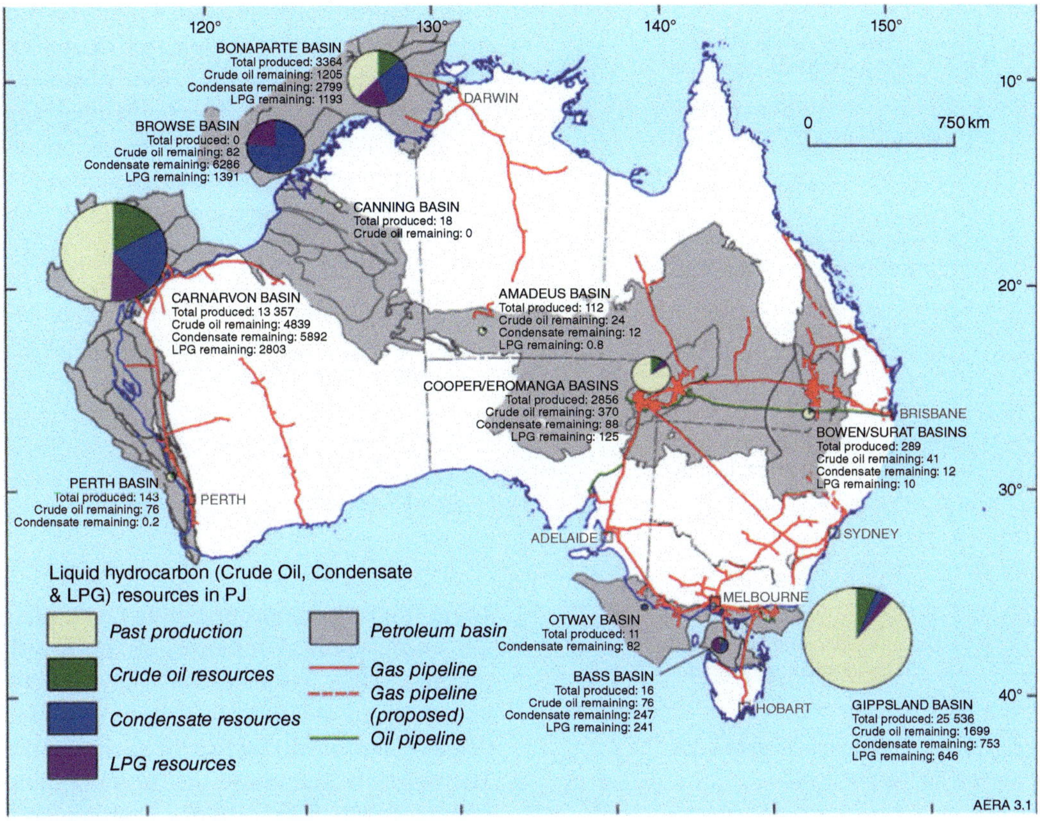

Figure 1.28 Basins of unconventional resources in Australia (Kuuskraa et al., 2013).

1.5 Unconventional Resources Interpretation Workflow

1.5.1 Workflow of Unconventional Reservoirs

The development of the unconventional reservoirs occurs in stages (Fig. 1.29), and each stage could take years to complete, hence, they are mega-multibillion-dollar projects. The first stage is to verify the existence of the unconventional resources. Analyzing geological data from existing wells could give clear signs if an unconventional resource exists near that area or not. Preliminary geological and geophysical surveys are then performed in this stage to verify the existence of the unconventional resources. This stage also involves acquiring the land in which the field will be established as well as obtaining the drilling permits. This stage could take more than a year to complete (CSUR, 2017b).

The second stage is the evaluation and estimation of the unconventional resource. In this stage, full seismic surveys are performed to map the reservoir and to locate the most favorable drilling spots. A vertical well is then drilled to evaluate the unconventional resource (oil or gas) by the use of a variety of well-logging suites as well as collecting core samples. A comprehensive Rock Eval pyrolysis is conducted in this stage to examine the total organic carbon (TOC) and the thermal maturity (Tmax) as well as other important parameters such as the hydrogen index (HI). The Rock Eval pyrolysis procedure will be discussed in detail in Chapter 3. The geomechanical properties (e.g., fracability index and brittleness index), which will be discussed in Chapter 5, are also evaluated

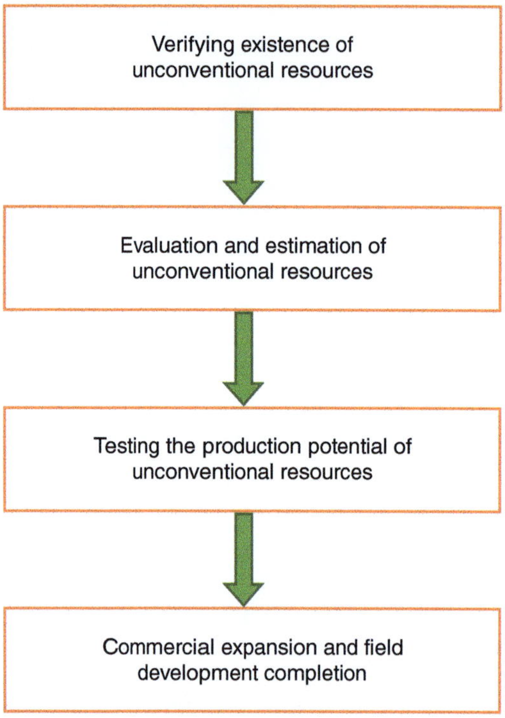

Figure 1.29 Development stages of unconventional resources.

in this stage to assist in identifying the optimum sweet spot locations for hydraulic fracturing. Furthermore, horizontal drilling and an initial hydraulic fracture are performed in this stage to estimate the productivity of the drilled well. This stage takes 2 to 3 years to complete (CSUR, 2017b).

The third stage is testing the production potential of the unconventional play. This stage involves drilling multiple horizontal wells and performing multistage fracturing. Geosteering is heavily involved in this stage to assist placing the horizontal wells in the targeted zones (Zee Ma, 2016). Microseismic surveys are performed in this stage to evaluate the hydraulic fracturing treatments and to optimize the completion techniques of the horizontal wells. Intensive hydraulic fracturing modeling is also practiced in this stage to optimize the conductivity of the hydraulic fractures and to minimize the treatment costs. This stage could take 4 years to complete.

The fourth stage is completing the field development for commercial production. It involves acquiring approvals for the construction of facilities and pipelines. Drilling more wells and sidetracks and refracturing are also practiced in this stage to further improve the production from the unconventional plays (CSUR, 2017b). EOR methods, such as CO_2 huff-n-puff, might also be practiced in this stage to maintain the production goal. This stage goes on until the play is no longer economically profitable.

1.6 Future Projection and Challenges

There is a wide controversy over the future of the unconventional resources. Some say the unconventional resources will have a huge impact on global oil supply, and the unconventional resources will redefine the energy market. The opposite side says the unconventional resources will make a minor contribution to the global oil supply and have a very limited effect on the oil market. It is important to note that the unconventional resources industry, shale sector in particular, is young; it is only over a decade old. Hence, it is hard to speculate whether or not the unconventional resources will completely change the game in the petroleum industry and consequently the energy market. However, the shale boom in the United States has significantly changed EIA projections in their yearly Annual Energy Outlook (AEO) reports. For instance, the EIA in its 2005 AEO report projected the oil production in the United States to be about 5.5 million bbl/day in 2015. However, because of the shale boom, U.S. oil production figures in 2015 were almost double that projected number at 9.4 million bbl/day (Ross, 2016).

Furthermore, everyone has witnessed the resilience of the shale sector, and how it survived the fall of oil prices. Nowadays, in certain countries, even the small local service companies have acquired knowledge in developing unconventional resources. Thus, with innovative technology and the competitiveness in the shale sector, the costs of developing the unconventional resources has gone down drastically. These signs clearly indicate that unconventional resources are going to play a big role in total oil and gas production in the near future (Ross, 2016). At least the unconventional resources will help certain nations (e.g., United States) to become more energy independent than ever before.

That said, there are challenges the unconventional resources industry has to tickle in order to influence the oil market. For instance, the geology, geomechanical, and geochemical properties, which will be discussed in the following chapters, of the unconventional resources vary significantly from one area to the other. Hence, a technique that works marvelously in one area might not be as effective in another area (Kennedy et al., 2016). It is even possible that it damages the formation in that area. For example, in some areas in the Eagle Ford shale play, operators have noticed a higher initial production from wells that were drilled in 2011 than the newly drilled wells in 2016 even though more advanced techniques (e.g., longer laterals) were implemented in the newly drilled wells. This is mainly due to the lithology in those spots, which are not great candidates for hydraulic fracturing (Kennedy et al., 2016).

There are also political and environmental issues that the unconventional resources industry has to face in order to maintain its drive and to further expand and grow (Neville, et al., 2017; Schade, 2017). Ultimately, the exploration and development of unconventional resources would become a standard practice.

1.7 General Remarks

1) There are several shale plays, in the United States and around the world, that are not mentioned in this chapter.
2) There are several shale plays that are recently disclosed; hence, there is no information on the active number of rigs or the cost of the wells. Also, in certain countries, information such as well costs, formation pressure, and daily production rates are considered extremely confidential.
3) The U.S. Energy Information Administration (EIA) and the Advanced Resources International, Inc. (ARI), reports are the best available sources to learn more about the shale plays.
4) Most of the technically recoverable resource estimates that are presented in this chapter are based on EIA and ARI reported estimates.
5) The technically recoverable resource estimates are the resources that can be produced with existing technology. It does not mean that extracting these resources is economically profitable.
6) The technically recoverable resources are estimated by multiplying the calculated original oil and gas in place by a recovery factor, which is based on the geology, geochemistry, and geomechanical properties of the play.
7) The cost of the wells could significantly vary based on the completion plan (e.g., number of hydraulic fracturing stages), formation depth, well location, as well as several other factors.

1.8 Problems

1 What are the two main reasons that led to the shale boom?

2 What are the three types of unconventional plays?

3 What is the current most productive shale gas play in the United States?

4 What are the three largest crude oil fields in the United States?

5 Discuss the difference between the upper Bakken, lower Bakken, and middle Bakken formation.

6 Discuss the obstacles that are limiting Russia and China from developing their massive unconventional resources by studying the location and the geomechanics of their unconventional reservoirs.

Additional Reading

U.S. Energy Information Administration (2011). *Review of emerging resources: U.S. shale gas and shale oil plays. U.S.* Department of Energy. Washington, DC.

Kuuskraa, V., Stevens, S., & Moodhe, K. (2013). *World shale gas and shale oil resource assessment.* Arlington, Advanced Resources International, Inc.

Ahmed, U., & Meehan, N. (2016). *Unconventional oil and gas resources: Exploitation and development.* Boca Raton: CRC Press Taylor & Francis Group.

References

Abogados, A. (2016). *Argentina: Chevron cleared to continue its joint venture agreement with YPF to exploit Vaca Muerta.* Retrieved from Mondaq: www.mondaq.com

Adams, M. (2018). *Cabot plans 85 wells in Marcellus shale in 2018.* Retrieved from *Oil & Gas Journal* /www.ogj.com

Ahmed, U., & Meehan, N. (2016). *Unconventional oil and gas resources: Exploitation and development.* Boca Raton: CRC Press Taylor & Francis Group.

Baker Hughes (2018). *North America Rotary Rig Count.* BHGE.

Bartis, J., LaTourrette, T., Dixon, L., Peterson, D., & Cecchine, G. (2005). *Oil shale development in the United States: Prospects and policy issues.* RAND Corporation.

Beattie, A. (2018). *How oil prices impact the U.S. economy.* Retrieved from Investopedia: www.investopedia.com

Blomquist, P. (2016). *Wolfcamp horizontal play, Midland basin, West Texas.* Las Vegas: IHS Markit: AAPG Pacific Section and Rocky Mountain Section Joint Meeting.

British Columbia Ministry of Natural Gas Development. (2015). *Unconventional natural gas assessment for the Cordova embayment in northeastern British Columbia; Oil and gas report.* British Columbia: Ministry of Natural Gas Development.

Brown, S., & Yücel, M. (2013). *The shale gas and tight oil boom: U.S. States' economic gains and vulnerabilities.* USA: the Council on Foreign Relations.

Chevron. (2017). *Business overview: American energy powerhouse the Permian basin.* Retrieved from Chevron: https://www.chevron.com/projects/permian

Coskun, O. (2014). *ExxonMobil in talks with Turkey over shale gas exploration.* Retrieved from Reuters: www.reuters.com

CSUR (2017a). *Understanding hydraulic fracturing.* Retrieved from http://www.csug.ca/resources/understanding-booklets

CSUR (2017b). *Understanding shale gas in Canada.* Retrieved from http://www.csug.ca/resources/understanding-booklets

CSUR (2017c). *Understanding tight oil.* Retrieved from http://www.csug.ca/resources/understanding-booklets

CSUR (2017d).*Understanding water and unconventional resources.* Retrieved from http://www.csug.ca/resources/understanding-booklets

Cudjoe, S., Vinassa, M., Henrique Bessa Gomes, J., & Barati, R. (2016). A comprehensive approach to sweet-spot mapping for hydraulic fracturing and CO_2 huff-n-puff injection in Chattanooga shale formation. *Journal of Natural Gas Science and Engineering*, 1201–1218. doi: 10.1016/j.jngse.2016.03.042

Cunningham, N. (2017). *Shale growth hides underlying problems*. Retrieved from Oilprice: https://oilprice.com/Energy/Crude-Oil/Shale-Growth-Hides-Underlying-Problems.html

Dong, D., Zou, C., Dai, J., Huang, S., Zheng, J., Gong, J., et al. (2016). Suggestions on the development strategy of shale gas in China. *Journal of Natural Gas Geoscience*, 413–423. doi:10.1016/j.jnggs.2016.11.011

Duncan, D., & Swanson, V. (1965). *Organic-rich shale of the United States and world land areas. Geological Survey Circular*. Washington, DC: United States Department of the Interior.

Egan, M. (2016). *Oil! Massive shale discovery in Texas*. Retrieved from CNN Money: http://money.cnn.com/2016/09/08/investing/apache-huge-oil-discovery/index.html.

Gaswirth, S. (2017). *Assessment of continuous oil resources in the Wolfcamp shale of the Midland basin, Permian basin province, Texas, 2016*. Reston, Virginia: U.S. Geological Survey.

Granados-Hernandez, J., Muñoz-Cisneros, R., Caraveo-Miranda, L., Guerrero-Tristán, M., García-Ortega, M., & Padilla-Bastida, R. (2017). *The emerging unconventional Upper Jurassic oil play in Mexico*. SPE Unconventional Resources Conference. Calgary, Alberta: Society of Petroleum Engineers. doi:10.2118/185024-MS

Gray, J. (1977). *Future gas reserve potential Western Canadian sedimentary basin*. 3rd National Technical Confrence. Canadian Gas Association.

Hameed, A., Bacciarelli, M., & Williams, P. J. (2014). *Unconventional reservoir formation evaluation challenges addressed with deployment-optimized open hole logging solutions*. SPE Eastern Regional Meeting. Charleston, WV, USA.

Hanania, J., Stenhouse, K., & Donev, J. (2013). Conventional vs. unconventional resources. Retrieved 03 29, 2018, from http://www.oilandgasinfo.ca/wp-content/uploads/Nov_2013_conv_vs_unconv.pdf

Higley, D., Rouse, W., Enomoto, C., Trippi, M., Klett, T., Mercier, T., et al. (2016). Assessment of undiscovered oil and gas resources of the Mississippian Sunbury shale and Devonian-Mississippian Chattanooga shale in the Appalachian basin province, 2016. Retrieved from U.S. Geological Survey. doi: 10.3133/fs20163085

Hiller, J. (2017). The Eagle Ford shale is back, kind of. Retrieved from *San Antonio Express News*: https://www-expressnews-com.cdn.ampproject.org/c/www.expressnews.com/business/eagle-ford-energy/amp/The-Eagle-Ford-Shale-is-back-kind-of-12366998.php

Hogg, J. (2016). An unconventional energy future. Retrieved from CSUR Technical Presentations: http://www.csur.com/csur-technical-luncheon-series-presentations.

Holditch, S., & Ayers, W. (2009). How technology transfer will expand the development of unconventional gas, worldwide. Houston: 29th Annual Gulf Coast Section SEPM Foundation Bob F. Perkins Research Conference. Retrieved from http://www.bergan.com/Downloads/HoldAyrs.pdf

IHS Markit (2017). Permian basin. Retrieved from IHS Markit: https://ihsmarkit.com/solutions/permian-basin.html.

Information on the Monterey shale. (n.d.). Retrieved from Natural Gas Intelligence: http://www.naturalgasintel.com/montereyinfo

International Energy Agency (2013). *Resources to reserves*. Paris: IEA.

Jackson, G. (1998). *Whaling*. Retrieved from Britannica: https://www.britannica.com/technology/whaling

Jolly, D. (2013). France upholds ban on hydraulic fracturing. Retrieved from *The New York Times*: http://www.nytimes.com/2013/10/12/business/international/france-upholds-fracking-ban.html

Katona, V. (2017). Will Russia ditch its ambitious oil tax reform? Retrieved from Oilprice: https://

oilprice.com/Energy/Energy-General/Will-Russia-Ditch-Its-Ambitious-Oil-Tax-Reform.html

Kelly, D. (2012). DJ basin. Retrieved from Noble Energy Inc.: www.nblenergy.com

Kennedy, R., Luo, L., & Kusskra, V. (2016). The Unconventional Basins—North America, the Rest of the World, and Emerging Basins. In U. Ahmed, & N. Meehan, *Unconventional oil and gas resources: Exploitation and development*. CRC Press Taylor & Francis Group.

King, H. (2017). Utica shale:The natural gas giant below the Marcellus. Retrieved from Geology and Earth Science News and Information: www.geology.com

Kirschbaum, M., Schenk, C., Cook, T., Ryder, R., Charpentier, R., Klett, T., et al. (2012). Assessment of undiscovered oil and gas resources of the Ordovician Utica shale of the Appalachian basin province, 2012. Retrieved from *U.S. Geological Survey*: https://pubs.usgs.gov/fs/2012/3116/

Kuuskraa, V., Stevens, S., & Moodhe, K. (2013). *World shale gas and shale oil resource assessment*. Arlington,: Advanced Resources International, Inc. Retrieved from http://www.adv-res.com/

Lesniewicz, J. (2012). *Unconventional gas and oil in the USA and Poland*. Göteborg: University of Gothenburg. doi:ISSN 1400-3821

Masters, J. (1979). Deep basin gas trap, Western Canada. *The American Association of Petroleum Geologists Bulletin*, *63*, 152–181.

Maugeri, L. (2013). *The shale oil boom: A U.S. phenomenon*. Cambridge: Harvard Kennedy School.

Natgas (2013). Unconventional natural gas resources. Retrieved from NaturalGas: http://naturalgas.org

Neville, K., Baka, J., Gamper-Rabindran, S., Bakker, K., Andreasson, S., Vengosh, A., ... Weinthal, E. (2017). Debating unconventional energy: Social, political, and economic implications. *Annual Review of Environment and Resources*, *42*(1). doi:10.1146/annurev-environ-102016-061102

NGI (2017). Information on the Monterey shale. Retrieved from Natural Gas Intelligence: http://www.naturalgasintel.com/montereyinfo

Ning, L., Liu, J., & Wu, X. (2015). *Research on the incentive policy of shale gas in China and America*. International Conference on Economics, Social Science, Arts, Education and Management Engineering. Atlantis Press.

OGJ (2018). Bahrain reports large offshore oil find. Retrieved from *Oil & Gas Journal*: https://www.ogj.com/articles/2018/04/bahrain-reports-large-offshore-oil-find.html

Railroad Commission of Texas (2017). Drilling permits issued. Retrieved from Eagle Ford Shale Information: http://www.rrc.state.tx.us/

Rapoza, K. (2017). When Russia finally hops on shale bandwagon, OPEC is finished. Retrieved from Forbes: https://www.forbes.com/sites/kenrapoza/2017/08/29/when-russia-finally-hops-on-shale-bandwagon-opec-is-finished/#641f91151af0

Ratledge, N., & Zachary, L. (2017). Impacts of unconventional oil and gas booms on public education. Retrieved from Resources for the Future: http://www.rff.org

Reinsalu, E., & Aarna, I. (2015). About technical terms of oil shale and shale oil. *Oil Shale: A Scientific-Technical Journal*, *32*(4), 291–292.

Ross, G. (2016). *The remarkable resilience of shale*. Platts: Insight Magazine.

Schade, G. (2017). How has the US fracking boom affected air pollution in shale areas? Retrieved from Oil and Gas Facilities: https://www.spe.org/en

Shale Experts. (2017). Wolfcamp shale overview. Retrieved from Shale Experts: http://www.shaleexperts.com

Slav, I. (2017). Can Mexico replicate the U.S. shale boom? Retrieved from Oilprice: https://oilprice.com/Energy/Natural-Gas/Can-Mexico-Replicate-The-US-Shale-Boom.html

Sonnenberg, S. (2012). *The Niobrara petroleum system, Rocky Mountain region*. Tulsa: Tulsa Geological Society dinner meeting.

Stevens, S., & Moodhe, K. (2015). *Evaluation of Mexico's shale oil and gas potential.* SPE Latin American and Caribbean Petroleum Engineering Conference. Quito: Society of Petroleum Engineers. doi:10.2118/177139-MS

U.S. Department of the Interior (2012). Oil shale/tar sands guide. Retrieved from Oil Shale and Tar Sands Programmatic EIS Information Center: http://ostseis.anl.gov/guide/oilshale/index.cfm

U.S. Energy Information Administration (2011). *Review of emerging resources: U.S. shale gas and shale oil plays.* Washington, DC: U.S. Department of Energy.

U.S. Energy Information Administration (2014). *Updates to the EIA Eagle Ford play maps.* Washington, DC: U.S. Department of Energy.

U.S. Energy Information Administration (2015). Shale gas development in China aided by government investment and decreasing well cost. Retrieved from U.S. Energy Information Administration: /www.eia.gov

U.S. Energy Information Administration (2015). Technically recoverable shale oil and shale gas resources: Argentina. U.S. Department of Energy.

U.S. Energy Information Administration (2015). Technically recoverable shale oil and shale gas resources: Australia. U.S. Department of Energy.

U.S. Energy Information Administration (2015). Technically recoverable shale oil and shale gas resources: Canada. U.S. Department of Energy.

U.S. Energy Information Administration (2015). *Technically recoverable shale oil and shale gas resources: Jordan.* Washington: U.S. Department of Energy.

U.S. Energy Information Administration (2015). *Technically recoverable shale oil and shale gas resources: Oman.* Washington: U.S. Department of Energy.

U.S. Energy Information Administration (2015). *Technically recoverable shale oil and shale gas resources: United Arab Emirates.* Washington: U.S. Department of Energy.

U.S. Energy Information Administration (2015). *Top 100 U.S. oil and gas fields.* Washington, DC: U.S. Department of Energy.

U.S. Energy Information Administration (2016). Maps: Exploration, resources, reserves, and production. Retrieved from EIA: https://www.eia.gov/maps/maps.htm

U.S. Energy Information Administration (2017). *Drilling productivity report for key tight oil and shale gas regions.* Washington, DC: U.S. Department of Energy.

U.S. Energy Information Administration (2017). International energy source. U.S. Energy Information Administration. Retrieved from https://www.eia.gov/beta/international

U.S. Energy Information Administration (2017). *Utica shale play: Geology review.* Washington, DC: U.S. Department of Energy.

U.S. Energy Information Administration. (2018). *How much shale gas is produced in the United States?* U.S. Energy Information Administration. Retrieved from https://www.eia.gov/tools/faqs/faq.php?id=907&t=8

Wang, Q., & Wang, L. (2016). Comparative study and analysis of the development of shale gas between China and the USA. *International Journal of Geosciences.* doi:10.4236/ijg.2016.72016

Warden, S. (2016). Unconventional resources the future of Mexican gas extraction? Retrieved from oilandgasmexico.com: http://www.oilandgasmexico.com/2016/01/13/unconventional-resources-the-future-of-mexican-gas-extraction/

Whitfield, S. (2017). After drop in activity, Haynesville shale sees resurgence. Retrieved from SPE: https://www.spe.org/en/jpt/jpt-article-detail/?art=3591

Wilkinson, R. (2015). Chevron exits Cooper basin unconventional gas project. Retrieved from *Oil and Gas Jounral*: https://www.ogj.com/articles/2015/03/chevron-exits-cooper-basin-unconventional-gas-project.html

Wolfcamp Shale Overview (n.d.). Retrieved from Shale Experts: http://www.shaleexperts.com/plays/wolfcamp-shale/Overview

World Energy Council (2016). World energy resources.

Xie, F., Zhu, J., Liang, H., & Liu, G. (1993). Basic characteristics of recent tectonic stress field in southwest China. *Acta Seismologica Sinica.* doi:10.1007/BF02651820

Yong, T., Fan, Y., Lv, Q., Tang, W., Wen, J., & Wang, H. (2018). Analysis of the tectonic stress field of SE Sichuan and its impact on the preservation of shale gas in lower Silurian Longmaxi formation of the Dingshan region, *China. Journal of the Geological Society of India.* doi:10.1007/s12594-018-0957-z

Zborowski, M. (2018). How does Vaca Muerta stack up vs. US shale? Data tell the tale. *Journal of Petroleum Technology*: https://www.spe.org/en/jpt/jpt-article-detail/?art=4918.

Zee Ma, Y. (2016). Unconventional resources from exploration to production. In Y. Zee Ma & S. Holditch, *Unconventional oil and gas resources handbook evaluation and development.* Gulf Professional Publishing Elsevier.

2

Petrophysical Properties of Unconventional Reservoirs

Negar Nazari[1], Mustafa M. Alhubail[2], Sherifa E. Cudjoe[2], and Reza Barati Ghahfarokhi[2]

[1] *Stanford University, Stanford, California, USA*
[2] *The University of Kansas, Lawrence, Kansas, USA*

2.1 Background

Due to the continuous worldwide demand for hydrocarbons as well as the decline in conventional resources, the unconventional reservoirs (e.g., shale formations) have become more attractive. Since 2005, production from shale formations has increased significantly in the United States (Jiang et al., 2016). For example, the shale oil production is more than 6 million barrels per day as of March 2018, and the shale gas production has reached record high figures (more than 2 Tcf per month; EIA, 2018). Despite the complexities involved in development of unconventional reservoirs, these resources must be developed in order to meet the worldwide increasing demand for energy. Studying the petrophysical properties of the unconventional reservoirs is a crucial step that must be accurately done to properly characterize and develop the unconventional reservoirs.

2.2 Petrophysics

Petrophysics is defined as studying the physical and chemical properties of the rock and their interaction with fluids (i.e., gases, liquid hydrocarbons, and aqueous solutions). The hydrocarbons are trapped within the void space in the rocks, and the nature of the reservoir rocks dictates their quantity, their ability to flow through the porous media, and their interaction with the surrounding environment. Studying the petrophysical properties of the rocks and the pore connectivity and tortuosity plays an important role in characterizing and understanding the production mechanisms of unconventional resources. Valuable information about lithology, porosity, permeability, water saturation, wettability, and other petrophysical properties of the porous media is attained by acquiring well log measurements, core analysis, and seismic data interpretations (Tiabb & Donaldson, 2004). Petrophysical studies are applicable in petroleum engineering, geology, mineralogy, exploration geophysics, and other related studies.

Unconventional resources, such as ultratight shale gas and shale oil plays, were not specifically well characterized due to their historical lack of productivity (Bratovich & Walles, 2016) until the industry started new developments in the late 1990s. The petrophysical properties of several unconventional plays are provided later in Tables 2.26–2.29.

Nanosized pores connected to microsized fractures, organic pores interacting with inorganic pores, complex lithology with significant clay content in some cases, in addition to

Unconventional Hydrocarbon Resources: Techniques for Reservoir Engineering Analysis, Advanced Textbook Series, First Edition. Edited by Reza Barati Ghahfarokhi and Mustafa M. Alhubail.

presence of gas, water, and oil in different stages of development make these systems' petrophysics very complex. It is only by detailed understanding of these interactions that one can develop comprehensive petrophysical models that can capture the physics underlying the fluid flow through these porous systems.

2.2.1 Evaluation of Rock Properties

There are different methods of analyzing rock properties. Cutting samples, core sample, and well log measurements are the major rock evaluation methods used in the industry.

Mud Logging

Mud properties such as mud weight, salinity, temperature, and viscosity in addition to the observation of gaseous phases in the returned mud are good representatives of the formation properties. This, in addition to the cutting analysis, gives a comprehensive, and most of the time reasonably accurate, description of the rock lithology in addition to the type of hydrocarbon to be produced. Mud logging information can be used to give a good estimation of the type of hydrocarbon using gas chromatography as well as the hydrocarbon window using carbon isotope measurements.

Coring

Coring is defined as obtaining actual samples of different formation rocks. Coring is classified into two different categories: conventional coring and side wall coring. In the conventional coring process, a rock sample is obtained using a specialized drill bit. During the sidewall coring process, multiple samples of rock are taken from the sides of the borehole after penetrating the formation (Halliburton, 2011). Direct measurement of the petrophysical properties, which is an expensive and time-consuming process, is feasible by analyzing the cores, cuttings, and crushed samples. Due to the heterogeneous nature of the unconventional reservoirs, properties obtained from small sections must be scaled up to measurements with larger radius of investigation (Sinha et al., 2012). Luckily, there exist well-logging measurements that can be used to scale up the petrophysical properties.

Well Logging Measurements

Well Logging is a method to obtain petrophysical properties downhole. Different readings and measurements can be obtained by using measurement while drilling (MWD), logging while drilling (LWD), and wireline methods (Sinha et al., 2012; Asquith & Krygowski, 2004). For example, during LWD, the wellogging operation is performed during the drilling process, and it provides real-time data. In MWD, pressure, temperature, and the wellbore trajectory are evaluated in three-dimensional spaces. Wireline logging includes spectral gamma, cross dipole acoustic, high resolution resistivity imager, neutron, and density logs on the vertical section of the well. Some of the well loggings readings must be corrected before they are used in evaluating the petrophysical properties. Logs measured only on the vertical section are used to develop correlations using a common lithology-representative log that is measured on the lateral section. Figure 2.1 shows the different components of conventional and unconventional reservoirs, which will be discussed in the following sections.

2.2.2 Shale Volume

The gamma ray logging tool is utilized commonly in unconventional reservoirs to calculate the shale volume. The shale volume is the volume of shale (i.e., the volume of silt, dry clay, and bound water) to the bulk volume. The calculation of the gamma ray index, which is also called the shale index (i.e., linear shale response), is the first step in determining the volume of shale. The linear shale response can be calculated by the use of equation (2.1). This method of shale volume calculation is

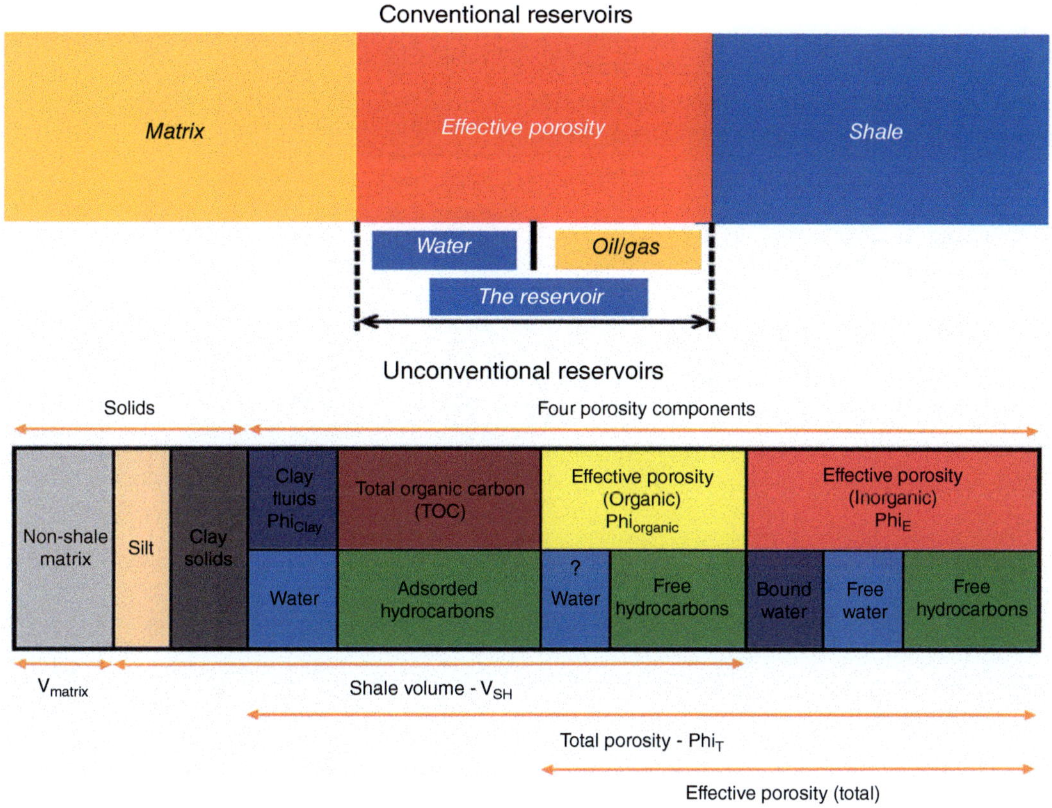

Figure 2.1 Different components of conventional and unconventional reservoirs modified after (Holmes et al., 2014, 2017).

applicable in most plays and the only exceptions are radioactive dolomites and sandstones as well as zones that contain uranium or feldspar (Crain, 2015).

$$I_{sh} = \frac{\gamma_{log} - \gamma_{min}}{\gamma_{max} - \gamma_{min}}, \tag{2.1}$$

where I_{sh} is the linear shale index, γ_{log} is the gamma ray reading in zone of interest, γ_{max} is the gamma ray reading in shale zone, and γ_{min} is the gamma ray reading in clean zone. Shale volume is sometimes simplified to be equal to the shale index calculated from the gamma ray log. This assumption overestimates the shale volume.

There also exist different models that can be used to calculate the shale volume. Those models are usually referred to as the nonlinear shale response models. Examples of these models are the Larionov model for older rocks (equation (2.2)), Larionov equation for tertiary rocks (equation (2.3)), the Steiber model for tertiary rocks (equation (2.4)), and the Clavier model for tertiary rocks (equation (2.5)).

$$V_{sh} = 0.33\left(2^{2\,I_{sh}} - 1.0\right), \tag{2.2}$$

$$V_{sh} = 0.083\left(2^{3.7\,I_{sh}} - 1.0\right), \tag{2.3}$$

$$V_{sh} = \frac{I_{sh}}{3 - 2I_{sh}}, \tag{2.4}$$

and

$$V_{sh} = 1.7 - \left[3.38 - \left(I_{sh} + 0.7\right)^2\right]^{1/2}, \tag{2.5}$$

where V_{sh} is the estimated shale volume.

Example 2.1 Find the Shale Volume

The gamma ray readings for a zone from the Bakken play is provided in Table 2.1. Estimate the shale volume for this specific zone using the Larionov's older rocks method knowing that the gamma ray readings in a clean zone and a shale zone are 12.7 and 131.4, respectively.

Solution

First, the linear shale index should be calculated. Then, the shale volume using the Larionov's older rocks method. A sample calculation for the depth 9,890 is shown here:

$$I_{sh} = \frac{\gamma_{log} - \gamma_{min}}{\gamma_{max} - \gamma_{min}} = \frac{29.12 - 12.7}{131.4 - 12.7} = 0.138$$

and

$$V_{sh} = 0.33\left(2^{(2 \times 0.138)} - 1.0\right) = 0.07.$$

The final results are summarized in Table 2.1.

Table 2.1 Summarized Shale Volume for Example 2.1

Depth (ft)	GR (API)	Ish	Vsh
9,890	29.12	0.138	0.070
9,891	28.14	0.130	0.065
9,892	28.37	0.132	0.066
9,893	28.00	0.129	0.065
9,894	27.22	0.122	0.061
9,895	27.47	0.124	0.062
9,896	33.92	0.179	0.093
9,897	53.84	0.347	0.204
9,898	62.95	0.423	0.263
9,899	61.50	0.411	0.253
9,900	49.59	0.311	0.178
9,901	52.91	0.339	0.198
9,902	58.36	0.385	0.232
9,903	48.44	0.301	0.171
9,904	51.73	0.329	0.191
9,905	57.88	0.381	0.229
9,906	52.39	0.334	0.195
9,907	47.83	0.296	0.167
9,908	49.97	0.314	0.180
9,909	62.22	0.417	0.258
9,910	59.68	0.396	0.241

2.2.3 Gamma Ray Spectroscopy

The spectral gamma wireline measures the total natural radioactivity of the formation regardless of its energy level or energy spectrum. The gamma ray log measures the potassium, thorium, and uranium concentration in the rock, as well as the total gamma. The spectral gamma ray log also detects the naturally occurring gamma ray and defines the energy spectrum of the radiations using equation (2.6).

$$\gamma_{API} = 4T_h + 8U + 16K, \qquad (2.6)$$

where γ_{API} is the total gamma ray in API, T_h is thorium in ppm, U is uranium in ppm, and K is potassium in %. Spectral gamma ray data is occasionally measured for the vertical section of the lateral wells.

The Natural Gamma Ray Tool (NGT) resolves the detected spectrum into the three most common components of the naturally occurring radiation (potassium, thorium, and uranium) by measuring the natural gamma ray radiation of the formation and five-window spectroscopy (International Ocean Discovery Program, 2014). Figure 2.2 shows the natural gamma ray energy spectrum.

Gamma ray spectroscopy is applicable in the following areas.

Clay Typing

Clay minerals are certain elements that decrease the porosity and permeability by filling the voids and blocking pores and canals, and they need to be quantified (Klaja & Dudek,

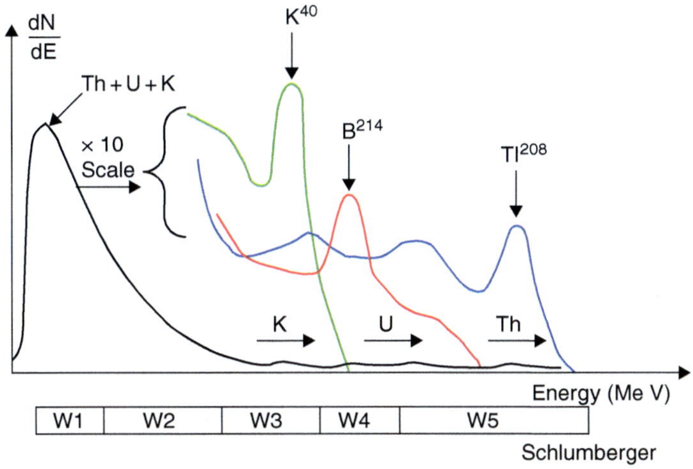

Figure 2.2 Natural gamma ray energy spectrum determined with sodium iodide crystal detector (courtesy of Schlumberger).

2016). Clay mineral content determination is restricted to five important minerals: chlorite, glauconite, illite, kaolinite, and smectite (Schlumberger, 1985). Potassium and thorium are the primary radioactive elements present in clays that are used to identify clay in the formation. A combination of potassium and thorium with the photoelectric effect of the lithodensity tool is used to determine the clay intervals in the formation (Hassan et al., 1976). Thorium is a good indicator of shale in the formation. In order to count for potassium and thorium, the uranium counts are subtracted from the total counts to give a uranium corrected gamma ray curve. If the uranium is reported in ppm, the total gamma ray can be corrected for uranium (uranium-free gamma ray). The uranium part of the gamma ray measurement is directly correlated to organic matter content of the organic-rich shales (Swanson, 1960). Therefore, calculating the uranium-free gamma ray is very important and can be done using equation (2.7).

$$\gamma_{uf} = \gamma_{total} - 8U, \tag{2.7}$$

where γ_{uf} is the uranium-free gamma ray

Mineralogy

Concentrations of the three main radioactive elements in the formation can often be used to give an indication of the mineralogy and/or geochemistry (Hassan et al.,1976). Carbonates usually display a low gamma ray; an increase of potassium can be related to an algal origin or to the presence of glauconite, while the presence of uranium is often associated with organic matter (Adams & Weaver, 1958). In sandstones, high K values may be caused by the presence of potassium feldspars or micas, and glauconite produces a very distinctive diagnostic spike in the K log (Humphreys & Lott, 1990).

Ash Layer Detection

Thorium is a good indicator of the heavy minerals, particularly in channel sand deposits overlying an erosional unconformity, and it is frequently found in ash layers. The ratio of Th/U can also help detect these ash layers. Increasing the Th values may also be associated with an increased input of terrigenous clays (Hassan et al., 1976).

Since hydraulic fracturing is an essential process in the development of the unconventional reservoirs, identifying different types of clays is very crucial for determining mechanical properties as well as designing appropriate fracturing fluids.

Example 2.2 Find the Total and Uranium-Free Gamma Ray

The spectral gamma was run in a well in the Bakken play. The readings of potassium, thorium, and uranium are provided in Table 2.2. Calculated the total gamma ray.

Solution

The total gamma ray can be calculated by the use of equation (2.6). A sample calculation for the depth 9,925 ft is shown below, and the final results are summarized in Table 2.2.

$$\gamma_{API} = 4T_h + 8U + 16K = 4 \times (7) + 8 \times (2.41) + 16 * (2.03) = 79.76\ (API).$$

The uranium-free gamma ray can be calculated by the use of equation (2.7) as follows:

$$\gamma_{uf} = \gamma_{total} - 8U = 79.76 - 8 \times 2.41 = 60.47\ (API).$$

Table 2.2 Gamma Ray Spectroscopy Data for Example 2.2

Depth (ft)	HFK (%)	HTHO (ppm)	HURA (ppm)	Calculated GR (API)	U-Free GR (API)
9,925	2.03	6.998	2.41	79.76	60.47
9,926	2.02	8.14	2.14	82.01	64.86
9,927	2.13	8.40	2.30	86.04	67.67
9,928	2.15	8.79	2.09	86.26	69.54
9,929	1.93	8.41	1.97	80.26	64.52
9,930	1.83	8.26	1.87	77.30	62.30
9,931	1.91	8.40	2.12	81.11	64.17
9,929	1.93	8.41	1.97	80.26	64.52
9,930	1.83	8.26	1.87	77.30	62.30
9,931	1.91	8.40	2.12	81.11	64.17
9,932	1.63	7.05	2.55	74.63	54.27
9,933	1.12	4.39	4.98	75.29	35.47
9,934	1.03	3.86	8.49	99.90	31.93
9,935	2.45	8.78	29.00	306.33	74.3
9,936	3.57	12.47	58.04	571.35	107.00
9,937	3.17	11.70	64.81	616.04	97.52
9,938	3.26	11.97	77.02	716.20	100.05
9,939	2.85	10.45	74.14	680.51	87.40
9,940	2.27	8.57	63.68	580.04	70.61

The different types of clay can also be identified by the use of the thorium- potassium plot. Figure 2.3 presents a plot of thorium versus potassium with representative clay types marked based on the Th/K ratios.

Information about the Th, U, and K concentrations is used to identify the mineral composition and analyze the sedimentary conditions. Thorium is usually combined with shales and heavy minerals in igneous rock and provides valuable information about the clay content. Uranium concentration is not linked to the clay contamination of rocks, but it can be used to analyze the organic matter content of the system. Potassium is a basic component of shale rocks and an indicator of feldspar

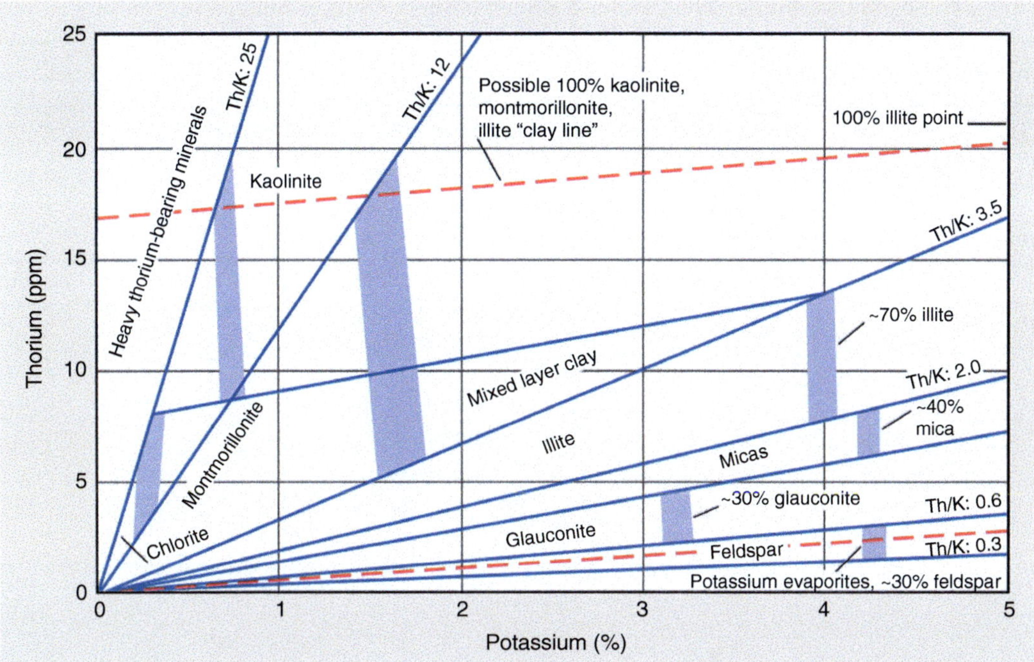

Figure 2.3 Thorium-potassium plot to identify different types of clay (courtesy of Schlumberger).

Table 2.3 Spectral Gamma Ray Applications

Ratio	Application
Th/U	Sedimentary condition analysis
	Estimation of organic matter content in clays
	Discontinuity detection
	Stratigraphic correlation evaluation using oxidizing-reducing conditions
U/K	Organic matter content evaluation in clay sediments
	Stratigraphic and natural fissure systems correlation analysis
	Diagenetic changes detection
Th/K	Sediments typing recognition
	Sedimentary condition evaluation
	Diagenetic changes evaluation in clay sediments
	Clay-type determination in clay minerals

Source: Klaja & Dudek (2016).

and mica presence. The potassium presence mainly occurs in oxidized form (K_2O). Basically, different ratios of uranium (U), thorium (Th), and potassium (K) with respect to each other is translated to different mineralogy and depositional environments. For example, increasing the TOC content of the clay-rich sediments associates to increase in the U and Th concentrations and decrease in the K content (Vigh, et al., 2013). Klaja and Dudek (2016) summarized the application and interpretation of spectral gamma ray. Tables 2.3 and 2.4 show the applications of the spectral gamma ray logs in geological analysis and depositional environment interpretation (Klaja & Dudek, 2016).

2.3 Lithology Evaluation

An example of unconventional reservoirs is shale gas and shale oil reservoirs, which are usually the source rocks. Shales are extremely heterogeneous, fine-grained sedimentary rocks. Presence of two or more minerals significantly

Table 2.4 Spectral Gamma Ray Interpretation Based on Different Ratios of Radioactive Material

Ratio	Interpretation
Th/U	Th/U > 7: Oxidizing conditions, continental environment, weathered soil
	Th/U < 7: Marine sediments, grey and green shales, greywackes
	Th/U < 2: Marine black shales, phosphorites, reducing conditions
Th/K	Order of Th/K increasing:
	glauconite→ muscovite→ illite→ mixed-layer minerals→ kaolinites→ chlorite→ bauxite

Source: Klaja & Dudek (2016).

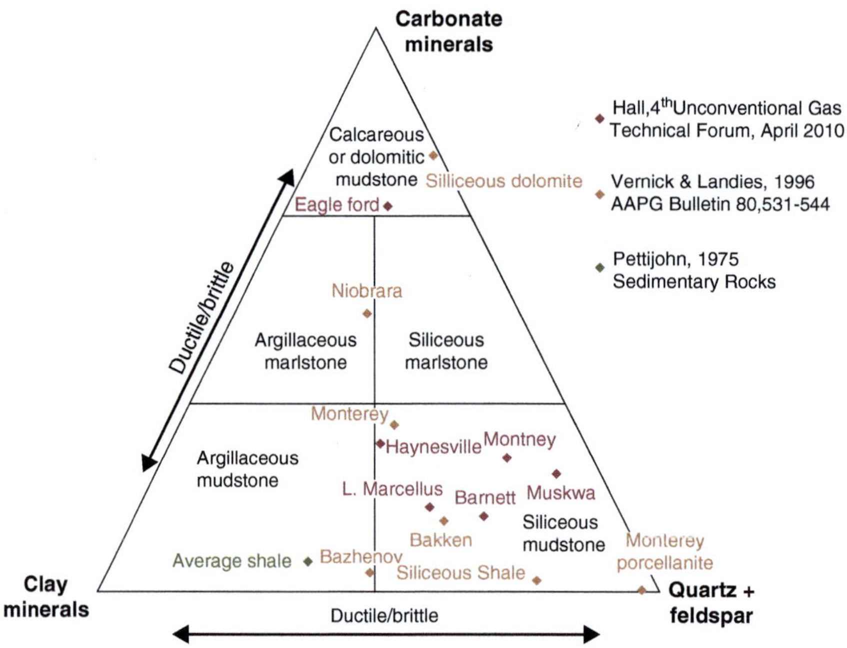

Figure 2.4 Mineral composition for unconventional plays modified after (Allix et al., 2010; courtesy of Schlumberger).

increases the difficulty of determining shale's complex lithology from wireline logs (Passey et al., 2010). Hence, the unconventional reservoirs require different evaluation strategies during their exploration and production phases (Hameed et al., 2014).

Figure 2.4 shows the main minerals composition for several unconventional plays. Knowing the formation lithology is crucial in the unconventional reservoirs because the minerals and their percentages play a decisive role in the success of the hydraulic fracturing treatments as they dictate the rock's brittleness. The rock brittleness is an important parameter that assists in locating the sweet spots for hydraulic fracturing. The mineral-based brittleness index will be discussed in detail in Chapter 5.

2.3.1 Lithology Measurements Using Cross-Plots

In this method, if the matrix is composed of three major minerals, the zone will be plotted within the triangle described by three points. Different cross-plot techniques are commonly used to identify the occurrence of mineralogy

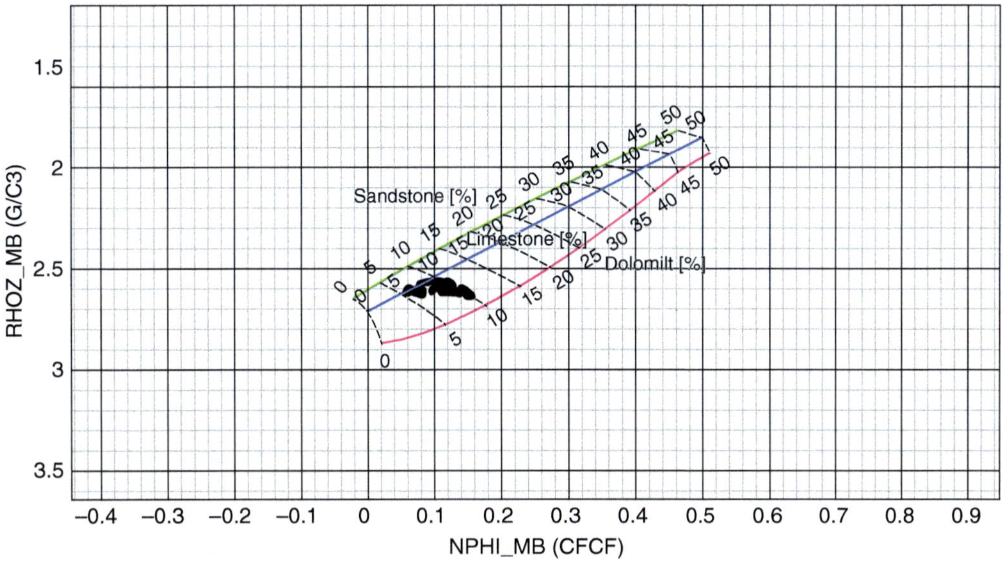

Figure 2.5 Neutron-density cross-plot (Techlog).

including the M-N cross-plot and the MID cross-plot. In these techniques, the ratio of the different lengths is used to calculate the mineral composition of different samples. A combination of neutron-density, sonic-density, or neutron-sonic cross-plot is typically used to determine the lithology. Figure 2.5 shows a neutron-density cross-plot of a sample from the middle Bakken formation. The two-component cross-plots simplify the lithology to two major components. This is typically not the case in unconventional shale oil/gas reservoirs.

2.3.2 Lithology Measurements Using a Combination of Logs

The multimineral log problems have been solved using a classical method known as the Chaveroo method (Burke et al., 1967). The method was developed in the 1960s during the drilling of the Chaveroo field in New Mexico. This method, which is also referred to as the "Martini problem," was further developed in 1979 by John Doveton at the Kansas Geological Survey. The strategy behind this method is to reduce the problem to a simple inversion of a response matrix, and then the formation lithology can be found by the use of neutron, density, and gamma ray logs, simultaneously. The default values of several log readings for different rock types are provided in Table 2.5.

Litho-Density Tools and Their Application

The new generation of density logs are the litho-density logs that have been typified by Schlumberger's Litho-Density Tool (LDT). This tool is composed of a source and two detectors. Two receivers will measure the high-energy and low-energy gamma rays emitted from the source. High- and low-energy gamma rays are used to measure the formation density and lithology of the formations in conventional reservoirs, respectively (Glover, 2001). The bulk density, electron density index, and effective photoelectric index of the formation are measured by the litho-density tool. However, the most important application of the litho-density tool is to determine the lithology of the formation. Table 2.6 shows an example of the Bakken lithology using the litho-density tool.

Table 2.5 Default Values for Pure Rock Types

	Bulk density (g/cm3)	DTC - Sonic (us/ft)	Gamma ray (API)	PEF - Photoelectric factor (B/E)	U_{photo} (B/cm^3)
Sandstone	2.65	56	15–35	1.8	4.8
Limestone	2.71	49	10–40	5.1	13.8
Dolomite	2.85	44	15–40	3.1	9
Anhydrite	2.98	50	10–15	5.1	15
Shale	2.45	100	60–150	4.0	9.4

Table 2.6 Lithology of Bakken Formation from Litho-Density Tool

Anhydrite	Calcite	Dolomite	Illite	Orthoclase	Pyrite	Quartz
0.0000	0.0000	0.3234	0.2757	0.0472	0.0001	0.2747
0.0003	0.0321	0.3340	0.2600	0.0000	0.0004	0.3024
0.0002	0.0468	0.3104	0.2357	0.0000	0.0000	0.3411
0.0002	0.0000	0.3256	0.2768	0.0000	0.0000	0.3358
0.0002	0.0000	0.2562	0.3049	0.0141	0.0000	0.3615
0.0000	0.0000	0.2325	0.3166	0.0000	0.0142	0.3587
0.0000	0.0556	0.1826	0.2403	0.0000	0.0011	0.4511
0.0000	0.1025	0.2174	0.1751	0.0237	0.0002	0.4179
0.0000	0.1254	0.2966	0.1432	0.0257	0.0001	0.3480
0.0000	0.0761	0.2821	0.1994	0.0204	0.0001	0.3615
0.0003	0.0228	0.2213	0.2754	0.0501	0.0000	0.3703

Dual-Mineral Analysis

In the case of a mixture of two minerals, for example a sandy limestone, a cross-plot technique is applied to calculate the relative proportions of the two minerals. To generate the P_e-ρ_b cross-plot, the P_e value is plotted against bulk density. P_e can be used in combination with bulk density to analyze two-mineral matrices and to determine the porosity. The lithology may then be checked against the other tool readings for consistency (Glover, 2001). The combination of equation (2.8) to equation (2.11) is used to obtain apparent total porosity and apparent matrix density (Gardner & Dumanoir, 1980).

$$\rho_b = \Sigma\,(V_i\,\rho_i) = V_1\,\rho_1 + V_2\,\rho_2 + \,\varnothing_{ta}\,\rho_f, \tag{2.8}$$

$$P_e\rho_e = \Sigma(V_iU_i) = V_1U_1 + V_2U_2 + \,\varnothing_{ta}U_f, \tag{2.9}$$

$$1 = \Sigma\,V_i = V_1 + V_2 + \,\varnothing_{ta}, \tag{2.10}$$

and

$$\rho_e = \frac{\rho_b + 0.1883}{1.0704}, \tag{2.11}$$

where ρ_b is the measured bulk density from the log (gr/cm^3), $\rho_{1,\,2}$ is the density of minerals 1, 2 (gr/cm3), ρ_f is the density of fluid (gr/cm3), $V_{1,\,2}$ is the fraction of the bulk volume of minerals

1 and 2, P_e is the measured effective photoelectric absorption cross-section index from the log, $U_{1,\ 2}$ is the volumetric cross-section of minerals 1, 2, U_f is the fluid volumetric cross-section, $\emptyset_{ta}$ is the apparent total porosity (fraction), and ρ_e is the electron density index.

Example 2.3 Determine the Lithology

Figure 2.6 shows a cross-plot of the photoelectric response and matrix density of a sample from the Bakken play. Estimate the composition (i.e., quartz, dolomite, and calcite) of the green point that is located on the cross-plot.

Solution

Based on the lithology identification chart, the composition of the green point is 17% dolomite, 41% calcite, and 42% quartz.

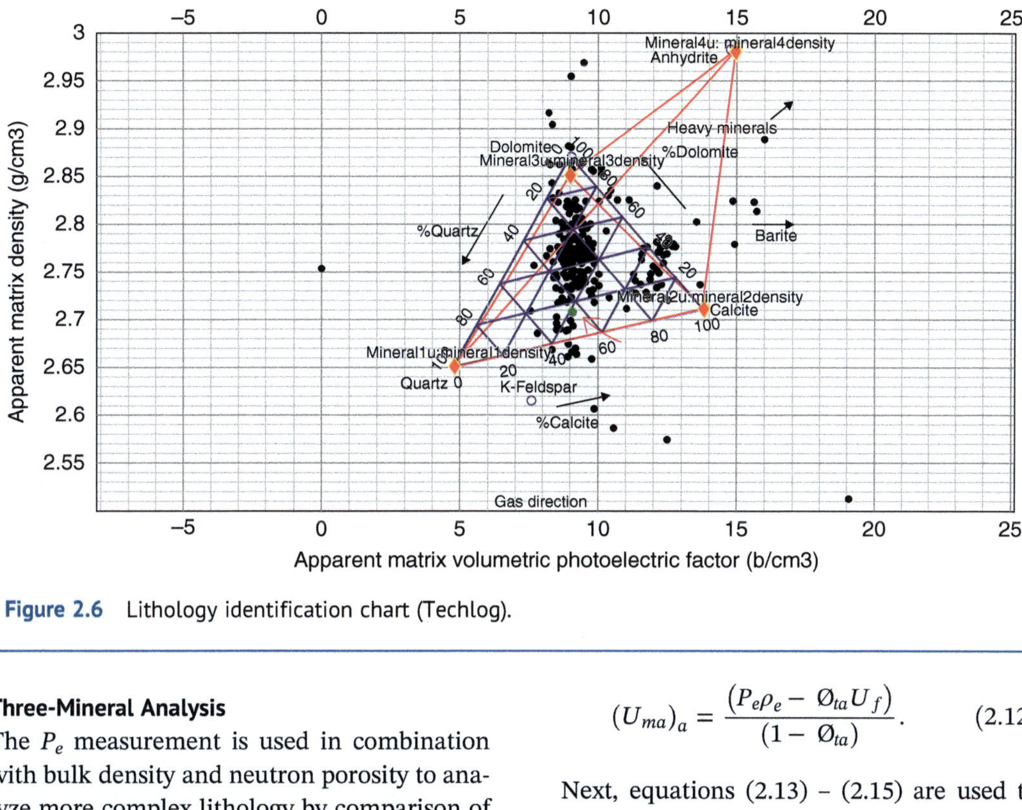

Figure 2.6 Lithology identification chart (Techlog).

Three-Mineral Analysis

The P_e measurement is used in combination with bulk density and neutron porosity to analyze more complex lithology by comparison of apparent matrix grain density and apparent volumetric cross-section. The relative proportions of three lithology types is read from the point lies between lines. Determining the relative proportion of three minerals in a matrix requires defining the apparent matrix volumetric cross-section $(U_{ma})_a$ and applying the matrix identification plot. $(U_{ma})_a$ is obtained from equation (2.12).

$$(U_{ma})_a = \frac{(P_e\rho_e - \emptyset_{ta} U_f)}{(1 - \emptyset_{ta})}. \tag{2.12}$$

Next, equations (2.13) – (2.15) are used to obtain the matrix composition made of any three minerals (Gardner & Dumanoir, 1980).

$$(\rho_{ma})_a = V_1\rho_1 + V_2\rho_2 + V_3\rho_3, \tag{2.13}$$

$$(U_{ma})_a = V_1 U_1 + V_2 U_2 + V_3 U_3, \tag{2.14}$$

and

$$1 = V_1 + V_2 + V_3, \tag{2.15}$$

where $(U_{ma})_a$ is the apparent matrix volumetric cross-section, $V_{1,\ 2,\ 3}$ is the proportion of minerals 1, 2, and 3 in the formation matrix, $\rho_{1,\ 2,\ 3}$ is the density of minerals 1, 2, and 3 (gr/cm3), and $U_{1,\ 2,\ 3}$ is the volumetric cross-section of minerals 1, 2, and 3.

Example 2.4 Determine the Lithology of the Formation

The log readings of an unconventional play (Chattanooga shale) are shown in Table 2.7, and the log readings for each mineral are shown in Table 2.8. Determine the lithology of the formation.

Table 2.7 Log Readings for Example 2.4

GR	RHOZ	PEFZ	NPHI
254.8	2.580	8.6	19.3
224.1	2.570	7.7	20.7
205.8	2.597	7.6	20.6
201.3	2.602	7.7	23.0
199.5	2.590	7.9	26.0
205.3	2.591	7.9	25.9
206.8	2.601	7.9	27.0
213.8	2.577	7.9	26.6
223.8	2.565	7.9	27.1
216.8	2.579	8.0	26.5
200.3	2.602	7.9	22.0

Source: Data are provided by John Doveton and are being used by his permission.

Table 2.8 Log Readings of Minerals for Example 2.4

	Clay	Quartz	Pyrite	Kerogen	PHIeff
GR	216	32	32	1755	32
RHOB	2.8	2.65	5	1.3	1
Uphoto	11.2	4.78	82.2	0.312	0.5
NPHI	43	−2.1	−1.9	60	100
UNITY	1	1	1	1	1

Solution

The first step is to find the inverse matrix of the log readings of the minerals. Hence, the inverse of Table 2.8 is Table 2.9.

The next step is to calculate the lithology of each mineral. The results are summarized in Figure 2.7 and Table 2.10. A sample calculation for the Clay is shown below:

$$\begin{aligned} Clay &= (0.0003\times 254.8)+(1.5216\times 2.58)+(-0.0462\times 8.6)+(0.0227\times 19.3)\\ &\quad +(-3.7718)=0.2574 \end{aligned}$$

Table 2.9 Inverse Matrix of Log Readings of Minerals

	GR	RHOZ	PEFZ	NPHI	UNITY
Clay	0.0003	1.5216	−0.0462	0.0227	−3.7718
Quartz	−0.0003	−0.6189	0.0059	−0.0195	2.5818
Pyrite	0.0000	−0.1672	0.0180	−0.0019	0.3535
Kerogen	0.0006	−0.1625	0.0049	−0.0024	0.3842
PHIeff	−0.0004	−0.5730	0.0174	0.0013	1.4523

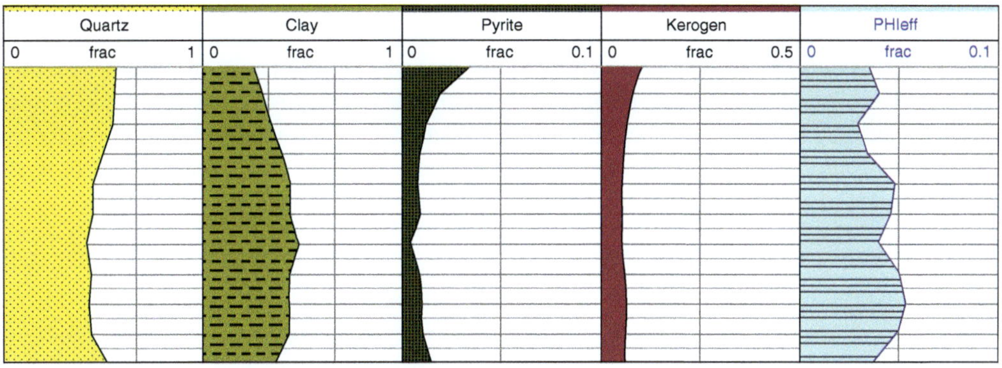

Figure 2.7 Display of lithology for Example 2.4 (Techlog).

Table 2.10 Lithology of Formation for Example 2.4

Clay	Quartz	Pyrite	Kerogen	PHIeff
0.2574	0.5707	0.0362	0.1018	0.0339
0.3075	0.5545	0.0194	0.0786	0.0399
0.3490	0.5461	0.0122	0.0636	0.0291
0.4040	0.4979	0.0092	0.0551	0.0338
0.4466	0.4475	0.0083	0.0495	0.0481
0.4430	0.4483	0.0094	0.0533	0.0460
0.4866	0.4199	0.0044	0.0495	0.0396
0.4433	0.4392	0.0091	0.0582	0.0502
0.4384	0.4333	0.0105	0.0645	0.0533
0.4408	0.4394	0.0106	0.0602	0.0490
0.3719	0.5190	0.0148	0.0579	0.0363

2.3.3 Lithology Measurements Using the Diffuse Reflectance Infrared Fourier Transform Spectroscopy Techniques

The diffuse reflectance infrared Fourier transform spectroscopy (DRIFTS) is used to analyze the mineralogy, total organic carbon (TOC), organic thermal maturity, and matrix density. A large number of cuttings from a vertical pilot well and a horizontal well in the Permian basin were analyzed using this technique (Loan et al., 2017). The samples were cleaned and after grinding to a fine powder, they were transferred to a sample cup. The prepared samples were studied using a vertically mounted infrared (IR) spectrometer. A Bruker Alpha-R spectrometer fitted with a reflection front-end module was used.

The intensity of the absorption spectrum, in Kubelka Munk units, formats the spectrum such that absorption is proportional to concentration (Kubelka & Munk, 1931). Each mineral has a unique set of absorption peaks and peak shapes. Mineral concentrations are quantified in an unknown sample by a weighted least-squares regression of the unknown measured spectrum against a library of known mineral-standard spectra. It was observed that the major changes in lithology with depth can be captured using DRIFTS technique. Comparing the results of the DRIFTS cuttings analysis with the high-definition spectroscopy log shows a good agreement between DRIFTS and log results throughout most of the sampled interval, even though the cuttings provide lower vertical resolution (Loan et al., 2017). Furthermore, this method in addition to mineralogy and organic matter concentration, can simultaneously provide an estimate of organic thermal maturity.

2.4 Porosity

The particles making up the inorganic parts of shale rocks usually never fit together perfectly due to a high degree of irregularity in their shapes. There is always void space between the grains, which is filled by the reservoir fluids including water, oil, or gas. The ratio of this void space, which is the volume not occupied by solid materials, to the total bulk volume of the rock is called *porosity*, which is shown in equation (21.6). Hence, porosity can be defined as the fraction of the reservoir bulk volume that is not occupied by the solid materials forming the reservoir framework.

$$\phi = \frac{V_p}{V_b} = \frac{V_b - V_{gr}}{V_b}, \tag{2.16}$$

where ϕ is porosity (fraction), V_b is the bulk volume of the reservoir rock (cm^3), V_p is the pore volume or the volume not occupied by rigid rock framework (cm^3), V_{gr} is the grain volume or the volume occupied by rigid rock framework (cm^3).

The combination of extremely small interparticle and intraparticle pores, both organic and inorganic, forms the pore system in the unconventional reservoirs. The effective porosity is the combination of both organic and inorganic porosity elements. The natural fractures present in the unconventional source rocks are also considered as a part of their complex porosity (Loucks et al., 2012). The pore-size distribution of the shale reservoirs is close to viruses and proteins. A combination of focused ion argon mill beam (FIB) sample preparation technique and advanced SEM technology (FIB-SEM), with the accuracy of 1-2 nm, is used for imaging the nanoscale porosity of unconventional reservoirs (Bratovich & Walles, 2016). Based on Holmes definition (Holmes et al., 2017) about unconventional reservoirs, the porosity components are described in four categories as follows:

1) Standard effective porosity (inorganic) including clay-free porosity in nonshale fractions
2) Effective organic porosity ($\varphi_{organic}$) contains formation fluids and is normally less than 5% of the total volume of the rock. The organic porosity is associated with TOC and degree of maturity of the organic matter.
3) Clay mineral porosity (φ_{clay})
4) Total organic carbon (TOC), which includes kerogen, bitumen, and adsorbed hydrocarbons.

In another definition, Loucks et al. (2010) divided the mudrock nonfracture hydrocarbon storage porosity into three categories: kerogen pores, interparticle pores, and intraparticle pores.

Figure 2.8 represents the following different geometries, wettability, roughness, and interconnection properties (Bratovich & Walles, 2016): (1) interparticle pores, Upper Cretaceous Eagle Ford shale; (2) combination of organic matter and interparticle pores, Mississippian Barnett shale; (3) organic matter pores, Mississippian Barnett shale; (4) combination of organic matter and intraparticle pores, Pearsall formation; (5) intraparticle pores, Upper Cretaceous Austin Chalk; (6) combination of interparticle and intraparticle pores, Upper Jurassic Haynesville formation; (7) combination of organic matter, interparticle, and intraparticle pores, Lower Cretaceous (lower Bexar shale member) Pearsall formation.

Furthermore, Figure 2.9 by Glorioso and Rattis (2012) show the fracture flow paths and an organic porosity example. In Figure 2.9 left image (Passey et al., 2010), dark red bubbles represent free gasses in the organic pores and

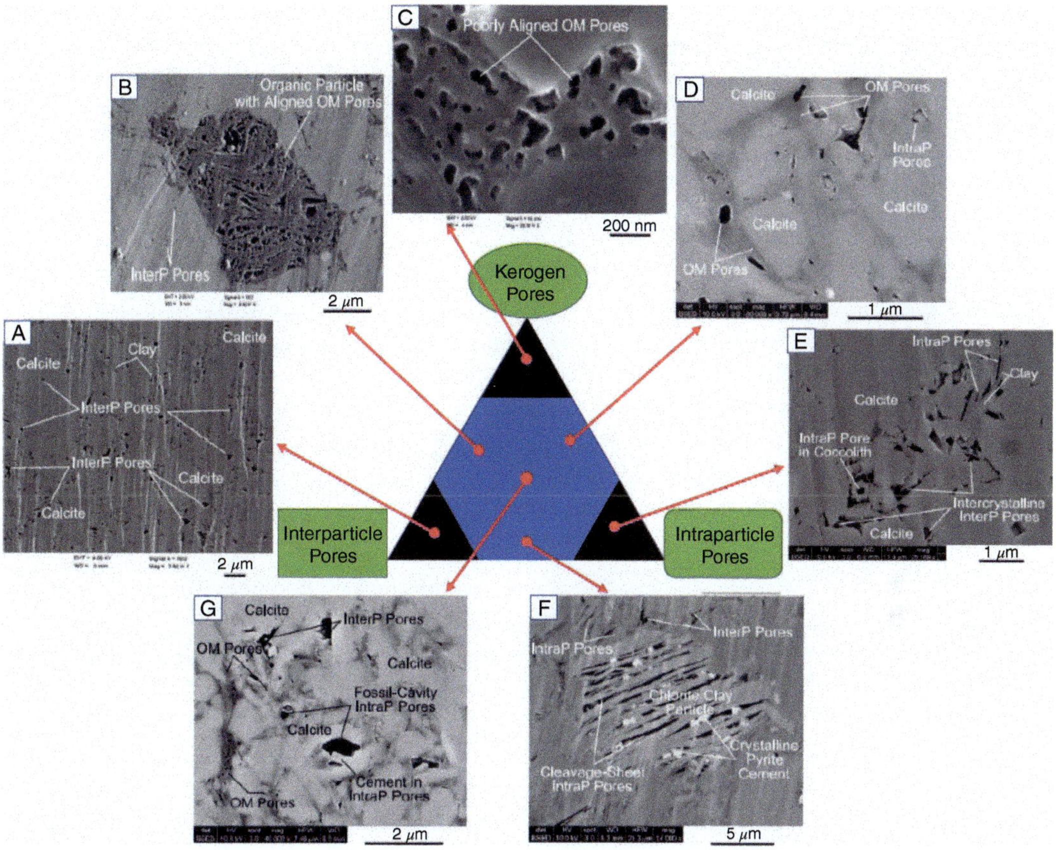

Figure 2.8 Division of nonfracture HC storage porosity systems (Bratovich & Walles, 2016, and Loucks et al., 2012, republished by permission of the Gulf Coast Association of Geological Societies, whose permission is required for further publication use).

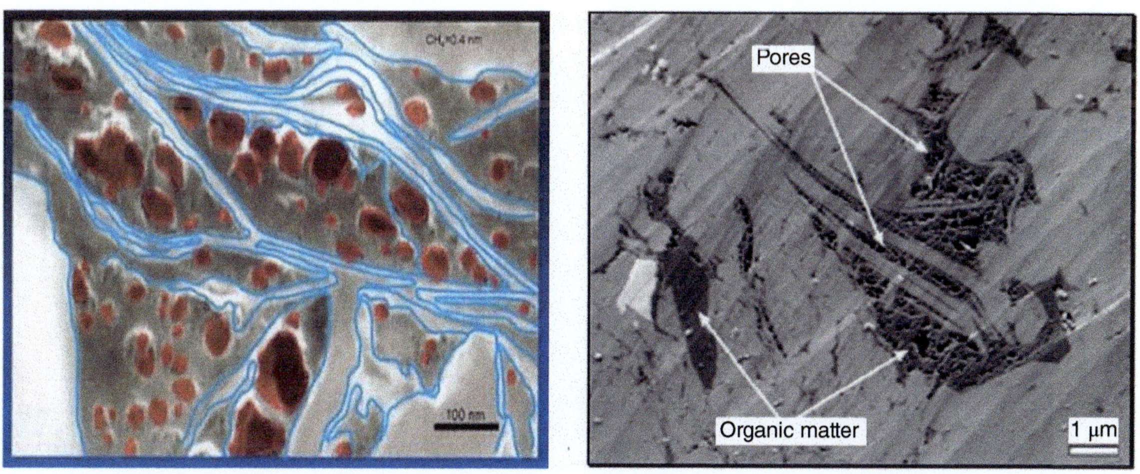

Figure 2.9 SEM images show (left) the fracture flow paths and (right) the organic matters porosity.

the microfractures are shown in blue lines. In Figure 2.9 right image (Loucks et al., 2010), the maturation of kerogen results in creating secondary porosity.

2.4.1 Porosity Measurement

Despite the commercial importance of unconventional plays in the industry, accurate porosity measurement of shale samples is still challenging. This complexity is due to the extremely low permeability and complex pore structure of the shale plays, which lead to a significant uncertainty in the economic assessment of these reserves (Kuila et al., 2014). The different techniques that are being used in the industry to calculate shale porosity are reported in the following sections.

Porosity Evaluation Using Well Logs

Refer to Table 2.11 for a porosity measurements comparison of the Bakken play based on different log evaluation methods.

Density Logs

Density is defined as the mass of a substance per unit volume. The common units of density are g/cm^3 or kg/m^3. The densities of rock and minerals are typically expressed as specific gravity. Specific gravity is defined as the ratio of rock density to the density of water. Considering that the density of water is 1 g/cm^3, therefore, the numbers of specific gravity are simply translated to the density of the rock in g/cm^3 (Alden, 2017). The minerals that encompass a particular rock type, have a significant effect on the rock density. The porous media are made of matrix, which is the rock and fluids such as oil, gas, and water. The bulk density is calculated using equation (2.17),

$$\rho_b = \O\rho_f + (1 - \O)\rho_{ma}, \tag{2.17}$$

where ρ_b is the formation bulk density (g/cm^3), ϕ is porosity, fraction/0 ρ_{ma} is the matrix density (g/cm^3), and ρ_f is the density of the fluid saturating the rock immediately surrounding the borehole-usually mud filtrate (g/cm^3).

There are different methods from which the porosity can be calculated using the density logs. The electron density of the formation is measured and used to analyze the density of the formation. In this method, the gamma rays emitted from a specific tool collide with the formation and lose their energy after collision with the electrons of the formation. The returning gamma rays, which are an indicator of formation bulk density, are counted by a detector

Table 2.11 Comparison of Different Porosity Evaluation Methods

Neutron-density method	Neutron-sonic method	Sonic method	NMR method	DMR
0.087	0.098	0.078	0.065	0.065
0.082	0.103	0.089	0.055	0.055
0.095	0.109	0.089	0.043	0.043
0.088	0.099	0.078	0.0478	0.0478
0.053	0.065	0.049	0.0536	0.0536
0.053	0.054	0.036	0.0473	0.0473
0.081	0.084	0.062	0.0435	0.0435
0.088	0.104	0.085	0.0654	0.0654
0.093	0.109	0.089	0.0758	0.0758
0.087	0.102	0.084	0.0698	0.0698
0.084	0.102	0.085	0.071	0.071

located at a fixed distance from the tool source (Crain, 2006). The low-energy gamma rays are absorbed by atoms and the number of gamma rays capable of photoelectric absorption indicates the atomic number and are used to obtain the photoelectric effect. This information is used to differentiate the lithology of the formation. Porosity, matrix density, and fluid contents characterize the bulk density of a formation. Formation bulk density measured by the log must be corrected for borehole irregularities. The total effective porosity of the formation for a clay-free rock is calculated as a function of measured densities by the use of equation (2.18) (Zhang et al., 2015).

$$\varnothing = \frac{\rho_{ma} - \rho_b + V_{TOC} \times (\rho_{TOC} - \rho_{ma})}{\rho_{ma} - \rho_f}, \quad (2.18)$$

where ϕ is porosity (fraction), ρ_{ma} is the matrix density (g/cm^3), ρ_b is the formation bulk density (log value) (g/cm^3), V_{TOC} is the total volume of organic matter, ρ_{TOC} is the organic matter density, ρ_f is the density of the fluid saturating the rock immediately surrounding the borehole, typically the density of the mud filtrate (g/cm^3).

In a different approach introduced by Bust et al. (2011) after defining the different parts of the medium similar to Figure 2.10, the total porosity of the shale gas formation can be calculated using equation (2.19).

$$\varnothing_T = \frac{\rho_b - \rho_m - w_k \rho_b \left(1 - \frac{\rho_m}{\rho_k}\right)}{\left(\rho_g - \rho_m\right) + \left(\rho_w - \rho_g\right) S_{wT}}, \quad (2.19)$$

where ϕ_T is defined as equation (2.20),

$$\varnothing_T = \varnothing_m + \varnothing_k, \quad (2.20)$$

where ρ is the density g/cm^3; subscripts, b, m, k, g, and w correspond to bulk, matrix, kerogen, gas, and water; S_{wT} is the water saturation relative to total porosity, fraction; and w_k is the weight fraction of kerogen, and it is defined using equation (2.21),

$$w_k = \frac{TOC}{c_k}, \quad (2.21)$$

where TOC is the total organic carbon and c_k is the weight fraction of carbon in kerogen. This parameter depends on thermal maturity condition of the reservoir and the kerogen source (Luffel et al.,1992).

Freedman et al. (2018) conducted a new technique to evaluate shale reservoirs. They used the density log porosities, NMR total porosities, TOC dry weight fraction, and dry weight matrix densities as the inputs of their developed method. The total dry weight TOC (Radtke et al., 2012) and dry weight matrix densities (Freedman et al., 2015) are obtained from the gamma ray spectroscopy tool. Freedman et al. (2015) developed different schematics to

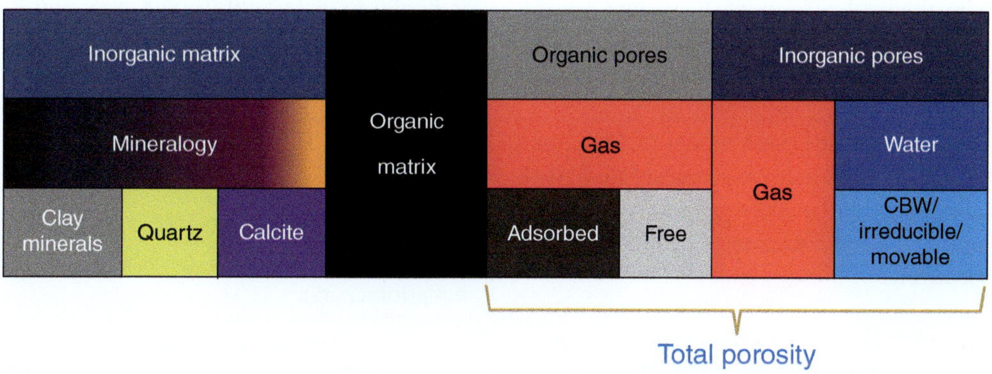

Figure 2.10 Petrophysical model developed for shale gas resources (Bust et al., 2013).

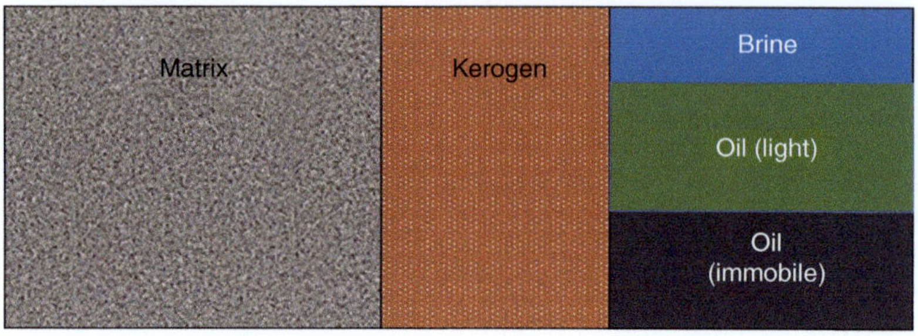

Figure 2.11 Schematic of an organic shale tight oil reservoir with nanopores in the kerogen phase being ignored (after Freedman et al., 2018).

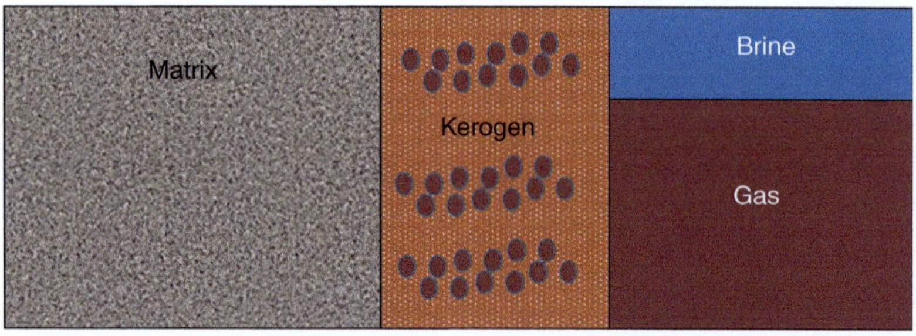

Figure 2.12 Schematic of a mature shale gas reservoir (after Freedman et al., 2018).

depict organic shale tight oil and mature shale gas reservoir, which are shown in Figures 2.11 and 2.12, respectively. Figure 2.11 shows that brine, high API gravity producible oil, and bitumen and/or immobile oil occupy the total porosity of the system, while Figure 2.12 shows that the mature shale gas model does not contain bitumen and/or immobile oil. In this model, the total porosity of the system is occupied by gas and brine and the organic matter porosity is shown by solid red dots (Loucks et al., 2012).

By using the proposed model of Freedman et al. (2018), the shale bulk density and NMR total porosity of the shale gas reservoir can be calculated using equations (2.22) and (2.23), respectively.

$$\rho_b = (1 - \emptyset - V_k)\rho_{ma} + \rho_k V_k + \rho_f(\emptyset - \emptyset_g) + \rho_g \emptyset_g, \quad (2.22)$$

and

$$\emptyset_{nmr} = \emptyset_g HI_g + \rho_k V_k + (\emptyset - \emptyset_g) HI_f, \quad (2.23)$$

where ρ is the density; $\emptyset_g$ is the gas porosity (fraction); $\emptyset_{nmr}$ is the NMR total porosity (fraction); HI_g and HI_f are the gas and fluid hydrogen index, respectively; V_k is the fractional volume of all solid organic matter; subscripts b, g, and f correspond to bulk, gas, and fluid; and $\emptyset$ is the total porosity (fraction), which is obtained from equation (2.24).

$$\emptyset = \emptyset_w + \emptyset_{im} + \emptyset_g, \quad (2.24)$$

where $\emptyset_w$ is the fractional water volume and $\emptyset_{im}$ is the fractional volume of all immobile fluids.

The gas porosity, $\emptyset_g$, is obtained by using equation (2.25),

$$\emptyset_g = \frac{(\emptyset_D - \lambda_2 V_k) - \frac{\emptyset_{nmr}}{HI_f}}{\left(1 - \frac{HI_g}{HI_f}\right) + \lambda}, \quad (2.25)$$

where $\emptyset_D$ is the density porosity. It is obtained using

$$\emptyset_D = \frac{\rho_{ma} - \rho_b}{\rho_{ma} - \rho_f}. \quad (2.26)$$

The λ and λ_2 parameters provide a light hydrocarbon correction for gas and a kerogen correction to $\emptyset_D$ and are defined as equations (2.27) and (2.28), respectively.

$$\lambda = \frac{\rho_f - \rho_g}{\rho_{ma} - \rho_f} \quad (2.27)$$

and

$$\lambda_2 = \frac{\rho_{ma} - \rho_k}{\rho_{ma} - \rho_f}\ . \quad (2.28)$$

Similarly, the shale bulk density and NMR total porosity of the shale tight shale oil reservoir can be calculated using equations (2.29) and (2.30), respectively.

$$\rho_b = (1 - \emptyset - V_k)\rho_{ma} + \rho_k V_k + \rho_f(\emptyset - \emptyset_o) + \rho_o \emptyset_o \quad (2.29)$$

and

$$\emptyset_{nmr} = \emptyset_o HI_o + \rho_k V_k + (\emptyset - \emptyset_o) HI_f, \quad (2.30)$$

where $\emptyset_o$is the porosity (fraction); HI_o is the oil hydrogen index; the subscript o corresponds to oil.

Estimating the fraction of the original organic carbon that can be converted to hydrocarbons in each source rock is necessary. This is because the transformation of kerogen with increasing thermal maturity and resulting kerogen porosity development will proceed until the available hydrogen is exhausted (Daly & Edman, 1987).

The porosity development within kerogen is defined as a function of different variables using a mass-balance relation (Daly & Edman, 1987; Modica, 2012) and is shown in equation (2.31).

$$\emptyset_{Kerogen} = ([TOC_0 \times C_c]k) T_R \times \left(\frac{\rho_b}{\rho_k}\right), \quad (2.31)$$

where $[TOC_0 \times C_c]$ represents the labile carbon mass, and C_c is the labile organic carbon fraction of the rock and is calculated using equation (2.32).

$$C_c = \left(\frac{1}{T_R}\right) * \left(1 - \frac{TOC}{TOC_0}\right), \quad (2.32)$$

where k term is a dimensionless scale factor used to estimate a kerogen mass from a labile carbon mass, ρ_b is the average bulk rock density, which can be obtained from density logs, ρ_k is an average kerogen density and T_R is the transformation ratio and is obtained from equation (2.33) (Bordenave et al., 1993).

$$T_R = \frac{S2_o - S2}{S2_o}, \quad (2.33)$$

where S_2 is the average measured partially spent kerogen yield from the more mature well, which can be obtained from Rock-Eval pyrolysis and S_{2o} is the original S_2 estimated from both original TOC (TOC_0) and original hydrogen index (HI_0). The density porosity should be corrected for shale and kerogen as it is shown in equation (2.34).

$$(\emptyset_D)_c = \emptyset_D - \left[V_{ker} \times (\emptyset_D)_{ker}\right]\ , \quad (2.34)$$

where $\emptyset_D$ is the density porosity log, $(\emptyset_D)_{sh}$ is the density porosity reading within the shale formation, V_{ker} is the kerogen volume, and $(\emptyset_D)_{\mathrm{ker}}$ is the density porosity of kerogen, defaulted to 0.65.

Work done by Mendelson (1985) for a system composed of just rock matrix and organic matter increasing the organic matter volume by 10% in the rock content, results in 0.17 g/cc decrease in true bulk density of limestone. Calculating TOC in terms of weight percentage results in equation (2.35).

$$TOC = a\left[\frac{1}{1 - \frac{\rho_k\left(\frac{\rho_m \rho_b}{\rho_k} + \rho_m V_m - \rho_w V_m - \rho_m\right)}{\rho_b + \rho_m V_m - \rho_w V_m - \rho_m}}\right], \quad (2.35)$$

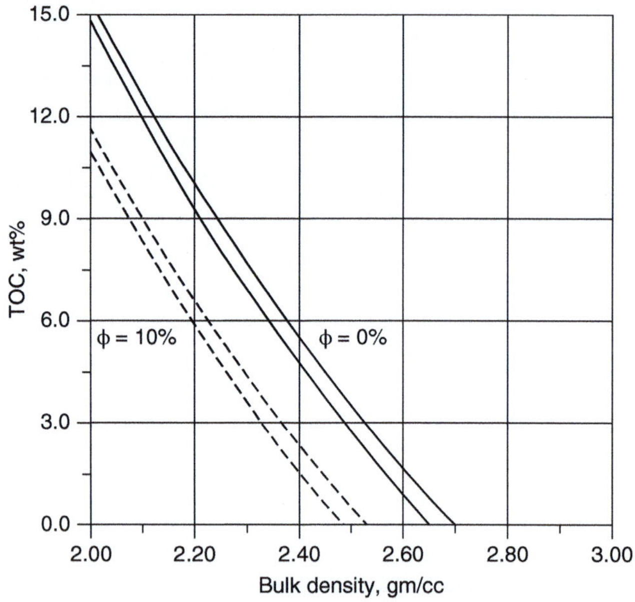

Figure 2.13 Bulk density versus TOC for two different porosities (Mendelson, 1985).

where *TOC* is total organic carbon, ρ is density, g/cm^3, V is volume fraction, a is a constant that relates weight percentage of TOC to weight percentage of kerogen with a common value of 0.7–0.8. Subscripts k, m, b, and w correspond to kerogen, matrix, bulk, and water.

Considering equation (2.35), adding 10 weight percent of TOC results in 0.50 g/cm^3 reduction in density. Figure 2.13 shows TOC vs. bulk density content for two different porosities. It can be observed that filling some of the pores by liquid shifts these curves to the left (lower densities).

Neutron Logs

The neutron log measures the hydrogen content of the formation. The energy loss of the emitted neutrons corresponds to the hydrogen content of the formation (Bassiouni, 1994; Hartman & Beaumont, 2000). Since hydrogen atoms are concentrated in the kerogen and fluid-filled pores, the formation porosity is obtained using the relationship between the energy loss of neutron particles and the hydrogen content of the formation. Reservoirs with gas-filled pores may show a lower porosity than the same pores filled with oil or water since gas has a lower concentration of hydrogen atoms than either oil or water (Crain, 2006). Due to this fact, the neutron log interpretation is related to the hydrogen content of the formation and more generally to the hydrogen index. Mendelson (1985) computed the porosity of the two-component model of Monterey formation using the neutron log by the use of equation (2.36).

$$\varnothing_N = HI_w V_w + 0.67 V_{ker}, \tag{2.36}$$

where $\varnothing_N$ is the neutron porosity (fraction);, HI_w is the hydrogen index of the water content of the formation, ignoring the water presented in clay minerals; V_w is the fractional volume of water; and V_{ker} is the fractional volume of kerogen.

In the presence of shale and kerogen, the neutron porosity is corrected for kerogen using equation (2.37) before it is corrected for the clay content.

$$(\varnothing_N)_c = \varnothing_N - \left[V_{ker} \times (\varnothing_N)_{ker}\right], \tag{2.37}$$

where $\varnothing_{Nker} = 0.45 - 0.75$ and the default is $\varnothing_{Nker} = 0.65$.

Acoustic Logs

The compressional wave velocity changes in different continuums, and this property is used in acoustic logging tools to evaluate the formation porosity. The acoustic energy takes longer to travel between the transmitter and the receiver in the fluid-filled pores (Tixier et al., 1959). The travel time of the acoustic waves is a function of the solid part or framework of the rock, rock lining of the pores, and the fluid-filled pore space. The Wyllie time-averaging equation (Wyllie et al., 1956) is an empirical relationship used to calculate the porosity of the formation. In terms of velocity, it is defined as equation (2.38), and in terms of transit time (Δt), it is defined as equation (2.39).

$$\frac{1}{v} = \frac{\phi}{v_f} + \frac{(1-\phi)}{v_{ma}}, \tag{2.38}$$

where ϕ is the fractional porosity of the rock, v is the velocity of the acoustic waves in the formation (ft/sec), v_f is the velocity of the acoustic waves in the fluid (ft/sec), and v_{ma} is the velocity of the acoustic waves in the rock matrix (ft/sec).

$$\phi_s = \frac{\Delta t_{log} - \Delta t_{ma}}{\Delta t_f - \Delta t_{ma}}, \tag{2.39}$$

where Δt is the acoustic transit time in the formation (μsec/ft), Δt_f is the acoustic transit time in the fluid (μsec/ft), and Δt_{ma} is the acoustic transit time in the rock matrix (μsec/ft).

The default value of 189 μsec/ft is used as fluid velocity in fresh drilling fluids and a slightly lower value, 185 μsec/ft, is used for salt muds (Carmichael, 1982). Ideally, one must iterate between saturation calculations in the invaded zone and porosity calculations to converge to the right porosity and saturation values.

The same analysis as the density log approach is used to analyze the acoustic logs. The kerogen effect on travel time depends on the shape of the inclusion. The stronger effect belongs to the more platelet forms (Toksoz & Cheng, 1978).

Because shale transit time varies greatly, failure to correct for the presence of shale may result in overly optimistic porosity calculations (Fertl, 1981). The sonic porosity must be corrected for kerogen using equation (2.40) before it is corrected for its clay content.

$$\phi_{Sc} = \phi_S - \left[V_{ker} \times (\emptyset_s)_{ker}\right], \tag{2.40}$$

where V_{ker} is the fractional volume of kerogen, $\phi_{Sker} = 105 - 160$ and the default value of kerogen porosity is $\phi_{Sker} = 125$.

Porosity Evaluation Using Combination of Different Logs and Cross-Plots

In a combination condition, a combination of two or three porosity measurement techniques is applied to estimate the formation porosity and lithology. In two-element cross-plots, one measurement is displayed along x-axis and another one is showed along y-axis (Sen et al., 2015). The same methodology as for lithology evaluation is used to interpret them for porosity measurements using cross-plots. To interpret them, the intersection of those values in cross-plot determines both the porosity and lithology of that point. If a point lies between two lithology lines, its composition is a mixture of both lithologies.

Neutron-Density Logs

Porosity calculated from neutron and density measurements is the most commonly used estimate of pore volume in rock formations (Ijasan et al., 2013). The combination of density and neutron logs is typically used simultaneously to measure the porosity effectively and help with lithology estimation of the rock (Hartman & Beaumont, 2000).

The difference in the porosity recorded by two logs can be due to (1) the error in the matrix density reading, which is used to calculate the porosity, (2) the presence of gas in the formation pore space; (3) the presence of shale/clay/organic matter in the formation.

To obtain correct porosities from density-neutron logs, the following approaches should be considered (Hartman & Beaumont, 2000):

(1) where density and neutron logs overlapped, both logs are presenting the true porosity; (2) if shale is present in the content of the formation, the density and neutron logs do not coincide.; in this condition, the density porosity is less than the neutron porosity, and the density log provides a reasonable approximation of formation porosity; (3) in the presence of gas in the formation, the density and neutron logs do not coincide. In this condition, the density porosity is greater than neutron porosity and the density porosity is recomputed using equation (2.41):

$$\phi = \frac{\rho_{ma} - \rho_b}{\rho_{ma} - \rho_g}; \tag{2.41}$$

(4) cross-plots are used to correct the porosity; and (5) in the absence of cross-plots, a combination of neutron and density porosities is used to obtain the formation porosity using equation (2.42) (Hartman & Beaumont, 2000; Gaymard & Poupon, 1968):

$$\phi = \left(\frac{\phi_N^2 + \phi_D^2}{2}\right)^{\left(\frac{1}{2}\right)}. \tag{2.42}$$

Porosity Evaluation by Helium Expansion and Helium Injection

This method is a laboratory method for accurate measurement of porosity for tight and ultratight rock. In this method, porosity is measured by helium expansion using the Gas Research Institute (GRI) method and helium injection under confining stress. In the helium porosimetry technique, porosity is measured on crushed rock samples. In the GRI technique, the bulk rock volume is measured by mercury immersion using Archimedes's principle, and crushed rock grain volume is measured by Boyle's law (Luffel & Guidry, 1992; Luffel et al., 1992). Helium expansion can be used for measuring the connected pores (i.e., accessible by helium gas) of crushed or intact samples. Luffel and Guidry suggested 30 min as the equilibration time for shales as opposed to 1 min for conventional rocks. Jia et al. (2017) used the pulse-decay setup shown in Figure 2.14 to measure the apparent porosity of shale samples. In this setup, the upstream and downstream reservoirs are installed before and after the core holder, respectively.

Based on the setup of Jia et al. (2017), equations (2.43) and (2.44) are used to calculate

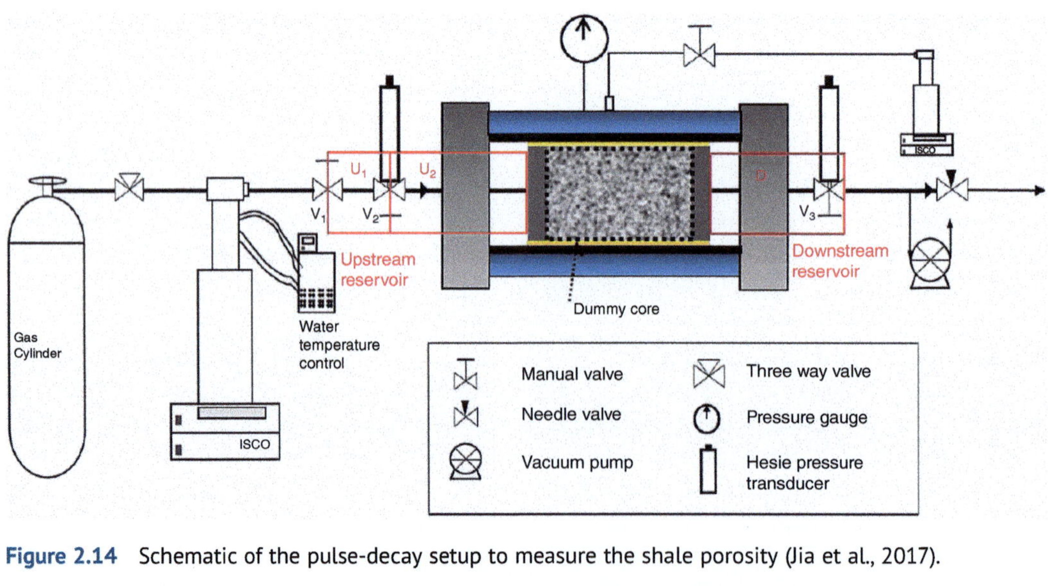

Figure 2.14 Schematic of the pulse-decay setup to measure the shale porosity (Jia et al., 2017).

the gas apparent porosity based on the pulse-decay pressure and volume data.

$$\frac{P_u V_{u1}}{Z_u} + \frac{P_d(V_{u2} + V_d + V_a)}{Z_a} = \frac{P_e(V_{u1} + V_{u2} + V_d + V_a)}{Z_e} \quad (2.43)$$

and

$$\emptyset_a = \frac{V_a}{V_b}, \quad (2.44)$$

where P_u, P_d, and P_e are initial upstream pressure, initial downstream pressure, and equilibrium pressure, respectively; V_{u1}, V_{u2}, and V_d are volumes of U_1, U_2, and D, respectively; U_1 is the part between valves V_1 and V_2V_2; and U_2 is the part between V_2 and the core inlet; Z_u, Z_d, and Z_e, are compressibility factors under pressures of P_u, P_d, and P_e, respectively; V_a is the estimated gas apparent pore volume; V_b is the core bulk volume; $\emptyset_a$ is the apparent porosity, which is defined as the ratio of gas apparent pore volume over core bulk volume. The main disadvantage of this method is that it will take into account only the pore volume connected to the pores that are at the entrance of the core. Tight shale samples are known to have different patches of connected pore network within the size of a core plug and some of those may not be taken into account during the typical pulse-decay measurements.

Anovitz and Cole (2015) applied the helium expansion technique (i.e., He-porosimetry) accompanying Boyle's law to calculate the porosity of low permeable rocks. In this method, the dry sample, which is either a core or crushed rocks, is placed in a container of known volume (V_a), which is connected to another container of known volume V_b. The bulk volume of the sample is known and referred to as V_{bulk}. The helium gas, which was stored in a reference cell, is introduced into the first container of volume and arbitrary pressure of V_a and P_1, respectively. The arbitrary pressure is typically around 100 psi. Thereafter, He is released into container V_b and the pressure is equilibrated between two chambers. Thereafter, the gas penetrates the pores and decreases to a new equilibrium pressure called P_2. Considering the ideal gas law, the pore volume of the sample is obtained by using equation (2.45) and the porosity of the sample is calculated using equation (2.46).

$$V_v = V_{bulk} - V_a - V_b\left(\frac{P_2}{P_2 - P_1}\right) \quad (2.45)$$

and

$$\phi = \frac{V_p}{V_b} = \frac{V_b - V_{gr}}{V_b}. \quad (2.46)$$

Porosity Evaluation Using Nitrogen Adsorption

Originally, this technique was used to measure total specific surface areas in conventional reservoirs. Kuila and Prasad (2011) conducted experiments to analyze the porosity of shale samples using the nitrogen adsorption technique. They used a volumetric method for measuring the amount of the adsorbed N_2 on the sample surface using the Micromeritics setup. In this technique, samples are crushed, sieved, and then degassed under vacuum. Thereafter, nitrogen is injected into the analysis chamber and N_2 is adsorbed on the crushed samples. The vacuum process is performed in high temperature to ensure the removal of free, capillary, and bound water adsorbed with the clays. In this condition, analyzing the thermogravimetric analysis (TGA) and derivative thermogravimetric (DTG) plots, which correspond to weight loss percentage and derivative weight loss percentage, provides valuable information about the water removal of the samples. Thereafter, the samples are exposed to nitrogen at a specific liquid nitrogen temperature and controlled pressures. Plotting the amount of adsorbed N_2 vs. P/P_0, which is known as t-plot, provides information about the pore types. The volume adsorbed at low P/P_0 ratios ($P/P_0 < 0.01$) is an indicator of micropores with the width smaller than 2 nm while the volume adsorbed at $P/P_0 = 1.0$ shows the total porosity of the samples. Therefore, the total pore volume of the sample is obtained by finding the N_2 adsorbed at $P/P_0 = 1.0$ and knowing the density of N_2 in its liquid state. Assuming that all the pores are filled with the liquid adsorbate, then the total pore volume is calculated by measuring the amount of adsorbed vapor N_2 on the sample at a relative pressure close to unity ($P/P_0 = 1.0$) using equation (2.47).

$$V_{ads} = \frac{V_{liq}RT}{P_a V_m}, \tag{2.47}$$

where V_{liq} is the volume of liquid N_2 in pores, cm^3; V_{ads} is the volume of gas adsorbed, cm^3; V_m is the molar volume of liquid adsorbate, cm^3/mol ($N_2 = 34.7\ cm^3/mol$); P_a is ambient pressure, atm; T is ambient temperature (K); and R is the gas constant, 82.057 $cm^3atm/(K\ mol)$.

Finally, the porosity of the samples is found by measuring the bulk volume of the samples and using equation (2.48):

$$\phi = \frac{V_p}{V_b} = \frac{V_b - V_{gr}}{V_b}. \tag{2.48}$$

The pore connection type is characterized by the size and shape of the isotherm and its hysteresis pattern in low-pressure gas adsorption technique. There are close relationship between the shape of the adsorption isotherms and the adsorbate and solid adsorbent properties and the geometry of the pore space. Hence, plotting the amount of the adsorbed gas vs. the relative pressure provides information about the size of the pores (Keller & Staudt, 2005). The adsorption isotherms are divided into six categories based on IUPAC classifications (Sing et al., 1985), which is shown in Figure 2.15. Type I isotherms are characterized by a horizontal plateau and can be described by the Langmuir equation. They are characteristics for micropores without multilayer adsorption. The surface area of the microporous materials cannot be described by BET method. Type II isotherms are characteristics for mesoporous materials that can be described by the BET equation or its generalization. This type can be observed in disperse, nanoporous, or only macroporous solids as well. Type III isotherms occur under the condition in which the adsorbate-sorbent (a-s) interaction is small compared with the adsorbate-adsorbate (a-a) interaction. Type IV isotherms show the

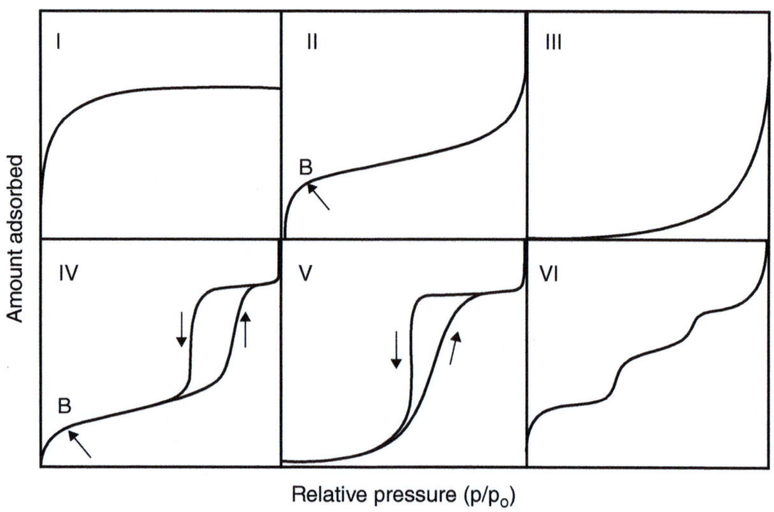

Figure 2.15 Main types of gas physiosorption isotherms (Keller & Staudt, 2005).

behavior of the adsorption for mesoporous materials that shows hysteresis behavior between the adsorption and desorption curves. Type V isotherms are similar to type IV except that the phase change may occur. The middle portion of the curve shows a near perpendicular behavior where the adsorption and desorption curves meet near the relative gas pressure of 0.5 ($P/P_0 = 0.5$). This indicates the occurrence of the phase change behavior in the mesoporous. Type VI isotherms are representative of stepwise multilayer adsorbates in which the layers become more noticeable (Keller & Staudt, 2005).

The first type, which is known as the classic Langmuir isotherm, is the most applied model in the unconventional gas reservoirs. This model assumes that there is only a single layer of molecules covering the solid surface, and it works based on the fact that there is a dynamic equilibrium between the adsorbed and the nonadsorbed gas at constant pressure and temperature (Langmuir, 1918). In this isotherm type, the model is represented by equation (2.49),

$$V(p) = \frac{V_L P}{P + P_L}, \tag{2.49}$$

where $V(p)$ is the gas volume of adsorption at pressure P; V_L is the maximum gas volume of adsorption at infinite pressure, which is called Langmuir volume; P_L corresponds to the one-half Langmuir volume, which is known as Langmuir pressure; and P is the pressure.

However, at high reservoir pressures, the natural gas forms multimolecular layers on the organic carbon surfaces due to sorption and the first type model cannot be used anymore. In this condition, the BET model, developed by Stephen Brunauer, Paul Hugh Emmett, and Edward Teller (Brunauer et al., 1983), is used as the generalized version of the Longmuir model to cover the multiple adsorbed layers. While using this model, the surface is considered homogeneous. Furthermore, it is assumed that there is no lateral interaction between molecules, and finally, the upper most layer is in equilibrium with the gas phase. This model is expressed as equation (2.50),

$$V(p) = \frac{V_m CP}{(P_0 - P)[1 + (C-1)P/P_0]}, \tag{2.50}$$

where P_0 is the saturation pressure of the gas, V_m is the maximum adsorption gas volume when the entire adsorbent surface is being covered with a complete monomolecular layer, and C is a constant that is related to the heat of adsorption. It is defined as equation (2.51),

$$C = \exp\left(\frac{E_1 - E_L}{RT}\right), \tag{2.51}$$

where E_1 is the heat of adsorption for the first layer and E_L is the heat of adsorption for the second or higher layers, which is considered to be the heat of liquefaction.

Figure 2.16, which has been done by Yu et al. (2016), shows the comparison between the Longmuir and BET isotherms. The extended Langmuir model (EML; Langmuir, 1918) is used to predict the multicomponent isotherms. This model works reasonably accurately for the binary mixtures. This model is expressed in equations (2.52) and (2.53):

$$V(p) = V_L \frac{bP}{1 + bP} \tag{2.52}$$

and

$$b = \frac{1}{P_L}. \tag{2.53}$$

Porosity Evaluation Using NMR

The principle of the NMR logs and its dependence on the hydrogen content of the formation implies that the total porosity can be calculated from NMR logs (Freedman et al., 1998). Total CMR (Schlumberger's NMR log tool) porosity is calculated using equation (2.54),

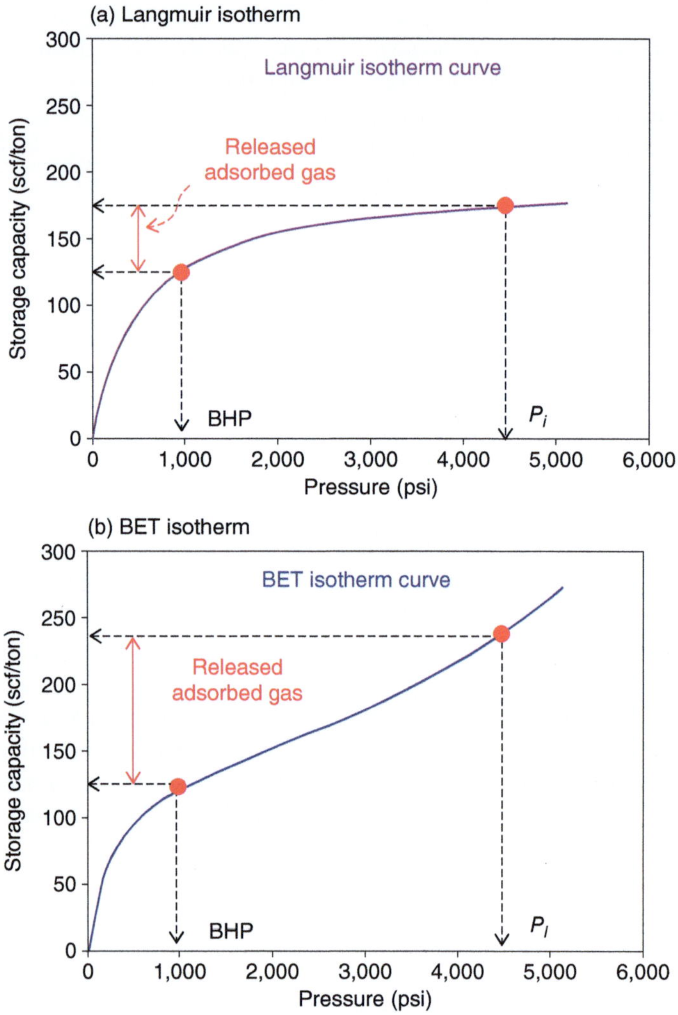

Figure 2.16 Comparison of the Longmuir and BET isotherms (Yu et al., 2016).

$$TCMR = \varphi \times \left[S_{g,xo} \times HI_g \times P_g + HI_f\left(1 - S_{g,xo}\right)\right], \quad (2.54)$$

where $TCMR$ is the total NMR porosity, fraction; φ is the total NMR formation porosity, fraction; $S_{g,\,xo}$ is the gas saturation at the flushed zone, fraction; HI_g is the hydrogen index of natural gas at reservoir conditions, fraction; HI_f is the hydrogen index of pore fluid in the flushed zone at reservoir conditions; and P_g is the gas polarization function and is obtained using:

$$P_g = 1 - exp\left[-\left(\frac{T_w}{T_{l,g}}\right)\right] \quad (2.55)$$

where T_w is the polarization time (s) and $T_{l,\,g}$ is the longitudinal relaxation time of natural gas (s).

In another approach by Freedman et al. (1998), the density-derived porosity is calculated by combining the responses of the total NMR porosity, equation (2.54), and formation bulk density, equation (2.56). The result is the DMR porosity. The DMR porosity after gas correction is presented in equation (2.57). Note that this corrected porosity is a function of saturation and lithology.

$$\rho_b = \rho_{ma}(1-\varphi) + \rho_f\varphi(1-S_{g,xo}) + \rho_g\varphi S_{g,xo}, \quad (2.56)$$

where ρ_b is the formation bulk density, gr/cm^3; ρ_{ma} is the formation matrix density, gr/cm^3; ρ_f is the fluid phase density in the flushed zone at reservoir conditions, gr/cm^3; and ρ_g is the gas density at reservoir conditions, gr/cm^3

$$\varnothing_{corr} = \frac{\varnothing_D\left(1-\frac{HI_gP_g}{HI_f}\right)+\left(\frac{\lambda\times TCMR}{HI_f}\right)}{\left(1-\frac{HI_gP_g}{HI_f}\right)+\lambda}, \quad (2.57)$$

where $\varnothing_D$ is the density porosity, fraction; and λ is calculated using equation (2.27).

An iterative solution may be required to estimate the saturations from resistivity logs and the lithology from cross-plots and then the corrected porosity is recalculated until convergence occurs.

2.4.2 NMR Core Porosity for Shales

Both low-field and high-field NMR instruments employed in the laboratory to determine petrophysical properties such as porosity, fluid saturation, wettability, as well as pore-size distribution, operate on a similar principle as the NMR logging tool outlined as follows (Ge, 2016; Coates et al., 1999):

1) Before the core sample is exposed to an external magnetic field, the hydrogen protons in the bound and movable fluids within the core sample are randomly oriented.
2) Upon exposing the sample to an external magnetic field, the protons are activated such that the NMR magnetic field polarizes the spin axes of the protons in a particular direction about a static field, B_0.
3) The polarization of the protons increases with a time constant, T_1 (longitudinal relaxation time). The longitudinal (T_1) relaxation time distribution is measured with the inversion recovery free induction decay (IRFID) (Dunn et al., 2002).
4) An oscillating magnetic field (B_1) of the NMR tool is then applied to tip the protons away from the longitudinal equilibrium position to a transverse plane; B_1 is perpendicular to B_0.
5) Turning off the oscillating field (B_1) causes the protons to dephase resulting in a decay signal detected in the transverse direction.
6) Finally, the NMR raw data made up of spin-echo trains are generated through the use of Carr-Purcell-Meiboom-Gill (CMPG) pulse sequence (Carr & Purcell, 1954) to excite the signal and inverse them to transverse relaxation time (T_2) by the Laplace nonnegative least square fitting (Lawson & Hanson, 1974).

The initial amplitude of the spin-echo can be calibrated to determine porosity since it is proportional to the number of polarized hydrogen nuclei in the pore fluids within a sensitive volume as shown in Figure 2.17 (Coates et al., 1999). The raw porosity provided by NMR is estimated as the ratio of the initial amplitude to the NMR response in a medium with 100% porosity. In conventional reservoirs, the NMR porosity is perceived to be independent of the rock lithology, but in unconventional shale reservoirs, the presence of organic matter (OM) and clay content elicit NMR responses (Mehana, 2016). Although, NMR is not sensitive to hydrogen in kerogen because it relaxes too fast, it responds to the presence of generated hydrocarbons from kerogen. The accuracy of the NMR raw porosity is dependent on a longer polarization time (TW) to achieve complete polarization of hydrogen nuclei in the fluids, a shorter interecho spacing (TE) to record decays of clay-bound fluids, and the hydrogen index (HI) being equal to one (Coates et al., 1999). In the presence of gas and light hydrocarbon fluids, longer TW and HI = 1 prove difficult to achieve, thus, affecting the accuracy of NMR porosity.

A bimodal T_2 spectrum of an Eagle Ford shale outcrop core (1 in. diameter by 3 in. length) saturated at 3,000 psi confining pressure to ensure a fully saturated core is shown

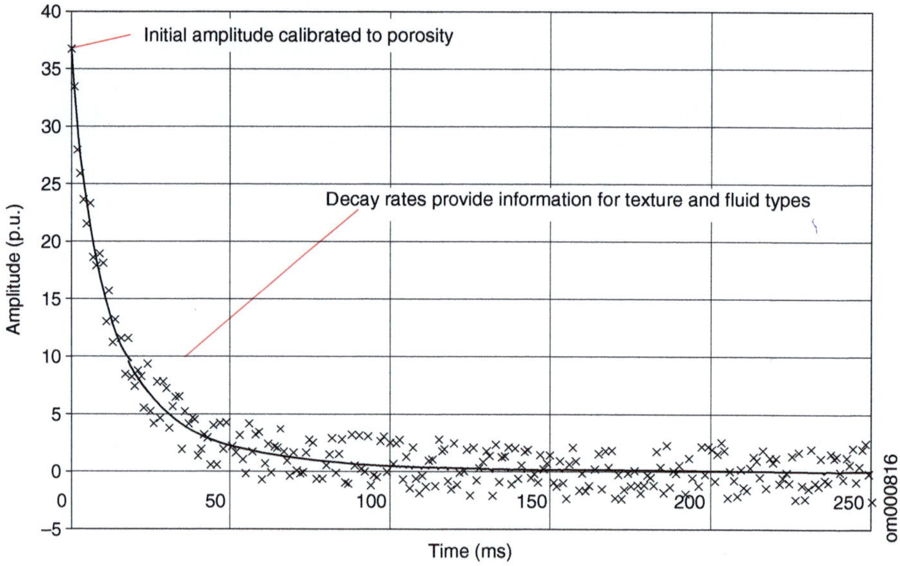

Figure 2.17 Spin-echo decay at 1 ms interecho spacing with the initial amplitude calibrated to porosity (Coates et al., 1999).

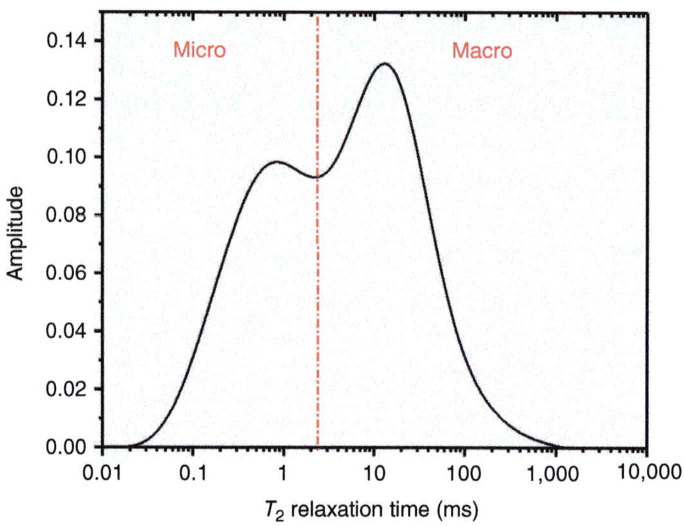

Figure 2.18 T_2 relaxation time spectrum of saturated Eagle Ford shale outcrop at confining pressure of 3,000 psi.

in Figure 2.18; the bimodal peak is representative of bound fluids (water and/ or bitumen) in the micropores and movable/ nonmovable hydrocarbons in the overlapped organic- and macropores, respectively. Small pores (micro-) have very short transverse relaxation time (T_2), while larger pores (including micro-fractures) have longer transverse relaxation time (T_2). NMR porosity in shales takes into account all other micropores, otherwise missed by routine laboratory measurements but underestimates the porosity in situations where the signals relax too fast before being detected, mostly

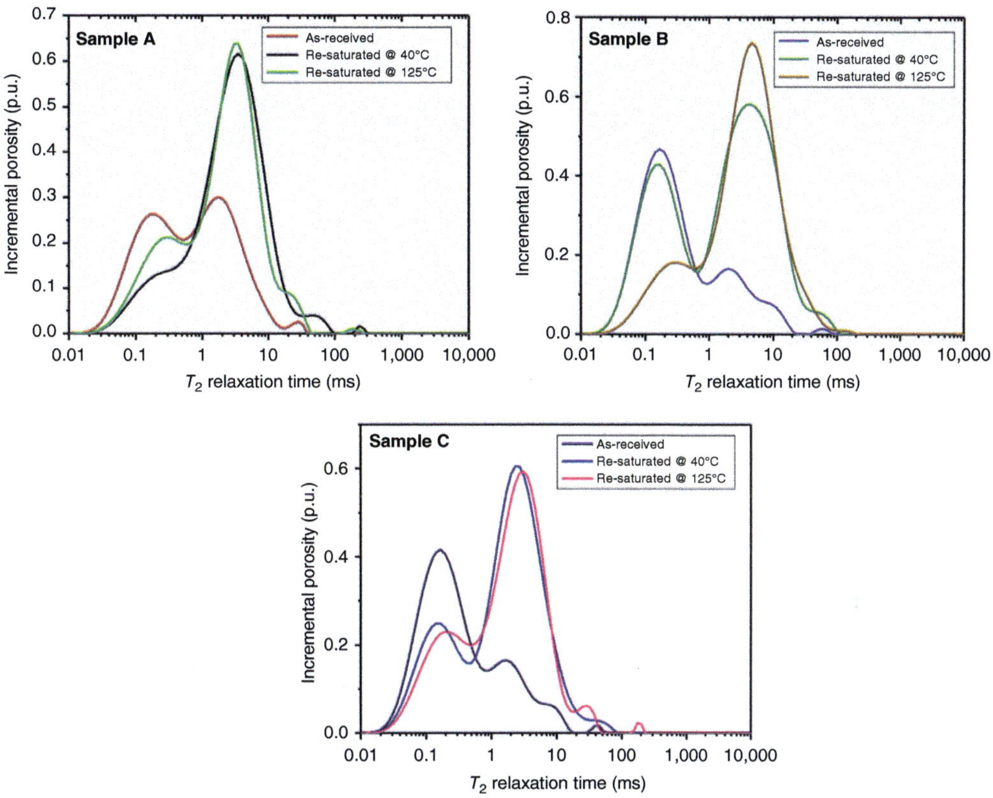

Figure 2.19 Different states of lower Eagle Ford (LEF) reservoir cores in the "as-received," "saturated at 40°C," and "saturated at 125°C," respectively (Cudjoe et al., 2019).

in micropores with bound fluids. The NMR porosity is reported to be influenced at a greater level by the echo time (TE) than the waiting time (TW). An ideal echo time of 0.3 ms and a waiting time of 9 ms is recommended to obtain a comparable NMR porosity with helium (He) porosity for low-field NMR porosity measurement in shales (Xu, 2014). The NMR measurements below were carried out with a 2 MHz instrument at an echo time of 100 μs with waiting time between 2,000 and 15,000 ms.

On the other hand, Figure 2.19 shows the NMR T_2 distribution of LEF reservoir core samples in the "as-received," "saturated at 40°C," and "saturated at 125°C," respectively. Different pore systems are highlighted under the different states. While, the as-received state is dominated by the bound fluids ($T_2 < 1$ ms) especially in Samples B and C, the saturated states are dominated by the oil in the organic pores ($1 < T_2 < 16$ ms), and signals in the inorganic pores ($T_2 >$ 16 ms) (Cudjoe et al., 2019). Because the T_2 distribution of a sample saturated with fluids reflects the signals of hydrogen nuclei from fluids within the pores, integrating the area under the T_2 spectrum provides a direct calculation of the rock porosity. Table 2.12 lists the cumulative porosity under the T_2 spectra of the fully saturated LEF reservoir cores with oil at 125°C.

Table 2.12 Cumulative NMR Porosity of LEF Reservoir Cores

LEF samples	NMR cumulative porosity (%)
A	12.47
B	14.57
C	12.38

Source: Cudjoe et al. (2019).

Example 2.5 Calculate the NMR Porosity

Given, the NMR T_2 measurement of an outcrop shale sample in Table 2.13. Plot the T_2 spectrum and determine the cumulative porosity of the outcrop shale sample.

Table 2.13 NMR Transverse (T_2) Relaxation Time Measurement of an Outcrop Shale Sample

	T_2 (ms)	Incremental porosity (p.u.)		T_2 (ms)	Incremental porosity (p.u.)		T_2 (ms)	Incremental porosity (p.u.)
1	0.010	0.000013	33	1.463	0.081994	65	214.115	0
2	0.012	0.000054	34	1.71	0.058948	66	250.215	0
3	0.014	0.000184	35	1.998	0.040122	67	292.402	0
4	0.016	0.00053	36	2.335	0.026416	68	341.701	0
5	0.019	0.001309	37	2.729	0.018232	69	399.312	0
6	0.022	0.002845	38	3.189	0.01541	70	466.637	0
7	0.025	0.005539	39	3.727	0.017229	71	545.312	0.001302
8	0.030	0.009826	40	4.355	0.022481	72	637.252	0.008972
9	0.035	0.01611	41	5.089	0.029622	73	744.694	0
10	0.041	0.024711	42	5.947	0.036988	74	870.25	0
11	0.047	0.035823	43	6.95	0.043033	75	1016.98	0
12	0.056	0.049491	44	8.122	0.046576	76	1188.44	0
13	0.065	0.065608	45	9.491	0.046996	77	1388.81	0
14	0.076	0.083917	46	11.091	0.04433	78	1622.97	0
15	0.089	0.10402	47	12.961	0.039267	79	1896.6	0
16	0.104	0.125389	48	15.147	0.033	80	2216.37	0
17	0.121	0.147383	49	17.701	0.026969	81	2590.05	0
18	0.141	0.169264	50	20.685	0.022536	82	3026.74	0
19	0.165	0.190213	51	24.172	0.020649	83	3537.05	0
20	0.193	0.209359	52	28.248	0.021587	84	4133.4	0
21	0.226	0.225809	53	33.01	0.024829	85	4830.3	0
22	0.264	0.2387	54	38.576	0.029113	86	5644.7	0
23	0.308	0.247244	55	45.08	0.032658	87	6596.4	0
24	0.360	0.250795	56	52.681	0.033531	88	7708.56	0
25	0.421	0.24891	57	61.563	0.030026	89	9008.23	0
26	0.492	0.241401	58	71.942	0.021	90	10527	0
27	0.575	0.228389	59	84.072	0.006058	91	12301.9	0
28	0.671	0.210328	60	98.246	0	92	14376	0
29	0.785	0.188016	61	114.81	0	93	16799.8	0
30	0.917	0.16257	62	134.168	0	94	19632.3	0
31	1.072	0.135371	63	156.789	0	95	22942.3	0
32	1.252	0.107975	64	183.223	0	96	26810.5	0

Solution

The T_2 spectrum is plotted by assigning the T_2 relaxation time (ms) column as the x-axis (in logarithmic scale) and the incremental porosity (p.u) as the y-axis. Next the cumulative porosity is calculated by integrating the area under the T_2 spectrum. This can be carried out in Excel.

The NMR cumulative porosity at the end is calculated as 4.61%.

2.5 Pore-Size Distribution

Characterization of pore structure depends on the size, shape, volume, surface area, connectivity, and spatial distribution (Sun et al., 2017). Pore-size distribution is of extreme importance in the evaluation and exploration of shale reservoirs. Understanding the pore-size distribution and bound water cutoff in shale is the first step in analyzing the reservoir capacity estimation and hydraulic fracturing (Cao-Minh et al., 2012; Yu et al., 2016). Pore-size distribution in shales varies from nanosized pores in the organic matters to microsized pores in the inorganic matter and clay, and microsize fractures along the beddings (Sondergeld et al., 2010). Nuclear magnetic resonance (NMR) and nitrogen adsorption are widely used to measure pore-size distribution.

2.5.1 Pore-Size Distribution Using NMR Logging

Yu et al. (2017) used NMR well logs to analyze different pore-size distributions. Mechanism of T_2 relaxation includes bulk, surface, and diffusion relaxation (Kenyon, 1992), which is shown in equation (2.58),

$$\frac{1}{T_2} = \frac{1}{T_{2,bulk}} + \frac{1}{T_{2,surface}} + \frac{1}{T_{2,diffusion}}. \tag{2.58}$$

In the uniform internal fields, the surface relaxation is the main contributor of T_2 relaxation and two other parameters ($\frac{1}{T_{2,surface}}$ and $\frac{1}{T_{2,bulk}}$) are ignored (Kleinberg et al., 1994). Therefore, the relaxation equation of pore distribution is simplified to equation (2.59),

$$\frac{1}{T_2} = \rho\frac{S}{V}, \tag{2.59}$$

where T_2 is the NMR relaxation time for fluids in a certain pore (s), ρ is the surface relaxivity, which is an intrinsic constant of rocks $\left(\frac{\mu m}{s}\right)$, and $\frac{S}{V}$ is the specific surface area of pores (μm^{-1}).

Therefore, for a system of spherical pores, the pore radius can be calculated as a function of T_2 as shown in equation (2.60) (Yu, et al., 2017; Sigal, 2015),

$$r = 3\,\rho T_2, \tag{2.60}$$

where r is the pore radius (μm).

An example NMR log is shown in Figure 2.20. In Figure 2.21, it can be observed that the T_2 T_2 distribution plot is divided into several logarithmic intervals with the same distances as shown in the header of the NMR log in Figure 2.20. The higher porosity percentage results in higher peaks as shown in Figure 2.20. Therefore, it is concluded that the micropores are the main contributors of the total porosity. The first peak corresponds to bound fluid volume, and the second and the third peaks correspond to free fluid volume (Yu et al., 2017). Since the third peak is not fully developed everywhere, it can be concluded that this peak is due to the presence of microfractures. Thereafter, these boundaries and the NMR well logs were used to recalculate the pore-size distribution from NMR well logs using the T_2 cutoff and the predicted boundaries. In can be observed that the large proportion of total

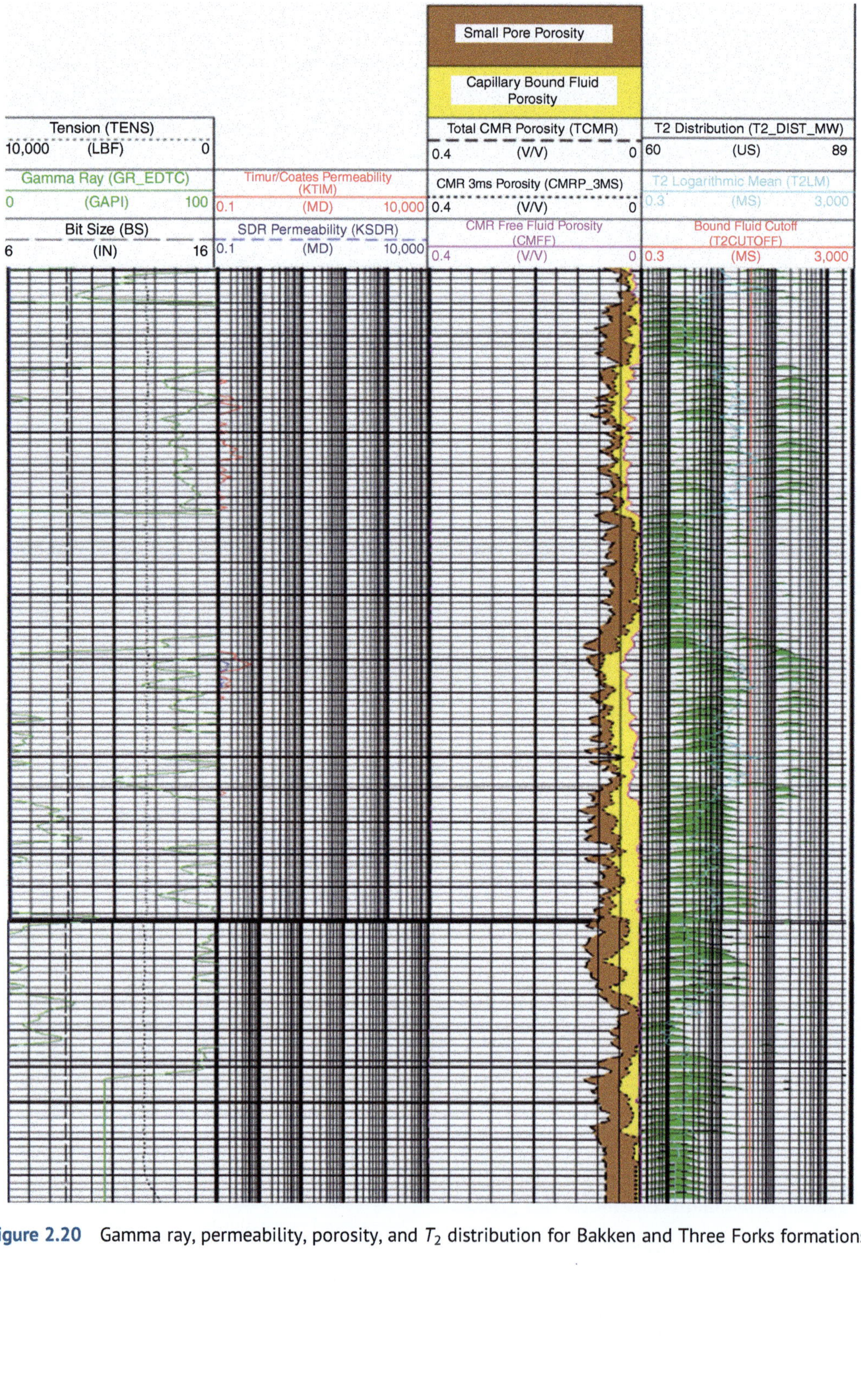

Figure 2.20 Gamma ray, permeability, porosity, and T_2 distribution for Bakken and Three Forks formations.

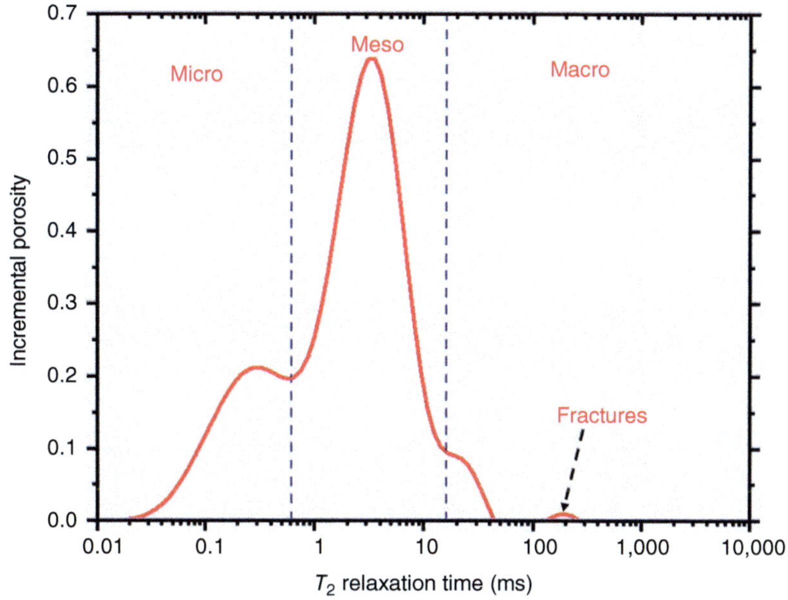

Figure 2.21 NMR T_2 pore-size distribution of an Eagle Ford shale sample; micropores (< 2 nm), mesopores (2–50 nm), and macropores (> 50 nm).

porosity comes from the bound-fluid volume. Although, microfractures can be found in most of the reservoirs, they are not the dominant portion of the pore space. Figure 2.20 shows the gamma ray, permeability, porosity, and T_2 distribution for the Bakken and Three Forks formation.

On the other hand, a LEF core sample analyzed with low-field (2.3 MHz) NMR is observed to be dominated by mesopores (organic pores) rather than micropores (bound fluid volume) on a T_2 spectrum as shown in Figure 2.21. Three types of pore systems are observed, notably, micropores (< 2 nm), mesopores (2–50 nm), and macropores (> 50 nm), including microfractures. This is confirmed with focused ion beam-scanning electron microscopy (FIB-SEM) organic pore analysis, where a FIB-SEM volume of 216 μm^3 (12 x 9 x 2 μm) shows organic pore distribution from 14 nm to 510 nm with a mean equivalent pore diameter of 32 nm as shown in Figure 2.22.

The relaxation time from NMR based on the surface relaxivity as expressed in equation (2.59) can be converted to estimate the pore diameters within a core sample knowing the surface-to-volume ratio (*S/V*) in terms of the pore diameter. For spherical pores, the surface-to-volume ratio (*S/V*) is assumed to be (6/*d*) leading to an expression of the pore diameter for transverse relaxation (Chen, 2017),

$$d_2 = 6\rho_2 T_2, \tag{2.61}$$

where, d_2 is the pore diameter (μm), ρ_2 is the surface relaxivity, and T_2 is the transverse relaxation time (ms).

The surface relaxivity (ρ) is required in order to determine the distribution of pore diameters from equation (2.61). Several surface relaxivity values have been reported in the literature for different shale plays either by calibrating pore-size distribution (PSD) of shales from SEM or FIB-SEM or in-situ measurements and some are listed below in Table 2.14.

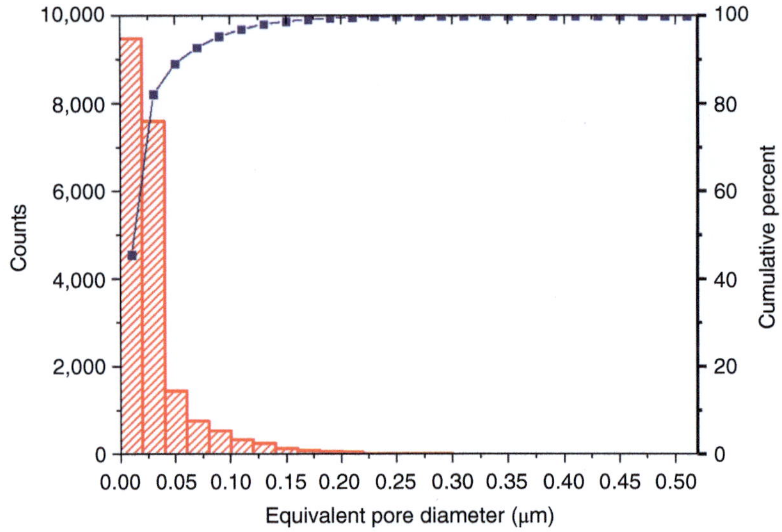

Figure 2.22 Equivalent organic pore diameter distribution of a lower Eagle Ford shale sample.

Table 2.14 Surface Relaxivity of Different Shale Formations

Surface relaxivity (ρ)	Method	Formation/rock component	Reference
4.4 nm/ms	Calibrated with PSD from SEM (large pores)	Eagle Ford shale	Jiang (2013)
18.9 ± 6.54 μm/s	Pulse-field gradient NMR	Kerogen (Barnett shale)	Zhang (2016)
16.6 ± 7.2 μm/s	Pulse-field gradient NMR	Kerogen (Rocky Mountains)	Zhang (2016)
23.4 μm/s	Pulse-field gradient NMR	Kerogen (Eagle Ford)	Zhang (2016)
0.05 μm/ms	Calibrated with PSD from SEM (nano pores)	Shales	Sondergeld (2010)
0.5 – 3.1 μm/s	Fluid imbibition experiments	Ordovician organic shales	Sulucarnain (2012)
5 μm/s	Assumed through NMR	Organic surfaces	Chi (2015)

2.5.2 Pore-Size Distribution Using Nitrogen Adsorption

The BJH method, which was developed by Barret-Joyner-Halenda, is known as the most useful technique to determine the pore-size distribution in the adsorbent (Dudek, 2016). In this method, the average pore radius, assuming cylindrical pore geometry, is estimated by equation (2.62). Before applying this technique to measure the surface area and pore-size distribution, the surface should be cleaned of water and organic vapors. The cleaning process can be performed under a flow of dry, inert gas, or with heating the samples under the vacuum (Leddy, 2012).

$$r_p = \frac{2V_{liq}}{S}, \tag{2.62}$$

where V_{liq} is calculated using equation (2.63) and S is calculated using equation (2.64),

$$V_{liq} = \frac{P_a V_{ads} V_m}{RT}, \tag{2.63}$$

where V_{liq} is the volume of liquid N_2 in pores (cm^3); V_{ads} is the volume of adsorbed gas (cm^3); V_m is the molar volume of liquid adsorbate (cm^3/mol; it is 34.7 cm^3/mol N_2); P_a is the ambient pressure (atm); T is the ambient temperature (K); R is the gas constant, 82.057 ($cm^3 atm/(K\ mol)$); and S is the specific surface area (cm^2).

$$S = \frac{S_t}{w}, \tag{2.64}$$

where w is the sample weight and S_t is the total surface area (cm^2). S_t is calculated using equation (2.65) as follow:

$$S_t = \frac{W_m N A_{cs}}{M}, \tag{2.65}$$

where N is Avogadro's number ($N = 6.023 \times 10^{23}$); M is the molecular weight of adsorbate; A_{cs} is the adsorbate cross-sectional area (16.2Å^2 for nitrogen); and W_m is the weight of monolayer. It is obtained from equation (2.66),

$$W_m = \frac{1}{s+i}, \tag{2.66}$$

where s (slope) and I (intercept) are found by the use of equation (2.67),

$$\frac{1}{W\left(\left(\frac{P_0}{P}\right)-1\right)} = \frac{1}{W_m C} + \frac{C-1}{W_m C}\left(\frac{P}{P_0}\right), \tag{2.67}$$

where W is the weight of gas adsorbed, $\frac{P_0}{P}$ is the relative pressure, W_m is the weight of adsorbate as monolayer, and C is the BET constant, and it can be found using equation (2.68),

$$C = \left(\frac{S}{i}\right) + 1 \tag{2.68}$$

Hence, plotting $\frac{1}{W\left(\left(\frac{P_0}{P}\right)-1\right)}$ vs. $\left(\frac{P}{P_0}\right)$ results in calculating s and i values. Therefore, s and i are the slope and the intercept of the curve, respectively. The values of s and i are defined using equation (2.69) and equation (2.70) as follow:

$$s = \frac{C-1}{W_m C}, \tag{2.69}$$

and

$$i = \frac{1}{W_m C}. \tag{2.70}$$

Example 2.6 Calculate the Average Pore Radius

For the data provided in Table 2.15 by Leddy (2012), determine the total surface area (S_t), and based on that, estimate the average pore radius knowing the volume of liquid N_2 (V_{liq}) in the pores of a sample is 7.5355 (cm^3/g). The nitrogen cross sectional area and molecular weight are 1.62×10^{-19} (m^2/mol) and 28.0134 (g/mol), respectively.

Table 2.15 BET Data for Example 2.6

Relative pressure (P/Po)	1/[w((Po/P)-1)]
1.11E-01	1.32E+01
1.53E-01	1.78E+01
1.99E-01	2.28E+01
2.48E-01	2.80E+01
2.97E-01	3.35E+01

Source: Leddy (2012).

Solution

The first step is to determine the slope (s) and intercept (i) to calculate the weight of monolayer (W_m). This can be accomplished by plotting the term 1/[w((Po/P)-1)] versus the relative pressure as shown in Figure 2.23.

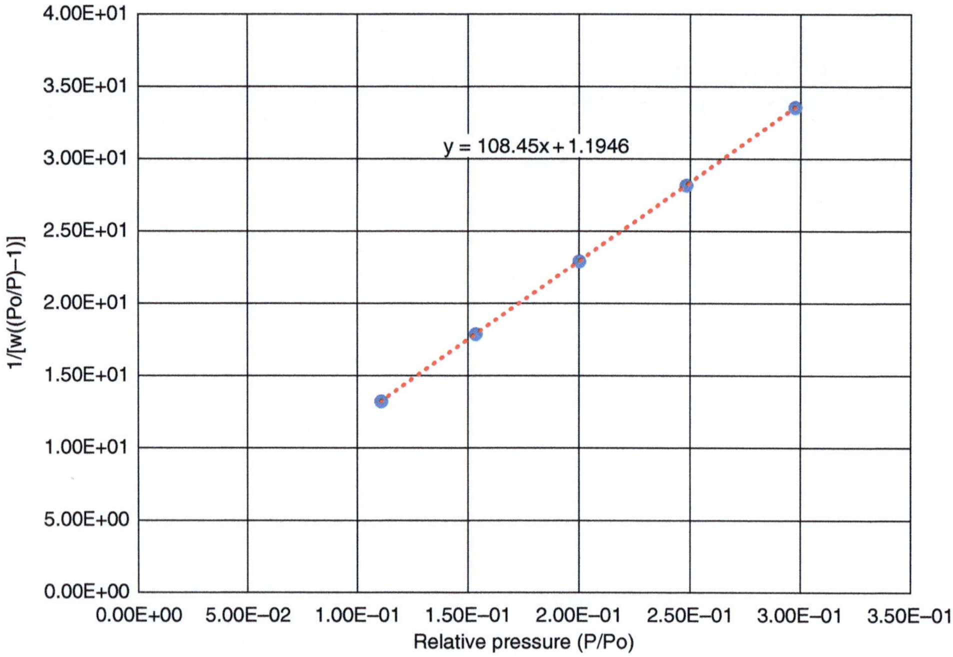

Figure 2.23 Slope and intercept for Example 2.6.

Hence, the slope (s) and the intercept (i) are 108.45 and 1.1946, respectively. Thus, the weight of monolayer (W_m) can be calculated by the use of equation (2.66),

$$W_m = \frac{1}{s+i} = \frac{1}{108.45 + 1.1946} = \mathbf{0.00912}.$$

The total surface area is then calculated by the use of equation (2.65),

$$S_t = \frac{W_m\, N\, A_{cs}}{M} = \frac{0.00912 \times 6.023 \times 10^{23} \times 1.62 \times 10^{-19}}{28.0134} = \mathbf{31.77} \left(\frac{\boldsymbol{m^2}}{\boldsymbol{g}}\right).$$

Finally, the average pore radius based on the total surface area is calculated by the use of equation (2.62),

$$r_p = \frac{2V_{liq}}{S_t} = \frac{2 \times 7.5355 \left(\frac{\text{cm}^3}{g}\right)}{31.77 \left(\frac{m^2}{g}\right)} = 4.74 \times 10^{-4}\,(cm) = \mathbf{4.74}\,(\boldsymbol{\mu m}).$$

2.6 Permeability

Permeability is defined as the capacity of a rock to transmit a fluid. Darcy's law, presented in equation (2.71), is used to measure the permeability of the rock in conventional reservoirs.

$$Q = \frac{kA}{\mu} \frac{dP}{dx}, \tag{2.71}$$

where Q is the measured flow rate, cc/sec; k is permeability; Darcy, A is surface area, cm^2; μ is fluid viscosity, cP; and $\frac{dP}{dx}$ is the pressure drop across the length of the sample.

Darcy's law is usually used to measure the permeability of different rocks under steady state conditions (Tinni et al., 2012). The permeability of the unconventional reservoirs plays an important role in the economic viability of the unconventional plays (Tinni et al., 2012). The unconventional gas and oil reservoirs have an extremely low permeability (e.g., nano-Darcy range; Bratovich & Walles, 2016).

The permeability measurements are also done on core plugs and crushed samples using unsteady state techniques (Tinni et al., 2012). The GRI method, pulse decay method, and oscillating pulse method are laboratory methods to determine the rock permeability (Passey et al.,2010; Sondergeld et al., 2010).

2.6.1 Unsteady-State Permeability Measurement Methods

The GRI Method

The Gas Research Institute (GRI) technique is known as the pressure decay method, and it was introduced by Luffel et al. (1993) to replace the plug permeability measurement. The reason for that is because shale samples are highly affected by the fractures contained within the core plug (Luffel et al., 1993). In this technique, in order to obtain matrix permeability, the rocks are crushed into small particles in the range of 0.7 (mm) (Luffel et al., 1993; Cui et al., 2009; Profice et al., 2011). This method is used to measure the shale permeability, but it cannot be performed at elevated confining pressures. Furthermore, the permeability measured by this technique is highly dependent on the particle size. The schematic of the pulse-decay experimental setup, drawn by He (1995) is shown in Figure 2.24.

For the purpose of this technique, the shale matrix gas permeability to helium or methane is measured in the presence of gas desorption (Schettler et al., 1989). In this technique, after holding a shale core plug in a rubber sleeve and applying 2,000 psi net pressure, the core sample was equilibrated at 500 psi for several hours. Thereafter, by exhausting the small dead volume at one core face to a vacuum, the cell was sealed. Finally, the pressure rebound was measured and used to measure the amount of the released gas into the dead volume. Finally, the gas permeability of the sample was derived from the slope of the isotherm (S) plot in using equation (2.72).

$$K = 3140\frac{zRT\mu G^2}{PS}, \tag{2.72}$$

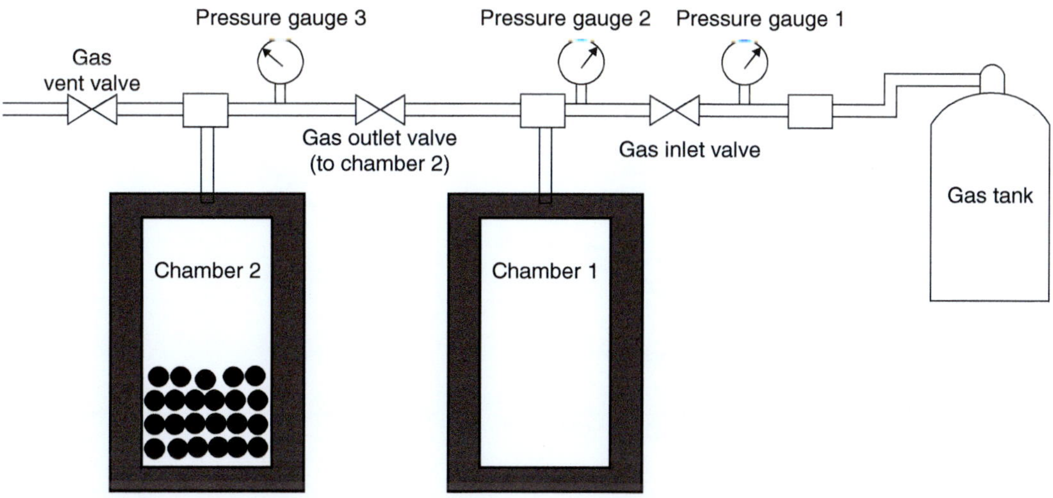

Figure 2.24 Schematic of experimental arrangement for pulse-decay technique (He, 1995).

where K is the core plug permeability, md; z is the gas compressibility factor; R is the gas constant, atm-cm^3/(gmol-K) ($R =$ 82.057); T is absolute temperature, in degrees Kelvin; μ is gas viscosity, cP; P is pressure, atm; S is the isotherm slope, gmol/(atm-cm^3), which is obtained from plotting gas content vs. the pressure; and G is the specific degassibility, moles/(cm-atm-sec$^{1/2}$), which is calculated from plotting released gas vs. the square root of time, Figure 2.25, using equation (2.73). Early time, straight-line slope of the produced gas is used to calculate G.

$$G = \frac{M}{2A\Delta P t^{1/2}}, \tag{2.73}$$

where M is the gas produced during degassibility, gmols; A is the degassing surface area, cm^2; ΔP is the pressure drop causing degassing, atm; and t is the time function, s.

This method is a fast technique for permeability measurement on low-permeability samples, but it has some limitations. The overburden pressure is missed and therefore, this technique cannot be performed under the reservoir conditions (Wilson, 2012). This method is conducted at pore pressures much lower than the reservoir pressure at which the gas flow through the sample as a free molecular flow regime. Furthermore, gas slip and Klinkenberg correction is not applied in this method. Therefore, significant errors from the measured permeability values are observed. Inconsistency in the interpreting techniques and lack of standard analytical expression results in different measured values of permeability for the same unconventional plays. In addition, the original permeability measurements using GRI method were conducted for Devonian shale, which had an original GRI crushed-particle size of 20/35 mesh (0.67 ± 0.17 mm) and permeability values in the nano-Darcy range, while the current commercial unconventional plays have permeability values in the micro-Darcy range. It implies that the permeability measurement of these shale samples should be performed 1,000 times faster. Furthermore, the size of these samples should be increased by a factor of $\sqrt{1,000}$ compared with the Devonian shale samples (Wilson, 2012).

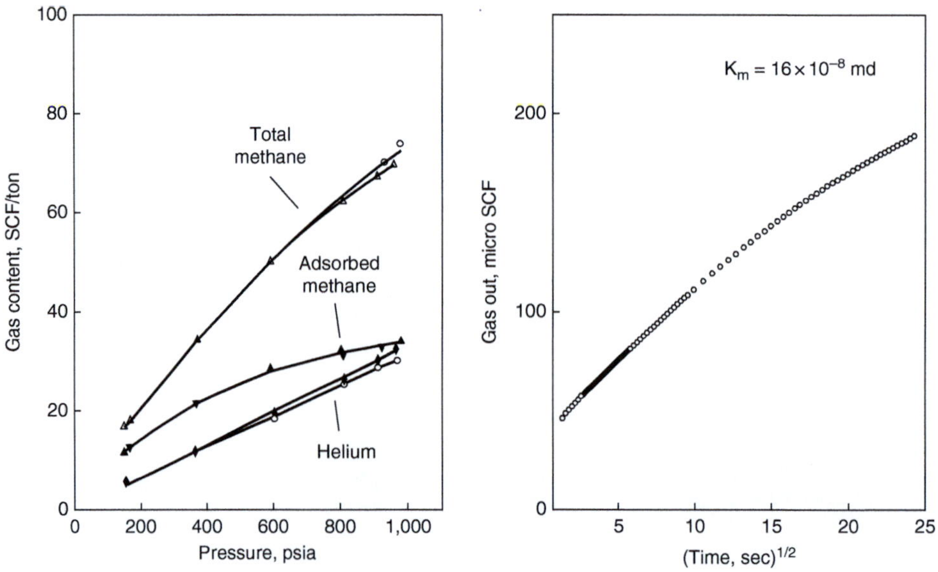

Figure 2.25 An example of desorption isotherms of (left) crushed shale samples and (right) methane degassibility test (Luffel et al., 1993).

Example 2.7 Determine the Permeability

Determine the permeability using methane for a core plug from the Devonian shale play at 2,000 psi and 100°C assuming the rubber sleeve pressure and the equilibrated pressure are 2,000 psi and 500 psi, respectively. The core has a diameter of 3/4 in., and the desorption isotherms of crushed shale samples and methane degassibility tests are shown in Figure 2.25.

Solution

First, the isotherm slope (S) and the specific degassibility (G) are found from Figure 2.25. After that, equation (2.72) is used to calculate the permeability. Therefore, the permeability of the core plug at the specified conditions is 14 nano-Darcy.

Pulse Decay Method

This method uses the transient pressure decay across a core plug to obtain the permeability of the rock under confining pressure, and it provides an accurate result for very low permeability cores down to approximately 1 nano-Darcy (Brace et al., 1968; Dicker & Smits, 1988; Jones, 1997). This test can be conducted using either gas or liquid flow. Injecting gas at high pore pressures reduces the Klinkenberg slippage effect. Jia et al. (2017) performed pulse decay experiments on a tight carbonate core by flowing helium in both forward and reverse directions under different pressures. The pore-size distribution of the core was analyzed using nuclear magnetic resonance (NMR) spectroscopy. Their studies indicated that the more permeable zones of the core are differentiable in both forward and reverse direction on pulse-decay experiments. The schematic of the pulse-decay experimental setup, shown in Figure 2.26, is drawn based on He applications (He, 1995). A transient flow suggested by Brace et al. (1968) was used to calculate the sample permeability using equation (2.74).

$$k = \frac{\alpha \mu c L}{A\left(\frac{1}{V_1} + \frac{1}{V_2}\right)}, \tag{2.74}$$

where k is the rock sample permeability, nD; μ is fluid viscosity, dyne-s/cm^2; c is fluid compressibility, bar^{-1}; L is length of sample, cm; A is cross-sectional area, cm^2; V_1 and V_2 are the upstream and downstream volumes, cm^3; and α is calculated as a function of pressure decay ($P_1 - P_f$) on a semilogarithm scale against time. It is obtained from equation (2.75),

$$(P_1 - P_f) = \Delta P\left(\frac{V_2}{V_1} + V_2\right).e^{-at}, \tag{2.75}$$

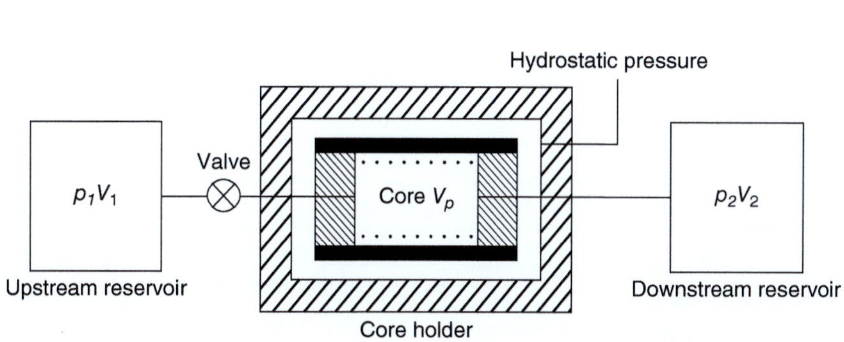

Figure 2.26 Schematic of experimental arrangement for pulse-decay technique (He, 1995).

where P_1 and P_2 are the upstream and downstream pressures, bar; P_f is the final pressure, bar; ΔP is the pressure difference between the upstream and the downstream pressures at $t = 0$, bar; V_1 and V_2 are the upstream and the downstream volumes, cm^3; and t is time, s.

Example 2.8 Estimate the Permeability Using the Pulse Decay Method

A core plug from an unconventional reservoir in the Green River formation in Colorado was chosen to test the permeability by using the pulse decay method. The core diameter and length are 1.48 in. and 1.46 in., respectively. The experimental data are provided in Table 2.16. Estimate the permeability of the core plug.

Solution

The first step is to convert all units to SI units to ease the calculation process. Then, equation (2.74) and the data in Table 2.16 can be used to calculate the permeability as follows:

$$k = \frac{\alpha\mu cL}{A\left(\frac{1}{V_1} + \frac{1}{V_2}\right)} = \frac{0.008\left(\frac{1}{s}\right) \times 1.63(poise) \times 9 \times 10^{-3}\left(\frac{1}{MPa}\right) \times 0.037(m)}{0.001(m^2) \times \left(\frac{1}{6.2 \times 10^{-4}(L)} + \frac{1}{0.004(L)}\right)} = 2.102 \times 10^{-20}\,(m^2)$$

and

$$k = 2.102 \times 10^{-20}\,(m^2) = 21.298\ nD.$$

Hence, the permeability is 21.298 nano-Darcy.

Table 2.16 Experimental Data for Example 2.8

Experimental data		
Upstream pressure	1,992	Psia
Sample volume	3.15	cm^3
Upstream volume 2	0.62	cm^3
Downstream pressure	1,786	Psia
Downstream volume	3.96	cm^3
Equilibrium pressure	1,885	Psia
Isothermal compressibility	9×10^{-3}	1/MPa
Viscosity	1.63×10^{-5}	Pa.s
The slope, α	0.008	1/s
Temperature	30.25	°C
Time	60	Sec

Modified Pulse Decay Method

Dicker and Smits (1988) presented a new pressure pulse decay technique to measure the permeability of tight rock samples. In this model, the upstream and downstream volumes were chosen to be equal and slightly higher than the sample pore volume. Figure 2.27 shows the schematic of the setup used by them to calculate the

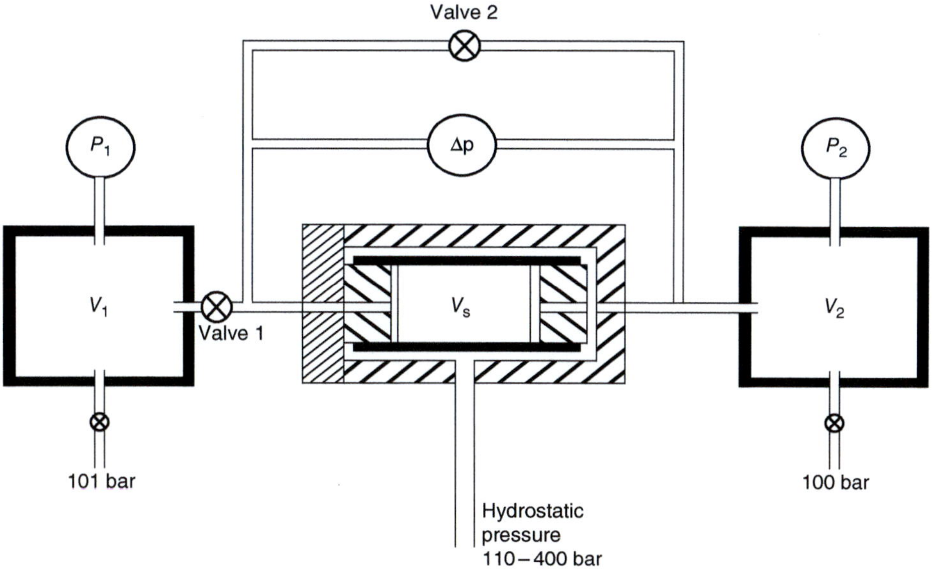

Figure 2.27 Schematic of pulse-decay experiment setup (Dicker & Smits, 1988).

rock sample permeability. In this method, the measured pressure difference is plotted on the semilog paper and the slope of the straight part of the plot is calculated. This straight part should be chosen from the part in which the pressure difference has decayed to 70% of the original pressure difference. Consequently, the permeability of the sample is calculated using equation (2.76) (Dicker & Smits, 1988).

$$k = \frac{c\mu\phi L^2 S}{f(a,b)}, \quad (2.76)$$

where k is the permeability of the core plug, m^2; c is the fluid compressibility, Pa^{-1}; μ is the fluid viscosity, Pa.s; ϕ is the core plug porosity, fraction; L is the core plug length, m; S is the pressure decay slope, which is obtained from plotting the normalized differential pressure $\left(\ln\left(\frac{\Delta P}{\Delta P_0}\right)\right)$ vs. time; and $f(a, b)$ is a function of chamber's and pore volumes and is defined as equation (2.77).

$$f(a,b) = (a + b + ab) - \frac{1}{3}(a + b + 0.4132ab)^2 + 0.0744(a + b + 0.0578ab)^3, \quad (2.77)$$

where a and b are defined using equations (2.78) and (2.79), respectively:

$$a = \frac{V_s}{V_1} \quad (2.78)$$

and

$$b = \frac{V_s}{V_2}. \quad (2.79)$$

This method is a simple and fast technique to evaluate the tight rock samples. This simple approximate solution is accurate to within 0.3% of the exact solution (Dicker & Smits, 1988).

Example 2.9 Estimate the Permeability Using the Modified Pulse Decay Method
Estimate the permeability of a core plug from the Bakken play using the modified pulse decay method. The core length is 1.41 in. and the porosity of the core is 0.073. The experimental data are provided in Table 2.17.

Table 2.17 Experimental data for Example 2.9

Experimental data		
Upstream volume	300	cm^3
Sample volume	0.3	cm^3
Downstream volume	30	cm^3
Isothermal compressibility	9×10^{-3}	1/MPa
The slope, s	0.426	1/s
Viscosity	1.63×10^{-5}	Pa.s

Solution

The first step is to convert all units to SI units. Then, the function of chamber's and pore volumes is calculated by the use of equation (2.77) to equation (2.79) as follows:

$$a = \frac{V_s}{V_1} = \frac{0.3}{300} = 0.001,$$

$$b = \frac{V_s}{V_2} = \frac{0.3}{30} = 0.01,$$

$$f(a,b) = (a + b + ab) - \frac{1}{3}(a + b + 0.4132ab)^2 + 0.0744(a + b + 0.0578ab)^3,$$

$$f(a,b) = (0.001 + 0.01 + 0.001 \times 0.01) - \frac{1}{3} \times (0.001 + 0.01 + 0.4132 \times 0.001 \times 0.01)^2$$
$$+ 0.0744 \times (0.001 + 0.01 + 0.0578 \times 0.001 \times 0.01)^3 = 0.011$$

Hence, the permeability can be calculated by the use of equation (2.76) as follows:

$$k = \frac{c\mu\phi L^2 s}{f(a,b)} = \frac{9 \times 10^{-3}\left(\frac{1}{MPa}\right) \times 1.63\,(poise) \times 0.073 \times 0.036^2(m) \times 0.426\left(\frac{1}{s}\right)}{0.011}$$
$$= 5.334 \times 10^{-17}(m^2),$$

$$k = 5.334 \times 10^{-17}(m^2) = 0.054\ mD.$$

Hence, the permeability is 0.054 milli-Darcy.

Oscillating Pulse Method

The oscillating pulse method is used to measure the tight rocks' permeability without destroying rock samples and in a relatively short time. This technique consists of applying a sinusoidal pressure wave of known amplitude and frequency to the upstream end of a rock sample (Suri et al., 1997) and measuring the phase shift of upstream and downstream pressure waves across a core plug (Kranz et al., 1990; Fischer & Peterson, 1992). Using in-depth interpretation of the amplitude attenuation and the phase retardation, while propagating through the sample, the sinusoidal oscillation of the pore pressure provides an accurate estimate of rock permeability (He,

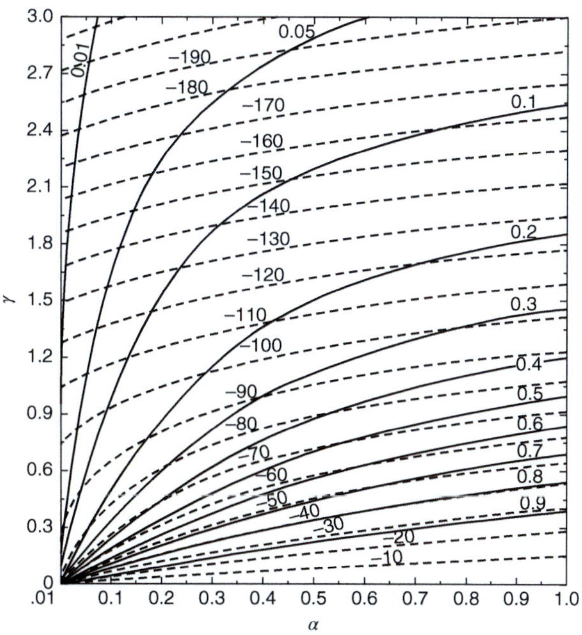

Figure 2.28 The dimensionless variables α and γ as functions of amplitude ratio (solid line) and phase lag (dashed lines) (Kranz et al., 1990).

1995). In this method, a sinusoidal pressure wave is generated in the upstream of the core sample after stabilizing the sample pore pressure in addition to upstream and downstream pressures. Thereafter, the generated wave is propagated through the core sample and the permeability of the core plug is obtained using the amplitude attenuation and phase shift between the upstream and downstream pressure wave at the downstream of the core plugs from equation (2.80) (Kranz et al., 1990).

$$k = \frac{\mu \beta V_2 \alpha \omega L}{A\gamma}, \tag{2.80}$$

where μ is the fluid viscosity, Pa • s; β is the fluid compressibility, 1/Mpa; V_2 is the volume of fluid between sample and outlet pump, cm^3; ω is the angular frequency, 1/sec; L is the sample length, cm; A is the cross-sectional area, cm^2; and α and γ are obtained from Figure 2.28 after calculating the amplitude ratio and the phase lag.

The amplitude ratio is defined as the ratio of the upstream peak to the downstream peak. The phase lag (offset) is defined as the phase difference between the upstream and the downstream core plug pressures. Figure 2.29 shows an example (Benson et al., 2017) for calculating the amplitude ratio and the phase offset in which amplitude ratio (R) is equal to 3.105/9 = 0.345; and phase offset (δ) is equal to 90°.

The oscillating pore pressure method is applicable for all permeability ranges, from low permeability values to high permeability values. In addition, permeability can be measured over time and, therefore, changes in the permeability can be monitored with time. The continuous measurements during slow changes in the state of the rock-environment system make this model useful for time-dependent diffusivity measurements during creep or consolidation of rock. Also, both diffusivity and permeability can be measured in the same time by knowing the amplitude ratio and the phase difference between the upstream and downstream reservoir pressures (Kranz et al., 1990).

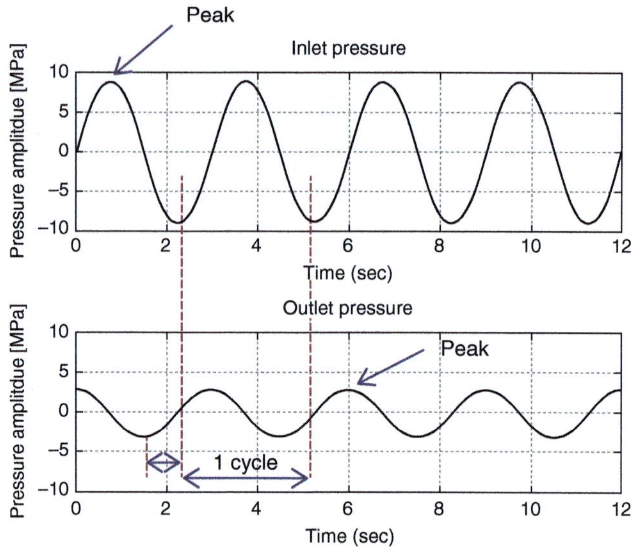

Figure 2.29 A representative plot for amplitude ratio and phase offset (Benson et al., 2017).

Example 2.10 Estimate the Permeability Using the Oscillating Pulse Method
Measure the permeability of a core plug using the oscillating pulse method. The inlet and outlet pressure values are provided in Figure 2.30. The core length and diameter are 6 cm and 5 cm, respectively. The volume of the fluid between the outlet and the pump valve is 177 cm^3. The fluid isothermal compressibility and viscosity are $4 \times 10^{-4}\left(\frac{1}{MPa}\right)$ and 1 (cP), respectively.

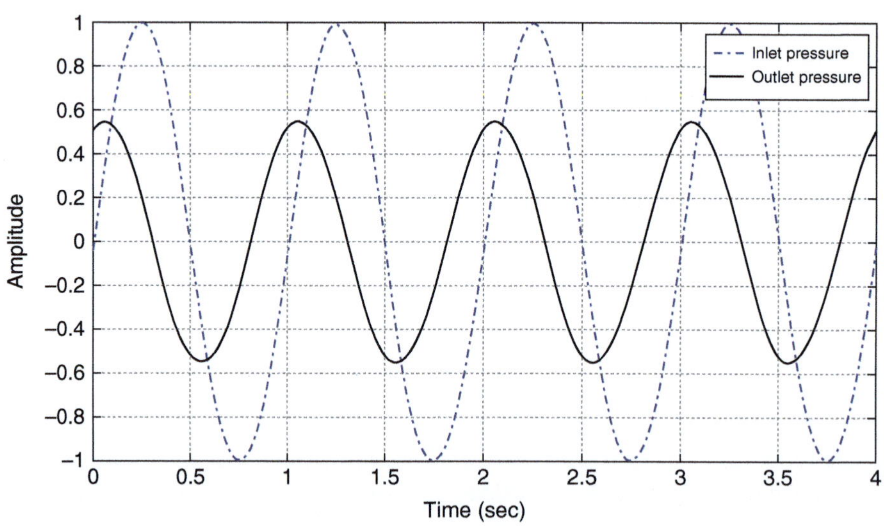

Figure 2.30 Inlet and outlet pressure values obtained from oscillating pore pressure experiment (Benson et al., 2017).

Solution

First, Figure 2.30 is used to determine the amplitude ratio and the phase offset. After that, Figure 2.28 is used to find the dimensionless variables α and γ. Then, equation (2.80) is used to calculate the permeability, accordingly. Therefore, the permeability of the core plus is 15.3 mD.

Example 2.11 Estimate Permeability Using the Oscillating Pulse Method

Measure the permeability of a core plug from the Barnett formation reported by Jin et al. (2015) using the oscillating pulse method. The inlet and outlet pressure values are provided in Figure 2.31. The core length and diameter are 1.21 cm and 2.53 cm, respectively. The oscillation period is 1,000 sec and the frequency is 0.001 (Hz) (Jin et al., 2015).

Solution

Based on the data, the permeability of the Barnett core sample using the oscillating pulse method is 13.7 nano-Darcy, which is equivalent to 0.0000137 mD (Jin et al., 2015).

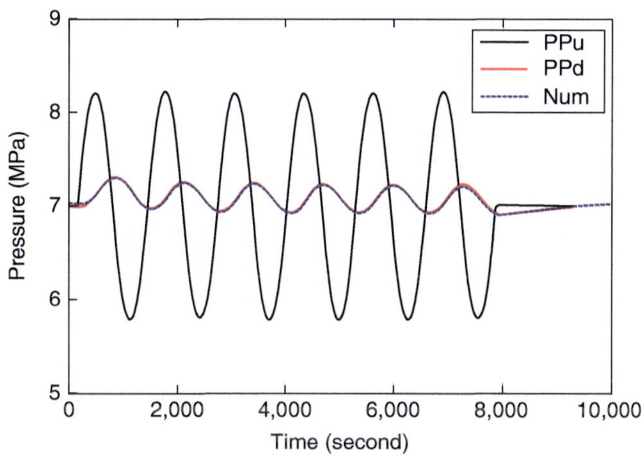

Figure 2.31 Inlet and outlet pressure values obtained from oscillating pore pressure experiment for Barnett core sample (Jin et al., 2015).

2.6.2 Single Phase Permeability Measurements

Due to the nonempirical apparent permeability (NAP) model proposed by Singh et al. (2014), a combination of Darcy components (represented by Kozeny-Carman equation) and a nonDarcy diffusive component is used to estimate the single-phase permeability of cylindrical shale cores using equation (2.81).

$$k = \frac{2\phi\mu w}{\pi\tau}\left(\frac{\pi wz}{64\mu} + \frac{\sqrt{2\pi MRT}}{3P_m M}\right), \quad (2.81)$$

where ϕ is the porosity, fraction; μ is the gas viscosity, Pa.s; w is the width of cylindrical cores, m; z is the gas compressibility factor; M is the molar mass of the gas, g/mol; R is the universal gas constant, $m^3\,Pa\,K^{-1}\,mol^{-1}$, ($R= 8.314$); T is the absolute temperature, K; τ is the tortuosity and defined as the square of the ratio of the actual path length to the shortest distance (equation (2.82)); and P_m is the mean pressure, Pa, which is calculated using equation (2.83),

$$\tau = \left(\frac{L_e}{L}\right)^2. \tag{2.82}$$

The presence of solid particles in any porous medium deviates the paths of species from straight lines and impose that the diffusion coefficients of species need to be corrected (Shen & Chen, 2007).

$$P_m = \frac{P_i + P_o}{2}, \tag{2.83}$$

where P_i is the inlet pressure of a capillary tube, Pa; and P_o is the outlet pressure of a capillary tube, Pa.

In definitions given by Sherwood et al. (1975), the diffusion coefficient is scaled up with tortuosity as equation (2.84),

$$d'_m = \frac{d_m \phi}{\tau}, \tag{2.84}$$

where d'_m is the scaled diffusion coefficient of the species; d_m is the diffusion coefficient of the species in question in the fluid (e.g., gas) without the presence of the sediment matrix; and ϕ is the porosity, fraction.

This equation is valid for flows at Knudsen numbers up to 1, which means that the mean free path of the gas molecules is greater than or equal to the average pore diameter (Daigle et al., 2015). The Knudsen number is defined as the ratio of the molecular mean free path length to a representative physical length scale, which is a dimensionless number. It can be obtained from equation (2.85),

$$K_n = \frac{\lambda}{L}, \tag{2.85}$$

where L is the representative physical length scale, m; and λ is the mean free path (Heidemann et al., 1984), m, which is defined as equation (2.86),

$$\lambda = \frac{K_B T}{\sqrt{2}\pi\delta^2 p}, \tag{2.86}$$

where K_B is the Boltzmann constant, J/K; ($K_B = 1.3805 \times 10^{-23}$); T is the temperature, K; p is the pressure, Pa; and δ is the collision diameter of gas molecules, m.

The division of different flow regimes based on the Knudsen number and the validity range of different models is shown in Table 2.18.

Table 2.18 Division of Different Flow Regimes Based on the Knudsen Number

	Knudsen number (kn)			
	Lower bound	**Upper bound**	**Flow regime**	
Validity of the NAP model	0	10^{-3}	Continuum/Darcy flow (no-slip flow)	Navier-Stokes equation
	10^{-3}	10^{-2}	Slip flow	
	10^{-2}	10^{-1}		
	10^{-1}	10^{0}	Transition flow	
	10^{0}	10^{1}		
	10^{1}	∞	Free-molecule flow	

Source: Singh et al. (2014).

2.6.3 NMR Permeability

Analyzing the NMR relaxation properties of rock samples shows a strong dependency on porosity, pore size, pore-fluid properties, and mineralogy of the rock samples (Coates et al., 1999). The rock permeability is estimated by NMR properties and different theoretical models are developed to show the dependency of rock permeability on porosity and pore-size distribution. Different studies showed that the permeability increases with both increasing porosity and increasing pore size (Timur, 1994, 1967, 1968; Ahmed et al., 1994; Allen, 1988; Thompson et al., 1987). Two different related models for permeability are developed. The free-fluid permeability model developed by Coates can be applied in formations containing water and/or hydrocarbons, while the average-T_2 model is applicable to pore systems containing only water (Marschall et al., 1997). In order to use these models for different reservoirs and different fields, measurements on core samples are necessary to refine these models and adjust them for specific locations (Coates et al., 1999).

The NMR logging tool targets the random oriented protons in the formation. Hence, the protons of the formation structure will be activated by applying the magnetic fields of the NMR tool on the randomly oriented protons. The protons are organized in a particular direction and subsequent removing of the oscillating field results in releasing the protons to tip back (Fukushima & Roeder, 1981). The amplitude of the spin echoes is measured by the NMR tool. The initial amplitude of the spin-echo train provides valuable information about the properties of the rocks and pore fluids in the measurement volumes. Relaxation time (T_2) is defined as an indication of the relaxation of the tipped protons in the field transversely relative to the axis of the static magnetic field. T_2 is obtained by summing of decaying exponentials (Coates et al., 1999). The T_2 values are proportional to the pore sizes, which means that smaller pores have smaller T_2 values and vice versa (Brownstein & Tarr, 1979). The three different ways of displaying the NMR T_2 distributions are shown in Figure 2.32.

$\boldsymbol{T_2 T_2 T_2}$

In another word, the T_2 wave is a distribution of porosity over time and assuming that there is only one hydrogen-containing fluid in the pores, the area under the curve gives the effective porosity. The cutoff value is different for different lithology types. Permeability is related to the area under the curve and to the right of the 33 ms cutoff line by Timur-Coates equation (2.87) (Coates et al., 1999).

$$k = \left(\frac{\phi}{C}\right)^m \times \left(\frac{FFI}{BVI}\right)^n, \qquad (2.87)$$

where k is permeability, md; ϕ is the effective porosity (%); C is the Coates constant, which is determined from capillary pressure for specific rock types, $md^{-0.25}$; a C value of 10 is used for younger (Tertiary/Miocene) rocks, but 23 is preferred for older clastic rocks; *FFI* or *MFFI* is free-fluid index and it is the estimation of the NMR log for producible porosity and *BVI* is the bulk volume irreducible. The default values for m and n are 4 and 2, respectively.

In this application, it is assumed that larger pores contain the producible fluids and the smaller pores are filled with the bound fluids. Therefore, by considering the relationship of T_2 values with pore-sized distribution, a T_2 value is selected as a cutoff value to divide the area under the T_2 distribution curves into FFI and BVI parts, Figure 2.33 (Coates et al., 1999). Green and Veselinovic (2010) performed a study and showed that shale reservoirs show bound/free NMR T_2 cutoff values of around 10 ms and clay bound cutoff values of around 1 ms.

In another method, the Schlumberger-Doll-Research (SDR) equation is used to calculate

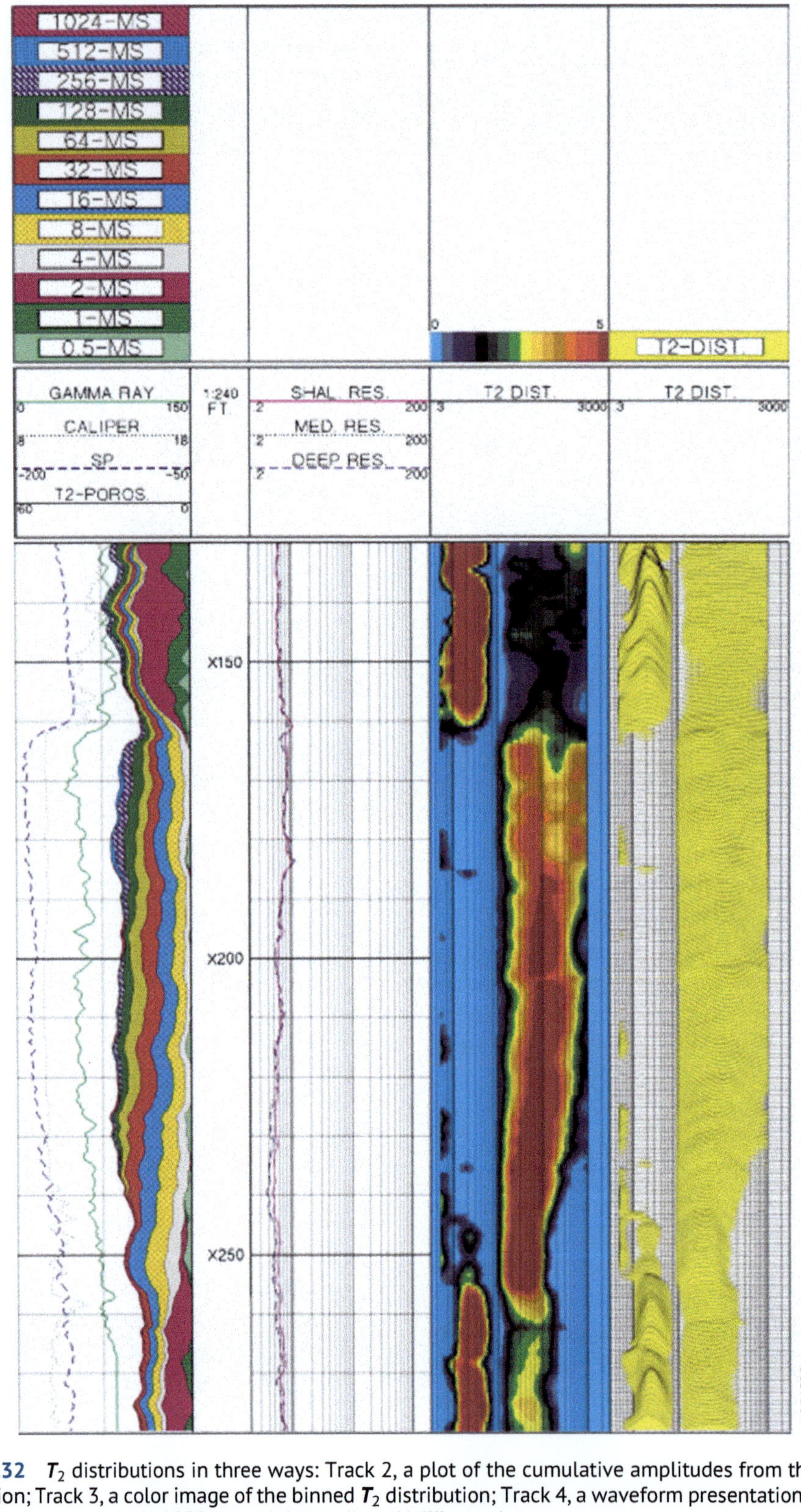

Figure 2.32 T_2 distributions in three ways: Track 2, a plot of the cumulative amplitudes from the binned T_2 distribution; Track 3, a color image of the binned T_2 distribution; Track 4, a waveform presentation of the same information (Coates et al., 1999; copyright credit to Halliburton).

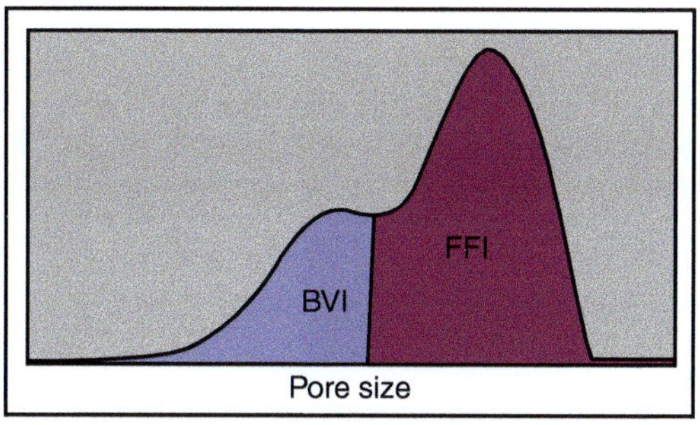

Figure 2.33 T_2 distribution showing the cutoff between bound and free fluid (Green & Veselinovic, 2010).

the permeability using equation (2.88) (Kenyon et al., 1988; Morriss et al., 1997).

$$k_{NMR} = c'\phi^a T_{lm}^b, \quad (2.88)$$

where k_{NMR} is calculated permeability (mD); ϕ is the total CMR porosity; T_{2lm} is logarithmic mean of T_2 distribution in ms; and a, b, and c' are formation-dependent constants; a, b, and c' have typical values of 4, 2, and approximately 4.6, respectively.

Example 2.12 Determine the Coefficients for the Timur-Coates Permeability Equation

The permeability, porosity, free fluid, and bulk volume irreducible data points for an interval in Bakken system from the NMR logging tool are provided in Table 2.19. Based on the data, determine the coefficients (i.e, C, m, and n) for the Timur-Coates equation. Are the values within the proposed range?

Solution

Since there are three unknowns (i.e., C, m, and n), three equations are needed to determine their values. Hence, the three equations can be formulated as follows:

$$0.0051 = \left(\frac{3.68}{C}\right)^m \times \left(\frac{0.0127}{0.024}\right)^n,$$

$$0.0045 = \left(\frac{4.01}{C}\right)^m \times \left(\frac{0.0118}{0.0283}\right)^n,$$

and

$$0.0022 = \left(\frac{3.45}{C}\right)^m \times \left(\frac{0.0098}{0.0247}\right)^n.$$

Next, the system of equations is solved to determine the three unknowns C, m, and n, which are found to be 9.781, 4.097, and 2.002, respectively. Thus, the values are within the proposed range.

Table 2.19 Data for Example 2.12

Permeability (mD)	Porosity (%)	FFI (cf/cf)	BVI (cf/cf)
0.0051	3.68	0.0127	0.024
0.0045	4.01	0.0118	0.0283
0.0022	3.45	0.0098	0.0247

Example 2.13 Evaluate the Permeability Using SDR Method
An NMR logging tool was run in a well in the Bakken play. The total NMR porosity (TCMR) and the logarithmic mean of T_2 distribution are provided in Table 2.20. Evaluate the permeability of the well, where the formation dependent constants a, b and c' are 4, 2, and 4.6, respectively.

Table 2.20 NMR Data for Example 2.13

Depth (ft)	T2LM (MS)	TCMR
9,880	5.5676	0.0848
9,882	14.7884	0.0668
9,884	14.8178	0.0759
9,886	21.1152	0.0949
9,888	30.3272	0.0592
9,890	72.2619	0.0497
9,892	196.2681	0.0449
9,894	37.5286	0.0583
9,896	53.5654	0.0540
9,898	30.6112	0.0801
9,900	7.2224	0.0586
9,902	23.6470	0.0720
9,904	34.0784	0.0605
9,906	32.6110	0.0564
9,908	22.6503	0.0628
9,910	23.5423	0.0723

Table 2.21 Summarized Results for Example 2.13

Depth (ft)	T2LM (MS)	TCMR	Permeability (mD)
9,880	5.5676	0.0848	0.0074
9,882	14.7884	0.0668	0.0200
9,884	14.8178	0.0759	0.0335
9,886	21.1152	0.0949	0.1663
9,888	30.3272	0.0592	0.0520
9,890	72.2619	0.0497	0.1466
9,892	196.2681	0.0449	0.7202
9,894	37.5286	0.0583	0.0748
9,896	53.5654	0.0540	0.1122
9,898	30.6112	0.0801	0.1774
9,900	7.2224	0.0586	0.0028
9,902	23.6470	0.0720	0.0691
9,904	34.0784	0.0605	0.0716
9,906	32.6110	0.0564	0.0495
9,908	22.6503	0.0628	0.0367
9,910	23.5423	0.0723	0.0697

Solution

The permeability can be calculated using the NMR technique by the use of equation (2.88). A sample calculation is shown below, and the final results are summarized in Table 2.21.

$$k_{NMR} = c'\phi^a T_{lm}^b = 4.6 \times (0.0848)^4 \times (5.5676)^2 = 0.0074\ mD.$$

Conversely, the Schlumberger-Doll Research (SDR) theoretical model for estimating permeability in formation using the NMR is modified to suit unconventional formations like shales. This is carried out by comparing the SDR equation with the Kozeny-Carman equation since both methods correlate the permeability to the surface area and pore geometry. The empirical coefficient is then expressed as (Daigle & Dugan, 2011):

$$A = \frac{\rho_2^2}{\varphi^3 \beta \tau^2}, \tag{2.89}$$

where ρ_2 is the transverse surface relaxivity (m/s), β is the shape factor of the pores, and τ is the tortuosity of the pore system. The surface relaxivity of different shale samples is reported in Table 2.14 The shape factor (β) and tortuosity (τ) values are obtained from digital rock analysis

Table 2.22 NMR Estimated Permeability Through a Modified Form of the SDR Model for the LEF Samples

LEF samples	Average β	Average τ	NMR permeability(mD)
A	1.35	1.58	0.001435
B	1.14	1.96	0.002404
C	0.78	1.79	0.000972

Source: Cudjoe et al. (2019).

of generated pore network models of the LEF samples, detailed in the work of Cudjoe et al. (2019). The estimated NMR permeability in Table 2.22 for the LEF samples is observed to range from 0.000972 mD to 0.002404 mD.

2.6.4 Relative Permeability

The multiphase flow in porous media is characterized by relative permeability of different fluids present in porous media. Relative permeability in a pore system is a function of fluid properties, fluid saturations, wettability, surface mineralogy, interfacial tension, viscosity contrast between the fluid phases, and geometry and connectivity of the pore system (Ojha et al., 2017; De-Prisco et al., 2012). Understanding the relative permeability of unconventional reservoirs provides insight into maximizing recovery of the reservoirs (Dacy, 2010).

Relative Permeability Measurements

The scope, reliability, and adoption of laboratory-based techniques to measure the relative permeability of unconventional reservoirs are extremely limited (Dacy, 2010). Due to the existence of nanopores, the current measurement techniques are not applicable in unconventional reservoirs (e.g., shale plays). Besides the high-pressure injection requirement to intrude the pores in shale, presence of slippage and Knudsen diffusion effects, and mixed wettability of shale pore systems due to the presence of both organic and inorganic surfaces, complicates the permeability estimation on shale samples (Dadmohammadi et al., 2016). There has been a significant amount of recent studies on quantifying the apparent permeability of shales by combining Poiseuille's law with consideration of nonzero velocity on pore walls (Javadpour, 2009; Civan, 2010; Darabi et al., 2012; Singh et al., 2014).

Relative Permeability Measurements Using Digital Rock Physics

Digital rock physics (DRP), or pore-scale modeling, has been introduced as a valuable technology for laboratory measurement of reservoir rock properties (Masalmeh et al., 2015). DRP creates digital representation of the rock by combining advanced 3D imaging techniques such as X-ray computed tomography (CT) scanning or focused ion beam scanning electron microscopy (FIB-SEM). Thereafter, advanced numerical models are used to evaluate different properties of the rock including relative permeability. In this method, the boundary conditions need to be defined appropriately to mimic the correct special core analysis (SCAL) experimental procedures. Thereafter, an improved lattice Boltzmann method (LBM) is used to simulate the complex fluid movement through the rock (De-Priscoet al., 2012). The DRP workflow can be summarized into three steps (Zahaf,et al., 2017):

Step 1 involves the selection and characterization of the core samples. This step is used to determine porosity and pore-size distribution using a combination of visual description, plug overview X-ray scanning, mineral evaluation, scanning electron microscopy, and backscattered electrons. Information obtained from this analysis is used to generate the micro-CT and nano-CT subsampling plan.

Step 2 is used to construct the rock model. The processing and segmentation of the 3D X-ray images were conducted prior to any simulations. The processing part includes image enhancement and generating a three-dimensional rock model. Thereafter, the rock properties are calculated after quality control. The quality control is performed to validate pore-size distribution and porosity of the digital rock model with laboratory data.

Step 3 consists of upscaling and multiphase flow modeling. The upscaling is used to average properties from pore-scale network modeling to the whole core plug (Zahaf et al., 2017). More details on this workflow are discussed in Chapter 4.

Relative Permeability Measurements Using Adsorption-Desorption Techniques

The size and the shape of the pores and their connections control mass and energy transfer through porous media. Relevant transport properties such as permeability is a function of pore connections and pore-size distribution (Katsube & Williamson, 1994; Sakhaee-Pour & Bryant, 2015). Daigle et al. (2015) proposed a model to estimate the relative permeability of wetting and nonwetting phases in shale rocks using a combination of percolation theory, critical path analysis (CPA), and effective medium theory. The percolation theory is a branch of probability theory that deals with the properties of random media (Berkowitz & Ewing, 1998). It was first introduced to describe polymerization (Flory, 1941) and penetration of fluids in porous media (Broadbent & Hammerslcy, 1957). The percolation theory provides universal laws to determine the geometrical and physical properties of a system (Berkowitz, 1993). This theory simply indicates the paths with the minimum pressure gradient and flow resistance in which the fluid can flow easily. The percolation properties related to permeability measurements were determined from N_2 adsorption-desorption (AD) measurements. In this method, after estimating the pore-size distribution and the fractal and percolation parameters, the relative permeability curves of aqueous and hydrocarbon phases are estimated (Ojha et al., 2017). Due to this method, the relative permeability of the wetting phase is expressed different θ_t conditions (Daigle et al., 2015). Equation (2.90) is used for $\theta_t > \varphi_1$, and equation (2.91) is used for $\theta_t < \varphi_1$. The relative permeability of the nonwetting phase is expressed as equation (2.92).

$$k_{rw}(\theta_w) = \left[\frac{\beta_2 - \varphi_2 + \theta_w - \theta_t}{\beta_1 - \theta_t}\right]^{\frac{3}{3-D_2}}, \tag{2.90}$$

$$k_{rw}(\theta_w) = \begin{cases} \left[\frac{\beta_1 - \varphi + \theta_w - \theta_t}{\beta_1 - \theta_t}\right]^{\left(\frac{3}{3-D_1}\right)}, & \theta_w - \theta_t > \varphi_2, \theta_w > \varphi_2 \\ \left(\frac{\beta_1^{\left(\frac{3}{3-D_1}\right)}}{\beta_2^{\left(\frac{3}{3-D_2}\right)}}\right) \left[\frac{\beta_1 - \varphi}{\beta_1}\right]^{\left(\frac{3}{3-D_1}\right)} \left[\frac{\beta_2 - \varphi_2 + \theta_w - \theta_t}{\beta_1 - \theta_t}\right]^{\left(\frac{3}{3-D_2}\right)}, & \theta_w - \theta_t < \varphi_2, \theta_w > \varphi_2 \end{cases}, \tag{2.91}$$

and

$$k_{rnw}(\theta_{nw}) = \begin{cases} \dfrac{(\varphi-\theta_t)(\theta_{nw}-\theta_t)^2}{(\theta_x-\theta_t)(\varphi-\theta_t)^2}, & \theta_t < \theta_{nw} < \theta_x \\ \dfrac{(\theta_{nw}-\theta_t)}{(\varphi-\theta_t)}, & \theta_x < \theta_{nw} < \varphi \end{cases}, \tag{2.92}$$

where k_{rw} is the permeability of the wetting phase; k_{rnw} is the permeability of the nonwetting phase; φ is the ADI porosity of the sample, which is obtained by multiplying the cumulative pore volume from calculated pore-size distribution at nitrogen saturation pressure with bulk density of the sample; β is the probability of drainage and it is defined as the ratio of pore volume to the sum of pore solid volume in a repeating fractal model (Perfect, 2005).

The parameter D is the fractal dimension and quantifies the complexity of a pattern in terms of the change in detail to the change in scale. In another word, D describes the self-similar scaling of a porous medium.

The variable $\theta(d)$ is the cumulative volume distribution of pore sizes and it is defined as equation (2.93). The variable θ_x is defined using equation (2.94), and φ_1 and φ_2 are calculated using equation (2.95) and (2.96), respectively.

$$\theta(d) = \begin{cases} \varphi_1(d), & d_{min} < d < d_x \\ \varphi_1(d_x) + \varphi_2(d), & d_x < d < d_{max} \end{cases}, \tag{2.93}$$

$$\theta_x = \theta_t + \frac{\varphi}{Z}, \tag{2.94}$$

$$\varphi_1(d) = \beta_1\left[\left(\frac{d}{d_x}\right)^{(3-D_1)} - \left(\frac{d_{min}}{d_x}\right)^{(3-D_1)}\right], \tag{2.95}$$

and

$$\varphi_2(d) = \beta_2\left[\left(\frac{d}{d_{max}}\right)^{(3-D_2)} - \left(\frac{d_x}{d_{max}}\right)^{(3-D_2)}\right], \tag{2.96}$$

where d_x is the pore diameter at which pore behavior changes from the fractal regime 1 to fractal regime 2 and D_1, β_1 to D_2, β_2; Z is the number of immediate neighbors of each pore in the pore structure; and d_{min} and d_{max} are minimum and maximum pore diameter in samples that influence the adsorption-desorption isotherm (Daigle et al., 2015).

Relative Permeability Measurements Using Modified Purcell Method

In this technique, the original Purcell's model, which was developed for uniform-wettability systems, is modified to be applied for two-phase flow of water and gas in a dual-porosity/dual-wettability medium. It means that the wettability and porosity of the system is not homogeneous and is varying along the core sample. Figure 2.34 shows the schematic of pore-throat (a), and the inverse-square of Pc (MICP) vs. water saturation (b) in a dual-porosity/dual-wettability system. Part I shows the small nano-sized pores, which are inside the organic matter, and part II shows the microscale intraparticle inorganic pores. In this model, pores within the range of $0 < r < r_{org}$ are strongly water repellant and are filled by gas. In this definition, r_{org} represents the largest pores in part I. Based on this condition, the pores within the range of $r_{org} < r < r_{Swa}$ in part II are filled with water and the pores in the range of $r > r_{Swa}$ in part II are filled with gas (Yassin et al., 2016). Therefore, water saturation and relative permeability values of wetting and nonwetting fluid phases, by considering the absence of water in part I, the wetting phase saturation (S_w) in MICP, is related to S_{org} and S_{wa} by equation (2.97), and the relative permeability of wetting (water) and nonwetting (gas) phases are defined using equations (2.98) and (2.99), respectively.

$$S_w = S_{org} + S_{wa}, \tag{2.97}$$

$$k_{rw}(S_{wa}) = \frac{\int_0^{(S_{org}+S_{wa})} \frac{dS_w}{P_c^2} - \int_0^{S_{org}} \frac{dS_w}{P_c^2}}{\int_0^1 \frac{dS_w}{P_c^2}} = \frac{\int_{S_{org}}^{(S_{org}+S_{wa})} \frac{dS_w}{P_c^2}}{\int_0^1 \frac{dS_w}{P_c^2}}, \tag{2.98}$$

and

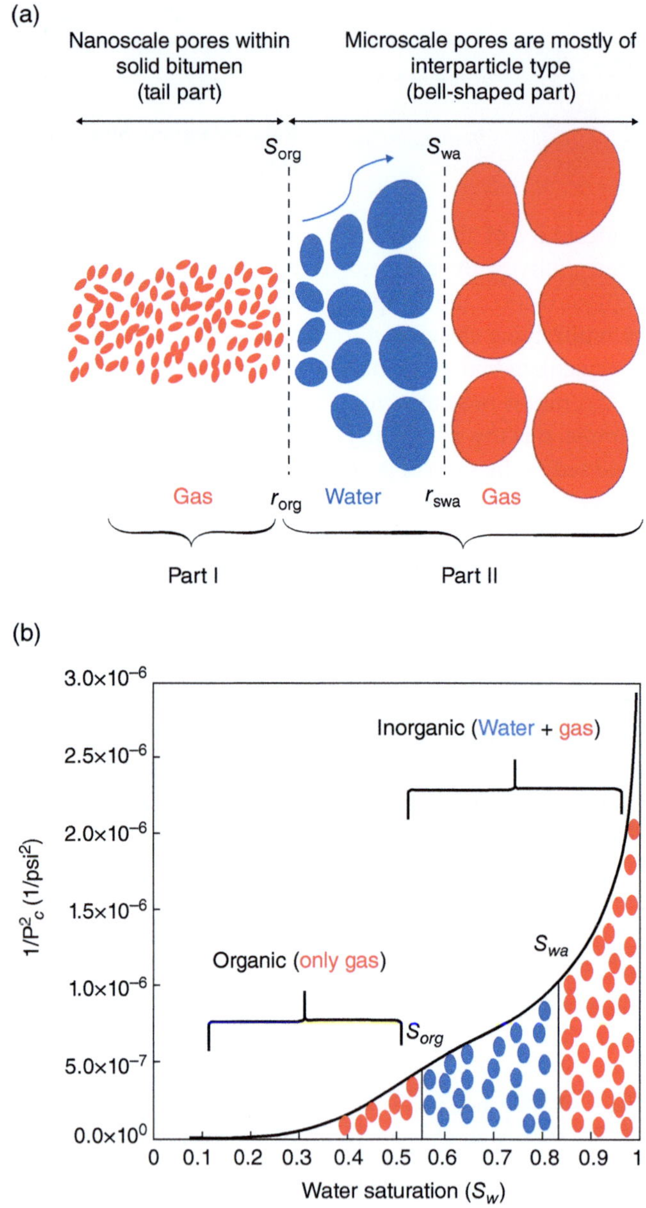

Figure 2.34 (a) Schematic of pore-throat and (b) inverse-square of Pc (MICP) versus water saturation in a dual-porosity/dual-wettability system (Yassin et al., 2016).

$$k_{rg}(S_{wa}) = \frac{\displaystyle\int_{(S_{org}+S_{wa})}^{1} \frac{dS_w}{P_c^2} + \int_0^{S_{org}} \frac{dS_w}{P_c^2}}{\displaystyle\int_0^1 \frac{dS_w}{P_c^2}}. \tag{2.99}$$

This model has two limitations. First, Part I was considered to have no water, but, it may contain clay particles, which have high affinity for water adsorption (Dehghanpour et al., 2013). Second, although the large pores within the range of $r_{org} < r < r_{Swa}$ in Part II are considered to be hydrophilic, some of them may be coated with organic matter (Yassin et al., 2016).

Yassin et al. (2016) observed that this definition can be extended to be applied on Wyllie and Gardner's equation (Wyllie & Gardner, 1958). In this model, which was proved by Nakornthap and Evans (1986), the relative permeability is obtained using the capillary pressure curve. Equations (2.100) and (2.101) are used to calculate the oil and gas relative permeability values, respectively.

$$k_{ro} = \left(S_o^*\right)^2 \frac{\int_0^{S_o^*} \frac{1}{P_c^2} dS_o^*}{\int_0^1 \frac{1}{P_c^2} dS_o^*}, \tag{2.100}$$

and

$$k_{rg} = \left(1 - S_o^*\right)^2 \frac{\int_{S_o^*}^1 \frac{1}{P_c^2} dS_o^*}{\int_0^1 \frac{1}{P_c^2} dS_o^*}, \tag{2.101}$$

where k_{ro} is the permeability of the oil phase, k_{rg} is the permeability of the gas phase, P_c is the capillary pressure, and S_o^* is the normalized oil saturation. It is defined by equation (2.102),

$$S_o^* = \frac{S_o - S_{or}}{1 - S_{or}}, \tag{2.102}$$

where S_o is the oil saturation, fraction; and S_{or} is the residual oil saturation, fraction.

2.6.5 NMR Capillary Pressure

As observed from the NMR T_2 spectra above for various shale formations, fluids in the micropore regions of shales with fast relaxation are challenging to extract during recovery due to contributing effects of capillary pressure. NMR T_2 spectra of shales show nonuniform pore shapes and sizes, which interact with fluids (both invading and evacuating) differently creating a disproportion in pressure known as the capillary pressure (P_c). Capillary pressure (P_c) is therefore a measure of the resistance of either the removal of the wetting phase or the injection of the nonwetting phase in a medium. It is known to vary with the interfacial tension between nonwetting and wetting phases, the advancing or receding contact angle between the fluid and the pore throat, and, inversely, as the pore radius (Altunbay, 2001; Martinez, 2000). It is expressed as

$$P_c = \frac{2\sigma \cos\theta}{r}, \tag{2.103}$$

where r is the pore radius, σ is the interfacial tension between the fluids, and θ is the contact between the fluid and rock surface.

A comparison between the NMR pore-size distribution equation (2.59), where the T_2 relaxation time is proportional to the surface-to-volume ratio (S/V), and that of the capillary pressure, equation (2.103), through the pore radius yields equation (2.104), where the surface-to-volume ratio (S/V) is expressed in terms of the pore size as the ratio of the shape factor (β) to the pore size (β/r). The shape factor (β) of round pores is equated to three, six for spherical pores, and two for cylindrical tubes. The radius in the capillary pressure equation is defined in terms of the pore throat (r_{pt}) since the nonequilibrium pressure between two phases is controlled by the pore throat, while the radius of the NMR pore-size distribution equation is defined in terms of the pore body/size (r_b) (Altunbay, 2001).

$$\frac{1}{P_c} = \frac{\rho}{2\sigma\cos\theta} \frac{r_{pt}}{r_b} \beta T_2. \tag{2.104}$$

A further simplification of equation (2.104) yields the correlation as shown in equation (2.105), which can be used to scale the fluid saturation versus T_2 spectrum to match fluid saturation with capillary pressure ($1/P_c$) curve (Altunbay, 2001; Martinez, 2000).

$$Log(A) = Log\left(\frac{1}{P_c}\right) - Log(T_2), \tag{2.105}$$

where A is

$$A = \frac{\rho}{2\sigma\cos\theta} \frac{r_{pt}}{r_b} \beta. \tag{2.106}$$

2.6.6 Relative Permeability from NMR Pseudocapillary Pressure

The irreducible fluid saturation or volume in NMR is defined by the BVI from T_2 pore-size distribution spectrum, while conventional methods use the fractional flow (F) defined as the relative flow rate of one fluid in the presence of other fluids. Thus, the fractional flow (F) is a function of both fluid viscosity (μ) and relative permeability (k_r) as expressed by equation (2.107) (Marschall, 1995),

$$F_w = \frac{1}{1 + \frac{k_{rnw}}{k_{rw}} \frac{\mu_w}{\mu_{nw}}}, \tag{2.107}$$

where, F_w is the fraction flow of the wetting phase; k_{rnw} and k_{rw} are the relative permeabilities of the nonwetting and wetting phases, respectively; and μ_w and μ_{nw} are the viscosities of the wetting and nonwetting phases, respectively. This method may still be applicable in some of the tight oil reservoirs.

For a fully saturated rock (i.e., $S_W = 100\%$), both the capillary pressure curve and the NMR T_2 spectrum are assumed to represent the pore-size distribution (Marschall, 1995). Relative permeability can be computed from the capillary pressure data using equations from Standing (1974) on the NMR pseudocapillary pressure curve for both drainage and imbibition. However, the equations (2.108) and (2.109) are for drainage.

$$K_{rw} = \left(S_w^*\right)^2 \frac{\int_0^{S_w^*} \frac{dS_w^*}{P_c^2}}{\int_0^1 \frac{dS_w^*}{P_c^2}} \tag{2.108}$$

and

$$K_{rnw} = \left(1 - S_w^*\right)^2 \frac{\int_{S_w^*}^1 \frac{dS_w^*}{P_c^2}}{\int_0^1 \frac{dS_w^*}{P_c^2}}, \tag{2.109}$$

where Sw*

$$S_w^* = \frac{(S_w - S_{wi})}{(1 - S_{wi})}. \tag{2.110}$$

2.7 Saturation

Saturation is the measure of the fluid volume occupying the pore volume of a porous medium. The saturation of each fluid is defined by the ratio of that fluid to the total pore volume (Helander, 1983). Therefore, considering three phases in the reservoir (i.e., water, oil, and gas), their saturations are defined in equations (2.111) to (2.113). Therefore, for a system containing three phases, the summation of all saturations is equal to unity, which is shown in equation (2.114).

$$S_w = \frac{V_w}{V_p}, \tag{2.111}$$

$$S_o = \frac{V_o}{V_p}, \tag{2.112}$$

$$S_g = \frac{V_g}{V_p}, \tag{2.113}$$

and

$$S_w + S_o + S_g = 1. \tag{2.114}$$

2.7.1 Techniques for Calculating Water Saturation

Water saturation can be determined by different methods and, hence, these different approaches lead to somewhat different S_w values that may result in considerable differences in the original oil in place (OOIP) or original gas in place (OGIP) volumes. These methods inclde (1) calculation of S_w using resistivity logs, (2) calculation of S_w from laboratory capillary pressure-saturation (P_c- S_w) measurements, (3) calculation of S_w using oil-based mud (OBM)-core-plug Dean-Stark water-volume, and (4) combinations of these methods.

2.7.2 Resistivity Logs

The most common technique for calculating S_w is the use of resistivity logs with a model (empirical or theoretical) that relates S_w to R_t, R_w, and porosity. The Archie's model is a well-known method to calculate water saturation.

Clean Sand–Archie's Model

In this model (Archie, 1942), the formation water is the only electrically conductive medium. This model can be applied in many sandstone and carbonate reservoirs with the low clay-mineral content, and the water saturation can be calculated using equations (2.115) and (2.116) (Archie, 1952, 1942).

$$S_w = \left(\frac{aR_w}{R_t * \phi^m}\right)^{1/n} \tag{2.115}$$

and

$$S_w = \left(\frac{R_o}{R_t}\right)^{1/n}, \tag{2.116}$$

where S_w is water saturation of the uninvaded zone; n is saturation exponent, which varies from 1.8 to 4.0 but normally is 2.0; R_w is formation water resistivity at formation temperature; φ is porosity; m is cementation exponent, which varies from 1.7 to 3.0 but normally is 2.0; R_t is true resistivity of the formation, corrected for invasion, borehole, thin bed, and other effects; and R_o is the resistivity of a water-saturated rock.

This model is mainly used for clean reservoirs. In the presence of a significant fraction of montmorillonite or finely laminated shale and sand sequences in the formation content, a shaly sand model must be used to compute the water saturation.

Shaly Sand Model

Due to the increase in the R_t values as a result of the clay-bound water, the calculated water-saturation values using the clean sand model are overestimated. Hence, a more advanced electrical model is developed to account for the effects of conductive clay minerals and shale on rock resistivity. All the shaly sand models are developed in a way that they cover the shale content effect, and they reduce to Archie's equation when the shale volume is zero.

Laminated Sand/Shale Model

A parallel resistor model is used to model the multiple thin parallel layers of 100% shale interbedded with the layers of clean sand. The width of the thin beds is within the vertical resolution of the resistivity-logging tool. Equation (2.117) is utilized for this purpose.

$$\frac{1}{R_t} = \frac{V_{sh}}{R_{sh}} + \frac{1-V_{sh}}{R_{sd}}, \tag{2.117}$$

where R_{sd} is the clean-sand resistivity and is computed using equation (2.118),

$$R_{sd} = \frac{R_w}{(\phi_{sd}^m * S_{wsd}^n)}, \tag{2.118}$$

in which, the effective porosity depends on the sand fraction of the bulk volume and it is expressed by equation (2.119). Here,

$$\phi_e = \phi_{sd}(1 - V_{sh}), \tag{2.119}$$

where ϕ_{sd} is assumed from neighboring thick sands; and S_{wsd} is the S_w of the sand, equation (2.120),

$$S_{wsd} = S_{we}\ . \tag{2.120}$$

Simandoux Model

The Simandoux (1982) model gave the algorithm for water saturation as shown in equation (2.121),

$$\frac{\phi_e^m}{a\,R_w} S_w^n + \frac{V_{sh}}{R_{sh}} S_w = \frac{1}{R_t}, \tag{2.121}$$

where a is tortuosity exponent; m is cementation exponent; n is saturation exponent; R_w is formation water resistivity; R_{sh} is shale resistivity; R_t is formation resistivity; S_w is water saturation, and ϕ_e is effective porosity.

Modified Simandoux Model

The Simandoux model was modified by inserting (1-V_{sh}) to give better accommodation for shaliness of the formation. The model came to be known as the modified Simandoux equation and is shown in equation (2.122):

$$\frac{\phi_e^m}{a\,R_w(1 - V_{sh})} S_w^n + \frac{V_{sh}}{R_{sh}} S_w = \frac{1}{R_t}\ . \tag{2.122}$$

Poupon-Leveaux (Indonesia) Model

The Indonesian equation introduced by Poupon and Leuveaux (1971) is also a modified Simandoux model to achieve a better evaluation of the fresh water formations (Hari-Kumar, 2010). This model is based on standard log analysis parameters, and it generates reliable results. The model is presented in equations (2.123) – (2.127). In this model, the effective porosity, $Ø_e$, shale volume and resistivity (V_{sh} and R_{sh}), and water and deep resistivity (R_w and R_t) are inputs; and S_w is an output that is usually taken to be the water saturation of the effective porosity (Poupon & Leveaux, 1971). However, recent studies suggested that the output is likely to estimate the water saturation of the total porosity, S_{wt}, as it is shown in the next model (Woodhouse & Warner, 2005).

$$A = \frac{a\,R_w}{\phi_e^m}, \tag{2.123}$$

$$B = 1 - (V_{sh} \times 0.5), \tag{2.124}$$

$$C = \frac{V_{sh}^B}{\left(\frac{R_{sh}}{R_t}\right)^{0.5}}, \tag{2.125}$$

$$D = \left(\frac{R_t}{A}\right)^{0.5}, \tag{2.126}$$

and

$$S_w = (C + D)^{\frac{-2}{n}} . \tag{2.127}$$

Waxman-Smits-Thomas and Dual-Water Models

The complex multiparameter Waxman-Smits-Thomas (WST) and dual-water (DW) equations are used iteratively to find S_{wt}. The WST model is based on laboratory measurements of resistivity, porosity, and saturation of real rocks (Waxman & Smits, 1968; Waxman & Thomas, 1974; Keelan & McGinley, 1979; Juhasz, 1979). The model is defined as equation (2.128),

$$\frac{1}{R_t} = Ø_t^{m^*} \times S_{wt}^{n^*}\left(\frac{1}{R_w} + \frac{BQ_V}{S_{wt}}\right), \tag{2.128}$$

where S_{wt} is water saturation of the total porosity; B is specific cation conductance in (1/ohm-m)/(meq/mL); Q_V is the cation-exchange capacity (CEC) per unit PV; and m^* and n^* apply to the total PV.

Considering the clay-bound water conductivity instead of WST's BQ_V factor results in the DW model (Woodhouse & Warner, 2005; Clavier et al., 1984; Best et al., 1979). This model uses S_{wb} as an alternative shale-volume descriptor, and the saturation of physically bound water in the total PV (Juhasz, 1979, 1989). The model is shown in equation (2.129). When V_{sh} is zero, S_{wb} is zero; and when V_{sh} is 100% BV, S_{wb} and S_{wt} are also 100% PV.

$$\frac{1}{R_t} = Ø_t^{m_o} S_{wt}^{n_o}\left[\frac{1}{R_{wf}} + \frac{S_{wb}}{S_{wt}}\left(\frac{1}{R_{wb}} - \frac{1}{R_{wf}}\right)\right], \tag{2.129}$$

where R_{wb} is resistivity of the clay-bound water (Morrison, 1984) in the shale and computed from equation (2.130); R_{wf} is resistivity of free formation water in the shale-free water zones and computed from equation (2.131); m_o and n_o must be smaller than the WST exponents (Clavier et al., 1984) and may have values similar to "clean" sand exponents.

$$R_{wb} = R_{sh}\phi_{sh}^m \tag{2.130}$$

and

$$R_{wf} = R_t\phi_t^m. \tag{2.131}$$

Alternatively, the water saturation using the dual-water method can be calculated using equation (2.132),

$$\frac{1}{R_w} S_w^n + \left(\frac{\phi_{sh}\,V_{sh}}{\phi_t}\right)\left(\frac{1}{\phi_{sh}^2\,R_{sh}} - \frac{1}{R_w}\right)S_w^{(n-1)} = \frac{a}{R_t\,\phi_t^m}, \tag{2.132}$$

where ϕ_{sh} is the porosity in 100% shale.

Liu et al. Approach

Liu et al. (2017) developed a new method for calculating water saturation in shale gas reservoirs by applying the ratio of compressional to shear wave velocity (V_p/V_s) and porosity. In this model, after estimating the ranges of the total water saturation, total porosity, permeability, fluid viscosity, and sonic frequency, an empirical equation was obtained to calculate the water saturation using total porosity and V_p/V_s ratio (Liu et al., 2017). In this model, the water saturation is defined as equation (2.133).

$$S_w = x_a \frac{\left(V_p/V_s/V_pV_{s0}\right)^{x_n}}{\varphi^{x_m}}, \tag{2.133}$$

where x_a, x_m, and x_n are empirical coefficients, which can be obtained using the correlation between experimental water saturation, experimental porosity and V_p/V_s ratio by dipole sonic logging; V_p/V_s is defined as equation (2.134),

$$\frac{V_p}{V_s} = a_1\varphi^{a_2}S_w^{a_3}, \tag{2.134}$$

where V_p is the compressional wave velocity, µft/s; V_s is the share wave velocity, µft/s; and a_1, a_2, and a_3 are the empirical coefficients to be determined.

Example 2.14 Determine the Water Saturation

Calculate the water saturation for a well targeting the Bakken play using the Simandoux model, the modified Simandoux model, the Indonesia model, and the dual-water method. The data are provided in Table 2.23. The shale resistivity (Rsh) and the shale total porosity (ϕ_{sh}) are 23.98 and 0.294, respectively. The cementation factor (m) is 1.4, the saturation exponent (n) is 1.74 and the tortuosity factor (a) is 1.

Solution

The water saturation using the Simandoux model can be found by the use of equation (2.121). A sample calculation for the depth of 9970 is shown below, and the final results are summarized in Table 2.24.

$$\frac{\phi_e^m}{a\,R_w}S_w^n + \frac{V_{sh}}{R_{sh}}S_w = \frac{1}{R_t} = \frac{0.1357^{1.4}}{1 \times 0.0297}S_w^{1.74} + \frac{0.7551}{23.98}S_w = \frac{1}{7.5025}.$$

Now, solve for the water saturation; thus $S_{w(simandoux)} = 0.202$.

The water saturation using the modified Simandoux model can be found by the use of equation (2.122). A sample calculation for the depth of 9970 is shown below:

$$\frac{\phi_e^m}{a\,R_w(1-V_{sh})}S_w^n + \frac{V_{sh}}{R_{sh}}S_w = \frac{1}{R_t} = \frac{0.1357^{1.4}}{1 \times 0.0297 \times (1-0.7751)}S_w^{1.74} + \frac{0.7551}{23.98}S_w = \frac{1}{7.5025}.$$

Now, solve for the water saturation; thus $S_{w(Modified\ simandoux)} = 0.091$.

Table 2.23 Data for Example 2.14

Depth	Total porosity	Effective porosity	RT	Rw	Vsh
9,962	0.2940	0.2646	258.8767	0.0175	0.1945
9,963	0.2932	0.2623	194.4657	0.0224	0.2039
9,964	0.3116	0.2688	128.1388	0.0268	0.2826
9,965	0.3572	0.3016	105.1738	0.0183	0.3671
9,966	0.3576	0.3133	42.8727	0.0191	0.2923
9,967	0.3510	0.3034	33.8783	0.0270	0.3143
9,968	0.3403	0.2493	30.7185	0.0246	0.6001
9,969	0.2939	0.1439	23.9789	0.0186	0.9900
9,970	0.2501	0.1357	7.5025	0.0297	0.7551

Table 2.24 Summarized Results for Example 2.14

Depth	Total porosity	Effective porosity	RT	Rw	Vsh	Sw (Simandoux)	Sw (Modified Simandoux)	Sw (Indonesia)	Sw (Dual-water method)
9,962	0.2940	0.2646	258.8767	0.0175	0.1945	0.012	0.010	0.011	0.194
9,963	0.2932	0.2623	194.4657	0.0224	0.2039	0.016	0.014	0.016	0.204
9,964	0.3116	0.2688	128.1388	0.0268	0.2826	0.022	0.018	0.021	0.266
9,965	0.3572	0.3016	105.1738	0.0183	0.3671	0.018	0.014	0.018	0.301
9,966	0.3576	0.3133	42.8727	0.0191	0.2923	0.030	0.025	0.029	0.243
9,967	0.3510	0.3034	33.8783	0.0270	0.3143	0.043	0.035	0.042	0.269
9,968	0.3403	0.2493	30.7185	0.0246	0.6001	0.050	0.030	0.048	0.518
9,969	0.2939	0.1439	23.9789	0.0186	0.9900	0.074	0.005	0.069	0.986
9,970	0.2501	0.1357	7.5025	0.0297	0.7551	0.202	0.091	0.182	0.905

The water saturation using the Indonesia model can be found by the use of the set of equations (2.123) – (2.127). A sample calculation for the depth of 9970 is shown below:

$$A = \frac{a\,R_w}{\phi_e^m} = 1 \times \frac{0.0297}{0.1357^{1.4}} = 0.485,$$

$$B = 1 - (V_{sh} \times 0.5) = 1 - (0.7551 \times 0.5) = 0.623,$$

$$C = \frac{V_{sh}^B}{\left(\frac{R_{sh}}{R_t}\right)^{0.5}} = \frac{0.7551^{0.623}}{\left(\frac{23.98}{7.5025}\right)^{0.5}} = 0.469,$$

$$D = \left(\frac{R_t}{A}\right)^{0.5} = \left(\frac{7.5025}{0.485}\right)^{0.5} = 3.932.$$

Thus, the water saturation is $S_{w(indonesia)} = (C + D)^{\frac{-2}{n}} = (0.469 + 3.932)^{\frac{-2}{1.74}} = 0.182$.

The water saturation using the dual-water method can be found by the use of equation (2.132). A sample calculation for the depth of 9970 is shown below:

$$\frac{1}{R_w} S_w^n + \left(\frac{\phi_{sh}\,V_{sh}}{\phi_t}\right)\left(\frac{1}{\phi_{sh}^2\,R_{sh}} - \frac{1}{R_w}\right) S_w^{(n-1)} = \frac{a}{R_t\,\phi_t^m},$$

$$\frac{1}{0.0297} S_w^{1.74} + \left(\frac{0.294 \times 0.7551}{0.2501}\right)\left(\frac{1}{0.294^2 \times 23.98} - \frac{1}{0.0297}\right) S_w^{(1.74-1)} = \frac{1}{7.5025 \times 0.2501^{1.4}}.$$

Thus, the water saturation is $S_{w(Dual\ water\ method)} = 0.905$.

2.7.3 NMR Saturation Estimation

The NMR T_2 distribution can be used to estimate the amount or percentage hydrocarbon recovered from a core after enhanced oil recovery by the integrated area difference between the remaining oil and that of the fully saturated distributions as detailed by Cudjoe et al. (2019), where their NMR recovery estimation after huff-n-puff recovery was comparable to the mass balance estimations. The area under the T_2 distribution curves encompasses the bound region ($T_2 < 1$ ms) to the inorganic pore region ($T_2 > 16$ ms).

2.8 Wettability

Wettability is defined as the tendency of one fluid to spread on a solid surface in the presence of other immiscible fluids (Craig, 1971). Wettability is a key property that has significant impact on the fluids distribution and fluid-flow behavior inside the reservoir rock (Buckley & Lord, 2003). In addition, wettability is important in the analysis of fluid loss in the hydraulic fracturing fluids in unconventional reservoirs (Roshan et al., 2016). Furthermore, wettability of the porous media affects the improved recovery performance using gas injection.

2.8.1 Wettability Measurement

Several methods have been proposed for rock wettability measurements, including contact angle and adhesion tests, spontaneous and forced imbibition (Amott and USBM), flotation, scanning electron microscopy (SEM), atomic force microscopy (AFM), capillary pressure, and nuclear magnetic resonance (NMR) (Anderson, 1986). Traditional techniques work well for homogeneous sandstones; however, they are not applicable in shale rock samples (Odusina et al., 2011). Due to the extremely low porosity and permeability of the unconventional reservoirs (e.g., shale plays) in addition to their water-sensitive clay content, Amott and USBM techniques are very difficult and time-consuming methods to apply (Thyne, 2013). Wang et al. (2012) used a modified Amott-Harvey technique to measure the wettability of thin slides of core plugs of Bakken play. Thereafter, Xu and Dehghanpour (2014) used contact angle measurement technique and imbibition tests to analyze the wettability of crushed shale packs.

Wettability Measurement Using NMR

Zhang et al. (2014) conducted different experiments to study the wettability of shale formations using NMR. Figure 2.35 shows the

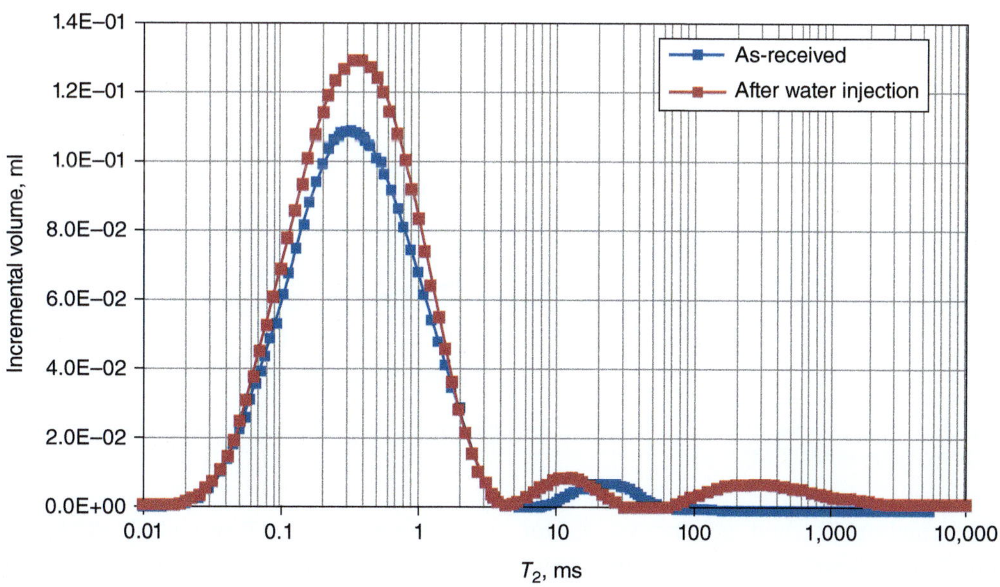

Figure 2.35 Incremental NMR fluid volume versus T_2 for Marcellus shale core before and after water injection (Zhang et al., 2014).

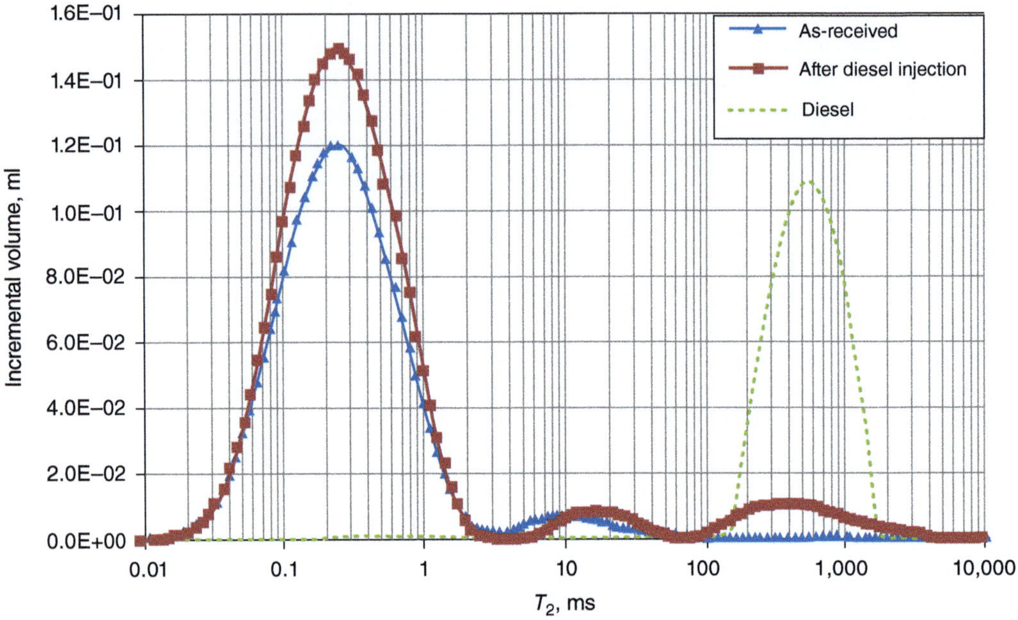

Figure 2.36 Incremental NMR fluid volume versus T_2 for Marcellus shale core before and after diesel injection (Zhang et al., 2014).

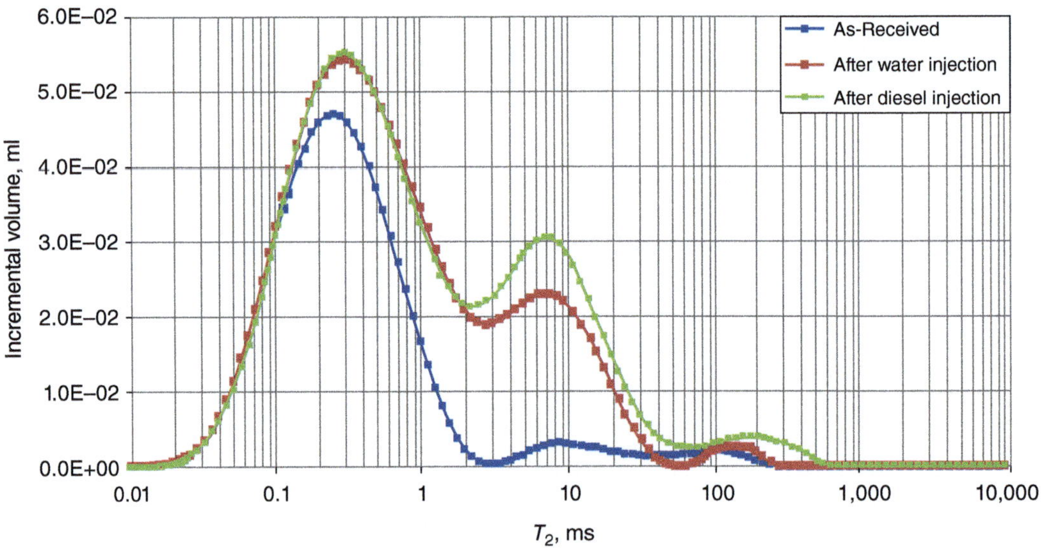

Figure 2.37 Incremental NMR fluid volume versus T_2 for Mancos shale core before and after water and diesel injection (Zhang et al., 2014).

Incremental NMR fluid volume vs. T_2 for Marcellus shale core before and after water injection. Figures 2.36 and 2.37 show the incremental NMR fluid volume versus the T_2 relaxation time for Marcellus and Mancos shale plays, respectively (Zhang et al., 2014). Comparing three plots in each figure provides information about the wettability of the cores. Since the first peak (before 1 ms) is correlated to the small pores, increasing the peak due to water and diesel injection corresponds to wetting properties of small pores. For small pores, with radii smaller than 60 nm ($T_2 < 5\ ms$), it can be observed that the NMR signal increases after water injection for both shale plays. However, after oil injection, Mancos shale core does not show any increase in NMR signal, while the NMR signal increases in Marcellus shale. Therefore, it can be concluded that the small pores in Marcellus shale are mixed-wet while they show water-wet behavior in Mancos shale core. For larger pores in the range of several hundred nanometers to micrometers, the Marcellus shale does not show any significant response, which can be due to not many pores with such a size existing in these core samples. For the Mancos shale core, a notable magnitude increase is observed after both water and oil injections, which is translated as the mixed-wet nature of the larger pores (Zhang et al., 2014).

Thus, a rock sample saturated with a fluid for NMR measurement prompts an increase in the T_2 amplitude and a shift to faster T_2 values when compared with the bulk fluids to indicate the dominance of fluid-pore surface interaction (surface relaxation) (Zhang et al., 2014). Figures 2.18 and 2.19, which show the NMR T_2 distributions of saturated and as-received Eagle Ford outcrop and LEF samples, are clear indications of this effect; the bulk relaxation of the hexadecane injected into the Eagle Ford outcrop core recorded a dominant peak at 614 ms and a secondary small peak at 0.24 ms as shown in Figure 2.38, while the oil injected into the reservoir cores recorded a sole dominant peak of 706 ms. Upon injecting hexadecane into the outcrop, assumed to have very low TOC and less-developed organic pores, the amplitude within the overlapped organic and inorganic pore region increased compared with that of the micropores in the bound region to indicate the affinity of hexadecane to the organic and inorganic pore region, due to the presence of organic pores (oil-wet) in the presence of TOC and mixed-wet pores, as opposed to the capillary-bound microregion. From Figure 2.38, the microregion coincides with partial bulk relaxation of hexadecane at ~0.24 ms. On the other hand, upon injecting oil into the LEF reservoir cores (Figure 2.19), the mesoregion identified to consist of organic pores increased in amplitude to suggest its affinity for oil (oil-wet). Both the micropores and the macropores including fractures had reduced amplitude but the macropores more so than the micropores; the micropores are

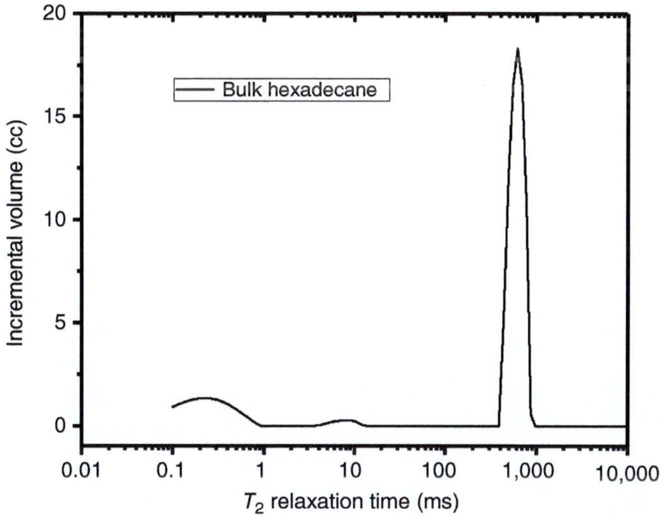

Figure 2.38 Bulk hexadecane (HD) relaxation time.

likely to be mixed-wet, while the macropores are likely be completely water-wet.

Wettability Measurement Using Modified Amott-Harvey and XPS Techniques

Modified Amott-Harvey and XPS techniques were used to measure surface composition, and were used to assess reservoir rock wettability (Mitchell et al., 1990; Araujo et al., 1995). X-ray photoelectron spectroscopy (XPS) is a surface-sensitive quantitative spectroscopic technique, which is used to measure the elemental composition, and chemical and electronic state of the elements. The scattered electrons with a characteristic energy collide to the surface, and the intensity of the peaks observed in the energy spectra is correlated to the element's concentration in the surface (Van der Heide, 2011). This technique has a lateral resolution of 5 mm (Araujo et al., 2017). Araujo de Itriago et al. conducted experiments to analyze the wettability of different shale samples. In their method, after the samples were put in a Hassler core holder, they were vacuumed, and the gas was removed from the lines and core. Thereafter, the imbibition tests were conducted by soaking the cores in crude oil for 24–32 hours. The injection process was started after the imbibition test under the same circumstances. The imbibition and injection volumes were obtained by reading the amount of oil in the burette, and the wettability condition was calculated using equations (2.135) - (2.137) (Araujo et al., 2017). In this method, the WI is translated to the wettability of the rock as follows: (1)The rock is oil wet if $-1.0 \leq WI \leq -0.3$; (2) the rock is intermediate wet if $-0.3 \leq WI \leq 0.3$; and (3) the rock is water wet if $0.3 \leq WI \leq 1.0$.

$$\delta_w = \frac{V_{w_{imbibition}}}{V_{w_{imbibition}} + V_{w_{injection}}}, \tag{2.135}$$

$$\delta_o = \frac{V_{o_{imbibition}}}{V_{o_{imbibition}} + V_{o_{injection}}}, \tag{2.136}$$

and

$$WI = \delta_w - \delta_o, \tag{2.137}$$

where δ_w is the ratio of spontaneous water imbibition to total water imbibition; δ_{wo} is the ratio of spontaneous oil imbibition to total oil imbibition; $V_{w_{imbibition}}$ is the water volume that imbibed spontaneously; $V_{o_{imbibition}}$ is the oil volume that imbibed spontaneously; $V_{w_{injection}}$ is the total volume of injected water; and $V_{o_{injection}}$ is the total volume of injected oil.

Thereafter, the XPS measurements were conducted for each sample in four different locations. The area of the corresponding peak in a high-resolution spectrum was used to calculate the elements' concentration. After the XPS measurements, the surface composition for the four locations was recorded, and the average of the four locations was used to represent the surface composition. Figure 2.39 shows atomic concentration plots obtained from XPS spectrum in four different location of the core used by Araujo et al. (2017). The XPS survey spectra results performed for different shale samples studied by Araujo et al. revealed three types of carbons and forms of silicon. The carbon bond types contained single bond carbon with hydrogen (C-H), single bond carbon with oxygen (C-O), and double bond carbon with oxygen (C=O). The silicon bond carbons contained silicon bonded to an oxygen atom (Si-O) and silicon single bonded to an H chain (Si-H). Other elements presented in the XPS survey included Al-O, Fe, Mg, Na, K, Ca, N, S, F, and Cl. The surface composition was translated to the wetting properties of the samples. Analyzing the mineral composition of their samples showed that the water-wet samples, which have similar bulk mineral compositions, exhibit a low carbon content of the type C-H, below 16 atomic percentage (at.%), and a low Si-H content (i.e., < 4 at.%). The samples in the intermediate-wet show carbon type C-H content in the range of 28.54–31.98 at.% and a relatively low Si-H content in the range of 6.06–7.26 at.%. These samples contain calcite, K-feldspar, quartz, and clay (illite) in different proportions and

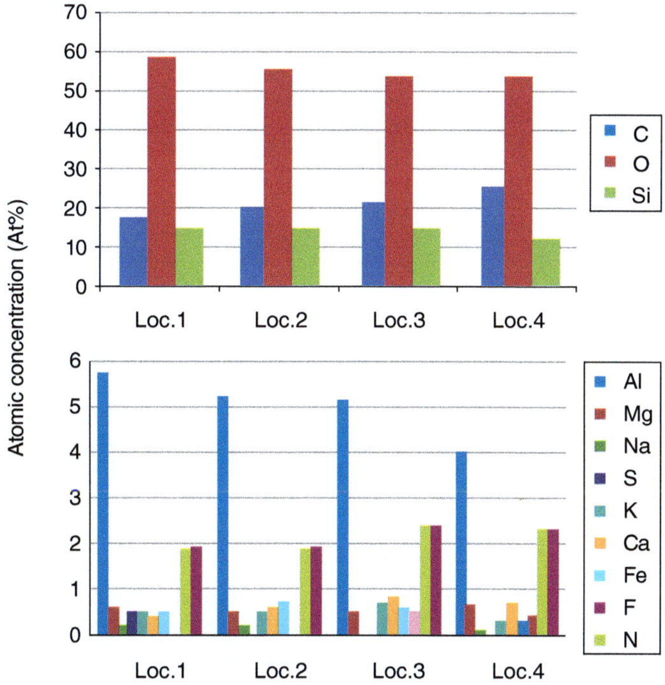

Figure 2.39 Atomic concentration plot in four different location of the core used by Araujo de Itriago (Araujo et al., 2017).

calcium content in the range of 1.12–1.45 at %. In this study, the oil-wet samples showed a high XPS organic carbon content of > 39.45 at.% and Si-H content of < 4 at.%. These samples show the highest calcium content (i.e., >2.53 at.%), which is correlated to the results of the XRD tests in the presence of calcite. The results of the XPS method confirm the result of the modified Amott-Harvey technique (Araujo et al., 2017).

Example 2.15 Determine the Wettability of the Samples

The XPS atomic concentration results of various minerals for 3 different shale samples are provided in Figure 2.40. Considering that the width of the bars are equal, comment on the wetting characteristics of the samples (i.e., water-wet, oil-wet or intermediate wet) based on the method introduced by Araujo de Itriago et al (Araujo et al., 2017).

Solution

According to the method by Araujo de Itriago et al., water-wet samples exhibit a low carbon content of the type C-H (below 16 atomic percentage, at.%) and a low Si-H content (<4 at.%), intermediate-wet samples show carbon type C-H content in the range of 28.54–31.98 at.% and a relatively low Si-H content in the range of 6.06–7.26 at.% and the oil-wet samples show a high XPS organic carbon content of >39.45 at.% and Si-H content of <4 at.%. Therefore, sample #1 is water-wet, sample #2 is oil-wet, and sample #3 is intermediate-wet.

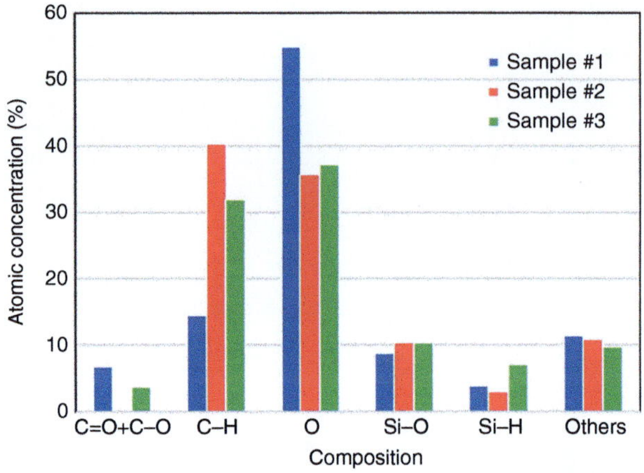

Figure 2.40 Atomic concentration for different compositions of three different samples.

2.9 Hydrocarbon Pore Volume and Reserve Estimation

In the absence of pertinent data, the reserve estimation based on analogy is the simplest approach to estimate possible reserves. The reservoir performance in a similar geologic setting is considered as a basis for reserves estimation. However, oil and gas estimation using volumetric method and decline curve analysis are the popular methodologies. While the volumetric method depends on static data, the material balance method and the decline curve analysis are conducted based on production data. McLean et al. (2014) determined the HCPV using GRI calibrated petrophysics. They correlated the petrophysical HCPV to estimated ultimate recovery (McLean et al., 2014). Another correlation was found by McLean et al. between porosity asserted organic matter (PAOM) HCPV and petrophysical HCPV (McLean et al., 2017). This correlation shows three times higher values for petrophysical HCPV compared to PAOM HCPV. This overestimation of HCPV corresponds to adding bulk volume water (BVW) and clay bound water to PAOM computed using GRI porosity estimation technique (Walls, et al., 2015) and results in significant under-estimation of recovery factor. Digital rock analysis (DRA) performed on SEM images was used to measure the organic matter content, porosity, and maturity of the samples (McLean et al., 2017). The GRI method provides the porosity values by indirect measurement techniques, which can be used to compute the oil and water saturation values. Scaling up process for such values measured at the microscale is a challenge and subject to more research.

2.9.1 Volumetric Analysis Theory

Volumetric analysis theory, which is known as static method, employs geological information to estimate original fluid in place. This information can come from core analysis, Rock-Eval, wireline logs, geological maps, and fracture extension maps. This analysis is performed prior to production for reserve estimation. In this method, the pore volume and the hydrocarbon pore volume are calculated using equations (2.138) and (2.139), respectively (Fekete Associates Inc., 2014).

$$V_p = A \times h \times \phi \tag{2.138}$$

and

$$V_{HCP} = V_p \times S_{HC} = A \times h \times \phi \times S_{HC}, \tag{2.139}$$

where V_p is the pore volume, m^3; V_{HCP} is the hydrocarbon pore volume, m^3; A is area corresponding to the stimulated reservoir volume, m^2; h is the reservoir thickness, m; ϕ is the reservoir porosity, fraction; and S_{HC} is the hydrocarbon saturation in the reservoir, fraction.

The original oil in place (OOIP) and the original gas in place (OGIP) of the reservoir in the field unit is obtained using equations (2.140) and (2.141), respectively. Historically, only free gas in place was considered for calculating the original gas in place (OGIP) in the reservoirs.

$$OOIP = 7758 \times A \times h \times \phi \times S_{oi} \times \frac{1}{B_{oi}} \tag{2.140}$$

and

$$\mathrm{OGIP} = 43560 \times A \times h \times \phi \times S_{gi} \times \frac{1}{B_{gi}}, \tag{2.141}$$

where A is the cross-sectional area open to the flow, acres; h is the reservoir thickness, ft; ϕ is the reservoir porosity, fraction; S_{oi} is the initial oil saturation in the reservoir, fraction; S_{gi} is the initial gas saturation in the reservoir, fraction; B_{oi} is the oil formation volume factor, bbl/STB; and B_{gi} is the gas formation volume factor, ft^3/SCF.

In certain unconventional reservoirs, such as shale gas and coalbed methane, the gas content of the reservoir is divided into free and adsorbed gas. In shales, the free gas content accounts for 15%–80% of the total gas content. This number can vary depending on reservoir pressure, saturation, and rock properties including porosity (Satter & Iqbal, 2016). The adsorbed gas remains in the matrix and miniscule pores of rock. In this condition, the adsorbed and free gas contents are calculated using equations (2.142) and (2.143), respectively, and the total gas in place can be calculated by equation (2.144). Figure 2.41 shows a typical plot of gas adsorption volume vs. pressure in a shale reservoir (Satter & Iqbal, 2016).

$$V_a = \frac{V_L P}{P_L + P}, \tag{2.142}$$

$$V_f = \frac{1}{B_g} C\phi_{eff}(1 - S_w)\frac{1}{\rho_b}, \tag{2.143}$$

and

$$GIP = V_a + V_f, \tag{2.144}$$

where V_a is the volume of adsorbed gas, SCF/ton; P is the reservoir pressure, psi; P_L is the Langmuir pressure, which is the pressure corresponding to half of Langmuir volume, psi; V_L is the Langmuir volume, which is the volume of gas adsorbed at infinite pressure, SCF/ton; V_f is the volume of free gas, SCF/ton; B_g is the gas formation volume factor; C is the conversion factor; ϕ_{eff} is the effective porosity, fraction; S_w is the connate water saturation, fraction; and ρ_b is the bulk density, g/cm^3.

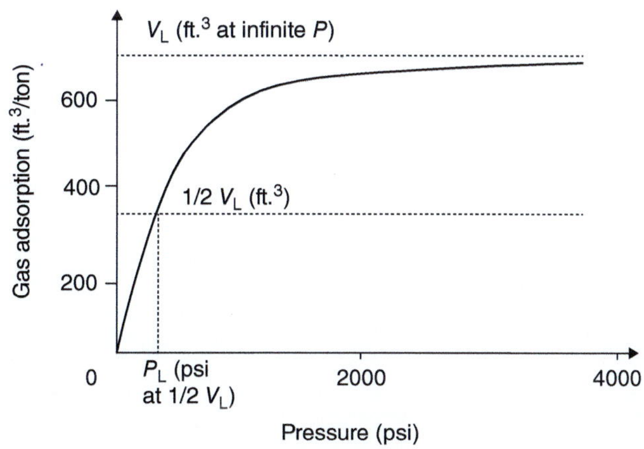

Figure 2.41 Typical plot of gas adsorption volume versus pressure in a shale reservoir (Satter & Iqbal, 2016).

Example 2.16 Determine the Adsorption Capacity

The Langmuir pressure and volume of three different shale samples from Longmaxi formation in Southern Sichuan basin in China are provided in Table 2.25 (Guo et al., 2017). Determine the adsorption capacity of the samples.

Solution

The adsorption capacity is calculated for all the samples over 20 MPa range of pressure using equation (2.142). A sample calculation for sample 1 and at a pressure of 5 (MPa) is shown below, and the summarized results are shown in Figure 2.42.

$$V_a = \frac{V_L P}{P_L + P} = \frac{1.15\left(\frac{m^3}{t}\right) \times 5\ (MPa)}{3.49\ (MPa) + 5\ (MPa)} = 0.677\left(\frac{m^3}{t}\right).$$

Downey et al. developed a model to measure the oil in place quantitatively considering the distillable oil in unconventional oil plays. In this method, the oil in place is obtained from the S_1 measurements of a standard Rock-Eval analysis. This method requires rock samples and determines volumes of distillable hydrocarbons in the rock in mg of hydrocarbon/g of dry rock. In this model, the volume of oil per section is calculated using equation (2.145) (Downey et al., 2011). The Rock-Eval analysis will be discussed further in Chapter 3.

$$V_{S_1HC} = \frac{A \times h \times \rho_{av} \times S_{1_{av}} \times 0.001}{(\rho_{oil})}, \tag{2.145}$$

where V_{S_1HC} is the volume of S_1 hydrocarbons per section, cm^3; A is the area of interest, cm^2; h is the reservoir height, cm; ρ_{av} is the average bulk density, gr/cm^3; S_{1_av} is the average S_1, mg/g; and ρ_{oil} is the oil density, gr/cm^3.

Table 2.25 Data for Example 2.16

	Sample #1	Sample #2	Sample #3
V_L (m^3/t)	1.15	1.17	1.2
P_L (Mpa)	3.49	2.99	3.18

Source: Guo et al. (2017).

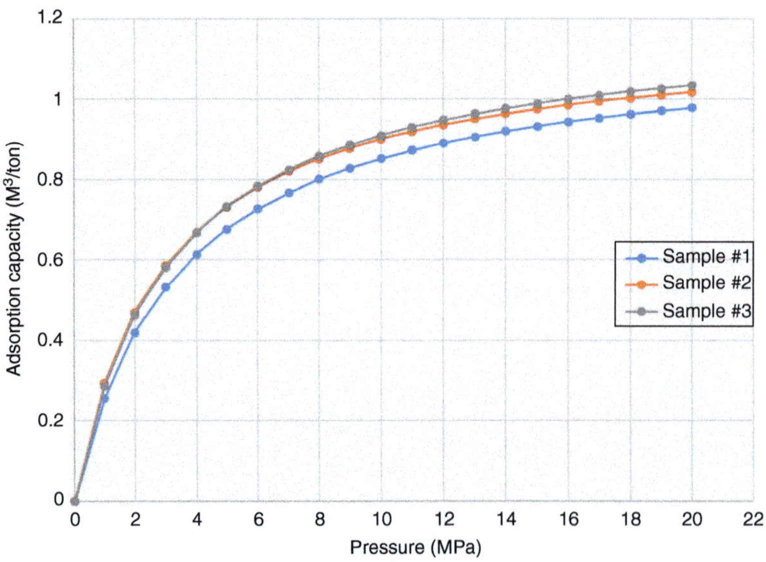

Figure 2.42 Results for Example 2.16 modified from the work of Guo et al. (Guo et al., 2017).

Table 2.26 Petrophysical Properties of Unconventional Plays in North America, Part 1

Unconventional play	Barnett	Bakken	Marcellus	Utica	Eagle Ford, US region	Wolfcamp	Niobrara	Woodford
Average porosity (%)	5–10	5–10	8	8	8–18	4–16	8–10	6–10
Average permeability	10–300 nD	100 nD–30 µD	20–300 nD	300 mD–0.1 mD	1–1,500 nD	<10 mD	<0.1 mD	3–274 nD
Average water saturation (%)	25–35	NA	12–35	35	7–31	37	NA	20–25
Dominant pore-size distribution (PSD)	10–300 nm	0.4-1.8 nm	<200 nm	<10 nm	0.1–300 nm	0.1–200 nm	1 nm–368 µm	0.38–5 nm
Dominant mineral composition	Quartz, clay, feldspar, calcite	Quartz, calcite, dolomite, clay, feldspar, pyrite	Quartz, clay, calcite, feldspar, pyrite	Quartz, calcite, clay, pyrite, dolomite	Quartz, calcite, pyrite clays	Mostly shale and argillaceous carbonates and sand intervals	Calcite, quartz, clay, pyrite, feldspar (chalk and marl)	Silica, clay, dolomite, pyrite

Table 2.27 Petrophysical Properties of Unconventional Plays in North America, Part 2

Unconventional play	Fayetteville	Haynesville	Monterey	Chattanooga	Horn River	Eagle Ford, Mexico region
Average porosity %	5–9	3–14	11	1.6–5.6	3–7	Up to 9.4
Average permeability	10s–100s nD	400 nD	8–14 µD	87–393 nD	Up to 0.2 nD	0.4 µD
Average water saturation %	17–-42	15–40	40	Up to 23	20–30	40
Dominant pore-size distribution (PSD)	5–100 nm	1–300 nm	20–100 nm	0.1–0.0005 µm	3 mn–100 µm	NA
Dominant mineral composition	Quartz, clay, dolomite, calcite, chlorite	Clay, quartz, feldspar, carbonate	Quartz, clay, feldspar, calcite, dolomite, pyrite	Clay, silica, carbonates, pyrite	Silicate dominated, carbonates, clay	Quartz, carbonates, clay

Table 2.28 Petrophysical Properties of Unconventional Plays in South America, Europe, and Australia

Unconventional play	Vaca Muerta	Llandovery	Goldwyer	Nappamerri	L. Kyalla
Location	Argentina	Poland	Australia	Australia	Australia
Average porosity %	6–22	1–14	<10	2–10	2–10
Average permeability	140–213 nD	200 nD–122 μD	NA	0.1–10 mD	10–100 nD
Average water saturation %	NA	NA	15–30	40-53	Up to 58
Dominant pore-size distribution (PSD)	NA	1 nm–10 μm	NA	2–50 nm	NA
Dominant mineral composition	Quartz, feldspar, clay, calcite, dolomite	Quartz, clay, carbonates	Quartz, calcite, dolomite, clay, anhydrite	Quartz, clay, pyrite, silicate	Silicate, clay, carbonate, chlorite

Table 2.29 Petrophysical Properties of Unconventional Plays in Asia, Middle East, and Africa

Unconventional play	Longmaxi	Qusaiba	Dadas	Whitehill	Ahnet basin	Sirte
Location	China	Saudi Arabia	Turkey	South Africa	Algeria	Libya
Average porosity %	10-Jan	Up to 10	Up to 6.8	Up to 7	7–9	Up to 22
Average permeability	1 nD–1 μD	Up to 100 μD	NA	NA	350–360 nD	Up to 87 mD
Average water saturation %	<45	NA	8–9	NA	18–24	25–30
Dominant pore-size distribution (PSD)	NA	NA	NA	NA	NA	NA
Dominant mineral composition	Clay, quartz, feldspar, carbonates	Quartz, mica, mudstone, silicate, clay	Quartz, feldspar, dolomite, calcite, kaolinite, illite, smectite	Quartz, clay, calcite, dolomite, feldspar, pyrite, gypsum, mica	Kaolinite, illite, chlorite	Limestone, shale, sandstone

** NA (not announced).

2.10 Problems

1 Use the data in Table 2.31 to calculate the shale volume of the provided interval using the Larionov tertiary rocks method, the Larionov older rocks method, the Clavier method, and the Steiber method. Compare the results and state the differences between each method. Assume that the GR readings in matrix and shale are 12.7 and 131.4, respectively.

2 Use the data in Table 2.31 and the results obtained from Problem 1 to calculate the total and effective porosity using the density method, the sonic method, the neutron-density method, and the neutron-sonic method. Compare the results and state the differences between each method.

3 Estimate the formation lithology based on the log readings that are provided in Table 2.30. Display the lithology on a stacked area plot.

4 Calculate the permeability for the interval provided in Table 2.30 using the SDR method. The values of a, b, and c' are 4, 2, and 4.6, respectively.

Table 2.30 Data for Problem 3

GR	RHOZ	PEFZ	NPHI
270.4	2.376	8.0	32.7
272.6	2.482	8.0	32.3
277.9	2.517	8.2	30.2
305.6	2.501	8.4	30.7
314.8	2.489	8.2	30.2
290.5	2.468	8.1	27.8
250.5	2.461	7.6	24.5
264.9	2.436	7.3	25.0
318.7	2.368	7.2	29.1
267.0	2.338	7.4	27.1
144.5	2.477	7.6	24.9

5 Calculate the water saturation for the interval provided in Table 2.31 and the results obtained from Problems 1 and 2. Use the Simandoux method, the modified Simandoux method, the Indonesia method, and the dual-water method. Assume that the shale resistivity (Rsh) and the shale total porosity (ϕ_{sh}) are 23.98 and 0.294, respectively. The cementation factor (m) is 1.4, the saturation exponent (n) is 1.74, and the tortuosity factor (a) is 1. Compare the results and state the differences between each method.

6 Figure 2.43 shows a cross-plot of the photoelectric response and matrix density. Estimate the composition (i.e., quartz, dolomite, and calcite) of the green point that is located on the cross-plot. What is the lithology-corrected porosity for this point?

7 Estimate the permeability using the Timur-Coates equation and SDR equation for the interval provided in Table 2.32. Compare the permeability from the two NMR methods.

8 Estimate the porosity for the interval provided in Table 2.33 using the density method, the neutron-density method, and the neutron-sonic method. Compare the results and state the differences between the methods.

9 Using the different surface relaxivity values listed in Table 2.14, determine the pore-size distribution (PSD) of the outcrop sample in Table 2.13 assuming spherical pores. Comment on the effect of the surface relaxivity on the pore-size distribution.

10 Deduce the wettability of the outcrop sample in Table 2.13 given the bulk relaxation of hexadecane to be 614 ms.

Table 2.31 Data for Problems 1, 2, 4, and 5

Depth (ft)	DT (us/ft)	GR (API)	Neutron porosity (fraction)	Bulk density (g/cm3)	Formation resistivity (OHMM)	Water resistivity (OHMM)	T2LM (MS)	TCMR (fraction)
9,866	106.3389	95.2922	0.3159	2.2862	60.0534	0.0274	2.9970	0.0454
9,866.5	114.0797	101.6151	0.3972	2.2067	93.9386	0.0231	2.2514	0.0482
9,867	116.7134	105.5544	0.4163	2.1582	143.3696	0.0211	1.4817	0.0670
9,867.5	116.3579	102.1358	0.4406	2.1529	160.0452	0.0237	1.1374	0.0860
9,868	115.4431	97.3090	0.4564	2.1570	219.8649	0.0329	1.0495	0.0785
9,868.5	115.6503	89.2616	0.4480	2.1642	333.0038	0.0395	1.1515	0.0578
9,869	116.0907	78.8318	0.4412	2.1726	497.2107	0.0387	1.2406	0.0492
9,869.5	115.9228	67.0508	0.4237	2.1711	335.1144	0.0362	0.9472	0.0619
9,870	115.9274	53.0536	0.4550	2.1540	171.4814	0.0371	1.0437	0.0643
9,870.5	114.7028	46.4266	0.4298	2.1545	103.1367	0.0388	0.9603	0.0642
9,871	112.7652	47.2203	0.4273	2.1496	106.4274	0.0387	1.2203	0.0593
9,871.5	111.0675	59.8692	0.3753	2.1439	102.4468	0.0353	2.4150	0.0557
9,872	110.2947	78.0412	0.3958	2.1195	46.7842	0.0316	3.7440	0.0496
9,872.5	107.8933	81.4154	0.4110	2.1140	26.0640	0.0296	3.0339	0.0587
9,873	104.6493	71.3291	0.4036	2.1458	15.1630	0.0282	2.5563	0.0595
9,873.5	102.7417	58.8489	0.3721	2.1878	28.2065	0.0292	2.6565	0.0553
9,874	102.0624	55.3381	0.3325	2.2227	42.5725	0.0325	2.6215	0.0520
9,874.5	102.6126	58.7204	0.3454	2.2154	40.8502	0.0310	2.2778	0.0460
9,875	105.1641	65.1456	0.3761	2.1938	48.0975	0.0237	2.1955	0.0366
9,875.5	107.3445	78.3421	0.3985	2.1609	37.9180	0.0193	2.2340	0.0343
9,876	110.7472	92.9670	0.4203	2.1282	20.8435	0.0235	2.8027	0.0260
9,876.5	110.1054	96.6526	0.4282	2.1077	17.8274	0.0345	5.8275	0.0301
9,877	104.7699	80.9965	0.3832	2.1490	15.4084	0.0399	10.9746	0.0420
9,877.5	94.9881	55.6945	0.2915	2.2738	9.3140	0.0361	15.1146	0.0522
9,878	83.8512	36.3251	0.1954	2.4412	6.1463	0.0318	23.4672	0.0549

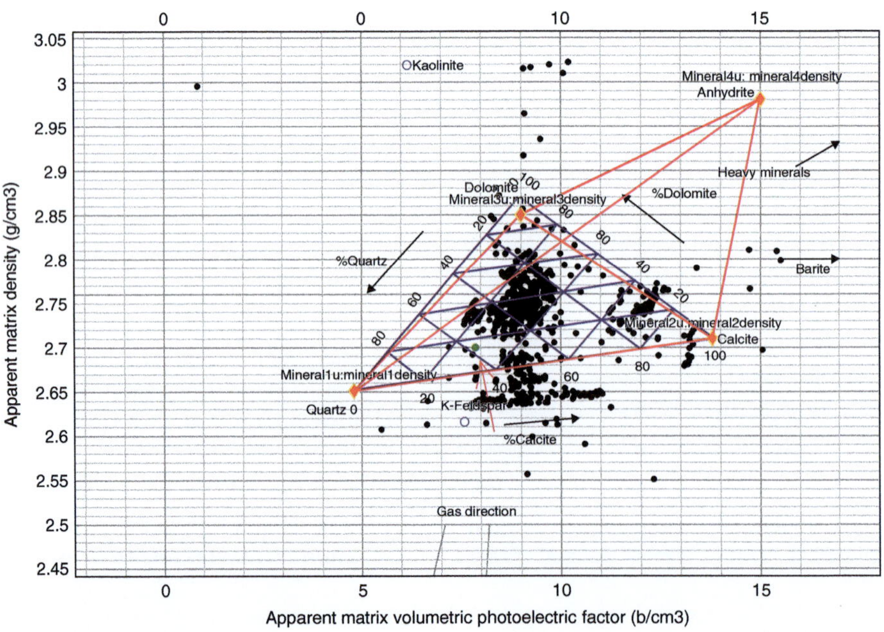

Figure 2.43 Cross-plot for Problem 6.

Table 2.32 Data for Problem 7

BVI (cf/cf)	FFI (cf/cf)	Logarithmic mean of T_2 distribution (MS)	Porosity (frac)
0.0668	0.034	14.8468	0.1008
0.0515	0.0344	24.0178	0.0858
0.0589	0.0241	17.2573	0.0829
0.0645	0.0223	15.2929	0.0869
0.0609	0.0258	16.7664	0.0867
0.0477	0.0312	29.4173	0.0789
0.0481	0.032	30.0356	0.0801
0.0478	0.0309	28.0734	0.0787
0.0451	0.0301	27.8911	0.0752
0.0345	0.0323	39.1652	0.0668

Table 2.33 Data for Problem 8

DT sonic (us/ft)	GR (gAPI)	Neutron porosity (cf/cf)	Bulk density (g/cc)
68.4152	128.0957	0.1516	2.6252
64.526	78.2845	0.1335	2.6263
64.3352	66.1453	0.1265	2.6554
64.6548	68.1817	0.1334	2.6652
62.6561	63.6282	0.1275	2.6603
64.4328	63.2743	0.1396	2.6364
62.4528	65.4292	0.1289	2.6485
60.5411	72.1801	0.1036	2.6638
62.4486	75.1541	0.1035	2.648
64.5073	87.9307	0.1104	2.6225
67.0517	96.2258	0.1422	2.6208

11 Estimate the NMR permeability of the outcrop core in Table 2.13 using the Coates formula using both the original constants and the modified constants listed under NMR permeability.

Additional Reading

Bassiouni. (1994). *Theory, measurement, and interpretation of well logs*. Richardson, TX: Society of Petroleum Engineers (SPE).

Coates, G. R., Xiao, L., & Prammer, M. G. (1999). *NMR logging: Principles and applications*. Halliburton Energy Services Publication.

Bateman, R.M. (2012) *Openhole Log Analysis and Formation Evaluation*. Society of Petroleum Engineers.

References

Adams, J. A., & Weaver, C. E. (1958). Thorium to uranium ratios as indicators of sedimentary processes: Example of concept of geochemical facies. *Bulletin of the American Association of Petroleum Geologists, 42*(2), 387–430.

Ahmad, M. (2014). *Petrophysical and mineralogical evaluation of shale gas reservoirs (A Cooper basin case study)*. Adelaid: The University of Adelaide.

Ahmed, U., Crary, S., & Coates, G. R. (1994). Permeability estimation: The various sources and their interrelationships. *Petrophysics, SPE, 39*, 118–127.

Alden, A. (2017). Densities of common rocks and minerals. (ThoughtCo) Retrieved March 28, 2018, from https://www.thoughtco.com/densities-of-common-rocks-and-minerals-1439119

Allen, D. (1988). Probing for permeability: An introduction of measurements. *The Technical Review, 36*(1), 6–20.

Allix, P., Burnham, A., Fowler, T., Herron, M., Kleinberg, R., & Symington, B. (2010). *Coaxing oil from shale*. Schlumberger.

Alshakhs, M. (2017). *Shale play assessment of the Goldwyer formation in the Canning basin using property modelling*. Bentley,

Western Australia, Australia: Curtin University.

Altunbay, M. M. (2001). *Capillary pressure data from NMR logs and its implications on field economics*. SPE Annual Technical Conference and Exhibition (1–11). New Orleans: SPE.

Anderson, G. W. (1986). Wettability literature survey, Part 2: Wettability measurement. *Journal of Petroleum Technology, 38*(11), 1246–1282.

Anovitz, L. M., & Cole, D. R. (2015). Characterization and analysis of porosity and pore structures. *Reviews in Mineralogy and Geochemistry, 80*, 61–164.

APERC,. (2015). *Pathways to shale gas development in Asia-Pacific*. Chuo-ku, Tokyo, Japan: Asia Pacific Energy Research Centre.

Araujo, Y. C., Albrecht, J., Möeller, A., & Dallmann, G. (2017). A different approach to evaluate wettability in shales. *Abu Dhabi International Petroleum Exhibition & Conference*. Abu Dhabi, UAE.

Araujo, Y. C., Toledo, P. G., Leon, V., & Gonzalez, H. Y. (1995). Wetting behavior of silane-treated silica surfaces as determined from XPS. *Journal of Colloid and Interface Science, 176*, 485–490.

Archie, G. E. (1942). The eletrical resistivity log as an aid in determining some reservoir characteristics. *Trans. of AIME, 146*(1), 54–62.

Archie, G. E. (1952). Classification of carbonate reservoir rocks and petrophysical considerations. *AAPG Bulletin, 36*(2), 218–298.

Asquith, G., & Krygowski, D. (2004). Basic well log analysis. *AAPG Methods, 16*, 31–35.

Aydin, E. (2017). *Pore characterization and FE-SEM analysis of the Niobrara formation in the Aristocrat Angus PC H11-07 core, Wattenberg field, Denver basin, Colorado*. Golden: Colorado School of Mines.

Bai, B., Elgmati, M., Zhang, H., & Wei, M. (2013). Rock characterization of Fayetteville shale gas plays. *Fuel, 105*, 645–652.

Banas, R. M. (2012). *Geophysical and geological analysis of the Collingwood member of the Trenton formation*. Michigan Technological University.

Bassiouni. (1994). *Theory, measurement, and interpretation of well logs*. Richardson, TX: Society of Petroleum Engineers (SPE).

Bednarczyk, S., Krzyżak, A. T., & Machowski, G. (2016). Comparative analysis of measurements and estimation of permeability of shales in selected well sections from Baltic basin (northern Poland). *International Multidisciplinary Scientific Geoconference and Expo*. Albena, Bulgaria.

Benson, S., Garing, C., Zahasky, C., Boon, M., Gran, M., & Ni, H. (2017). *Laboratory measurement of reservoir rock properties: Permeability*. Stanford, CA: Stanford University.

Berkowitz, B. (1993). Percolation theory and its application to groundwater hydrology. *Water Resources Research, 29*(4), 775–794.

Berkowitz, B., & Ewing, R. P. (1998). Percolation theory and network modeling applications in soil physics. *Surveys in Geophysics, 19*, 23–72.

Best, D. L., Gardner, J. S., & Dumanoir, J. L. (1979). A computer-processed wellsite log computation. *SPWLA Annual Symposium*. Casper, WY.

Blomquist, P. K. (2016). *Wolfcamp horizontal play, Midland basin, West Texas*. AAPG Pacific Section and Rocky Mountain Section Joint Meeting. Las Vegas, NV.

Bordenave, M. L., Espitalie, J., Leplat, P., Oudin, J. L., & Vandenbroucke, M. (1993). *Screening techniques for source rock evaluation*. Paris, France: Applied Petroleum Geochemistry.

Boreham, C., Edwards, D., Hall, L., Bernecker, T., Carr, L., Chen, J., et al. (2013). *Unconventional Middle Cambrian petroleum systems in the Georgina basin, Australia*. AAPG Hedberg Conference, Fundamental Controls on Petroleum Systems in Lower Paleozoic and Older Strata. Beijing, China.

Brace, W. F., Walsh, J. B., & Frangos, W. T. (1968). Permeability of granite under high pressure. *Journal of Geophysical Research, 73*(6), 2225–2236.

Bratovich, M., & Walles, F. (2016). Formation evaluation and reservoir characterization of source rock reservoirs. In *Unconventional Oil and Gas Resources: Exploitation and Development*, 8-1–8-6. CRC Press.

Broadbent, S. R., & Hammerslcy, J. M. (1957). Percolation processes. *I. Crystals and mazes. Proceedings of the Cambridge Philosophical Society*, *53*(3), 629–641.

Brownstein, K. R., & Tarr, C. E. (1979). Importance of classical diffusion in NMR studies of water in biological cells. *Physical Review Series A*, *19*, 2446–2453.

Brunauer, S., Emmett, P. H., & Teller, E. (1983). Adsorption of gases in multimolecular layers. *Journal of The American Chemical Society*, *60* (2), 309–319.

Buckley, J. S., & Lord, D. L. (2003). Wettability and morphology of mica surfaces after exposure to crude oil. *Journal of Petroleum Science and Engineering*, *39*, 261–273.

Burke, J., Curtis, M., & Cox, J. (1967). Computer processing of log data improves production in Chaveroo field. *Journal of Petroleum Technology*. doi:10.2118/1576-PA

Bush, B. L. (2013). *Rock physics characterization of organic rich shale formation to predict organic properties*. Colledge Station, TX: Texas A&M University.

Bust, V., Majid, A., Oletu, J., & Worthington, P. (2013). The petrophysics of shale gas reservoirs: Technical challenges and pragmatic solutions. *Petroleum Geoscience*. doi:10.1144/petgeo2012–031.

Bust, V. K., Majid, A. A., Oletu, J. U., & Worthington, P. F. (2011). *The petrophysics of shale gas reservoirs: Technical challenges and pragmatic solutions*. International Petroleum Technology Conference. Bangkok, Thailand.

Bustin, R., Bustin, A., Cui, X., Ross, D., & Pathi, V. M. (2008). *Impact of shale properties on pore structure and storage characteristics*. SPE Shale Gas Production Conference, SPE. Fort-Worth, TX.

Cao-Minh, C., Crary, S. F., Zielinski, L., Liu, C., Jones, S., & Jacobsen, S. J. (2012). *2D-NMR applications in unconventional reservoirs*. SPE Canadian Unconventional Resources Conference: Society of Petroleum Engineers. Calgary, Alberta, Canada.

Carmichael, R. S. (1982). *Handbook of physical properties of rocks* (vol. *2*). Boca Raton, Florida: CRC Press Inc.

Carr, H., & Purcell, E. (1954). Effects of diffusion on free precession in nuclear magnetic resonance experiments. *Physical Review*, *94*(3), 630–638.

Chen, Z. a. (2017, October). Effects of bitumen extraction on the 2D NMR response of saturated kerogen. *Petrophysics*, *58*(5), 470–484.

Cho, Y., Eker, L., Uzun, I., Yin, X., & Kazemi, H. (2016). *Rock characterization in unconventional reservoirs: A comparative study of Bakken, Eagle Ford, and Niobrara formations*. SPE Low Perm Symposium. Denver, CO.

Civan, F. (2010). Effective correlation of apparent gas permeability in tight porous media. *Transport in Porous Media*, *82*(2), 375–384.

Clavier, C., Coates, G., & Dumanoir, J. (1984). Theoretical and experimental bases for the dual-water model for interpretation of shaly sands. *SPE*, *24*(2), 153–168.

Close, D. I., Baruch, E. T., Altmann, C. M., Cote, A. J., Mohinudeen, F. M., Richards, B., & Stonier, S. (2016). *Unconventional gas potential in Proterozoic source rocks: Exploring the Beetaloo sub-basin*. Australia: AGES 2016 Proceedings, NT Geological Survey.

Coates, G. R., Xiao, L., & Prammer, M. G. (1999). *NMR logging: Principles and applications*. Halliburton Energy Services Publication.

Craig, F. F. (1971). *The reservoir engineering aspects of waterflooding*. Richardson, TX: SPE Monograph Series.

Crain, E. R. (2006). Crain's petrophysical handbook. https://spec2000.net/00-contactus.htm

Crain, E. R. (2015). Crain's petrophysical handbook. Retrieved June 28, 2018, from https://www.spec2000.net/11-vshgr.htm

Cudjoe, S., Barati, R., Goldstein, R., Tsau, J.-S., Nicoud, B., Bradford, K., et al. (2019). *An integrated pore-scale characterization workflow for hydrocarbon gas huff-n-puff injection into the lower Eagle Ford shale*. Unconventional

Resources Technology Conference (1–20). Denver: URTeC.

Cudjoe, S., Barati, R., Gomes, J. H., & Vinassa, M. (2016). A comprehensive approach to sweet-spot mapping for hydraulic fracturing and CO_2 huff-n-puff injection in Chattanooga shale formation. *Journal of Natural Gas Science and Engineering, 33*, 1201–1218.

Cudjoe, S., Oraki, I., Barati, R., Tsau, J.-S., Zhang, C., Nicoud, B., et al. (2019). Nuclear magnetic resonance estimation of petrophysical properties and evaluation of hydrocarbon huff-n-puff gas injection in lower Eagle Ford shale oil samples. Unconventional Resources Technology Conference (1–15). Denver: URTeC.

Cui, X., Bustin, A. M., & Bustin, R. M. (2009). Measurement of gas permeability and diffusivity of tight reservoir rocks: different approaches and their applications. *Geofluids, 9*, 208–223.

Dacy, J. M. (2010). *Core tests for relative permeability of unconventional gas reservoirs.* SPE Annual Technical Conference and Exhibition. Florence, Italy.

Dadmohammadi, Y., Misra, S., Sondergeld, C. H., & Rai, C. S. (2016). *Simultaneous estimation of intrinsic permeability, effective porosity, porevolume compressibility, and Klinkenberg-slip factor of ultra-tight rock samples based on laboratory pressure-step-decay method.* SPE Low Perm Symposium. Society of Petroleum Engineers.

Daigle, H., & Dugan, B. (2011). An improved technique for computing permeability from NMR measurements in mudstones. *Journal of Geophysical Research, 116*, 1–14.

Daigle, H., Ezidiegwu, S., & Turner, R. (2015). *Determining relative permeability in shales by including the effects of pore structure on unsaturated diffusion and advection.* SPE Annual Technical Conference and Exhibition. Houston, TX.

Daly, A. R., & Edman, J. D. (1987). Loss of organic carbon from source rocks during thermal maturation (abs.). *AAPG Bulletin, 71*(5), 546.

Dang, S. S. (2017). *Interpretation of NMR response to hydrocarbons: application of miscible EOR experiments.* International Symposium of the Society of Core Analysts (1–12). Vienna: SCA.

Daniels, J. J., Cole, D., Murphy, M., Welch, S., & Sheets, J. (2011). *Mineralogic and physical property characteristics of the Utica-Point Pleasant shale from core and geophysical log analysis.* Presented at Taking a Deeper Look at Shales: Geology and Potential of the Upper Ordovician Utica Shale in the Appalachian Basin, June 21, 2011, New Philadelphia, Ohio.

Darabi, H., Ettehad, A., Javadpour, F., & Sepehrnoori, K. (2012). Gas flow in ultra-tight shale strata. *Journal of Fluid Mechanics, 710*, 641–658.

Dehghanpour, H., Lan, Q., Saeed, Y., Fei, H., & Qi, Z. (2013). Spontaneous imbibition of brine and oil in gas shales: effect of water adsorption and resulting microfractures. *Energy Fuels, 27* (6), 3039–3049.

Density-neutron log porosity. (2016). (AAPG WIKI) Retrieved June 20, 2018, from http://wiki.aapg.org/Density-neutron_log_porosity

De-Prisco, G., Toelke, J., & Dernaika, M. R. (2012). Computation of relative permeability functions in 3d digital rocks by a fractional flow approach using the lattice Boltzmann method. *International Symposium of the Society of Core Analysts.* Aberdeen, Scotland.

DeWitt, H. (2008). Marcellus shale overview. Investor_and_Analyst_Meeting-Marcellus.

Dicker, A. I., & Smits, R. M. (1988). *A pratical approach for determining permeability from laboratory pressure-pulse decay measurements.* SPE International Meeting in Petroleum Engineering. Tianjin, China.

Downey, M. W., Garvin, J., Lagomarsino, R. C., & Micklin, D. F. (2011). *Quick look determination of oil-in-place in oil shale resource plays.* AAPG Annual Convention and Exhibition. Houston, TX.

Drake, W. R., Longman, M. W., & Moses, A. (2018). Assessment of the Woodford shale petroleum system within a deep subbasin on the central basin platform, Permian basin.

AAPG Annual Convention and Exhibition. Salt Lake City, Utah.

Dudek, L. (2016). Pore size distribution in shale gas deposits based on adsorption isotherm analyses. *Nafta-Gaz*, *8*, 603–609.

Dunn, K., Bergman, D., & LaTorraca, G. (2002). *Nuclear magnetic resonance: petrophysical and logging applications*. Amsterdam: Elsevier Science.

Echikh, K. (1998). Geology and hydrocarbon occurrences in the Ghadames Basin, Algeria, Tunisia, Lybia. *Geological Society, London, Special Publications*, *132*, 109–129.

EIA (2011). *Review of emerging resources*. Washington, DC: US Energy Information Administration.

EIA (2015). *Technically recoverable shale oil and shale gas resources: Australia*. US Energy Information Administration (EIA). Washington, DC: US Energy Information Administration.

EIA (2018). *Petroleum and other liquids-drilling productivity report*. US Energy Information Administration. Washington, DC: US Energy Information Administration.

El-Bakush, S., & Minas, H. A. (2008). Petrophysical analysis of Nubian reservoir, Rimal Oil Field-Sirte Basin. *Journal of Engineering Research*, *9*, 55.

Enno, R., & Indrek, A. (2015). About technical terms of oil shale and shale oil. *Oil Shale, A Scientific-Technical Journal*, *32*(4), 291–292.

Ezidiegwu, S. N. (2015). *Investigation of the pore size and structure in organic-rich shales*. Austin: The University of Texas at Austin.

Fauchille, A. L., Ma, L., Rutter, E., Chandler, M., Lee, P. D., & Taylor, K. G. (2017). An enhanced understanding of the basinal bowland shale in Lancashire (UK), through microtextural and mineralogical observations. *Marine and Petroleum Geology Journal*, *86*, 1374–1390.

Fekete Associates Inc. (2014). IHS Inc. Retrieved July 12, 2018, from http://www.fekete.com/SAN/WebHelp/FeketeHarmony/Harmony_WebHelp/Content/HTML_Files/Reference_Material/Analysis_Method_Theory/Volumetrics_Theory.htm

Fertl, W. H. (1981). Openhole crossplot concepts a powerful technique in well log analysis. *Journal of Petroleum Technology*, *33*(3), 535–549.

Finger, L. J., Godet, A., & Billingsley, L. (2017). Porosity evaluation in the lower Eagle Ford shale formation, Atascosa County, South Texas. *Gulf Coast Association of Geological Societies Transactions*, *67*, 587.

Fischer, G. J., & Peterson, M. S. (1992). Measurement of permeability and storage capacity in rocks during deformation at high temperature and pressure. *Fault Mechanics and Transport Properties of Rocks*, *51*, 213–252.

Fishman, N., Ellis, G., Paxton, S. T., Abbott, M., & Boehlke, A. R. (2010). Chert beds in the Upper Devonian-Lower Mississipian Woodford shale in Oklahoma, Possible reservoir rocks? Rocky Mountain, 62nd Annual Meeting. Rapid City, SD.

Fishman, N., Guthrie, J. M., & Honarpour, M. (2013). Development of organic and inorganic porosity in the Cretaceous Eagle Ford formation, *South Texas*. AAPG Annual Convention and Exhibition. Pittsburgh, PA.

Fleury, M. (2016). Characterization of shales using T1-T2 NMR maps. *Journal of Petroleum Science and Engineering*, 55–62.

Flory, P. J. (1941). Molecular size distribution in three dimensional polymers. *Journal of American Chemical Society*, *63*(11), 3083–3090.

Frantz, J. H., Williamson, J. J., Sawyer, W. K., Johnston, D., Waters, G., Moore, L. P., ... March, K. S. (2005). *Evaluating Barnett shale production performance using an integrated approach*. Society of Petroleum Engineers. Dallas, TX.

Freedman, R., Herron, S., Anand, V., Herron, M., May, D., & Rose, D. (2015). New method for determining mineralogy and matrix properties from elemental chemistry measured by gamma ray spectroscopy logging tools. *SPE Reservoir Evaluation & Engineering*, *18*(04), 599–608.

Freedman, R., Minh, C. C., Gubelin, G., Freeman, J. J., McGinness, T., Terry, B., & Rawlence, D. (1998). Combining NMR and density logs for petrophysical analysis in gas-bearing formations. *Society of Petrophysicists and Well-Log Analysts*. Keystone, CO.

Freedman, R., Rose, D., Sun, B., Brown, R. L., & Malizia, T. (2018). Novel method for evaluating shale gas and shale tight oil reservoirs using advanced well-log data. *SPE Reservoir Evaluation and Engineering*, 2–20.

Fukushima, E., & Roeder, S. B. (1981). *Experimental pulse NMR: A nuts and bolts approach*. Massachusetts: Addison-Wesley Publishing Company.

Gardner, J. S., & Dumanoir, J. L. (1980). *Litho-density log interpretation*. SPWLA 21 Annual Logging Symposium. Lafayette, LA.

Gaymard, R., & Poupon, A. (1968). Response of neutron and formation density logs in hydrocarbon bearing formations: The log analyst. *Society of Petrophysicists and Well-Log Analysts*, *9*(5), 3–12.

Ge, X. F. (2016). Investigation of organic related pores in unconventional reservoir and its quantitive evaluation. *Energy Fuels*, 4699–4709.

Geel, C., Bordy, E., Schulz, H.-M., & Krooss, B. (2017). *Geochemical and petrophysical analysis of the Permian lower Ecca group, Karoo basin, South Africa*. Cape Town, South Africa: AAPG Africa Region, Geoscience Technology Workshop, Exploration and Development of Unconventional Hydrocarbons.

Gillespie, J., & Sanford, R. M. (1967). The geology of the Sarir oilfield, Sirte basin, Libya. *World Petroleum Congress*, Mexico City, Mexico.

Glorioso, J. C., & Rattis, A. (2012). *Unconventional reservoirs: Basic petrophysical concepts for shale gas*. SPE/EAGE European Unconventional Resources Conference and Exhibition. Vienna, Austria.

Glover, P. (2001). The litho-density log. In Petrophysics MSc Course Notes (139–149).

Godinez, L. J. (2014). *Important mineralogical factors for fluid saturation, specific surface area and pore size distributions based on gas adsorption, cation exchange capacity and 2D dielectric microscopy: A case study of quartz phase porcelanites in the Miocene Monterey Fm*. Golden, CO: Colorado School of Mines.

Green, D. P., & Veselinovic, D. (2010). *Analysis of unconventional reservoirs using new and existing NMR methods*. Fredericton, New Brunswick, Canada: Green Imaging Technologies.

Guo, W., Hu, Z., Zhang, X., Yu, R., & Wang, L. (2017). Shale gas adsorption and desorption characteristics and its effect on shale permeability. *Energy Exploration & Exploitation*, *35*(4), 463–481.

Guoyo, L. (2009). *World atlas of oil and gas basins*. London: Wiley-Blackwell.

Hackley, P. (2012). Geological and geochemical characterization of the Lower Cretaceous Pearsall formation, Maverick basin, South Texas: A future shale gas resource? *AAPG Bulletin*, *96*(8), 1449–1482.

Halliburton. (2011). Sidewall coring. Retrieved March 29, 2018, from https://web.archive.org/web/20111011073101/http://www.halliburton.com/ps/default.aspx?navid=1831&pageid=1141&prodgrpid=PRG%3a%3aJCFUH1E8Z

Hameed, A., Bacciarelli, M., & Williams, P. J. (2014). Unconventional reservoir formation evaluation challenges addressed with deployment-optimized open hole logging solutions. *SPE Eastern Regional Meeting*. Charleston, WV: SPE Eastern Regional Meeting.

Hanania, J., Stenhouse, K., & Donev, J. (2013). Conventional vs. unconventional resources. Retrieved March 29, 2018, from http://www.oilandgasinfo.ca/wp-content/uploads/Nov_2013_conv_vs_unconv.pdf

Hari-Kumar, K. C. (2010). On the application of Simandoux and Indonesian shaly sand resistivity interpretation models in low and high rw regimes. 8th Biennial International Conference and Exhibition on Petroleum Geophysics.

Hartman, D. J., & Beaumont, E. A. (2000). Predicting reservoir system quality and performance. In *Treatise of Petroleum Geology / Handbook of Petroleum Geology: Exploring for Oil and Gas Traps*. AAPG.

Hassan, M., Hossin, A., & Combaz, A. (1976). *Fundamentals of the differential gamma ray log interpretation technique*. SPWLA 17th Annual. Symposium Transactions. Denver, CO.

He, J. (1995). *Research about low permeability measurement*. Grand Forks: University of North Dakota.

Heidemann, S. R., Hamborg, M. A., Thomas, S. J., Song, B., Lindley, S., & Chu, D. (1984). Spatial organization of axonal microtubules. *Journal of Cell Biology*, *99*, 1289–1295.

Helander, D. (1983). Fundamentals of formation evaluation. *OGCI Publications*. Tulsa, OK.

Heller, R., Vermylen, J., & Zoback, M. (2014). Experimental investigation of matrix permeability of gas shales. *The American Association of Petroleum Geologists (AAPG) Bulletin*, *98*(5), 975–995.

Holmes, M., Holmes, A., & Holmes, D. (2014). *A comprehensive deterministic petrophysical analysis procedure for reservoir characterization: Conventional and unconventional reservoirs*. Denver: AAPG Rocky Mountain Section Meeting.

Holmes, M., Holmes, A. M., & Holmes, D. I. (2017). *Mixed reservoir wetting in unconventional reservoirs and interpretation of porosity/resistivity cross plots, derived from triple-combo log data*. SPWLA 58th Annual Logging Symposium. Oklahoma City, OK.

Hull, D., Loucks, B., & Milliken, K. (2012). *Lower Cretaceous Pearsall oceanic anoxic events and associated development of shale-gas reservoirs in south Texas*. AAPG Annual Convention and Exhibition. Long Beach, CA.

Humphreys, B., & Lott, G. K. (1990). An investigation into nuclear log responses of North Sea Jurassic sandstones using mineralogical analysis. *Geological Society, London, Special Publications*, *48*, 223–240.

Ijasan, O., Torres-Verdin, C., & Preeg, W. E. (2013). Interpretation of porosity and fluid constituents from well logs using an interactive neutron-density matrix scale. *Society of Exploration Geophysicists and American Association of Petroleum Geologists*, *1*(2), T143–T155.

International Ocean Discovery Program (2014). (Schlumberger) Retrieved from http://iodp.ldeo.columbia.edu/TOOLS_LABS/LEGACY/ngt.html

Javadpour, F. (2009). Nanopores and apparent permeability of gas flow in mudrocks (shales and siltstone). *Journal of Canadian Petroleum Technology*, *48*(08), 16–21.

Jia, B., Tsau, J.-S., & Barati, R. (2017). *Different flow behaviors of low-pressure and high-pressure CO_2 in shales. Unconventional Resources Technology Conference. Austin, TX.*

Jia, B., Tsau, J.-S., & Barati, R. (2017). *Evaluation of core heterogeneity effect on pulse-decay experiment*. International Symposium of the Society of Core Analysis. Vienna, Austria.

Jiang, S., Peng, Y., Gao, B., Zhang, J., Cai, D., Xue, G., et al. (2016). Geology and shale gas resource potentials in the Sichuan basin, China. *Energy, Exploration & Exploitation*, *34*(5), 689–710.

Jiang, Z., Z. W., Liang, C., Wang, Y., Liu, H., & Chen, X. (2016). Basic characteristics and evaluation of shale oil reservoirs. *Petroleum Research*, *2*, 149–163.

Jin, G., Pérez, H., Al Dhamen, A., Ali, S., Nair, A., Agrawal, G.,et al. (2015). *Permeability measurement of organic-rich shale: Comparison of various unsteady-state methods*. SPE Annual Technical Conference and Exhibition. Houston: Society of Petroleum Engineers. doi:10.2118/175105-MS

Jones, S. C. (1997). Technique for faster pulse-decay permeability measurements in tight rocks. *SPEREE*, *12*, 19–25.

Juhasz, I. (1979). The central role of Qv and formation water salinity in the evaluation of shaly formations. *Society of Petrophysicists and Well-Log Analysts*. Tulsa, OK.

Jun, X. (2018). Rapid shale gas development accelerated by the progress in key technologies: a case study of the Changninge-Weiyuan National Shale GasDemonstration Zone. *Natural Gas Industry*, *B5*, 283–292.

Kadri, M. M., & Hacini, M. (2018). Preliminary reservoirs characterizations of Silurian shale, case of Ahnet basin, southern Algeria. *International Journal of Latest Research in*

Engineering and Management (IJLREM), 2(2), 29–33.

Katsube, T. J., & Williamson, M. A. (1994). Effect of diagenesis on shale nano-pore structure and implications for sealing capacity. *Clay Minor, 29*(4), 451–461.

Katsube, T. J., Mudford, B. S., & Best, M. E. (1991). Petrophysical characteristics of shales from the Scotian shelf. *Society of Exploration Geophysicists (SEG), 56*(10), 1681.

Kausik, R. F. (2014). *NMR petrophysics for tight oil shale enabled by core resaturation.* International Symposium of the Society of Core Analysts (4–9). Avignon: SCA.

Keelan, D. K., & McGinley, D. C. (1979). Application of cation exchange capacity in a study of the Shannon sand of Wyoming. *SPWLA Annual logging Symposium*. Tulsa, OK.

Keller, J. U., & Staudt, R. (2005). *Gas adsorption equilibria: Experimental methods and adsorption isotherms*. Siegen, Germany: Springer.

Kenyon, W. (1992). Nuclear magnetic resonance as a petrophysical measurement. *The International Journal of Radiation Applications and Instrumentation: Part E. Nuclear Geophysics, 6*(2), 153–171.

Kenyon, W., Day, P., Straley, C., & Willemsen, J. (1988). A three-part study of NMR longitudinal relaxation properties of water-saturated sandstones. *SPE Formation Evaluation.* doi:10.2118/15643-PA

Klaja, J., & Dudek, L. (2016). Geological interpretation of spectral gamma ray (SGR) logging in selected boreholes. *Nafta-Gaz, 1*, 3–14.

Kleinberg, R., Kenyon, W., & Mitra, P. (1994). Mechanism of NMR relaxation of fluids in rock. *Journal of Magnetic Resonance, 108*(2), 206–214.

Kosanke, T. H., & Warren, A. (2018). Geological controls on matrix permeability of the Eagle Ford shale (Cretaceous), south Texas, U.S.A. Houston, TX: The American Association of Petroleum Geologists.

Krakowska, P., Puskarczyk, E., Jędrychowski, M., Madejski, P., & Habrat, M. (2017). *Shale gas formations in Poland: New evaluation methods for processing and quantitative interpretation using computed X-ray tomography.* International Symposium of the Society of Core Analysts. Vienna, Austria.

Kranz, R. L., Saltzman, J. S., & Blacic, J. D. (1990). Hydraulic diffusivity measurements on laboratory rock samples using an oscillating pore pressure method. *International Journal of Rock Mechanics and Mining Sciences and Geomechanics Abstracts, 27*, 345–352.

Kubelka, P., & Munk, F. (1931). An article on optics of paint layers. *Ein Beitrag Zur Optik Der Farbanstriche, Zeitschrift für Technische Physik, 12*, 593–601.

Kuila, U., & Prasad, M. (2011). Surface area and pore-size distribution in clays and shales. *SPE Annual Technical Conference and Exhibition.* Denver, CO.

Kuila, U., McCarty, D. K., Derkowski, A., Fischer, T. B., & Prasad, M. (2014). Total porosity measurement in gas shales by the water immersion porosimetry (WIP) method. *Fuel, 117*, 1115–1129.

Langmuir, I. (1918). The adsorption of gasses on plane surfaces of glass, mica, and platinum. *Journal of the American Chemical Society, 40* (9), 1361–1403.

Law, B. E., & Spencer, C. W. (1993). Gas in tight reservoirs: An emerging major source of energy. *The Future of Energy Gasses, US Geological Survey, 1570*, 233–252.

Lawson, C., & Hanson, R. (1974). *Solving least square problems*. Englewood Cliffs, NJ: Prentice-Hall.

Leddy, N. (2012). *Surface area and porosity*. CMA Analytical Workshop.

Lehr, J. H., & Keeley, J. (n.d.). *Alternative energy and shale gas encyclopedia*. Wiley.

Lesniak, G., Cicha-Szot, R., Stadtmuller, M., Mroczkowska-Szerszen, M., Dudek, L., Tallec, G., et al. (2014). Possibilities of quantitative evaluation of Polish shale-gas rocks petrophysical and geomechanical parameters by laboratory analytical methods and well logs interpretation. *International Symposium of the Society of Core Analysts*. Avignon, France.

Lewis, R., Ingraham, D., Pearcy, M., Williamson, J., Sawyer, W., & Frantz, J. (2004). New evaluation techniques for gas shale reservoirs. Reservoir Symposium.

Lineback, J. (1970). Stratigraphy of the New Albany shale in Indiana. *Indiana Geological Survey Bulletin*, *44*(73), 18.

Liu, K., Ostadhassan, M., Zou, J., Gentzis, T., Rezaee, R., Bubach, B., & Carvajal-Ortiz, H. (2018). Multifractal analysis of gas adsorption isotherms for pore structure characterization of the Bakken shale. *Feuls*, *219*, 296–311.

Liu, K., Sun, J., Zhang, H., Liu, H., & Chen, X. (2017). A new method for calculation of water saturation in shale gas reservoirs using VP-to-VS ratio and porosity. *Journal of Geophysics and Engineering*, *15*, 224–233.

Loan, M. L., Herron, M. M., Craddock, P., Prioul, R., Prange, M., Schwartz, K. M., & Charsky, A. M. (2017). *Rapid quantification of mineralogy, organic matter, and thermal maturity of cuttings with diffuse reflectance infrared fourier transform spectroscopy (DRIFTS): A Permian basin case study.* Austin, TX: Unconventional Resources Technology Conference.

Log Interpretation Principles/Applications (1989). *Schlumberger*. Houston, TX: Schlumberger.

Loucks, R. G., Reed, R. M., Ruppel, S. C., & Hammes, U. (2010). Preliminary classification of matrix pores in mudrocks. *Gulf Coast Association of Geological Societies(GCAGS)*, *60*, 435–441.

Loucks, R. G., Reed, R. M., Ruppel, S. C., & Hammes, U. (2012). Spectrum of pore types and networks in mudrocks and a descriptive classification for matrix-related mudrock pores. *AAPG Bulletin*, *96*(6), 1071–1098.

Luffel, D. L., & Guidry, F. K. (1992). New core analysis-methods for measuring reservoir rock properties of Devonian shale. *Journal of Petroleum Technology*, *44*(11), 1184–1190.

Luffel, D. L., Guidry, F. K., & Curtis, J. B. (1992). Evaluation of Devonian shale with new core and log analysis methods. *Journal of Petroleum Technology*, *44*(11), 1192–1197.

Luffel, D. L., Hopkins, C. W., Holditch, S. A., & Schettler, P. D. (1993). Matrix permeability measurement of gas productive shales. *SPE Annual Technical Conference and Exhibition*. Houston, TX.

Luneau, B., Longman, M., Kaufman, P., & Landon, S. (2011). *Stratigraphy and petrophysical characteristics of the Niobrara formation in the Denver basin, Colorado and Wyoming*. Cheyenne, WY: AAPG.

Ma, L., Taylor, K. G., Dowey, P. J., Coutois, L., Golinia, A., & Lee, P. D. (2017). Multi-scale 3D characterisation of porosity and organic matter in shales with variable TOC content and thermal maturity: Examples from the Lublin and Baltic basins, Poland and Lithuania. *International Journal of Coal Geology*, *180*, 100–112.

Marschall, D., Gardner, J., & Curby, F. (1997). MR laboratory measurements-requirements to assure successful measurements that will enhance MRI log interpretation. *1997 International Symposium Proceedings, SPWLA*.

Marschall, D. G. (1995). Method for correlating NMR relaxometry and mercury injection data. *Society of Core Analysts* (1–12). San Francisco: SCA.

Martinez, G. (2000). *Petrophysical measurements on shales using NMR*. SPE/AAPG Western Regional Meeting (1–10). Long Beach: SPE.

Masalmeh, S. K., Jing, X., Roth, S., Wang, C., Dong, H., & Blunt, M. (2015). *Towards predicting multi-phase flow in porous media using digital rock physics: workflow to test the predictive capability of pore-scale modeling*. SPE Abu Dhabi International Petroleum Exhibition and Conference. Abu Dhabi, UAE.

McLean, R., Miertschin, D., Owen, K., Winters, N., & Whitman, N. (2014). *Technical aspects controlling the emerging Eagle Ford play east of the San Marcos arch*. URTEC Conference, Denver, CO.

McLean, R., Miller, C., Guzman, B., & Walls, J. (2017). *Quantifying organic porosity and predicting estimated ultimate recovery (EUR) in the Eagle Ford formation*. Unconventional Resources Technology Conference. Austin, TX.

Mehana, M. (2016). Shale characteristics impact on nuclear magnetic resonance (NMR) fluid typing methods and correlations. *Petroleum*, 138–147.

Mendelson, J. (1985). Petroleum source rock logging. *Massachusetts Institute of Technology. Earth Resources Laboratory*, 253–330.

Milliken, K. L., Rudnicki, M., Awwiller, D. N., & Zhang, T. (2013). Organic matter–hosted pore system, Marcellus formation (Devonian), Pennsylvania. *AAPG Bulletin*, *97*(2), 177–200.

Mills, R. M. (2008). *The myth of the oil crisis: overcoming the challenges of depletion, geopolitics, and global warming*. Greenwood Publishing Group.

Mitchell, A. G., Hazell, L. B., & Webb, K. J. (1990). *Wettability determination: Pore analysis*. SPE Annual Technical Conference and Exhibition. New Orleans, LA.

Modica, C. A. (2012). Estimation of Kerogen porosity in source rocks as a function of thermal transformation: Example from the Mowry shale in the Powder River basin of Wyoming. *American Association of Petrolium Geologists (AAPG)*, *96*(1), 87–108.

Monaghan, A. A. (2014). *The Carboniferous Shales of the Midland Valley of Scotland: Geology and Resource Estimation*. London, UK: British Geological Survey for Department of Energy and Climate Change.

Morrison, G. R. (1984). Bureau of Mineral Resources, Geology and Geophysics.

Morriss, C., Rossini, D., Straley, C., Tutunjian, P., & Vinegar, H. (1997). Core analysis by low-field Nmr. *The Log Analyst*.

Mukhopadhyay, P. K. (2008). *Mississippian Lacustrine Horton formation source rocks from Nova Scotia and New Brunswick, eastern Canada: Major shale gas and oil shale resource plays*. AAPG Annual Convention. San Antonio, Texas.

Murphy, E., Warner, M., & Sarmah, B. (2013). A workflow to evaluate porosity, mineralogy, and TOC in the Utica-Point Pleasant shale play. *SPE Eastern Regional Meeting*. Pittsburgh, PA.

Murphy, M., Daniels, J., Cole, D., Sheets, J., & Welch, S. (2012). *Pore distribution in the Ordovician shale of the Utica/Point Pleasant sub-basin*. AAPG Annual Convention and Exhibition. Long Beach, CA.

Mustafa, A., Shani, A., Abouelresh, M. O., & Hariri, M. M. (2014). *Geological characteristics of the Lower Silurian Qusaiba shale, Rub Al-Khali basin, Saudi Arabia*. Second EAGE/SPE/AAPG Shale Gas Workshop. Dubai, UAE.

Nakamura, K., Hatakeyama, T., & Hatakeyama, H. (1981). Studies on bound water of cellulose by differential scanning calorimetry. *Textile Research Journal*, *51*(9), 607–613.

Nakornthap, K., & Evans, R. D. (1986). Temperature-dependent relative permeability and its effect on oil displacement by thermal methods. *SPE Reservoir Engineering*, *1*(3), 230–242.

Odusina, E., Sondergeld, C., & Rai, C. (2011). *An NMR study on shale wettability*. SPE, Canadian Unconventional Resources Conference. Calgary, Alberta, Canada.

Oilfield Glossary. (n.d.). (Schlumberger) Retrieved March 28, 2018, from http://www.glossary.oilfield.slb.com/Terms/l/lwd.aspx

Ojha, S. P., Misra, S., Tinni, A., Sondergeld, C., & Rai, C. (2017). Pore connectivity and pore size distribution estimates for wolfcamp and eagle ford shale samples from oil, gas and condensate windows using a dsorption-desorption measurements. *Journal of Petroleum Science and Engineering*, *158*, 454–468.

Ojha, S. P., Misra, S., Tinni, A., Sondergeld, C., & Rai, C. (2017). Relative permeability estimates for Wolfcamp and Eagle Ford shale samples from oil, gas and condensate windows using adsorption-desorption measurements. *Fuel*, *208*, 52–64.

Ozyurtkan, M. H. (2016). Researchers test mud formulas for Turkey's Dadas shale. *Oil & Gas Journal*.

Papay, J. (2014). Exploitation of light tight oil plays. *NAFTA*, *65*(3), 231–237.

Pashin, J. C. (2011). *Geological foundation for production of natural gas from diverse shale formations*. Tuscaloosa, AL: Geological Survey of Alabama.

Pashin, J. C., Kopaska-Merkel, D. C., Arnold, A. C., McIntyre, M. R., & Thomas, W. A. (2012). Gigantic, gaseous mushwads in Cambrian shale: Conasauga formation, southern

Appalachians, USA. *International Journal of Coal Geology, 103*, 70–91.

Passey, Q. R., Bohacs, K. M., Esch, W., Klimentidis, R., & Sinha, S. (2010). From oil-prone source rock to gas-producing shale reservoir: geologic and petrophysical characterization of unconventional shale gas reservoirs. *International Oil and Gas Conference and Exhibition*. Beijing, China.

Pawlewicz, M. J., & Hatch, J. R. (2007). Petroleum assessment of the Chattanooga shale/Floyd shale: Paleozoic total petroleum system, Black Warrior basin, Alabama and Mississippi. In *Geologic assessment of undiscovered oil and gas resources of the Black Warrior basin province, Alabama and Mississippi*. Reston, VA: U.S. Department of the Interior U.S. Geological Survey.

Perfect, E. (2005). Modeling the primary drainage curve of prefractal porous media. *Vadpose Zone Journal, 4*(4), 959–966.

Perry, C. (2012). *Geology update of the Ohio Utica-Point Pleasant play*. Ohio EPA Brownfield Conference. Columbus, OH.

Piane, C. D., Almqvist, B. S., MacRae, C. M., Torpy, A., Mory, A. J., & Dewhurst, D. N. (2015). Texture and diagenesis of Ordovician shale from the Canning basin, Western Australia: Implications for elastic anisotropy and geomechanical properties. *Journal of Marine and Petroleum Geology, 59*, 56–71.

Poupon, A., & Leveaux, J. (1971). Evaluation of water saturations in shaly formations. *Society of Petrophysicists and Well-Log Analysts, 12*(4).

Profice, S., Lasseux, L., Jannot, Y., Jebara, N., & Hamon, G. (2011). Permeability, porosity, and Klinkenberg coefficient determination on crushed porous media. *International Symposium of the Society of Core Analyst*. Austin, TX.

Radtke, R. J., Lorente, M., Adolph, B., Berheide, M., Fricke, S., Grau, J., et al. (2012). *A new capture and inelastic spectroscopy tool takes geochemical logging to the next level*. SPWLA 53rd Annual Logging Symposium. Cartagena, Colombia.

Rathakrishnan, E. (2004). *Gas dynamics*. New Delhi: Practice Hall of India.

Reinsalu, E., & Aarna, I. (2015). About technical terms of oil shale and shale oil. *Estonian Academy Publishers, Oil Shale: A Scientific-Technical Journal, 32*(4), 291–292.

Romero-Sarmiento, M.-F., Ramiro-Ramirez, S., Berthe, G., Fleury, M., & Littkec, R. (2017). Geochemical and petrophysical source rock characterization of the Vaca Muerta formation, Argentina: Implications for unconventional petroleum resource estimations. *International Journal of Coal Geology, 184*, 27–41.

Roshan, H., Al-Yaser, A. Z., Sarmadivaleh, M., & Iglauer, S. (2016). On wettability of shale rocks. *Journal of Colloid and Interface Science, 475*, 104–111.

Ryan, B. E. (2017). *Petrophysical properties of the Woodford formation in the Ardmore basin in Oklahoma, U.S.A.* Arlington, TX: The University of Texas at Arlington.

Ryder, R. T. (1996). *Fracture patterns and their origin in the Upper Devonian Antrim shale gas reservoir of the Michigan basin: A review*. Reston, VA: U.S. Department of the Interior U.S. Geological Survey.

Sahin, A. (2013). *Unconventional natural gas potential in Saudi Arabia*. SPE Middle East Oil and Gas Show and Exhibition. Manama, Bahrain.

Saidian, M., Kuila, U., Prasad, M., Barraza, S. R., & Godinez, L. J. (2016). A comparison of measurement techniques for porosity and pore size distribution in shales (Mudrocks): A case study of Haynesville, eastern European Silurian, Niobrara, and Monterey formations. In *A comparison of measurement techniques for porosity and pore size distribution in shales (mudrocks)* (89–144). AAPG Memoir.

Saidian, M., Kuila, U., Rivera, S., Godinez, L. J., & Prasad, M. (2014). *Porosity and pore size distribution in mudrocks: A comparative study for Haynesville, Niobrara, Monterey and eastern European Silurian formations*. Unconventional Resources Technology Conference. Denver, CO.

Sakhaee-Pour, A., & Bryant, S. L. (2011). *Gas permeability of shale*. SPE Annual Technical Conference and Exhibition. Denver, CO.

Sakhaee-Pour, A., & Bryant, S. L. (2015). Pore structure of shale. *Fuel, 143*, 467–475.

Salehi, I. (2010). *New Albany shale gas project.* Des Plaines, IL: Gas Technology Institute.

Sarg, J. F. (2012). The Bakken: An unconventional petroleum and reservoir system. *Colorado School of Mines.* Golden, CO.

Satter, A., & Iqbal, G. M. (2016). *Reservoir engineering the fundamentals, simulation, and management of conventional and unconventional recoveries.* Houston, TX: Gulf Professional Publishing.

Schettler, P. D., Parmely, C. R., & Lee, W. J. (1989). Gas storage and transport in Devonian shales. *SPE Formation Evaluation Journal, 4*(3), 371–376.

Schlumberger. (1985). *Interpretation charts.* New York: Schumberger.

Sen, S., Ghosh, I., & Kumar, M. (2015). *Uncertainty in well log analyses and petrophysical interpretations.* 11th Biennial International Conference & Exposition, Society of Petrolum Geophysicists. JECC, Jaipur, India.

Shen, L., & Chen, Z. (2007). Critical review of the impact of tortuosity on diffusion. *Chemical Engineering Science, 62*, 3748–3755.

Sherwood, T. G., Pigford, R. L., & Wilke, C. R. (1975). *Mass transfer.* New York: McGraw-Hill.

Sigal, R. F. (2015). Pore-size distributions for organic-shale-reservoir rocks from nuclear-magnetic-resonance spectra combined with adsorption measurements. *SPE Journal, 20*(04), 824–830.

Simandoux, P. (1982). Dielectric measurements in porous media and application to shaly formation. *Revue del'Institut Francais du Petrole* (Translated text in SPWLA Reprint Volume Shaly Sand), 193–215.

Sing, K., Everett, D., Haul, R., Moscou, L., Pierotti, R., Rouquérol, J., & Siemieniewska, T. (1985). IUPAC commission on colloid and surface chemistry including catalysis:. Reporting physisorption data for gas/solid systems with special reference to the determination of surface area and porosity. *Pure Applied Chemistry, 57*, 603–619.

Singh, H., Javadpour, F., Ettehadtavakkol, A., & Darabi, H. (2014). Nonempirical apparent permeability of shale. *SPE Reservoir Evaluation and Engineering, 17*(3), 414–424.

Sinha, S., Braun, E., & Passy, Q. (2012). *Advances in measurement standards and flow properties measurements for tight rocks such as shales.* SPE/EAGE Conference. Vienna, Austria.

Soeder, D. J. (2011). *Petrophysical characterization of the Marcellus and other gas shales.* Arlington, VA: U. S. Department of Energy.

Sondergeld, C. H., Ambrose, R. J., Rai, C. S., & Moncrieff, J. (2010). *Micro-structural studies of gas shale.* SPE Unconventional Gas Conference. Pittsburg, PA.

Sonnenberg, S. A. (2012). The Niobrara petroleum system, Rocky Mountain region. *Tulsa Geological Society Meeting.* Tulsa, OK.

Standing, M. (1974). *Notes on relative permeability relationships.* Thesis, Division of Petroleum Engineering and Applied Geophysics. The Norwegian Institute of Technology, University of Trondheim.

Stevens, P. (2012, 08 15). The shale gas revolution: developments and changes. Retrieved March 30, 2018, from *The Shale Gas Revolution: Developments and Changes.* Chatham House

Stidham, J. (2011). *Chesapeake energy Eagle Ford shale overview.* Oklahoma City: Chesapeake Energy.

Stoneburner, R. (2009). *The Haynesville shale: A look back at the first year.* 8th Gas Shales Summit. Dallas, TX.

Sun, S., Liang, F., Tang, L., Wu, J., & Ma, C. (2017). Microstructural investigation of gas shale in Longmaxi formation, Lower Silurian, NE Sichuan basin, China. *Energy Exploration and Exploitation, 35*(4), 406–429.

Suri, P., Azeemuddin, M., Zaman, M., Kukreti, A. R., & Roegiers, J.-C. (1997). Stress-dependent permeability measurement using the oscillating pulse technique. *Journal of Petroleum Science and Engineering, 17*, 247–264.

Swanson, V. E. (1960). *Oil yield and uranium content of black shales: Uranium in carbonaceous rocks.* Washington, DC: United States Government Printing Office.

Tan, J., Horsfield, B., Fink, R., Krooss, B., Schulz, N. M., Rybacki, E., et al. (2014). Shale gas potential of the major marine shale formations in the upper Yangtze platform, south China, Part III: Mineralogical, lithofacial, petrophysical, and rock mechanical properties. *Energy & Fuels*, *28*, 2322–2342.

Thompson, A., Katz, A., & Krohn, C. (1987). The microgeometry and transport properties of sedimentary rock. *Advances in Physics*, *36*.

Thyne, G. (2013). A review of the measurement of wettability. Science Based Solution Report based on a report for Sandia National Laboratory.

Tiabb, D., & Donaldson, E. C. (2004). *Petrophysics (1)*. Oxford: Elsevier.

Timur, A. (1967). Pulsed nuclear magnetic resonance studies of porosity, movable fluid and permeability of sandstones. *Journal of Petroleum Technology*, *21*(6), 775–786.

Timur, A. (1968). Effective porosity and permeability of sandstones investigated through nuclear magnetic principles. *The Log Analyst*, *10*(1), 3–11.

Timur, A. (1994). An investigation of permeability, porosity, and residual water saturation relationships for sandstone reservoirs. *The Log Analyst*, *9*(4), 128–137.

Tinni, A., Fathi, E., Agarwal, R., Sondergeld, C. H., Akkutlu, I. Y., & Rai, C. S. (2012). *Shale permeability measurements on plugs and crushed samples*. SPE Canadian Unconventional Resources Conference. Calgary, Alberta, Canada.

Tixier, M. P., Alger, R. P., & Doh, C. A. (1959). Sonic logging. *Journal of Petroleum Technolology, Transactions, AIME*, *216*(106).

Toksoz, M. N., & Cheng, C. H. (1978). Modeling of seismic velocities in porous rocks and its application to seismic exploration. *Arabian Journal of Science and Engineering*, 253–330.

Topcu, G. Y. (2013). *Quantification of the uncertainties in shale gas reservoirs, a case study for Dadas shale formation*. Middle East Technical University.

Torsaeter, M., Vullum, P. E., & Nes, O. M. (2012). Nanostructure vs. macroscopic properties of Mancos shale. *SPE Canadian Unconventional Resources*. Calgary, Alberta, Canada.

U.S. Energy Information Administration (2013). Retrieved March 30, 2018, from www.EIA. gov

Unconventional resources: Tight gas sands and shale gas (n.d.). http://www.energy.gov.dz/francais/uploads/slides/Unconventional_Resources_Tight_Gas_Sands_Shale_Gas.pdf.

Unlocking Turkey's tight gas potential. (2013). Valeura Energy Inc.

Utica Shale Appalachian Basin Exploration Consortium (2012–2015). *A geologic play book for Utica shale Appalachian basin exploration*. West Virginia University.

Valori, A. V. (2017). Permeability estimation from NMR time dependent methane saturation monitoring in shales. *Energy Fuels*, 5913–5925.

Van der Heide, P. (2011). *X-ray photoelectron spectroscopy: An introduction to principles and practices*. Hoboken, NJ: John Wiley and Sons.

Viamontes, J. (n.d.). A case study into the successful evaluation and completion nonconventional wells in Mexico. NUTECH.

Vigh, T., Kovacs, T., Somlai, J., Kavasi, N., Polgari, M., & Biro, L. (2013). Terrestrial radioisotopes in black shale hosted mn-carbonate deposit (Úrkút, Hungary). *Acta Geophysica*, *61*(4), 831–847.

Walls, J., Anderson, E., & Ceron, M. (2013). Results and methodology from ANH (Colombia) unconventional resources core project, search and discovery applying the "Geo" into organic geochemistry of shale gas systems. *SPE ATW Unconventional Gas Reservoirs*. Keystone, CO.

Walls, J., Guzman, B., Anderson, J., Foster, M., McLean, R., & Cicciotelli, M. (2015). Eagle Ford lateral wellbore analysis using drill cuttings. *2015 International AAPG Convention Poster Session and Exhibition*. Denver, CO.

Walter, Y. (1998). *Shale oil: The elusive energy*. Colorado, CO: Hubbert Center Newsletter, Colorado School of Mines.

Wang, D., Butler, R., & Zhang, J. (2012). Wettability survey in Bakken shale with surfactant-formulation imbibition. *SPE Reservoir Evaluations and Engineering*, *15*(6), 695–705.

Waxman, M. H., & Smits, L. J. (1968). Electrical conductivities in oil-bearing shaly sands. *SPE, 8* (2), 107–122.

Waxman, M. H., & Thomas, E. C. (1974). Electrical conductivities in shaly sands. I. The Relation between hydrocarbon saturation and resistivity index. II. The temperature coefficient of electrical conductivity. *Journal of Petroleum Technology, 26*(2), 213–225.

Wilson, A. (2012). Advances in measuring permeability address shortcomings of crushed-rock technique. *JPT*, 57–60.

Wilson, T. K., & Bustin, R. M. (2017). *Unconventional petroleum systems analysis of Upper Devonian organic-rich shale units in the Horn River and Liard basins, northeastern British Columbia and adjacent western Alberta: Preliminary report.* Alberta, Canada: Geoscience BC.

Woodhouse, R., & Warner, H. R. (2005). *Sw and hydrocarbon pore volume estimates in shaly sands - routine oil-based-mud core measurements compared with several log analysis models.* SPE Annual Technical Conference and Exhibition. Dallas, TX.

World Energy Resources (2016).World Energy Council.

Wyllie, M., & Gardner, G. (1958). The generalized Kozeny-Carman equation. *World Oil, 146*(4), 121–128.

Wyllie, M. R., Gregory, A. R., & Gardner, L. W. (1956). Elastic wave velocities in heterogeneous and porous media. *Geophysics, 21*(1), 41–70.

Xiao, Y., Bosch, R. V.-D., Liu, F., & Steffen, K. (2012). *Performance evaluation in data Rich Fayetteville shale gas plays: Integrating physics-based reservoir simulations with data driven approaches for uncertainty reduction.* International Petroleum Technology Conference (IPTC). Bangkok, Thailand.

Xu, H. T. (2014). A precise measurement method for shale porosity with low-field nuclear magnetic resonance: A case study of the Carboniferous-Permian strata in Linxing area, eastern Ordos basin, China. *Fuel*, 47–54.

Xu, M., & Dehghanpour, H. (2014). Advances in understanding wettability of gas shales. *Energy Fuels, 28*(7), 4362–4375.

Yassin, M. R., Dehghanpour, H., Wood, J., & Lan, Q. (2016). A theory for relative permeability of unconventional rocks with dual-wettability pore network. *SPE Journal, 21*(6), 1970–1980.

Ybarra, M. (2014). *Well log and core-derived reservoir properties of Barnett shale of Fort Worth basin.* Arlington, TX: The University of Texas at Arlington.

Yu, H., Wang, Z., Rezaee, R., Arif, M., & Xiao, L. (2016). *Characterization of elastic properties of lacustrine shale reservoir using well logging and core analysis.* SPE Asia Pacific Oil & Gas Conference and Exhibition: Society of Petroleum Engineers. Perth, Australia.

Yu, H., Wang, Z., Rezaee, R., Su, Y., Tan, W., Yuan, Y., et al. (2017). *Applications of nuclear magnetic resonance (NMR) logs in shale gas reservoirs for pore size distribution evaluation.* Unconventional Resources Technology Conference. Austin, Texas.

Yu, W., Sepehrnoori, K., & Patzek, T. W. (2016). Modeling gas adsorption in Marcellus shale with Langmuir and BET isotherms. *SPE Journal, 21*(02), 589–600.

Yu, Y. (n.d.). *PMI pulse decay permeameter for shale rock characterization.* New York: Porous Materials Inc.

Zahaf, K., Lecoq, T. F., Al-Badi, B. S., Roth, S., Dong, H., & Blunt, M. J. (2017). *Prediction of relative permeability and capillary pressure using digital rock physics: Case study on two giant Middle Eastern carbonate reservoirs.* International Symposium of the Society of Core Analysts. Vienna, Austria.

Zhang, B., Gomaa, A. M., Sun, H., Qu, Q., & Chen, J. H. (2014). *A study of shale wettability using nmr measurements.* International Symposium of the Society of Core Analysts. Avignon, France.

Zhang, Y., Jin, S., Jiang, H., Wang, Y., & Jia, P. (2015). Review of well logs and petrophysical approaches for shale gas in Sichuan basin, China. *Petroleum Engineering Journal, 11*, 316–324.

3

Petroleum Geochemistry in Organic-Rich Shale Reservoirs

Sherifa E. Cudjoe, Mustafa M. Alhubail, and Reza Barati Ghahfarokhi

The University of Kansas, Lawrence, Kansas, USA

3.1 Background

Petroleum geochemistry deals with the application of chemical principles to study the origin, generation, migration, accumulation, and alteration of petroleum in order to improve hydrocarbon exploration and recovery (Brooks & Welte, 1984; Tissot & Welte, 1984). The petroleum exploration industry is mostly concerned with where the petroleum accumulations are located, whereas, petroleum geochemistry deals with the "when and how" petroleum is formed in addition to its migration and accumulation. Thus, organic geochemistry starts with the preserved organic matter and ends with the generation and accumulation of hydrocarbons after considerable depletion of hydrogen under thermal degradation. Moreover, petroleum geochemistry in unconventional resources helps in identifying and locating the sweet spots for perforation during hydraulic fracturing. This chapter details the generation and expulsion of petroleum, highlighting some of the methods used to determine and evaluate the geochemical parameters.

3.2 Evolution of Organic Matter

Substantial evidence supports the organic origin of petroleum and natural gas rather than the opposed claims of inorganic origins (Brooks &Welte, 1984; Waples, 1984). Therefore, this chapter is written with the assumption of organic origin of petroleum in mind. The organic material is composed of organic molecules (monomeric and/or polymeric) derived from the organic parts of organisms deposited in sediments either directly or indirectly. Blue-green algae and photosynthetic bacteria coupled with a variety of marine phytoplanktonic organisms were the primary producers of organic matter in the Precambrian era until the emergence of land plants by the middle Devonian (Tissot & Welte, 1984) as shown in Figure 3.1 (Peters et al., 2005). A potential source rock is shown in anoxic settings (Fig. 3.1), which eventually transforms into an effective source rock under considerable burial and heat for both conventional and unconventional reservoirs (Peters et al. 2005). In recent times, marine phytoplankton and

Unconventional Hydrocarbon Resources: Techniques for Reservoir Engineering Analysis, Advanced Textbook Series, First Edition. Edited by Reza Barati Ghahfarokhi and Mustafa M. Alhubail.

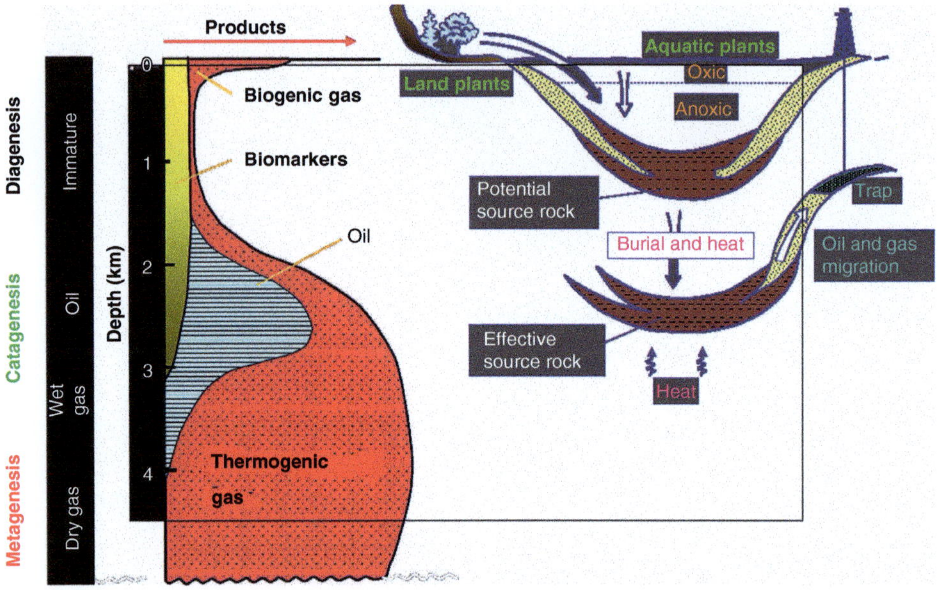

Figure 3.1 Evolution of organic matter (modified after Peters et al., 2005).

bacteria are responsible for a large percentage (50%–60%) of the world's organic carbon production (Tissot & Welte, 1984). Organic matter is only preserved and fossilized in subaquatic sediments due to the inactivity of aerobic bacteria as a result of limited access to dissolved molecular oxygen. Only about 0.01% to 0.1% of the primary productivity of organic matter (OM) is preserved in sediments (Tissot & Welte, 1984). During the evolution of organic matter, biopolymers (i.e., proteins and carbohydrates) are transformed by microorganisms, and the resulting constituents undergo polycondensation and polymerization to eventually form an insoluble kerogen (Sato, 1990). This process constitutes the diagenetic stage, which is one of the three main stages of organic matter transformation. The three stages are (1) diagenesis, (2) catagenesis, and (3) metagenesis.

3.2.1 Diagenesis

Diagenesis involves the deposition of sediments at shallow burial depths in subaquatic environments under equilibrium (100–1,000 m) resulting in the consolidation of the sediments. Microbial activity controls sediment transformations at low temperatures and pressures during diagenesis (Tissot & Welte, 1984). This biochemical degradation stage corresponds to a thermal maturity of ~0.5 Ro%, and the most important hydrocarbon formed during this stage is methane.

3.2.2 Catagenesis

Catagenesis occurs at greater burial depths due to continuous deposition of sediments at considerable increase in temperature (~50–150°C) and pressure (~4,351–14,503 psi) as well as the effect of tectonics. Kerogen from diagenesis produces first liquid petroleum, which is further transformed to produce wet gas and condensate together with significant amounts of methane as illustrated in Figure 3.2 (Jarvie et al., 2007). The decomposition of kerogen into bitumen and thermogenic gas, marked as step 1 in Figure 3.2, follows primary cracking (catagenesis), in addition to the decomposition of bitumen into oil and thermogenic gas, marked as step 2. Biodegradation of oil into biogenic gas (methane) is marked as step 3 (Jarvie et al.,

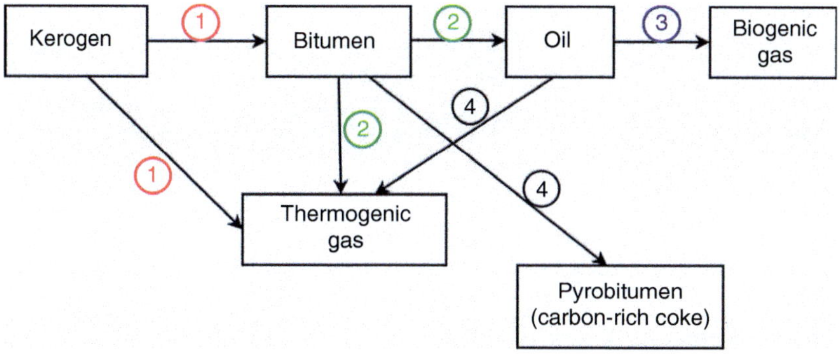

Figure 3.2 Transformational stages of kerogen developed under diagenesis (modified after Jarvie et al., 2007).

2007). The end of catagenesis is marked by the onset of the disappearance of aliphatic carbon chains in kerogen and the development of ordered kerogen units, corresponding to vitrinite reflectance of ~2 Ro% (Tissot & Welte, 1984).

3.2.3 Metagenesis

Metagenesis occurs in deep zones at much higher temperatures and pressures in addition to hydrothermal effects. Organic matter at this stage is made up of only methane and carbon residue with the development of crystalline ordering reflecting a thermal maturity of ~ 4 Ro% (Hunt, 1996; Tissot & Welte, 1984). Figure 3.1 (left) summarizes the degradation of kerogen from diagenesis (before thermal alteration) to metagenesis (thermal destruction). Prior to thermal alteration, biogenic methane is generated by microbial activity on organic matter at shallow depths. Diagenesis marks the onset of the oil window. At greater depths of burial and heating, the cracking of kerogen produces crude oil and wet gas in the catagenesis stage and thermogenic gas in the metagenesis stage (McCarthy, et al., 2011; Jarvie et al., 2007; Peters & Casa, 1994). The dry gas (i.e., thermogenic gas) is produced either directly from the degradation of kerogen or through secondary cracking of retained oil as shown in Figure 3.2. The cracking of kerogen creates organic porosity in the nanometer range, and the extent of kerogen conversion to petroleum is quantified by the transformation ratio (Jarvie et al., 2007). Many molecular fossils (biomarkers) survive thermal degradation by showing little or no change in structure from their parent organic molecules mainly in the diagenesis and catagenesis stages (Peters & Moldowan, 1993).

3.3 Total Organic Carbon (TOC)

Total organic carbon (TOC) measures the concentration of organic carbon in a rock or sediment expressed in weight percent (wt. %), and it doubles as a reliable parameter to screen for generative potential of a source rock (Miles, 1994). TOC includes both insoluble kerogen and soluble bitumen. Kerogen is referred to as the insoluble fraction of organic matter remaining after extraction of crushed rock sample with organic solvents (Hunt, 1961). On the other hand, bitumen is the fraction of organic matter soluble in organic solvents (5%–10% of TOC), and it is composed of three main fractions: asphaltenes, resins (higher molecular weight), and hydrocarbons (lower molecular weight) (Hunt, 1996; Forsman & Hunt, 1958).The main components of what is referred to as bitumen today is high viscosity asphaltenes and wax materials.

TOC can be measured by direct combustion, pyrolysis, and geophysical methods, each of

Table 3.1 Guidelines for Source Rock Organic-Richness

Quality	TOC (wt. %)	S_1 (mg HC/g)	S_2 (mg HC/g)	Bitumen (ppm)	Hydrocarbons (ppm)
Poor	< 0.5	< 0.5	< 2.5	< 500	< 300
Fair	0.5–1.0	0.5–1.0	2.5–5.0	500–1,000	300–600
Good	1.0–2.0 (possible threshold of shale oil)	1.0–2.0	5.0–10.0	1,000–2,000	600–1,200
Very good	2.0–4.0 (possible threshold to shale gas)	2.0–4.0	10.0–20.0	2,000–4,000	1,200–2,400
Excellent	> 4.0	> 4.0	> 20.0	> 4,000	> 2,400

Source: Peters and Cassa (1994).

which presents a limitation. Moreover, TOC is not a clear indicator of petroleum potential; the presence of substantial hydrogen is required to be associated with the carbon to generate more hydrocarbons (Dembicki, 2009). Table 3.1 lists some TOC guidelines for potential source rock organic richness. TOC in the range of 1–2 wt. % is found to be the possible threshold for oil expulsion, while TOC above 2 wt. % is determined to be the possible threshold for gas in shales (Ruble et al., 2015; Lewan, 1987). In practice, one would have to consider TOC, bitumen content, hydrogen index (HI), oxygen index (OI), S_1, and S_2 to be able to classify the shale resources accurately.

As previously stated, TOC is not a clear indicator of petroleum potential, however, considering TOC for geochemical and petrophysical investigation of shale reservoirs is relevant for several reasons. First, TOC provides the carbon for the hydrocarbons, and it also provides the adsorptive sites for the hydrocarbons (mostly, kerogen for storage of adsorbed gas and to retain oil in nanopores for cracking to gas). In addition, TOC provides increased porosity in accordance with increasing thermal maturation. Finally, the estimated ultimate recovery (EUR) of shale gas reservoirs is directly proportional to TOC and formation thickness (Ruble et al., 2015).

3.4 Kerogen, Bitumen, and/or Pyrobitumen

Chemically, kerogen is defined as the insoluble organic component of the sedimentary rocks, while bitumen is described as the extractable fraction that is soluble in organic solvents. Petrographically, kerogen is described as particulate debris pervading intergranular spaces between silt and clay-sized grains, whereas migrated bitumen or solid bitumen/pyrobitumen, depending on the thermal maturity of the formation, are identified as void-filling organic material within chambers of foraminifera bodies, microfractures as well as matrix pores (Camp, 2015; Milliken et al., 2014). The term *bitumen* refers to hydrocarbons with varying viscosities including petroleum, resins, and asphaltenes, while solid bitumen and pyrobitumen is described as a solid, nonvolatile residue of oil made up of hydrocarbon complexes, according to the American Geological Institute (AGI) Glossary of Geology. Kerogen undergoes compaction and thermal degradation to produce bitumen and/or oil at a maturity of ~ 0.75 %Ro (Loucks & Reed, 2014). The produced oil/bitumen migrates into surrounding mineral pores and fossil cavities/chambers (foraminifera bodies) and undergoes further thermal decomposition to transform into solid bitumen (oil window) or pyrobitumen (gas window) (> 1.3 %Ro) with developed organic pores (Bernard et al., 2012).

Foraminifera bodies are made up of outer calcite linings and overgrowths with significantly larger chambers than matrix pores. The calcite overgrowths suggest an open pore prior to migration of bitumen/oil (Camp, 2015; Loucks & Reed, 2014). In so doing, the dimensions of the migrated organic matter (OM) is controlled by the size of the hosting pore (Loucks & Reed, 2014). Figure 3.3 shows back-scattered electron (BSE) image of an oil-matured lower Eagle Ford (LEF) shale sample with distributions of depositional kerogen embedded in mineral grains without calcite overgrowths and migrated organic matter (OM) within foraminifera cavities. In addition to the organic material, the foraminifera chambers also house other minerals, such as kaolinite, quartz, calcite, and pyrite, which predate the migrated OM.

As previously stated, the migrated oil/bitumen transforms into solid bitumen, which preserves organic pores and/ or pyrobitumen at much higher maturity with organic pores. Two types of migrated OM are observed within the foraminifera chamber in Figure 3.3, namely, meniscus-type OM (nonporous) and solid bitumen (porous). The meniscus-type OM, which appears to be partially filling the chamber, can be interpreted as pre-oil-generation bitumen or another form of solid bitumen not prone to organic porosity within the oil window. Solid bitumen is defined as *a very viscous bitumen capable of preserving pores,* according to the AGI Glossary (Loucks & Reed, 2014). This lower Eagle Ford (LEF) shale sample has an estimated maturity of 1.01 %Ro, which falls within the oil window. Pyrobitumen is presumed to form in gas windows also with secondary (organic) porosity (Bernard et al., 2012). Thus, organic porosity can occur not only in pyrobitumen in shale gas formations as stated in some literature findings but also in solid bitumen of shale oil formations. Kerogen also develops organic pores, but they are presumed to be mostly covered by produced oil or bitumen. Therefore, organic porosity in shale reservoirs are due to contributions by both the depositional kerogen and solid bitumen or pyrobitumen (Camp, 2015; Loucks & Reed, 2014). Although, the migrated bitumen (solid bitumen/pyrobitumen) is reported to contribute about 80% of the organic pores, while depositional kerogen contributes ~2%; the remaining 18% is attributed

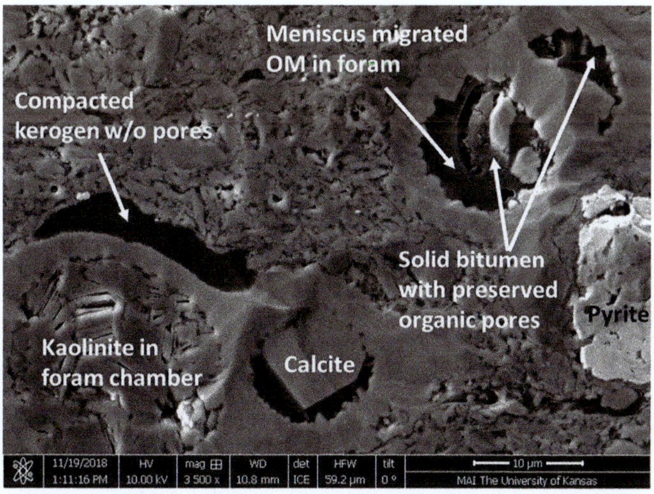

Figure 3.3 BSE image of an Eagle Ford shale sample with distributions of kerogen and migrated bitumen/solid bitumen at a maturity level of 1.01 %Ro.

to contributions by both types of organic matter, which are difficult to distinguish especially within the matrix (Cudjoe et al., 2019a,b).

3.4.1 Classification of Kerogen

Different types of kerogen experience different kinetics of hydrocarbon generation and can be recognized by optically and/ or physicochemically. Three (3) main kerogen types are widely accepted and characterized by means of their respective evolution pathways in the van Krevelen (H/C vs. O/C) diagram in addition to a secondary type IV kerogen. A modified van Krevelen diagram of hydrogen index (HI) versus oxygen index (OI) shown in Figure 3.4 is used to characterize the kerogen type and quality. Moreover, the kerogen quality is evaluated by the use of the S_2 versus TOC plot in Figure 3.5. In addition to characterizing kerogen types based on their

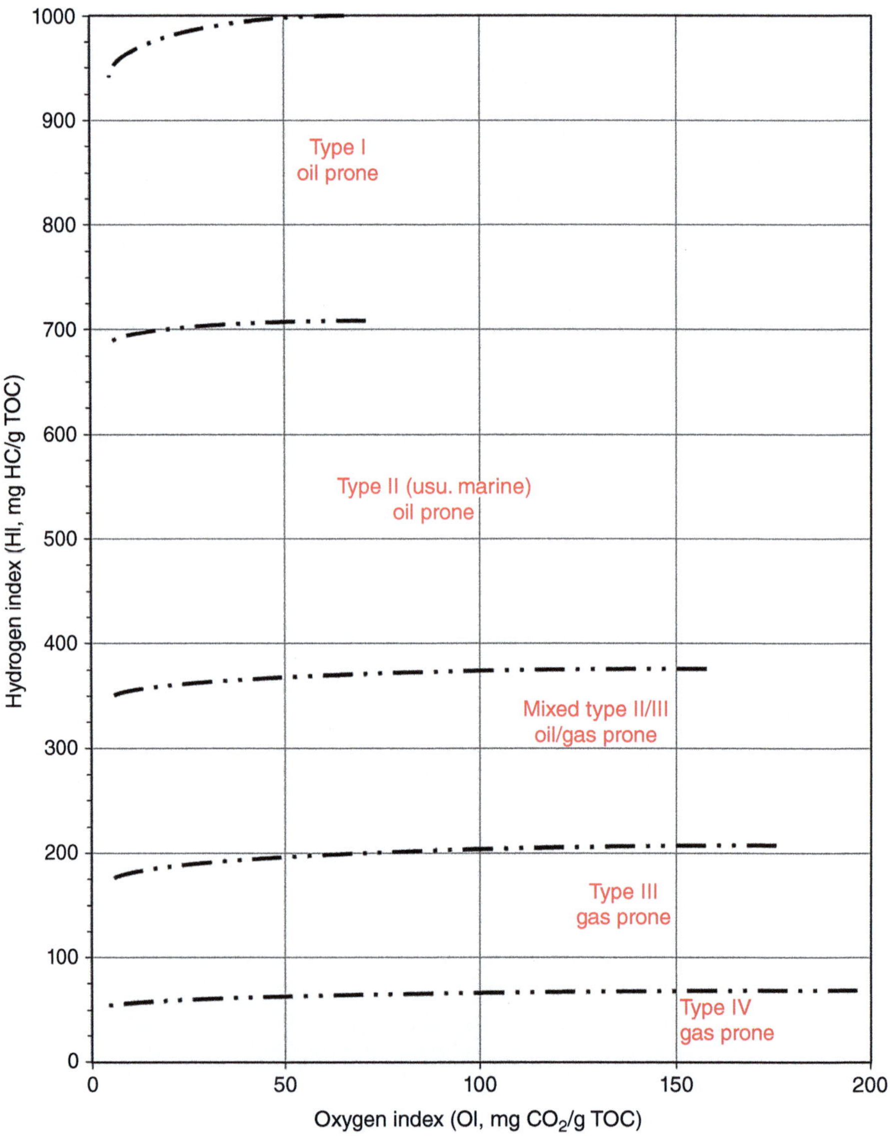

Figure 3.4 HI-OI plot to determine kerogen type.

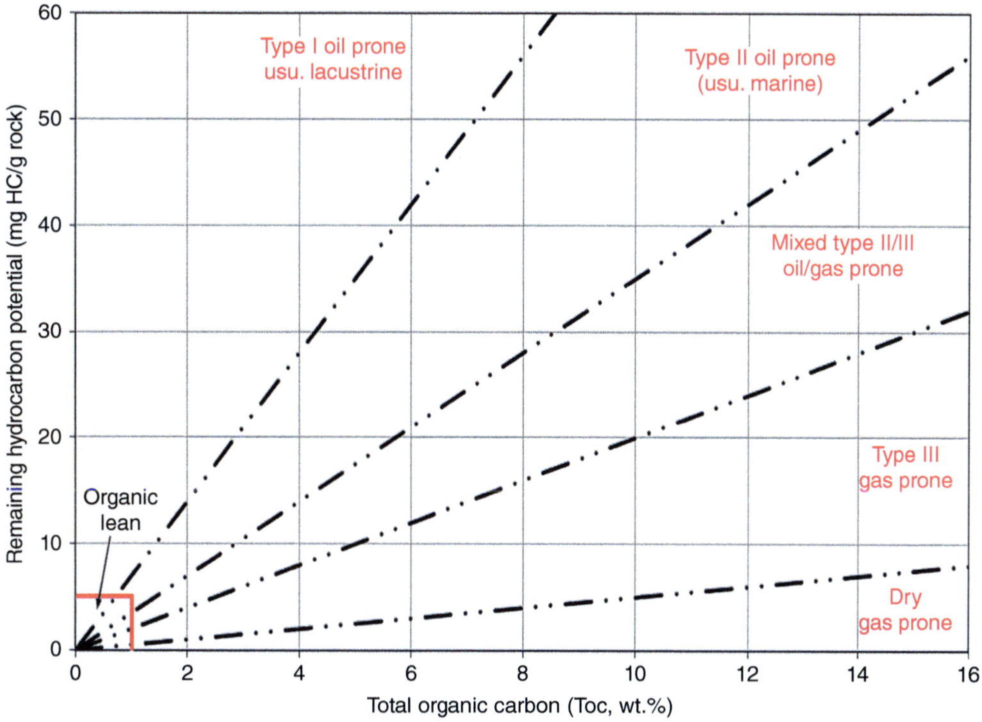

Figure 3.5 S2-TOC plot to determine kerogen quality.

Table 3.2 Kerogen Type Characterization

Kerogen type	Maceral group	Macerals	Organic precursors
I	Liptinite	Alginite I	Fresh water algae
II		Alginite II	Marine algae
II		Exinite	Spores, pollen
II		Cutinite	Leaf cuticle
II		Resinite	Resin, tree sap
III	Vitrinite	Vitrinite, pseudovitrinite	Woody tissue
IV	Inertinite	Semifusinite, fusinite, sclerotonite	Reworked/oxidized material, charcoal

Source: Hutton et al. (1994).

respective evolution pathways, maceral groups can be assigned to the kerogen types to broaden the characterization process. Macerals are defined as *naturally occurring homogeneous optic organic materials that show unique physical and chemical properties varying with increasing thermal maturity* (Cardott, 2012; Mukhopadhyay, 1994). Table 3.2 lists the various maceral groups and their corresponding macerals correlated with kerogen type as well as organic precursors. The three primary maceral groups are identified as liptinite (hydrogen-rich), vitrinite

Table 3.3 KerogenType and Hydrocarbon Products

Kerogen type	Hydrogen index (mg HC/g TOC)	S_2/S_3	Main hydrocarbon product at peak maturity
I	> 600	> 15	Oil
II	300–600	10–15	Oil
II/III	200–300	5–10	Mixed oil and gas
III	50–200	1–5	Gas
IV	< 50	< 1	None

Source: Peters and Casa (1994).

(oxygen-rich), and inertinite (carbon-rich). The change in the macerals as a result of thermal maturity can be quantified by either the vitrinite reflectance (%Ro) or the thermal alteration index (TAI) (Fishman et al., 2012; Teerman et al., 1995). Moreover, Table 3.3 lists the kerogen types with their corresponding hydrocarbon products to be expected when characterized by Rock-Eval parameters.

Type I Kerogen

Type I kerogen is characterized by alginate macerals enriched in lipids and is predominantly made up of aliphatic chains. It possesses the highest average pseudo activation energy and requires strong C–C bonds to produce bitumen (Waples, 1984). It has a high H/C atomic ratio and low O/C ratio, therefore its potential to generate hydrocarbons is high (Graciano et al., 2012).

Type II Kerogen

Type II kerogen is characterized by exinite macerals in marine-reducing environment and contains more aromatic and naphthenic rings with low O/C ratio and relatively high H/C ratio (Vandenbroucke, 2003). It also has a high sulphur content and begins to generate both oil and gas at a thermal maturity value of ~0.5 %Ro (Tissot & Welte, 1984).

Type III Kerogen

Type III kerogen is characterized by vitrinite and psuedovitrinite macerals of terrestrial higher plant origin. It is made up of condensed polyaromatics and oxygenated functional groups, with very few aliphatic chains (Tissot & Welte, 1984). It has lower H/C ratio with higher O/C ratio and yields the highest amounts of gas at a vitrinite reflectance > 1.3 %Ro (Jarvie et al., 2007; Hutton et al., 1994).

Type IV Kerogen

Type IV kerogen is residual kerogen characterized by inertinite macerals and consists of transformed oxidized organic material from subaerial weathering with no potential for hydrocarbon (i.e., oil or gas) generation (Tissot & Welte, 1984).

3.5 Vitrinite Reflectance

Vitrinite reflectance (%Ro) is a thermal maturity indicator, which measures the percentage of light reflected off the presence of commonly occurring vitrinite maceral. Thermal alteration index (TAI) and the spore coloration index (SCI) are also maturity indicators that measure the color changes observed in pollen or spores microfossils; TAI/SCI measurements are comparable to vitrinite reflectance measurements (Cardott, 2012; Mukhopadhyay, 1994). The maceral or spore colorization is determined through optical analysis of either thin sections or pellet-sized samples. Vitrinite reflectance (% Ro) of a sample increases during thermal maturation due to complex aromatization reactions that are irreversible (Peters, 1986). Table 3.4 lists some assigned vitrinite reflectance values from the onset to postmature periods of hydrocarbon generation. However, vitrinite reflectance measurements can be affected by inclusions (e.g., pyrite, bitumen, and other macerals), weathered or fractured vitrinite textures, oxidized vitrinite and oily vitrinite, among other factors (Fang & Jianyu,

Table 3.4 Guidelines for Vitrinite Reflectance (% Ro) Thermal Maturity

Thermal maturity	Vitrinite reflectance (%Ro)
Immature	0.20–0.60
Mature	0.60–0.65 (Early)
	0.65–0.90 (Peak)
	0.90–1.35 (Late)
Pomature	> 1.35

Source: Jarvie et al. (2007).

2007). In general, the vitrinite reflectance range of 0.5–0.6 (%Ro) is an indicator of the onset of the oil generation, and the end of the oil generation window is typically in the range of 0.85–1.1 (% Ro). The onset of the gas generation window is typically in the range of 1.0–1.3 (%Ro). It is important to note that in the absence of vitrinite macerals (e.g., prior to the Devonian geological period), the vitrinite reflectance method is not useful in determining the thermal maturity (Mukhopadhyay & Dow, 1994). Thus, other methods (e.g., Rock-Eval pyrolysis T_{max}) have to be used to determine the thermal maturity.

Example 3.1 Determine the Thermal Maturity

The calculated vitrinite reflectance (%Ro) for the upper Bakken samples is provided in Table 3.5. Determine the thermal maturity based on the average values of the data.

Table 3.5 Data for Example 3.1

Depth	Formation	Cal. %Ro
9,864	Upper Bakken	0.688
9,865	Upper Bakken	0.724
9,866	Upper Bakken	0.616
9,867	Upper Bakken	0.634
9,868	Upper Bakken	0.616
9,869	Upper Bakken	0.526
9,870	Upper Bakken	0.652

Solution

The average calculated vitrinite reflectance (%Ro) is 0.637. Hence, by the use of the guidelines provided in Table 3.4, the average value shows that the kerogen is in the range of early thermal maturity for this zone.

3.6 Solid Bitumen Reflectance

Bitumen is genetically classified into pre-oil-solid bitumen and postoil bitumen. Pre-oil bitumen consists of early diagenetic products that are expelled from the organic source as very viscous fluids and have migrated a short distance into fractures and other mineral pores. Whereas, postoil bitumen is made up of thermally altered (cracked) products of retained crude oil, such as solid bitumen and pyrobitumen, migrated from source into fossil cavities and mineral pores/fractures (Landis & Castano, 1994; Curiale, 1986). A vitrinite reflectance equivalent regression equation is developed for solid bitumen (migra-bitumen) when the pre-oil bitumen is used as a thermal maturity indicator upon solid bitumen reflectance (BRo) and given by Schoenherr et al.(2007), Landis and Castano (1994), and Jacob (1989) (Schoenherr et al., 2007; Landis & Castano, 1994; Jacob, 1989), respectively, for different formations.

$$\%R_o = \frac{(BR_o + 0.2443)}{1.0495} \tag{3.1}$$

$$\%R_o = \frac{(BR_o + 0.41)}{1.09} \tag{3.2}$$

$$\%R_o = 0.618(BR_o) + 0.40 \tag{3.3}$$

Equation (3.3) describes a linear relationship between the vitrinite and bitumen at concentrations between < 0.1 %Ro and 3.0 %BRo, using only the lowest reflectance data to represent the degree of hydrocarbon maturity (Jacob, 1989). Equation (3.2) is derived based on measurements carried out on locally derived solid hydrocarbons. Both vitrinite and bitumen reflectances are measured on the same rock samples (tar sands of western Canada and South America). At vitrinite concentrations of < 4.0 %Ro, the measured % BRo significantly differed from the measured %Ro. However, at concentrations of %Ro > 4.0, both vitrinite and bitumen measurements were found to be similar (Landis & Castano, 1994). On the other hand, equation (3.1) is developed to include higher maturity samples (%BRo > 2.7) from the Infra-Cambrian Ara Group (South Oman salt basin). However, equation (3.1) does not correlate with equation (3.2), also applicable to higher maturity samples. This points to the variation of bitumen and vitrinite in different lithologies.

3.7 Organic Porosity

Organic porosity refers to the nanoscale pores observed in either kerogen or solid bitumen and pyrobitumen after undergoing thermal maturation to produce hydrocarbons as explained under the kerogen and bitumen section. Organic porosity in shales is a contribution of both depositional and migrated organic matter (OM). Depositional OM porosity can be found in kerogen, solid bitumen, and pyrobitumen, while migrated OM porosity is found in only solid bitumen and pyrobitumen (Camp, 2015; Loucks & Reed, 2014; Bernard et al., 2012). Using focused ion beam-scanning electron microscopy (FIB-SEM), the organic pores of lower Eagle Ford shale samples with thermal maturity in terms of vitrinite reflectance (%Ro) are given as 0.98 %Ro (Fig. 3.6), 1.01 %Ro (Fig. 3.7), and 1.03 %Ro (Fig. 3.8), respectively. The range of vitrinite reflectance (%Ro) values for the lower Eagle Ford shale samples puts them in the late mature oil window. Thermal maturity (%Ro) shows to slightly increase with increasing depth as these samples are at different intervals within 10 to 20 ft of each other. The presence of mainly large bubble pores and some small spongy pores in the FIB-SEM images in figures below are indications of produced hydrocarbon as a result of thermal cracking of retained oil (solid organic matter plugs) and/or kerogen.

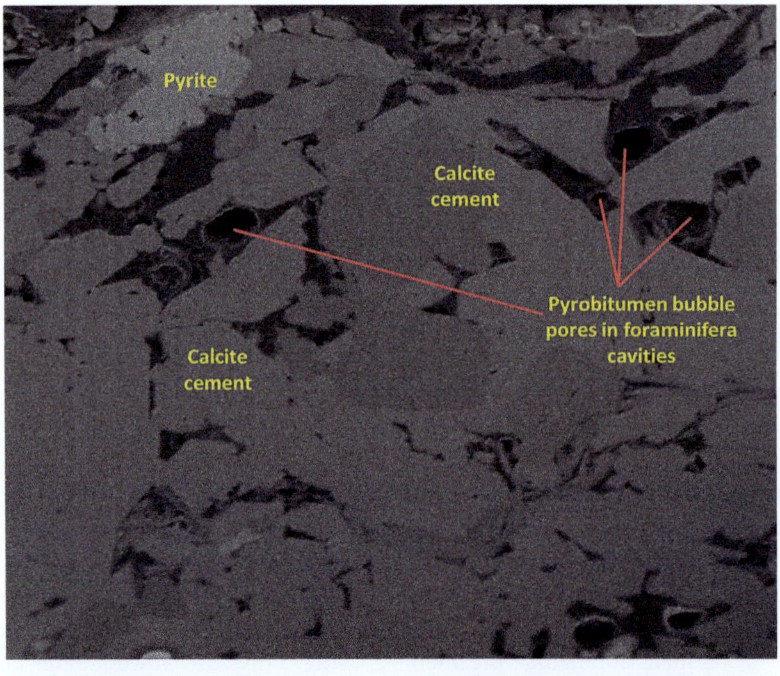

Figure 3.6 FIB-SEM image of a lower Eagle Ford shale sample with vitrinite reflectance of 0.98 %Ro.

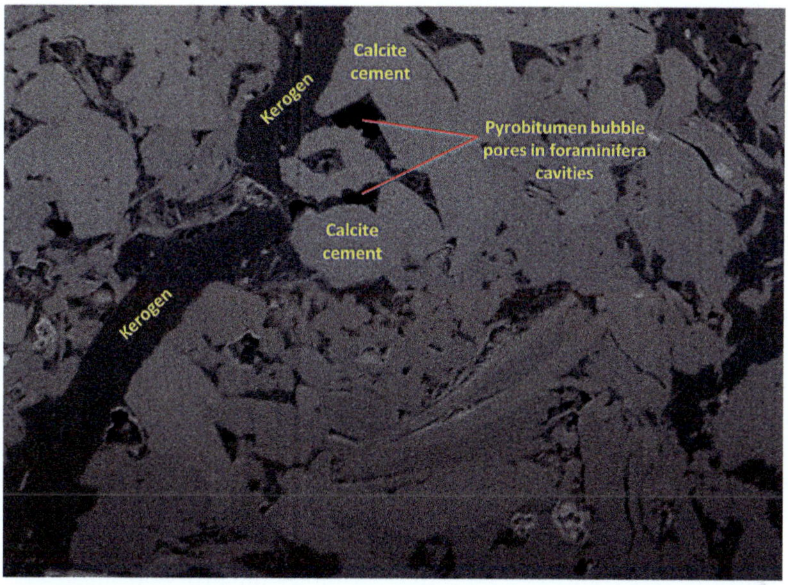

Figure 3.7 FIB-SEM image of a lower Eagle Ford shale sample with vitrinite reflectance of 1.01 %Ro.

Whereas, Figure 3.6 and Figure 3.8 show mostly solid bitumen or pyrobitumen porosity with calcite crystal lining, Figure 3.7 shows both kerogen (with no crystal face lining) and solid bitumen/pyrobitumen porosities (with crystal face lining). The kerogen and solid bitumen/pyrobitumen in Figure 3.7 lie very close (<1 μm) to each other suggesting similar thermal histories. But the pores within the solid bitumen/pyrobitumen are more developed than those in the depositional kerogen, indicating the presence of different organic macerals in both features, which exhibit different levels of porosity even though thermal maturity is the same. Organic porosity is postulated to increase with increasing thermal

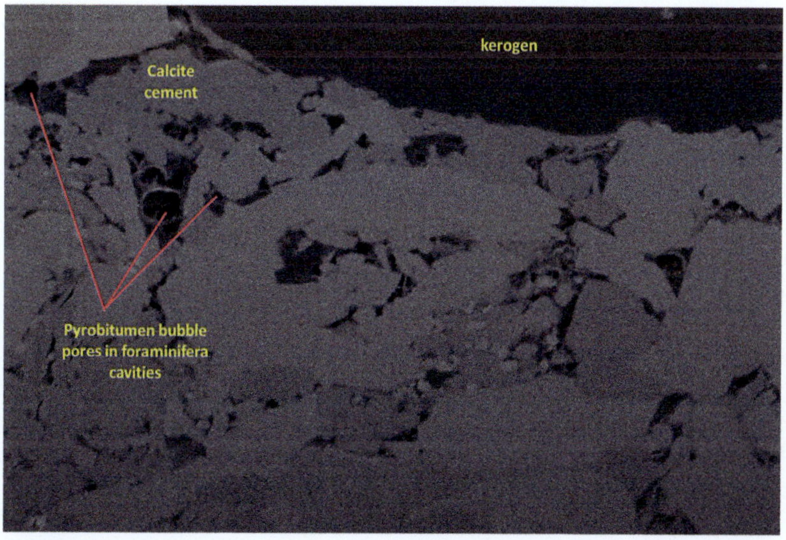

Figure 3.8 FIB-SEM image of a lower Eagle Ford shale sample with vitrinite reflectance of 1.03 %Ro.

maturity and decreasing TOC (Tissot & Welte, 1984). The TOC values obtained from Rock-Eval pyrolysis for the lower Eagle Ford shale samples are presented as 5.75 wt. % (Fig. 3.6), 5.06 wt. % (Fig. 3.7), and 5.29 wt.% (Fig. 3.8), respectively, and the corresponding organic porosity estimated through digital rock analysis with PerGeos for each sample is given as follows: (1) 2.6% with pore sizes ranging from 13nm to 360 nm in diameter (Fig. 3.6); (2) 1.4% with pore size ranging from 14 nm to 510 nm in diameter (Fig. 3.7); and (3) 0.57% with pore size ranging from 13 nm to 310 nm in diameter (Fig. 3.8), respectively.

Curtis et al. (2011) investigated the relationship between organic porosity and thermal maturity in the Marcellus shale by comparing samples with vitrinite reflectance of 1.1+ %Ro (toward the end of the oil window and beginning of the gas window) and >>3.1 %Ro (gas window), respectively. The Marcellus samples in the late mature oil window towards the onset of the gas window (1.1 %Ro) recorded an estimated organic porosity of 18.5% for pore sizes ranging from 10 nm to 140 nm in diameter. Whereas, the Marcellus samples in the gas window (>>3.1 %Ro) had an organic porosity estimated at 15.4% with pore sizes ranging from 5 nm to 20 nm in diameter. The porosity estimation was carried out by drawing ellipses around the organic pores in grayscale BSE images of the Marcellus samples and calculating the total area of all ellipses compared with the total area of the organic material (pyrobitumen or kerogen). It was found that increase in thermal maturity does not correspond to an increase in organic porosity in the selected Marcellus shale samples.

The role played by different organic macerals in the development of organic porosity is observed by comparing Rock-Eval and porosity data for three shale plays, the Ordovician Viola group (type II kerogen), the Cretaceous Niobrara formation (types II and III kerogen), and the Silurian mudrocks (types II and III to type III kerogen) (Keel, 2015). The Viola group samples have calculated vitrinite reflectance (% Ro) ranging from 0.63% to 0.89% and TOC between 1.9 and 6 wt. %. On the other hand, the Niobrara samples have calculated vitrinite reflectance (%Ro) ranging from 0.58 to 0.94 % and TOC between 2 and 7 wt. %. Finally, the Silurian hot shale have calculated %Ro ranging from 0.67 to 0.92% and TOC between 0.86 and 5.93 wt. %. Measured helium porosity for the Niobrara formation varies from 0.2 to 2.5 % (Keel, 2015). No organic porosity was observed in the Viola samples but the Silurian hot shale samples showed macroscale variation of organic porosity, while the Niobrara samples showed microscale variation of organic porosity (Keel, 2015). Two populations of organic matter were observed coexisting within the samples on FIB-SEM images of each sample group, namely, lamellar and isolated organic matter. The lamellar organic matter (depositional kerogen) showed no indications of organic pores, while the isolated organic matter (migra-bitumen or pyrobitumen) showed evidence of organic pores within the samples. Thus, this phenomenon points to the presence of two types of organic macerals, aliphatics and aromatics, and their response to organic porosity. The aliphatics in lamellar organic matter did not generate organic pores at a given maturity level but the aromatics generated organic porosity.

Subsequent sections below outline some correlations to estimate organic porosity from Rock-Eval data (Kuchinskiy, 2013; Modica & Lapierre, 2012) as well as wireline logs quantification of organic porosity (Galford et al., 2013).

3.8 Methods of Determining Source Rock Potential

The TOC can be measured by several methods including direct combustion, pyrolysis, and geophysical methods, each of which presents a limitation.

3.8.1 Direct Combustion

Direct combustion is the most commonly used method for determining TOC (Peters, 1986). It involves acidifying the ground rock sample with a concentration of 6 N HCl to dissolve most carbonates, sulphides (ZnS, FeS_2), and basic or amphoteric oxides and hydroxides as shown in the balanced chemical equation below (Saxby, 1970).

$$CaCO_3 + 2H^+ \longrightarrow Ca^{2+} + H_2O + CO_2(\text{carbonates})$$
$$ZnS + 2H^+ \longrightarrow Zn^{2+} + H_2S(\text{sulphides})$$
$$Fe(OH)_3 + 3H^+ \longrightarrow Fe^{3+} + 3H_2O(\text{hydroxides})$$

The filtrate/residue is then removed by washing or decantation and drying at ~55°C. Eventually, the dried sample is combusted with metallic oxide accelerator at ~1,000°C. As a result, carbon dioxide (CO_2) is generated during combustion and analyzed with either infrared (IR) or thermal conductivity detectors (TCD). This method is limited in its application in organic-lean, thermally immature rocks and carbonate-rich rocks (Peters & Casa, 1994).

3.8.2 Indirect Method

The indirect method of determining TOC is more accurate for organic-lean (< 0.5 wt. % TOC) and carbonate-rich samples. In the indirect method, two weighted portions of a pulverized homogeneous sample are treated separately. One portion is mixed with acceler ator and combusted to produce total carbon, equation (3.4), while the other portion is treated drop-wise with HCl and the evolved CO2 is measured. TOC is then estimated as the difference between the total carbon (TC) and $C_{carbonate}$ (Peters & Casa, 1994).

$$TC = C_{organic} + C_{carbonate} \tag{3.4}$$

3.8.3 Rock-Eval Pyrolysis Method

Rock-Eval pyrolysis evaluates the source-rock potential (e.g., TOC) and thermal maturity of organic-rich rocks. Pyrolysis involves the heating of organic matter in the absence of oxygen to yield organic compounds. It makes use of ground samples of whole rock of about 100 mg, pyrolyzed at 300°C initial temperature isotherm, and then programmed at a rate of 25°C/min to 600°C final temperature isotherm for 1 minute and 580°C oxidation oven temperature (Peters, 1986). The heating process distills the free organic compounds (bitumen), then cracks pyrolytic products from kerogen. Parameters obtained from the Rock-Eval pyrolysis include S_1, S_2, S_3 as illustrated in Figure 3.9 as well as S_4, T_{max}, TOC, hydrogen index (HI), oxidation index (OI), and productivity index (PI) to be discussed in subsequent sections.

S_1

S_1 refers to the amount of free hydrocarbons (i.e., gas and oil) initially present during kerogen deposition or generated after its deposition; it increases with depth (Graciano et al., 2012). The free hydrocarbons under S_1 peak typically include hydrocarbons up to C_{20+} depending on the S_1 hold time as shown in Figure 3.9. However, heavier free hydrocarbons and non-hydrocarbons (resins, asphaltenes) vaporize at higher temperatures and are included in the S_2 peak instead (Peters, 1986; Jiang et al., 2015). Moreover, contamination of samples by drilling fluids and mud can yield an abnormally high value for S_1. A low S_1 does not necessarily indicate little or no oil as volatile components (gas and light condensate) may have already evaporated away leaving the rock with a lower potential. Therefore, knowing the maturity of the sample is key to evaluating S_1, S_2, and T_{max} parameters, respectively.

S_2

S_2 represents the amount of hydrocarbons generated through thermal cracking of insoluble organic matter or kerogen (Lafargue et al., 1998) also shown in Figure 3.9. The S_2 parameter represents the hydrocarbon quantity that the rock can potentially produce under a continuous burial and thermal maturation (Ramachandran et al., 2013). Therefore, a low

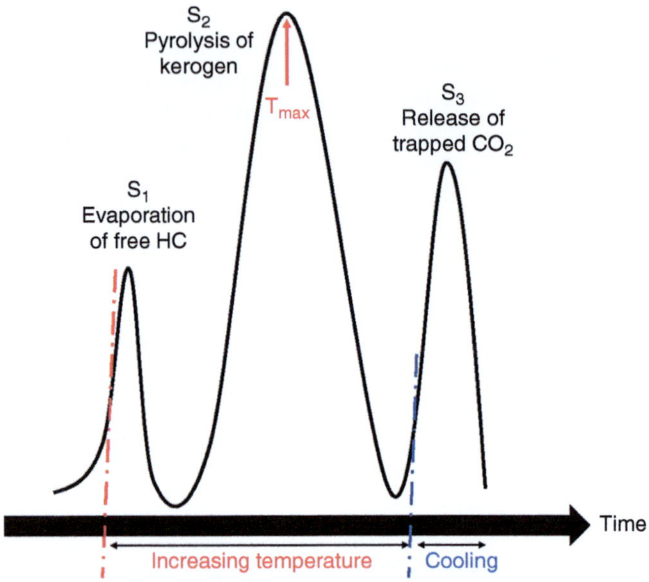

Figure 3.9 Measured parameters from programmed pyrolysis (modified from Ruble et al., 2015).

S_2 does not always point to low potential of the rock as the rock may have already generated its hydrocarbons especially in high-maturity formations. However, low organic content in samples usually results in an erroneous and broad S_2 peak.

S_3

S_3 is the amount of CO_2 (in mg CO_2/g of rock) produced as a result of breaking the carboxyl groups and oxygen-rich compounds in kerogen (El Nady et al., 2015; Graciano et al., 2012); the S_3 peak is shown in Figure 3.9. Therefore, S_3 is an indicator of the amount of organic carbon dioxide (Horsfield & Jurgen, 1994).

S_4

S_4 is the remaining carbon content (in mg carbon/gram of rock) in the sample. The remaining carbon content in the sample has little or no potential of generating any hydrocarbons due to the nonexistence of hydrogen (Baudin et al., 2015).

T_{max}

T_{max} is the temperature (°C) at which the maximum of S_2 hydrocarbons is generated as marked in Figure 3.9. A correlation between the maximum temperature (T_{max}) and vitrinite reflectance (%Ro) is derived for a collection of shales with low-sulfur type II and type III kerogen as shown in equation (3.5) (Jarvie et al., 2007).

$$R_o\% = 0.0180 * T_{max} - 7.16 \tag{3.5}$$

Table 3.6 shows some interpreted T_{max} results in terms of maturity from immature, mature, and postmature samples. T_{max} is unreliable when $S_2 < 0.50$ and the S_2 curve is broad and flat with no definite peak. The presence of

Table 3.6 T_{max} Results Interpreted in Terms of Maturity

Immature	T_{max} < 435°C
Mature	$T_{max:}$ 435°–445°C (Early)
	$T_{max:}$ 445°–450°C (Peak)
	T_{max}: 450°–470°C (Late)
Post-mature	T_{max} > 470°C

Source: Jarvie et al. (2007).

heavy and free hydrocarbons (bitumen) in S_2 peak results in abnormally low T_{max} values (< 400°C), while the presence of inert carbon results in abnormally high T_{max} values (> 550°C) (Peters, 1986).

TOC

The total organic carbon (TOC) can be calculated from the Rock-Eval pyrolysis method by the use of pyrolysis carbon (PC) and residual carbon (RC) (Lafargue et al., 1998). PC, which is the sum of the carbon content in both the free hydrocarbon generated (S_1) and the remaining potential hydrocarbon (S_2), can be calculated by the use of equation (3.6).

$$PC = 0.83\frac{(S1 + S2)}{10} \tag{3.6}$$

The residual carbon (RC), which is the remaining carbon, is the carbon found in kerogen, and it is calculated by the use of equation (3.7):

$$RC = \frac{S4}{10}. \tag{3.7}$$

Therefore, TOC can be calculated by equation (3.8) or equation (3.9) (Lafargue et al., 1998).

$$TOC(wt.\%) = PC + RC \tag{3.8}$$

$$TOC(wt.\%) = \left[\frac{0.83(S1 + S2)}{10} + \frac{S4}{10}\right] \tag{3.9}$$

Hydrogen Index

The hydrogen index (HI) corresponds to the quantity of pyrolyzable hydrocarbons from S_2 relative to the total organic carbon (TOC) in the sample expressed in equation (3.10). HI is also inferred from the slope of the S_2 versus TOC plot (Tobey & Campbell, 2016). HI is proportional to the amount of hydrogen in the kerogen and, therefore, indicates the potential of the rock to generate oil but falls short due to nonuniform nature of source rock mineralogy and inconsistency of the amount of TOC and quality of organic matter (Tobey & Campbell, 2016; Dargahi & Rezaee, 2013).

$$HI = (S2/TOC)100 \tag{3.10}$$

Oxygen Index

The oxygen index (OI) is a measure of the amount of oxygen in the kerogen expressed in terms of S_3 relative to the TOC (El Nady et al., 2015) and given in equation (3.11). The OI component is not as reliable as the other Rock-Eval parameters due to the interference of carbonate minerals and kerogen oxidation (Peters, 1986). However, OI is proved to be reliable during diagenesis and early catagenesis stages (Hetényi, 1998).

$$OI = (S3/TOC)100 \tag{3.11}$$

Productivity Index

The productivity index (PI) is identical to the transformation ratio (TR) except in cases without recorded S_1, thus, indicative of thermal maturation. PI is the ratio of hydrocarbons already generated in the sample (S_1) to the potential hydrocarbons that could still be generated in the sample (Stanley et al., 2014; Peters & Moldowan, 1993) expressed in equation (3.12). Thus, PI increases with conversion of thermally altered components in kerogen (S_2) to free hydrocarbons. Immature samples have a PI value of 0.1 or less, while mature samples have a PI from 0.1 to 0.4 (Peters, 1986).

$$PI = S1/(S1 + S2) \tag{3.12}$$

High Oil Saturation Index

High oil saturation index (OSI) is the normalized oil content and it serves as the basis of the oil crossover effect. Oil crossover effect occurs when free oil in the reservoir exceeds TOC (Jarvie et al., 2011). The oil saturation index can be calculated by the use of equation (3.13) (Jarvie et al., 2011).

$$OSI = \left(\frac{S1}{TOC}\right)100 \tag{3.13}$$

Rock-Eval pyrolysis results for three rock samples from the Bakken system are shown in Table 3.7 and Figures 3.10–3.12, respectively.

Table 3.7 Summarized Geochemical Data for the Three Samples from the Bakken System

Depth (ft)	TOC (Wt. %)	S1 (mg/gm)	S2 (mg/gm)	S3 (mg/gm)	PI	T_{max} (°C)	HI	OI	cal. Ro%
9,770.00	8.21	0.14	33.82	0.55	0.004	445	412	7	0.85
9,773.25	3.54	0.08	10.8	0.22	0.007	450	305	6	0.94
9,777.00	13.9	0.24	59.57	0.42	0.004	445	429	3	0.85

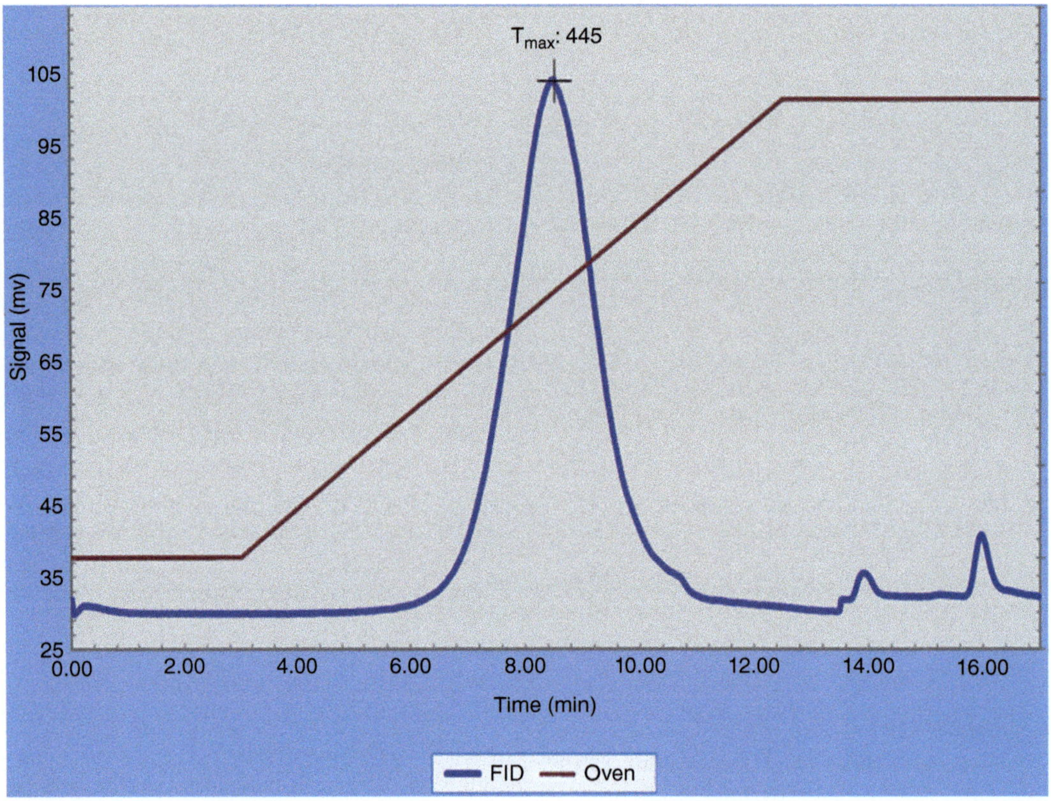

Figure 3.10 Rock-Eval test for a sample from the Bakken system at depth 9,770 ft.

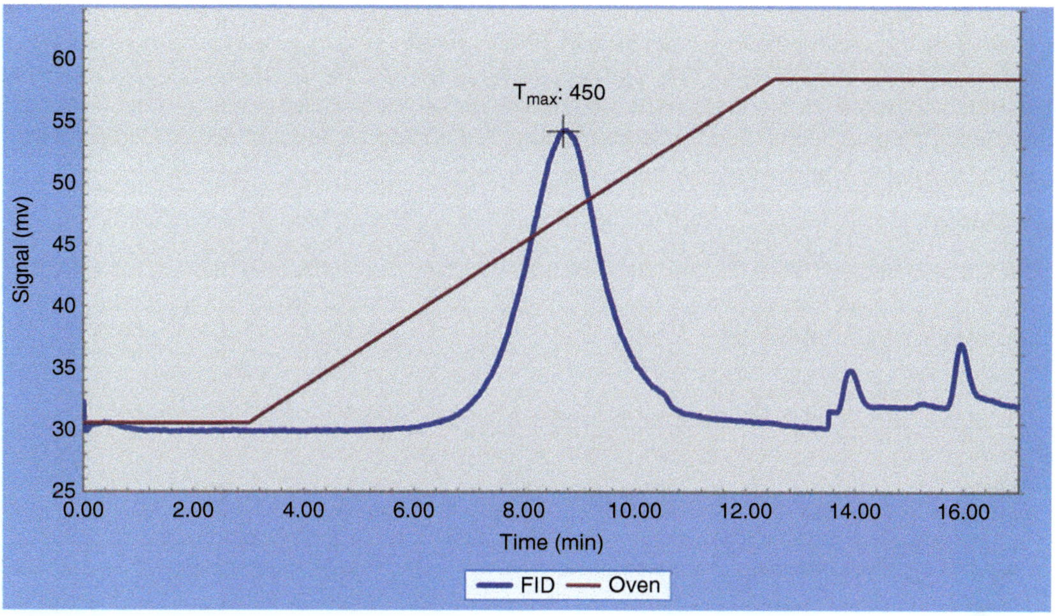

Figure 3.11 Rock-Eval test for a sample from Bakken system at depth 9,773.25 ft.

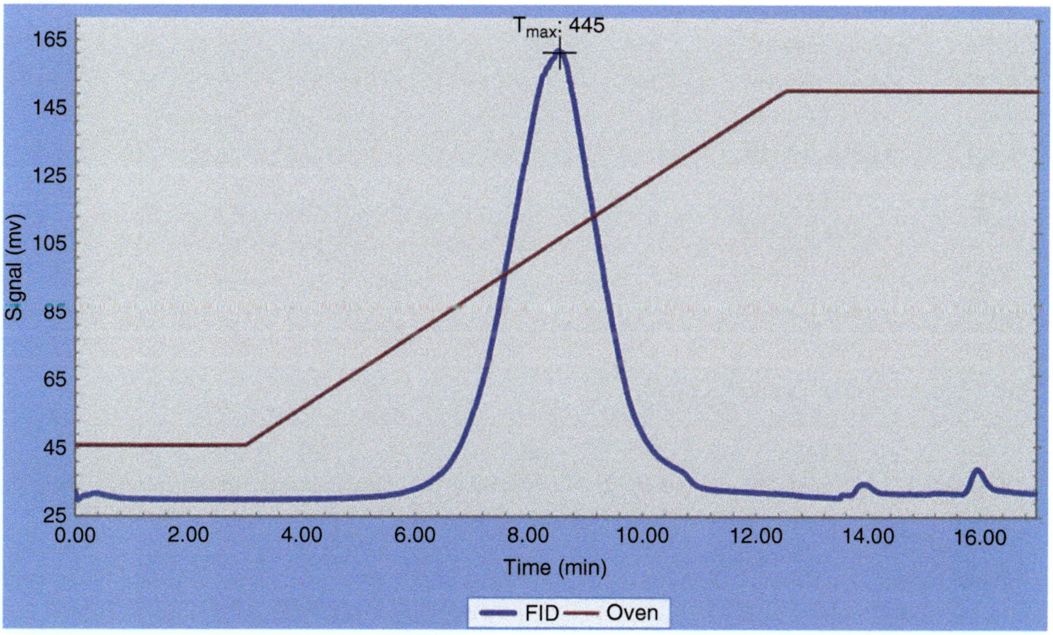

Figure 3.12 Rock-Eval test for a sample from Bakken system at depth 9,777 ft.

Example 3.2 Determine the Quality of Source Rock

Rock-Eval pyrolysis test was conducted on core samples from the upper and middle members of the Bakken system. The test results are provided in Table 3.8. Based on the average values of the reported parameters, determine the quality of the rocks in terms of organic richness for each (the upper and the middle) members.

Solution

First, the average of TOC, S_1, and S_2 is calculated. Then, the guidelines that are provided in Table 3.1 are used to determine the quality of each member of the Bakken system. The results are summarized in Table 3.9.

Table 3.8 Data for Example 3.2

Depth (ft)	Formation	TOC (wt. %)	S_1 (mg HC/g)	S_2 (mg HC/g)	S_3 (mg HC/g)	T_{max} (°C)
9,864	Upper Bakken	13.9	7.29	74.23	1.38	436
9,865	Upper. Bakken	14.2	7.77	74.68	1.01	438
9,866	Upper Bakken	16.1	8.59	76.34	1.29	432
9,867	Upper Bakken	14.8	8.18	83.62	1.11	433
9,868	Upper Bakken	18.3	9.96	94.63	0.58	432
9,869	Upper Bakken	16.8	10.27	76.06	1.44	427
9,870	Upper Bakken	12.3	8.11	74.33	1.20	434
9,875	Middle Bakken	0.39	1.87	0.68	0.48	420
9,880	Middle Bakken	0.84	6.93	1.68	0.46	411
9,885	Middle Bakken	0.30	1.89	0.55	0.26	434
9,890	Middle Bakken	0.46	3.37	0.82	0.32	416
9,895	Middle Bakken	0.45	2.82	0.62	0.34	410
9,900	Middle Bakken	0.48	3.09	0.93	0.33	421
9,905	Middle Bakken	0.50	3.19	0.87	0.31	418

Table 3.9 Results for Example 3.2

Formation	TOC (wt. %)	S_1 (mg HC/g)	S_2 (mg HC/g)	Overall source rock organic-rich quality
Upper Bakken	15.2	8.596	79.127	Excellent
Middle Bakken	0.489	3.309	0.879	Poor

The average TOC, S_1, and S_2 for the upper Bakken are 15.2, 8.596, and 79.127, respectively. Since TOC > 4.0, $S_1 > 4$, and $S_2 > 20$, then the upper Bakken can be considered as an excellent source rock organic-rich formation. On the other hand, the average TOC, S_1, and S_2 for the middle Bakken are 0.489, 3.309, and 0.879, respectively. The high S_1 value in the middle Bakken is solely an indicator of the amount of free hydrocarbon that can be thermally distilled from the rock. However, since TOC < 0.5 and $S_2 < 2.5$, the middle Bakken is considered as a poor organic-rich formation, and this confirms that the middle Bakken is neither a source rock nor an organic-rich shale formation but rather a fine-grained carbonate formation.

Example 3.3 Determine the Kerogen Type and Quality

For the provided data in Example 3.2, determine the kerogen type and quality for the upper Bakken organic-rich formation. The hydrogen index and oxygen index for the upper Bakken are provided in Table 3.10.

Table 3.10 Data for Example 3.3

Depth (ft)	Formation	TOC (wt. %)	S_2 (mg HC/g)	HI (mg HC/g TOC)	OI (mg HC/g TOC)
9,864	Upper Bakken	13.90	74.23	534.03	9.93
9,865	Upper Bakken	14.20	74.68	525.92	7.11
9,866	Upper Bakken	16.10	76.34	474.16	8.01
9,867	Upper Bakken	14.80	83.62	565.00	7.50
9,868	Upper Bakken	18.30	94.63	517.10	3.17
9,869	Upper Bakken	16.80	76.06	452.74	8.57
9,870	Upper Bakken	12.30	74.33	604.31	9.76

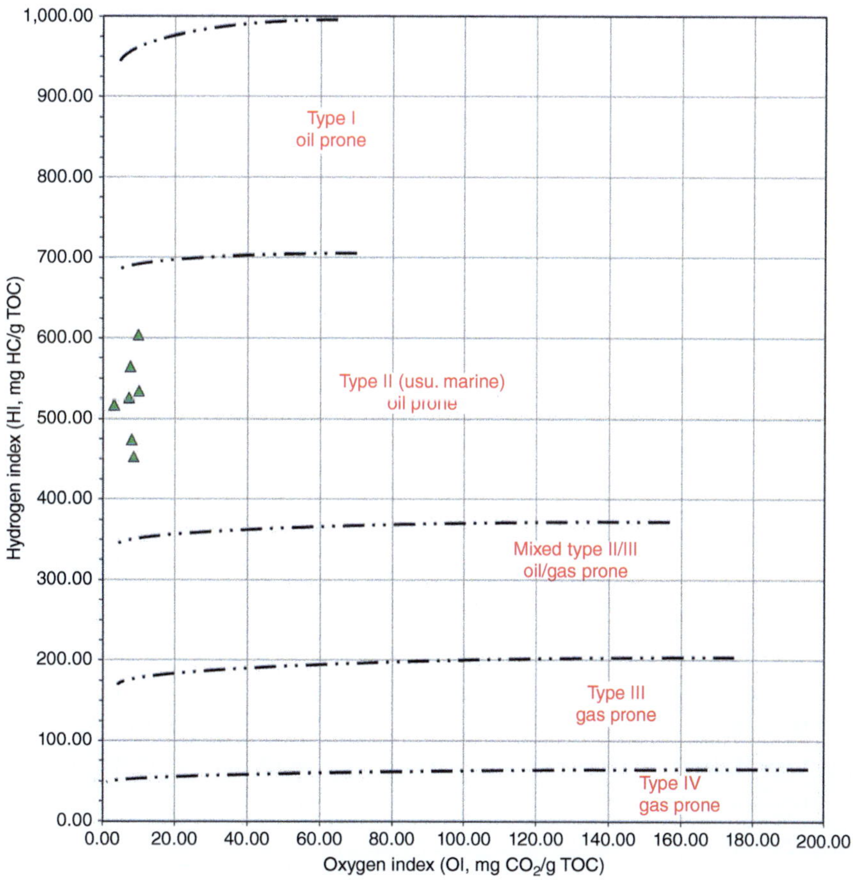

Figure 3.13 Kerogen type for Example 3.2.

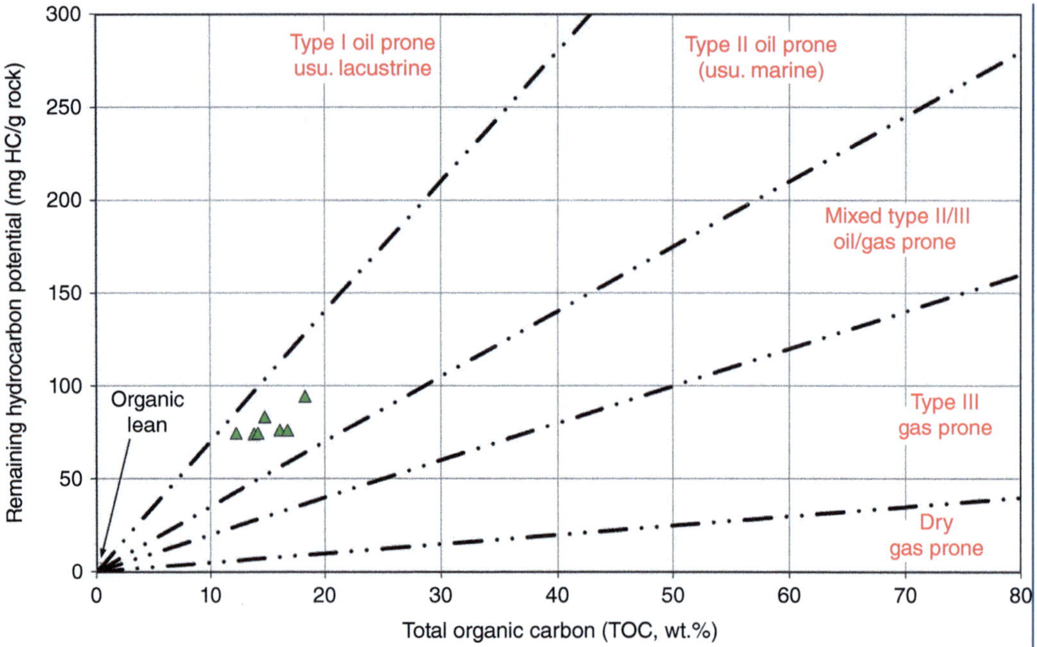

Figure 3.14 Kerogen type and quality for Example 3.2.

Solution

The kerogen type and quality can be assisted by the use of Figures 3.4 and 3.5. The plotted results are shown in Figures 3.13 and 3.14. Both figures confirm that the provided data are of type II kerogen and in the oil window.

Example 3.4 Calculate HI, OI, and PI

Rock-Eval pyrolysis test was conducted on core samples from the upper Bakken formation. The results of the test are provided in Table 3.11. Calculate the hydrogen index, oxygen index, productivity index, and vitrinite reflectance (%Ro).

Table 3.11 Data for Example 3.4

Depth	Formation	TOC (Wt. %)	S_1 (mg/gm)	S_2 (mg/gm)	S_3 (mg/gm)	T_{max} (°C)
9,864	Upper Bakken	13.90	7.29	74.23	1.38	436.00
9,865	Upper Bakken	14.20	7.77	74.68	1.01	438.00
9,866	Upper Bakken	16.10	8.59	76.34	1.29	432.00
9,867	Upper Bakken	14.80	8.18	83.62	1.11	433.00
9,868	Upper Bakken	18.30	9.96	94.63	0.58	432.00
9,869	Upper Bakken	16.80	10.27	76.06	1.44	427.00
9,870	Upper Bakken	12.30	8.11	74.33	1.20	434.00

Table 3.12 Summarized Results for Example 3.4

Depth	Formation	TOC	S_1	S_2	S_3	T_{max}	HI	OI	PI	%Ro
9,864	Upper Bakken	13.90	7.29	74.23	1.38	436.00	534.03	9.93	0.09	0.69
9,865	Upper Bakken	14.20	7.77	74.68	1.01	438.00	525.92	7.11	0.094	0.72
9,866	Upper Bakken	16.10	8.59	76.34	1.29	432.00	474.16	8.01	0.10	0.62
9,867	Upper Bakken	14.80	8.18	83.62	1.11	433.00	565.00	7.50	0.09	0.63
9,868	Upper Bakken	18.30	9.96	94.63	0.58	432.00	517.10	3.17	0.095	0.62
9,869	Upper Bakken	16.80	10.27	76.06	1.44	427.00	452.74	8.57	0.12	0.53
9,870	Upper Bakken	12.30	8.11	74.33	1.20	434.00	604.31	9.76	0.098	0.65

Solution

A sample calculation for the depth of 9,864 ft will be shown for all calculations, and the summarized results can be seen in Table 3.12.

The hydrogen index (HI) can be calculated by the use of equation (3.10) as follows:

$$HI = \left(\frac{S2}{TOC}\right)100 = \left(\frac{74.23}{13.9}\right) \times 100 = 534.$$

The oxygen index (OI) is calculated by equation (3.11):

$$OI = \left(\frac{S3}{TOC}\right)100 = \left(\frac{1.38}{13.9}\right) \times 100 = 9.93.$$

The productivity index (PI) is calculated by equation (3.12):

$$PI = \left(\frac{S1}{S1 + S2}\right) = \left(\frac{7.29}{(7.29 + 74.23)}\right) = 0.09.$$

Finally, the vitrinite reflectance (%Ro) can be estimated from T_{max} by the use of equation (3.5):

$$(\%Ro) = 0.018\ T\,max\ - 7.16 = 0.018 \times 436 - 7.16 = 0.69.$$

3.8.4 In-Situ Measurements

Well logging methods, in the form of wireline logs, can also be used to determine TOC, specifically, the bulk density (RHOB), sonic transit time (DT), and resistivity (RT) logs. Lower bulk density log readings coupled with lower sonic transit time (slowness) and higher resistivity readings give indications of source rock potential as listed in Table 3.13 for thermally matured formations with sufficient TOC and thickness within the resolution of the logs (Ruble et al., 2015). Furthermore, an overlay

Table 3.13 Geophysical Properties of Kerogen

Parameter	Value
Bulk density (RHOB)	1.1–1.4 g/cm^3
Low photoelectric absorption (PE)	0.24
Uranium	0.18–0.28
Neutron porosity (NPHI)	50–65 pu
Gamma ray (GR)	500–4000 gAPI
Sonic velocity (DT)	80–160 μs.ft^{-1}

Source: Ruble et al. (2015).

of the DT curve with the RT curve yields the delta logR (ΔlogR) method to estimate TOC.

Schmoker Method

Schmoker and Hester presented a method, which is usually referred to as the Schmoker method, to determine TOC based on the formation bulk density log readings (Schmoker & Hester, 1983). This correlation was originally derived for the Appalachian Devonian shales and later modified for the upper and lower members of the Bakken formation (Schmoker & Hester, 1983); it could as well be extended to other formations by modifying the assumptions made (i.e., changing the two coefficients) to suit the formation in question. In developing the TOC - ρ_b correlation, the shale formation is treated as a four-component system made up of rock matrix (m), interstitial pores (i), pyrite (p), and organic matter (o), where the formation density is a function of the densities (ρ) and volume fractions (φ) of each of the four components as expressed in equation (3.14):

$$\rho_b = \varphi_o\rho_o + \varphi_p\rho_p + \varphi_i\rho_i + \left(1-\varphi_o-\varphi_p-\varphi_i\right)\rho_m. \tag{3.14}$$

The presence of pyrite shifts the density log significantly as such pyrite-organic matter ($\varphi_p - \varphi_o$) correlations are developed for different formations. A simplified two-variable expression is obtained in equation (3.14) by assuming a constant mineral composition and porosity throughout the formation. Therefore, the Schmoker equation (i.e., TOC-ρb correlation) for the Bakken system is expressed as equation (3.15) based on the low-density organic matter measurements from the bulk density log (Charsky & Herron, 2013; Schmoker & Hester, 1983).

$$TOC = \frac{154.497}{\rho b} - 57.261, \tag{3.15}$$

where TOC is the total organic carbon (wt. %) and ρ_b is the formation density log reading (g/cc).

Example 3.5 Estimate the TOC from Bulk Density

The bulk density log reading from a well targeting the lower Bakken formation is provided in Table 3.14. Calculate the TOC for the provided data using the Schmoker method. The TOC from the Rock-Eval pyrolysis test is also provided at the same depth of the log readings. Compare the TOC from Schmoker method and TOC from the Rock-Eval pyrolysis test by plotting both TOC values. Also, find the relative error assuming the Rock-Eval TOC is the true TOC.

Table 3.14 Data for Example 3.5

Depth (ft)	Formation	Rock-Eval-TOC (wt. %)	Bulk density (g/cc)
9,940	Lower Bakken	16.2	2.10
9,941	Lower Bakken	13.4	2.15
9,942	Lower Bakken	13.9	2.16
9,945	Lower Bakken	12.0	2.19
9,946	Lower Bakken	12.0	2.20
9,947	Lower Bakken	14.0	2.18
9,953	Lower Bakken	15.3	2.15
9,954	Lower Bakken	14.0	2.18
9,955	Lower Bakken	13.5	2.21
9,956	Lower Bakken	11.6	2.20
9,957	Lower Bakken	13.0	2.18

Solution

A sample calculation for the depth of 9,940 ft will be shown. First, the TOC can be calculated from the bulk density by the use of the Schmoker equation, equation (3.15), as follows:

$$TOC = \frac{154.49}{bulk\ density} - 57.261$$
$$= \frac{154.49}{2.10} - 57.26 = 16.15.$$

Now, the results can be compared with the Rock-Eval TOC by finding the relative error as follows:

$$Relative\ Error(\%) = \frac{(16.2-16.15)}{16.2} \times 100 = 0.32(\%).$$

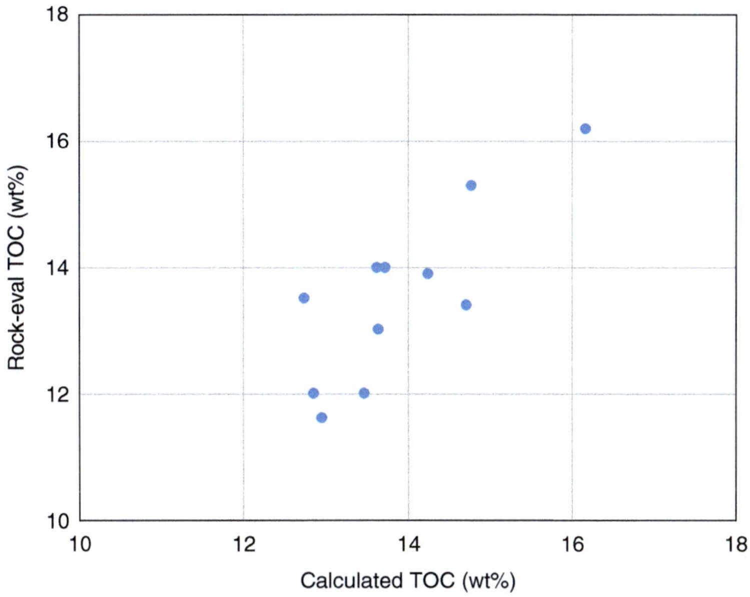

Figure 3.15 Comparison of TOC.

Table 3.15 Summarized Results for Example 5.5

Depth (ft)	Formation	Rock-Eval–TOC (wt. %)	Bulk density (g/cc)	Calculated TOC (wt. %)	Relative error (%)
9,940	Lower Bakken	16.2	2.11	16.15	0.32
9,941	Lower Bakken	13.4	2.15	14.69	9.64
9,942	Lower Bakken	13.9	2.16	14.23	2.39
9,945	Lower Bakken	12.0	2.19	13.46	12.17
9,946	Lower Bakken	12.0	2.20	12.84	7.01
9,947	Lower Bakken	14.0	2.18	13.71	2.07
9,953	Lower Bakken	15.3	2.15	14.76	3.56
9,954	Lower Bakken	14.0	2.18	13.61	2.81
9,955	Lower Bakken	13.5	2.21	12.72	5.75
9,956	Lower Bakken	11.6	2.20	12.95	11.60
9,957	Lower Bakken	13.0	2.18	13.63	4.81

Figure 3.15 shows a plot of Rock-Eval TOC versus calculated TOC from the Schmoker method, and the final results are summarized in Table 3.15.

GR-TOC Method

Another empirical correlation from logs for quantitative estimation of TOC is the uranium component of the gamma ray. Gamma ray (GR) is capable of differentiating between kerogen and pore fluids, where organic-rich shales induce high gamma ray readings associated with the concentration of uranium in kerogen (Gonzalez, et al., 2013; Heslop, 2010). Moreover, the GR-TOC is advantageous over other correlations since GR readings are acquired for almost all wellbores (Gonzalez, et al., 2013). A GR-TOC correlation is developed for the Chattanooga shale formation in Barber

County, Kansas. The correlation is also applicable in other shale formations of similar properties. Ten core plugs from different depths of the Chattanooga shale interval were analyzed through Rock-Eval pyrolysis to obtain TOC. The GR log is known to identify shale formations with high readings due to the presence of radioactive uranium associated with the organic matter (Conant & Swanson, 1961). The TOC values at each depth is then matched with the GR reading at the same depth and a correlation is developed as shown in equation (3.16) (Cudjoe et al., 2016):

$$TOC = 0.022{*}GR - 3.037 \quad (3.16)$$

where GR is the gamma ray log reading (API).

Example 3.6 Calculate the TOC from Gamma Ray Readings

The gamma ray readings for the same well in Example 3.5 are provided in Table 3.16. Calculate the TOC using the derived gamma ray correlation. Compare the results against the Rock-Eval TOC by calculating the relative error.

Solution

First, it is important to note that the derived correlation to calculate the TOC from the GR readings was based on Chattanooga shale interval. Thus, the correlation might not yield accurate results when applied in Bakken formation. A sample calculation for the depth of 9,940 ft is shown below, and the final results are summarized in Table 3.17. The TOC can be calculated from the GR readings by the use of equation (3.16):

$$TOC = 0.022\,GR - 3.037 = 0.22 \times 624.33 - 3.037 = 10.7.$$

The Rock-Eval TOC, from Example 3.5, is 16.2. Hence, the relative error is

Table 3.16 Data for Example 3.6

Depth (ft)	GR (API)
9,940	624.33
9,941	554.26
9,942	579.29
9,945	542.65
9,946	496.06
9,947	536.96
9,953	603.41
9,954	500.50
9,955	465.06
9,956	514.10
9,957	608.89

Table 3.17 Summarized Results for Example 3.6

Depth (ft)	Formation	Rock-Eval-TOC (wt. %)	GR (API)	Calculated TOC (wt. %)	Relative Error (%)
9,940	Lower Bakken	16.2	624.33	10.70	33.96
9,941	Lower Bakken	13.4	554.26	9.16	31.67
9,942	Lower Bakken	13.9	579.29	9.71	30.16
9,945	Lower Bakken	12.0	542.65	8.90	25.82
9,946	Lower Bakken	12.0	496.06	7.88	34.36
9,947	Lower Bakken	14.0	536.96	8.78	37.31
9,953	Lower Bakken	15.3	603.41	10.24	33.08
9,954	Lower Bakken	14.0	500.50	7.97	43.04
9,955	Lower Bakken	13.5	465.06	7.19	46.71
9,956	Lower Bakken	11.6	514.10	8.27	28.68
9,957	Lower Bakken	13.0	608.89	10.36	20.32

$$Re\,lative\ Error(\%) = \frac{(16.2-10.7)}{16.2} \times 100 = 33.96(\%).$$

It is clear that there is a substantial difference between the two TOC values. Hence, the derived GR-TOC correlation is not valid in the Bakken formation. Therefore, a new correlation should be derived for the Bakken formation before the GR readings can be used to estimate the TOC.

Delta LogR Method

Passey and Stroud (1990) introduced the Δ log R method for identifying and calculating TOC in organic-rich shales using the deep resistivity (RT) and sonic transit time (DT) logs. In water-saturated, organic-lean rocks, the DT and RT curves, when properly scaled, are parallel to each other and so are overlain, while in hydrocarbon, organic-rich non-reservoir rocks, a separation (Δ log R) is observed between the curves. The separation (Δ log R) is explained in terms of the DT curve responding to the presence of low-density, low-velocity kerogen, whereas the RT curve responds to the presence of formation fluid. However, in immature organic-rich rock with no hydrocarbon, a separation between the RT and DT curve is solely attributed to the porosity (DT) curve's response to the presence of kerogen. The magnitude of the curve separation in organic-rich formation is calibrated to TOC and maturity, applicable to other shale formations with varying lithologies and maturities. In applying the Δ log R method, the DT and RT curves are scaled such that their relative scaling is -100 µsec/ft per two logarithmic resistivity cycles and a baseline is defined, where the curves overlie each other in an organic-lean rock over a significant depth range as shown in Figure 3.17. The Δ log R separation is calculated as shown in equation (3.17) (Passey & Stroud, 1990).

$$\Delta \log R = \log\left(\frac{RT}{RT_{baseline}}\right) + 0.02*(DT - DT_{baseline}), \quad (3.17)$$

where RT is the measured deep resistivity (ohm-m), DT is the measured transit time (µsec/ft.), $RT_{baseline}$ is the resistivity corresponding to the $DT_{baseline}$ value when the curves are overlain in organic-lean, clay-rich rocks, and 0.02 is obtained from the ratio of -50 µsec/ft per one resistivity cycle

Subsequently, the Δ log R separation is linearly related to TOC and level of maturity (LOM) as shown in Figure 3.18 (a); LOM is determined through either vitrinite reflectance (%Ro) or thermal alteration index (TAI). Figure 3.16 shows a plot of vitrinite reflectance (%Ro) versus the level of maturity (LOM). An increase in the vitrinite reflectance (%Ro) corresponds to an increase in the level of maturity (LOM). An inaccurate estimation of LOM results in erroneous TOC values. The empirical equation for estimating TOC is therefore given as shown in equation (3.18) (Passey & Stroud, 1990):

$$TOC = (\Delta \log R)*10^{(2.297 - 0.1688*LOM)}. \quad (3.18)$$

Likewise, if the kerogen type is known, the S_2 peak derived through Rock-Eval analysis can be correlated with the TOC via maturity (LOM) for different kerogen types; Figure 3.18(b) shows the TOC-S_2 correlations for kerogen type II (left) and type III (right), respectively (Passey & Stroud, 1990). LOM can also be estimated by the use of equation (3.19) (Techlog, 2018):

$$\begin{aligned} LOM &= 0.0989\,Ro^5 - 2.1587\,Ro^4 \\ &+ 12.392\,Ro^3 - 29.032\,Ro^2 \\ &+ 32.53\,Ro - 3.0338. \end{aligned} \quad (3.19)$$

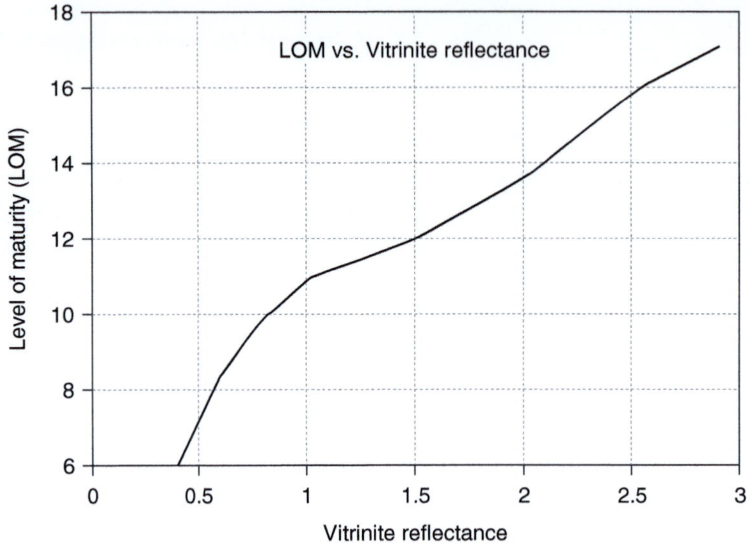

Figure 3.16 LOM relationship with vitrinite reflectance (Ro%) (Passey & Stroud, 1990).

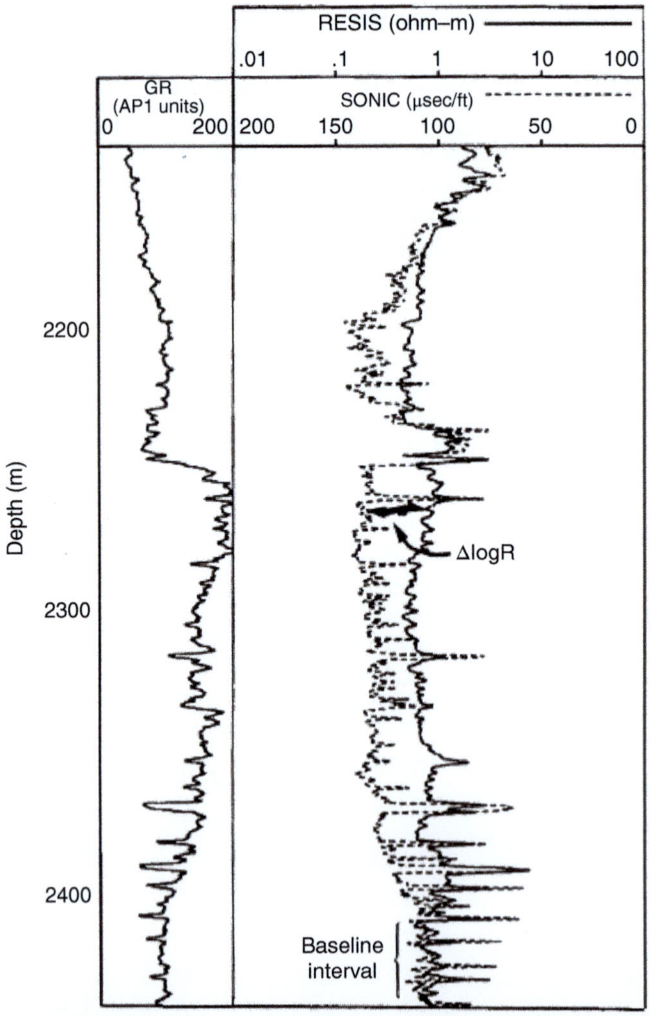

Figure 3.17 DT-RT overlay showing ΔlogR separation in an organic-rich interval (Passey & Stroud, 1990).

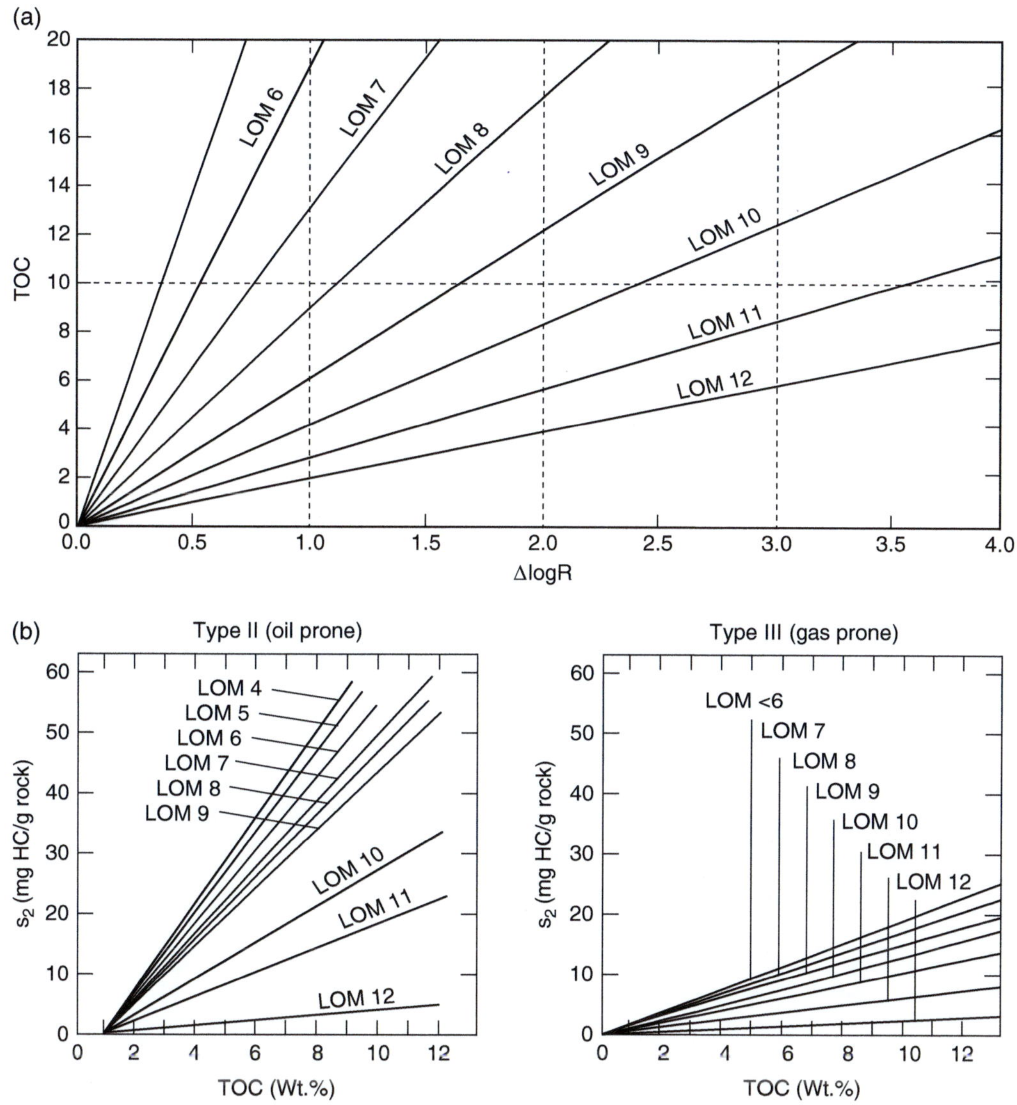

Figure 3.18 (a) ΔlogR separation correlated to TOC through LOM; (b) TOC-S2 correlation via maturity for kerogen types II and III, respectively (Passey & Stroud, 1990).

Example 3.7 Estimate the TOC Using ΔlogR Method

Estimate the TOC using the Δ log R method for the lower Bakken core samples provided in Table 3.18. Compare the results against the Rock-Eval pyrolysis TOC. The baseline RT and DT are found to be 6.5245 and 60.3636, respectively. The average vitrinite reflectance (R_o%) Ro% for the samples is 0.59.

Solution

The first step is to find the level of maturity (LOM). LOM can be estimated from the vitrinite reflectance ((R_o%)Ro%) either by Figure 3.19 or equation (3.19). LOM from equation (3.19) is

Table 3.18 Data for Example 3.7

Depth (ft)	Formation	Rock-Eval–TOC (wt. %)	RT (ohm)	DT Sonic (us/ft)	Neutron porosit (frac.)	Bulk density (g/cc)
9,940	Lower Bakken	16.2	100.98	111.64	0.40	2.11
9,941	Lower Bakken	13.4	103.65	110.36	0.45	2.15
9,942	Lower Bakken	13.9	107.99	113.27	0.39	2.16
9,945	Lower Bakken	12.0	145.82	104.36	0.35	2.19
9,946	Lower Bakken	12.0	125.02	106.73	0.41	2.20
9,947	Lower Bakken	14.0	109.82	108.73	0.44	2.18
9,953	Lower Bakken	15.3	108.21	106.73	0.39	2.15
9,954	Lower Bakken	14.0	114.87	106.00	0.33	2.18
9,955	Lower Bakken	13.5	112.89	103.27	0.33	2.21
9,956	Lower Bakken	11.6	101.51	104.00	0.33	2.20
9,957	Lower Bakken	13.0	93.10	106.36	0.41	2.18

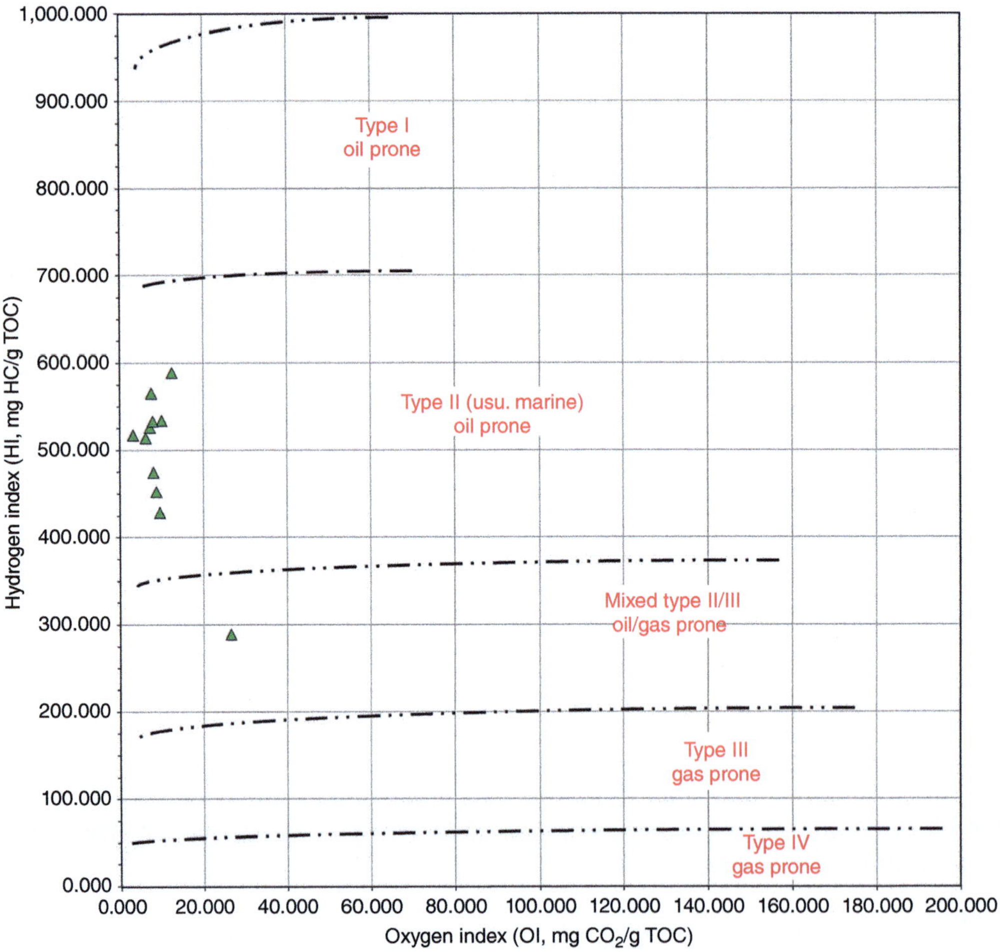

Figure 3.19 Kerogen type for Example 3.8.

$$LOM = 0.0989\,Ro^5 - 2.1587\,Ro^4 + 12.392\,Ro^3 - 29.032\,Ro^2 + 32.53\,Ro - 3.0338,$$

$$LOM = 0.0989 \times 0.59^5 - 2.1587 \times 0.59^4 + 12.392 \times 0.59^3 - 29.032 \times 0.59^2 + 32.53 \times 0.59 - 3.0338 = 8.34.$$

Next, is the calculation of the Δ log R by using equation (3.17). A sample calculation for the depth of 9,940 ft will be shown, and the final results are summarized in Table 3.19.

$$\Delta\, log\, R = \log\left(\frac{RT}{RTbaseline}\right) + 0.02(DT - DTbaseline)$$

and

$$\Delta\, log\, R = \log\left(\frac{RT}{RTbaseline}\right) + 0.02(DT - DTbaseline) = \log\left(\frac{100.981}{6.5245}\right) + 0.02 \times (111.636 - 60.3636) = 2.22.$$

Finally, the TOC is calculated by equation (3.18):

$$TOC_{\Delta\, log\, R} = (\Delta\, log\, R)10^{(2.297 - 0.1688LOM)}$$

and

$$TOC_{\Delta\, log\, R} = (2.215) \times 10^{(2.297 - 0.1688 \times 8.343)} = 17.14.$$

Thus, the relative error would be

$$Re\,lative\,Error(\%) = \frac{(16.2 - 17.141)}{16.2} \times 100 = 5.81(\%).$$

Table 3.19 Summarized Results for Example 3.7

Depth (ft)	Formation	Rock-Eval-TOC	RT	DT	ΔlogR	$TOC_{\Delta logR}$	Relative error (%)
9,940	Lower Bakken	16.2	100.98	111.64	2.22	17.14	5.81
9,941	Lower Bakken	13.4	103.65	110.36	2.20	17.03	27.10
9,942	Lower Bakken	13.9	107.99	113.27	2.28	17.62	26.76
9,945	Lower Bakken	12.0	145.82	104.36	2.23	17.25	43.75
9,946	Lower Bakken	12.0	125.02	106.73	2.21	17.1	42.49
9,947	Lower Bakken	14.0	109.82	108.73	2.19	16.97	21.23
9,953	Lower Bakken	15.3	108.21	106.73	2.15	16.61	8.59
9,954	Lower Bakken	14.0	114.87	106.00	2.16	16.70	19.30
9,955	Lower Bakken	13.5	112.89	103.27	2.10	16.22	20.16
9,956	Lower Bakken	11.6	101.51	104.00	2.07	15.98	37.73
9,957	Lower Bakken	13.0	93.10	106.36	2.07	16.05	23.48

Table 3.20 Comparison of ΔlogR Methods

Depth	Rock-Eval TOC	DT- TOC	NP- TOC	BD- TOC	DT- error (%)	NP-error (%)	BD-error (%)
9,940	16.2	17.14	14.16	19.57	5.81	12.61	20.80
9,941	13.4	17.03	14.05	18.83	27.10	4.83	40.54
9,942	13.9	17.62	14.64	18.70	26.76	5.29	34.56
9,945	12.0	17.25	14.27	19.26	43.75	18.89	60.47
9,946	12.0	17.1	14.12	18.37	42.49	17.62	53.05
9,947	14.0	16.97	13.99	18.45	21.23	0.08	31.80
9,953	15.3	16.61	13.63	19.01	8.59	10.92	24.28
9,954	14.0	16.70	13.72	18.54	19.20	2.02	32.44
9,955	13.5	16.22	13.24	17.95	20.16	1.95	32.97
9,956	11.6	15.98	12.99	17.73	37.73	12.01	52.84
9,957	13.0	16.05	13.07	17.85	23.48	0.52	37.29

It is clear from the relative error that the sonic Δ log R method is not the best method to estimate the TOC for the lower Bakken formation.

Table 3.20 shows the estimated TOC values for the different Δ log R method versions (i.e., sonic, neutron porosity, and bulk density) and their relative errors. Overall, the neutron porosity Δ log R version gave the best results compared with the other Δ log R versions. Note that, the difference in the scale of measurement between the Rock-Eval experiments versus well logging tools is another important reason for the geophysical methods to show a different result compared with the Rock-Eval data.

3.9 Original TOC and Hydrocarbon Yield Determinations

The original TOC (TOC_o) and the original components of TOC (i.e., S_2 original and hydrogen index original) can be back calculated on high maturity samples to determine the hydrocarbon yield. First, the original hydrogen index (HI_o) is calculated using the visual kerogen data as shown in equation (3.20) (Jarvie et al., 2007):

$$HI_o = \left(\frac{\%TypeI}{100}750\right) + \left(\frac{\%TypeII}{100}450\right) + \left(\frac{\%TypeIII}{100}125\right) + \left(\frac{\%TypeIV}{100}50\right). \quad (3.20)$$

The present-day HI (HI_{pd}) is less than the original HI (HI_o) at increasing thermal maturity; hence, the HI_{pd} will not reflect the original kerogen type but will indicate the type of hydrocarbon to be generated (i.e., oil, wet-gas, and/or dry gas) (Jarvie et al., 2007).

Next step involves the calculation of the transformation ratio (TR_{HI}) or the fractional conversion parameter, which determines the extent of organic matter conversion from the change in the original HI (HI_o) to the present-day HI (HI_{pd}). The transformation ratio (TR_{HI}) can be calculated by the use of equation (3.21) (Jarvie et al., 2007; Peters et al., 2006),

$$TR_{HI} = 1 - \frac{HI_{pd}[1200 - HI_o(1 - PI_o)]}{HI_o[1200 - HI_{pd}(1 - PI_{pd})]}, \quad (3.21)$$

where, 1,200 is assumed to be the maximum amount of hydrocarbons that could be formed assuming 83.33% carbon in hydrocarbons (Pelet, 1985).

The original TOC (TOC_o) is then back calculated from the present-day TOC (TOC_{pd}) and its components as shown in equation (3.22) (Jarvie et al., 2007):

$$TOC_o = \frac{HI_{pd}\left(\frac{TOC_{pd}}{1+k}\right)83.33}{\left[HI_o(1-TR_{HI})\left(83.33-\left(\frac{TOC_{pd}}{1+k}\right)\right)\right]+\left[HI_{pd}\left(\frac{TOC_{pd}}{1+k}\right)\right]}, \quad (3.22)$$

where, 83.33 is the average carbon content in hydrocarbons, k is a correction factor based on residual organic carbon (C_R) (C_R) with an increase in carbon over original values at high maturity (Burnham, 1989).

The increase in residual carbon (C_R)(C_R) at high maturity for type II kerogen is assigned a value of 15%, while type I kerogen and type III kerogen are given the values of 50% and 0%, respectively (Burnham, 1989). The correction factor (k) is calculated as shown in equation (3.23) (Jarvie et al., 2007):

$$k = TR_{HI} \times C_R. \quad (3.23)$$

A similar plot of the generative potential (S_2) versus organic richness (TOC) for immature to early mature samples can be used to obtain HI_o and C_R from the slope and x-intercept, respectively (Jarvie et al., 2007). The original hydrocarbon generation potential (S2o) is calculated using both the original hydrogen index (HI_o) and the original TOC (TOC_o) as shown in equation (3.24):

$$S2_o = \left(\frac{HI_o \times TOC_o}{100}\right). \quad (3.24)$$

Finally, the hydrocarbon (HC) yield can be estimated by equation (3.25). Although the hydrocarbon yield is useful for evaluation, it does not represent in-place reserves.

$$Generated\_HCs = Original(S2_o) - \text{Re}\,maining(S2). \quad (3.25)$$

Example 3.8 Determine the Hydrocarbon Yield

A Rock-Eval pyrolysis test was conducted on core samples from the upper Bakken formation, and the test results are provided in Table 3.21. Determine the hydrocarbon yield for the provided data knowing that the original hydrogen index and the original productivity index are 575 and 0.02, respectively.

Table 3.21 Data for Example 3.8

Depth	Formation	TOC (wt. %)	S_1 (mg HC/g)	S_2 (mgHC/g)	S_3 (mg HC/g)	C_R (%) of TOC
9,861	Upper Bakken	4.49	4.39	12.99	1.20	68
9,862	Upper Bakken	15.4	7.23	66.00	1.45	61
9,864	Upper Bakken	13.9	7.29	74.23	1.38	51
9,865	Upper Bakken	14.2	7.77	74.68	1.01	52
9,866	Upper Bakken	16.1	8.59	76.34	1.29	56
9,867	Upper Bakken	14.8	8.18	83.62	1.11	49
9,868	Upper Bakken	18.3	9.96	94.63	0.58	53
9,869	Upper Bakken	16.8	10.27	76.06	1.44	57
9,871	Upper Bakken	10.8	7.40	63.60	1.34	45
9,872	Upper Bakken	16.6	10.80	88.48	1.28	50
9,873	Upper Bakken	19.1	11.71	98.19	1.18	52

Solution

The first step is to confirm that the original hydrogen index is within the range of the given value. The original hydrogen index (HI_o) can be estimated by knowing the kerogen type of the samples. Thus, the hydrogen index (HI) and the oxygen index (OI) should be calculated and plotted to determine the kerogen type. A sample calculation for the depth of 9,861 ft will be shown, and the final results are summarized in Table 3.22.

The hydrogen index, the oxygen index and productivity index are calculated by the use of equation (3.10), equation (3.11), and equation (3.12), respectively.

$$HI = \left(\frac{S2}{TOC}\right)100 = \left(\frac{12.99}{4.49}\right) \times 100 = 289.31,$$

$$OI = \left(\frac{S3}{TOC}\right)100 = \left(\frac{1.2}{4.49}\right) \times 100 = 26.73,$$

and

$$PI = \left(\frac{S1}{S1 + S2}\right) = \left(\frac{4.39}{4.39 + 12.99}\right) = 0.25.$$

After all the HI and OI points are calculated, the results are plotted on Figure 3.4 to determine the kerogen type.

Figure 3.19 shows that the data points are of kerogen types II and III. Hence, the plot and equation (3.20) confirm that the value of HI_o is in the range of 575; thus, it is reasonable to use that value in estimating the hydrocarbon yield.

The next step is to calculate the transformation ratio (TR_{HI}) by the use of equation (3.21):

$$TR_{HI} = 1 - \frac{HI_{pd}[1200 - HI_o(1 - PI_o)]}{HI_o[1200 - HI_{pd}(1 - PI_{pd})]} = 1 - \frac{289.31[1200 - 575(1 - 0.02)]}{575[1200 - 289.31(1 - 0.253)]} = 0.67.$$

Next is the calculation of the original TOC, which also requires the calculation of the k value. The k value is calculated by the use of equation (3.23), where C_R is given in terms of % of TOC value:

Table 3.22 Summarized Results for Example 3.8

Depth	TOC	S_1	S_2	S_3	Cr	HI	OI	PI	TR	k	TOC_o	$S2_o$	HC Yield
9,861	4.49	4.39	12.99	1.2	68	289.31	26.73	0.25	0.67	0.02	6.61	38.00	25.01
9,862	15.4	7.23	66.00	1.45	61	428.57	9.42	0.10	0.42	0.04	18.05	103.80	37.80
9,864	13.9	7.29	74.23	1.38	51	534.03	9.93	0.09	0.17	0.01	15.10	86.81	12.58
9,865	14.2	7.77	74.68	1.01	52	525.92	7.11	0.09	0.2	0.01	15.56	89.45	14.77
9,866	16.1	8.59	76.34	1.29	56	474.16	8.01	0.10	0.32	0.03	18.28	105.11	28.77
9,867	14.8	8.18	83.62	1.11	49	565.00	7.50	0.09	0.09	0.01	15.62	89.84	6.22
9,868	18.3	9.96	94.63	0.58	53	517.10	3.17	0.10	0.22	0.02	19.97	114.82	20.19
9,869	16.8	10.27	76.06	1.44	57	452.74	8.57	0.12	0.37	0.04	19.43	111.75	35.69
9,871	10.8	7.40	63.60	1.34	45	588.89	12.41	0.10	0.03	0.001	11.31	65.04	1.44
9,872	16.6	10.80	88.48	1.28	50	533.01	7.71	0.12	0.19	0.02	18.13	104.22	15.74
9,873	19.1	11.71	98.19	1.18	52	514.08	6.18	0.11	0.23	0.02	20.96	120.51	22.32

$$k = TR_{HI}\, C_r = 0.674 \times (0.68 \times 0.0449) = 0.02.$$

Now, the original TOC can be calculated by the use of equation (3.22):

$$TOC_o = \frac{HI_{pd}\left(\frac{TOC_{pd}}{(1+k)}\right)83.33}{\left[HI_o(1-TR_{HI})\left(83.33-\left(\frac{TOC_{pd}}{(1+k)}\right)\right)\right]+\left[HI_{pd}\left(\frac{TOC_{pd}}{(1+k)}\right)\right]},$$

$$TOC_o = \frac{289.31 \times \left(\frac{4.49}{(1+0.021)}\right) \times 83.33}{\left[575 \times (1-0.674) \times \left(83.33-\left(\frac{4.49}{(1+0.021)}\right)\right)\right]+\left[289.31 \times \left(\frac{4.49}{(1+0.021)}\right)\right]} = 6.609,$$

$$TOC_o = 6.61(wt.\%).$$

The next step is to calculate the original S2 (S2o) by the use of equation (3.24):

$$S2_o = \frac{HI_o\, TOC_o}{100} = \frac{575 \times 6.609}{100} = 38.00.$$

Finally, the hydrocarbon yield can be estimated by the use of equation (3.25):

$$Generated\ HCs = S2_o - S2_{pd} = 38.003 - 12.99$$
$$= 25.01.$$

The final results for all the samples are summarized in Table 3.22.

3.9.1 Organic Porosity from Rock-Eval Parameters

Aside from estimating organic porosity from either FIB-SEM or BSE/SEM grayscale images through segmentation/thresholding as carried out in the lower Eagle Ford shale samples, organic porosity can also be calculated directly from Rock-Eval data as observed in studies by Matyasik et al. (2017), Chen and Jiang (2016), and Kuchinskiy (2013).

Kuchinskiy (2013) proposed an organic porosity method based on an empirical relationship between Rock-Eval parameters (e.g., original HI and present-day HI) and measured porosity. It calculates the organic porosity directly from present-day hydrogen index (HI_{pd}) without considering any geological processes and other geochemical characteristics of the shale formation. This method is limited in its application as organic porosity is not just a function of thermal maturity but also a function of the original organic composition of the organic material.

Furthermore, Modica and Lapierre (2012) proposed a method for estimating organic porosity based on hydrocarbon generation kinetics and maturity applied in a gas shale formation (upper Albian Mowry shale in the Powder River basin of Wyoming). It assumes the remaining free hydrocarbons in the shale do not affect the initial TOC (TOC_o) estimation and this may lead to an overestimation of the organic porosity. The organic porosity is derived as shown in equation (3.26) (Modica & Lapierre, 2012):

$$\varphi_{org} = [(TOC_o \times C_c)k]TR\left(\frac{\rho_b}{\rho_k}\right), \tag{3.26}$$

where, ρ_b is the bulk rock density, and ρ_k is the kerogen density. The k is a dimensionless scale factor used to estimate a kerogen mass from convertible carbon mass. An assigned value of 1.118 assumes 95% of the TOCo is from kerogen and a convertible carbon mass (Cc) represents ~85% of a convertible kerogen mass (i.e., k = 0.95/0.85 = 1.118). *TR* is the transformation ratio defined in terms of the hydrocarbon generative potential ($TR = (S2_o - S2)/S2_o$), and C_c is the fraction of original organic carbon that can be converted to hydrocarbons in a given source rock and is given as shown in equation (3.27) (Modica & Lapierre, 2012):

$$C_c = \left(\frac{1}{TR}\right)\left(1 - \frac{TOC}{TOC_o}\right). \quad (3.27)$$

Similarly, Chen and Jiang (2016) provide a method based on Rock-Eval data but take into account the effect of the remaining hydrocarbons within the source rock on the restoration of the initial organic carbon content (TOCo). This method is restricted to nanopores formed due to thermal maturation (decomposition) of organic matter during primary hydrocarbon generation. Using the transformation ratio (TR_{HI}) calculated above, the amount of hydrocarbon expelled (ε) is calculated by equation (3.28) (Chen & Jiang, 2016):

$$\varepsilon = \frac{S2 \times TR_{HI}}{1 - TR_{HI}} - S1. \quad (3.28)$$

From the amount expelled (ε), the expulsion efficiency (f) is estimated as the ratio of the amount expelled to the total hydrocarbons generated, and it is expressed by equation (3.29) (Chen & Jiang, 2016):

$$f = \frac{\varepsilon}{\varepsilon + S1}. \quad (3.29)$$

The expulsion efficiency is deemed reliable upon assuming no hydrocarbon expulsion at $T_{max} < 435°C$ and that the expelled hydrocarbon does not exceed 85% (Chen & Jiang, 2016). Bearing in mind that within the oil window, about 60% to 90% of generated hydrocarbon is expelled (Cooles et al., 1986) and the expulsion efficiency can be up to 80% of the extract (Leythaeuser et al., 1984).

Next, the initial TOC (TOC_o) is estimated from the current TOC and a mass conversion factor (y) is used to correct the carbon equivalent of expelled hydrocarbons. The mass conversion factor (y) is defined as the ratio of present-day rock mass after hydrocarbon generation and expulsion over initial rock mass and given by equation (3.30) (Chen & Jiang, 2016):

$$y = 1 - 0.833\frac{TOC_o}{100}. \quad (3.30)$$

Therefore, the initial TOC (TOC_o) is expressed as shown in equation (3.31) (Chen & Jiang, 2016):

$$TOC_o = y\frac{TOC}{1 - \alpha f TR_{HI}}, \quad (3.31)$$

where, $\alpha = (HI_o/1200)$ defined by Chen and Jiang (2016) and Kuchinskiy (2013) as a scaling parameter associated with the kerogen type. However, assuming full expulsion ($f = 1$ & $y = 1$) ($f = 1$ and $y = 1$) of generated hydrocarbons and neglecting a possible difference between the present-day rock mass and initial rock mass, an overestimation of the TOC_o will ensue.

Finally, the organic porosity is estimated by converting the amount of carbon transformed to hydrocarbon from weight percentage to a volume percentage, and it is expressed by equation (3.32) (Chen & Jiang, 2016):

$$\varphi_{org} = \gamma\left[TOC_o \alpha f TR_{HI}\left(1 - \frac{0.833TOC}{100}\right)\right]\frac{\rho_b}{\rho_k} \quad (3.32)$$

where, ρ_b is the bulk rock density, ρ_k is the kerogen density, and γ represents the carbon equivalent mass of kerogen (~1.2).

Both organic porosity equations may prove inaccurate when used with logs since both equations depend on the original TOC (TOC_o) weight fractions, while the log readings are a function of present-day organic content. Galford et al. (2013) developed an expression compatible with wireline logs by correlating the organic porosity to present-day solid kerogen volume leading to the equation below, where the organic porosity (ϕ_k) is related to solid kerogen volume (V_{sk}):

$$\varphi_k = \frac{\varphi}{1 - \varphi}V_{sk}. \quad (3.33)$$

By normalizing the organic porosity equation from Modica and Lapierre (2012) with the initial kerogen volume (V_{ki}), the organic porosity is estimated by equation (3.34) (Galford et al., 2013):

$$\varphi_k = C_c \times k \times TR \times TOC_k. \quad (3.34)$$

3.10 Thermal Maturity and Source Rock Assessment

Oil from different conventional reservoirs have different fingerprints and this is attributed to many factors including the change in composition of oil from a source rock as kerogen maturity increases. Moreover, within the same source rock, different organic facies (macerals) of kerogen exist and undergo different thermal maturities with each facies generating a slightly different oil, which becomes increasingly difficult to distinguish (Curiale, 1986; Dahl et al., 1994). Stable molecular markers such as biomarkers and diamondoids can be used to provide molecular and isotopic fingerprints leading to the source. Biomarkers decrease in abundance at the onset of oil cracking, while diamondoids stay constant during maturation and increase during oil cracking (Zinniker, et al., 2011).

3.10.1 Biological Markers (Biomarkers)

A biomarker is described as an organic compound in generated hydrocarbons (bitumen and crude oil) and source rocks, with a basic skeleton linked to its known natural material (Philp & Lewis, 1987; Mackenzie & Mackenzie, 1984). Thus, biomarkers are able to retain all or most of the original carbon skeleton of the original product and can be categorized based on the following characteristics (Peters et al., 2005; Seifert et al., 1980): (1) Source input marker; (2) age diagnostic marker; (3) depositional environment marker; and (4) maturity marker.

The source input marker is a common biomarker in many organisms, but some markers are found only in specific types of organisms. The age diagnostic marker is based on the fact that some biomarkers do not appear until their parent organism has evolved. The depositional environment is of the principle that some biomarker structures are more stable in certain depositional environments than others. Finally, the maturity marker is inferred from the subtle chemical transformation from one chemical structure into another due to thermal stress.

Biomarkers are analyzed through gas chromatography–mass spectrometry (GC-MS) by separating a crude petroleum sample into saturate and aromatic fractions prior to biomarker analysis as outlined in Table 3.23 (Peters & Moldowan, 1993). A reliable biomarker ratio of pristane/phytane (C_{19})/(C_{20}) isoprenoids is commonly used along with other parameters to determine depositional environments based on the fact that pristane (C_{19}) is produced in an oxidizing environment, while phytane (C_{20}) is produced in a reducing environment (Moldowan et al., 1985). High pristane (Pr) to phytane (Ph) ratio (>1) indicates a shale source rock, while low Pr to Ph ratio (<1) indicates a carbonate source (Huang et al., 2015). Thus, for a biomarker analysis in different basins, different correlations exist for the same ratios due to the nonlinear relation between source rock and biomarker maturities (Philp & Lewis, 1987). Moreover, a given biomarker parameter is applicable only over a specific maturity range as specific indicators reach their terminal values (Mackenzie & Mackenzie, 1984).

Furthermore, isoprenoid to normal alkane (n-alkane) ratios such as Pr/nC_{17} and Ph/nC_{18} are used together to determine depositional environment and maturation of source rocks. The two (Pr/nC_{17} and Ph/nC_{18}) ratios decrease with increasing thermal maturity (Huang, Zhang, & Su, 2015). Isoprenoids are classified into three classes; acyclics and cyclics, polycyclic terpenoids, and tetraterpenoids (Gomez, 2016). Saturated forms of terpenoids form terpanes composed of tricyclics, tetracyclics, hopanes, and steranes widely used to characterize petroleum oils and source rocks (Jarvie et al., 2011).

Some notable biomarkers used to indicate migration of petroleum in shale-sandstone formations is the monitoring of C_{15+} alkanes (lower molecular weight components), presumed to undergo migration more readily than higher molecular weight components. In addition, n-alkanes migrate faster than branched alkanes (Moldowan et al., 1985). The maturity of a source can be inferred from thermally stable isomers as their concentrations are influenced in the process (Philp & Lewis, 1987).

Table 3.23 Source Rock Maturity Petroleum Biomarkers

Petroleum fraction	Biomarker parameter	Effect of increasing maturity	Observations
Saturated hydrocarbons	C_{29} steranes [20S/(20S + 20R)]	Increase	Useful in early to mid oil window, decreases at very high thermal maturity levels
	C_{29} steranes [abb/(abb + aaa)]	Increase	Useful in early to mid oil window
	Moretane/hopane	Decrease	Useful in early oil window
	C_{31} hopane [22S/(22S + 22R)]	Increase	Useful in immature rocks to onset of early oil window
	Tricyclic terpane/ hopanes	Increase	Useful in late oil window; also increases at high levels of biodegradation
	Diasteranes/steranes	Increase	Useful in late oil window; also affected by source lithology (low in carbonates, high in shales); increases at high levels of biodegradation
Aromatic hydrocarbons	Monoaromatic steroids $(C_{21}+C_{22})/[(C_{21}+C_{22}+C_{27}+C_{28}+C_{29})]$	Increase	Useful in early to late oil window; resistant to effects of biodegradation
	Triaromatic steroids $(C_{20}+C_{21})/[(C_{20}+C_{21}+C_{26}+C_{27}+C_{28})]$	Increase	Useful in early to late oil window; resistant to effects of biodegradation
	Triaromatic/ (monoaromatic + triaromatic steroids)	increase	Useful in early to late oil window; resistant to effects of biodegradation

Source: Weatherford Labs (2018); Peters et al. (1993).

Tricyclic Terpanes (TT) (C19–C30)

Tricyclic terpanes (TT) (C_{19}–C_{30}) are aliphatic cyclic hydrocarbons that make up the terpenoid biomarker group with a precursor of C_{30} tricyclohexaprenol (Neto et al., 1982; Ourisson et al., 1982). High concentration of TT and their aromatic counterparts are correlated with Tasmanite algae, to suggest a source (Aquino Neto et al., 1992). However, TT may not be solely found in Tasmanites as there exists a possibility of other sources or factors. Other forms of TT include tricyclic diterpanes (C_{19}–C_{20}) used as terrigenous indicators (Jarvie et al., 2011). An abundance of C_{19} and C_{20} tricyclic terpanes indicate a terrestrial source of organic matter (Zumberge, 1987). In addition to determining the source, TT can also be used to assess thermal maturity, depositional environment (C_{19}–C_{23} diterpanes), and correlation of crude oils (Peters et al, 2005). TT ratios include: C_{22}TT/C_{21}TT, C_{24}TT/C_{23}TT, and C_{26}TT/C_{25}TT. Terrestrial oils are commonly enriched with C_{19} and C_{20} TT, while C_{23} is most dominant in marine source oils (Philp & Lewis, 1987).

Tetracyclic Terpanes (TeT)

Tetracyclic terpanes (TeT) commonly occur in petroleum samples and are assumed to originate from microbial openings of hopanes (C_{30}) (Peters & Moldowan, 1991; Mackenzie & Mackenzie, 1984). Combining both tetracyclic and tricyclic terpanes can be used to differentiate carbonate from evaporite depositional environments. An abundance of C_{24} tetracyclic terpanes (TeT) is most indicative of carbonate environments, while depleted C_{24} TeT suggests a shale source (Peters et al., 2005). Commonly used tetracyclic terpane (TeT) ratios include, C_{24}TeT/

C_{30}Hopane, C_{24}TeT/C_{23}TT, and C_{24}TeT/C_{26}TT to differentiate and identify oils. On the other hand, an increase in the C24TeT/C_{30} Hopane ratio is indicative of increasing thermal maturity and biodegradation (Gomez, 2016). TeT are more resistant to biodegradation than hopanes, enabling the differentiation between carbonate and evaporitic source rock settings. High C_{24}TeT/C_{26}TT values (>1) indicate carbonate organic matter (Moldowan et al., 2001), while low to medium values suggest marine shale source and low C_{24}TeT/C_{26}TT values and have been observed in deltaic shale sources (Peters & Moldowan, 1993).

Hopanes and Steranes (C27+)

Hopanes and steranes (C_{27+}) are pentacyclic terpenoids in higher plants commonly used to identify hydrocarbon sources (Ourisson et al., 1982). Carbonate source rocks are often dominated by C_{29} hopanes (Peters & Moldowan, 1993). On the other hand, steranes are thought to originate from steroids with eukaryotic organisms (Mackenzie & Mackenzie, 1984). The C_{29} and C_{27} sterane biomarkers are used to differentiate organic matter sources from higher plants (terrigenous origin) and zooplankton, respectively (Peters et al., 2005). Figure 3.20 shows a sterane ternary diagram highlighting the relationships between oils and source rocks (Moldowan et al., 1985). High values of diasterane/sterane (>1) are found in marine shale oils (Ourisson et al., 1982).

Both hopane and sterane biomarkers are thermodynamically stable and resistant to biodegradation. In less biodegraded oils, hopanes are removed before steranes, while steranes are removed before hopanes in heavily biodegraded oils (Mackenzie & Mackenzie, 1984). Some commonly used hopane ratios as source parameters include, C_{24}TeT/C_{30}H, C_{29}H/C_{30}H, C_{35}/C_{34} homohopane, and C_{33}/C_{32} homohopane, (C_{28}+C_{29}TT)/C_{30}H. Maturity-related hopane and sterane ratios include Tm/C_{30}H, Ts/C_{30}H, and C_{29}Ts/C_{29} norH (Michael et al., 1995). Table 3.23 outlines some selected biomarkers for characterizing the source rock maturity from the analysis of migrated oil

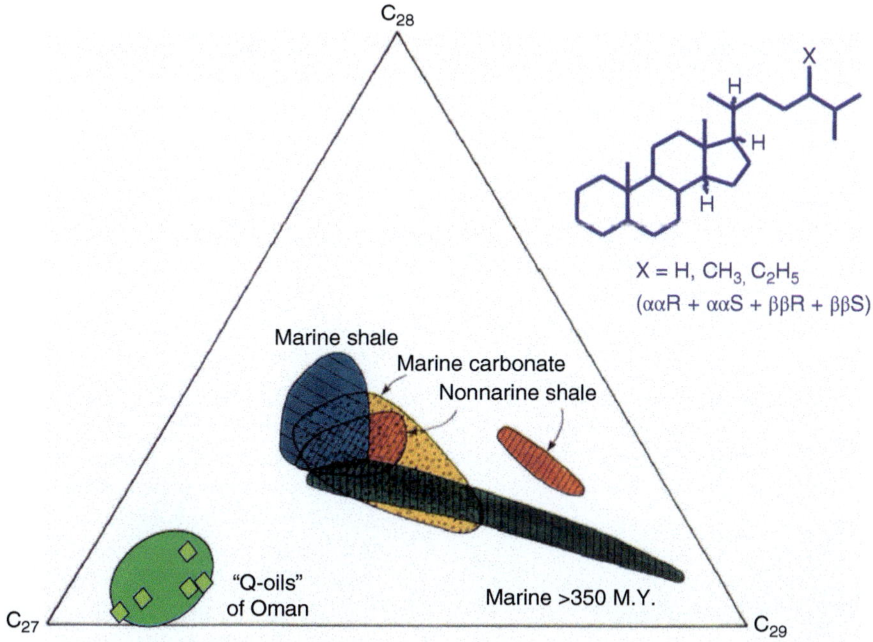

Figure 3.20 Sterane distribution. Ternary diagram showing the relative abundances of C_{27}, C_{28}, and C_{29} regular steranes (5α, 14α, 17α(H) 20S + 20R and 5α, 14β, 17β(H) 20S + 20R) in the saturate fractions of crude oils determined by GCMSMS (M+ → 217) (Moldowan et al., 1985).

(Peters & Moldowan, 1993; Peters et al., 1993). The outlined biomarker maturity parameters make use of processes that occur during source rock maturation, such as cracking (large molecules break into smaller molecules), isomerization (changes in the three-dimensional arrangements of atoms in molecules), and aromatization (loss of hydrogen from naphthenes-formation of aromatic rings). On the other hand, Table 3.24 outlines a few examples of oil biomarker parameters from which depositional environment of source rock and origin of the organic matter in the source rock can be inferred (Peters et al., 2005; Moldowan et al., 1995; Seifertet al., 1980).

3.10.2 Diamondoids

Diamondoids are nanometer-sized (0.5–2 nm) caged hydrocarbons ($C_{4n+6}H_{4n+12}$) with a basic repeated unit of adamantane found in all liquid petroleum and source rocks (Ma, 2016; Moldowan et al., 2015). Diamondoids are extremely stable both thermally and biologically and can be used as effective indicators of oil cracking and mixed oils, and also to assess the degree of thermal alteration in unconventional shales (Zinniker, et al., 2011).

Figure 3.21 shows a molecular structure of diamondoid built by face-fusing of adamantane units. Diamondoids increase in molecular size depending on specific conditions in source

Table 3.24 Source Rock and Depositional Conditions From Selected Biomarker Parameters

Source information	Biomarker parameter	Observations
Marine source rock	24-n-propylcholestanes	Universal in oils from marine source rocks (Seifert et al., 1980; Moldowan et al., 1990)
	C_{42}–C_{46} cyclopentyalkanes with odd/even carbon preference	Carlson et al., 1993; Hsieh & Philip, 2001
Lacustrine source rock	Sterane/Hopanes	Low in oils derived from lacustrine source rocks (Seifert et al., 1980; Moldowan et al., 1985)
	C_{26}/C_{25} tricyclic terpanes	> 1 in many lacustrine-shale-sourced oils (Zumberge, 1987)
	Tetracyclic polyprenoids	High in oils from lacustrine sources (Holba et al., 2000)
	C_{42}–C_{46} cyclopentyalkanes with even/odd carbon preference or with no preference	Carlson et al., 1993
Coal source rock	Pristane/phytane	Very high in coal-sourced oils (> 3) (Hughes et al., 1995)
	C_{31} homohopanes	High relative to total C_{31}–C_{35} in some coal-sourced oils
Anoxic depositional environment	C_{35} homohopanes	High relative to total hopanes in oils derived from source rocks under anoxic conditions (Peters & Moldowan, 1991)
		Abundance in oils is correlated with source rock hydrogen index (HI) (Dahl et al., 1994)
	Pristane/phytane	1.0 can indicate anoxic conditions but the ratio is affected by many other factors
Carbonate source rock	Diasteranes/steranes	Low in carbonate-sourced oils (Rubinstein et al., 1975; Hughes, 1984)

Source: Peters and Moldowan (1993).

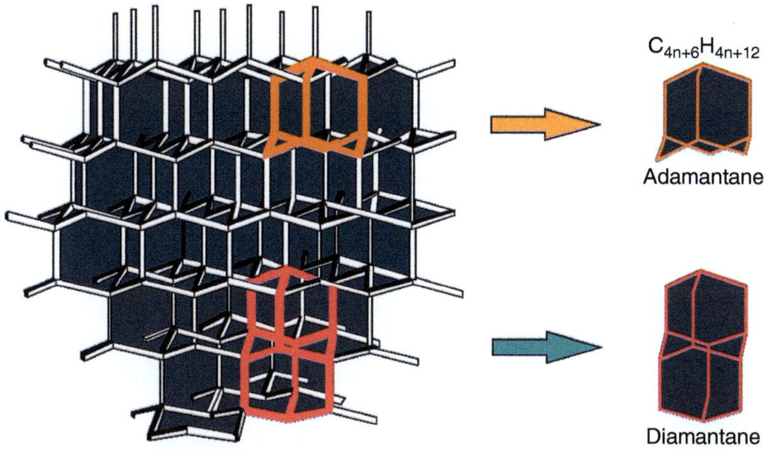

Figure 3.21 Diamondoid lattice made up of adamantanes, diamantanes, and other larger members (Moldowan et al., 2015).

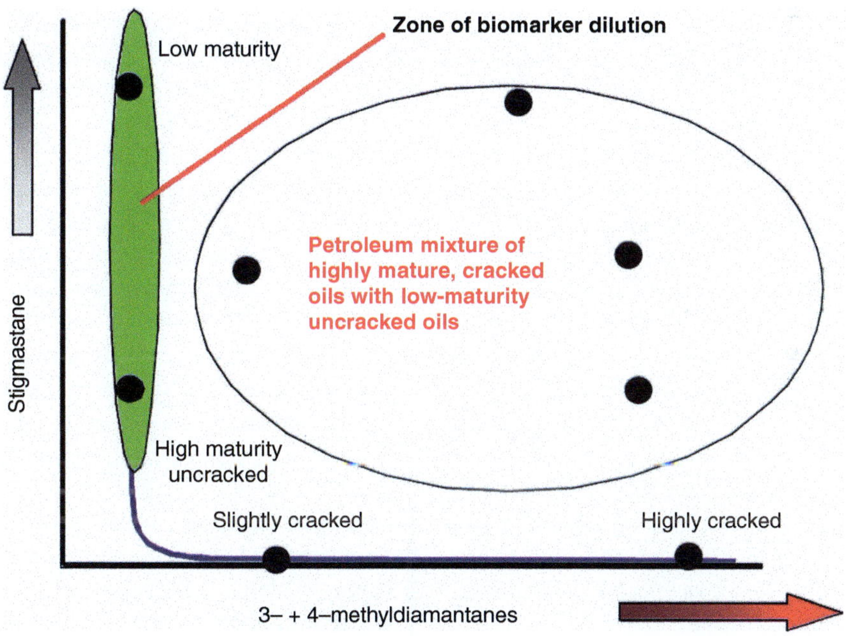

Figure 3.22 A plot of biomarkers versus diamondoids to determine oil-cracking value (modified after Zinniker et al., 1999).

rocks. Some polyhomologous members of adamantanes include diamantanes, triamantanes, tetramantanes, pentamantanes, and hexamantanes (Ma, 2016; Dahl et al., 1999). Adamantanes, diamantanes, and triamantanes make up lower diamondoids, while tetramantanes, pentamantanes, and hexamantanes belong to higher diamondoids (Gunawan et al., 2014). Lower diamondoids like adamantane are easily transported with dry gas, while higher diamondoids like cyclohexamantane are not. Moreover, higher diamondoids could be used to indicate a source signature (Yang et al., 2012; Zinniker et al., 2011).

Figure 3.22 shows a plot of concentrations of stigmastanes (biomarkers) (ppm) versus concentrations of methyldiamantanes (diamondoids) (ppm) to determine the value of oil cracking in oils of different maturities (Dahl et al., 1999). It can be observed that cracking oil has high contents of diamondoids and low contents of C_{29} steranes (biomarkers). When high contents of diamondoids and C_{29} steranes (biomarkers) are encountered, a mixed oil (petroleum) is formed from both a low-maturity (biomarker-rich) source and a high-maturity (diamondoid-rich) source. The zone of biomarker dilution consists of (1) late cracking of kerogen in the source, producing low-biomarker light oil; and (2) cracking of biomarkers in the reservoir or source (Dahl et al., 1999). The percentage of oil cracking into gas and pyrobitumen is therefore expressed in equation (3.35) (Dahl et al., 1999):

$$CR = \left(1 - \frac{C_o}{C_c}\right) 100, \qquad (3.35)$$

where, *CR* is the cracked ratio, C_o is the concentration of methyldiamantanes in the uncracked oil samples (diamondoid baseline), and C_c is methyldiamantanes concentration of any cracked oil sample obtained from the same starting oil. The accurate determination of the extent of oil cracking depends largely on the diamondoid baseline, proposed to be within 4 to5 ppm (Moldowan et al., 2015).

Although, oil cracking can also be estimated from gas-oil ratios (GORs), the GORs are affected by fractionation and diffusion processes within the reservoirs and during petroleum migration by displacement of oil by gas. Hence, GORs are not as effective as diamondoids (Dahl et *al.*, 1994).

3.11 Raman Spectroscopy Analysis of Thermal Maturity in Kerogen

Standard thermal maturity techniques such as vitrinite reflectance (%Ro) and Rock-Eval pyrolysis (T_{max}) fall short in the absence of the vitrinite maceral in small amounts of organic matter and T_{max} becomes unreliable in the presence of free hydrocarbons (bitumen) and inert carbon. On the other hand, Raman spectroscopy investigates the microscale heterogeneity of crystalline, nanocrystalline, and amorphous carbons within a sample without relying on a maceral and is nondestructive. In collecting Raman spectra, laser light is impinged on a sample at a chosen wavelength to excite the molecules within the sample; scattered light is detected and analyzed as a result. The incident light interacts with the vibrational modes of the molecules, causing a shift in frequency in some of the scattered light, and new frequencies known as Raman scattering make up the Raman spectra with information on the molecular structure of the sample (Long, 1977). Therefore, Raman spectroscopy serves as a useful tool for determining the composition of materials and providing thermal maturity information about catagenesis in carbonaceous materials (Kelemen & Fang, 2001).

Figure 3.23 shows a first-order Raman spectra of kerogen made up of a D-band around 1,350 cm^{-1} and a graphite-like G-band between 1,580 and 1,600 cm^{-1}; natural graphite shows a single first-order Raman band at 1,578 cm^{-1} (Kelemen & Fang, 2001). The D-band occurs as a result of crystalline disorder and is assigned to a A_{1g} vibrational mode associated with structural defects of carbonaceous material and decreasing crystallite size, while the G-band is indicative of ordered carbon structures assigned to an in-phase, doubly degenerate C-C stretching of E_{2g2} crystal symmetry (Kelemen & Fang, 2001; Ferrari & Robertson, 2000).

The second-order Raman spectra occur between 2,000 and 3,500 cm^{-1} and are attributed to overtones and combinations of the first-order with much weaker intensity and broad bandwidths, hidden by fluorescence for low mature samples (Schito et al., 2017; Brolly et al., 2016).

Thermal maturity indicators are identified as the labeled Raman parameters (Fig. 3.23) consisting of the full width at half maximum (FWHM) of both the G and D bands, the

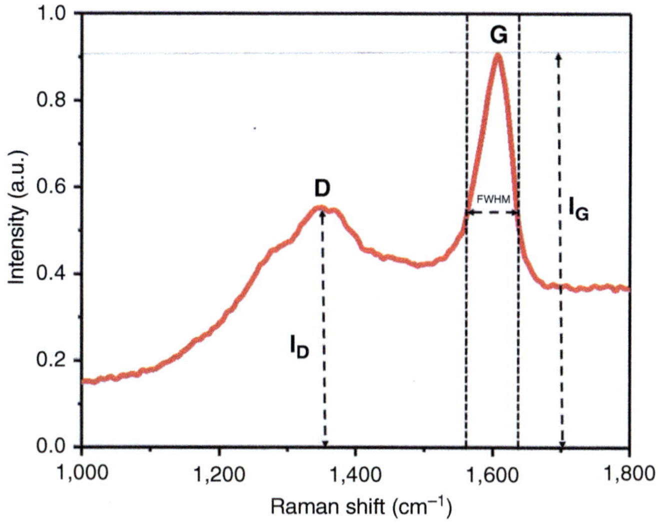

Figure 3.23 First-order Raman spectra of a LEF sample at a thermal maturity of 1.03% with labeled Raman parameters (Cudjoe et al., 2019)

intensity ratio of the D band (ID) to that of the G band (IG), among others. The FWHM of the G band is reported to be proportional to the bond-angle disorder at sp^2 sites (Ferrari & Robertson, 2000). In so doing, the narrowing of the G band and an increase in the G band intensity relative to the D band can be correlated to thermal maturity. On the other hand, the intensity ratio of the D band to that of the G band is defined to be inversely proportional to the in-plane crystallite size in disordered graphite (Tuinstra & Koenig, 1970), where an increase in the ratio will result in a corresponding increase in the Tuinstra-Koenig (TK) relationship (Ferrari & Robertson, 2000).

3.11.1 Thermal Maturity Controls of Organic Matter Types in LEF Samples

Thin sections prepared from LEF samples at different maturity levels (0.98 %Ro and 1.01 %Ro,) but all within the oil window were studied under the Renishaw inVia Reflex Spectrometer at the University of Kansas. An excitation wavelength of 514.5 nm, 2,400 l/mm grating, 10s exposure time, and 1 accumulation was used. The system was calibrated with a 514.5 nm laser using an internal silicon (Si) standard and a low-laser power between 0.5% and 5% to prevent alteration of samples. The first-order spectral range was set from 100 to 1,800 /cm to collect Raman spectra. Samples were first examined under reflected light microscopy with objective lens of 10x and 20x, respectively, to aid in the characterization of different organic maceral morphologies.

Based on the morphological criteria of Loucks and Reed (2014), the depositional kerogen (DK) is identified to be embedded in the matrix, while the migrated bitumen (MB) is observed in the infills of foraminifera chambers. As a result, different DK and MB features were selected for lower Eagle Ford (LEF) samples at different thermal maturities of 0.98% and 1.01%, respectively, and analyzed with the Raman spectrometer. Figure 3.24 shows

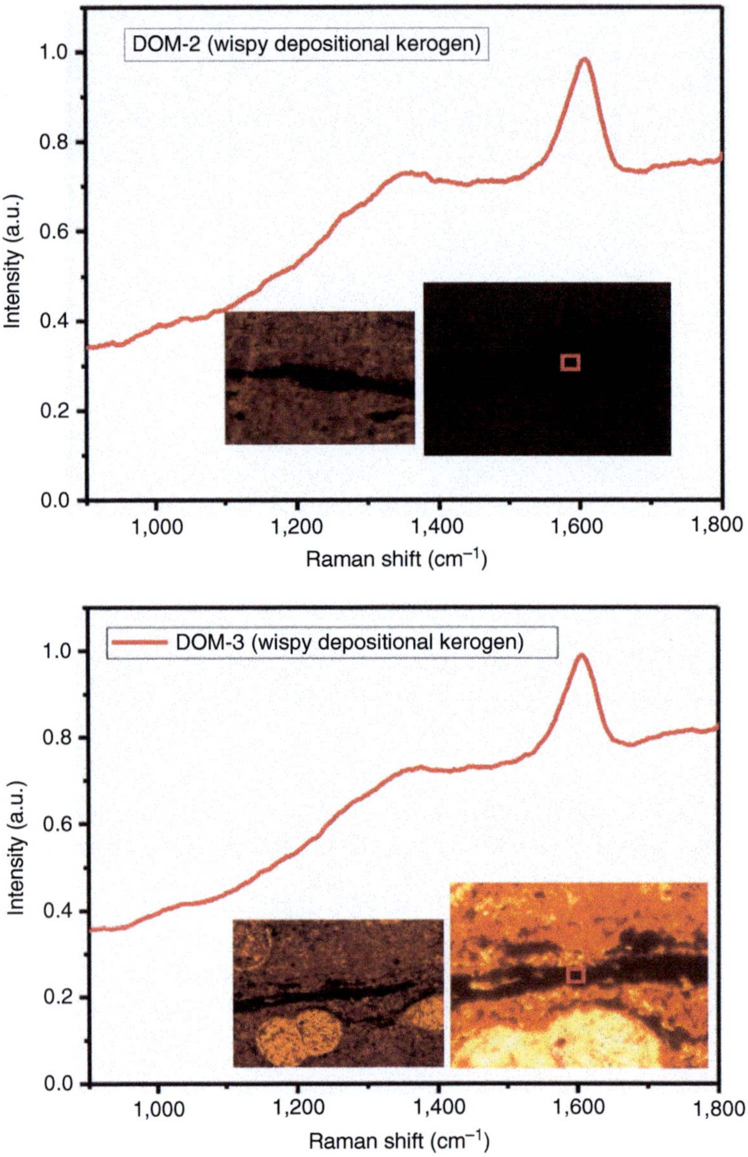

Figure 3.24 Selected depositional kerogen features within a LEF sample at thermal maturity (VR_o %) of 0.98% (Cudjoe et al., 2019).

the depositional kerogen features, and Figure 3.25 shows the migrated bitumen features for a LEF sample at thermal maturity of 0.98% as detailed in the work of Cudjoe et al. (2019a,b).

On the other hand, Figure 3.26 shows both the DK and MB features selected for another LEF sample at a thermal maturity of 1.01%. The Raman measurements for both LEF samples at different thermal maturities with different organic matter (OM) features are compared in Table 3.25. The FWHM of the G bands and D bands (Table 3.25) indicate that the MB features appear more

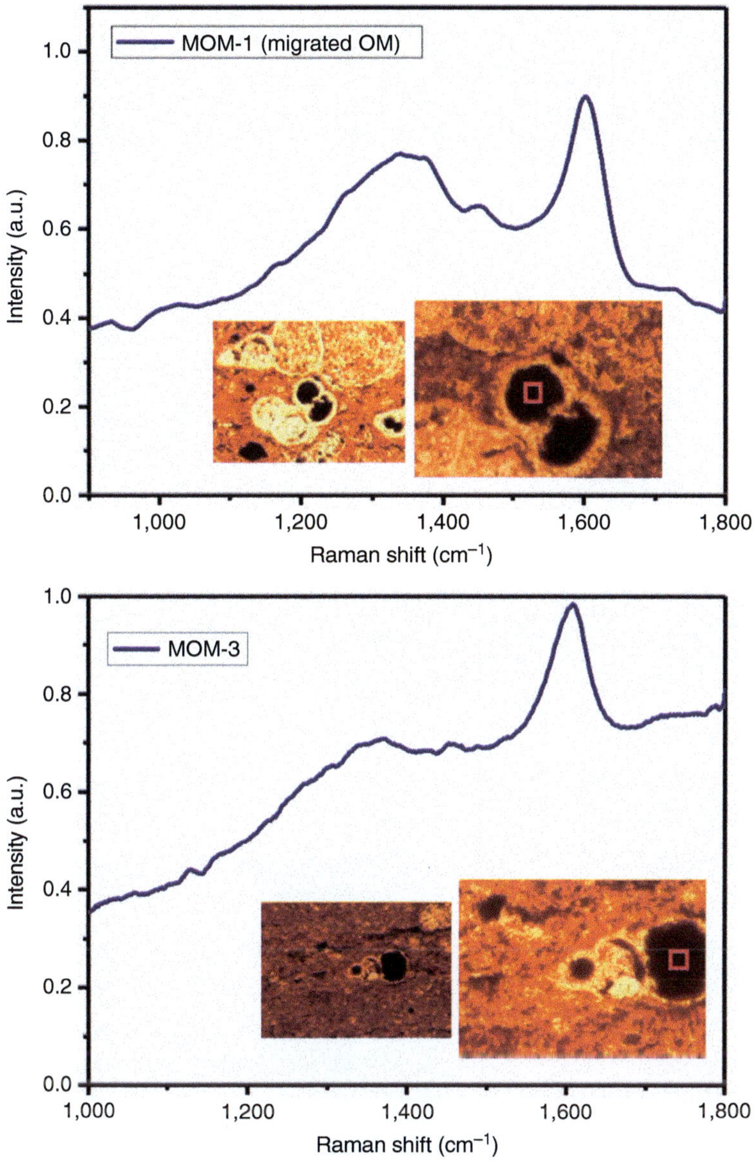

Figure 3.25 Selected migrated bitumen features within a LEF sample at a thermal maturity (VR$_o$ %) of 0.98% (Cudjoe et al., 2019).

thermally matured than the DK features. However, this is not feasible since both features occur within the same sample at the same maturity window. Cudjoe et al. (2019a,b) attributed this discrepancy to the influence of the maceral composition of the MB features playing a significant role. Thus, the maceral composition of the MB features may be more prone to develop organic pores than the DK features, which goes further to add to literature findings not just as temperature but also, most important, as the source influencing thermal maturity indicators with Raman.

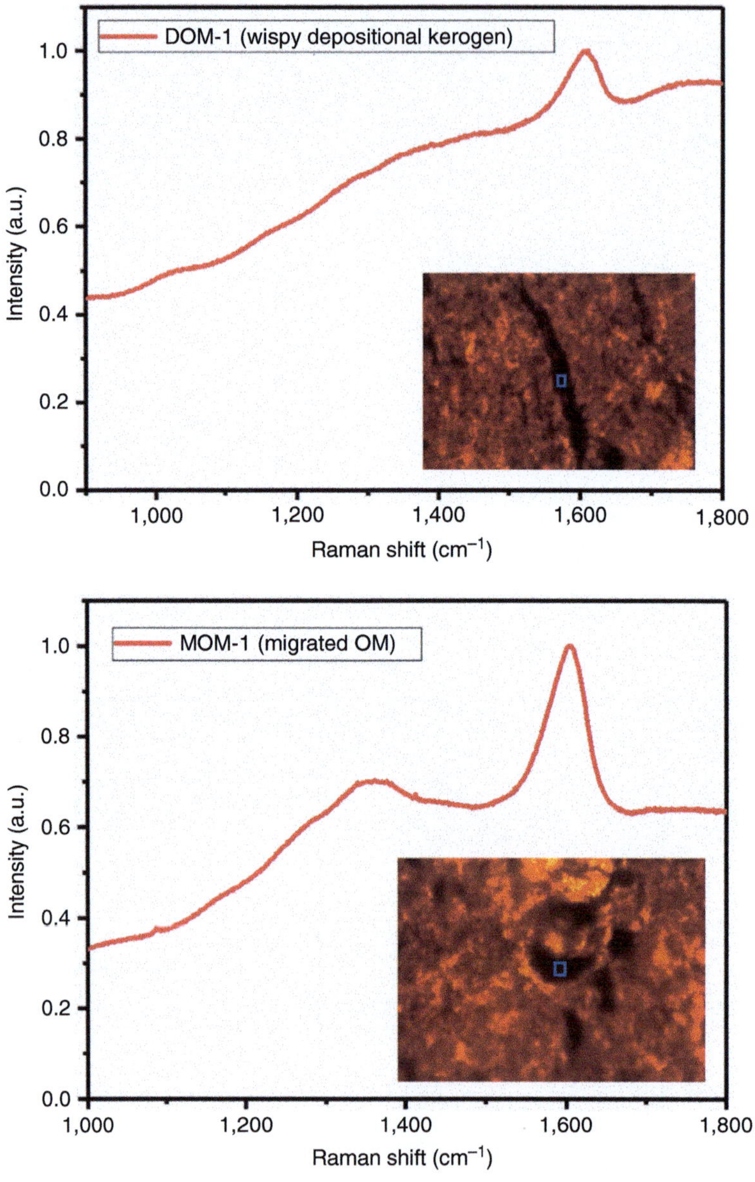

Figure 3.26 Selected depositional kerogen (left) and migrated bitumen (right) features within a LEF sample at a thermal maturity (VR_o %) of 1.01% (Cudjoe et al., 2019).

To further prove the hypothesis that the migrated bitumen (MB) undergoes further thermal decomposition in the chambers of the forams after migration to generated solid bitumen/pyrobitumen with developed organic pores, Raman measurements were acquired for Gilsonite (Cudjoe et al., 2019a,b). Gilsonite represents a form of migrated bitumen believed to be thermally immature and sourced from the Green River formation in the Uinta basin, Utah (Johnson et al., 2017). Figure 3.27 shows the corresponding Raman spectra of selected regions (bright and dark) on the Gilsonite sample and no discernable Raman bands are

Table 3.25 Thermal Maturity Indicators Inferred From Raman Parameters of LEF Samples at 0.98% and 1.01% Thermal Maturities, Respectively

Type of OM	G-band (cm^{-1})	D-band (cm^{-1})	FWHM of G-band (cm^{-1})	FWHM of D-band (cm^{-1})
DK-2 (VRo = 0.98%)	1,607	1,348	55	157
DK-3 (VRo = 0.98%)	1,606	1,349	55	168
DK-1 (VRo = 1.01%)	1,604	Not discernable	55	--
MB-1 (VRo = 0.98%)	1,600	1,347	40	43
MB-3 (VRo = 0.98%)	1,608	1,337	53	152
MB-1 (VRo = 1.01%)	1,605	1,340	55	150

Source: Cudjoe et al. (2019).

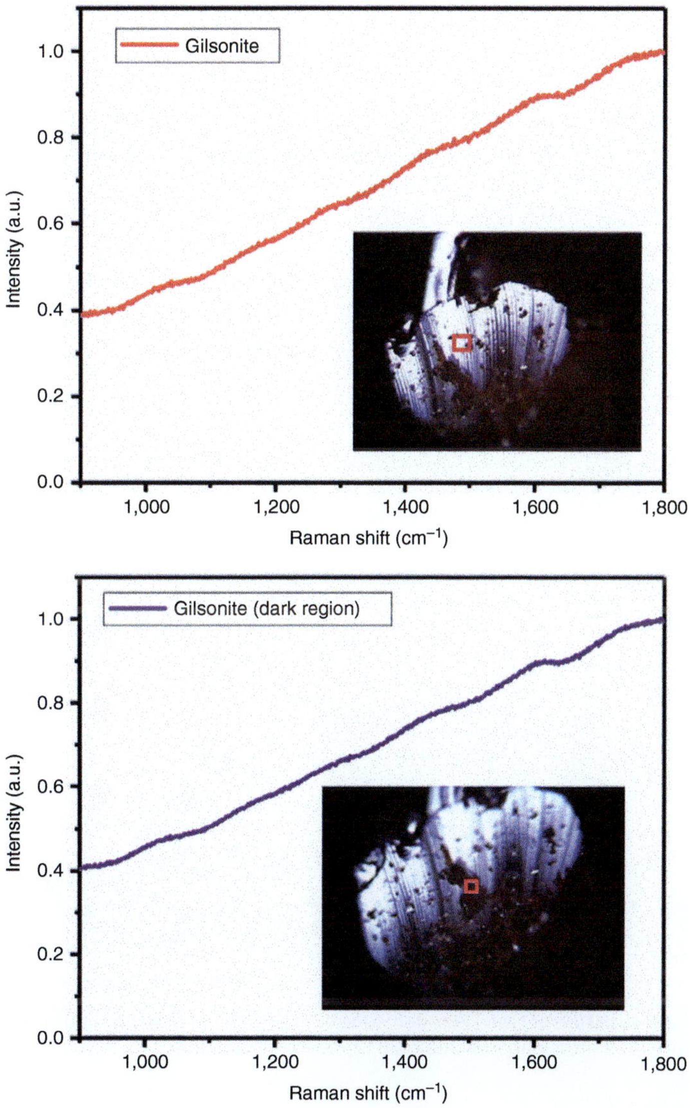

Figure 3.27 Raman spectra of Gilsonite from the Uinta basin of Utah (Cudjoe et al., 2019).

observed due the fluorescence of the C-H bonds (Cudjoe et al., 2019a,b).

3.11.2 Maturity-Related Changes

Chemical structure of organic matter is significantly altered during maturation (catagenesis and metagenesis) and is reflected in the distribution of C-H bonds and aromatic skeletal backbones of the maceral at each stage. Figure 3.28 shows a highly aliphatic maceral structure in type I kerogen with low aromatic contents. This characteristic is reflected in Raman spectroscopy with the absence of the D band but the presence of a G band. On the other hand, Figures 3.29 and 3.30 show the effect of increasing thermal maturity on the molecular structure of kerogen, with a more prominent influence on type III kerogen (Fig. 3.30) than type II kerogen (Fig. 3.29) (Ferrari & Robertson, 2000). Thus, as thermal maturity increases, the macerals begin to shrink due to devolatilization reactions and hydrocarbon expulsion. This results in the increase in aromaticity and the development of void spaces/ organic pores within the maceral. Aromatization increases during catagenesis, while graphitization begins at later stages of metagenesis. This phenomenon is reflected in Raman spectra with the presence of both D bands and G bands. The D band narrows, intensifies, and shifts to relatively higher wave numbers. The G band also narrows and intensifies for kerogen in rocks with maturity < 2.0 %Ro, above which the change in the G band is negligible (Kelemen & Fang, 2001; Ferrari & Robertson, 2000).

3.12 Drifts Analysis of Kerogen Maturity

Unlike labor-intensive Fourier transform infrared radiation (FTIR), diffuse reflectance infrared Fourier transform spectroscopy (DRIFTS) is an easier and a faster mode of infrared spectral acquisition for determining kerogen maturity especially in drill cuttings. It involves shining an infrared laser onto a sample (cuttings or powder) and measuring the corresponding reflected light intensity by a

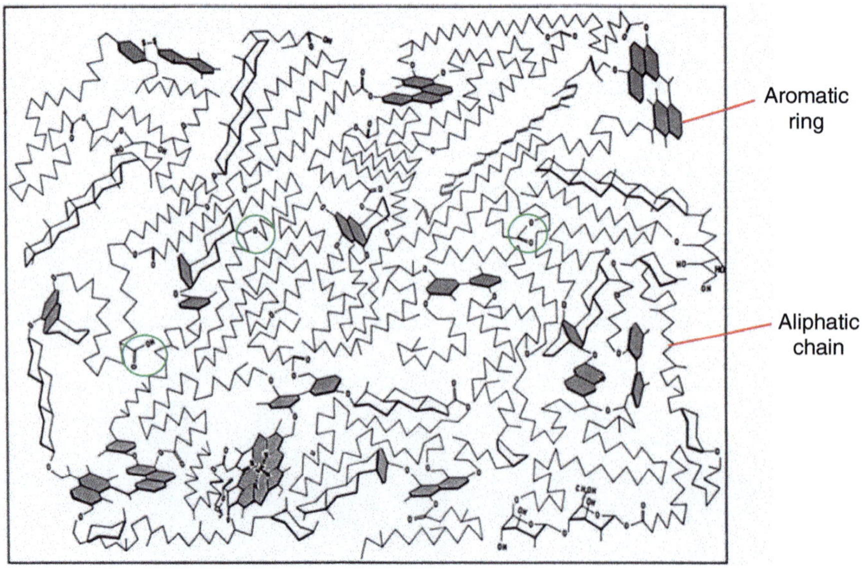

Figure 3.28 Chemical structure of type I kerogen; long aliphatic chains, low aromatic content, and presence of ester-acids-ether (modified after Behar & Allix, 2012; courtesy of Total).

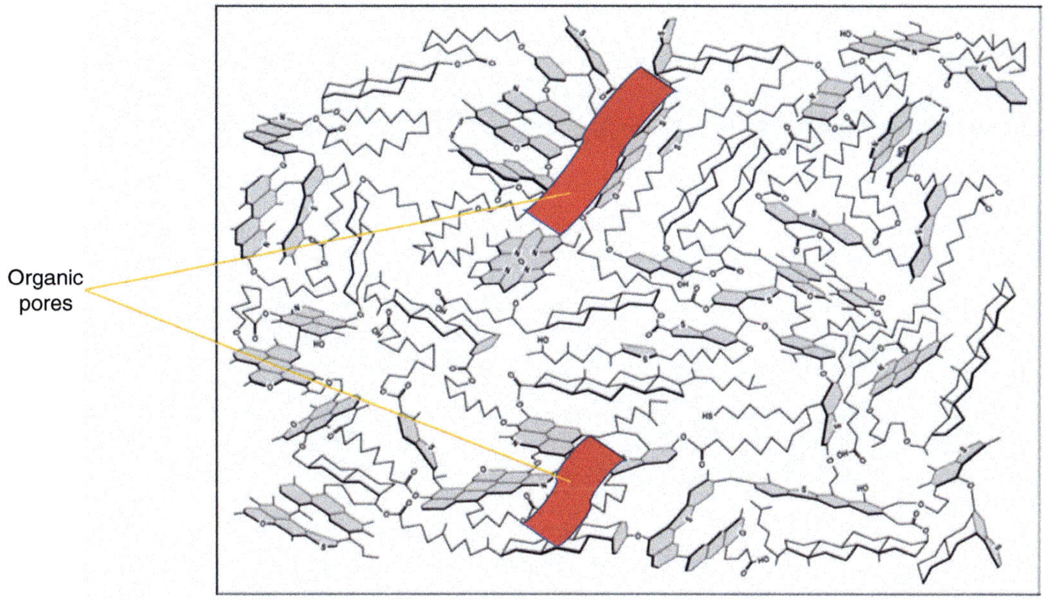

Figure 3.29 Chemical structure of type II kerogen showing the effect of increasing thermal maturity (modified after Behar & Allix, 2012; courtesy of Total).

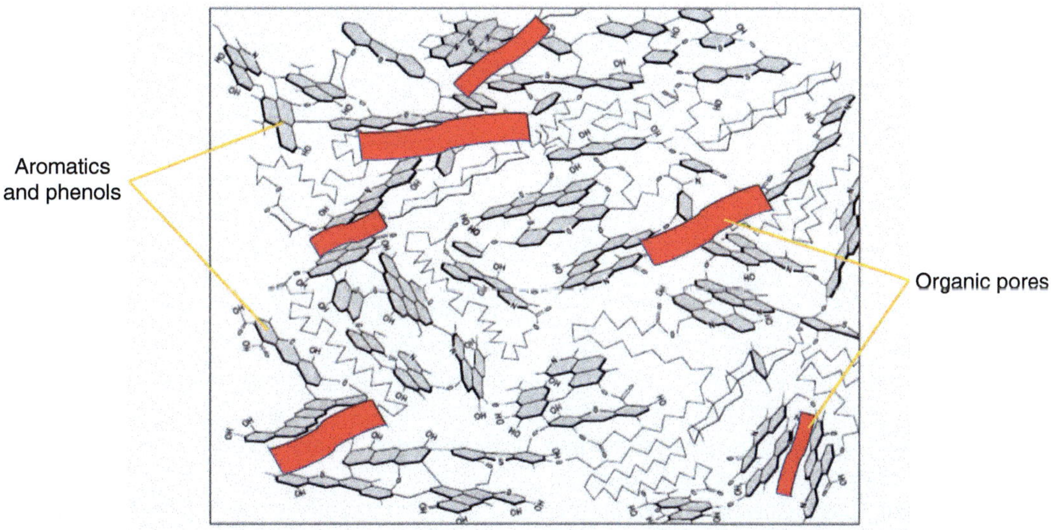

Figure 3.30 Chemical structure of type III kerogen showing increase in aromaticity (modified after Behar & Allix, 2012; courtesy of Total).

detector as a function of frequency (Herron et al., 2014). The infrared light could be specularly or diffusely reflected or transmitted through the sample and measured as a function of the wave number (Charsky & Herron, 2012). Each mineral component of the sample gives off molecular vibrations. In so doing, DRIFTS responds to the different modes of vibrations,

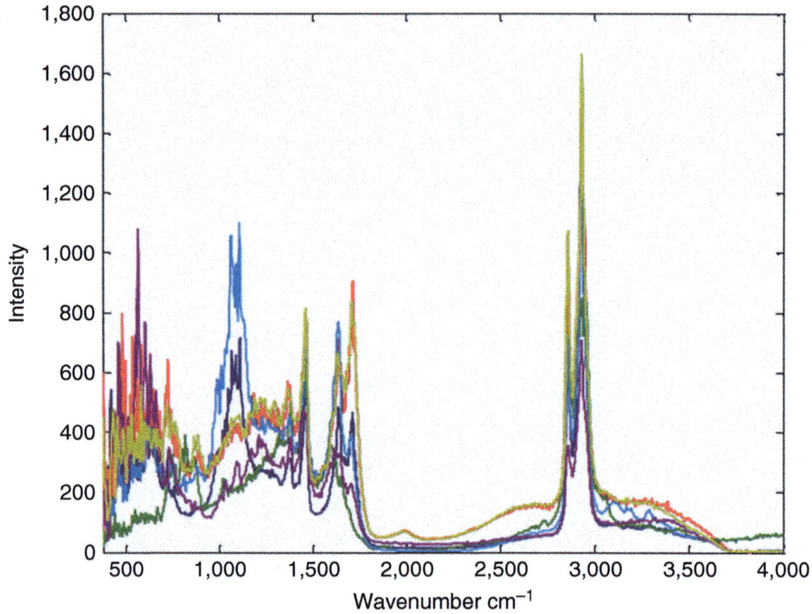

Figure 3.31 Standard isolated kerogen DRIFTS spectra; different colors represent different kerogen types at different maturity levels (Charsky & Herron, 2012).

which corresponds to a specific absorbance band leading to the extraction of components from DRIFTS spectrum through linear combination of known standard minerals (Craddock et al., 2015; Charsky & Herron, 2012). Figure 3.31 shows a kerogen spectra of five demineralized rocks collected to build a library of standard organic spectra, which may include bitumen (Charsky & Herron, 2012; Durand, 1980).

DRIFTS scans are collected over mid-IR range between 375 and 4,000 cm^{-1} at 4 cm^{-1} resolution; the region of the spectrum between 2,800 cm^{-1} and 3,100 cm^{-1} is most susceptible to kerogen characterization and quantification (Craddock et al., 2015; Chen et al., 2012). It is worth noting that, this range (2,800 cm^{-1} - 3,100 cm^{-1}) corresponds to the second-order Raman spectra. Oil based mud and bitumen (if present) are usually removed from cuttings with organic solvents before DRIFTS analysis leaving behind insoluble kerogen and other minerals in order not to interfere with the accuracy of the method. Moreover, spectral effects are minimized by reducing the particle size for a minimum sample amount of ~1 g (Charsky & Herron, 2012).

The DRIFTS spectrum is made up of an average number of scans collected over a certain period (e.g., 24 scans within 20 seconds) in Kubelka-Munk (KM) intensity and assumed to obey Beer-Lambert law (Charsky & Herron, 2012; Lis et al., 2005). Thus, the KM theory relates the concentration of the absorbing material/component in KM intensity to the reflectance (R) in percentage (%) of the sample expressed as shown in equation (3.36) (Kubelka & Munk, 1931):

$$KM = \frac{(1-R)^2}{2R}. \tag{3.36}$$

According to Craddock (2015) and Herron et al. (2014), in the kerogen region, six IR absorption bands resulting from aliphatic C-H chains and aromatic C-H rings stretching modes are identified. In addition to kerogen characteristics, absorption bands of carbonate minerals can be observed, which can be eliminated by solving for the carbonate content of the sample.

Changes in kerogen due to thermal maturity are reflected in the IR spectrum of kerogen and expressed in terms of several IR structural parameters such as CH_3/CH_2 ratio, aromaticity, A-factor, and C-factor, which can all be individually correlated to thermal maturity (Chen et al. 2012). The IR structural parameters can be estimated in transmission FTIR spectra given the following principal absorption bands (Craddock et al., 2015):

1) Aliphatic groups (CH_2 and CH_3 deformations ~1,375 and 1,450–1,460 cm^{-1}; CH, CH_2, and CH_3 stretches ~2,800–3,000 cm^{-1})
2) Aromatic groups (C = C stretch ~1,600 cm^{-1}; CH stretch ~3,000–3,100 cm^{-1})
3) Oxygenated groups (carboxyl and carbonyl C = O stretches ~1,650–1,800 cm^{-1})

Craddock (2015) identified the structure of bitumen to lie between kerogen and generated petroleum through infrared spectroscopy (IR) and exhibits oxygen-rich tendencies at low artificial maturity (EASY%Ro < 0.8%) but oxygen deficient at higher maturity. Kerogen is identified by the depletion of oxygenated functional groups at a low artificial maturity (EASY%Ro < 0.9 %) but becomes oxygen-rich at higher artificial maturity (EASY%Ro >1.2 %). Although both kerogen and bitumen lose their aliphatic carbon content during thermal maturation and become more aromatic, the aliphatic content of bitumen is greater than that of kerogen.

3.13 Problems

1 What controls the transformation of organic matter during diagenesis and what final product(s) is produced at the end of diagenesis?

2 What characterizes catagenesis and how is kerogen transformed?

3 What constituents make up organic matter at the metagenesis stage?

4 What percentage of depositional organic matter is preserved in sediments and what depositional environment favors preservation?

5 What single hydrocarbon is generated in abundance over the diagenetic stage of organic matter evolution?

6 The results of a Rock-Eval pyrolysis test are provided in Table 3.31. Based on the provided, calculate:
   a Hydrogen index
   b Oxygen index
   c Productivity index
   d Vitrinite reflectance (%Ro)

7 Determine graphically the kerogen type and quality for the data provided in Table 3.31. Determine the overall source rock quality based on the average values.

8 Determine the hydrocarbon yield for the data provided in Table 3.26 assuming the original hydrogen index is 575 and the original productivity index is 0.05.

Table 3.26 Data for Problem 6

Depth (ft)	TOC (wt. %)	S_1 (mg HC/g)	S_2 (mg HC/g)	S_3 (mg CO2/g)	T_{max} (deg. C)
4,894	10.09	4.21	61.47	0.25	439
4,899	2.44	2.82	13.91	0.18	440
4,903	2.00	1.31	9.48	0.18	442
4,909	1.88	1.47	9.57	0.18	441
4,914	2.23	1.38	10.23	0.18	442
4,919	2.43	1.75	10.76	0.18	442
4,925	2.96	2.14	16.39	0.17	441
4,929	3.91	2.70	23.54	0.15	441
4,934	3.70	2.75	22.70	0.19	439
4,939	0.49	0.42	1.34	0.27	437

9 Three intervals are provided in Table 3.27. Calculate the HI, OI, PI, and normalized oil ratio for each interval. Based on the calculated results, determine the kerogen type and the overall source rock potential for each interval.

Table 3.27 Data for Problem 8

TOC (wt. %)	S_1 (mg HC/g)	S_2 (mg HC/g)	S_3 (mg CO2/g)	T_{max} (deg. C)	C_r %
2.44	2.82	13.91	0.18	440	8.20
2.00	1.31	9.48	0.18	442	7.45
1.88	1.47	9.57	0.18	441	9.40
2.23	1.38	10.23	0.18	442	7.00
2.43	1.75	10.76	0.18	442	8.64
2.96	2.14	16.39	0.17	441	5.40

10 Develop correlations for the data provided in Table 3.28 to estimate the TOC, S_1, and S_2 from the gamma ray log readings provided in Table 3.29. Find the correlations for TOC and S_1 and TOC and S_2. Also, find the best correlation for the GR log readings with either S_1 or S_2. Compare the TOC, S_1, and S_2 values that are found from the correlations for the data provided in Table 3.29 against the Rock-Eval pyrolysis values provided in Table 3.30.

Table 3.28 Data for Problem 9

Interval	TOC (wt. %)	S_1 (mg HC/g)	S_2 (mg HC/g)	S_3 (mg CO2/g)	T_{max} (deg. C)
Interval 1	10.80	7.40	63.60	1.34	434
Interval 1	18.30	11.86	93.19	1.33	429
Interval 1	16.60	10.8	88.48	1.28	431
Interval 1	15.20	9.64	87.70	1.22	436
Interval 1	19.10	11.71	98.19	1.18	429
Interval 2	0.53	2.85	0.85	0.37	430
Interval 2	0.49	2.51	0.87	0.36	426
Interval 2	0.45	2.43	0.58	0.37	433
Interval 2	0.31	1.33	0.29	0.32	428
Interval 2	0.54	2.17	0.82	0.40	430
Interval 3	23.50	13.8	107.90	0.98	429
Interval 3	17.20	11.05	74.96	1.83	432
Interval 3	16.50	10.16	69.58	2.17	432
Interval 3	14.70	9.01	62.32	1.38	433
Interval 3	13.10	7.84	51.47	1.48	433

Table 3.29 Data for Problem 10

TOC (wt. %)	S_1 (mg HC/g)	S_2 (mg CO2/g)	GR (gAPI)
15.6	7.19	67.34	31.62
13.9	7.29	74.23	31.34
14.2	7.77	74.68	92.91
16.1	8.59	76.34	263.22
14.8	8.18	83.62	392.87
18.3	9.96	94.63	418.98
16.8	10.27	76.06	378.29
12.3	8.11	74.33	315.60
10.8	7.40	63.60	274.11
16.6	10.80	88.48	387.24
19.1	11.71	98.19	617.35

Table 3.30 GR Readings Data for Problem 10

Depth (ft)	GR (gAPI)
9,934	68.19
9,935	124.96
9,936	394.00
9,937	578.11
9,938	644.80
9,939	704.52
9,940	624.33

11 The G and D band positions for a group of samples from six different wells are provided in Table 3.32. Determine some maturity-related parameters and comment on their level of maturity. The Raman spectra representative of each sample are shown in Figure 3.32.

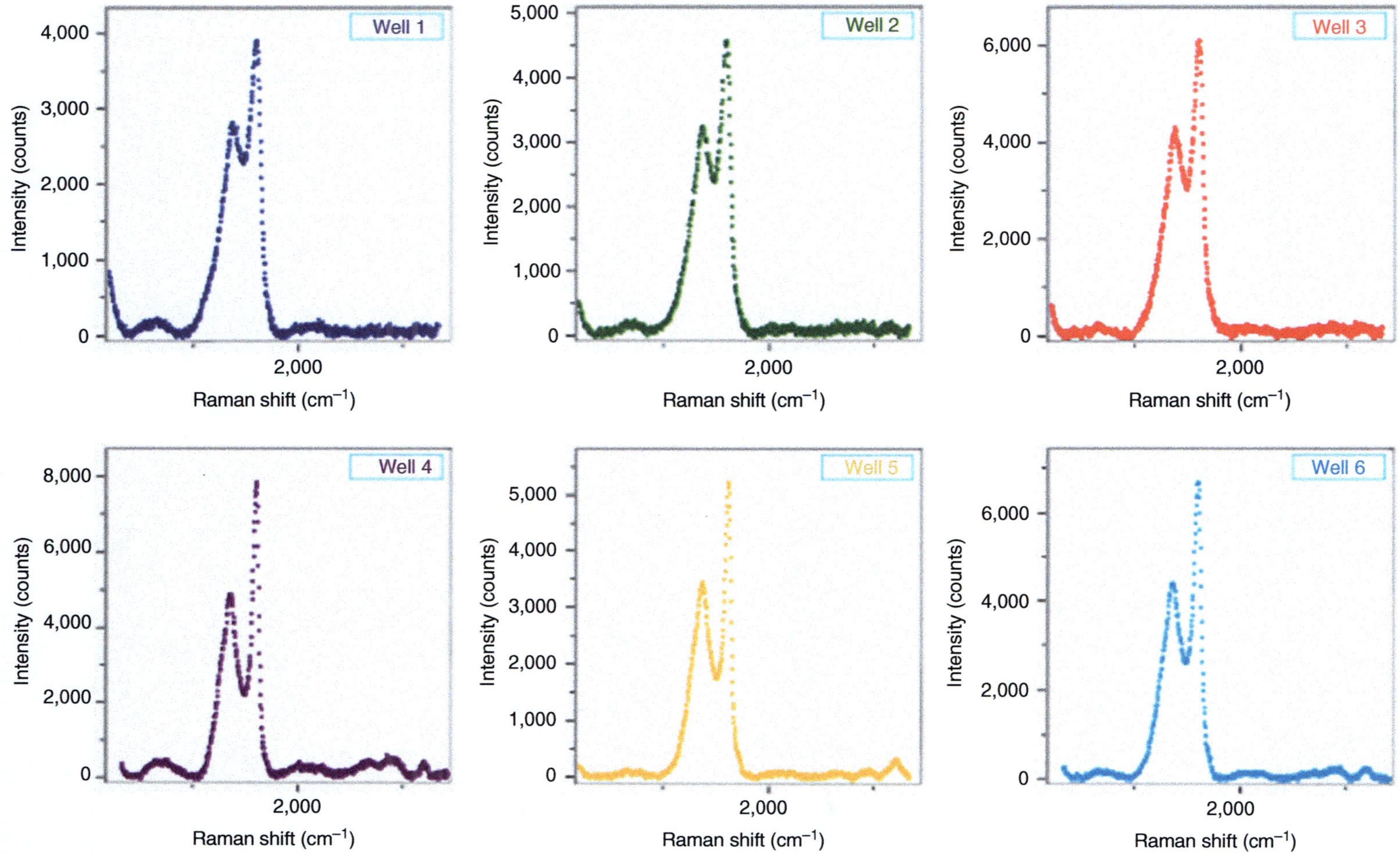

Figure 3.32 Raman spectra representative of each sample from six different wells (Khatibi et al., 2018).

Table 3.31 Rock-Eval Data

Depth (ft)	TOC (wt. %)	S_1 (mg HC/g)	S_2 (mg CO_2/g)
9,934	16.1	10.57	72.85
9,935	19.7	11.83	93.21
9,936	19.8	11.75	89.08
9,937	17.4	10.18	82.93
9,938	19.7	11.55	89.81
9,939	20.2	12.41	105.32
9,940	16.2	11.42	72.83

Table 3.32 D- and G- Band Positions for Each Sample

Well	D-band (cm^1)	G-band (cm^1)
1	1,367.0	1,585.00
2	1,367.0	1,587.00
3	1,361.0	1,588.50
4	1,351.0	1,595.00
5	1,357.5	1,600.29
6	1,354.0	1,591.00

Source: Khatibi et al. (2018).

Additional Reading

Cooles, G., Mackenzie, A., & Quigley, T. (1986). Calculation of petroleum masses generated and expelled from source rocks. *Organic Geochemistry*.

Passey, Q. & Stroud, J. D. (1990). A practical model for organic richness from porosity and resisitivity logs. *AAPG Bulletin*, *74*(12), 1777–1794.

Peters, K. E., Walters, C.C., & Moldowan, J. (2005). *The biomarker guide* (2nd ed.). Cambridge, UK: Cambridge University Press.

References

Aquino Neto, F., Trigüis, J., Azevedo, D., Rodriques, R., & Simoneit, B. (1992). Organic geochemistry of geographically unrelated tasmanites. *Organic Geochemistry*, *18*(6), 791–803.

Baudin, F., Disnar, J., Aboussou, A., & Savignac, F. (2015). Guidelines for Rock-Eval analysis of recent marine sediments. *Organic Geochemistry*, *86*, 71–80.

Behar, F., & Allix, P. (2012). Kinetic modelling of kerogen cracking during oil shale process: influence of organic matter source. *Oil shale symposium*,1–40. Golden, Colorado: Total.

Bernard, S., Horsfield, B., Schulz, H., Wirth, R., Schreiber, A., & Sherwood, N. (2012). Geochemical evolution of organic-rich shales with increasing maturity: A STXM and TEM study of the Posidonia shale (Lower Toarcian, northern Germany). *Marine and Petroleum Geology*, *31*(1), 70–89.

Bernard, S., Wirth, R., Schreiber, A., Schulz, H., & Horsfield, B. (2012). Formation of nanoporous pyrobitumen residues during maturation of the Barnett shale (Fort Worth basin). *International Journal of Coal Geology*, *103*, 3–11.

Brolly, C., Parnell, J., & Bowden, S. (2016). Raman spectroscopy: Caution when interpreting organic carbon from oxidising environments. *Planetary and Space Science*, *121*, 53–59.

Brooks, J. & Welte, D. (1984). Introduction. In D. Brooks, J. Welte, & D. Brooks (Ed.), *Advances in Petroleum Geochemistry*, 1–7. London: Academic Press.

Burnham, A. (1989). On the validity of the pristane formation index. *Geochimica Cosmochimica Acta*, *53*, 1693–1697.

Camp, W. (2015). Diagenetic evolution of organic matter cements in unconventional shale reservoirs. *AAPG Annual Convention & Exhibition*. Denver, CO: AAPG.

Cardott, B. (2012). *Introduction to vitrinite reflectance as a thermal maturity indicator.* Tulsa Geological Society Luncheon, 823–830. Tulsa: AAPG.

Carlson, R. M. K., Teerman, S. C., Moldowan, J. M., Jacobson, S. R., Chan, E. I., Dorrough, K. S., et al. (1993). *High temperature gas chromatography of high-wax oils. 22nd Annual Convention Proceedings*. Jakarta, Indonesia: Indonesian Petroleum Association.

Charsky, A., & Herron, M. (2012). *Quantitative analysis of kerogen content and mineralogy in shale cuttings by Diffuse Reflectance Infrared Fourier Transform Spectroscopy*. International Symposium of the Society of Core Analysts. Aberdeen, Scotland: SCA.

Charsky, A., & Herron, S. (2013). Accurate, direct total organic carbon (toc) log from a new advanced geochemical spectroscopy tool: Comparison with conventional approaches for TOC estimation. Adapted from extended abstract prepared in conjunction with poster presentation at AAPG Annual Convention and Exhibition, Pittsburg, Pennsylvania, May 19–22, 2013.

Chen, Y., Mastalerz, M., & Schimmelmann, A. (2012). Characterization of chemical functional groups in macerals across different coal ranks via micro-FTIR spectroscopy. *International Journal of Coal Geology*, *104*, 22–33.

Chen, Z., & Jiang, C. (2016). A revised method for organic porosity estimation in shale reservoirs using Rock-Eval data: Example from Duvernay Formation in the Western Canada Sedimentary Basin. *AAPG Bulletin*, *100*(3), 405–422.

Conant, L., & Swanson, V. (1961). Chattanooga Shale and related rocks of Central Tennessee and nearby areas. *USGS Professional Paper*, *357*, 91.

Cooles, G., Mackenzie, A., & Quigley, T. (1986). Calculation of petroleum masses generated and expelled from source rocks. *Organic Geochemistry*, *10*(1–3), 235–245.

Craddock, P., Le Doan, T., Bake, K., Polyakov, M., Charsky, A., & Pomerantz, A. (2015). Evolution of kerogen and bitumen during thermal maturation via semi-open pyrolysis investigated by infrared spectroscopy. *Energy and Fuels*, *29*(4), 2197–2210.

Cudjoe, S., Barati, R., Goldstein, R., Tsau, J.-S., Nicoud, B., Bradford, K.,et al. (2019a). An Integrated Pore-Scale Characterization Workflow for Hydrocarbon Gas Huff-n-Puff Injection into the Lower Eagle Ford Shale. *Unconventional Resources Technology Conference* (pp. 1–20). Denver: URTeC.

Cudjoe, S., Barati, R., Marshall, C., Goldstein, R., Tsau, J.-S., Nicoud, B., et al. (2019b). Application of Raman Spectroscopy in Investigating the Effect of Source and Temperature on the Maturity of the Organic Matter Exposed to Hydrocarbon Gas Injection. *Unconventional Resources Technology Conference*. Denver.

Cudjoe, S., Vinassa, M., Henrique Bessa Gomes, J., & Barati, R. (2016). A comprehensive approach to sweet-spot mapping for hydraulic fracturing and CO2 huff-n-puff injection in Chattanooga shale formation. *Journal of Natural Gas Science and Engineering*, *33*, 1201–1218.

Curiale, J. (1986). Origin of solid bitumens, with emphasis on biological marker results. *Organic Geochemistry*, *10 (1–3)*, 559–580.

Curtis, M., Ambrose, R., Sondergeld, C., & Rai, C. (2011). Investigation of the Relationship Between Organic Porosity and Thermal Maturity in The Marcellus Shale. *North American Unconventional Gas Conference and Exhibition*, 14–16 June, The Woodlands, Texas. doi.org/10.2118/144370-MS

Dahl, J., Moldowan, J., Teerman, S., McCaffrey, M., Sundararaman, P., & Stelting, C. (1994). Source rock quality determination from oil biomarkers I: A new geochemical technique. *American Association of Petroleum Geologists Bulletin*, *78*(1)), 1507–1526.

Dahl, J. E., Moldowan, J.M., Peters, K. E., Claypool, G.E., Rooney, M.A., Michael, G.E., Mello, M. R., and Kohnen, M. (1999).

Diamondoid hydrocarbons as indicators of natural oil cracking. *Nature*, *399*(5), 54–56.

Dargahi, H., & Rezaee, R. (2013). Geochemical Analyses on Prospective Gas Shale Reservoirs at Perth Basin, Western Australia. Adapted from extended abstract prepared in conjunction with poster presentation at AAPG Annual Convention and Exhibition, Pittsburgh, Pennsylvania, May 19–22, 2013.

Dembicki, H. (2009). Three common source rock evaluation errors made by geologists during prospect or play appraisals. *AAPG Bulletin*, *93* (3), 341–356.

Durand, B. (1980). *Kerogen: Insoluble organic matter from sedimentary rocks*. Paris: Technip.

El Nady, M., Ramadan, F., Hammad, M., & Lotfy, N. (2015). Evaluation of organic matters, hydrocarbon potential and thermal maturity of source rocks based on geochemical and statistical methods: Case study of source rocks in Ras Gharib oilfield, central Gulf of Suez, Egypt. *Egyptian Journal of Petroleum*, *24*(2), 201–211.

Fang, H., & Jianyu, C. (2007, 12). The cause and mechanism of vitrinite reflectance anomalies. *Journal of Petroleum Geology*, *15*(3), 419 – 434.

Ferrari, A., & Robertson, J. (2000). Interpretation of Raman spectra of disordered and amorphous carbon. *Physical Review B*, *61*(20), 95–107.

Fishman, N., Hackley, P., Lowers, H., Hill, R., Egenhoff, S., Eberl, D., & Blum, A. (2012). The nature of porosity in organic-rich mudstones of the Upper Jurassic Kimmeridge Clay Formation, North Sea, offshore United Kingdom. *International Journal of Coal Geology*, *103*, 32–50.

Forsman, J., & Hunt, J. (1958). Insoluble organic matter (kerogen) in sedimentary rocks. *Geochimica et Cosmochimica Acta*, *15*(3), 170–182.

Galford, J., Quirein, J., Westacott, D., & Witkowsky, J. (2013). *Quantifying organic porosity from logs*. SPWLA 54th Annual Logging Symposium, New Orleans.

Gomez, K. (2016). *Application of Biomarker Data in Oil Characterization Kiara Gomez-Application of Biomarker Data in Oil Characterization*. Technical University of Crete.

Gonzalez, J., Lewis, R., Hemingway, J., Grau, J., Rylander, E., & Pirie, I. (2013). Determination of Formation Organic Carbon Content Using a New Neutron-Induced Gamma Ray Spectroscopy Service that Directly Measures Carbon. *Unconventional Resources Technology Conference, Denver, Colorado, 12–14 August 2013*.

Graciano, J., Filho, M., Menezes, T., Freitas, T., & Rondon, N. (2012). Organic Facies: Palynofacies and Organic Geochemistry Approaches. In J. Graciano, M. Filho, T. Menezes, T. Freitas, N. Rondon, & D. Panagiotaras (Ed.), *Geochemistry - Earth's System Processes* (pp. 211–248). InTech.

Gunawan, M., Hierso, J., Poinsot, D., Fokin, A., Fokina, N., Tkachenko, B., & Schreiner, P. (2014). Diamondoids: Functionalization and subsequent applications of perfectly defined molecular cage hydrocarbons. *New Journal of Chemistry*, *38*, 28–41.

Herron, M., Loan, M., Charsky, A., Herron, S., & Pomerantz, A. (2014). Kerogen Content And Maturity, Mineralogy And Clay Typing From Drifts Analysis Of Cuttings Or Core. *SPWLA 55th Annual Logging Symposium* (pp. 1–14). Abu Dhabi, UAE: SPWLA.

Heslop, K. (2010). PS generalized method for the estimation of TOC from GR and Rt. Adapted from poster presentation at AAPG Annual Convention and Exhibition, New Orleans, Louisiana, April 11–14, 2010.

Hetényi, M. (1998). Oxygen index as an indicator of early organic maturity. *Organic Geochemistry*, *29*(1–3), 63–77.

Holba, A. G., Tegelaar, E., Ellis, L., Singletary, S. M., & Albrecht, P. (2000). Tetracyclic polyprenoids: Indicators of freshwater (lacustrine) algal input. *Geology*, *28*(3), 251–254.

Horsfield, B., & Rullkotter Jurgen. (1994). Diagenesis, Catagenesis, and Metagenesis of Organic Matter. In B. Horsfield, Rullkotter Jurgen, L. Magoon, & Dow W.G. (Eds.), *The Petroleum System-From Source to Trap*. AAPG Memoir 60.

Hsieh, M. & Philip, R. (2001). Ubiquitous occurrence of high molecular weight hydrocarbons in crude oils. *Organic Geochemistry*, *32*, 955–966.

Huang, H., Zhang, S., & Su, J. (2015). Geochemistry of Tri- and Tetracyclic Terpanes in the Palaeozoic Oils from the Tarim Basin, Northwest China. *Energy and Fuels*, *29*(11), 7014–7025.

Hughes, W. (1984). Use of Thiophenic Organosulfur Compounds in Characterizing Crude Oils Derived from Carbonate Versus Siliciclastic Sources. *Petroleum Geochemistry and Source Rock Potential of Carbonate Rocks*, *30*, 181–196.

Hughes, W., Holba, A., & Dzou, L. (1995). The ratios of dibenzothiophene to phenanthrene and pristane to phytane as indicators of depositional environment and lithology of petroleum source rocks. *Geochimica et Cosmochimica Acta*, *59*(17), 3581–3598.

Hunt, J. (1961). Distribution of hydrocarbons in sedimentary rocks. *Geochimica et Cosmochimica Acta*, *22*(1), 37–49.

Hunt, J. (1996). *Petroleum Geochemistry and Geology* (2nd Edition ed.). New York: W.H. Freeman&Company.

Hutton, A., Bharati, S., & Robl, T. (1994). Chemical and Petrographic Classification of Kerogen/Macerals. *Energy and Fuels*, *8*(6), 1478–1488.

Jacob, H. (1989). Classification, structure, genesis and practical importance of natural solid oil bitumen ("migrabitumen"). *International Journal of Coal Geology*, *11*(1), 65–79.

Jarvie, D., Coskey, R., Johnson, M., & Leonard, J. (2011). Chapter 9: The Geology and Geochemistry of the Parshall Area, *Mountrail County, North Dakota*.

Jarvie, D., Hill, R., Ruble, T., & Pollastro, R. (2007). Unconventional shale-gas systems: The Mississippian Barnett Shale of north-central Texas as one model for thermogenic shale-gas assessment. *AAPG Bulletin*, *91*(4), 475–499.

Jiang, C., Mort, A., Sanei, H., & Chen, Z. (2015). S1 peak of Rock-Eval analysis: What does it represent for unconventional hydrocarbon resource assessment? *Geoconvention 2015: New Horizons*, (pp. 1–4). Calgary.

Keel, M. (2015). Organic Porosity Distribution: A Function of Aromaticity in Organic-Rich Mudrocks, MS thesis, Deartment of Geology. Lawrence: University of Kansas.

Kelemen, S., & Fang, H. (2001). Maturity trends in Raman spectra from kerogen and coal. *Energy and Fuels*, *15*(3), 653–658.

Khatibi, S., Ostadhassan, M., Tuschel, D., Gentzis, T., & Carvajal-Ortiz, H. (2018). Evaluating molecular evolution of kerogen by raman spectroscopy: Correlation with optical microscopy and rock-eval pyrolysis. *Energies*, *11*(6), 1406.

Kubelka, P., & Munk, F. (1931). *An Article on Optics of Paint Layers*, *12*, 593–609.

Kuchinskiy, V. (2013). Organic Porosity Study: Porosity Development within Organic Matter of the Lower Silurian and Ordovician Source Rocks of the Poland Shale Gas Trend. Adapted from oral presentation given at AAPG Annual Convention and Exhibition, Pittsburgh, Pennsylvania, May 19–22, 2013.

Lafargue, E., Marquis, F., & Pillot, D. (1998). Rock-Eval 6 Applications in Hydrocarbon Exploration, Production, and Soil Contamination Studies. *Revue de l'Institut Français du Pétrole*, *53*(4), 421–437.

Landis, C., & Castano J.R. (1994). Maturation and bulk chemical properties of a suite of solid hydrocarbons. *Organic Geochemistry*, *22*(1), 137–149.

Lewan, M. (1987). Petrographic study of primary petroleum migration in the Woodford Shale and related rock units.

Leythaeuser, D., Mackenzie, A., Schaefer, R., & Bjoroy, M. (1984). Novel Approach for Recognition and Quantification of Hydrocarbon Migration Effects in Shale-Sandstone Sequences. *American Association of Petroleum Geologists Bulletin*, *68*(2), 196–219.

Lis, G., Mastalerz, M., Schimmelmann, A., Lewan, M., & Stankiewicz, B. (2005). FTIR absorption indices for thermal maturity in camparison with virtinite reflectance R 0 in type-II kerogens from Devonian black shales. *Organic Geochemistry, 36*(11), 1533–1552. doi:1.1016/j.orggeochem.2005.07.001

Long, D. (1977). *Raman spectroscopy.* New York: McGraw-Hill.

Loucks, R., & Reed, R. (2014). Scanning-electron-microscope petrographic evidence for distinguishing organic-matter pores associated with depositional organic matter versus migrated organic matter in mudrocks. *GCAGS Journal, 3*, 51–60.

Ma, A. (2016). Advancement in application of diamondoids on organic geochemistry. *Journal of Natural Gas Geoscience, 1*(4), 257–265.

Mackenzie, A., & Mackenzie, A. (1984). Applications of Biological Markers in Petroleum Geochemistry. In A. Mackenzie, & A. Mackenzie, *Advances in Petroleum Geochemistry.*

MacKenzie, K., Marshall, C., & Walter, M. (2008). Regional organic geochemistry of host sediments of Palaeoproterozoic McArthur River Ore deposit, Australia. *Theoretical Chemistry Accounts.*

Matyasik, M., Kierat, M., Bieleń, W., Brzuszek, P., Kania, M., & Janiga, M. (2017). EXPERIMENTS TO EVALUATE OF ORGANIC POROSITY IN SHALE GAS EXPLORATION.

McCarthy, K., Rojas, K., Niemann, M., Palmowski, D., Peters, K., & Stankiewicz, A. (2011). Basic Petroleum Geochemistry for Source Rock Evaluation. *Oilfield Review, 23*(2), 32–43.

Miles, J. (1994). *Illustrated Glossary of Petroleum Geochemistry.* Oxford: Oxford University Press.

Milliken, K., Ko, L., Pommer, M., & Marsaglia, K. (2014). Sem Petrography of Eastern Mediterranean Sapropels: Analogue Data For Assessing Organic Matter In Oil and Gas Shales. *Journal of Sedimentary Research.*

Modica, C., & Lapierre, S. (2012). Estimation of kerogen porosity in source rocks as a function of thermal transformation: Example from the Mowry Shale in the Powder River Basin of Wyoming. *AAPG Bulletin.*

Moldowan, J., Dahl, J., Zinniker, D., & Barbanti, S. (2015). Underutilized advanced geochemical technologies for oil and gas exploration and production-1. The diamondoids. *Journal of Petroleum Science and Engineering.*

Moldowan, J., Fago, F., Lee, C., Jacobson, S., Watt, D., Slougui, N., ... Young, D. (1990). Sedimentary 24-n-propylcholestanes, molecular fossils diagnostic of marine algae. *Science.*

Moldowan, J. M., Dahl, J., McCaffrey, M., Smith, W., & Fetzer, J. (1995). *Application of biological marker technology to bioremediation of refinery byproducts* (Vol. 9).

Moldowan, J. M., Dahl, J., Zinniker, D., Talyzina, N., Winship Taylor, D., Li, H., Barbanti, S. M., et al. (2001). Expanding evolutionary knowledge using molecular fossils. *20th International Meeting on Organic Chemistry,* (pp. 241–242). Nancy, France.

Moldowan, J.M., Seifert, W.K., and Gallegos, E. (1985). Relationship between petroleum composition and depositional environment of petroleum source rocks. *AAPG Bulletin, 69*(8), 1255–1268.

Mukhopadhyay, P. (1994). Vitrinite reflectance as maturity parameter petrographic and molecular characterization and its applications to basin modeling.

Mukhopadhyay, P., & Dow, W. (1994). *Vitrinite reflectance as a maturity parameter: applications and limitations.* American Chemical Society. doi:10.1021/bk-1994-0570

Neto, F., Restle, A., Connan, J., Albrecht, P., & Ourisson, G. (1982). Novel tricyclic terpandes (C19, C20) in sediments and petroleums. *Tetrahedron Letters, 23*(19), 2027–2030.

Ourisson, G., Albrecht, P., & Rohmer, M. (1982). *Predictive microbial biochemistry — from molecular fossils to procaryotic membranes* (Vol. 7).

Passey, Q. & Stroud, J. D. (1990). A practical model for organic richness from porosity and resisitivity logs. *AAPG Bulletin, 74*(12), 1777–1794.

Pelet, R. (1985). Evaluation quantitative des produits formes lors de l'evolution geochimique de la matiere organique. *Revue Institut Francais du Petrole, 40*(5), 551–562.

Peters, K. (1986). Guidelines for evaluating petroleum source rock using programmed pyrolysis.

Peters, K., & Moldowan, J. (1991). Effects of source, thermal maturity, and biodegradation on the distribution and isomerization of homohopanes in petroleum. *Organic Geochemistry.*

Peters, K., & Moldowan, J. (1993). *The biomarker guide: Interpreting molecular fossils in petroleum and ancient sediments.* Englewood Cliffs, NJ: Prentice Hall.

Peters, K., Kontorovich, A., Andrusevich, V., Moldowan, J., Demaison, G., & Stasova, O. (1993). Geochemistry of selected oils and rocks from the central portion of the West Siberian basin, Russia.

Peters, K., Walters, C., & Mankiewicz, P. (2006). Evaluation of kinetic uncertainty in numerical models of petroleum generation. *AAPG Bulletin.*

Peters, K. E., & Casa, M. (1994). Applied source rock geochemistry. In L. Magoon & W. Dow (Eds.), *The petroleum system: From source to trap* (93–120). Tulsa: American Association of Petroleum Geologists.

Peters, K. E., Walters, C.C., & Moldowan, J. (2005). *The biomarker guide* (2nd ed.). Cambridge, UK: Cambridge University Press.

Philp, R., & Lewis, C. (1987). Organic geochemistry of biomarkers. *Annual Review of Earth and Planetary Sciences.*

Ramachandran, K., Babu, V., Behera, B., & Harinarayana, T. (2013). Source rock analysis, thermal maturation and hydrocarbon generation using rock-eval pyrolysis in parts of Krishna-Godavari basin, India: A case study. *Journal of Petroleum Exploration and Production Technology.*

Rubinstein, I., Sieskind, O., & Albrecht, P. (1975). Rearranged sterenes in a shale: occurrence and simulated formation. *Journal of the Chemical Society, Perkin Transactions 1.*

Ruble, T., Drozd, R., & Heck, W. (2015). Practical geochemical methods to assess unconventional reservoirs a case study from the Permian basin, Texas. Retrieved from WTGS Fall Symposium 2012: http://www.wtgs.org/media/files/None/2012_Ruble_WFTL_WTGS_Presentation.pdf

Sato, M. (1990). Thermochemistry of the formation of fossil fuels. *The Geochemical Society*(2), 271–283.

Saxby, J. (1970). Isolation of kerogen in sediments by chemical methods. *Chemical Geology, 6,* 173.184.

Schito, A., Romano, C., Corrado, S., Grigo, D., & Poe, B. (2017). Diagenetic thermal evolution of organic matter by Raman spectroscopy. *Organic Geochemistry.*

Schmoker, J., & Hester, T. (1983). Organic Carbon In Bakken Formation, United States Portion Of Williston Basin. *American Association of Petroleum Geologists Bulletin.*

Schoenherr, J., Littke, R., Urai, J., Kukla, P., & Rawahi, Z. (2007). Polyphase thermal evolution in the Infra-Cambrian Ara Group (South Oman Salt Basin) as deduced by maturity of solid reservoir bitumen. *Organic Geochemistry.*

Seifert, W., Moldowan, J., & Jones, R. (1980). Application of biological marker chemistry to petroleum exploration. *World Oil.*

Stanley, R., Lillis, P., Pawlewicz, M., & Haeussler, P. (2014). Rock-Eval pyrolysis and vitrinite reflectance results from the Sheep Creek 1 well, Susitna basin, south-central Alaska. U.S. Geological Survey Open-File Report 2013.2017, Reston, Virginia.

Techlog (2018). *Techlog Manual.* Schlumberger.

Teerman, S., Cardott, B., Harding, R., Lemos De Sousa, M., Logan, D., Pinheiro, H., et al. (1995). Source rock/dispersed organic matter characterization: TSOP research subcommittee results. *Organic Geochemistry.*

Tissot, B., & Welte, D. (1984). *Petroleum Formation and Occurrence* (Vol. *66*).

Tobey, M., & Campbell, C. (2016). *Hydrogen Index as a maturity proxy: Some pitfalls and how to overcome them.* AAPG 2016 Pacific Section and Rocky Mountain Section Joint Meeting. 41964. Las Vegas: AAPG.

Tuinstra, F., & Koenig, J. (1970). Raman spectrum of graphite. *J. Phys. Chem., 1126–1130.*

Vandenbroucke, M. (2003). Kerogen: From types to models of chemical structure. *Oil & Gas Science and Technology*, *58*(2), 243–269.

Waples, D. (1984). Thermal models for oil generation. In D. Waples, D. Brooks, & J. Welte (Eds.), *Advances in petroleum geochemistry* (Vol.*1*, pp. 7–61). London: Academic Press.

Weatherford Labs. (2018). Geochemistry for exploration using oil biomarkers in petroleum. *Geochemistry for Exploration*, 1–8.

Yang, C., Wang, Z., Hollebone, B., Brown, C., Landriault, M., Fieldhouse, B., & Yang, Z. (2012). Application of light petroleum biomarkers for forensic characterization and source identification of spilled light refined oils. *Environmental Forensics.*

Zinniker, D., Dahl, J., Moldowan, J., He, M., Mesdakom, M., Magoon, L., et al. (2011). New techniques for recognizing (and determining the source of) high-maturity hydrocarbons. California Petroleum Provinces 4th Annual Stanford BPSM Industrial Affiliates Meeting.

Zumberge, J. (1987). Prediction of source rock characteristics based on terpane biomarkers in crude oils: A multivariate statistical approach. *Geochimica et Cosmochimica Acta.*

4

Application of Imaging Techniques in the Characterization of Organic-Rich Shales

Sherifa E. Cudjoe and Reza Barati Ghahfarokhi

The University of Kansas, Lawrence, Kansas, USA

4.1 Background

Unconventional shale resources remain an important energy breakthrough in recent times, which calls for improved understanding of this complex medium. Shales are mostly made of an inorganic mineral matrix interspersed with dark spots of organic matter (kerogen). Whereas, the inorganic matrix of shales consists of numerous interparticle and intraparticle micropores and microfractures, the organic material (kerogen) is made up of nanopores. Thus, as the degree of kerogen maturation increases to a maximum, the kerogen nanopores are formed, which adds to the complexity of shales (Chen et al., 2014; Chen et al., 2013; Bernard et al. 2012; Loucks et al. 2010). On the other hand, conventional analytical and experimental methods fail to accurately characterize shale reservoirs due to the presence of ultratight (microdarcy and nanodarcy permeability) and multiscale micropore to nanopore size distributions. However, recent advances in both X-ray microcomputed tomography (X-ray micro-CT) and electron microscopy (SEM, FIB-SEM, BIB-SEM) provide detailed information of the internal structure for better understanding of these shale hydrocarbon reservoirs, while being able to scale up the measurements from nanoscale/microscale to macroscale (Lemmens & Richards, 2013; Elgmati et al., 2011; Curtis et al., 2010; Sondergeld et al., 2010; Javadpour et al. 2007). This section of the book details the application of various imaging techniques capable of providing porescale information of organic-rich shales from nanoscale to core scale, while taking full advantage of shale's potential as a reservoir as well as a caprock for CO_2 storage.

Frequently used terminologies in this chapter include

Resolution, which is defined as the distance between objects at which they can still be identified as independent from each other in an image; higher resolution implies more image detail, while lower resolution means less image detail.

Pixel represents the smallest sampled 2D element in an image.

Voxel is the smallest distinguishable box-shaped part of a 3D image.

Field of view (FOV) is the part of the sample that is visible through the imaging equipment (micro-CT, nano-CT, SEM, etc.) lens at a particular position and orientation in space. Therefore, regions outside the FOV are not recorded as part of the acquired CT images (radiographs or tomographs).

Magnification is the ratio between the apparent size of an object and its true size.

CT image reconstruction is defined as the use of a mathematical process to generate CT images from X-ray projection data obtained at different angles of a sample.

Unconventional Hydrocarbon Resources: Techniques for Reservoir Engineering Analysis, Advanced Textbook Series, First Edition. Edited by Reza Barati Ghahfarokhi and Mustafa M. Alhubail.

True resolution is less than the reconstructed voxel resolution and it is the maximum resolution at which the two boundary voxels can be identified and differentiated by an image processing software (Hsieh, 2012).

X-ray spot is the focused X-ray beam from the source.

Attenuation coefficient characterizes how easily a material/sample can be penetrated by an X-ray beam. Large attenuation coefficient means the beam is quickly absorbed/weakened as it passes through the medium, while a small attenuation coefficient means the medium is relatively transparent to the beam.

Contrast is the difference between high and low attenuation coefficent materials on an image quality.

Center shift is the clearest image in the range of pixel values and the spot where the pores are reversed

4.2 X-ray Microcomputed Tomography (X-ray Micro-CT)

X-ray micro-CT is a nondestructive technique used for acquiring images of the internal structure of a sample by measuring the X-ray disseminated from the sample, in this case, an organic-rich shale sample. This results in the visualization of pores as well as the distribution of minerals and eventually in the estimation of transport parameters that control flow and storage potential in shales (Rodriguez et al., 2014). The X-ray micro-CT operates in a similar manner as CT/CAT scans used in health care centers and was introduced to the petroleum industry in the early 1980s (Wildenschild & Sheppard, 2013; Fujii & Uyama, 2004). In the medical CT, the X-ray source and detector system rotate around a stationary patient, while the most common circular X-ray micro-CT possesses fixed X-ray source and a detector around a rotating sample. The circular micro-CT captures good images from mostly smaller samples in a single scan but requires multiple scans for taller samples. The advantage of the rotating stage with the sample in X-ray micro-CT allows for the adjustment of image resolution. Thus, smaller samples result in higher image resolution, while larger samples produce lower image resolution as depicted in Figure 4.1. Smaller sample sizes (1–10 mm) correspond to high- medium CT resolutions, while larger sample sizes (20 mm–50 mm) correspond to low CT

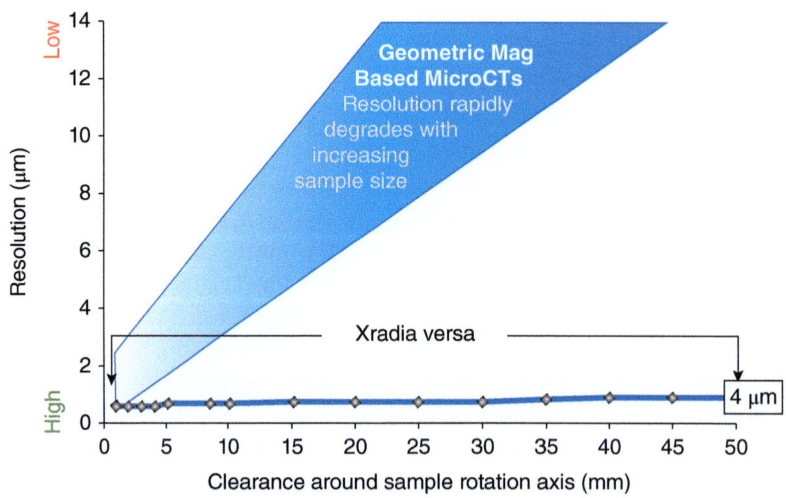

Figure 4.1 Image resolution variation with sample size (*Zeiss*, 2013).

resolution. Therefore, there is the need to compromise between the resolution and size of the sample, where smaller objects would result in a reduced field of view (FOV) but produce high resolution images and larger objects would provide a larger field of view (FOV) but with limited resolution (Hampton et al., 2014). In addition, the application of micro-CT facilitates set up modification for a range of materials, where higher voltage is applied to denser materials, and lower voltage for less-dense materials (*Zeiss*, 2013). Therefore, CT scanning resolves at a micrometer scale (pore scale) and beyond, where, individual grains and fluid interfaces could be resolved and identified. A submicron resolution constitutes a nano-CT with resolutions as low as 0.5 μm (Nia et al., 2016; Curtis et al., 2010; Neuser & Suppes, 2007).

Besides the circular CT scan system with a rotating stage, a helical micro-CT system is also applied in the geoscience industry, where the sample moves in a helical trajectory, as the name implies, to capture precise images from tall samples using single continuous scan at high cone-angles of approximately 50° (Johnston et al., 2014; Varslot et al., 2011). In helical scans, both the sample stage and the detector are mounted on rails for independent adjustment of both sample distance and camera length, which results in flexibility in sample size and acquisition geometry. It can image samples from 1 mm to over 100 mm in diameter, and full height of approximately 20 mm or more (Varslot et al., 2011). While, the circular cone beam trajectories are better suited for smaller size samples, the helical trajectory systems are fitted for larger and taller samples (Johnston et al., 2014; Brink et al., 1994).

Based on the setup of the CT scanner (either the source and detector rotate around the object or the object rotates 360°), tens to thousands of X-ray images (tomographs) are captured. The perecentage of X-ray passing through a detector is determined by the density, thickness, and material composition of the sample. As a result, the linear attenuation coefficient (μ) defined by Beer's law in equation (4.1) at each point in the 2D X-ray image is measured by the detector (Rodriguez et al., 2014).

$$\frac{I}{I_o} = \exp(-\mu h) \tag{4.1}$$

where, I_o is the incident X-ray intensity and I is the intensity remaining after the X-ray passes through a thickness, h of the sample.

At each angle of rotation from 0° to 360°, the spatial variation in the attenuation coefficient (μ) is detected, which results in a 2D CT image (tomograph or radiograph) of the sample. Thus, the CT scan operates on the principle that an image on a closed plane can be reconstructed if projections in all directions through any points on the plane are given (Fujii & Uyama, 2004). Therefore, resulting 2D CT images can be reconstructed with the proprietary XMReconstructor software applying the filtered back-projection (FBP) algorithms, where representative volume elements/pixels (voxels) are a function of the scan geometry (distance between X-ray source, detector, and sample) and the grayscale values are a function of the attenuation coefficient (μ) value or the CT number (CTN) (Rodriguez et al., 2014). Figure 4.2 shows a reconstructed 3D volume of a Chattanooga shale sample on a grayscale, where low-density organic matter (kerogen) appears darker and high-density pyrite presents as white (or brighter). Other matrix minerals with densities in between the extremes (kerogen and pyrite) are presented as different shades of gray; these minerals include quartz, carbonate, clay, and so on. This is explained with the CTN values, which are often displayed with grayscale color coding such that low-density materials are dark and high-density materials are bright. However, upon postprocessing of the CTN values, a wide range of color schemes can be obtained for reconstructed CT scans (Rodriguez et al., 2014).

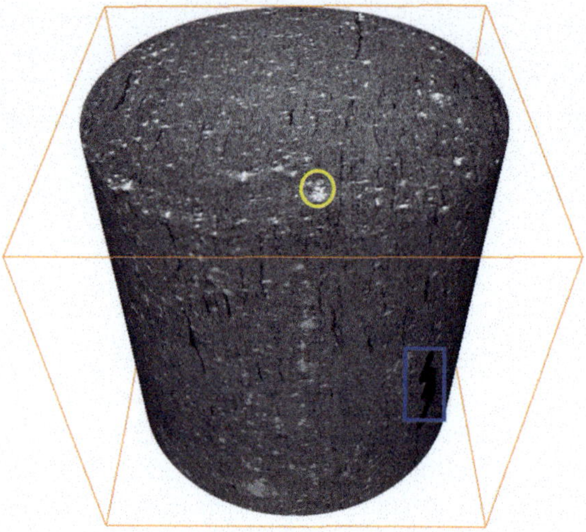

Figure 4.2 A reconstructed 3D view of Chattanooga shale sample (1 µm voxel size resolution with total image size: 948 x 995 x 974 voxels, 16 bit unsigned grayscale value integers in the range of 0 to 65,535) imaged with Xradia MicroXCT-400 highlighting varied mineral composition on a grayscale. Dark spot within blue rectangle is organic matter (kerogen) and bright spot within yellow circle is pyrite. The varying degree of gray background makes up the mineral matrix (quartz or carbonate, clay, etc.).

4.2.1 Operation of X-Ray Micro-CT

The Xradia MicroXCT-400 is capable of taking sample sizes up to 100 mm and revealing the internal morphology (structure) with full 3D imaging of features to approximately 5 to 150 µm resolution and below (Plessis et al., 2017). In so doing, most of the examples outlined in this book would be tailored to the Xradia MicroXCT-400 and Chattanooga shale sample from Barber County, Kansas, although, shale samples like Eagle Ford and Bakken, among others, would be featured as an example. An X-ray micro-CT is similar to the design of industrial scanners but allows small samples (e.g., thin section of rock samples) to be scanned very close to the X-ray source and designed in enclosures for laboratory use (Rodriguez et al., 2014). The micro-CT is made up of a source, which generates X-rays to illuminate the sample placed on a rotating stage, and a detector on the opposite end, which detects the attenuated X-rays as illustrated in Figure 4.3.

The X-rays are generated by an interaction between fast moving electrons at a voltage of approximately 240 kV in vacuum and a tungsten metal target. As shown in Figure 4.3, the sample is placed in the path of the X-ray beam and rotates through a specific angle (180° or 360°), while the detector transforms the attenuated radiation through the sample into 2D images made up of thousands of pixels. In the case of a parallel beam (synchrotron), usually 180° angle of rotation is used, while 360° is used in the case of a cone-beam (Plessis et al., 2017; Schoeman et al., 2016; Lemmens & Richards, 2013; Baker et al., 2012; Brunke, 2010). The sample can move from the source to detector (*x*-axis), upward and downward (*y*-axis) as well as toward the door and to the inner side (*z*-axis) of the XRadia MicroCT-400 in Figure 4.4. In addition, the lead-coated chamber of XRadia's MicroXCT-400 protects against radiation while a terminal panel provides access to an operator to monitor and manipulate the sample through a camera and computer screen.

The resulting CT images are created by the sample attenuating the incoming X-ray through two main mechanisms: (1) Photoelectric absorption (dominant for 8keV nano-CT) and (2) Compton scattering (hard X-rays) (Babu et al., 2015). As previously mentioned, X-ray attenuation is a function of X-ray energy, the density and composition of the material under investigation (Phillion & Milani, 2001). In so doing, denser materials have greater X-ray absorption, and appear brighter, while

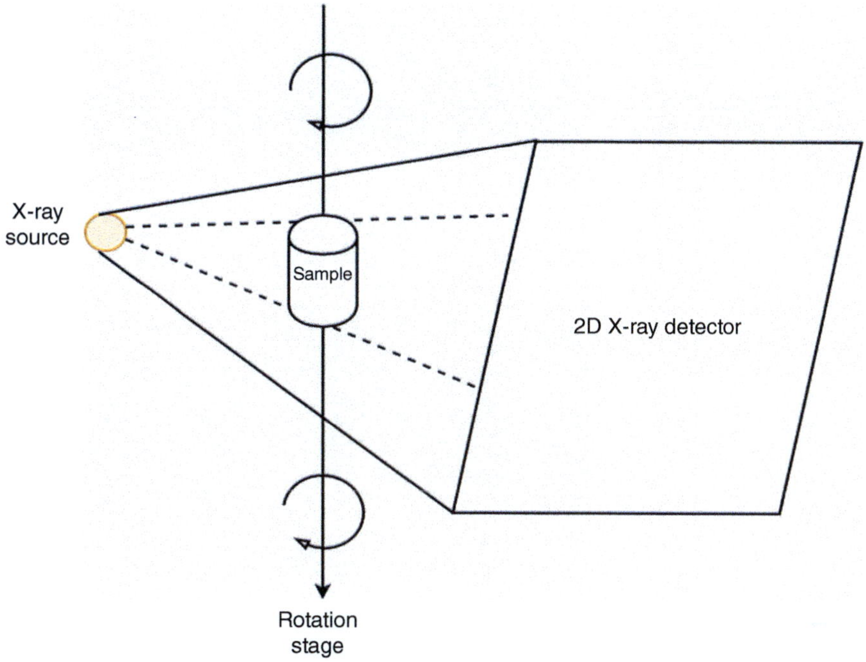

Figure 4.3 Schematic illustration of X-ray micro-CT.

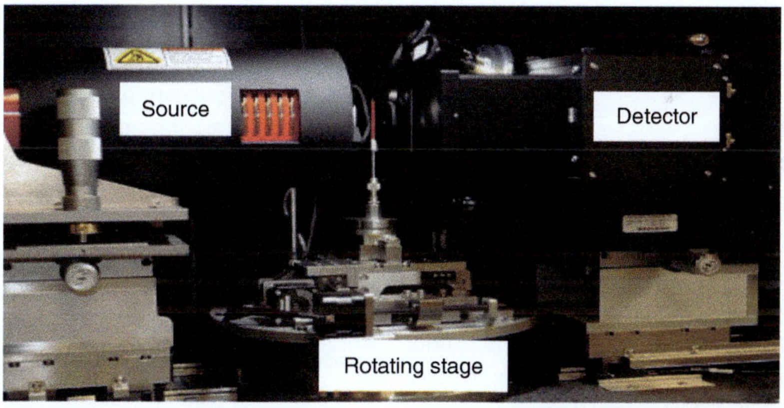

Figure 4.4 XRadia MicroCT-400 scanner. Close-up of source (left), sample holder (middle), and detector (right) (Rodriguez et al., 2014).

less dense materials have less X-ray absorption and appear darker. This imaging technique combines various 2D radiographs (CT images) at different rotation angles to produce 3D reconstructions of a sample through proprietary Xradia XMReconstructed software, which makes use of the filtered back-projection (FBP) algorithms. The 2D tomographs (images) are composed of voxels (volume elements) on gray scales, which correspond to each X-ray attenuation value. The grayscale levels in a CT slice correspond to X-ray attenuation, which reflects

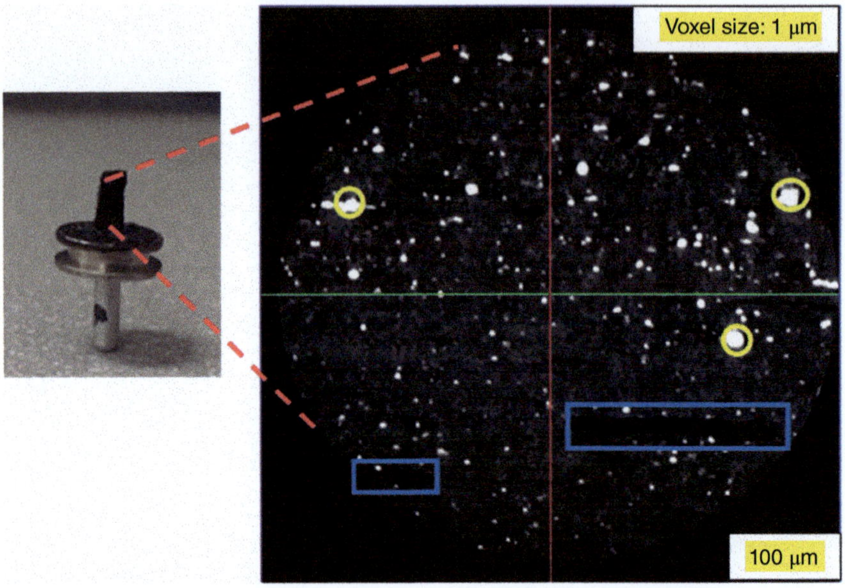

Figure 4.5 Single reconstructed slice (XY-plane) of a Chattanooga shale cuboid sample (5.62 mm x 3.18 mm x 1.2 mm). Dark spots in blue rectangle are organic matter (kerogen), bright spots in yellow circles are pyrite, and gray background represents mineral matrix (e.g., carbonate, quartz, clay, etc.).

the proportion of X-rays scattered or absorbed as they pass through each voxel. Figure 4.5 shows a 2D (XY-plane) CT image of a Chattanooga shale sample with an image resolution of 1 µm/voxel. The darkest spots represent organic matter (less dense), while the brightest spots represent pyrite (denser); quartz appears as light gray, together with other minerals that can be accurately distinquished by either the X-ray diffraction (XRD) method or the quantitative evaluation of minerals by scanning electron microscopy (QEMSCAN). The interparticle and intraparticle micropores of the shale matrix as well as nanopores of kerogen (if present) could not be resolved at this scale of the micro-CT.

A 3D volume made up of a series of reconstructed CT images can be obtained with various processing softwares (PerGeos, Image J, and Mango, among others) to visualize and quantify the distribution of minerals and fractures, as well as to create a pore network model in the presence of resolved pores (Plessis et al., 2017; Lemmens & Richards, 2013; Kalender, 2006).

4.2.2 Sample Preparation

Most rock samples can be scanned as received in X-ray micro-CT without much sample preparation. Samples are presented either as cylindrical cores (≥1 mm in diameter and ≥3 mm in height) cut along directions perpendicular to and parallel to the bedding plane (Zhou et al., 2018) or microchips (≥1 mm thick) cut from cylindrical cores. Depending on the type of experiment conducted with the X-ray micro-CT, some microchips are embedded in epoxy for stability since loading the sample to avoid movement during scanning is crucial to minimize the occurrence of artifacts (errors). Figure 4.6 shows different means of sample preparation for X-ray micro-CT imaging: (a) shows a core plug of Chattanooga shale with diameter of ~1 inch; (b) shows a microchip of ~1.5 mm thickness (red square) cut from the Chattanooga core plug and mounted on a sample holder with carbon tape; (c) shows five microplugs (~4 mm diameter) of Longmaxi shale cores and a microchip (yellow arrow) (Zhou et al., 2018); (d) shows a thin section of

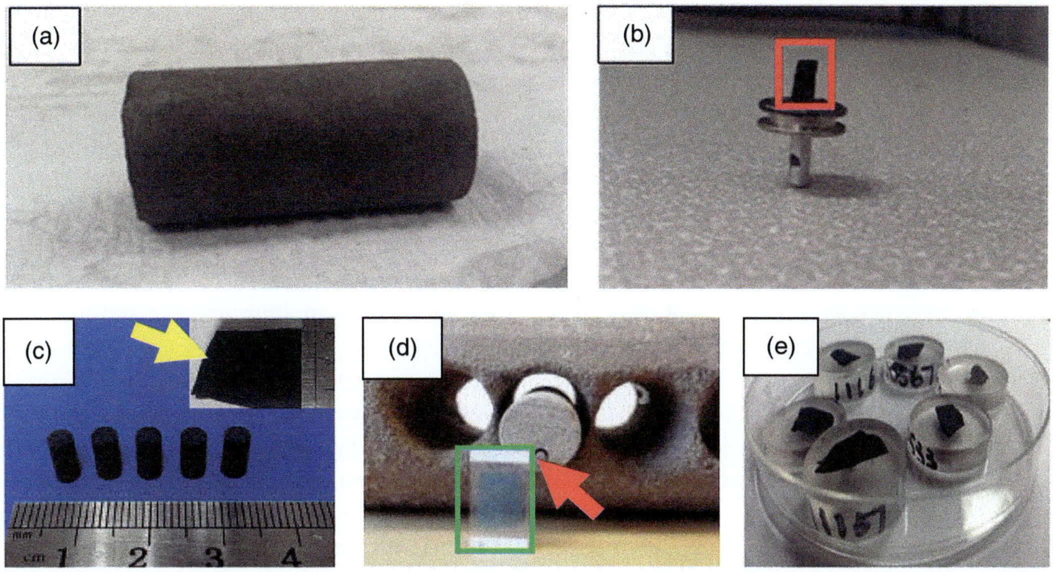

Figure 4.6 (a) Chattanooga whole core (~1 in. diameter), (b) Chattanooga microchip (~1.5 mm thick), (c) five microplugs (~4 mm diameter by 8 mm height) with microchip pointed out by the yellow arrow (Zhou et al., 2018), (d) thin section enclosed in green rectangular box, microplug with red arrow within a whole core (Olson 2015), (e) shale microchips embedded in epoxy (Liu et al., 2016).

Figure 4.7 Different sample holders for X-ray micro-CT scanning (Hsieh, 2012).

rock sample held onto a glass surface with epoxy (green box), and microplug (red arrow) cut from a whole core (Olson, 2015); and (e) shows Bakken shale samples embedded in epoxy (Li et al., 2019). However, epoxy and thick tube containers are not recommended since epoxy has an attenuation coefficient similar to quartz and thick tubes absorb X-ray photons (Hsieh, 2012). Thus, shale samples are recommended to be scanned as either miniplugs (diameter ~1.5 mm) or as rock chips cut from whole cores. Figure 4.7 shows multiple adjustable sample holders for different types of samples, which hold tube or rock chips on the sample stage and keep the sample perpendicular during scanning (Hsieh, 2012). Microchips are usually mounted on less-dense aluminum or glass rods varying in size (thin to thick), secured with glue or double-sided tapes to be placed on a platform. For most samples placed in core holders, the core holder can be placed directly on the platform within the scanner if it is made of aluminum, fiberglass, peek, or titanium. More important, the sample holders should be such that they hold the sample in place, but also separate the sample from the more dense rotating stage (Plessis et al. 2017; Kalender, 2006).

Samples are recommended to be loaded at a slight angle since samples loaded at parallel to the X-ray beam lead to artifacts and sample surfaces not being properly penetrated by the beam (Wildenschild & Sheppard, 2013; Brunke, 2010). In the geoscience industry, most rocks are scanned either as dry/cleaned cores or as fluid-satured samples. In the case of fluid-saturated samples, the saturated samples are dried first then wrapped in plastic film (parafilm) to keep the sample moist and avoid evaporation. Moroever, saturated fluids could be doped with contrasting fluids like iodine and xenon to increase contrast of a fluid from their surroundings (Fogden et al., 2014).

4.2.3 X-ray Micro-CT Scanning Procedure

Sample size plays a major role in image resolution (the detail an image holds) as such careful selection of sample size is required for X-ray MCT scans. For a sample of known dimensions, a useful guideline is that the optimal resolution is a factor of 1,000x smaller than the width of the sample. However, this guideline is based on a standard practice of making use of only 1,000 of the 2,000 available pixels of the detector to curtail edge effects (Plessis et al., 2017; Kalender, 2006; Brunke, 2010). The voxel size of a CT image is dependent on both the magnification and size of the object. But regardless of a good voxel size, a CT image could still have a lower resolution. In so doing, the size of the X-ray spot is restrained to the required voxel size in most systems (Plessis et al., 2017; Schoeman et al., 2016a,b).

On the other hand, the aquistion time of single projection images varies from system to system and also is based on the size of the sample and the differences in distance from source to detector (Singhal et al., 2013). For example, the distance between the Chattanooga sample and the X-ray source, as well as the detector was recorded to be 19 mm and 7 mm, respectively. As such, a total image size of 948 x 995 x 974 voxels (2-GB file) of the Chattanooga shale sample was acquired after 8 hours of scanning and, subsequently, processing. While, the acquistion time varies for different equipment designs, a typical acquisition time for a walk-in cabinet system like the Xradia MicroXCT-400 with a 16-bit flat-panel detector is 500 ms per image (*Zeiss*, 2013). Moreover, image contrast is enhanced by increasing the image acquisition time to near saturation of the detector, as very low acquisition time results in poor contrast (Plessis et al., 2017). The CT image of the Chattanooga shale sample in Figure 4.5 was acquired with a 20x magnification lens at a field of view (FOV) of 1.3 mm after processing with a proprietary software into a 16-bit unsigned grayscale image. In addition to the the 20x magnification lens, the Xradia MicroXCT-400 equipment has other magnification lenses (0.5x, 4x, 10x, and 40x) as shown in Figure 4.8. Xradia's micro-CT has a four megapixel (2,048 x 2,048), 16-bit digital CCD camera incoporated with five microscope objective lenses. The X-ray lens converts the incoming X-rays into optical light and magnifies the field of view (FOV) (Hsieh, 2012). Only one lens is applied at a time for a single scan and each of the outlined lenses for the Xradia MicroXCT-400 has its own magnification scale and resolution (Hsieh, 2012; Kalender, 2006). Table 4.1 lists a range of resolutions, FOV, and source/detector distance for the different lenses (Boas & Fleischmann, 2012; Hsieh, 2012; Kalender, 2006). In order to acquire a high-resolution image, a smaller field of view (FOV) together with a lower magnification is compromised on. For example, the 0.5x lens produces the largest FOV yet its image resolution is the least of all the other lenses in Table 4.1. The higher the magnification, and therefore number of pixels used on the detector, the larger the number of images required for a good reconstruction. A useful guideline is that the number of pixels covered by the sample on the detector in width (pixels) multipled by 1.6 equals the number of projection steps required (Plessis et al., 2017).

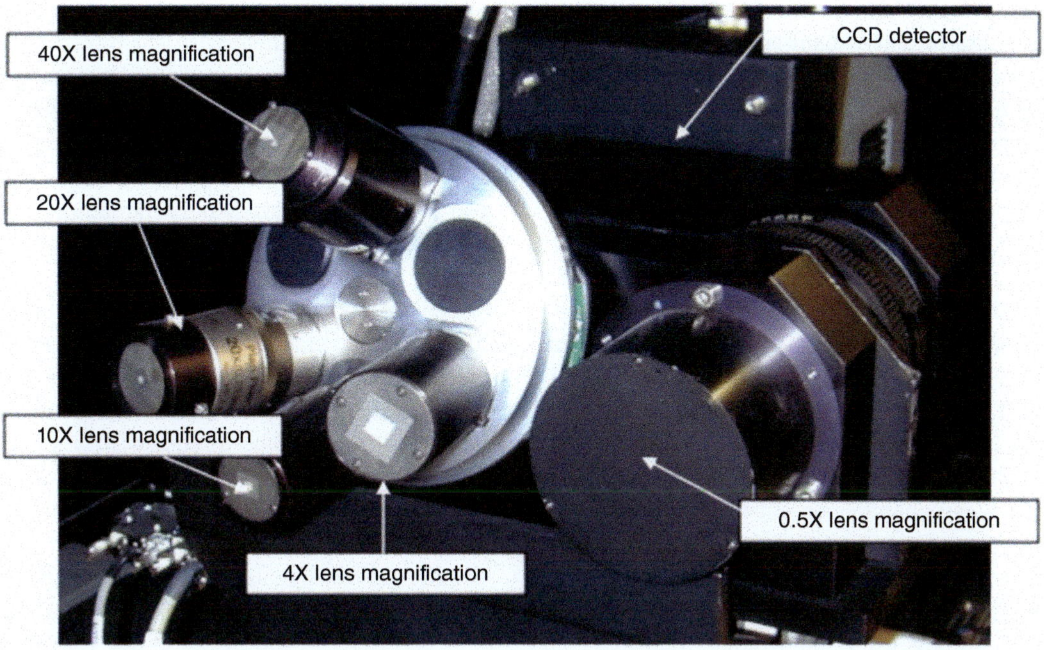

Figure 4.8 A set of magnification lenses and a detector of the XRadia MicroCT-400 system (Hsieh, 2012).

Table 4.1 A List of XRadia's Magnification Lens With Image Resolution, FOV, and Distance

Lens	Spatial resolution	Field of view (FOV)
0.5x	~30–50 microns	24 mm
4x	~5 microns	6 mm
10x	~3 microns	2.4 mm
20x	~1.5 microns	1.2 mm
40x	~1.0 microns	0.6 mm

Source: Hsieh (2012); Xradia (2010).

Scanning parameters include applied voltage and filter selection. The X-ray voltage depends on the material composition and type of material. Therefore, the X-ray source voltage determines the linear attenuation coefficient (μ), which is a function of energy and atomic number through Beer's law (Plessis et al., 2017; Zhang et al., 2012; Kalender, 2006). While, higher X-ray energy levels are required for materials with higher atomic number, low- density and low atomic materials transmit too many photons and saturate the dector at higher voltages (Hsieh, 2012). Therefore, samples like shales, which possess a mixture of materials with both high and low atomic numbers, require an appropriate energy level, strategically selected for the source. The X-ray source is mostly stabilized after heating to generate photons by cooling down to -60°C for about 15 min (Hsieh, 2012). On the other hand, a filter is commonly selected to reduce the polychromacity of the X-ray beam from the source, which in effect reduces the artifacts; this process is also referred to as beam filtering (Plessis et al., 2017). In filter selection, the generated X-rays pass through a chosen filter based on a calculated transmission value. The transmission value is obtained from the ratio of the image with the sample in view to the image without the sample in view (reference image). The Chattanooga shale sample was scanned with a low-energy filter (LE#3) for a calculated transmission value in the range of 0.45 to 0.50 for the 20x lens. Moreover, a filter could be placed between the detector and the sample, mostly in the case of dense materials and/or nanostructured samples with large amounts of scattering (detector filtration), which is also used to protect the detector from low-energy X-ray emissions and minimize noise. The general guideline for voltage selection for small rocks is between 60 and 150 kV and 160 and 240 kV for large rocks (Plessis et al., 2017). A voltage of 90 kV was used for the Chattanooga shale sample.

Before scanning with the micro-CT, the background of the resulting image is normalized by creating a reference image to fit all intensity variations across the detector. This is carried out by removing the sample from the X-ray chamber or out of view, and using the X-ray beam from the source at the set parameters to acquire reference image(s) with corrected intensity. Basic steps for MicroXCT tomography from the Xradia MicroXCT-400 manual are listed as (1) Find the region of interest (ROI) at low magnification; (2) fine-tune ROI at high magnification; (3) set source and detector positions; (4) select proper source filter and current (kV); (5) determine proper acquisition parameters (number of projections, acquisition time,etc.); and (6) set up recipe and start tomography.

CT images (radiographs) are then acquired through automatic scanning after the necessary measures have been put in place. However, frequent supervision is recommended for source instability or filament burn errors that may occur during the process. For shorter scanning time, a reduction in the exposure time is recommended (Plessis et al., 2017; Schoeman et al., 2016; Kalender, 2006). The exposure time for the Chattanooga shale sample was 7 s.

Three major artifacts reported in the literature for micro-CT imaging are: beam hardening, the cone-beam effect and phase-contrast (Schoeman et al., 2016; Singhal et al., 2013). Beam hardening occurs when X-ray beam encounters differences in absorption from different angles and along different paths through the object due to either a dense object or a dense part of the object. Cone-beam artifacts affect the edges of materials near the edges of the detector and can be corrected by the use of less magnification to fill fewer pixels on detector (Plessis et al., 2017). In addition to the three major effects, ring artifacts and noise also occur due to movement of the sample during acquisition, and statistical error of photon counts, respectively (Schoeman et al., 2016; Cnudde & Boone, 2013; Boas & Fleischmann, 2012). These artifacts can be corrected to some extent using preprocessing median and nonlocal means filters in most visualization software tools, which would be discussed in a subsequent subsection.

The Heliscan micro-CT is specifically designed for digital rock applications and delivers the highest signal-to-noise ratio compared with the circular micro-CT. It has a spatial resolution of 800 nm with X-ray source voltage in the range of 20 to 160 kV. The maximum image size acquired on the X-ray detector is 3,072 x 3,072 pixels; a sample diameter varies up to 240 nm (ThermoFisher Scientific, 2018).

4.2.4 Image Reconstruction

A 3D reconstructed volume of the acquired CT images is obtained by mapping each voxel with projection image representations of a particular voxel from many angles, while making use of the mathematical filtered back-projection (FBP) algorithm, also referred to as the reverse of the Fourier transform (Plessis et al., 2017; Hsieh, 2012) as shown in Figure 4.9. In the FBP algorithm, CT images are computationally back projected into sample space and a high pass filter is applied to remove low-frequency blurring. Typically, 400 to 900 projections are obtained for high resolution (Babu et al., 2015).

Linear attenuation coefficient (μ) associated with reconstructed CT image is displayed as a CT number (CTN). CT number (CTN) is defined as a normalized value obtained from converting the attenuation coefficients of a reconstructed voxel by calibrating with

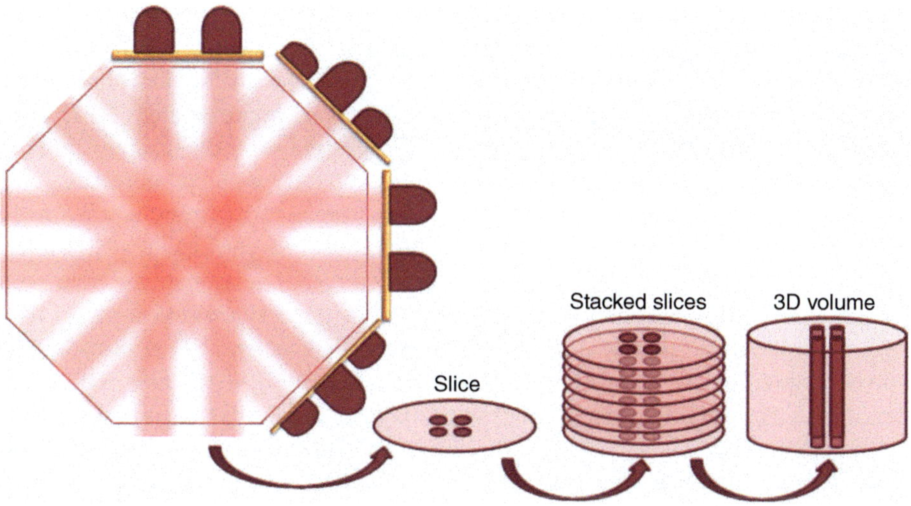

Figure 4.9 Filtered back projection reconstruction (Litster et al., 2015).

reference to water. It is measured in Hounsfield unit (HU) scale, which is a linear conversion of the linear attenuation coefficient into one in which the radiodensity of distilled water at standard pressure and temperature (STP) is defined as zero (0) HU, and that of air at STP is defined as -1,000 HU. The Hounsfield unit (HU) is given as (Goldman, 2007)

$$HU = 1,000\frac{\mu - \mu_{water}}{\mu_{water} - \mu_{air}}, \tag{4.2}$$

where, μ_{water} and μ_{air} are the linear attenuation coefficients of water and air, respectively.

The CTN can be used to estimate petrophysical properties such as porosity and saturation of a rock sample, which is discussed in section 4.2.6 regarding visualization and processing of CT images. The Xradia MicroXCT-400 system among other commercial systems has built-in reconstruction software (such as XMReconstructor for Xradia MicroXCT-400; www.zeiss.com) with different settings but based on the same algorithms (Xradia, 2010). The output file after reconstruction is usually a 16-bit file but an 8-bit file can also be selected if storage space is limited. Output files come in *.txm* or *.tiff* images, and *.dat* or *.nc* files for Heliscan CT, which can be opened in any visualization software for further analysis. The tomography file is not saved in a geometry of vectors but rather in a geometry of voxels. A single projection image in 2D is captured by the detector and visualized by a proprietary software of the system and saved with the extension *.xrm*. The *.xrm* file contains coordinate information, source voltages, exposure time, and angles, among others. On the other hand, information on projection images are saved with the extension *.txrm*. The *.txrm* data set can be reconstructed by a proprietary software of the system to form 2D slices oriented in the XZ-plane and stored with the extension *.txm*. In addition, the *.txm* file can be exported as an 8-bit tiff, bitmap, jpeg, 16-bit raw tiff, or binary file (.bin).

The rotation axis in each projection is determined by locating the central pixels in all 2D radiographs as it improves the quality of reconstruction. Guidelines for image reconstruction in the XMReconstructor software for the Xradia MicroXCT-400 system include the following:

1) Use the nondrifter version of the tomography acquisition file.
2) Check for the "center shift" with a pixel value range. Example: start at -10 and end at 10.

3) Each pixel number would have a reconstructed image.
4) Scroll through the images to find the center shift. If no such shift is observed, then the center shift is outside the specified range and the pixel range must be widened. This often occurs in the case of a high-magnification lens.
5) Once the shift is found, double check by making the pixel range and step size smaller. For example, if the shift is at 3, assign the pixel range from 2 to 4 and then make the step size 0.25. This will give four images for each pixel.
6) Record the pixel number in the center shift interface that corresponds to the center shift.
7) Set the reconstruction settings by entering the pixel number for the center shift, choosing the correction option for "bright and dark spots," and then choosing a kernel size between 0.5 and 0.7.
8) Begin the reconstruction process by clicking on its start icon. It usually takes between 20 to 30 min.

As a result of rotational movement, the imaged volume appears as a cylinder (Fig. 4.2) with a diameter dependent on the cone angle of the X-ray beam, the number of detector pixels, and the geometrical magnification (Cnudde & Boone, 2013). The sample size of Chattanooga shale sample used in the X-ray micro-CT imaging was 1.2 mm (thick), which resulted in a spatial resolution of 1 μm/voxel after image processing.

4.2.5 Application of X-ray Micro-CT on Shale Samples

Chattanooga shale (CS) is the name given to Upper Devonian-Mississippian black shales distributed in the Appalachian and Ozark regions with varying thicknesses and not necessarily in communication with each other (Conant & Swanson, 1961). The Chattanooga shale sample used in this application is from #1-34 Spriggs drilled by Metro Energy group in 2005 to investigate shale gas production located at a depth of approximately 4,900 ft (Newell & Doveton, 2010). The Chattanooga shale interval has a thickness of approximately 50 ft and records high gamma ray (GR) readings in the range of 120 to 300 API with no information on the uranium content accompanying this reading. The high GR counts correspond to high total organic carbon (TOC) measurements from source rock analysis of 10 sidewall cores from the Spriggs well. The average TOC is measured to be 3.2 wt.%, matching a vitrinite reflectance (%R_O) maturity level of 0.77 (gas window) at a maximum temperature (Tmax) of 440°C. Therefore, the high TOC content (~3.2 wt.%) coupled with a reasonable formation thickness (~50 ft) presents a potential formation that is worth investigating.

Figure 4.10 shows a 3D volume of a Chattanooga shale sample cut with thickness of ~1.2 mm and imaged at a resolution of 1 μm/voxel with the Xradia micro-CT. The resulting 3D volume is created from the 2D reconstructed CT images (tomographs) using processing and analysis software such as Image J, Avizo, and/or PerGeos. The X-ray micro-CT could not resolve the micropores and nanopores for the sample at 1 μm/voxel size resolution. As a result, the only analysis that can be carried out through rock digital physics is the estimation of the volume fraction of the different phases (organic matter, pyrite, etc.) observed on the grayscale image for this particular sample. Micro-CT will be used during the process of upscaling of rock properties from images with nanometer resolution to images with micrometer to centimeter resolution.

Saif et al. (2017) also applied the X-ray micro-CT to characterize the heterogeneity and mineral distribution of an immature oil shale sample from the Green River (Mahogany zone) formation in Utah. The oil shale sample of size 1 mm diameter by 5 mm length was cut and prepared normal to the bedding for X-ray micro-CT imaging. Figure 4.11 shows a 2D CT image of the oil shale sample at 0.8 μm/voxel resolution. On the grayscale image

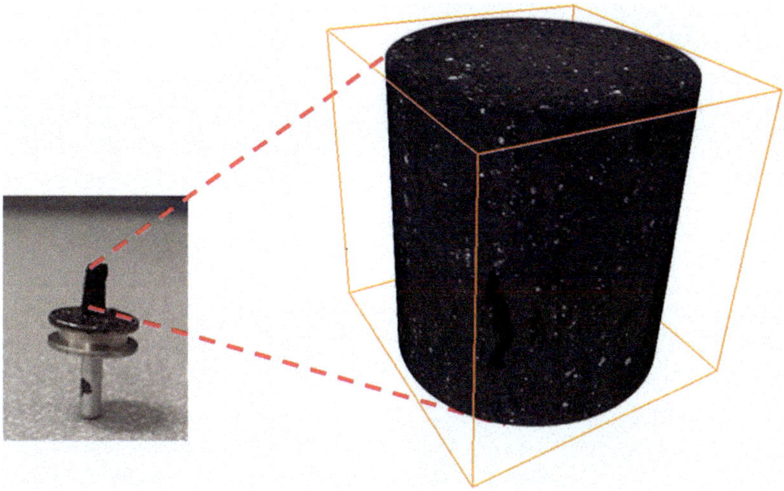

Figure 4.10 3D volume of Chattanooga shale sample in grayscale. Resolution is 1 μm/voxel. White = pyrite, gray = mineral matrix, and black = organic matter.

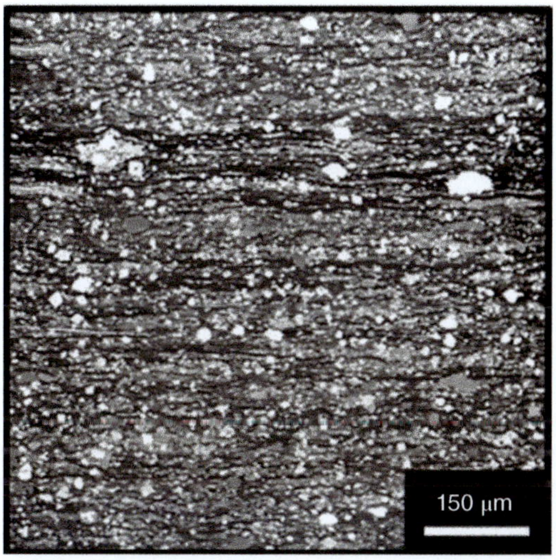

Figure 4.11 2D micro-CT image of immature Green River (Mahogany zone) oil shale sample at 0.8 μm voxel size resolution (Saif et al., 2017).

below, denser materials appear brighter, while less dense materials like organic matter (kerogen) appear darker. However, other materials with densities between the two extremes (highest and lowest) appear gray.

Another example of the use of the X-ray micro-CT for core plugs is the use of a cylindrical plug of North West Shelf shale (Australia) acquired with a medical CT scanner operating at 120 kV (Josh et al., 2012). Figure 4.12 shows a grayscale volume of a North West Shelf Shale after geomechanical testing. Although, the mechanical testing would not be discussed here, the phase contrast due to the density variation of the different minerals is observed. Unlike the previous shale samples, where organic matter (kerogen) was present, and identified as darker spots, the image below shows little or no signs of kerogen but the presence of higher density minerals (pyrite, carbonate, etc.), and gray mineral matrix.

In addition to viewing the mineral distribution and heterogeneous pore space with the use of X-ray micro-CT, microfractures can also be investigated in the process. Both naturally occurring fractures and hydraulically induced fractures in shales enhance the storage and propagation of reservoir fluids in ultra-tight shale formations. As such several investigations have been carried out to study the mechanical properties and fracture propagation in shales making use of the X-ray micro-CT (Liu et al., 2018; Zhou et al., 2018).

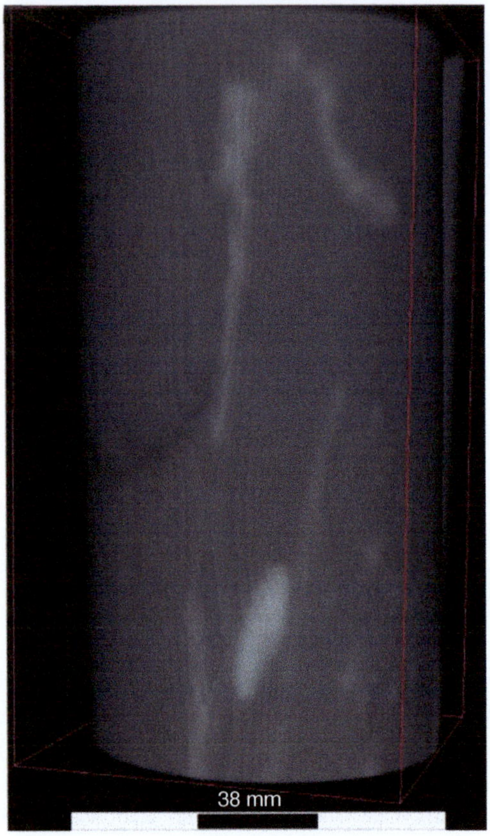

Figure 4.12 3D volume of a North West Shelf sample from X-ray medical CT scanner (Josh et al., 2012).

The resulting CT images show empty space fractures as black lines due to density close to zero, as such fractures show the lowest intensity and hence appear the darkest just as pores (Hampton et al., 2014). Figure 4.13 shows a fracture propagation induced in a Longmaxi shale sample (4 mm in diameter by 8 mm in height) at a load stress (σ_1) of 103.52 MPa to measure its mechanical properties at different bedding orientations (Zhou et al., 2018).

Moreover, CT scans were obtained from Bakken, Marcellus, and Woodford shale subcores (~3 mm) to investigate the internal structures (Verba et al., 2016). Figure 4.14 shows the acquired CT images of the Bakken, Marcellus, and Woodford shale samples, respectively. Since the micropores and nanopores could not be resolved with the micro-CT's resolution, dark spots in the CT images are interpreted as organic matter (kerogen) because of their less-dense nature compared with other minerals present. However, microfractures are represented as dark lines oriented along the bedding plane of all three shale samples (Verba et al., 2016). Pyrite appears brighter due to its high density, while the color gray represents other matrix minerals with densities either higher than that of kerogen or lower than that of pyrite.

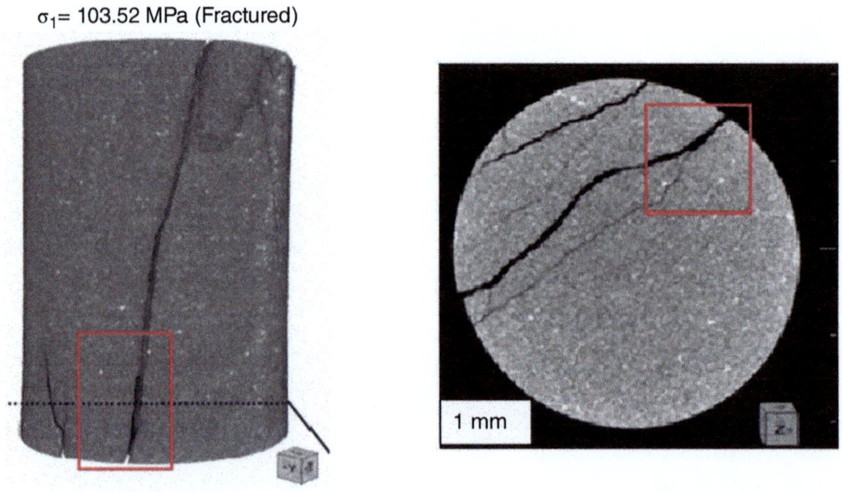

Figure 4.13 3D view (left) and 2D view (right) CT images of Longmaxi shale sample showing induced fractures as dark lines (Zhou et al., 2018).

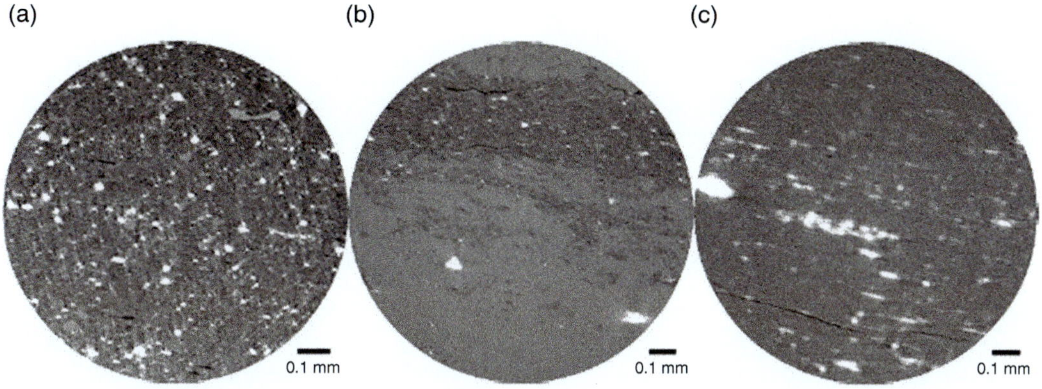

Figure 4.14 CT images of the (a) Bakken, (b) Marcellus, and (c) Woodford shale samples (Verba et al., 2016).

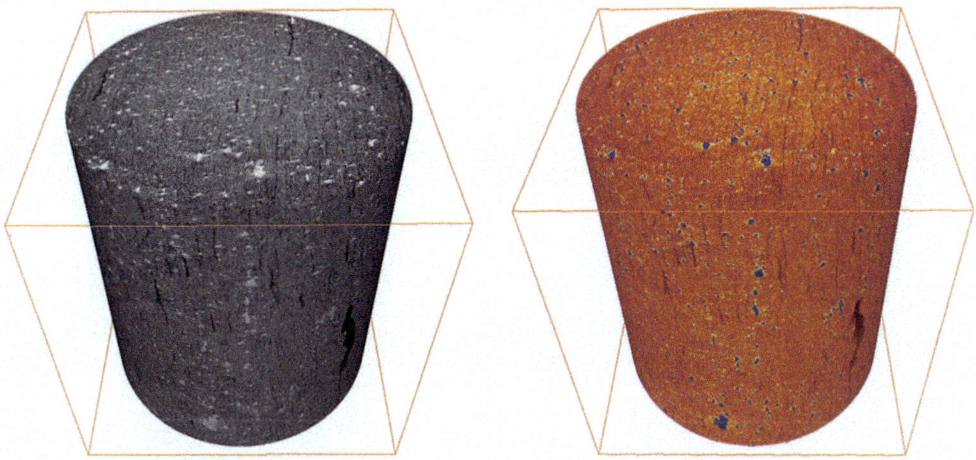

Figure 4.15 3D rendered volumes: (left) grayscale image and (right) colored mapscale. Black/red = organic matter, white/blue = pyrite, and gray/orange = matrix.

The X-ray micro-CT operates on a larger volume than electron microscopy, as such its resolution is not refined enough to directly visualize pores (micro- and nano-) in shales.

4.2.6 Image Visualization and Processing

The reconstructed file (*.txm*) or image slices (*.tiff*) can be loaded into proprietary visualization software (ImageJ, Mango, Avizo Fire, and/or PerGeos) for further processing and analysis (see Figure 4.19). Most visualization softwares automatically assign a data category, matrix and voxel dimensions, as well as an option to enter or change assigned coordinate units of the loaded data. Imported data can be visualized as either 3D rendered volume (Fig. 4.15) or 2D orthogonal slices (Fig. 4.16) in different planes. Volume rendering is a 3D view, which involves user-defined grayscale gradient or threshold value. Three main phases with varying intensities were identified in the CT images as shown in Figure 4.18. The lowest

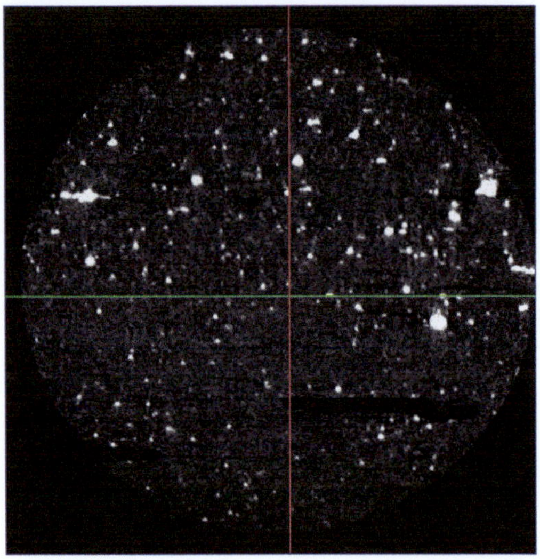

Figure 4.16 2D single slice (XY) of a Chattanooga shale sample.

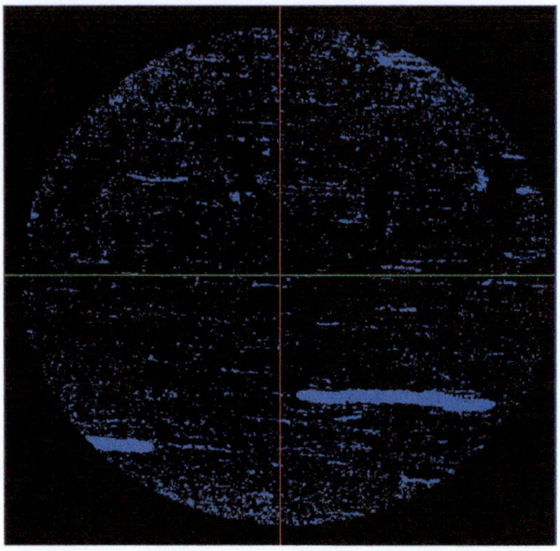

Figure 4.17 Binary (segmented) image of a Chattanooga shale sample; selected organic matter in blue and nonselected regions in black.

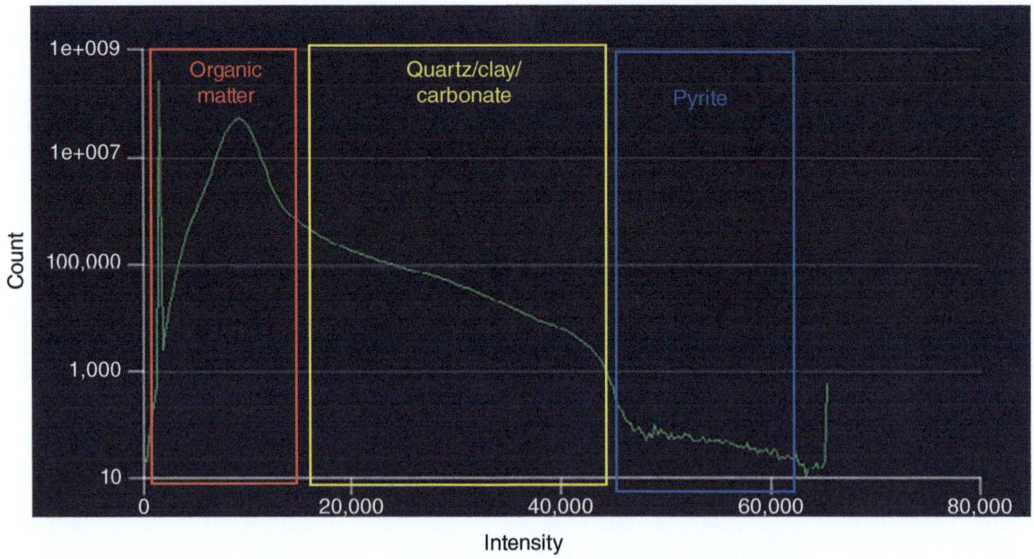

Figure 4.18 Histogram of the intensity range of the different mineral compositions.

intensity area under the curve represents the less-dense organic matter (OM), while the highest intensity represents the most-dense pyrite mineral. The area in between the lowest and highest intensities under the curve represents other mineral phases (quartz/clay/carbonate) with densities between the most-dense and less-dense minerals. Thus, the visible internal features of a sample are segregated into individual groups based on their intensities or CT numbers. The CT images of the Chattanooga shale sample are at 1 μm/voxel resolution with volume dimensions of 948 x 995 x 974 voxels, grayscale 16-bit unsigned data.

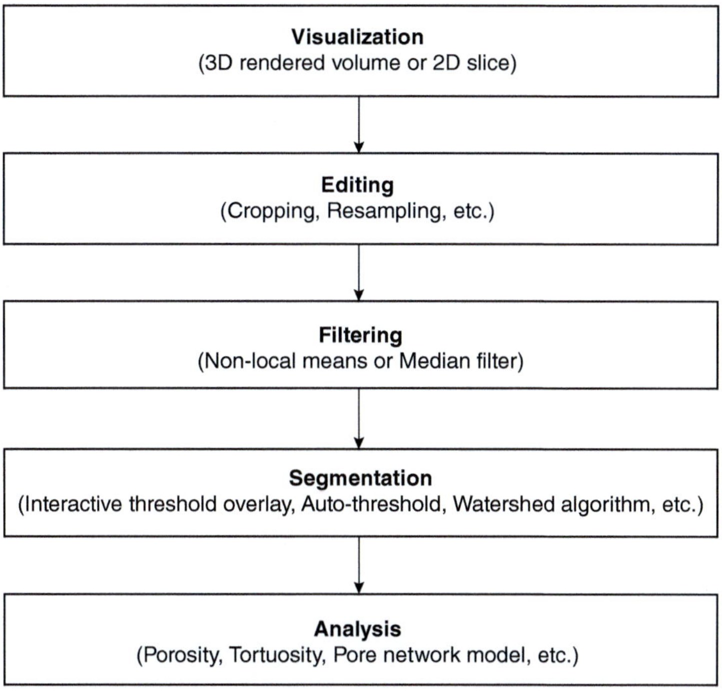

Figure 4.19 Flowchart outlining the various processes a reconstructed CT image should undergo once loaded in a visualization software (e.g., PerGeos, Avizo, ImageJ, Mango, etc.). The chart starts off with preprocessing steps, such as filtering and editing, and ends with generating a pore network model (PNM) for permeability estimation and fluid flow simulation.

A 16-bit detector is able to identify color depth levels from 0 to 65,535 pixels in an unsigned short format; as such the maximum reading of X-ray intensity cannot exceed 65,535 pixels. Moreover, a CT number (CTN) can be assigned in this range in line with the light intensity measured on the detector. The measured intensity can be expressed according to Beer's law, where an emerging photon intensity (I) of a monochromatic beam with energy (E) and incident photon intensity (I_0), upon passing through different materials that absorb X-ray photons with thickness (x) is expressed as equation (4.3) (McCullough, 1975).

$$I = I_0 \exp(-\mu(\rho, Z, E)x), \tag{4.3}$$

where μ is the linear attenuation coefficient, dependent on density (ρ) and atomic number (Z) and defined as equation (4.4) (McCullough, 1975).

$$\mu = \rho\left(a + \frac{bZ^{3.8}}{E^{3.2}}\right), \tag{4.4}$$

where a = Klen-Nishina coefficient (nearly energy independent coefficient), b=a constant, and E is the energy of a monochromatic beam.

The 3D rendered volume or the 2D slices as part of preprocessing can be edited by cropping to remove edge artifacts and to also obtain a desirable size or focus on a region of interest (ROI) to reduce processing time. Another preprocessing step is filtering to remove noise and other artifacts. To accurately estimate petrophysical properties of the CT images through digital rock analysis, pores and mineral phases are segmented. Segmentation involves the segregation of the gray-level voxels of the CT images into distinct phases by assigning user-defined threshold values to pixels belonging to internal features based on their representative intensities or CT numbers (gray color or brightness) into individual groups (Adeleye & Akanji, 2017). As a result, grayscale images are converted to binary images, where selected phases (pores or organic matter) are assigned the value of 1 (represented as blue) and nonselected features are assigned the value of zero as shown in Figure 4.17. The resulting binary (labeled) image can be used for quantitative analysis such as volume fraction estimation of pores (porosity) or minerals, pore and throat size distribution, for permeability calculation. Finally, a 3D pore network model (PNM) can be constructed for computational fluid flow purposes.

Regardless of the spatial resolution obtained from CT imaging of different shale sample sizes, the question remains whether the acquired images would adequately represent the microstructural characteristics of the shale sample. In so doing, a representative elementary volume (REV) is defined and considered a good representative volume to capture heterogeneity. The representative elementary volume (REV) is defined as the minimum volume of the sample field large enough to capture a representative amount of heterogeneity but varies for different internal features (Lin et *al.*, 2010) (Table 4.2).

Filtering

The filtering process transforms an image to highlight a specific feature of its structure by reducing contrast and noise. To estimate a signal mixed with noise as part of the filtering theory, different solutions are implemented based on the proprietary software in question. For example, the PerGeos software makes use of the moving average solution, where the value of each pixel is replaced by the average of its neighbors based on the assumption that small variations result from noise. The filtered output can then be viewed as the main trend of the function. Both 3D median filters and nonlocal means filters are commonly applied to CT images but the nonlocal means filter is preferred, since it preserves edges, while removing noise, hence, time consuming. Figure 4.20 shows an unfiltered image of the Chattanooga shale sample (left) and its corresponding filtered (nonlocal means filter) image on the right. The filtered image on the right shows less noise (smoother) or less coarse background compared with the unfiltered image on the left. While at it, an intensity correction tool can also be applied to compensate for nonuniform illumination so as to make way for image segmentation.

Table 4.2 Imaging Scale and Corresponding Representative Elementary Volume (REV)

Scale	Phase	REV
Macroscale	Fractures	9 mm
Mesoscale	Granular minerals	400 μm
Microscale	Organic matter/clay minerals	40 μm
Nanoscale	Pores	6 μm

Source: Lin et al. (2010).

Segmentation

Figure 4.22 shows a segmented image of the Chattanooga shale sample highlighting the different mineral distributions, from kerogen (brown) to pyrite (green) and matrix (turquoise), while Figure 4.23 shows a segmented slice of the Green River oil shale with organic matter (kerogen) in blue and a 3D rendered volume with resolved organic matter in blue and mineral matrix in gray (Saif et al., 2017). The segmentation step involves the extraction of pore spaces or minerals at a given threshold value, which corresponds to the CT image resolution. Particle segmentation is based on 3D watershed algorithms, where an image is thought of as a geological landscape made up of catchment basins and watershed lines. The watershed lines determine boundaries, which separate regions within an image. Bright areas are represented as peaks, while dark areas appear as troughs (Ruparelia, 2012). Upon

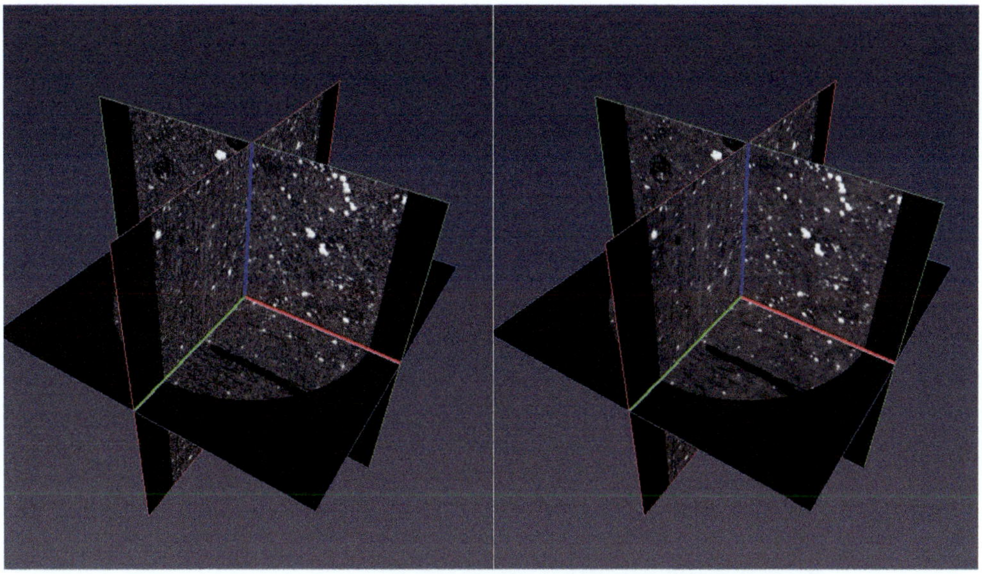

Figure 4.20 Unfiltered (left) versus filtered (right) images of the Chattanooga shale sample.

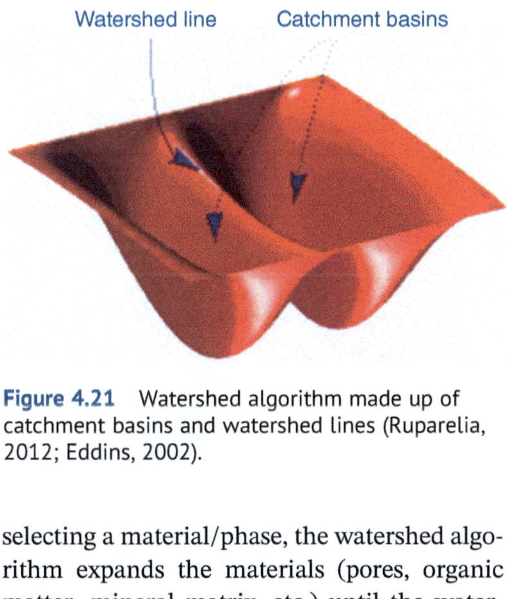

Figure 4.21 Watershed algorithm made up of catchment basins and watershed lines (Ruparelia, 2012; Eddins, 2002).

selecting a material/phase, the watershed algorithm expands the materials (pores, organic matter, mineral matrix, etc.) until the watershed line is reached. Subsequently, catchment basins and watershed lines for any grayscale image are computed (Eddins, 2002). Two input images, a label (segmented image) and a grayscale image, are required to serve as seed areas for flooding and catchment basin, respectively. Figure 4.21 shows an example of a watershed line and catchment basin with two minima points. Each regional minimum is imagined to be flooded with water from the bottom at a constant rate for the flooding based watershed algorithms. In effect, the water level will rise uniformly in different catchment basins but is prevented from merging with nearby catchment basins in the form of the watershed lines (Ruparelia, 2012). Thus, watershed algorithm changes ones image into another image whose catchment basins are the phases/materials to identify (Eddins, 2002).

Different threshold for binarization of CT images shows that the X-ray micro-CT, although, unable to resolve micro- and nanopores, can be used after proper calibration for upscaling from micro-/nano-scale to macroscale. (Adeleye & Akanji, 2017).

Analysis

Unlike the statistical models, recent advances in X-ray micro-CT enables image analysis for complex 3D characterization. Porosity, pore-size distribution, pore geometry, and fracture intensity are among the important properties determined through characterization of organic-rich shales since they influence the

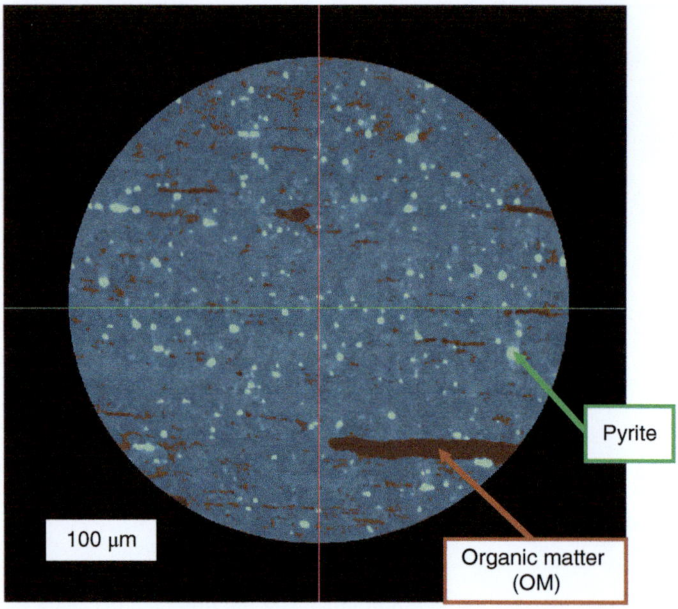

Figure 4.22 Labeled image with organic matter (brown), quartz/clay/carbonate (turquoise), and pyrite mineral (green).

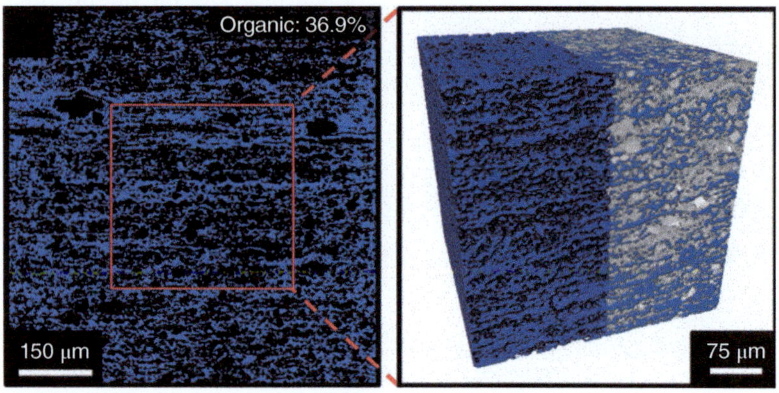

Figure 4.23 Segmented slice (left) with OM in blue and 3D rendered volume with OM in blue and mineral matrix in gray (Saif et al., 2017)

permeability of fluid transport through porous media. The labeled image acquired through segmentation in Figure 4.22 is processed through pore space analysis. From the 2D slice of the Chattanooga shale (Figure 4.16), it is observed that the background (dark) contributes to the volume of organic matter (also dark) to some extent; as such, the background is corrected and subtracted from the CT image. Therefore, the estimated volume percentage of labeled organic matter (OM) within the CT image of Chattanooga shale is 6.39%. Moreover, the estimated volume fraction of OM through imaging analysis can be converted to equivalent total organic carbon (TOC) as equation (4.5) (Peng et al., 2014),

$$TOC = \left(\frac{f_{OM + pores} - \varphi}{1 - \varphi}\right)\left(\frac{\rho_{OM}}{\rho_R}\right), \tag{4.5}$$

where $f_{OM + pores}$ is the fraction of organic matter and pores, φ is the organic porosity, ρ_{OM} is the density of OM, and ρ_R is the bulk density of the shale rock. The density of OM (kerogen) is given as 1.3 g/cc (Kelly et al., 2016) and the bulk density of the Chattanooga shale from density log is ~2.56 g/cc. This estimates the equivalent TOC to be 3.25% without organic pores.

On the other hand, LECO CR-412 TOC analysis of the Chattanooga shale samples from the Spriggs well, Barber County, Kansas, reported the total organic carbon (TOC) to be 3.2 wt. %, which is close to the estimated TOC from image analysis. LECO CR-412 is a carbon analyzer used to determine the total organic carbon (TOC) content in crushed, isolated, and acidified organic sediments (Gonzalez et al. 2013; Jarvie, 2012; Bernard et al., 2004).

4.2.7 Estimating Porosity from CT number (CTN) of CT Images

Aside from rock digital analysis, correlations can also be used to estimate petrophysical properties from either a single CT scan or multiple scans making use of their CT numbers (CTNs). Having acquired the relative attenuation coefficient values for each pixel after image reconstruction, the attenuation coefficients are next converted into corresponding CT numbers by normalizing with the linear attenuation coefficient of water (Akin & Kovscek, 2003). Thus, CT numbers are assigned according to mineral attenuation coefficient, and distributed in the range of highest to lowest readings based on a global scale. A global scale assigns CT number of each voxel depending on the maximum and minimum attenuation coefficients, making it easier to distinguish between different mineral phases (Aljamaan et al., 2017; Bryant et al., 2012). As previously mentioned, the resolution of the X-ray micro-CT systems makes it practically impossible to visualize micropores/nanopores. Therefore, porosity and in-situ saturation of CT images can be estimated locally once CT numbers are measured (Akin & Kovscek 2003). A standard CT number is measured in Hounsfield units (HU) defined in equation (4.2) (Aljamaan et al., 2017; Josh et al., 2012; Goldman, 2007). Each Hounsfield unit represents a 0.1% change in density with respect to calibration density scale (Akin & Kovscek, 2003). In effect, materials other than water have energy and atomic number-dependent CT numbers as given in equation (4.4). Gases like krypton (Kr) provide sufficient contrast when introduced into a rock sample for storage or porosity estimation compared with carbon dioxide (CO_2). In addition to the normalized and known CT numbers of water and air, Aljamaan et al.(2017) established the CT numbers of pure CO_2 and Kr as a function of pressure and constant room temperature of 23.8°C inside a core holder without the sample, shown in Figure 4.24. It is observed that injecting CO_2 at a higher pressure causes an increase in gas density to obtain a greater attenuation coefficient in relation to equation (4.4), while Kr records high attenuation coefficient even at low pressures. The CT values were acquired at an X-ray tube energy of 140 kV and a current of 120 mA (Aljamaan et al., 2017).

Furthermore, CT numbers for common materials are listed in Table 4.3. Most of the listed CT values were acquired at different currents of the X-ray system and different fields of view (FOV) of the material in question (Aljamaan et al., 2017).

Therefore, to estimate the porosity of a rock sample through CTN empirical correlations, a single scan or dual and/or multiple scans can be performed on the sample in question. In the case of a single scan, the porosity at each location is determined directly through pixel by pixel subtraction of the dry CT image from the fluid-saturated CT image and dividing by a calibration constant given below (Saadat et al., 2011),

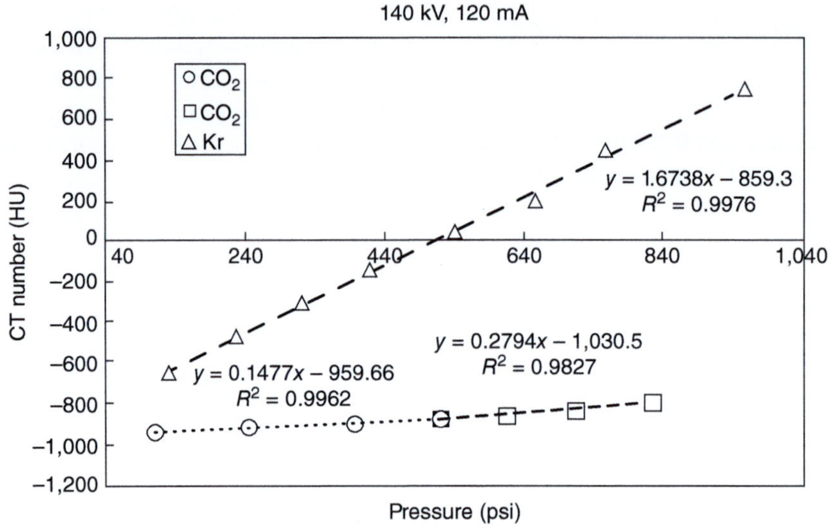

Figure 4.24 CT numbers of pure CO_2 and Kr as a function of pressure (Aljamaan et al., 2017).

Table 4.3 CTN of common materials

Material	CT (HU) - 130 keV	CT (HU) - 100 keV	Density (kg/m^3)
Air*	−1,000		1.82
Water*	0		1,000
N-hexane*	−285		660
N-decanc*	−283		730
1 wt.% KBr in water*	70	91	1,005
Quartz**	1,589	1,836	2,190
Berea sandstone**	1,608	1,835	2,120
Indiana limestone**	1,531	1,750	2,220
Alumina**	2,478	2,866	2,820

Source: Akin and Kovscek (2003).
* CTN obtained on a Picker 1200 SX instrument with a current of 65 mA and a FOV of 16 cm
** CTN for minerals and rocks were measured on a Philips Tomoscan instrument with a tube current of 250 mA and a 160 mm field of view (FOV).

$$\varphi = \frac{CTN_m^f - CTN_m^{vac}}{CTN_f - CTN_{vac}} \tag{4.6}$$

where CTN_m^f is the measured CTN of the fluid-filled core (100%), CTN_m^{vac} is the measured CTN of the dry and evacuated core, CTN_f is the CTN of the fluid in question, and CTN_{vac} is the CTN of air/vacuum.

As observed in equation (4.6), the mineralogy information of the sample is of no importance here but the saturating fluid (gas, water, oil, etc.) must have a CT number larger than the CT number of vacuum (-1,000 HU) (Saadat et al., 2011). If brine is the saturating fluid, it should be doped with a contrast agent (e.g., iodine) to enhance the attenuation coefficient of the brine versus oil. However, if using brine in shales leads to clay swelling, a contrasting gas (krypton, xenon, etc.) can be used instead. A close agreement of ±1 porosity (%) has been reported for CT-derived and volumetrically derived porosities (Akin & Kovscek, 2003).

On the other hand, dual/multiple scans are carried out at an exact location selected to be saturated with fluids for single/dual energy levels. It is recommended for direct measurement of porosity for shales with multiple porosities (organic and inorganic porosity, microfractures). Such a dual/multiple scan requires a contrasting agent (argon, krypton, iodine), time-sequenced scanning, and image subtraction to measure dual/multiple porosity (Moss et al., 1992; Withjack, 1988). First, a dry and evacuated core contained in a core holder is scanned with preferred CT parameters to determine the CTN of the dry core. A region of interest (ROI) is then selected from the resulting CT image, which is then subjected to time-sequenced scanning after a contrast fluid is introduced. At equilibrium, the saturated core sample is rescanned and evaluated by subtracting the dry images from the saturated images after scaling (Moss et al., 1992). Figure 4.25 shows a single dry CT image of a Monterey shale sample on the left with a corresponding CT image of the porosity obtained by subtracting the dry image from the xenon saturated CT image. Regions of high CTN (bright spots) on the dry CT image (left) correspond to regions of low porosity (dark spots) on the porosity CT image (right). On the contrary, there appear to be regions with higher CTN, where the porosity is higher than in regions with lower CTN. This discrepancy could otherwise have been lost in a single-scan porosity estimation (Moss et al., 1992). Figure 4.26 shows some marked regions of interest (ROIs) from 0 to 4 on the dry core to determine the rate of xenon gas injection into the Monterey shale sample. The entire slice is marked as ROI 0, ROI 3 has the lowest CTN with high permeability and high porosity (Moss et al., 1992).

Essentially, by subtracting the CT image of a saturated (oil and water) rock sample from the CT image of water-saturated rock, the contribution of rock is neglected from the resultant image. In so doing, a normalized difference in CT numbers between fully saturated rocks with water-and-oil provides the fraction of the pore space filled with water and scales quantities between 0 and 1 (Akin & Kovscek, 2003).

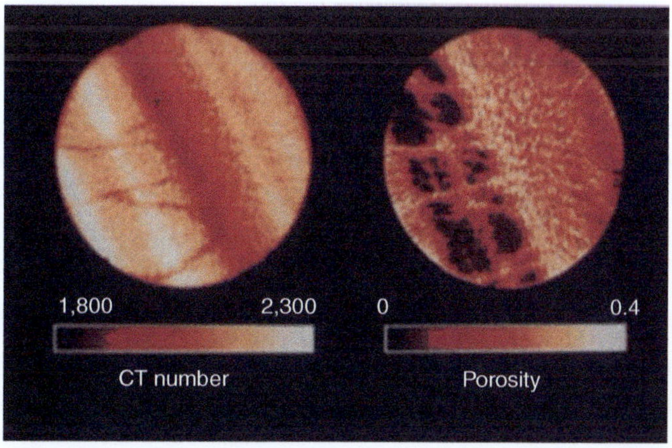

Figure 4.25 CTN image of an evacuated Monterey shale sample (left) and a corresponding porosity image (right) of fully saturated sample (Moss et al., 1992).

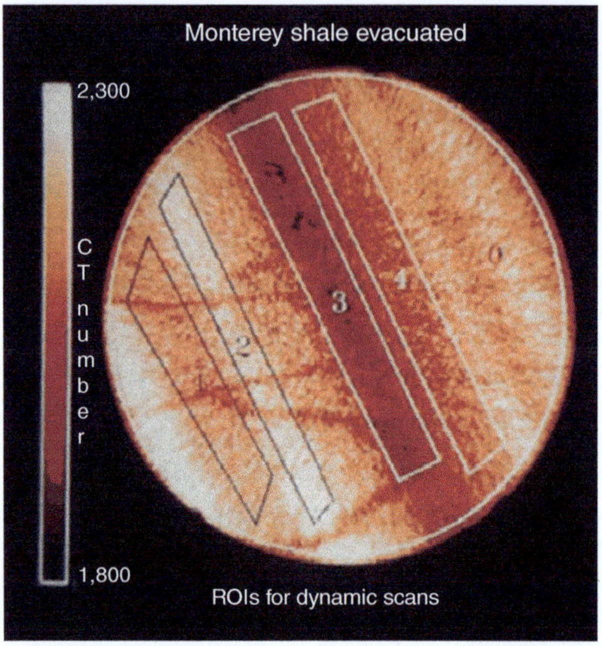

Figure 4.26 CT image of Monterey shale sample; ROIs for multiple scans with tube current of 100 mA (Moss et al., 1992).

4.2.8 Permeability Estimation from CT scanner

Although permeability cannot be measured directly, its distribution may be estimated from measuring in-situ flood-front velocities by X-ray micro-CT scan as well as through pore network modeling with the lattice Boltzmann method (LBM) (Chen et al., 2015; Zhang et al., 2015; Withjack et al., 1991). However, since kerogen micropores and nanopores cannot be resolved with the micro-CT for pore network modeling, this section will focus on the in-situ flood-front velocities (Peng et al., 2017; Xiong et al., 2016; Zhang et al., 2015; Withjack et al., 1991). For permeability estimation through in-situ flood, the core is assumed to be formed by a bundle of stream tubes with negligible variation of cross-sectional area along the length of each tube. Therefore, the permeability of each voxel of the stream tube can be calculated from equation (4.7) with known core permeability ($k_{average}$), individual stream tube porosities (φ_i), and the time for the solvent to flow along a given length of the core (t_i).

$$k_1 = \frac{k_{average} A_T}{A\left(1 + \frac{\phi_3 t_1}{\phi_1 t_2} + \frac{\phi_3 t_1}{\phi_1 t_3} + \ldots + \frac{\phi_n t_1}{\phi_1 t_n}\right)}, \tag{4.7}$$

where, A= area and subscript T= tube.

4.2.9 Two-Phase Fluid Saturations

Fluid saturation during multiphase, conventional coreflood experiments can be determined by CT measurements through either the linear interpolation between pure states (Akin et al., 1996) or linear regression (Akin & Kovscek, 2003). The linear interpolation between pure states assumes that the CT number of the core lies on the straight line connecting complete saturation by phase 1 to complete saturation by phase 2. In so doing, the saturation of either water or oil can therefore be estimated in each voxel as (Akin & Kovscek, 2003; Sharma et al., 1997)

$$S_o = \frac{CT_{wr} - CT_{owr}}{CT_{wr} - CT_{or}}, \tag{4.8}$$

where CT_{owr} is the CTN of rock containing both oil and water phases, CT_{wr} is the CTN of rock fully saturated with water, and CT_{or} is the CTN of rock fully saturated with oil.

From equation (4.8), the CT_{wr} is obtained by scanning a water-saturated core, while CT_{or} is obtained by saturating the core with oil (Sharma et al., 1997) after cleaning and drying to get rid of water or other fluids. On the other hand, if porosity distribution of the rock in its dry state is known, water saturation can be calculated without the need for complete water-saturated and oil-saturated images as

$$S_w = \frac{CT_{ow} - CT_{or}}{\phi(CT_w - CT_o)}, \tag{4.9}$$

where, CT_{ow} is the CTN of oil and water, CT_{or} is the CTN of rock fully saturated with oil, CT_w is the CTN of water, and CT_o is the CTN of oil.

To obtain accurate CT readings for the CTN of water and oil, the core holder is filled solely with water and oil without the sample (Akin & Kovscek, 2003; Sharma et al., 1997).

4.3 X-Ray Nano-CT

The nano-CT resolves images at a submicron level and higher resolution than the micro-CT. The resolution per pixel of nanofocus-CT ranges from 500 nm to approximately 50 nm, about an order of magnitude improvement from the best available micro-CT systems (Rodriguez et al., 2014). In a manner similar to the X-ray micro-CT, the nano-CT utilizes an X-ray source in addition to a Freznel Zone Plate objective (FZP) and a Zernike phase contrast ring to enhance images on the detector (Singhal et al., 2013; Withers, 2007) as shown in Figure 4.27. The Fresnel Zone Plate objective (FZP) is made up of diffractive X-ray lenses, which focus X-rays by constructive interference, while the Zernike phase contrast is used for low absorption contrast materials such as polymers, porous carbon, and tissue/cells (Babu et al., 2015). The source of X-ray is a rotating copper (Cu) anode and the X-ray energy depends on the target material's emission line. The copper (Cu) anode balances absorption contrast and X-ray penetration depth for high resolution in between soft and hard X-rays (Akbarabadi et al., 2017; Litster, et al., 2015; Withers, 2007). The nano-CT has two different fields of view made up of (1) a high resolution (HRES) with a field of view of 16.3 μm and (2) a large field of view (LFOV) with a field of view of 65.6 μm (Akbarabadi et al., 2017; Withers, 2007).

Figure 4.28 shows a 3D volume of a mudstone without organic matter (OM) acquired with a nano-CT; dark features enclosed in the

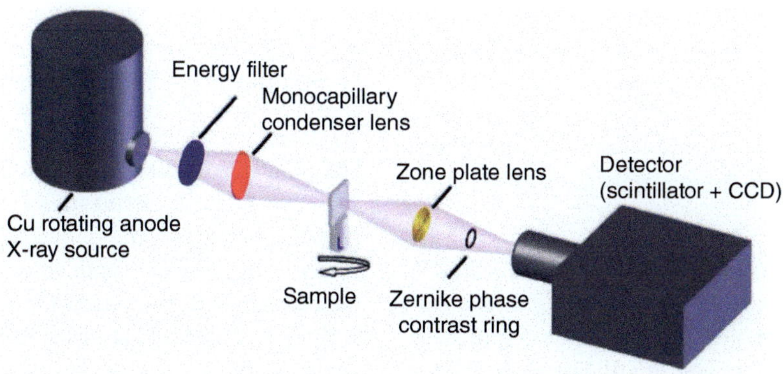

Figure 4.27 Schematic diagram of a nano-CT with X-ray optics for resolutions better than 500 nm in X-ray CT (Litster et al., 2015).

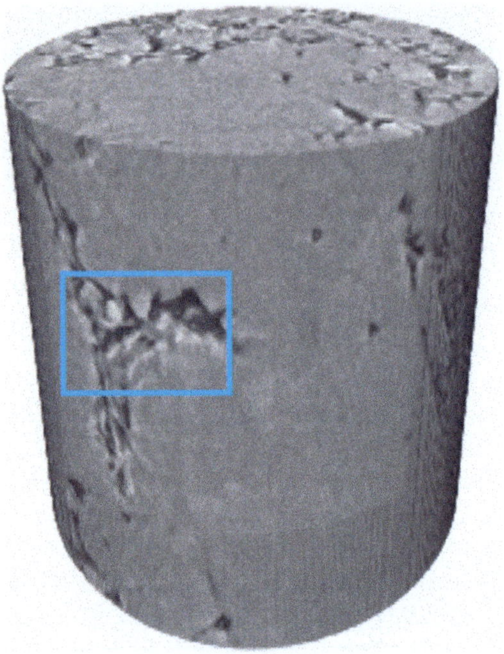

Figure 4.28 Nano-CT 3D volume of a mudstone sample; dark pores enclosed in blue rectangular box and gray mineral matrix (Akbarabadi et al., 2017).

blue rectangular box represent pores, while the gray background represents the mineral matrix (carbonate, quartz, and/or clay). The dimensions of the nano-CT image is 25 μm by length and 16.5 μm in diameter with a resolution of 16 nm (Akbarabadi et al., 2017). Unlike micro-CT, nano-CT systems have the potential to visualize the sharp contrast between minerals and pore space of organic-rich shales but are still practically limited to resolve nanopores compared with electron microscopy imaging techniques (Rodriguez et al., 2014).

4.3.1 Sample Preparation for X-Ray Nano-CT

In a manner similar to the X-ray micro-CT, sample sizes vary with the required image resolution, where smaller (thinner) sample sizes produce better image quality with the X-ray nano-CT. As a result, several methods including mechanical crushing, laser milling, and ion milling are employed to produce symmetrical and thin samples for better resolution. While, the first two methods (mechanical crushing and laser milling) mostly produce asymmetrical samples with severe damage to the pore structure of the samples, ion milling, on the other hand, reduces damage to the pore structure and is proven to be more reliable (Akbarabadi et al., 2017). In the ion milling technique, according to a method reported by Akbarabadi et al. (2017), a region of interest (ROI) of $\sim 100 \times 100\ \mu m^2$ is first identified from a scanning electron microscopy (SEM) image of the sample. The selected ROI is then coated with a protective platinum (~1 μm thick) and, subsequently, the volume around the platinum layer is removed with focused ion beam (FIB) at 30 kV Gallium (Ga) ion and 65 nA beam current as shown in Figure 4.29a. Next, the top of the sample is attached to the tip of a micromanipulator through platinum welding, having cut the cylindrical miniature core with FIB, shown in Figure 4.29b. The bottom of the cylindrical miniature core (height is twice the diameter) is then attached to a needle also by means of platinum welding and the micromanipulator is cut free as shown in Figure 4.29c. Finally, the platinum layer is removed and the sample is trimmed and polished, ready for X-ray nano-CT scan. Figure 4.29d shows a miniature core (15–60 μm diameter) on the tip of a needle, mounted on the sample holder for the X-ray nano-CT (Akbarabadi et al., 2017).

4.3.2 In-Situ Wettability and Spontaneous Imbibition at Nanoscale

Wettability is an important reservoir rock property, which controls multiphase flow properties such as relative permeability and capillary pressure leading to the optimization of hydrocarbon recovery in both conventional and unconventional reservoirs (Anderson, 1986). Although wettability is easily evaluated in conventional reservoir rocks through experimental methods (contact angle measurement, Amott test, and spontaneous imbibition), it becomes

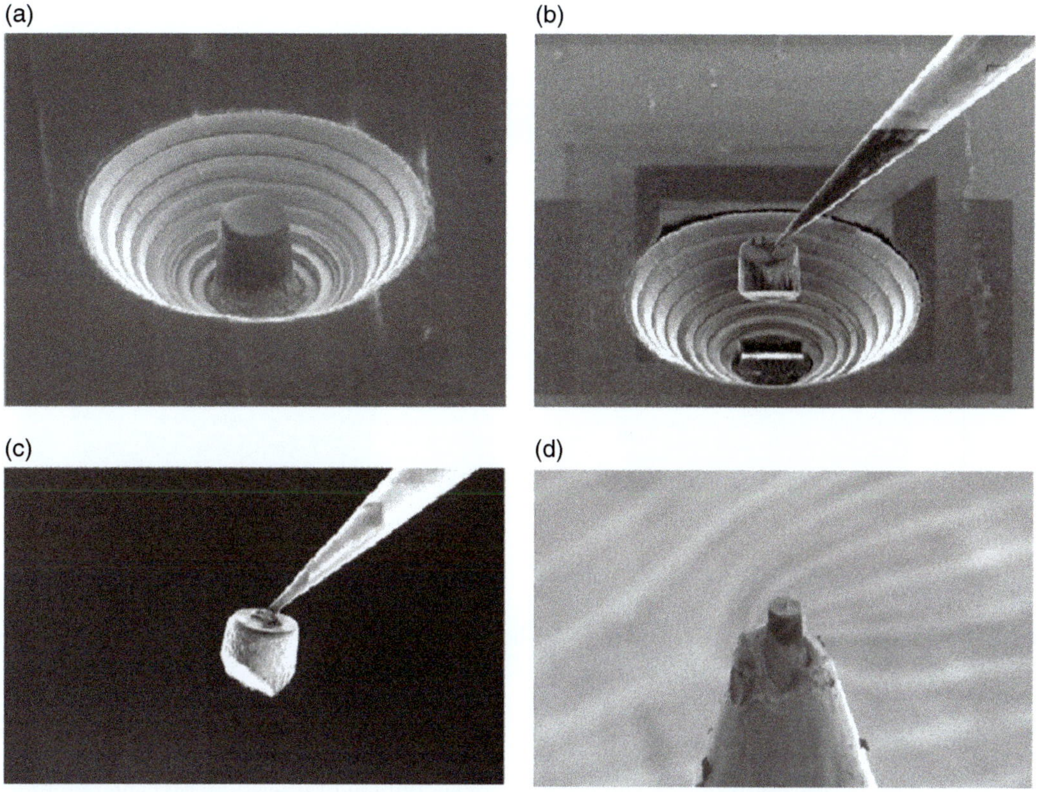

Figure 4.29 Ion milling sample preparation. (a) The volume around the platinum layer is removed with focused ion beam (FIB); (b) the top of the sample is attached to the tip of a micromanipulator through platinum welding, having cut the cylindrical miniature core with FIB; (c) the bottom of the cylindrical miniature core (height is twice the diameter) is then attached to a needle and the micromanipulator is cut free; (d) a miniature core is mounted on the sample holder for the X-ray nano-CT (Akbarabadi et al., 2017).

more complex in unconventional shale reservoirs due to very low permeability and poor pore-scale connectivity (Fogden et al., 2015; Borysenko et al., 2009). Akbarabadi et al. (2017) investigated spontaneous imbibition behavior and in-situ wettability at nanoscale in shale reservoir rocks and ultratight rock samples by integrating a core-flooding apparatus with imaging techniques such as X-ray nano-CT. In addition, the focused ion beam-scanning electron microscopy (FIB-SEM) was used to investigate the pore structures as well as the associated flow paths. The petrophysical properties, such as porosity and permeability, were estimated through image analysis of the resulting nano-CT and FIB-SEM images. The samples used were symmetrical and thin enough to produce high-quality images using the X-ray nano-CT. The imbibition of both oil and brine phases into ultratight rock samples revealed fluid inside individual nanopores in addition to a complex wettability identified in three main types: water-wet, oil-wet, and fractionally wet (Akbarabadi et al., 2017). Oil and brine phases are observed in Figure 4.30 to be spontaneously imbibed into the pore network of an ultratight rock sample D; red represents oil, blue represents brine, yellow represents gas (air), and green represents organic matter (OM). The intensity of the oil phase in a grayscale CT or FIB-SEM image lies between the intensity of the gas and brine

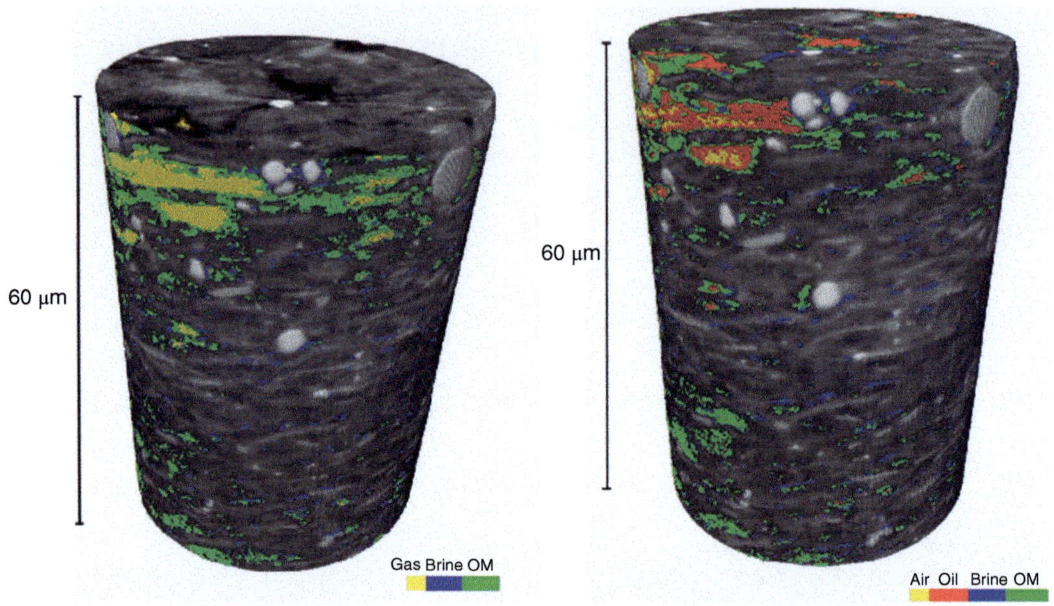

Figure 4.30 3D nano-CT volume of a shale sample showing the distribution of fluids during brine imbibition (left) and oil imbibition (right) (Akbarabadi et al., 2017).

phases, respectively. As such, lower threshold value defines the gas-oil interface, while upper threshold determines the oil-brine interface. Pores within the organic matter (green) during brine imbibition (Fig. 4.30, left) are observed to be hydrophobic (oil-wet); as such, during oil imbibition (Fig. 4.30, right) the organic pores are seen to be filled with oil and trapped gas (intermediate-wetting phase). On the other hand, some of the mineral pores within the shale matrix show strongly water-wet conditions, where water was spontaneously imbibed rather than oil. The remaining pore minerals exhibited partial trapping of gas upon water imbibition at the contact with organic matter. Therefore, upon a second imbibition test, oil was imbibed into these matrix pores to replace the trapped gas indicating a fractionally wet system (Akbarabadi et al., 2017).

Text Box 4.1 X-Ray Imaging Techniques, What to Expect and What Can Be Analyzed

X-ray Micro-CT

- Unresolved micro- or nano- pores
- Resolved micro-fractures
- Resolved organic matter, if present and shale mineral matrix
- Volume fraction of OM and mineral matrix can be estimated/analyzed

Nano-CT

- Resolved micro-pores and micro-fractures
- Resolved organic matter, if present and shale mineral matrix
- Volume fraction of pores, OM, and mineral matrix can be analyzed

4.4 Electron Microscopy

Unlike X-ray micro-CT and nano-CT, electron microscopy utilizes electron beam instead of X-ray to provide internal and surface (topographic) as well as elemental analysis of materials/samples. It includes scanning electron microscopy (SEM), focused ion beam-scanning electron microscopy (FIB-SEM), and quantitative evaluation of minerals by scanning electron microscopy (QEMSCAN).

4.4.1 Scanning Electron Microscopy (SEM)

Scanning electron microscopy (SEM) provides topographical information after the surface of the sample under investigation is bombarded with electron beam. SEM produces images by recording various signals generated as a result of the interaction between the electron beam and the atoms of a sample in a raster pattern (Huang et al., 2013). Figure 4.31 shows the main components of the SEM. The electron gun in the form of a tungsten filament or otherwise depending on the manufacturer, emits electrons, which are propagated down the electron column to the two-part lens (condenser lens and objective lens) system, where the electron beam is focused. First, the condenser lens demagnifies the electron while the objective lens focuses the beam down to the desirable spot size for detection (Ameruddin, 2015; Huang et al., 2013; Camp et al., 2013).

As a result of electron beam bombardment on the surface of a sample, several electrons are generated, which distinctively provide structural information about the sample and differ from one another in origin, energy, and traveling direction as shown in Figure 4.32 (Huang et al., 2013; Lemmens & Richards, 2013). In Figure 4.32, the type I secondary electrons (SE1) are emitted in close proximity from the impact point and are known to carry high-resolution surface sensitive (topographic) information. On the other hand, the type II

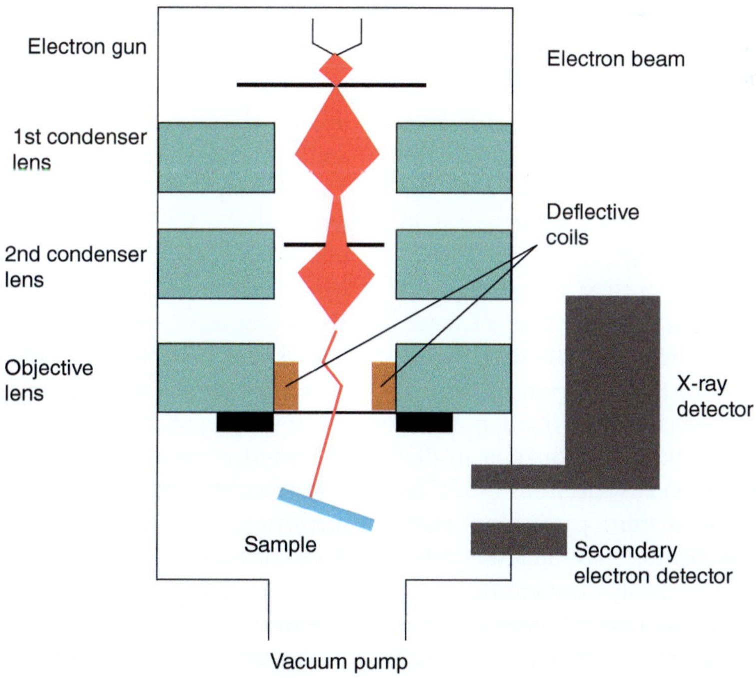

Figure 4.31 SEM main components (Ameruddin, 2015).

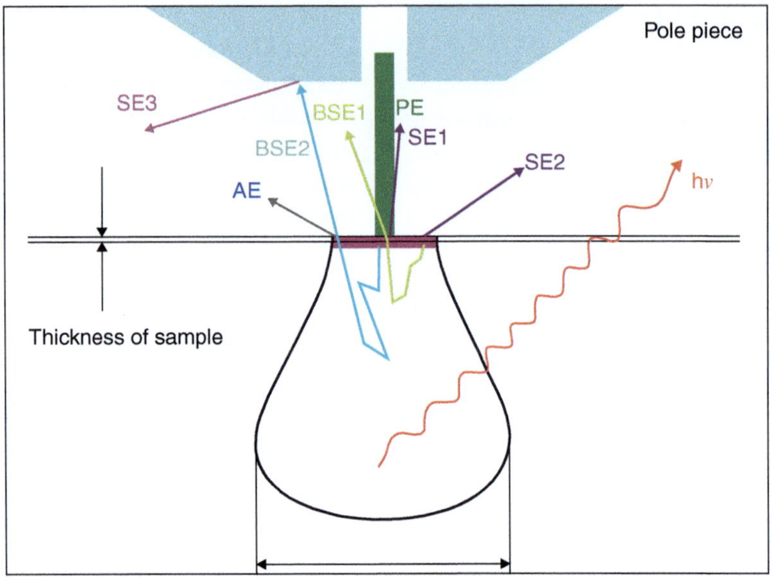

Figure 4.32 Illustration of typical signals from specimen-electron interaction (Huang et al., 2013).

secondary electrons (SE2) are generated much deeper at a wider volume than the SE1 and provide intrinsically lower-resolution topographical information (Huang et al., 2013; Camp et al., 2013). There is also the singly scattered backscattered electrons (BSE1) discharged at a high angle, which is associated with compositional contrast, while the multiple-scattered backscattered electrons (BSE2) at a lower angle are used to characterize composition and crystalline structures of a sample. In addition, the electron excitation generates photons in the form of X-ray characteristic lines useful in chemical analysis with SEM.

Figure 4.33 shows an SEM image acquired with the FEI Versa 3D DualBeam of a Chattanooga shale sample from McGrath well, Barber County, Kansas. The SEM image reveals more of topographical (surface) information at the expense of compositional information in line with the type I secondary electrons (SE1). As such, the organic matter, pores, and other matrix minerals (pyrite, clay, quartz, carbonate, etc.) cannot be easily distinguished.

On the contrary, BSE image (Figure 4.34) of a Chattanooga shale sample acquired with the FEI Helios NanoLab DualBeam provides compositional contrast of minerals within the sample in line with the singly scattered backscattered electrons (BSE1). Another example is a BSE image of an Eagle Ford shale sample (courtesy of Thermo Fisher Scientific) shown in Figure 4.35. The grayscale level of an electron image is composed of both high secondary electrons (SE), which appear brighter, and low secondary electrons (SE), darker in appearance and similar to the grayscale image of a CT image. This is to say that less dense pores and fractures would appear the darkest, followed by organic matter (OM), while denser minerals like pyrite would appear the brighter. Other matrix minerals (quartz, clays, etc.) with densities in between the less-dense pores and denser pyrite would appear in different shades of gray. This clear distinction of in-situ minerals and pores enables the characterization shales on a microscale.

Furthermore, the BSE image highlights the maturity level of organic matter found in different shale formations as well as the interparticle and intraparticle pores. A knowledge of the maturity level of the organic matter in a shale formation contributes to the development and

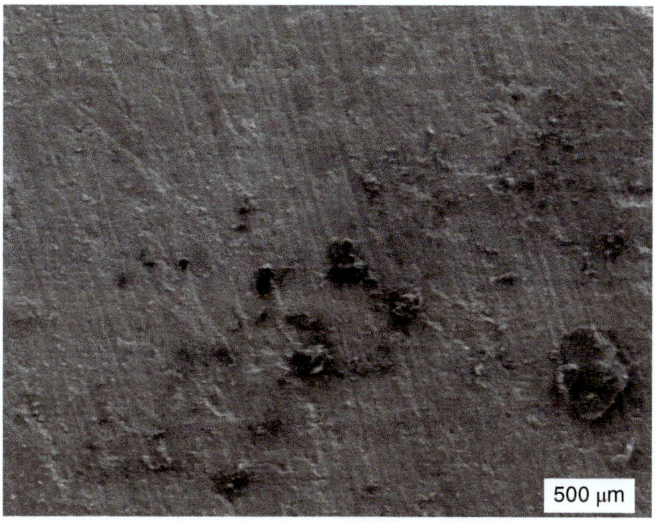

Figure 4.33 SEM image of a Chattanooga shale sample from McGrath well, Kansas.

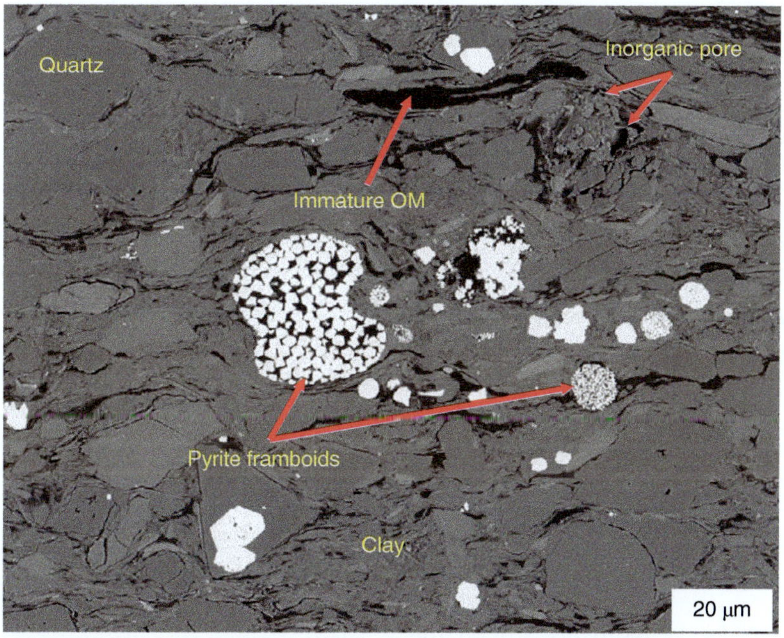

Figure 4.34 BSE image of a Chattanooga shale sample from McGrath well, Barber County, Kansas.

hydrocarbon potential of that particular shale formation. A majority of marine shale oil reservoirs in the United States are predominantly type II kerogen with relatively high TOC (> 3%) and thermal maturity (R_o) of 0.6% to 1.5%, while shale oil is developed in semideep and deep lake sedimentary environments, possesses both type I and type II kerogen with TOC above 2%, and thermal maturity (R_o) in the range of 0.5% to 1.1% (Jiang et al., 2016). In

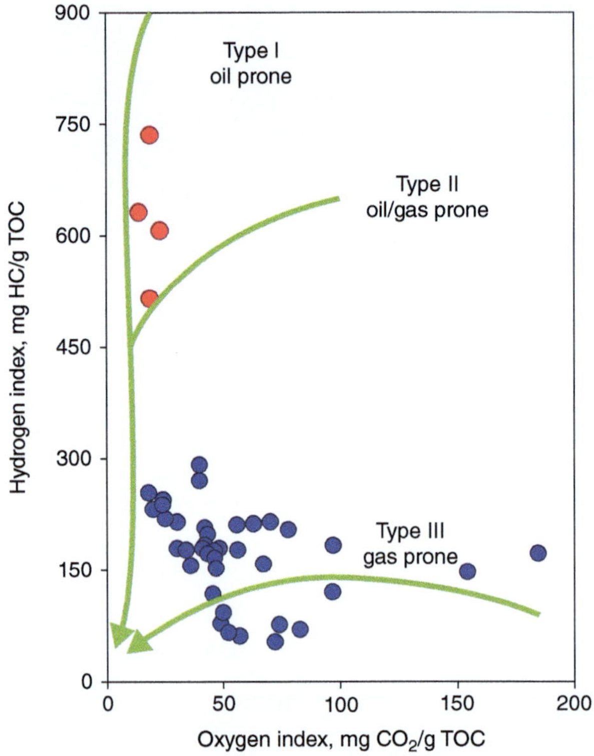

Figure 4.35 Modified van Krevelen plot (Alimi, 2018; Hill et al., 2007).

Table 4.4 Thermal Maturity Guidelines

Vitrinite reflectance (R_o)	Maturity guidelines
< 0.6%	Immature
0.6–1.1%	Oil window
1.1–1.4%	Wet gas window
1.4–3.2%	Dry gas window

Source: Alimi (2018); Jarvie et al. (2007); Pollastro et al. (2007).

comparison, shale gas reservoirs around the world are predominantly type III kerogen with thermal maturity (>1.40%). Figure 4.35 shows a modified van Krevelen plot classified into kerogen types and the corresponding hydrocarbon fluids to expect is listed in Table 4.4 with different ranges of the thermal maturity of a shale formation measured by the vitrine maceral reflectance through optical microscopy. However, in electron microscopy, the thermal maturity of a shale sample is inferred from the organic pores present within the organic material of the sample. Shale plays with varying porosity from organic to inorganic are shown in section 4.4.2 highlighting the mineral distribution as well.

4.4.2 SEM/BSE Images of Various Ultratight, Organic-Rich Formations

SEM/BSE images collected for different shale plays at different magnifications with resolved organic matter and grain types in addition to organic and matrix pores (interparticle and intraparticle) are covered in this section. Figure 4.36 shows a BSE image of a Chattanooga shale sample from Barber County, Kansas, of late Devonian origin at 5 μm scale. The BSE image highlights the presence of OM (dark), pyrite (white), mineral matrix (gray),

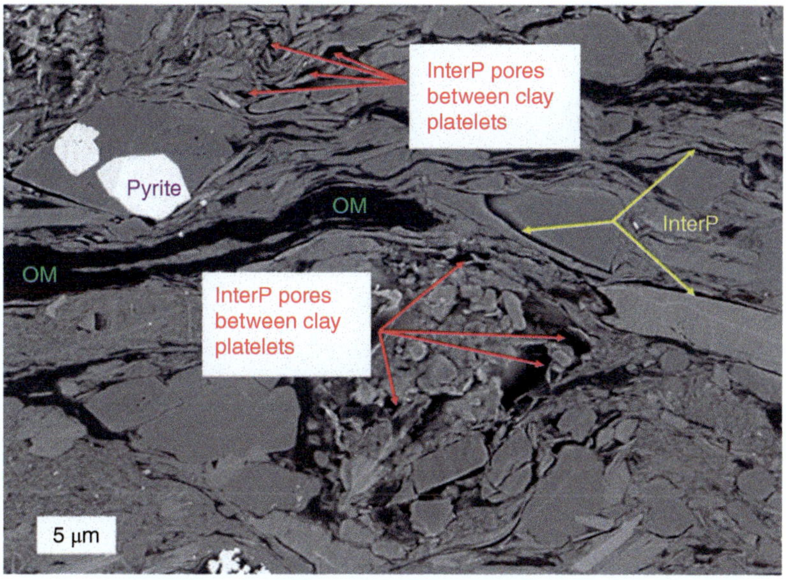

Figure 4.36 BSE image of Chattanooga shale, Barber County, Kansas.

and the absence of organic pores (OMP). The nonporous OM (Tasmanites) in Chattanooga shale are commonly found in Woodford shale formations, which belong to the Devonian period, and the Chattanooga shale is regarded as an extension of the Woodford shale in southern Kansas. Tasmanites almost always appear nonporous on SEM images either their pores are not large enough to be resolved or have not undergone considerable thermal decomposition to develop organic pores (Löhr et al., 2015). On the other hand, there exist inorganic pores in the Chattanooga shale sample, namely, interparticle pores (interP) and intraparticle pores (intraP). Interparticle pores (interP) are defined as well-connected mineral pores existing between mineral grains and crystals; and intraparticle pores are defined to occur within grain and crystal boundaries, and are relatively isolated (Ko et al., 2017; Loucks et al., 2010). The isolated interparticle pores in Figure 4.36 are located between clay platelets. Selecting four different categories (OM, matrix, InterP, and pyrite) of the distributed internal features on the BSE image, 1,000 points were counted using the JMicroVision software (Roduit, 2008) to estimate the percentage of each phase as shown in Figure 4.37. The results of the point counting were given as OM = 17.82%, matrix = 71.61%, InterP = 9.15%, and pyrite = 1.42%. The interP make up the matrix porosity of 9.15% for the Chattanooga shale sample.

Figure 4.38 shows a BSE image of an Eagle Ford shale sample at 39 nm pixel resolution with the presence of organic pores to indicate a thermally matured shale reservoir. An OM body is embedded in the mineral matrix of the shale sample synonymous to shale reservoirs in addition to intraparticle pores (intraP) within the pyrite crystals and interparticle pores (interP) between the organic matter body and the matrix as well as between other minerals in the matrix. Accordingly, the BSE image was processed through point counting to estimate the percentage of each mineral phase as shown in Figure 4.39.

Five categories of choice were defined as organic matter (OM), mineral matrix (matrix), mineral pores (both intraP and interP), pyrite,

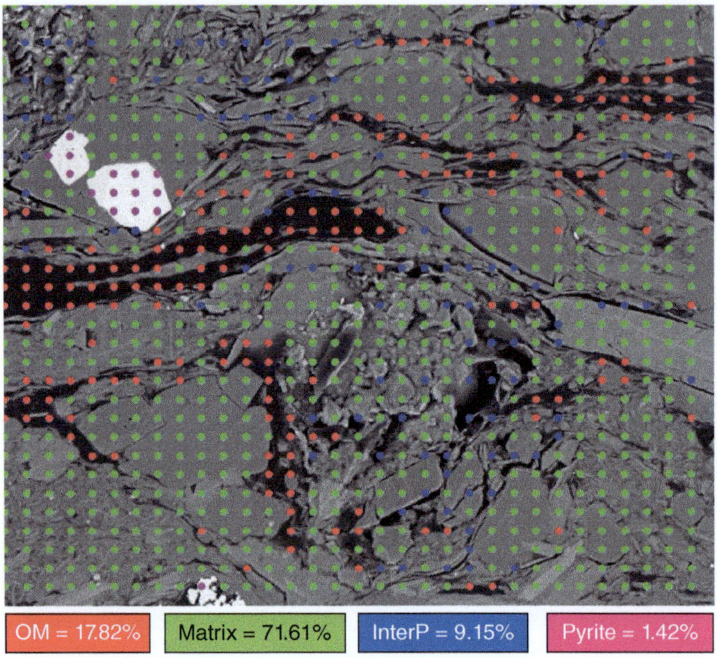

Figure 4.37 Point counting percentage of each mineral phase in the Chattanooga shale sample.

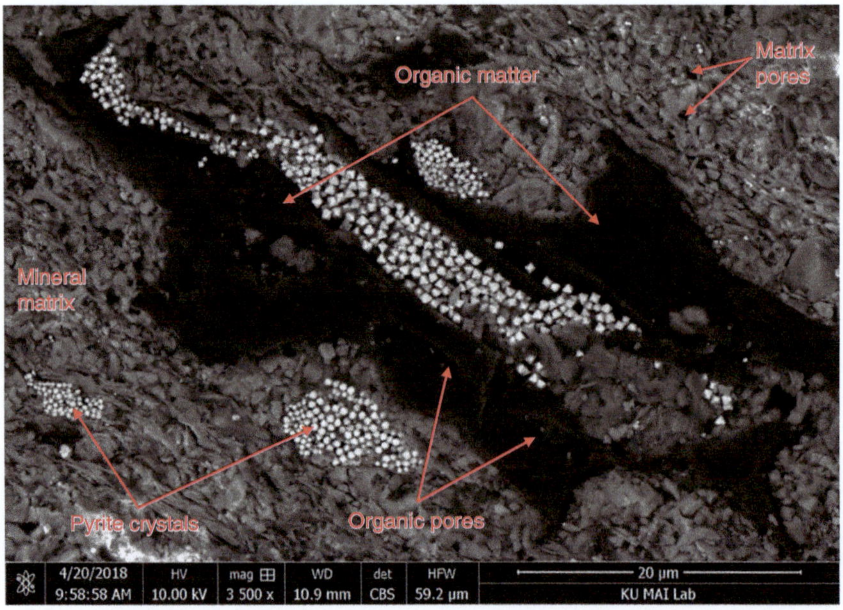

Figure 4.38 BSE image of a lower Eagle Ford shale sample at a thermal maturity of 1.03% (Cudjoe, et al., 2019).

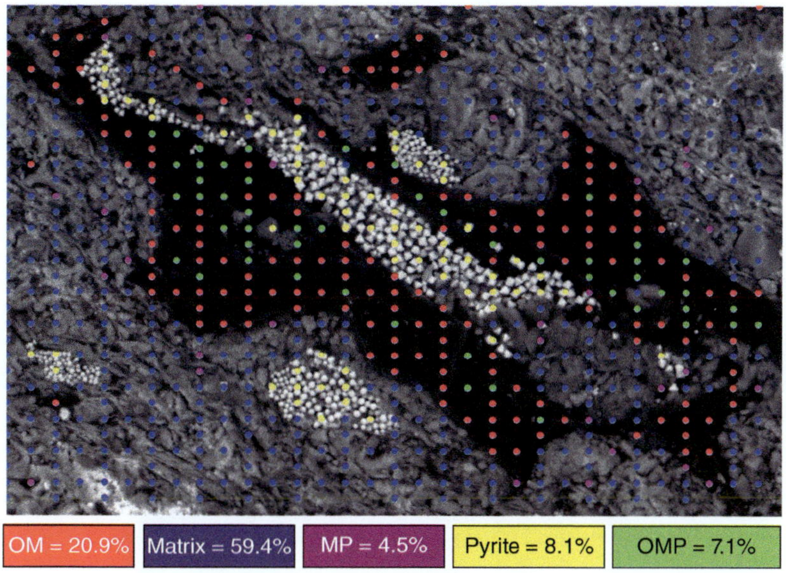

Figure 4.39 Point counting analysis on an Eagle Ford shale sample.

and organic matter pore (OMP). They were analyzed with 1,000 points and their percentages given as 20.9%, 59.4%, 4.5%, 8.1%, and 7.1%, respectively. Note that the porosity of this particular shale medium increases due to additional organic porosity, which is otherwise missed on large-scale wireline logging characterization and bulk laboratory measurements.

BSE images of a lower Eagle Ford (LEF) sample are shown in Figure 4.40 with underlying grain composition of quartz, clay minerals, and depositional kerogen, in addition to diagenetic phases such as pyrite, calcite cement, kaolinite, and migrated bitumen (Cudjoe, et al., 2019). A higher magnification view of the dark feature enclosed in the red box (Fig. 4.40a) shows resolvable organic pores estimated at a volume fraction of ~11 vol.% with equivalent pore diameter ranging from 17 nm to 4 μm (Cudjoe, et al., 2019). A similar mineralogical distribution of OM, quartz, clay minerals, pyrite, among other unidentified matrix minerals is observed in a BSE image of a Barnett shale sample in Figure 4.41 (courtesy of Thermo Fisher Scientific). Applying the point counting analysis with 1,000 points for five categories namely, OM (organic matter), OMP (organic matter pores), pyrite, and interP, the resulting surface percentages of the mineral phases are estimated at 73.2%, 22.02%, 0.39%, 2.82%, and 1.57%, respectively (see figure 4.42). Although SEM image-based porosity underestimates bulk porosity since pores below 5 nm are not resolved, SEM provides pore and mineral distribution information otherwise not provided by other methods.

Figure 4.43 shows an SEM image of another LEF sample at a thermal maturity of 1.03%. At about 5 μm scale, the carbonate microfossil (foraminifer test) is observed to be occupied by diagenetic minerals such as kaolinite, quartz, pyrite, and migrated bitumen. The migrated bitumen is described according to the morphological criteria of Loucks and Reed (2014), where the migrated bitumen is perceived to fill existing chambers of the forams.

Figure 4.44c and d, respectively show BSE images of the Demerara Rise (equatorial North Atlantic) black shales, where discrete OM is embedded in clay minerals and characterized by abundant, circular to angular pores (Löhr et al., 2015). The Permian Stuart ranges formation of South Australia is shown in Figure 4.45 enriched with pyrite and organic matter.

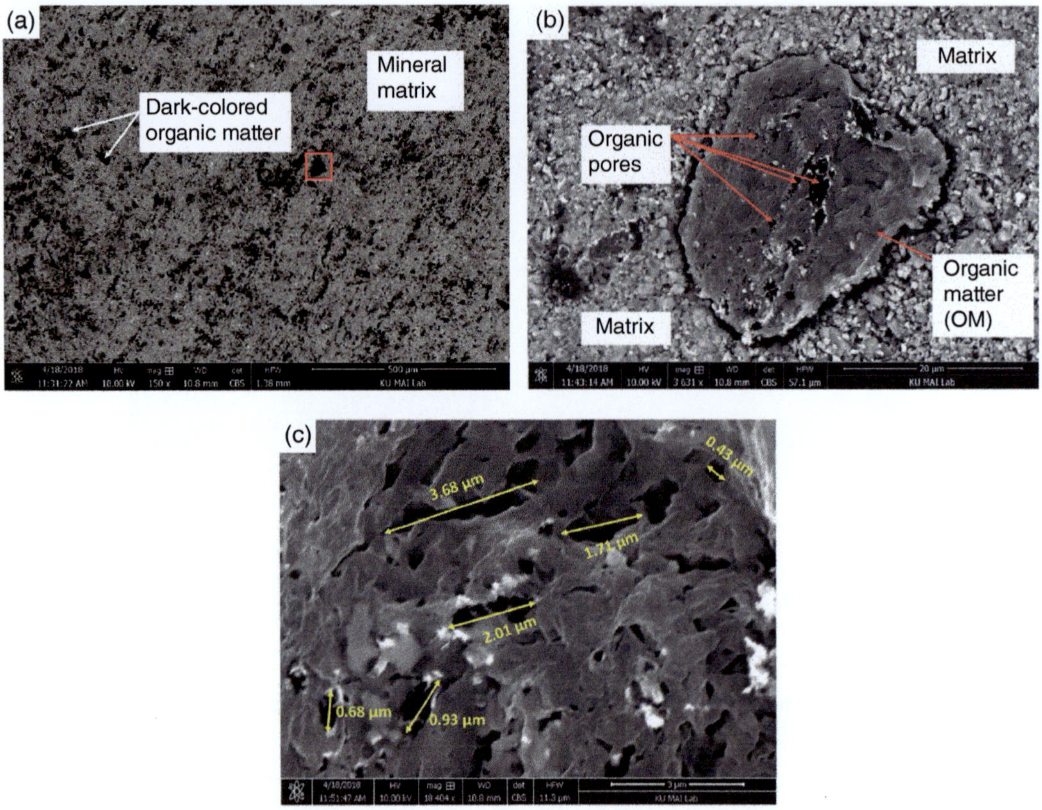

Figure 4.40 BSE image of a lower Eagle Ford (LEF) sample at thermal maturity of 0.98% (Cudjoe et al., 2019).

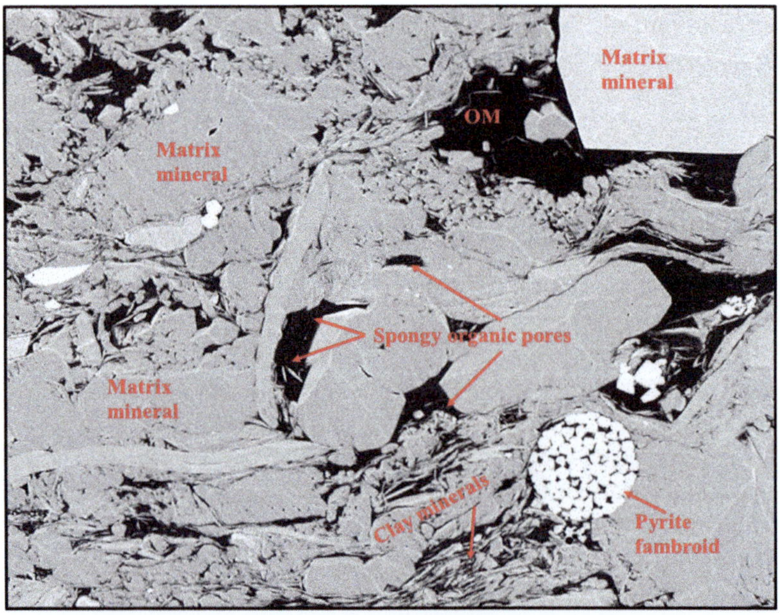

Figure 4.41 BSE image of a Barnett shale sample (courtesy of Thermo Fisher Scientific).

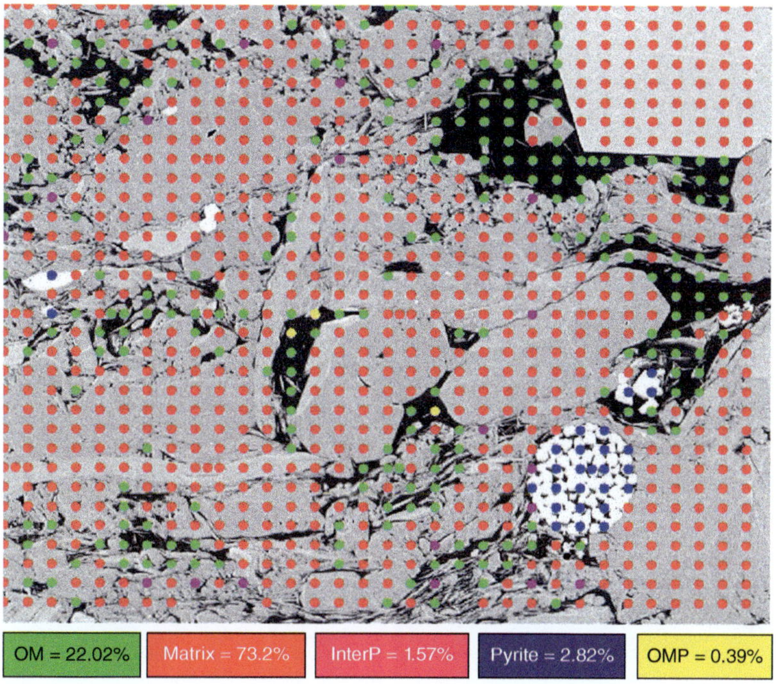

Figure 4.42 Point counting analysis of a Barnett shale sample.

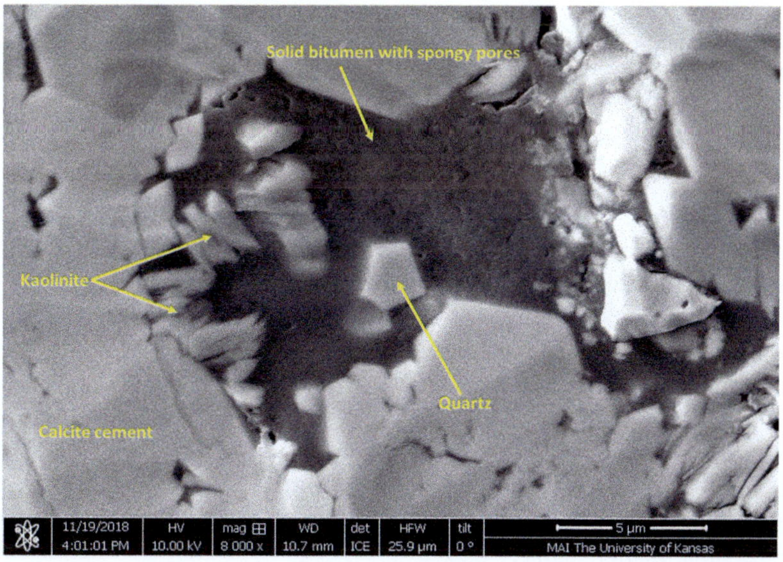

Figure 4.43 SEM image of a LEF sample with migrated bitumen infilling carbonate microfossil (Cudjoe et al., 2019).

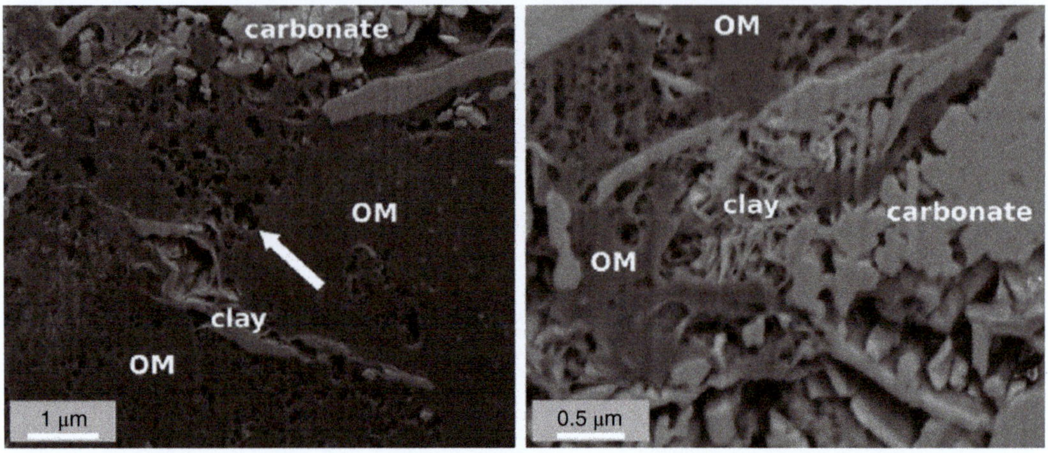

Figure 4.44 BSE image of Demerara Rise black shale formation (equatorial North Atlantic) (Löhr et al., 2015).

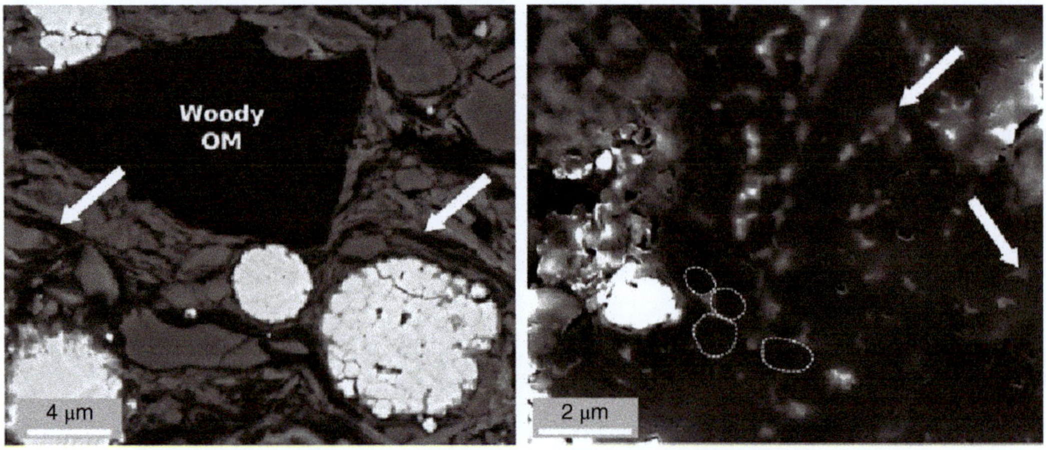

Figure 4.45 BSE image of Stuart Ranges Fm. (South Australia) (Löhr et al., 2015).

Figure 4.45c identifies nonporous woody OM of terrigenous origin, in addition to discrete OM laminae (white arrows), interpreted as lamalginite, while Figure 4.45d shows spherical to ovoid OM bodies (examples outline) with pores between adjacent bodies (Löhr et al., 2015).

Depositional Kerogen (DK) Versus Migrated Bitumen (MB)

Two types of organic matter, namely depositional kerogen (DK) and migrated bitumen (MB) are identified in the LEF shale samples by adopting Loucks and Reed (2014) morphological criteria. DK is described to be embedded within the inter- and intra-particle pores within the matrix, while the MB is perceived as the organic matter infilling existing chambers of foraminifera tests (carbonate microfossils). However, since both features appear as dark features on a grayscale SEM image, it becomes very challenging to accurately distinguish between them within the matrix without specific features. The significance of distinguishing between depositional and migrated (bitumen) OM results in a more predictive understanding of pore generation potential in organic-rich shales (Milliken et al., 2014). Primary DK is mainly kerogen, which can be transformed into diagenetic MB, oil, and gas during thermal maturation. Chemically, kerogen is defined as an insoluble OM in neither alkaline solvents nor organic solvents, while bitumen is soluble in these solvents (Milliken et al., 2014; Tissot & Welte, 1984). Petrographically, kerogen is described as minute debris deposited as part of detrital grain accumulation, whereas bitumen is described as a diagenetic product released as a result of kerogen maturation and occupies the interparticle pores (Loucks & Reed, 2014; Milliken et al., 2014; Durand, 1980).

Figure 4.46 shows a SEM (left) of a LEF sample at a thermal maturity of 1.01%, where the distribution of foraminifera fossils with primary intraparticle pores are filled with either kaolinite, calcite, pyrite, or MB transformed into solid bitumen/pyrobitumen; the transformed solid bitumen/pyrobitumen within the forams is observed to develop organic pores. The DK pointed out on the SEM image (left) is observed to be embedded within the matrix with no developed organic pores. Fishman et al. (2013) show the paragenetic sequence of diagenetic events within the chambers of the forams to begin with pyrite, followed by calcite, quartz, kaolinite, and then the MB. The MB, thus, predates all the other diagenetic materials, and together they preserve the integrity of the primary foram pores from undergoing compaction. The MB undergoes further thermal decomposition to develop organic pores, while the DK observed on the SEM images of LEF samples rarely developed organic pores under the same thermal maturity. Cudjoe et al. (2019) attributed this effect to a possible difference in maceral type, where that of the MB is more prone to develop organic pores. The FIB-SEM image (right) also shows little or no pores developed within the labeled depositional kerogen embedded within the matrix compared with MB transformed into solid bitumen/pyrobitumen with both spongy and bubble pores. It was found that, although both types of organic matter contribute to the organic porosity, about 80% are contributions from the MB, while about 2% are from DK; the remaining 18% are contributions from both OM types not easily differentiated (Cudjoe, et al., 2019).

Whereas, organic pores of DK are classified as not very well connected, OM pores in MB converted into solid bitumen/pyrobitumen are thought to form better connected pore pathways (Loucks & Reed, 2014). On the other hand, Löhr (2015) pointed out that detrital OM or DK pores are common in thermally immature (<0.4%Ro) shales but absent in thermally

Figure 4.46 SEM (left) and FIB-SEM (right) images of a LEF sample at thermal maturity of 1.01% (Cudjoe et al., 2019).

matured samples (0.5–1.1%Ro) due to infilling of pores by generated and retained bitumen, but are present in gas-matured samples (≥1.5 %Ro) and classified as secondary pores (Löhr et al., 2015). This may not be true for all thermally matured shales, since the LEF sample with thermal maturity of 1.03% in Figure 4.38 shows the presence of bubble pores within the detrital OM or DK.

Another mechanism to explain the absence of detrital pores in oil-matured samples is the compaction associated with burial and thermal maturation of the sediment (Milliken et al., 2014; Fishman et al., 2012).

The work of Loucks and Reed (2014) outlines some SEM petrographic criteria for distinguishing detrital OM from migrated OM. These criteria include

1) Fossil body-cavity voids filled with migrated OM. As previously mentioned, the generated bitumen migrates to fill available voids in the matrix and one such void is the fossil bodies (yellow circles) shown in Figure 4.46.
2) Migrated bitumen, occurring after cementation in mineral pores, shows a mix of spongy OM and large bubble pores. These pores are generated during thermal decomposition of bitumen into solid bitumen/pyrobitumen with developed pores to create more connected pathways.

On the other hand, migrated OM/bitumen is visualized in light microscopy (white and ultraviolet lights) for a LEF sample and it shows migrated OM/bitumen to fill foraminifera chambers in Figure 4.47a and b, respectively. Both OM and bitumen appear dark in the white light image (Fig. 4.47a) but the migrated OM (bitumen) in Figure 4.47b gives off spots of red fluorescence in certain foraminifera fillings, while the dispersed OM within the matrix appear as nonfluorescing. In addition, abundant yellow fluorescing particles within the matrix are observed.

Table 4.5 lists some prominent shale oil/gas resources characterized according to the kerogen type, TOC, thermal maturity, and lithology.

4.4.3 Energy-Dispersive X-Ray Spectrometry (EDS/EDX)

Energy-dispersive X-ray spectrometry (EDS) is an auxiliary technique, which involves the characterization of the elemental composition of the formation. While BSE and diffraction, undergo elastic scattering after electron bombardment, characteristic X-ray and secondary electron (SE) undergo inelastic scattering. During inelastic scattering, an incident electron ionizes the sample atom by ejecting an inner-shell electron. The vacancy is then filled by another electron (higher shell), which results in excess energy emission of X-ray photons. The de-excitation process produces characteristic X-rays. The accumulated counts from a sample produce an X-ray spectrum, typically made up of major spectral peaks superimposed on a broad background. Each element supplies characteristic spectral peaks detected by the EDS for qualitative and quantitative analysis. Figure 4.48 shows selected regions of interest (ROIs) labeled in images Figure 4.48a and b of two Chattanooga shale samples, while Figure 4.49 provides the quantitative elemental composition of the selected regions of interest (ROIs).

Dark spots in a selected region in the absences of pores prompt peaks in elemental carbon (C), deduced to be organic matter (kerogen). Meanwhile, gray areas representing shale mineral matrix show peak dominance of silicon (Si), aluminum (Al), magnesium (Mg), among other elements, deduced to be quartz (SiO_2), clay [$Al_2Si_2O_5(OH)_4$], and dolomite [$CaMg(CO_3)_2$], respectively.

Another EDS analysis performed on an Eagle Ford shale sample is shown in Figure 4.50, with the elemental composition dominated by calcium (Ca) - 22 wt.% of the matrix, followed by the carbon (C) element corresponding partially to dark organic matter patches. Other elements present include silicon (Si), aluminum

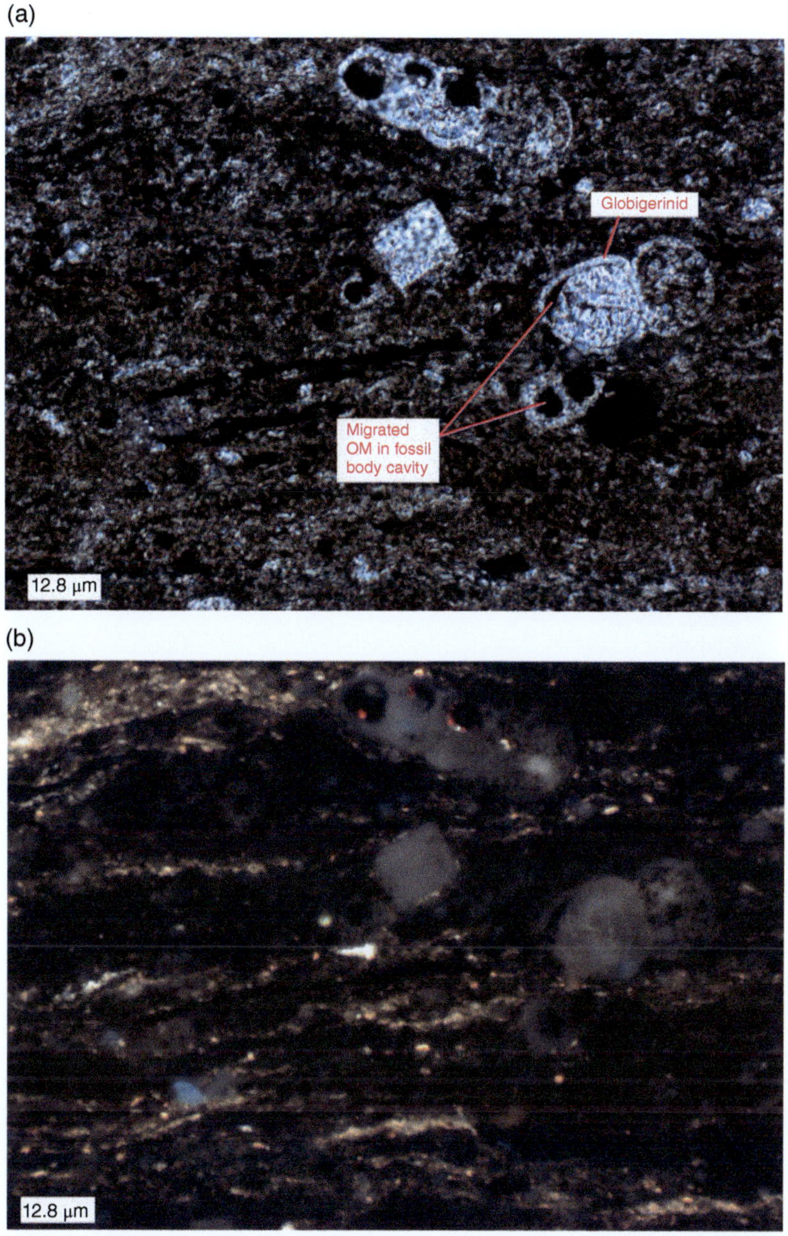

Figure 4.47 Light microscopy images of an Eagle Ford shale sample from an oil-matured formation: (a) white light image, (b) UV image.

(Al), magnesium (Mg), and sulphur (S). These identified elements are indicative of the following corresponding minerals: quartz, clay, dolomite, and pyrite, respectively, but not limited to these outlined minerals.

It is worth noting that in most cases, the recorded percentage weight of elemental carbon (C) in a particular ROI may not be solely contributed by organic matter (kerogen) but also from the matrix especially in the presence

Table 4.5 Characteristics of Some Shale Oil/Gas Resource Systems Around the World

Formation (period)	Mineral composition	Kerogen type	TOC (wt. %)	Thermal maturity (%Ro)	Porosity (%)	Reference
Monterey (Miocene)	Quartz, calcite, illite, mixed-layer illite/smectite, kaolinite, albite, apatite, pyrite, zeolite, OM	Type II S[a]	9.6–18.8	< 0.35[b]		Löhr et al., 2015
Hetaoyuan (Paleogene)	Calcareous shale, silty shale, OM	Type I–II	~2.61	0.52–0.87		Jiang et al., 2016
Demerara Rise black shales (Cretaceous)	Calcite, smectite, illite, quartz, biogenic opal, zeolite, apatite, pyrite, OM	Type II	10.4–12.0	< 0.35[b]		Löhr et al., 2015
Lower Eagle Ford shale (Upper Cretaceous)	OM, calcite, dolomite, quartz, pyrite, clay, K-Feldspar	Type II	5.06–5.75	0.98–1.03[c]	0.05–2.55 (organic)	
Eagle Ford (Upper Cretaceous)	Calcareous mudrock to calcite-rich siliceous mudrock	Type I/ II	4–8	0.6–1.4[c]		Ko et al., 2017
Bossier (Late Jurassic)	Silica, clay, carbonate, OM, chlorite	Type III/IV (gas)	0.5–4.0	1.10–2.40[c]		Jarvie, 2012
Haynesville (Late Jurassic)	Silica, clay, carbonate, OM	Type III/IV (gas)	0.5–4.0	1.20–2.40 (1.5)[c]		Jarvie, 2012
Kimmeridge Clay (Late Jurassic)	Quartz, K-feldspar, OM, illite, smectite, chlorite, muscovite, dolomite	Type I/II Type III/IV	0.15–9.6 < 1.42	< 0.5–1.3		
Montney (Triassic)	Silica, carbonate, clay, chlorite, OM, mica, pyrite	Type II/III	0.2–11.0	0.90–2.50[c]		Jarvie, 2012
Stuart Range (Early Permian)	Quartz, pyrite, Na & K-feldspars, mica, illite, kaolinite, smectite, OM	Type II	5.8–7.7	0.46–0.48[b]		Löhr et al., 2015
Barnett shale, Forth Worth Basin (Mississippian)	Siliceous shale, calcareous shale, limestone, OM	Type II/III	3–12	0.85–2.1[c]		Jiang et al., 2016
Barnett - Williston Basin (Mississippian)	Limestone, shale, dolomite siltstone, OM	Type II	7.2–13	0.65–0.9		Jiang et al., 2016

Table 4.5 (Continued)

Formation (period)	Mineral composition	Kerogen type	TOC (wt. %)	Thermal maturity (%Ro)	Porosity (%)	Reference
Fayetteville (Mississippian)	Clay, carbonate, silica, chlorite, OM	Type III	2–10	2–4.5 (2.5)[c]		Jarvie, 2012
Chattanooga shale[d] (Devonian)	OM, quartz, pyrite, calcite, clay	Type II	3.2	0.63–0.77		Cudjoe et al., 2016 Newell & Doveton, 2010
Marcellus (Devonian)	Silica, clay, carbonate, chlorite, OM	Type II/III	2–13	0.9–5.0 (1.5)[c]		Goodman et al., 2018 Jarvie, 2012
Woodford (Devonian)	Quartz, illite, pyrite, apatite, K-feldspars, chlorite, dolomite, OM	Type II	3–12	0.7–4.0 (1.5)[c]		Jarvie, 2012 Löhr et al., 2015
Muskwa (Devonian)	Carbonate, clay, OM, dominated by silica	Type III	1–10	1.4–2.2 (2.0)[c]		
Longmaxi shale (Lower Silurian)	OM, clay minerals, quartz, feldspar, minor carbonate and pyrite	Type III/IV	0.27–4.25 (1.90)	2.97–3.23[c]	0.08–2.73[e] (organic) 0.06–2.65 (inorganic)	Yang & Bao, 2017 Chen et al., 2014 Chen et al., 2011
Llandovery-Wenlock series (Lower Silurian)	Clay minerals (chlorite & muscovite), quartz, pyrite, fine-grained limestones, marine carbonates, OM, albite	Baltic basin Type I/II Lublin basin Type III/IV	1–10 (3.9)	Baltic samples (0.8) Lublin samples (2.2)[c]	1–20*[e] (organic) 10–63 (InterP) 1–2 (IntraP)	Ma et al., 2017
Utica (Ordovician)	Mostly chlorite, silica, clay, carbonate, OM	Type II /III	0.8–5.0	0.8–3.0[c]		Jarvie, 2012

Note: (TOC) or (Ro) values in parenthesis imply an average value.

[a] Denotes kerogen type II but sulfur-rich (8–14 wt. % organic sulfur) (http://wiki.aapg.org/Type_IIS_kerogen).

[b] Calculated %Ro based on $\%Ro = 0.00180{*}T_{max}$ (°C) – 7.16.

[c] Organic pores are mostly observed in samples from late oil window (> 0.9 %Ro) to abundance in gas and over-mature samples (≥ 1.5 %Ro) (Löhr et al., 2015).

[d] Reported gas production for a brief period from this formation in Barber County, Kansas. Limited gas production was attributed to the presence of oil molecules plugging the pores (Newell & Doveton, 2010).

[e] Bubble-like pores form in gas-matured samples.

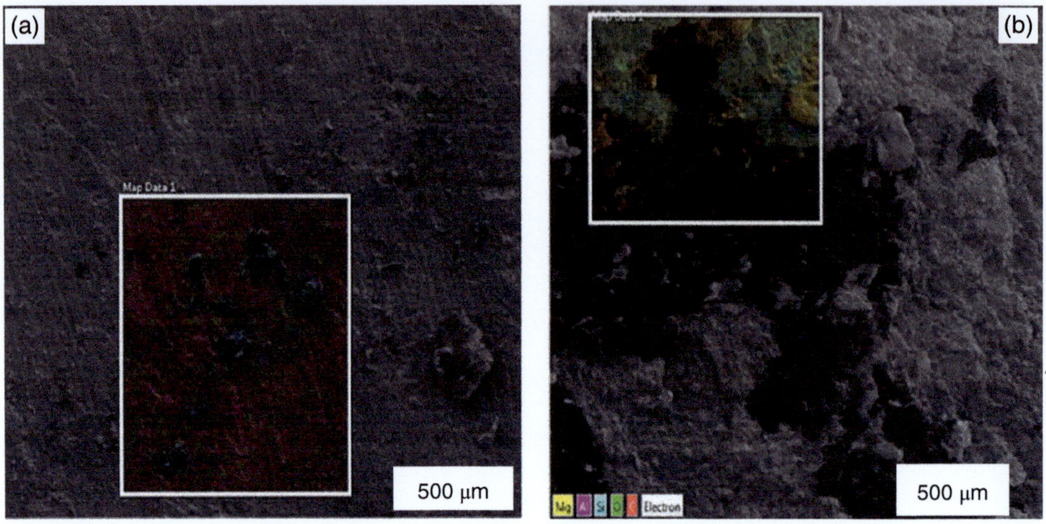

Figure 4.48 Regions of interest in both images (a) and (b), respectively of Chattanooga shale samples for EDS.

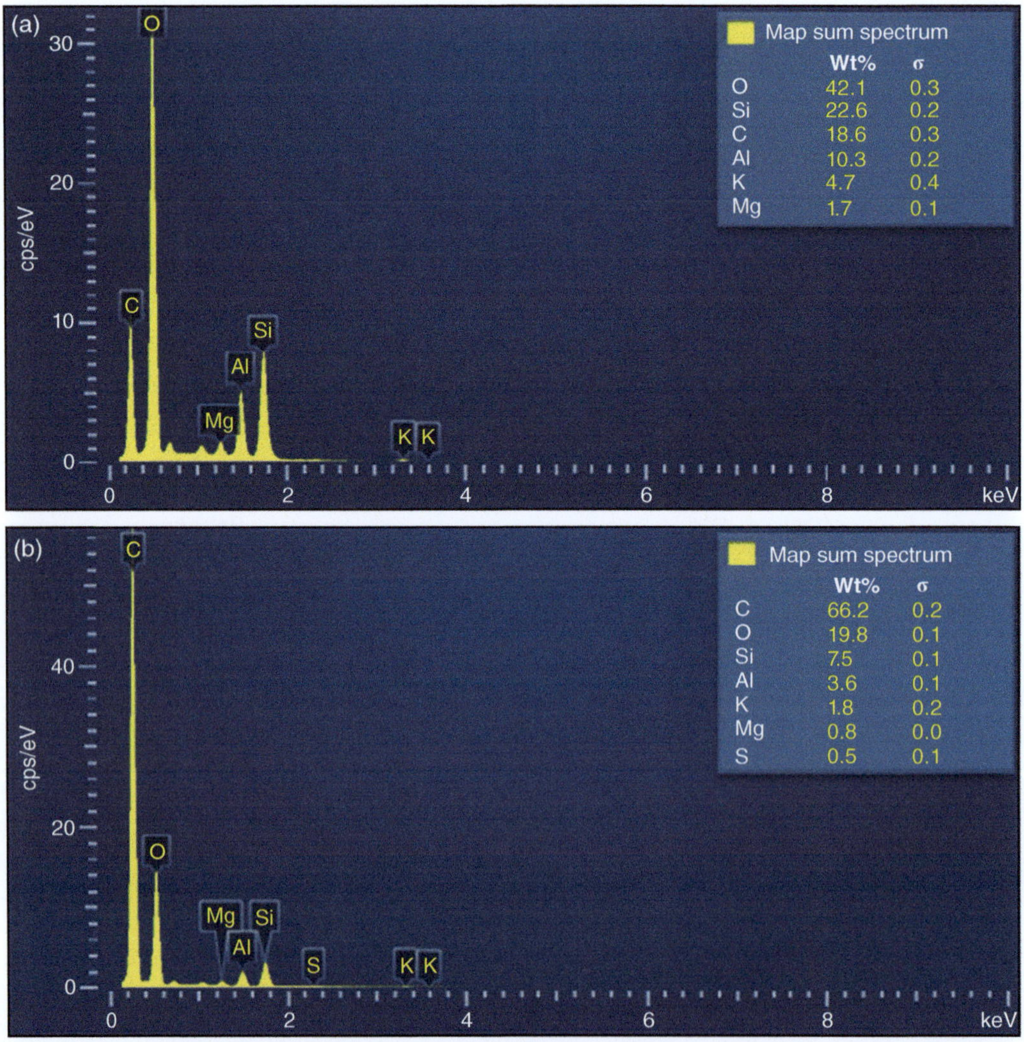

Figure 4.49 Elemental compositions of selected regions of Chattanooga shale samples for EDS analysis.

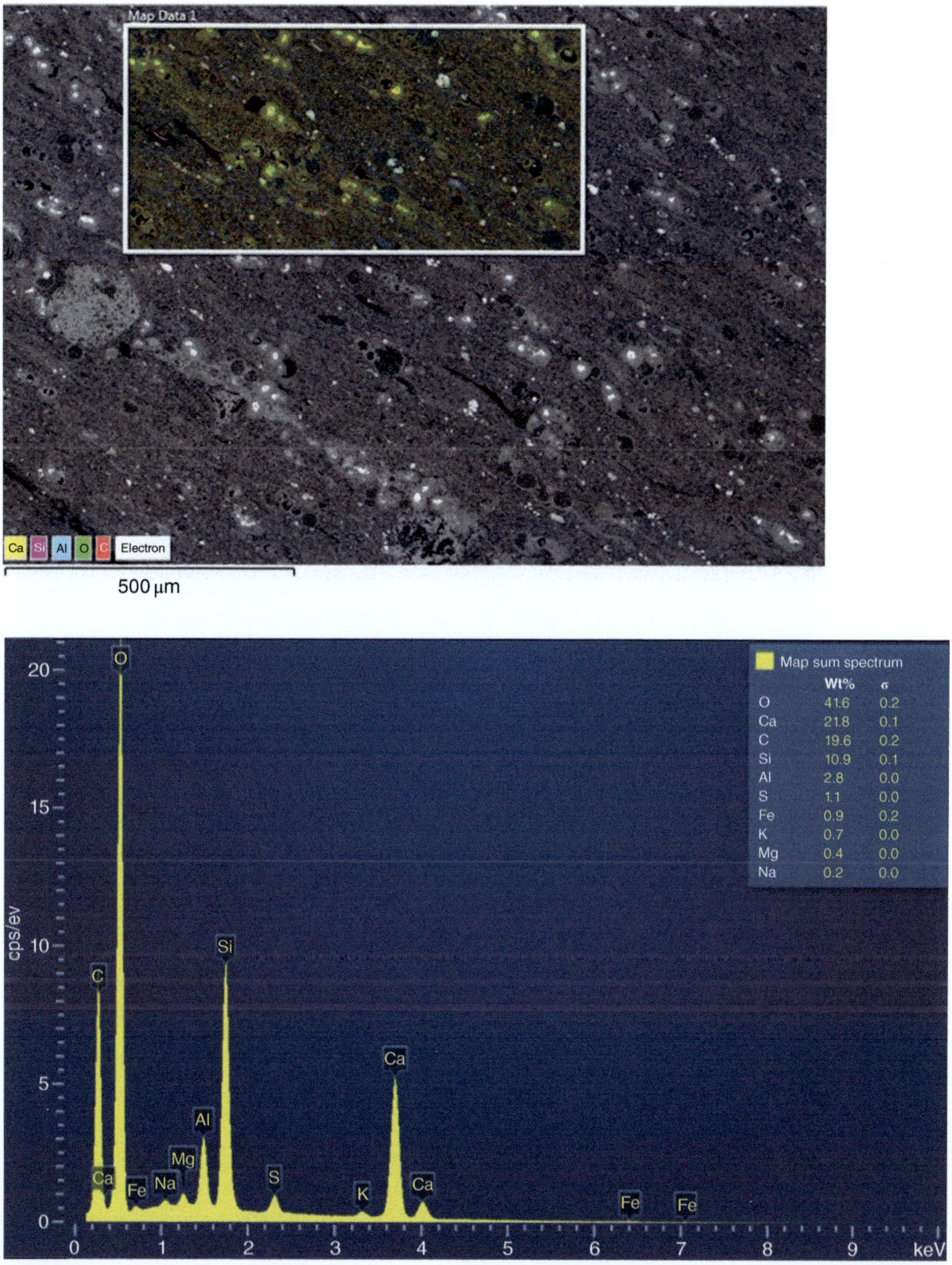

Figure 4.50 BSE image (top) of an Eagle Ford shale sample with selected ROI (white box) and (bottom) corresponding elemental composition.

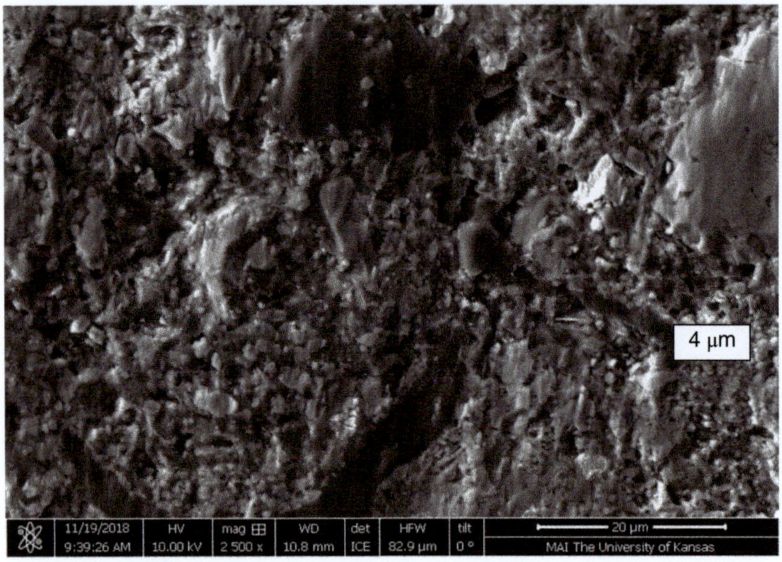

Figure 4.51 SEM image of an ion-milled Eagle Ford shale sample.

of carbonate minerals. Figure 4.51 shows an SEM image of an ion-milled Eagle Ford shale sample; organic matter/kerogen/pores are represented as dark/black, while matrix minerals are represented as gray.

The entire surface is analyzed through EDS and the resulting elemental map is shown in Figure 4.52. It can be noted that the carbon (C) map in turquoise extends beyond the dark patches into the matrix. Hence, the estimated carbon composition of 48.1% is due not only to the presence of organic matter, although dominant, but also to the presence of other matrix minerals.

Another observation in the elemental analysis maps is the absence of oxygen in almost all the regions containing organic matter. Figure 4.53 shows an SEM image of another Eagle Ford shale sample on a grayscale image.

The corresponding elemental maps are generated through EDS and shown in Figure 4.54. The carbon map (C-map) in turquoise outlines the shape of the OM, while the oxygen (O) map in red, next to the C-map, shows the absence of oxygen in the OM region. This indicates dominate carbon and hydrogen elements with the OM region with negligible oxygen.

An Eagle Ford shale sample at a much deeper depth than the one discussed above is analyzed (Fig. 4.55) with a selected ROI as shown in Figure 4.56 and a corresponding elemental composition plot in Figure 4.57. The elemental composition plot records a high carbon composition of ~46 wt.% in the presence of silicon (Si), calcium (Ca), and sulphur (S) elements. These elements are interpreted as OM (C), quartz (Si), calcite (Ca), and pyrite (S).

Figure 4.58a and b shows EDS maps of an Eagle Ford shale formation in the south Texas Upper Cretaceous shelf. The dominant elements of Si, Ca, Na, Mg, and Al correspond to mineral constituents in Figure 4.59. The identified minerals are quartz, calcite, albite, dolomite, and clay, respectively (Ko et al., 2017).

EDS is an advantageous elemental/mineral quantification method, which does not constrain the number of minerals per pixel for a single scan and also avoids complex mineral combinations (Saif et al., 2017).

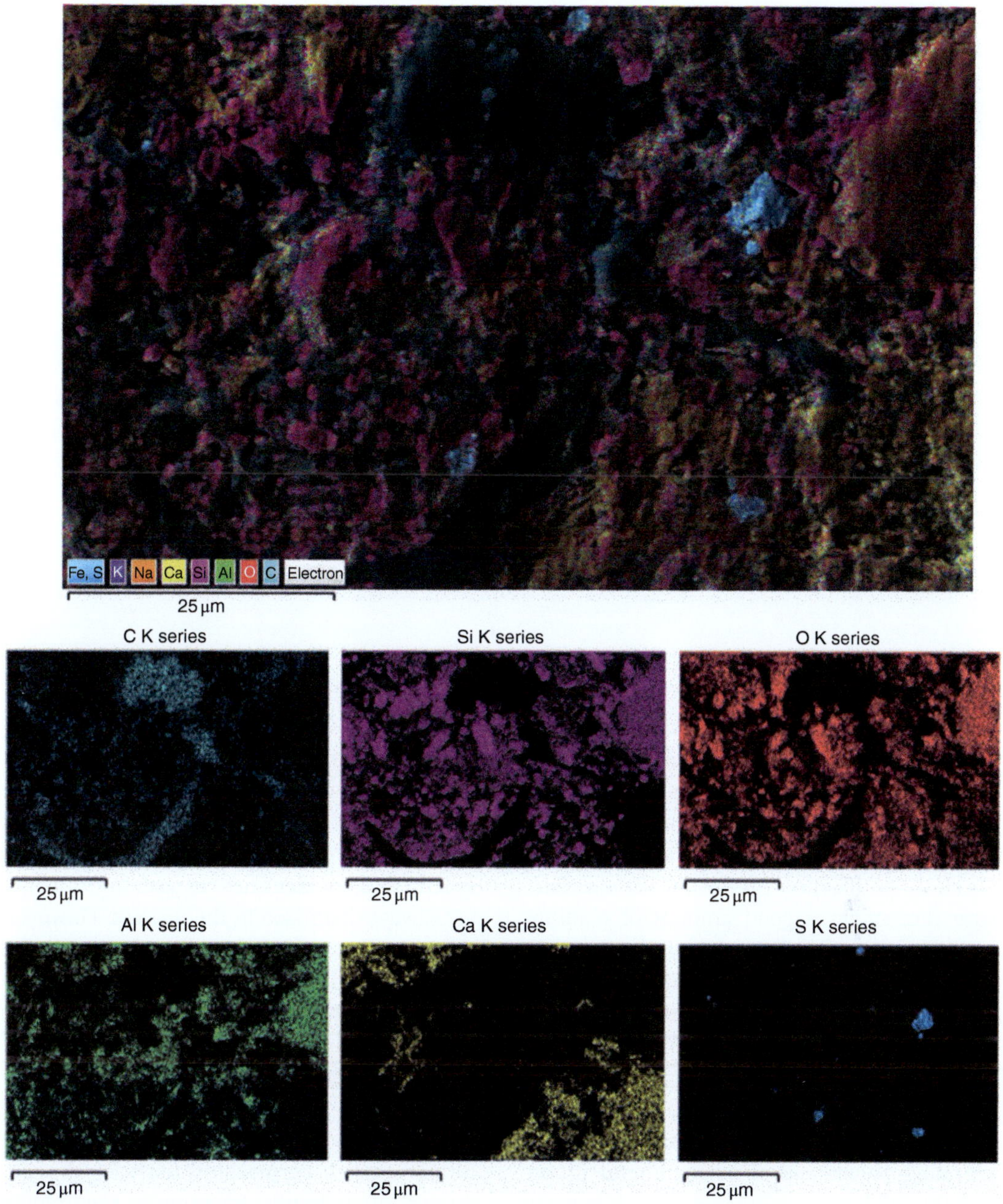

Figure 4.52 EDS analysis on an Eagle Ford shale with corresponding elemental maps.

4.4.4 Quantitative Evaluation of Minerals by Scanning Electron Microscopy (QEMSCAN)

QEMSCAN is a frequently used SEM microanalysis system for investigating quantitative mineral composition and chemical properties of samples in addition to generating high-resolution mineral maps (Ayling et al., 2012). The QEMSCAN makes use of SEM in addition to four EDS systems. As previously described,

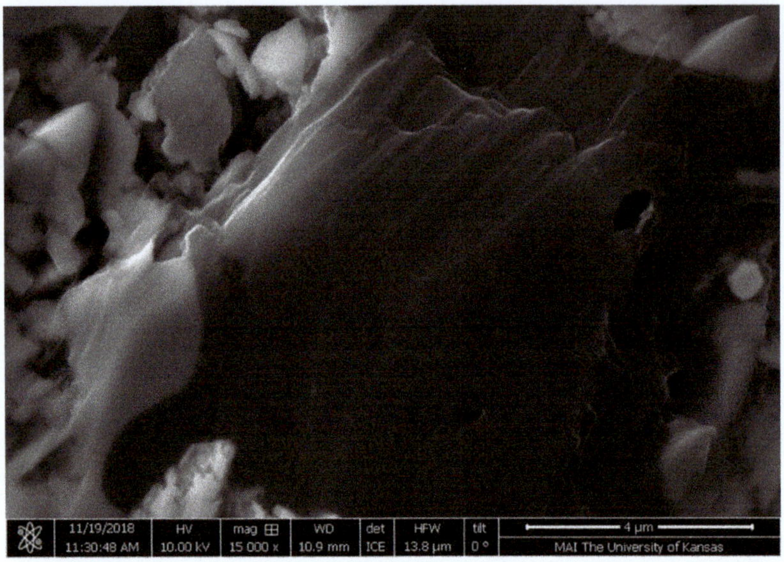

Figure 4.53 SEM image of an Eagle Ford shale sample.

SEM imaging is achieved by the subsequent interaction of emitted electrons with the surface of the material to produce signals at different intervals. Based on the X-ray characteristic signals, QEMSCAN provides mineral distribution maps for the sample surface utilizing SIP. Species Identification Protocol (SIP) is a database that stocks a good amount of readings for different minerals under different conditions to identify the mineral properties from previous scans of similar materials (Josh et al., 2012). A QEMSCAN mineral distribution map is shown in Figure 4.60 for a Chattanooga shale sample, where every color on the map represents a specific mineral and the corresponding percentage of each mineral in the sample under investigation is shown in Table 4.6.

The tabulated mineral composition shows that this particular selected region of the Chattanooga shale matrix is dominated by quartz (silica), pyrite, and clay minerals, weight percentage-wise. Organics are not typically measured with EDS because they are composed mostly of carbon (C), which would take a lot longer to acquire a 2,000 count spectra compared to quartz or calcite minerals. Maps mineralogy works on an acquisition time, where each spot must collect at least 2,000 counts within 8 ms, or it is classified as a low-count spectra and not identified. In so doing, organics are always dismissed in the process. However, the organics can be factored in afterward with BSE thresholding as they are not as dark as the background but usually much darker than the surrounding mineral phases.

Mineral-based brittleness index (MBI) can be estimated from the percentage weight of the minerals provided by QEMSCAN. Due to the ultralow matrix permeability of shales, the generation of hydraulic fractures is required to enhance fluid flow, and identifying sweet spots is the key to a successful fracture network creation. MBI serves as a means to delineate best spots to fracture and place horizontal wells based on its capability to differentiate ductile zones from brittle zones (Cudjoe et al., 2016). The MBI is given as (Wang & Gale, 2009)

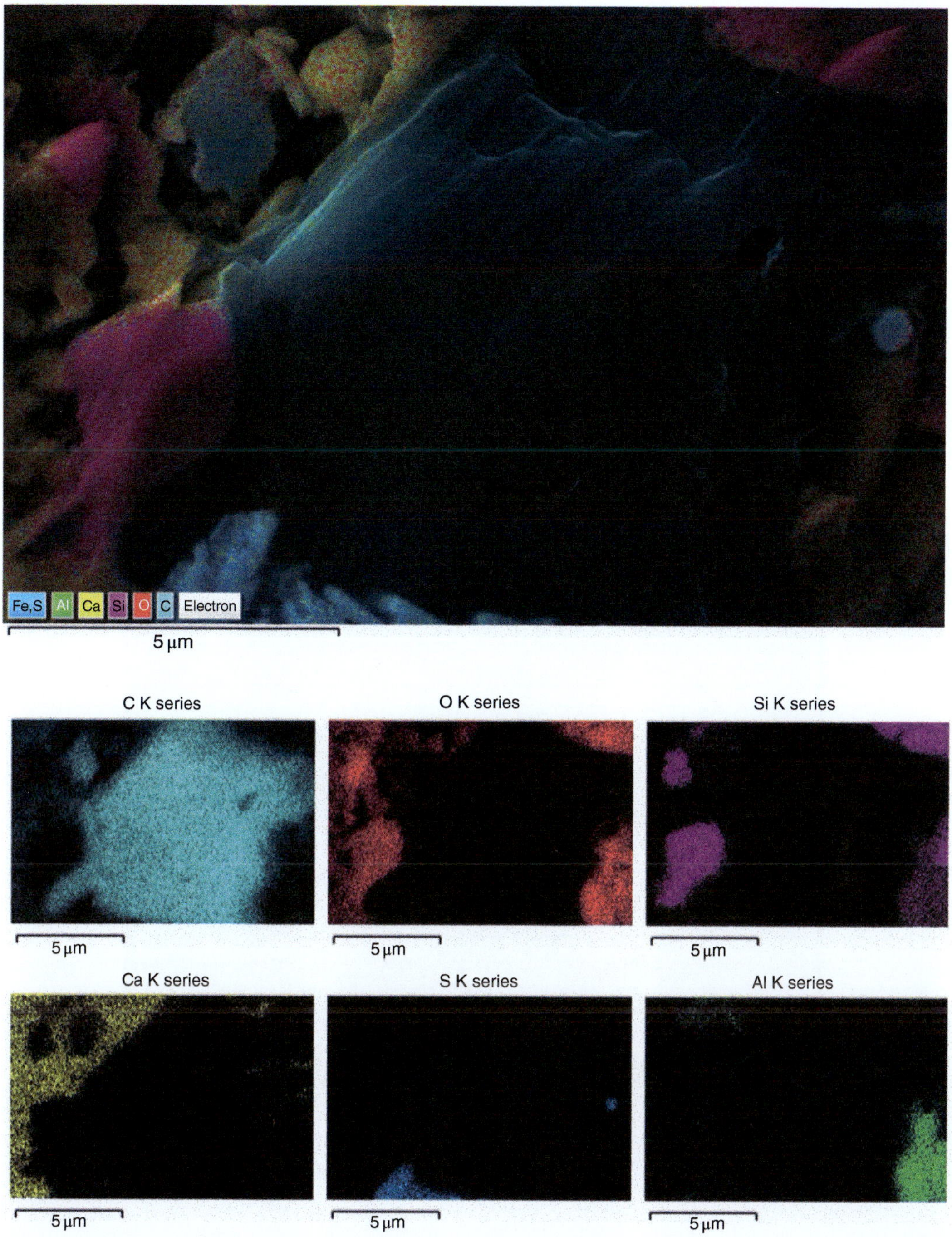

Figure 4.54 Elemental map showing the different distribution of elements in an Eagle Ford shale sample.

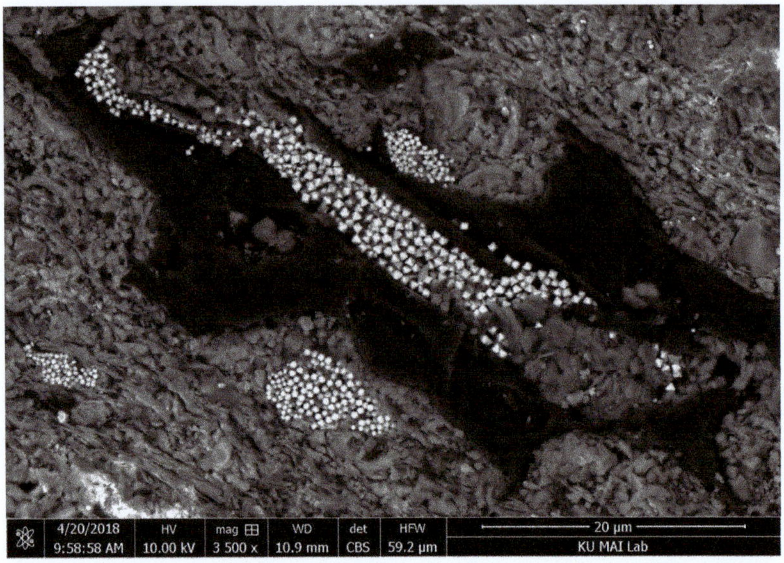

Figure 4.55 BSE image of an Eagle Ford shale sample.

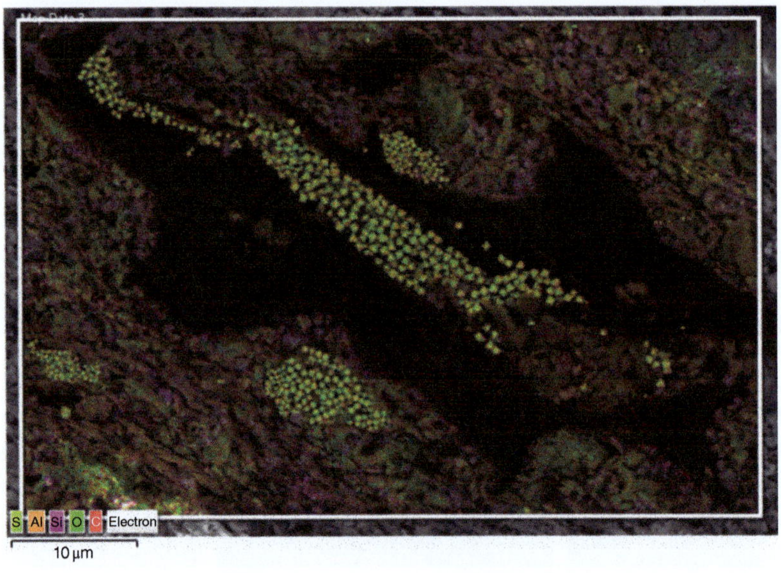

Figure 4.56 Selected ROI on BSE image of an Eagle Ford shale sample for EDS analysis.

$$MBI = \frac{QFM + CAR}{CAR + QFM + CL + TOC}, \tag{4.10}$$

where, QFM is the weight percentage of quartz, feldspar, and mica; CAR is the weight percentage of carbonate (calcite and dolomite); CL is clay minerals; and TOC is the total organic carbon.

The calculated MBI for the Chattanooga shale sample is 0.68. An MBI greater than 0.6 is classified as the most brittle spot suitable

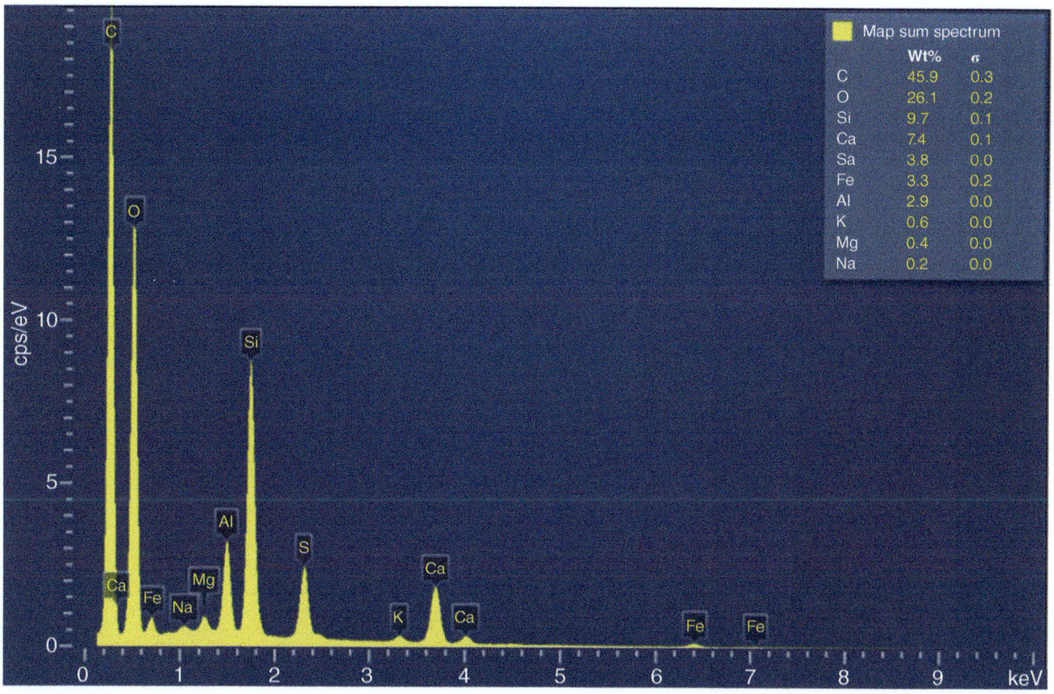

Figure 4.57 Elemental composition of corresponding ROI in Figure 4.56.

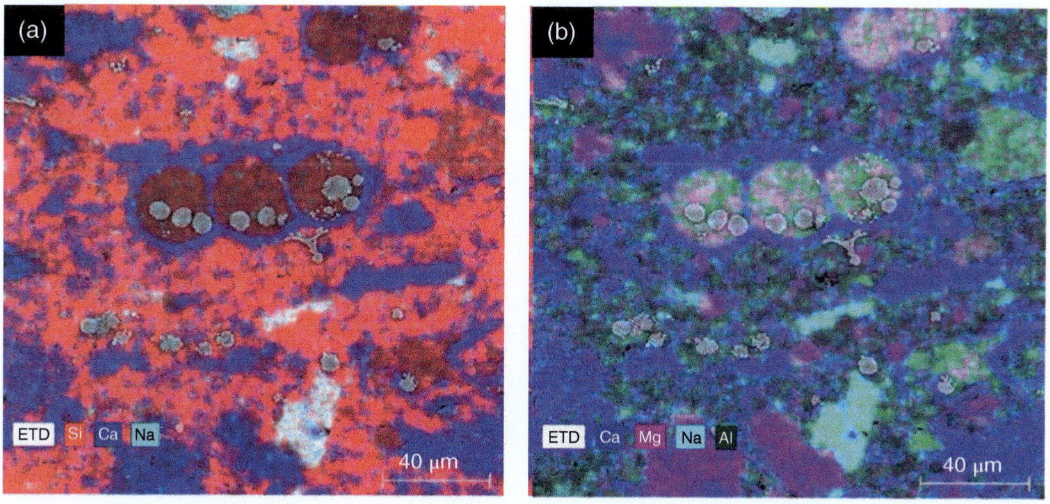

Figure 4.58 EDS elemental maps of an Eagle Ford sample at a depth of 11,901 ft (Ko et al., 2017).

for fracturing, while an MBI less than 0.6 is considered the most ductile zone not favorable for fracturing based on well-log analysis (Alzahabi et al., 2015). Figure 4.61 shows the spatial distribution of minerals on an overlay of two BSE mosaics into a large mineral map of an oil shale sample from the Green River dominated by carbonate (dolomite and calcite), and K-feldspar, among other minerals (Saif et al., 2017).

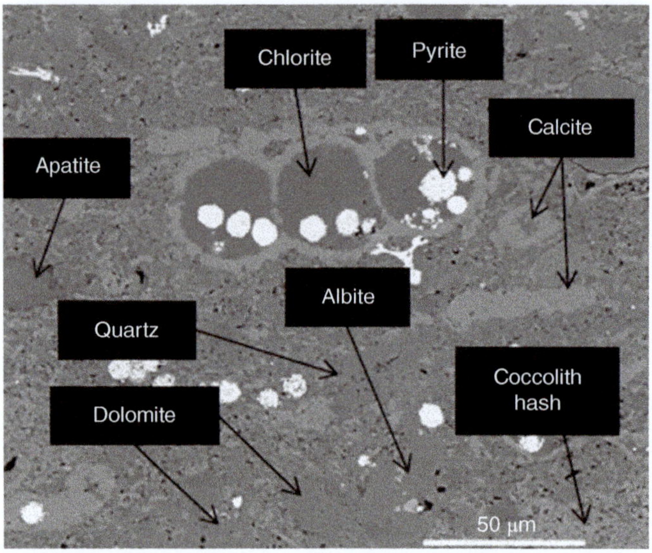

Figure 4.59 Composite map showing Eagle Ford minerals identified through elemental maps (Ko et al., 2017).

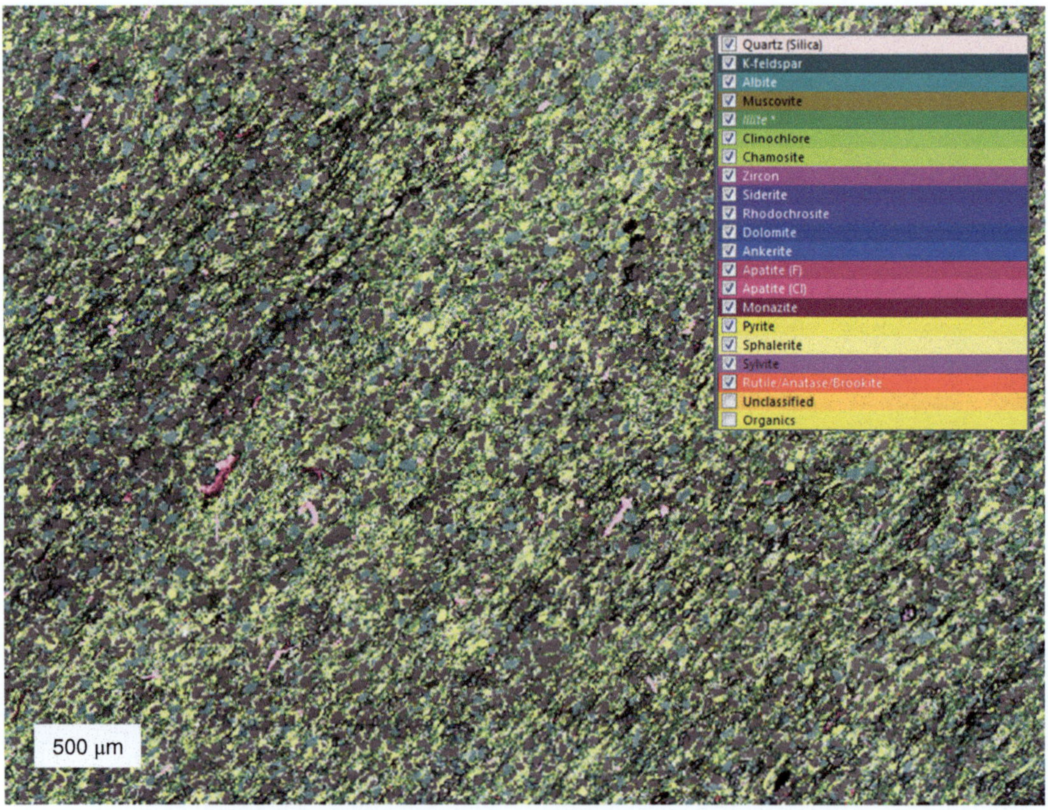

Figure 4.60 Mineral distribution map of a Chattanooga shale sample with the QEMSCAN.

Table 4.6 Measurement of Mineral Composition in a Chattanooga Shale Sample

Sequence	Group	Measurement (area %)	Measurement (wt. %)
0	Quartz (Silica)	37.14%	32.38%
1	Feldspars	17.13%	14.57%
3	Micas	4.01%	3.78%
4	Clay minerals	22.94%	20.93%
8	Zircon	0.03%	0.05%
9	Calcites	0.05%	0.06%
12	Dolomites	0.03%	0.03%
14	Apatites	1.04%	1.10%
16	Monazite	0.00%	0.00%
17	Pyrite	16.61%	26.77%

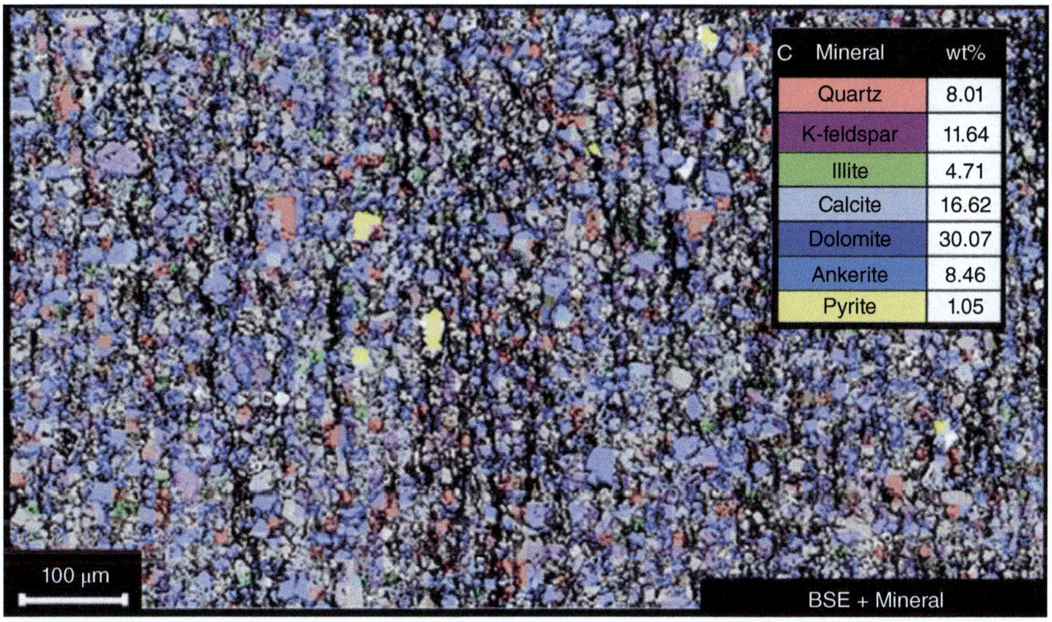

Figure 4.61 Quantification of minerals present in the Green River oil shale sample (Saif et al., 2017).

QEMSCAN provides less scan time and a reliable chemical composition with attractive graphical results to the mineral industry.

4.4.5 Focused Ion Beam-Scanning Electron Microscopy (FIB-SEM)

The focused ion beam-scanning electron microscopy (FIB-SEM) consists of a vacuum system, liquid metal ion source, ion column, stage, detectors, gas inlets, and a computer. The system exists as a stand-alone beam instrument or is incorporated into a scanning electron microscope (SEM). Combining the SEM column with a gallium (Ga) focused ion beam (FIB) forms a FIB-SEM system where a Ga beam is used to mill away thin serial slices of

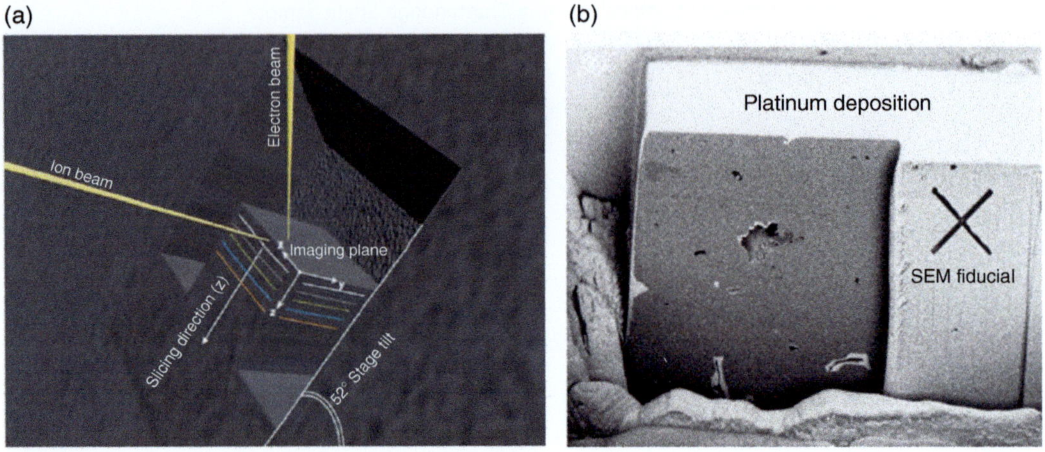

Figure 4.62 (a) FIB-SEM imaging profile and (b) SEM image of a cross-sectioned shale (Camp et al., 2013; Lemmens & Richards, 2013).

material and consecutively acquire SEM or BSE images of the exposed surface. A stack of 2D images is produced as a result, from which a 3D volume can be created. The constructed 3D volume reveals 3D distribution of mineral types, organic matter, porosity (organic and inorganic), and pore connectivity in shale samples (Camp et al., 2013; Lemmens & Richards, 2013). Figure 4.62a shows a typical FIB-SEM sample imaging profile where the ion beam and electron beam are oriented at an angle of 52° to allow for sequential sputtering of material and simultaneous SEM imaging. Figure 4.62b shows the cross-section of a shale sample for FIB-SEM imaging after digging out trenches as part of the sample preparation. The FIB-SEM acquistion process starts with preparing a site of interest by depositing a layer (strip) of platinum (Pt) (~1 μm) to prevent the target area from damage during ion milling. This is followed by digging out trenches around the coated region of interest to aid the serial sectioning and imaging (Lemmens & Richards, 2013; Curtis et al., 2011; Elgmati et al., 2011).

The focused ion beam (FIB) is then used to sputter away thick layers (~20 nm) off the cross-sectioned face, and either SEM or BSE images of the exposed surface are obtained; BSE images are preferrred for clear distinction of mineral phases. The total time required to mill a 10 x 10 x 25 μm^3 volume of a rock sample is between 15 to 20 hr (Elgmati et al., 2011). An automated program (auto slice and view software) is implemented for ease and the fiducial marks created in the preparation stage facilitates in directing the software for sectioning and subsequent imaging (Camp et al., 2013; Lemmens & Richards, 2013; Curtis et al., 2011; Elgmati et al., 2011; Holzer et al., 2004). Samples, especially shale rocks, usually require polishing to provide a flat, smooth surface for better milling rate and time and to eliminate artifacts. Most physical artifacts associated with FIB-SEM images of shale samples are due to the ion beam on the sample. Moreover, different materials (carbonate, pyrite, quartz, organic material, etc.), which constitute a shale rock, mill at different rates resulting in unevenness of the imaged surface (Kelly et al., 2016). Charging occurs when electrons accumulate on the sample surface due to lack of a conductive path and present as extremely bright areas, warping of the image, and even streaks. Curtaining, also known as waterfall effect, is caused by the spatial variation of the sputter rate of the sample due to porous materials, rough surfaces, and composite of hard and soft materials. Milled faces show lines and/or irregular nonflat surface due to the curtaining effect (Reuteler, 2017). Figure 4.63 highlights some common artifacts encountered during FIB-SEM acquisition of shale rocks as curtaining and charging effects. Coating the shale sample with carbon (thickness ~10 nm) reduces the charging effects to a minimum. Another challenge is distinguishing organic material from pores as can be seen in Figure 4.63 in blue square boxes, and in some cases the matrix from the organic material through SEM imaging (Kelly et al., 2016). Therefore, Argon ion (Ar) milling is recommended as opposed to mechanical milling to smooth the shale surfaces, since ion milling preserves the pore structure unlike the latter. Mechanical polishing of shale samples is not ideal because of the different hardnesses of the organic and inorganic components resulting in an uneven surface.

Argon (Ar) ion milling is a noncontact polishing technique for sample preparation for FIB-SEM, which uses Ar ions to obtain a smooth surface. The approach for preparing shale oil and gas samples with Ar ion mill uses a mask to create a cross-section resulting in a flat section limited to 1 x 1 mm. The Fischione 1060 SEM Mill (Fig. 4.64) is commonly used by first attaching samples to a 25 mm diameter pin-type stub using carbon tape and then loading into the SEM mill. A polished surface is then produced after 1.5 hr for 5 kV ions under a 5° incident angle and 360° sample rotation (Lemmens & Richards, 2013). Figure 4.65 shows (a) a SEM image of a Marcellus shale sample before ion polishing with a very rough surface compared to (b) a BSE image of polished Marcellus shale sample with a smooth surface (Erdman et al., 2012).

4.4.6 Three-Dimensional (3D) Rock Model

The iterative milling of thin slices (~20 nm) and subsequently acquiring SEM/BSE images of each exposed surface results in a stack of 2D SEM/BSE images, which can be reconstructed into a digital 3D volume consisting of mineral matrix, organic matter, and pores/fractures. Developing a 3D volume from the sequential 2D FIB-SEM images requires image processing, which includes shearing of the 2D images to identify bad slices, cropping and alignment, filtering and intensity correction attainable in

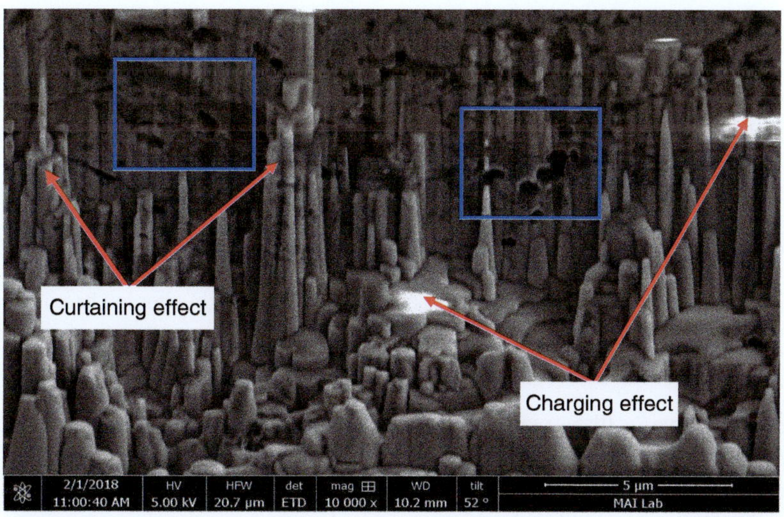

Figure 4.63 FIB-SEM image of an unpolished shale sample showing curtaining effect and charging effect.

Figure 4.64 Schematic diagram of the Argon Ion Mill (Lemmens & Richards, 2013; with permission from Fischione Instruments).

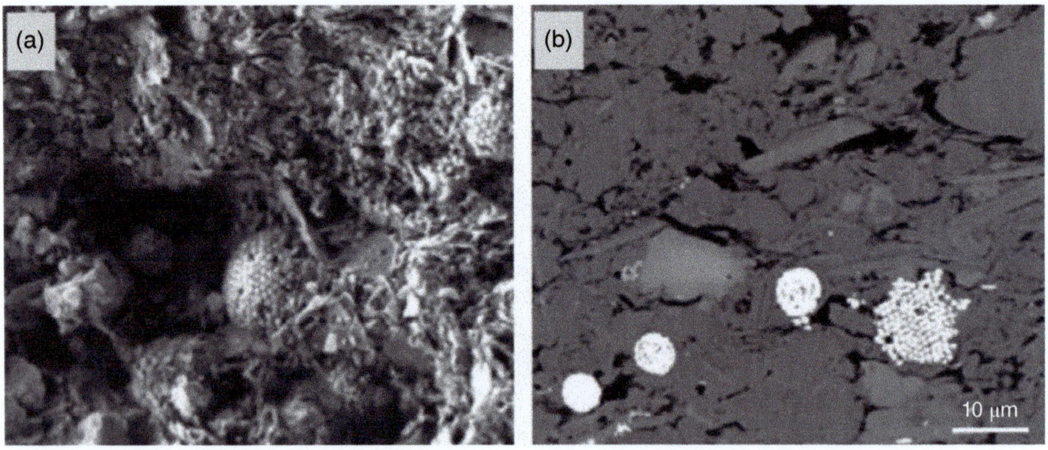

Figure 4.65 FE-SEM images of a Marcellus shale sample (a) before and (b) after ion beam polishing (Erdman et al., 2012).

image-processing softwares like Avizo, ImageJ, and PerGeos. The shear tool compensates for geometric effects of the stage tilt by shifting each XY-slice in Y-direction. Also, bad slices from acquisition can be identified and replaced by interpolation between two good slices during alignment. A filtering tool can also be applied to minimize curtaining effects and reduce noise reduction while preserving edges. These preprocessing steps are taken to get rid of

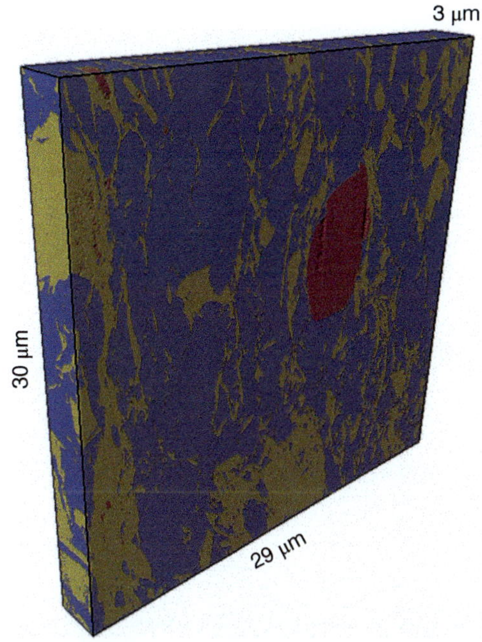

Figure 4.66 3D segmented volume outlining kerogen network (blue), pyrite (red), and mineral matrix (yellow).

artifact and obtain an accurate 3D model for further quantitative analysis.

Figure 4.66 shows a generated 3D volume of the Chattanooga shale sample from sequential 2D FIB-SEM images, segmented to subjective threshold values, which correspond to each distinct and visible material. Due to the absence of pores in both the inorganic crystalline-matrix and organic material, pore size distribution and a pore network model could not be generated. The 3D volume consists of 148 slices; the dimensions of the resulting volume is given as 29.2 µm x 30 µm x 2.94 µm with a resolution of 15 nm/voxel. The extracted volumes of kerogen and crystalline-matrix minerals through segmentation as previously explained are estimated as 68.05% of organic material (OM), 2.05% of pyrite, and 29.58% of crystalline matrix as listed in Table 4.7. There were no signs of organic pores at the 15 nm resolution which could be mainly attributed to the immature nature of the Chattanooga shale sample (Ro% = 0.63).

Table 4.7 Volume Fraction Estimation of Chattanooga Shale FIB-SEM Volume

Materials	Volume fraction (%)
Organic material (OM)	68.05
Mineral matrix	29.58
Pyrite	2.05

On the other hand, Figure 4.67 shows a segmented 3D volume of LEF samples at different depths with varying lithofacies. The FIB-SEM volumes of samples A, B, and C were focused on the organic content, while that of sample B (inorganic) was focused on the inorganic section of sample B described in detail by the work of Cudjoe et al. (2019). Table 4.8 shows the calculated volume fractions of the various materials/components selected based on the watershed segmentation method. The organic porosity is highest in sample A and these extracted pores would be further analyzed to obtain pore and throat size distribution to generate a pore network model for fluid flow simulations and bulk transport estimations.

Although structural detail of sample is lost with the FIB-SEM technique, it is immensely compensated by the detailed level of sample heterogeneity (Camp et al., 2013; Lemmens & Richards, 2013).

4.4.7 Pore Network Model (PNM) and Pore Size Distribution (PSD)

To accurately predict the distribution of the organic pores and develop a representative pore network model to bulk transport properties and eventually simulate fluid flow, a connected pore space tool is applied to the segmented (binary) image. The axis connectivity tool found in most visualization softwares (Avizo, PerGeos), generates a binary image containing all paths connecting two planes (XY, XZ, or YZ parallel faces). In so doing, the estimated volume fraction of the connected pores is always less than the volume fraction of

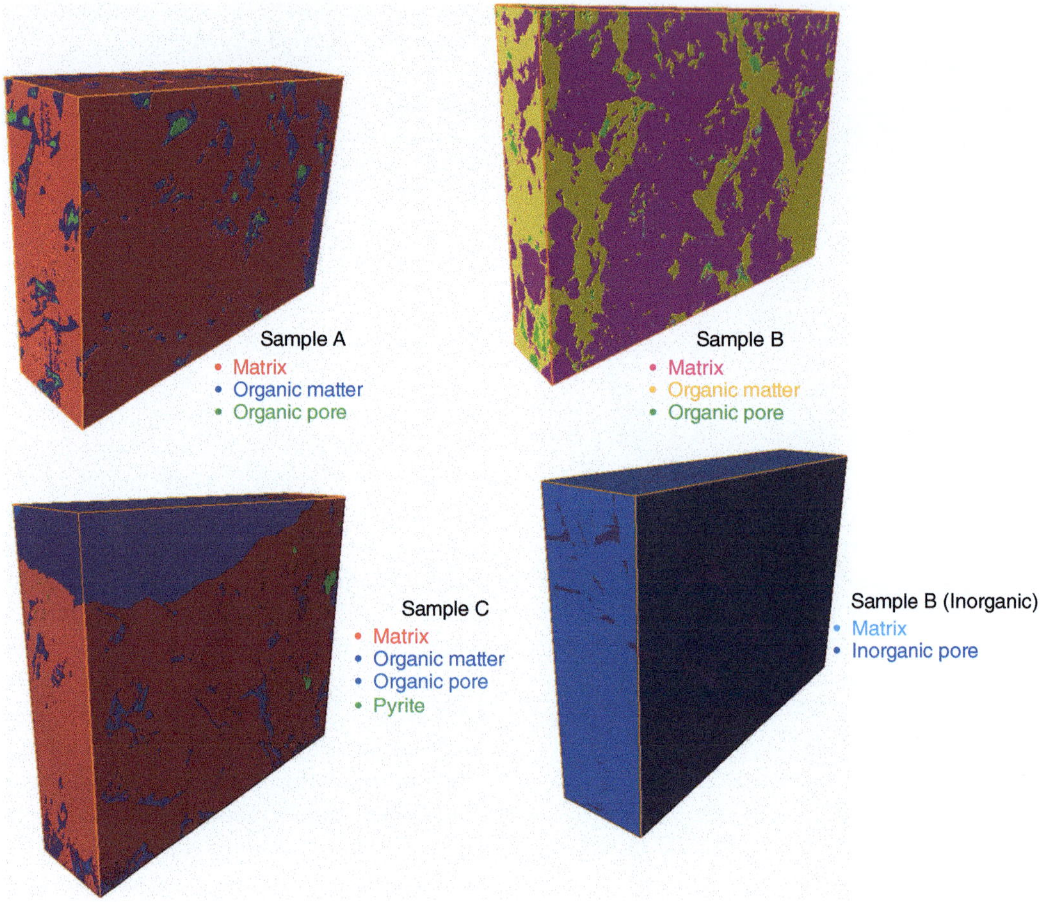

Figure 4.67 3D segmented FIB-SEM volumes LEF samples, where the different colors represent different phases within the grayscale image (Cudjoe et al., 2019).

Table 4.8 Volume Fraction Estimation Each Component Within the 3D Volumes of the LEF Samples

Materials	Sample A	Sample B	Sample C	Sample B (inorganic)
Organic material (OM)	13.2	27.2	17.8	-
Pores	2.11	1.65	0.87	1.46
Matrix	84.69	71.15	80.95	98.54

Source: Cudjoe et al. (2019).

all pores including isolated pores. The resulting connected pores are separated into groups of particles for further analysis of pore morphology and generation of a pore network model. Figure 4.68 shows a generated pore network model based on the segmented pores from Figure 4.67. The pore nodes are scaled to the estimated volume of each pore and the pore

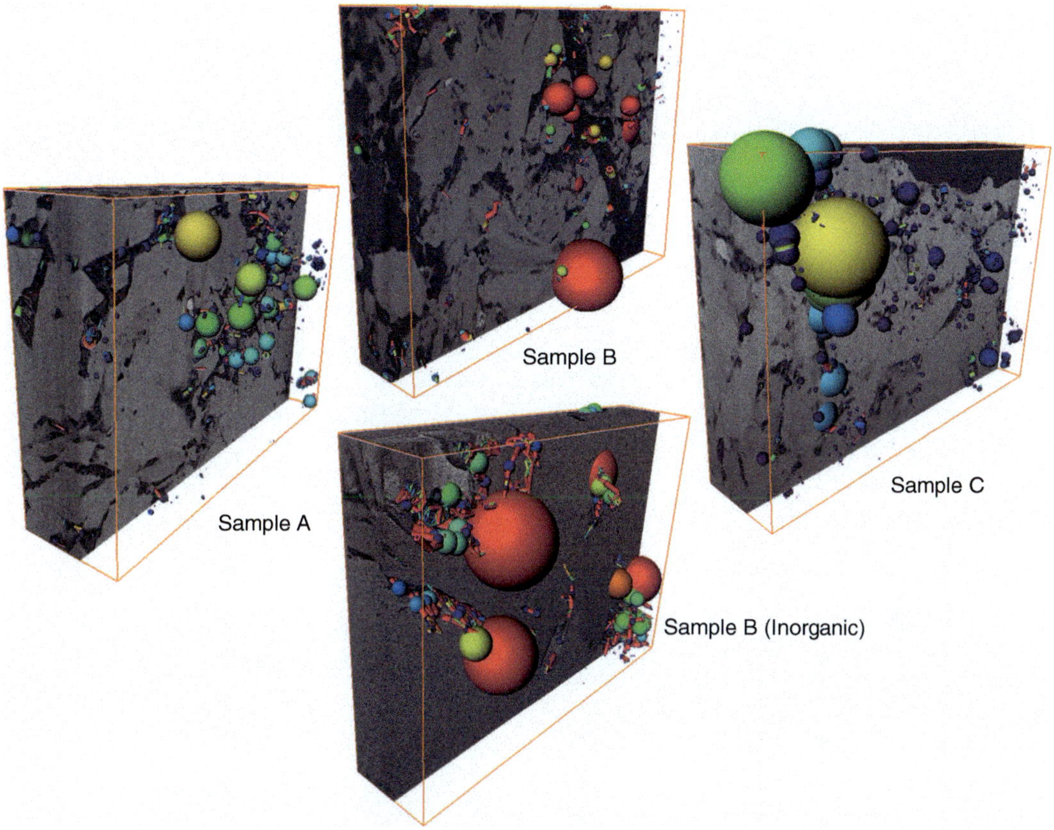

Figure 4.68 Generated PNMs of LEF samples with varying pore and throat sizes. Pores are represented as spheres, while throats are represented as rods (Cudjoe et al., 2019).

color is scaled to the equivalent radius of each pore. On the other hand, the throats and the color of the throats are both scaled to the equivalent radius of each throat. The equivalent diameter of the organic pores ranges from 13 nm to 500 nm, whereas the equivalent organic throat diameter ranges from 5 nm to 300 nm as shown in Figure 4.69 for samples A and B. The inorganic pores were observed to range from 13 nm to 400 nm with throat size ranging from 5 nm to 330 nm (Cudjoe et al., 2019).

Tortuosity

Tortuosity is a fundamental rock property capable of measuring the geometric complexity of a porous medium and characterizing the convoluted pathways of fluid diffusion and electrical conduction through the medium. It is therefore defined as the ratio of the actual length of the flow path to a straight line length in the direction of flow (Backeberg et al., 2017). Moreover, tortuosity is used to relate effective diffusivity to the Knudsen diffusion coefficient in an attempt to describe Knudsen transport in shale oil/gas reservoirs as given in equation (4.11) (Chen et al., 2015; Holt & Smith, 1988),

$$D_e - \varphi \frac{D_k}{\tau}, \tag{4.11}$$

where D_e is effective diffusivity, D_k is the Knudsen diffusion coefficient, τ is the tortuosity, and φ is the sample porosity.

Shale gas reservoir rocks are governed by different transport mechanisms during extraction

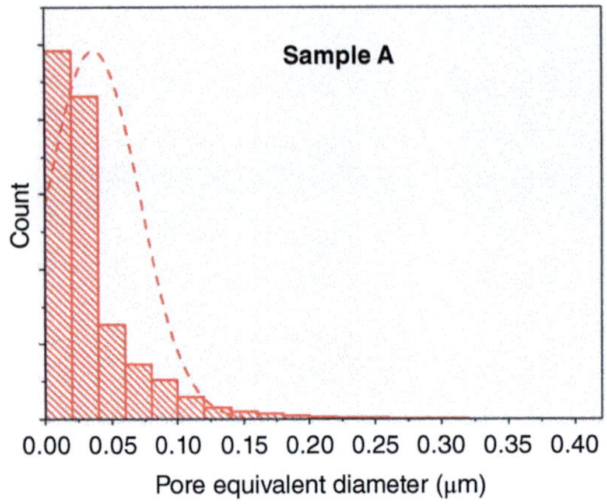

Figure 4.69 Pore size distribution of a LEF shale sample ranging from 13 nm to 500 nm (Cudjoe et al., 2019).

of gas due to the multiscale characteristics of the medium moving from macroscale in the fractures obeying Darcy's flow to molecular scale in the organic pores introducing Knudsen diffusion, as well as Fickian gas diffusion from the matrix pores to the fractures (Wang & Marongiu-Porcu, 2015; Javadpour et al., 2007). Thus, the transport mechanisms are dominated by advective-dispersive transport along the fractures into the surrounding rock matrix as well as Knudsen diffusion of adsorbed gas onto the surface of the organic material and into the matrix. Therefore, knowing the tortuosity of the shale medium, the effective diffusivity can be accurately estimated given the porosity and diffusion coefficient of the medium.

The centroid path tortuosity (CPT) in Figure 4.70 through image analysis is estimated by dividing the sum of the distance between the centroids (d_n) of each image plane to the total distance along the z-axis (H) (PerGeos Tortuosity Module). The resulting organic and inorganic tortuosity values for the LEF samples estimated using the CPT ranged from 13 to 35, which are quite high (Cudjoe et al., 2019).

The hydraulic tortuosity (HT) was implemented to compare the results. The HT is based on velocities obtained from the absolute permeability estimation using Darcy's law and is

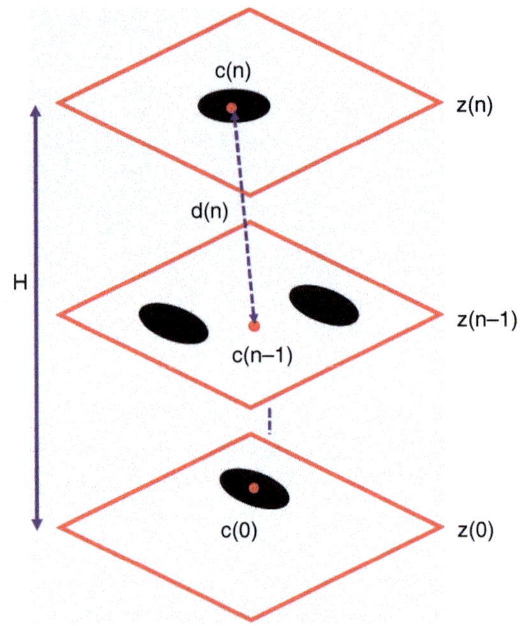

Figure 4.70 Centroid path tortuosity module (modified illustration from PerGeos Tortuosity Module).

defined as the ratio of the sum of the length of all velocities to the sum of the projection of the velocities along the flow direction. This is based on the assumption that the velocities of each throat are known. By simulating a single phase methane at an inlet pressure of 3,500 psi and outlet pressure of 3,000 psi in the

Table 4.9 Estimated HT for the LEF samples

LEF Samples	HT in the *x*-direction
A	1.84
B	2.34
C	1.33
B-Inorganic	1.57

Source: Cudjoe et al. (2019).

x-direction, the estimated HT values are listed in Table 4.9.

4.4.8 Permeability Estimation

Image analysis makes permeability estimation possible as it gives insights into pore networks including pore sizes, shape factors, and tortuosity, while serving as a good alternative for non-representative laboratory measurements (Peng et al., 2017). As a result, permeability based on image analysis for highly heterogeneous formations like the Eagle Ford shale sample can be estimated with the classical Kozeny-Carman (K-C) equation given as (Backeberg et al., 2017; Peng et al., 2015, 2017; Berg, 2014; Matyka et al., 2008; Costa, 2006)

$$K_{total} = \sum_{i=1}^{N}(K_i) = \sum_{i=1}^{N}\frac{r_{hyd_i}^2}{\beta_i\tau^2}\varphi, \quad (4.12)$$

where, K_{total} is the total permeability (m^2), φ is the porosity of the FIB-SEM volume, r_{hyd_i} is the hydraulic radius (μm) of individual pores within the FIB-SEM volume, β_i is the individual shape factor of each pore within the FIB-SEM volume, N is the total number of individual labeled pores within the FIB-SEM volume, and τ is the tortuosity of the FIB-SEM volume. The hydraulic radius is defined as the ratio of the volume of a pore to the solid-fluid interfacial area (Borst, 1983; Costa, 2006), while the shape factor is a dimensionless quantity used in image analysis to numerically describe the shape of the pores.

The K-C equation relates the permeability (K) to the porosity through well-defined parameters of the pore structure. Through pore morphology analysis of the connected and separated pores, the average shape factor for the LEF samples ranges from 0.98 (inorganic) to 1.35 (organic) with an average surface area of ~164.5 μm^2 for Sample B (organic) and 78.3 μm^2 for the inorganic pores of Sample B. The estimated geometric average permeability from the modified K-C equation ranged from 0.000633 to 0.0048 mD for the organic PNMs and 0.00433 mD for the inorganic PNM (Cudjoe et al., 2019). A plot of permeability against the average specific surface area of the three FIB-SEM volumes of the LEF samples yielded a linear correlation, where a higher average specific surface area (SSA) resulted in a higher permeability and a lower SSA resulted in lower permeability.

On the other hand, the lattice Boltzmann method (LBM) based on the kinetic theory can be applied to the generated pore network model (PNM) from FIB-SEM imaging to estimate both the intrinsic permeability without pore-wall interactions and the permeability incorporating the pore-wall interactions (slip flow, adsorption, Knudsen diffusion, and surface diffusion), which is dominate at the pore-scale for kerogen nanopores (Chen et al., 2015; Zhang et al., 2015; Ning et al., 2016; Wang et al., 2016).

4.5 Broad Ion Beam-Scanning Electron Microscopy (BIB-SEM)

BIB-SEM is a new field of pore investigation in shales and clays (fine-grained rocks) using argon (Ar) broad ion beam (BIB) milling in combination with SEM. Unlike the previously discussed electron imaging techniques, BIB-SEM enables imaging of relatively large sample areas (approximately 1–2 mm^2) at a high resolution to be viewed with the SEM. Thus, while

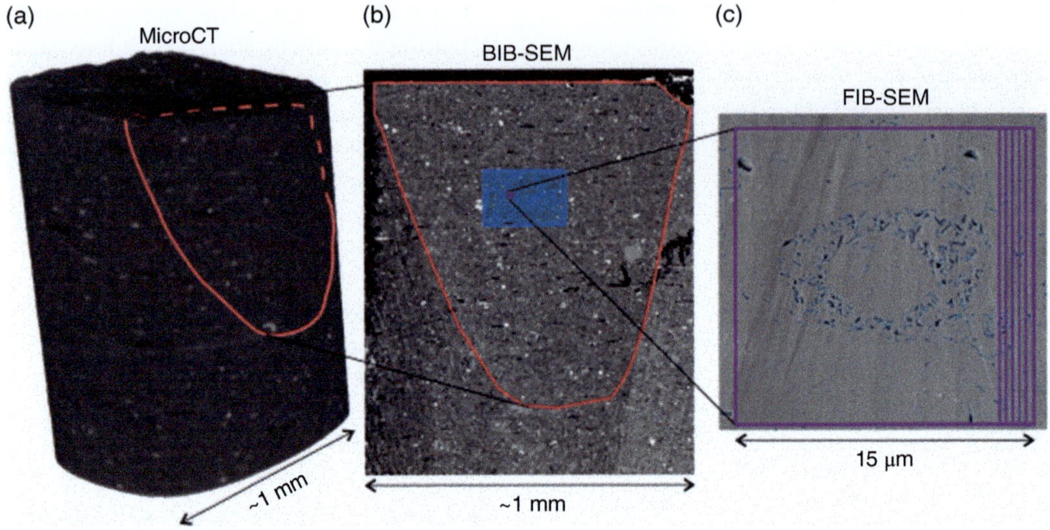

Figure 4.71 A link between micro-CT, BIB-SEM, and FIB-SEM: (a) shows a micro-CT volume with the outline of the BIB cross section; (b) shows a BIB-SEM cross section with segmented pores (blue) and ROI (purple) for FIB-SEM; and (c) shows the outline of a FIB-SEM image with some slices in purple and segmented pore edges in blue on the BIB cross section (Klaver et al., 2014).

BIB milling is commonly used to analyze millimeter-sized 2D sections, FIB is used for micrometer-sized volumes (Sinn et al., 2017; Klaver et al., 2014; Giffin et al., 2013).

Figure 4.71 shows the link between micro-CT, BIB-SEM, and FIB-SEM, respectively. The large area obtained through BIB-SEM imaging reveals information about pore sizes, pore morphology and spatial distribution, rock microstructure, and anisotropy (a measure of the pores' deviation from spherical shape) (Sinn et al., 2017; Klaver et al., 2014). To easily distinguish between mineral phases within an imaged sample, the BSE detector can be used in place of the SEM for clear and accurate distinction between the pores, organic material, and mineral phase as previously mentioned. In addition to EDS, BIB-SEM/BSE enables the identification of individual grains (Sinn et al., 2017; Schmatz et al., 2016). Moreover, unlike the FIB-SEM, where an imaged area is permanently damaged, an ion-milled BIB surface can be revisited for more detailed study (Kelly et al., 2016; Hemes et al., 2015; Klaver et al., 2014). However, the relatively large slice thickness makes BIB-SEM serial sectioning for 3D pore network development irrelevant for shales (Klaver et al., 2014).

4.5.1 Sample Preparation, BIB-SEM Acquisition and Processing

Subsamples of shales for BIB-SEM imaging can be obtained as rectangular blocks of suitable dimensions from end pieces of core plugs using a low-speed microdiamond saw (Sinn et al., 2017). Sample size is determined by the capacity of the BIB equipment, which polishes a cross-section of about 2 mm^2 using an argon beam for about 8 hr of milling under a vacuum (10^{-3}–10^{-4}, 6 kV and 150 μA) (Giffin et al., 2013).The sides of the cut sample is prepolished with silicon carbide (SiC) paper from very fine grit to ultrafine grit with an average particle diameter of ~10 μm before using a polisher (Sinn et al., 2017; Klaver et al., 2014). Wet and fresh samples are first dried by means of vacuum, oven, room temperature, or freeze drying methods. The edges of the sample are recommended to be perpendicular to each other to ensure the optimum sample setup for the BIB sample holder and to provide better ion-polished surfaces (Klaver et al., 2014; Houben et al., 2013).

Milling locations are checked on the subsample through the use of a polisher to produce planar, Gaussian-shaped cross-sections of approximately 2 mm^2. Most often than not, the side perpendicular to the sample bedding is ion milled. The samples are then coated with either gold or carbon to prevent charging during imaging since rocks are generally poor conductors (Sinn et al., 2017; Klaver et al., 2014). Polishing mostly removes about 100 μm of material and creates a cross-section with a topography of about 5 nm, which allows easy selection of porous areas to be studied within the polished cross-section (Klaver et al., 2014; Giffin et al., 2013). The final image of selected areas imaged with either SEM or BSE is a mosaic, created by taking hundreds of pictures at magnifications of either 2,000, 5,000, or 30,000x using 10%–20% overlap as shown in Figure 4.72 (Klaver et al., 2014; Giffin et al., 2013). A conventional SEM image (1,024 x 768) represents an area of about 50,000 nm^2 at a magnification of 20,000x, which is a fraction of the entire BIB cross-section (Klaver et al., 2014).

Hence, making use of the stage scan function of the SEM, large areas at high magnifications can be mapped as the sample stage moves laterally to an adjacent location after each scanned SEM/BSE image while maintaining some overlap. Therefore, combining all SEM images form one large mosaic as seen in Figure 4.72; dashed boxes indicate the overlap, while arrows indicate the mapping trajectory (Klaver et al., 2014). In a manner similar to FIB-SEM images, the mosaics obtained via BIB-SEM can be studied further in detail using image software for rock digital physics.

4.6 Acknowledgment

This chapter was made possible by the immense contributions from Thermo Fisher Scientific. Specifically, Erin Honse at Thermo Fisher Scientific with expertise on SEM, FIB-SEM, and QEMSCAN as well as Arash Aghaei, also at Thermo Fisher Scientific, an expert on the PerGeos software.

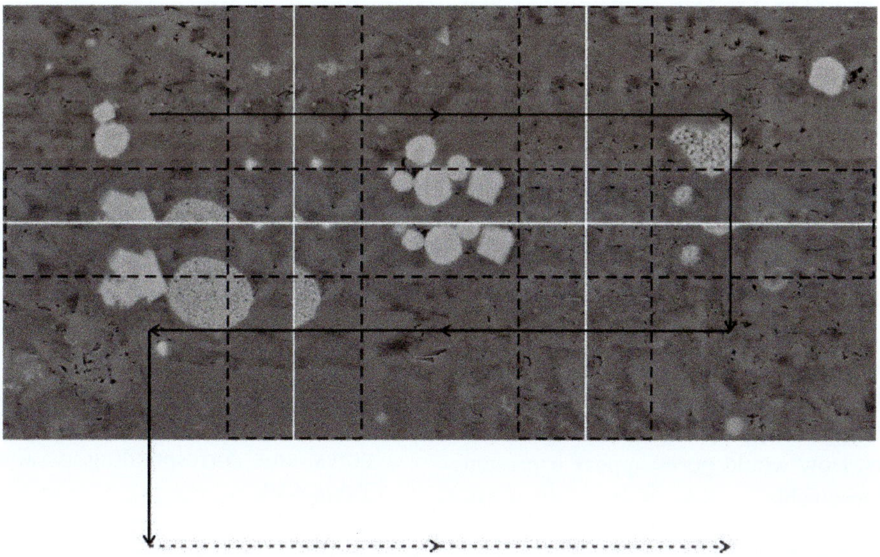

Figure 4.72 Mosaic mapping of SEM/BSE images with 25% of overlap. Dashed boxes indicate the overlap and arrows indicate the mapping trajectory (Klaver et al., 2014).

4.7 Problems on Micro-CT and Nano-CT

1 For a sample of known dimensions, a useful guideline is that the optimal resolution is a factor of 1,000 times smaller than the width of the sample assuming only 1,000 of the 2,000 available pixels of the detector are used. What would be the optimal resolution of the Chattanooga shale chip with a width of 3 mm?

2 What ideal sample size (small or large) would produce a better CT image resolution?

3 What is the difference between a circular micro-CT and a helical micro-CT?

4 Comment on the resolution of both micro-CT and nano-CT tools and their applicability for imaging of ultratight, organic-rich shale samples that are of interest to the oil and gas industry.

5 Estimate the CTN of a material with mass attenuation coefficient (μ/ρ) of 5,200 cm^2/g given the mass attenuation coefficient (μ/ρ) of water to be 4,078 cm^2/g at 1,000 eV and that of air at 1,000 eV to be 3,606 cm^2/g (https://physics.nist.gov/PhysRefData/XrayMassCoef/ComTab/water.html; https://physics.nist.gov/PhysRefData/XrayMassCoeff/ComTab/air.html).

6 How would both less-dense and denser materials, respectively, appear on a micro-CT or nano-CT image?

7 From the CT image of the Chattanooga shale sample, organic matter appears as dark spots, while pyrite appear as bright spots. How would pores appear if present and resolvable?

8 What would low CTN and high CTN, respectively, represent in a CT image in terms of porosity?

9 Calculate the porosity of a core sample given the measured CTN of the fluid-filled core to be 300 HU and the CTN of the clean, dry core to be 200 HU. The CTN of the fluid is -500 HU.

10 Estimate the saturation of oil (S_o) for a shale core given the CTN of rock saturated with both oil and water to be 150 HU, the CTN of rock saturated with only water to be 250 HU, and that of the rock saturated with only oil to be 50 HU.

4.8 Problems on Electron Microscopy

1 Identify the mineral distribution as well as the type of pores on the grayscale images in Figure 4.73, Figure 4.74, and Figure 4.75.

2 Figure 4.76 shows a selected ROI on a BSE image of a shale sample with a corresponding elemental composition plot in Figure 4.77. What minerals can be inferred from the identified elements?

3 In a similar manner, Figure 4.78 shows a selected ROI on a BSE image of a shale sample, while Figure 4.79 shows the elemental composition. What minerals do the identified elements represent?

4 Estimate the total permeability in millidarcy (mD) for a 3D volume with porosity of 2.4% and pore tortuosity of 25 using the K-C equation for the given number of pores and corresponding parameters in Table 4.10.

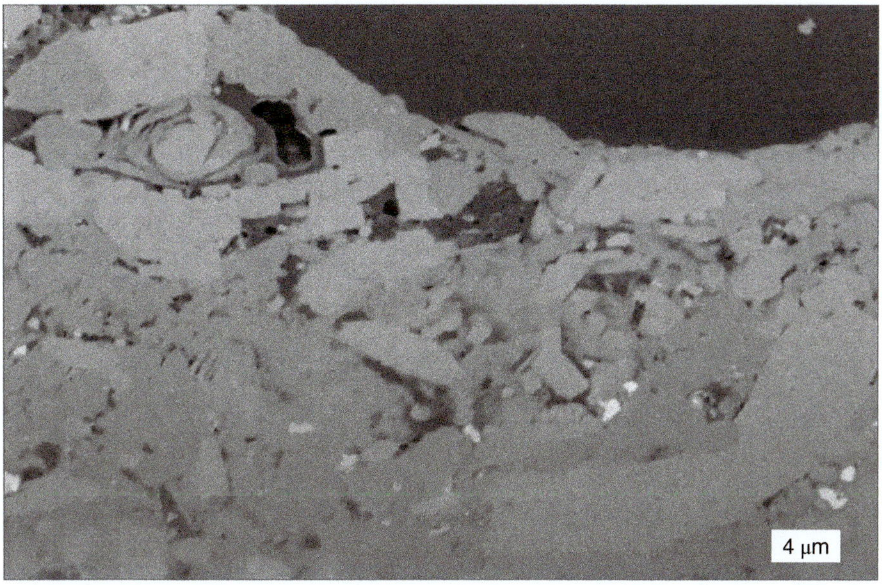

Figure 4.73 BSE image of an Eagle Ford shale sample.

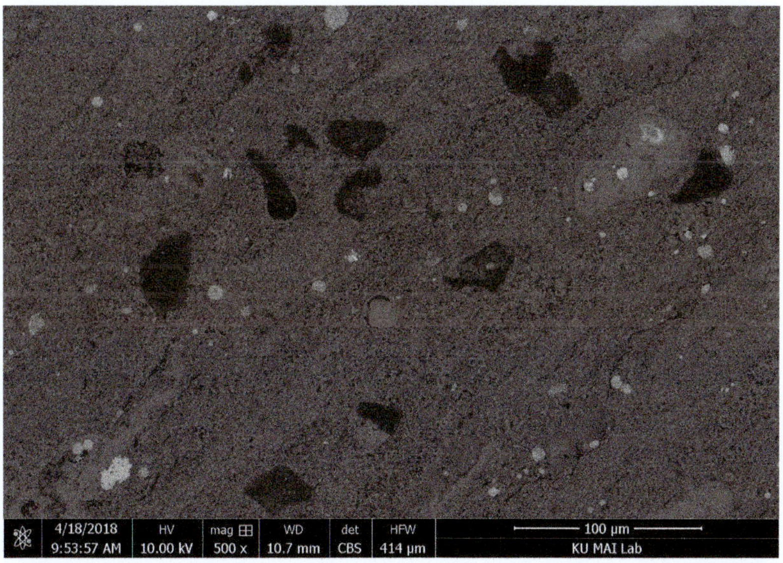

Figure 4.74 BSE image of an Eagle Ford shale sample.

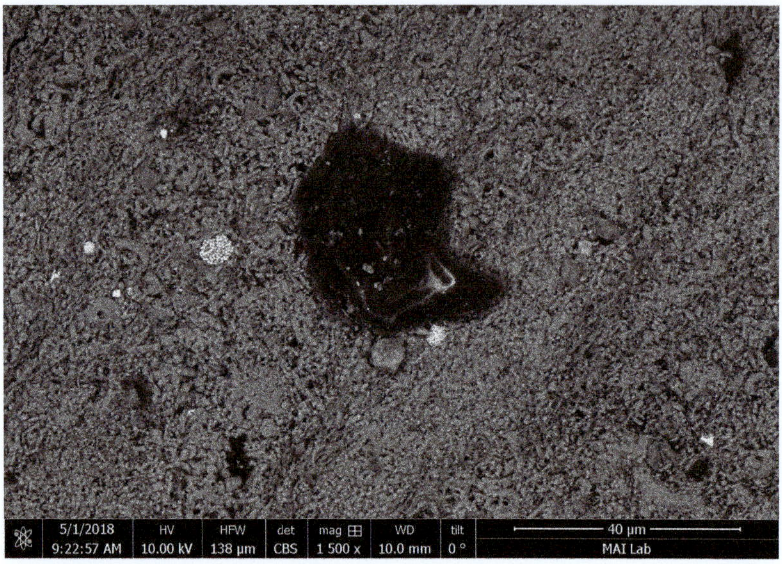

Figure 4.75 BSE image of an Eagle Ford shale sample.

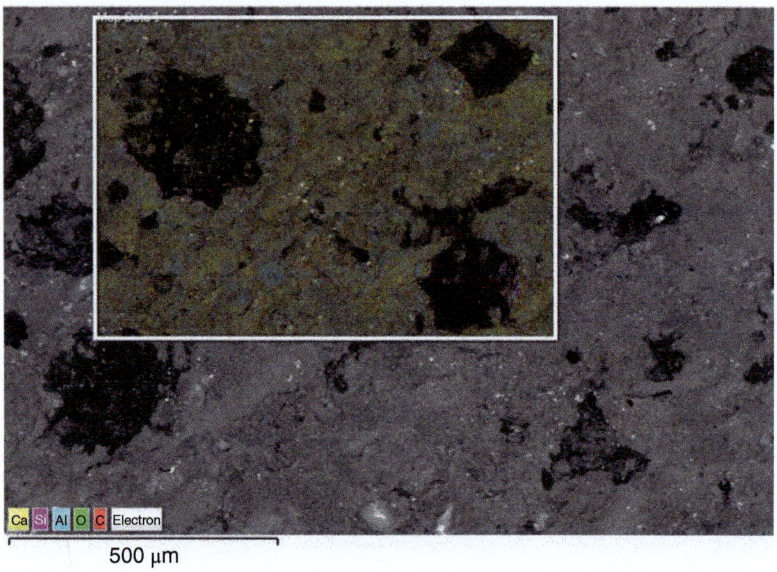

Figure 4.76 Selected ROI on a BSE image of an Eagle Ford shale for EDS.

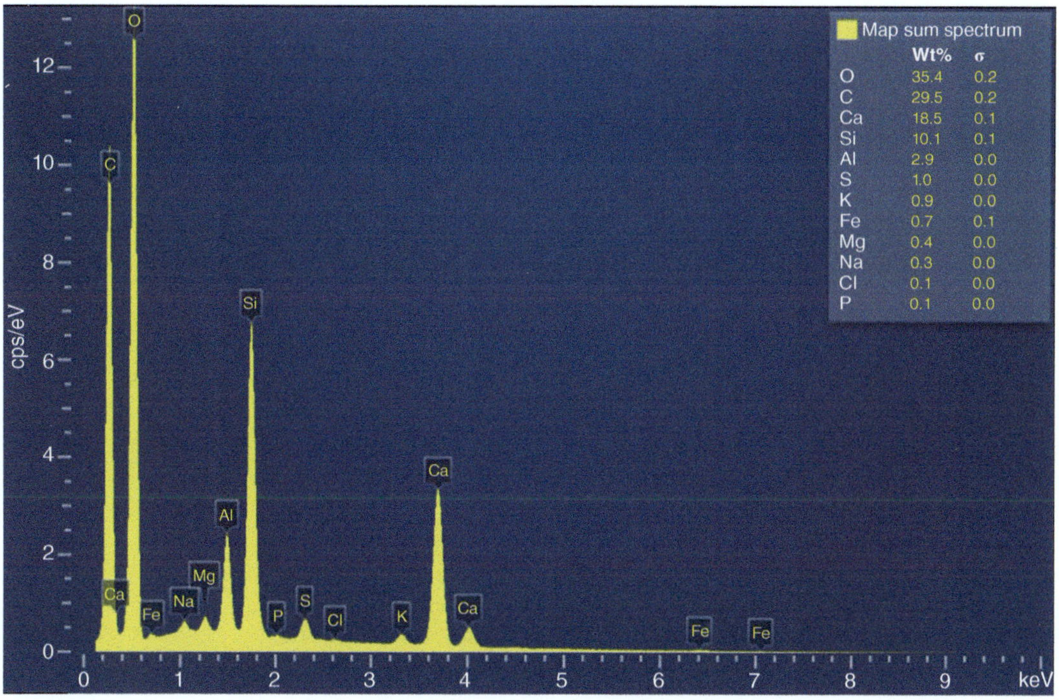

Figure 4.77 Elemental composition of the selected ROI in Figure 4.76.

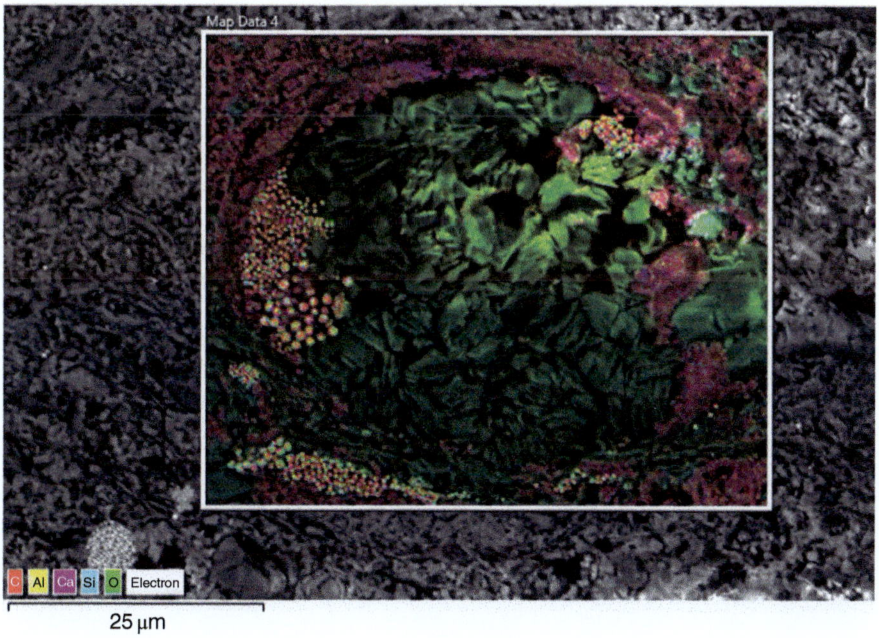

Figure 4.78 Selected ROI on a BSE image of an Eagle Ford shale sample.

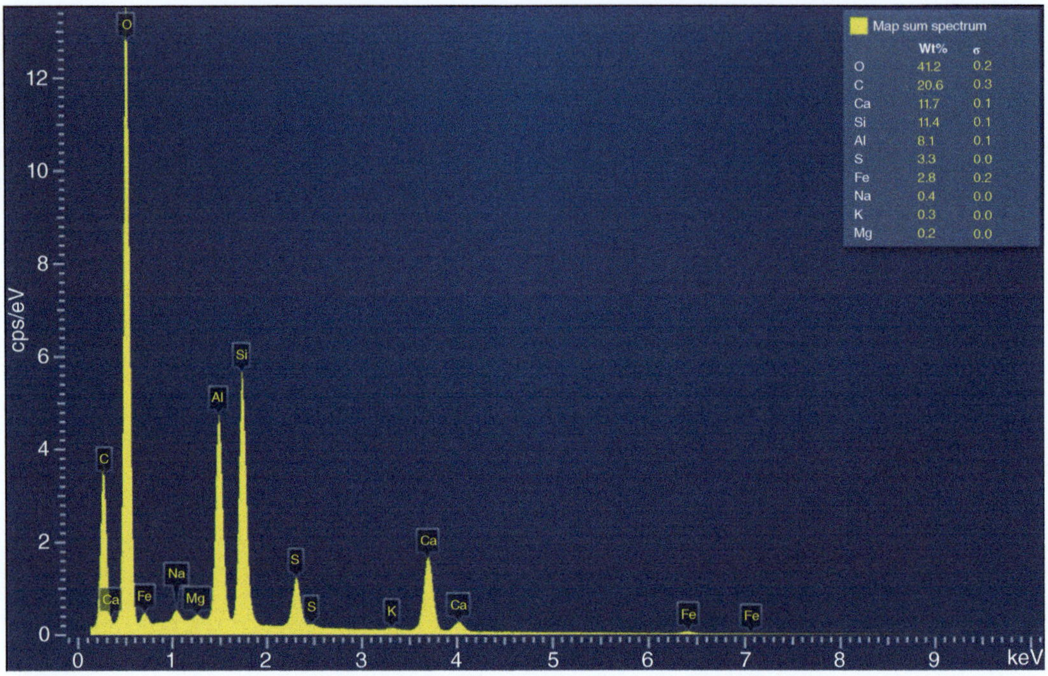

Figure 4.79 Corresponding elemental composition of the selected ROI in Figure 4.78.

Table 4.10 Pore Analysis Parameters For A 3D Shale Volume

Number of pores	Volume (μm^3)	Area (μm^2)	Shape factor
1	0.00015576	0.0175525	1.97084
2	0.000420401	0.0515524	6.85435
3	0.00296096	0.283187	22.9036
4	5.75E-05	0.00736542	1.06988
5	1.66E-05	0.00372211	1.64775
6	4.54E-06	0.00106565	0.519891
7	3.18E-05	0.00614296	2.03237
8	1.97E-05	0.00387744	1.3337
9	0.00051416	0.0785676	16.2212
10	8.47E-05	0.0106168	1.47541

Additional Reading

Hemes, S., et al. (2015). Multi-scale characterization of porosity in Boom Clay (HADES-level, Mol, Belgium) using a combination of X-ray micro-CT, 2D BIB-SEM and FIB-SEM tomography. *Microporous and Mesoporous Materials*, *208*, 1–20.

Kelly, S., El-Sobky, H., Torres-Verdín, C., et al. (2016). Assessing the utility of FIB-SEM images for shale digital rock physics. *Advances in Water Resources*, *95*, 302–316.

Ostadhassan, M., Liu, K., Li, C., & Khatibi, S. (2018). *Fine scale characterization of shale reservoirs: Methods and challenges*. Switzerland: Springer.

References

Adeleye, J. O., & Akanji, L. T. (2017). Pore-scale analyses of heterogeneity and representative elementary volume for unconventional shale rocks using statistical tools. *Journal of Petroleum Exploration and Production Technology*. Available at http://link.springer.com/10.1007/s13202-017-0377-4

Akbarabadi, M., et al. (2017). Nano-scale experimental investigation of in-situ wettability and spontaneous imbibition in ultra-tight reservoir rocks. *Advances in Water Resources*, *107*, 60–179.

Akin, S., & Kovscek, A. R. (2003). Computed tomography in petroleum engineering research. *Geological Society of London*, *215*, 23–38.

Akin, S., Demiral, M. R. B., & Okandan, E. (1996). *A novel method of porosity measurement utilizing computerized tomography*.

Alimi, H. (2018). *Short course on organic geochemistry and application of advanced geochemical methods in exploration and production*. Department of Chemical Engineering and Material Science, University of Southern California (USC).

Aljamaan, H., Ross, C. M., & Kovscek, A. R. (2017). Multiscale imaging of gas storage in shales. *SPE Journal*, *22*(6), 1760–1777.

Alzahabi, A., et al. (2015). *Fracturability index is a mineralogical index: A new approach for fracturing decision*. SPE Saudi Arabia Section Annual Technical Symposium and Exhibition (June). Available at https://www.onepetro.org/conference-paper/SPE-178033-MS%5Cnhttp://www.onepetro.org/doi/10.2118/178033-MS

Ameruddin, A. S. (2015). *Growth and characterisation of gold-seeded Indium gallium arsenide nanowires for optoelectronic applications*. The Australian National University. Available at http://hdl.handle.net/1885/101762

Anderson, W. G. (1986). Wettability literature survey, Part 1: Rock/oil/brine interactions and the effects of core handling on wettability. *Journal of Petroleum Technology*, *38*(10), 1125–1144.

Ayling, B., Rose, P., Petty, S., Zemach, E., & Drakos, P. (2012). *QEMSCAN® Quantitative evaluation of minerals by scanning electron microscopy: Capability and application to fracture characterization in geothermal systems*. Proceedings, Thirty-Seventh Workshop on Geothermal Reservoir Engineering Stanford University, Standford, California, 30 January - 1 February. Available at https://pangea.stanford.edu/ERE/pdf/IGAstandard/SGW/2012/Ayling.pdf

Backeberg, N. R., et al. (2017). Quantifying the anisotropy and tortuosity of permeable pathways in clay-rich mudstones using models based on X-ray tomography. *Scientific Reports*, *7*(1), 1–12.

Berg, C. F. (2014). Permeability description by characteristic length, tortuosity, constriction and porosity. *Transport in Porous Media*, *103*(3), 381–400.

Bernard, B., Bernard, H., & Brooks, J. (2004). Determination of total carbon, total organic carbon and inorganic carbon in sediments. *TDI-Brooks International/B&B Lab*,1–5. Available at http://www.tdi-bi.com/analytical_services/environmental/NOAA_methods/TOC.pdf

Bernard, S., et al. (2012). Geochemical evolution of organic-rich shales with increasing maturity: A STXM and TEM study of the Posidonia shale (lower Toarcian, northern Germany). *Marine and Petroleum Geology*, *31*(1), 70–89.

Boas, F. E., & Fleischmann, D. (2012). CT artifacts: Causes and reduction techniques. *Imaging in Medicine*, *4*(2), 229–240.

Borst, R. L. (1983). *Methods for calculating shale permeability*, *23*. Society of Petroleum Engineers.

Borysenko, A., et al. (2009). Experimental investigations of the wettability of clays and shales. *Journal of Geophysical Research: Solid Earth*, *114*(7),1–11.

Brink, J., et al. (1994). Helical CT: Principles and technical considerations. *Radiographics: A review publication of the Radiological Society of North America, Inc*, *14*(4),887–893.

Brunke, O. (2010). High-resolution CT-based defect analysis and dimensional measurement. *Insight: Non-Destructive Testing and Condition Monitoring*, *52*(2), 91–93.

Bryant, J. A., Drage, N.A., & Richmond, S. (2012). CT number definition. *Radiation Physics and Chemistry*, *81*(4), 358–361.

Chen, C., et al. (2013). Nanometer-scale characterization of microscopic pores in shale kerogen by image analysis and pore-scale modeling. *Geochemistry, Geophysics, Geosystems*, *14*(10), 4066–4075.

Chen, F., Lu, S., & Ding, X. (2014). Organoporosity evaluation of shale: A case study of the lower Silurian Longmaxi shale in southeast Chongqing, China. *Scientific World Journal*.

Chen, L., et al. (2015). Nanoscale simulation of shale transport properties using the lattice Boltzmann method: permeability and diffusivity. *Scientific Reports*, *5*, 8089.

Cnudde, V., & Boone, M. N. (2013). High-resolution X-ray computed tomography in geosciences: A review of the current technology and applications. *Earth-Science Reviews*, *123*, 1–17. doi: 10.1016/j.earscirev.2013.04.003

Conant, L. C., & Swanson, V. E. (1961). Chattanooga shale and related rocks of central Tennessee and nearby areas. *USGS Professional Paper*, *357*, 91.

Costa, A. (2006). Permeability-porosity relationship: A reexamination of the Kozeny-Carman equation based on a fractal pore-space geometry assumption. *Geophysical Research Letters*, *33*(2), 1–5.

Curtis, M. E., Ambrose, R. J., & Rai, C. S. (2011). Investigating the microstructure of gas shales by FIB/SEM tomography & STEM imaging. The Society for Organic Petrology 28th Annual Meeting, 31 July to 4 August 2011, Halifax, Nova Scotia, Canada.

Curtis, M. E., Ambrose, R. J., & Sondergeld, C. H. (2010). Structural characterization of gas shales on the micro- and nano-scales. Canadian Unconventional Resources and International Petroleum Conference, October.

Camp, W. (2015). *Diagenetic evolution of organic matter cements in unconventional shale reservoirs*. AAPG Annual Convention & Exhibition, Denver.

Cudjoe, S., Barati, R., Goldstein, R., Tsau, J., Nicoud, B., Bradford, K., et al. (2019). *An integrated pore-scale characterization workflow for hydrocarbon gas huff-n-puff injection into the lower Eagle Ford shale*. Unconventional Resources Technology Conference (URTeC), Denver.

Cudjoe, S., Barati, R., Marshall, C., Goldstein, R., Tsau, J.-S., Nicoud, B., et al. (2019). Application of Raman spectroscopy in investigating the effect of source and temperature on the maturity of the organic matter exposed to hydrocarbon gas injection. Unconventional Resources Technology Conference, Denver.

Deirieh, A. (2016). From clay slurries to mudrocks: a cryo-sem investigation of the development of the porosity and microstructure of mudrocks. Massachusetts Institute of Technology.

Durand, B. (1980). *Kerogen: Insoluble organic matter from sedimentary rocks.* Paris: Technip.

Eddins, S. (2002). The watershed transform: Strategies for image segmentation. *MathWorks.* Available at https://www.mathworks.com/company/newsletters/articles/the-watershed-transform-strategies-for-image-segmentation.html

Elgmati, M., et al. (2011). Submicron-pore characterization of shale gas plays. SPE North American Unconventional Conference and Exhibition.

Erdman, N., Nielsen, C., & Robertson, V. E. (2012). Shedding new light on cathodoluminescence, a low voltage perspective. *Microscopy and Microanalysis, 18* (6), 1246–1252.

Fishman, N., Guthrie, J., & Honarpor, M. (2013). *The stratigraphic distribution of hydrocarbon storage and its effect on producible hydrocarbons in the Eagle Ford formation, south Texas.* Unconventional Resources Technology Conference. Denver.

Fishman, N. S., et al. (2012). The nature of porosity in organic-rich mudstones of the upper Jurassic Kimmeridge clay formation, North Sea, offshore United Kingdom. *International Journal of Coal Geology, 103*, 32–50.

Fogden, A., et al. (2014). Micro-CT analysis of pores and organics in unconventionals using novel contrast strategies. Proceedings of the Second Unconventional Resources Technology Conference. doi: 10.15530/urtec-2014.1922195

Fogden, A., et al. (2015). Dynamic micro-CT imaging of diffusion in unconventionals. Unconventional Resources Technology Conference, 20–22 July, San Antonio, Texas. doi: 10.15530/URTEC-2015-2154822

Fujii, M., & Uyama, K. (2004). Recent advance on X-ray CT. In J. Otani & Y. Obara (Eds.), *X-ray CT for Geomaterials: Soils, concrete, rocks* (pp. 3–12). Lisse: Swets & Zwitlinger.

Giffin, S., et al. (2013). Application of BIB-SEM technology to characterize macropore morphology in coal. *International Journal of Coal Geology, 114*, 85–95.

Goldman, L.W. (2007). Principles of CT and CT technology. *Journal of Nuclear Medicine Technology, 35*(3), 115–128. Available at http://tech.snmjournals.org/cgi/doi/10.2967/jnmt.107.042978

Gonzalez, J., et al. (2013). Determination of formation organic carbon content using a new neutron-induced gamma ray spectroscopy service that directly measures carbon. Unconventional Resources Technology Conference, 1100–1109. Available at http://library.seg.org/doi/abs/10.1190/urtec2013-112.

Hampton, J. C., et al. (2014). Cumulative volumetric deformation of a hydraulic fracture using acoustic emission and micro-CT imaging. 48th US Rock Mechanics / Geomechanics Symposium, 1–15.

Hemes, S., et al. (2015). Multi-scale characterization of porosity in Boom Clay (HADES-level, Mol, Belgium) using a combination of X-ray micro-CT, 2D BIB-SEM and FIB-SEM tomography. *Microporous and Mesoporous Materials, 208*, 1–20.

Hill, R. J., et al. (2007). Oil and gas geochemistry and petroleum systems of the Fort Worth Basin. *AAPG Bulletin, 91*(4), 445–473.

Holt, T. E., and Smith, D. M. (1988). Surface roughness effects on Knudsen diffusion. *Chemical Engineering Science, 44*(3), 780–781.

Houben, M., Urai, J. L., & Drury, M. R. (2013). In situ characterization of the microstructure and porosity of Opalinus clay (Mont Terri Rock Laboratory, Switzerland). *Fakultät für Georessourcen und Materialtechnik, Dissertati, 192.*

Hsieh, C.-H. (2012). *Procedure and analysis of mineral samples using high resolution x-ray micro tomography.* The University of Utah.

Huang, J., Cavanaugh, T., & Nur, B. (2013). An introduction to SEM operational principles and geologic applications for shale hydrocarbon reservoirs. *Electron Microscopy of Shale Hydrocarbon Reservoirs: AAPG Memoir, 102*, 1–6.

Jarvie, D. M., et al. (2007). Unconventional shale-gas systems: The Mississippian Barnett shale of north-central Texas as one model for thermogenic shale-gas assessment. *AAPG Bulletin*, *91*(4), 475–499.

Javadpour, F., Fisher, D. & Unsworth, M. (2007). Nanoscale gas flow in shale gas sediments. *Journal of Canadian Petroleum Technology*, *46* (10), 55–61.

Jiang, Z., et al. (2016). Basic characteristics and evaluation of shale oil reservoirs. *Petroleum Research*, *1*(2), 149–163. Available at http://linkinghub.elsevier.com/retrieve/pii/S209624951730039X

Josh, M., et al. (2012). Laboratory characterisation of shale properties. *Journal of Petroleum Science and Engineering*, *88–89*, 107–124. Available at: http://dx.doi.org/10.1016/j.petrol.2012.01.023

Kalender, W. A. (2006). X-ray computed tomography. *Physics in Medicine and Biology*, *51*(13).

Kelly, S., El-Sobky, H., Torres-Verdín, C., et al. (2016). Assessing the utility of FIB-SEM images for shale digital rock physics. *Advances in Water Resources*, *95*, 302–316.

Klaver, J., et al. (2014). *Pore space characterization of organic-rich shales using BIB-SEM*. Fakultät für Georessourcen und Materialtechnik, PhD project.

Ko, L.T., et al. (2017). Origin and characterization of Eagle Ford pore networks in the south Texas Upper Cretaceous shelf. *AAPG Bulletin*, *101*(3), 387–418.

Komini Babu, S., et al. (2015). *Nano-scale resolution X-ray computed tomography*. Workshop on 3D Microstructural Studies. Carnegie Mellon University. Available at http://muri.materials.cmu.edu/wp-content/uploads/2015/06/Litster-3D_Summer_School.pdf

Kutchko, B., et al. (2018). *Characterization of the CO_2-fluid-shale interface via feature relocation using field-emission scanning electron microscopy, in situ infrared spectroscopy, and pore size analysis shale samples. Pittsburgh, SPE/AAPG. doi: 10.1130/abs/2017NE-291244*

Lemmens, H., & Richards, D. (2013). Multiscale imaging of shale samples in the scanning electron microscope. *Electron microscopy of shale hydrocarbon reservoirs: AAPG Memoir*, *102*, 27–36.

Li, C., et al. (2019). Multi-scale assessment of mechanical properties of organic-rich shales: a coupled nanoidentation, deconvolution analysis and homogenization method. *Journal of Petroleum Science and Engineering*, *174*, 80–91.

Litster, S., Babu, S., Frisco, S., Mandal, P., Epting, W., Kumar, A., et al. (2015). Nano-scale resolution X-ray computed tomography. Pittsburgh, Pennsylvania.

Liu, K., et al. (2016). Bakken formation shales nano-scale analysis understand mechanical parameters. 50th US Rock Mechanics / Geomechanics Symposium, 1–8.

Liu, S., Wang, Z., & Zhang, L. (2018). Experimental study on the cracking process of layered shale using X-ray microCT. *Energy Exploration and Exploitation*, *36*(2), 297–313.

Löhr, S. C., et al. (2015). Is organic pore development in gas shales influenced by the primary porosity and structure of thermally immature organic matter? *Organic Geochemistry*, *87*, 119–132.

Loucks, R. G., & Reed, R. M. (2014). Scanning-electron-microscope petrographic evidence for distinguishing organic-matter pores associated with depositional organic matter versus migrated organic matter in mudrocks. *GCAGS Journal*, *3*, 51–60.

Loucks, R. G., et al. (2010). Preliminary classification of matrix pores in mudrocks. *Gulf Coast Association of Geological Societies Transactions*, *60*(2), 435–441.

Ma, L., et al. (2017). Multi-scale 3D characterisation of porosity and organic matter in shales with variable TOC content and thermal maturity: Examples from the Lublin and Baltic Basins, Poland and Lithuania. *International Journal of Coal Geology*, *180*, 100–112.

Matyka, M., Khalili, A. & Koza, Z. (2008). Tortuosity-porosity relation in porous media flow. *Physical Review E: Statistical, Nonlinear, and Soft Matter Physics*, *78*(2), 1–8.

McCullough, E. (1975. Photon attenuation in computed tomography. *Medical Physics (Lancaster)*, *2*(6), 307.

Milliken, K. L., et al. (2014). SEM petrography of eastern mediterranean sapropels: Analogue data for assessing organic matter in oil and gas shales. *Journal of Sedimentary Research*, *84*(11), 961–974. Available at http://jsedres.sepmonline.org/cgi/doi/10.2110/jsr.2014.75

Moss, R. M., Pepin, G. P., & Davis, L. A. (1992). Direct measurement of the constituent porosities in a dual-porosity matrix. *The Log Analyst*, 126–136.

Neuser, E., & Suppes, A. (2007). Visualizing internal 3D structures with submicrometer resolution, 18.

Newell, K. D., & Doveton, J. H. (2010). *Gas production from the Chattanooga shale in southern Kansas: First tries and initial data for possible future reconnaissance studies*, Lawrence. Available at https://gsa.confex.com/gsa/2010NC/finalprogram/abstract_171401.htm

Nia, S. F., et al. (2016). An integrated approach for the characterization of shales and other unconventional resource materials. *Industrial and Engineering Chemistry Research*, *55*(12), 3718–3728.

Ning, Y., et al. (2016). Permeability prediction considering surface diffusion for gas shales by lattice Boltzmann simulations on multi-scale reconstructed digital rocks. *International Petroleum Technology Conference*. Available at http://www.onepetro.org/doi/10.2523/IPTC-18845-MS

Olson, T. (2015). Digital rock technology and applications to unconventionals, March, 47. Available at http://dwls.spwla.org/2015-03-17 DWLS Presentation.pdf.

Ostadhassan, M., Liu, K., Li, C., & Khatibi, S. (2018). *Fine scale characterization of shale reservoirs: Methods and challenges*. Switzerland: Springer. doi: 10.1007/978-3-319-76087-2

Peng, S., et al. (2015). An integrated method for upscaling pore-network characterization and permeability estimation: Example from the Mississippian Barnett shale. *Transport in Porous Media*, *109*(2), 359–376.

Peng, S., Hassan, A. & Loucks, R. G. (2017). Permeability estimation based on pore characterization and flow modeling from thin-section image analysis of grain-dominated carbonates problem statement and objectives K2D / K3D ratio examination of the K2D / K3D relationship, 42117, 5–7.

Peng, S., Zhang, T., & Ruppel, S. C. (2014). Upscaling of pore network and permeability from micron to millimeter scale in organic-pore dominated mudstones, 41429. Available at http://www.searchanddiscovery.com/documents/2014/41429peng/ndx_peng

Plessis, A., et al. (2017). Laboratory x-ray micro-computed tomography: A user guideline for biological samples. *GigaScience*, *6*, 1–11.

Pollastro, R. M., et al. (2007). Geologic framework of the Mississippian Barnett shale, Barnett-Paleozoic total petroleum system, Bend Arch-Fort Worth Basin, Texas. *AAPG Bulletin*, *91*(4), 405–436.

Reuteler, J. (2017). FIB artifacts and how to overcome them. *ScopeM*, 1–25.

Rodriguez, R., Crandall, D., Song, X., Verba, C., & Soeder, D. (2014). *Imaging techniques for analyzing shale pores and minerals*, 40. NETL Technical Report Series, U.S. Department of Energy, National Energy Technology Laboratory, Morgantown, West Virginia. doi: 10.18141/1432658

Roduit, N. (2008). JMICROVISION version 1.2.7: Image analysis toolbox for measuring and quantifying components of high-definition images. Available at http://www.jmicrovision.com

Ruparelia, S. (2012). Implementation of watershed based image segmentation algorithm in FPGA. Universitat Stuttgart. Available at ftp://ftp.informatik.uni-stuttgart.de/pub/library/medoc.ustuttgart_fi/MSTR-3256/MSTR-3256.pdf

Saadat, K., et al. (2011). Empirical correlation for porosity deduction from X-ray computed tomography (CT), *1*(2), 47–54.

Saif, T., et al. (2017). Multi-scale multi-dimensional microstructure imaging of oil shale pyrolysis using X-ray micro-tomography, automated ultra-high resolution SEM, MAPS Mineralogy and FIB-SEM. *Applied Energy*, *202*, 628–647. Available at http://dx.doi.org/10.1016/j.apenergy.2017.05.039

Schmatz, J., et al. (2016). Nanoscale morphology of brine-oil-mineral contacts in connected pores of carbonate reservoirs: Insights on wettability from Cryo-BIB-SEM. SPE Bergen One Day Seminar, 1–11. Available at: http://www.onepetro.org/doi/10.2118/180049-MS

Schoeman, L., et al. (2016a). X-ray micro-computed tomography (μCT) for non-destructive characterisation of food microstructure. *Trends in Food Science and Technology*, *47*, 10–24. Available at http://dx.doi.org/10.1016/j.tifs.2015.10.016

Schoeman, L., et al. (2016b). X-ray micro-computed tomography (μCT) for non-destructive characterisation of food microstructure. *Trends in Food Science and Technology*, *47*, 10–24.

Scientific, T. (2018). HeliScan microCT. *ThermoFisher Scientific*, *2*. Available at https://assets.thermofisher.com/TFS-Assets/MSD/Datasheets/heliscan-microct-materials-science-datasheet.pdf

Sharma, B. C., Brigham, W. E., & Castanier, L. M. (1997). *CT imaging techniques for two-phase and three-phase in-situ saturation measurements*. California. Available at https://geo.stanford.edu/ERE/research/supria/publications/public/tr107.pdf

Singhal, A., Grande, J. C., & Zhou, Y. (2013). Micro/nano-CT for visualization of internal structures. *Microscopy Today*, *21*(02), 16–22. Available at http://www.journals.cambridge.org/abstract_S1551929513000035

Sinn, C. J. A., et al. (2017). Using BIB-SEM imaging for permeability prediction in heterogeneous shales. *Geofluids*, 2017.

Slatt, R. M., & O'Brien, N. R. (2011). Pore types in the Barnett and Woodford gas shales: Contribution to understanding gas storage and migration pathways in fine-grained rocks. *AAPG Bulletin*, *95*(12), 2017–2030.

Sondergeld, C. H., et al. (2010). Micro-structural studies of gas shales. Society of Petroleum Engineers Unconventional Gas Conference, SPE Paper 131771.

Tissot, B. P., & Welte, D. H. (1984). *Petroleum formation and occurrence*. Berlin: Springer-Verlag. Available at https://raregeologybooks.files.wordpress.com/2014/12/b-p-tissot-and-d-h-welte-petroleum-formation-and-occurence.pdf

Varslot, T., et al. (2011). High-resolution helical cone-beam micro-CT with theoretically exact reconstruction from experimental data. *Medical Physics*, *38*(10), 5459–5476.

Verba, C., Crandall, D., & Moore, J. (2016). *Multiscale shale pore characterization*. San Antonio, Texas: Unconventional Resources Technology Conference (URTeC).

Wang, F. P., & Gale, J. F. W. (2009). Screening criteria for shale-gas systems. *Gulf Coast Association of Geological Societies Transactions*, *59*, 779–793.

Wang, H., & Marongiu-Porcu, M. (2015). Impact of shale-gas apparent permeability on production: Combined effects of non-darcy flow/gas-slippage, desorption, and geomechanics. SPE Reservoir Evaluation & Engineering, October, 1–13. Available at http://www.onepetro.org/doi/10.2118/173196-PA

Wang, J., et al. (2016). Apparent permeability prediction of organic shale with generalized lattice Boltzmann model considering surface diffusion effect. *Fuel*, *181*, 478–490. Available at http://dx.doi.org/10.1016/j.fuel.2016.05.032

Wildenschild, D., & Sheppard, A. P. (2013). X-ray imaging and analysis techniques for quantifying pore-scale structure and processes in subsurface porous medium systems. *Advances in Water Resources*, *51*, 217–246. Available at http://dx.doi.org/10.1016/j.advwatres.2012.07.018

Withers, P. J. (2007). X-ray nanotomography. *Materials Today*, *10*(12), 26–34.

Withjack, E. M. (1988). Computed tomography for rock-property determination and fluid-flow visualization. *SPE Formation Evaluation*, December, 696–704.

Withjack, E. M., Orsham, S. K., & Yang, C. T. (1991). CT determination of heterogeneities and miscible displacement characteristics. *SPE Formation Evaluation*, *6*(4), 3–8.

Xiong, Q., Baychev, T. G., & Jivkov, A. P. (2016). Review of pore network modelling of porous media: Experimental characterisations, network constructions and applications to reactive transport. *Journal of Contaminant Hydrology*, *192*, 101–117. Available at http://dx.doi.org/10.1016/j.jconhyd.2016.07.002

Xradia (2010). XRADIA microXCT Manual. *XRadia, MicroXCT-400*, 1–10. Available at http://www.imaging.sbes.vt.edu/BIDLib/SAMCT/equipment-docs/Xradia_manual_v4.pdf

Yang, Y., & Bao, F. (2017). Characteristics of shale nanopore system and its internal gas flow: A case study of the lower Silurian Longmaxi formation shale from Sichuan Basin, China. *Journal of Natural Gas Geoscience*, *2*(5–6), 303–311.

Zeiss (2013). ZEISS Xradia 520 Versa submicron X-ray imaging: Extending the limits of exploration. *Zeiss*, 1–20. Available at https://www.zeiss.com/content/dam/Microscopy/us/download/pdf/Products/xradia520versa/xradia-520-versa-product-information.pdf

Zhang, P., et al. (2015). Micro/nano-pore network analysis of gas flow in shale matrix. *Scientific Reports*, *5*(August), 13501. Available at http://www.pubmedcentral.nih.gov/articlerender.fcgi?artid=4642512&tool=pmcentrez&rendertype=abstract

Zhang, S., Klimentidis, R. E., & Barthelemy, P. (2012). *Micron to millimeter upscaling of shale rock properties based on 3D imaging and modeling*. International Symposium of the Society of Core Analysts, SCA2012-20.

Zhang, X. L., et al. (2015). Investigation of shale gas microflow with the lattice Boltzmann method. *Petroleum Science*, *12*(1), 96–103.

Zhou, M., et al. (2018). Mechanical property measurements and fracture propagation analysis of longmaxi shale by micro-CT uniaxial compression. *Energies*, *11*(6).

5

Geomechanical Properties of Unconventional Reservoirs

Mustafa M. Alhubail, Anil Misra, and Reza Barati Ghahfarokhi

The University of Kansas, Lawrence, Kansas, USA

5.1 Background

Designing a hydraulic fracturing treatment requires a comprehensive understanding of the formation mechanical properties. Hence, the mechanical properties (e.g., minimum principal stress, fracture gradient, and fracture closure pressure) have to be assessed in order to apprehend the process of fracturing. It is important to understand that fractures initiate perpendicular to the least minimum principal stress. In the early 1950s, the hydraulic fracturing treatments were conducted at shallow depths, which generated horizontal fractures. Consequently, all induced hydraulic fractures were thought to be horizontal (Smith & Montgomery, 2015). This is a consequence of the limited knowledge of how the formation mechanical properties affected the fracture initiation and propagation back then. With the advancement in technology and the ability to acquire and assess the formation mechanical properties, it has become a well-known fact that the induced hydraulic fractures initiate perpendicular to the least minimum principal stress, and they are mostly vertical at the depths that are being pursued in unconventional reservoirs.

5.2 Basic Concepts and Definitions

Before discussing the mechanical properties of the unconventional reservoirs, it is important to keep in mind that the two main reasons the mechanical properties are studied in this chapter are to better understand the geometry of the hydraulic fractures and to further improve the hydraulic fracturing development process.

5.2.1 Stress

The stress can be defined simply by equation (5.1) as force per unit area. The SI unit of stress is (N/m^2), which is equivalent to Pascal (Pa), and in field units, it is pound-force per square inch (lb_f/in^2). In short it is (psi).

$$\sigma = \frac{F}{A} \tag{5.1}$$

where σ is stress, F is force (N), and A is surface area (m^2).

The stress is a vector, which means it has both a magnitude and a direction. Stresses that are normal to the plane are either compressive stresses or tensile stresses as shown

Unconventional Hydrocarbon Resources: Techniques for Reservoir Engineering Analysis, Advanced Textbook Series, First Edition. Edited by Reza Barati Ghahfarokhi and Mustafa M. Alhubail.

in Figures 5.1 and 5.2, respectively. Squeezing, which is due to the compressive stress, results in positive stress, whereas stretching, which is due to the tensile stress, results in negative stress. Stresses that are parallel to the plane are called shear stresses as show in Figure 5.3. Tearing, which is due to the shear stress, results in positive shear stress when the implied rotation is anticlockwise (Reeves, 2017).

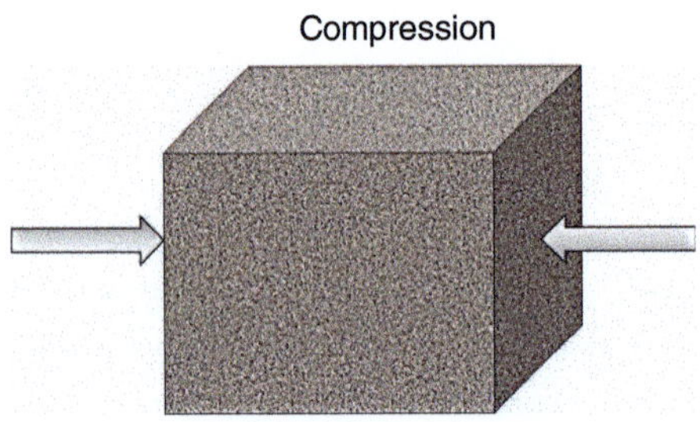

Figure 5.1 Compressive stress.

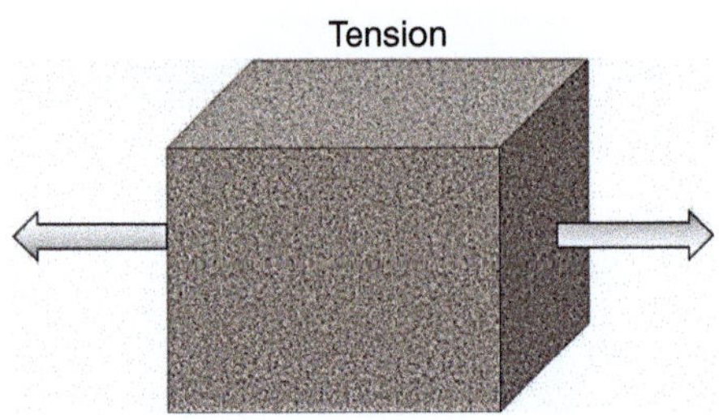

Figure 5.2 Tensile stress.

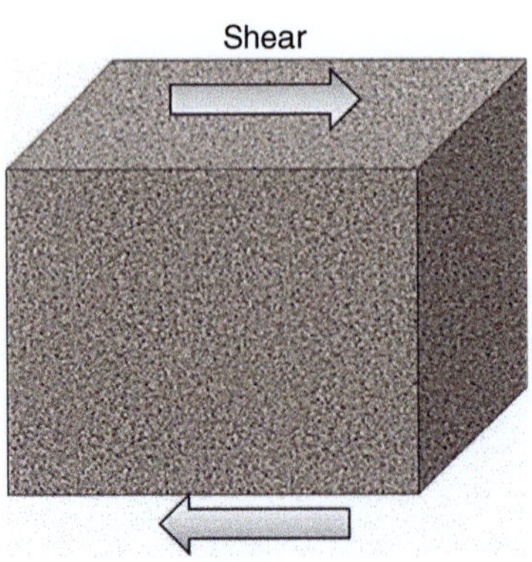

Figure 5.3 Shear stress.

5.2.2 Strain

Strain can be defined as the change in dimension over the original dimension of the deformed body. There are two main types of strain, and they are dimensionless. The first type is typically referred to as the normal strains, which are due to the compressive and tensile stresses. The normal strains are the longitudinal strain and the lateral strain. The longitudinal strain is the change of length due to stress over the original length of the body, and it is represented by equation (5.2). The lateral strain, which is also called the transverse strain, is the change of diameter over the original diameter of the body as shown in equation (5.3). The second type of strain is the shear strain, which is due to the shear stress. The shear strain is defined as the angle change between lines in the deformed body as shown in equation (5.4). For small strains, equation (5.5) is usually used to obtain the shear strain.

$$\varepsilon_l = \frac{\Delta L}{L}. \tag{5.2}$$

In this equation, ε_l is the longitudinal strain, ΔL is the change in length due to the stress, and L is the original length of the body.

$$\varepsilon_t = \frac{\Delta D}{D} \tag{5.3}$$

and

$$\varepsilon_s = \tan\theta, \tag{5.4}$$

where θ is the deformation angel, and it is in radian.

$$\varepsilon_s = \theta. \tag{5.5}$$

5.2.3 Elastic Constants

The elastic constants are defined within the elastic limit, which means the body returns to its original condition if the applied stress is removed. There are three elastic constants: modulus of elasticity (Young's modulus,) modulus of rigidity (shear modulus), and bulk modulus. The elastic constants are measured

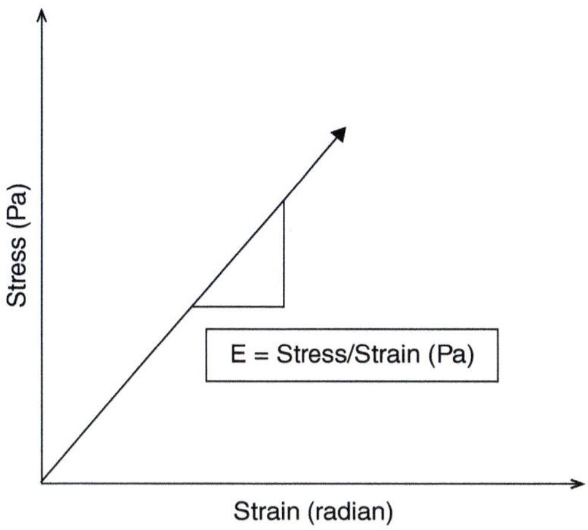

Figure 5.4 Elastic modulus.

by the use of true triaxial tests, and they can be estimated from well-logging measurements as will be demonstrated later in this chapter.

Elastic Modulus

The elastic modulus (E), which is also called the Young's modulus, is the ratio, within the elastic limit, of the compressive stress or the tensile stress to the corresponding strain. Most rocks behave in an elastic manner; therefore, the plot of stress versus strain yields a linear curve. Thus, the slope of that linear curve is the elastic modulus as shown in Figure 5.4. Consequently, the elastic modulus can be calculated using equation (5.6). The elastic modulus has the same unit as the stress,

$$E = \frac{\frac{F}{A}}{\frac{\Delta L}{L}} = \frac{\sigma}{\varepsilon_l}. \quad (5.6)$$

Shear Modulus

The shear modulus (G) is also called the modulus of rigidity. It is the ratio, within the elastic limit, of the shear stress to the corresponding shear strain (Ramasamy & Purushothama Raj, 2010). The shear modulus is the slope of the linear curve from the shear stress versus shear strain plot as shown in Figure 5.5, and equation (5.7) can be used to calculate it.

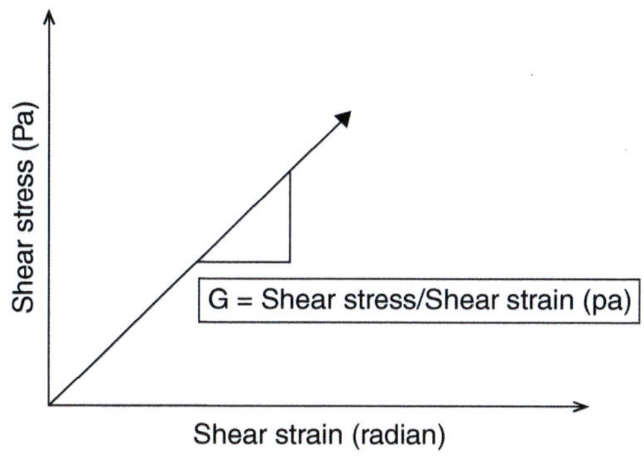

Figure 5.5 Shear modulus.

Alternatively, equation (5.8) can be used to calculate the shear modulus for isotropic materials.

$$G = \frac{Shear\ stress}{Shear\ strain} = \frac{\sigma}{\varepsilon_s} \quad (5.7)$$

and

$$G = \frac{E}{2(1+\nu)}, \quad (5.8)$$

where, ν is the dimensionless Poisson's ratio.

Bulk Modulus

The bulk modulus (K), which is the reciprocal of the rock compressibility, is the ratio of stress to the volumetric strain, and it can be calculated by equation (5.9). Alternatively, equation (5.10) can be used to calculate the bulk modulus for isotropic materials.

$$K = \frac{Stress}{Volumetric\ Strain} = \frac{\sigma}{\left(\frac{\Delta V}{V}\right)} \quad (5.9)$$

and

$$K = \frac{E}{3(1-2\nu)}. \quad (5.10)$$

5.2.4 Poisson's Ratio

Poisson's ratio can be defined as the ratio of the transverse strain to the longitudinal strain. It is a measure of the geometric change of shape under stress. Poisson's ratio is usually denoted as ν, which is a lower-case Greek letter nu. The Greek letter mu (μ) is also used to represent the Poisson's ratio in some rock mechanics books. Poisson's ratio can be calculated by the use of equation (5.11),

$$\nu = \frac{\varepsilon_t}{\varepsilon_l}. \tag{5.11}$$

Example 5.1 Calculate the Static Elastic Moduli

A stress of 1.68×10^7(Pa) was applied on a core sample from the Bakken formation. The original core length and diameter are 0.023 (m) and 0.02534 (m), respectively. The change of length and diameter due to the applied force are 4.19×10^{-5} (m) and 1.2×10^{-5} (m), respectively.

a) Calculate the axial and lateral strains.
b) Calculate Poisson's ratio.
c) Calculate the elastic constants:
   1) The elastic modulus (E)
   2) The shear modulus (G)
   3) The bulk modulus (K)

Solution

a) The axial strain, which is the change of length divided by the original length, can be calculated by the use of equation (5.2),

$$\varepsilon_l = \frac{\Delta L}{L} = \frac{4.19 \times 10^{-5}}{0.023} = 1.82 \times 10^{-3}.$$

The lateral strain, which is the change of diameter divided by original diameter, can be calculated by the use of equation (5.3),

$$\varepsilon_t = \frac{\Delta D}{D} = \frac{1.2 \times 10^{-5}}{0.02534} = 4.74 \times 10^{-4}.$$

b) Poisson's ratio is the lateral strain divided by the axial strain, can be calculated by the use of equation (5.11),

$$\nu = \frac{\varepsilon_t}{\varepsilon_l} = \frac{4.74 \times 10^{-4}}{1.82 \times 10^{-3}} = 0.26.$$

c) The elastic constants are calculated as follows:
   1) The elastic modulus (E) can be calculated by the use of equation (5.6),

$$E = \frac{\sigma}{\varepsilon_l} = \frac{1.68 \times 10^7}{1.82 \times 10^{-3}} = 9.1967 \times 10^9 Pa \equiv 1.334 \times 10^6 psi.$$

   2) The shear modulus (G) can be calculated by the use of equation (5.8)

$$G = \frac{E}{2(1+\nu)} = \frac{9.1967 \times 10^9 Pa}{2(1+0.26)} = 3.6496 \times 10^9 Pa \equiv 5.29 \times 10^5 psi.$$

   3) The bulk modulus (K) can be calculated by the use of equation (5.10),

$$K = \frac{E}{3(1-2\nu)} = \frac{9.1967 \times 10^9 Pa}{3(1\text{-}2(0.26))} = 6.3852 \times 10^9 Pa \equiv 9.261 \times 10^5 psi.$$

5.3 Stresses and Pressure Gradients

There are three distinctive principal stresses in each rock block. They are the maximum, intermediate, and minimum principal stresses, as shown in Figure 5.6. In almost all the unconventional reservoirs as well as the conventional reservoirs that are not shallow, the maximum principal stress is the vertical stress σ_V (i.e., the overburden pressure) (Smith & Montgomery, 2015). The intermediate stress is $\sigma_{H_{max}}$ and the minimum stress is $\sigma_{H_{min}}$. Therefore, the generated fracture is going to be vertical since the least minimum principal stress (i.e., $\sigma_{H_{min}}$) is horizontal, as shown in Figure 5.6. Hence, estimating the least principal stress leads to knowing how the fracture initiation is going to occur as well as knowing how much pressure should be applied to crack the formation open and initiate a fracture. Note that, the stress field may be a local set of vectors and be dependent on the rock type along the propagation path of the fracture. This may result in fracture propagation not being on a straight line.

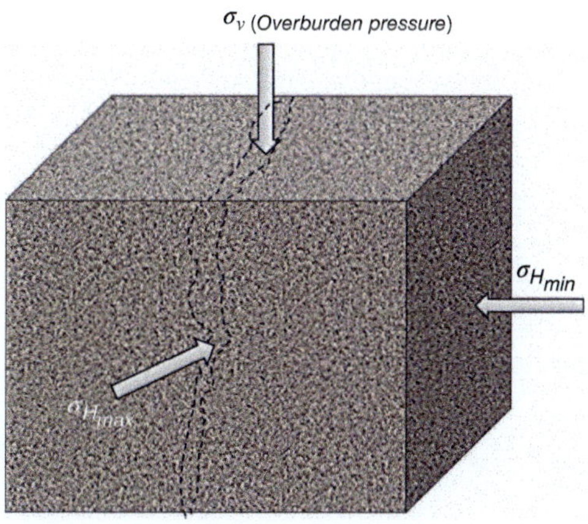

Figure 5.6 True triaxial stress.

5.3.1 Vertical Stress and Overburden Pressure

The overburden pressure is the pressure that is caused by the weight of the overlying formations above the formation of interest. The overburden pressure, which is also called the absolute vertical stress, can be initially estimated by equation (5.12) in SI units. Equation (5.13) can be used to estimate the overburden pressure in fields units.

$$P_{OB} = \rho_b\, g\, TVD, \tag{5.12}$$

where P_{OB} is the overburden pressure (Pa), ρ_b is formation bulk density (kg/m^3), g is the standard gravity (9.8 m/s^2), and TVD is the true vertical depth (m).

$$P_{OB} = 0.052\rho_b\, TVD, \tag{5.13}$$

where P_{OB} is the overburden pressure (psi), ρ_b is formation bulk density (lb/gal), and TVD is the true vertical depth (ft).

The overburden pressure gradient (G_{OB}) is estimated by equation (5.14) and equation (5.15) in SI units and field units, respectively. A typical value for the absolute overburden pressure gradient that is widely used in the petroleum industry is $2.49 \times 10^4 \frac{Pa}{m}$ in SI units and $1.1 \frac{Psi}{ft}$ in fields units (Valkó & Economides, 1995). Note that this value varies from well to well and from rock type to tock type in practice.

$$G_{OB} = \rho_b\, g \tag{5.14}$$

and

$$G_{OB} = 0.052\, \rho_b. \tag{5.15}$$

The formation bulk density, which is the density of the matrix and the fluid, can be found by well-logging measurements, or it can be estimated by equation (5.16).

$$\rho_b = (1-\varphi)\rho_m + \varphi\, \rho_f \tag{5.16}$$

where ρ_m is the matrix density, ρ_f is the average fluid density in the pores, and φ is the formation porosity.

Example 5.2 Estimating the Absolute Vertical Stress

Estimate the following for the Middle Bakken formation where the bulk density of the formation is 2,723 kg/m³, the oil density is 870 kg/m³, and the average formation porosity is 4.64%.

a) Estimate the pressure gradient.
b) Estimate the absolute vertical stress at 2,793.5 m.
c) Estimate the pore pressure gradient.
d) Estimate the pore pressure (minimum reservoir pressure).

Solution

a) The pressure gradient (G_{OB}) can be estimated directly by equation (5.14),

$$G_{OB} = 2{,}723 \frac{kg}{m^3} \times 9.8 \frac{m}{s^2} = 2.67 \times 10^4 \ \frac{Pa}{m} \equiv 1.18 \frac{psi}{ft}.$$

b) The overburden pressure is equivalent to the absolute vertical stress. Hence, the vertical stress can be estimated by multiplying the pressure gradient from part a by the given depth. Alternatively, equation (5.12) can be used directly to estimate the absolute vertical stress,

$$\sigma_v = P_{OB} = 2.67 \times 10^4 \ \frac{Pa}{m} \times 2{,}793.5 \ m = 7.45 \times 10^7 \ Pa$$

and

$$\sigma_v = P_{OB} = 2723 \ \frac{kg}{m^3} \times 9.8 \ \frac{m}{s^2} \times 2{,}793.5 \ m = 7.45 \times 10^7 Pa \equiv 10{,}805.3 \ psi.$$

c) The pore pressure gradient can be estimated by the use of equation (5.14) and the pore fluid density as follows:

$$G_p = 870 \ \frac{kg}{m^3} \times 9.8 \ \frac{m}{s^2} = 8.53 \times 10^3 \ \frac{Pa}{m} \equiv 0.377 \frac{psi}{ft}.$$

d) The pore pressure, which can be assumed to be the minimum reservoir pressure, can be estimated as a hydrostatic pressure by substituting the pore fluid density instead of the formation bulk density in equation (5.12).

$$P_p = 870 \ \frac{kg}{m^3} \times 9.8 \ \frac{m}{s^2} \times 2{,}793.5 \ m = 2.38 \times 10^7 \ Pa \equiv 3{,}454.42 \ psi.$$

5.3.2 Effective Vertical Stress

The vertical stress that was discussed in the previous section is the absolute vertical stress; neither production nor injection affect the absolute stresses. Hence, the absolute stresses do not change with time. However, the effective stresses do change with time, and producing or injecting a fluid affects the effective stress drastically. Basically, since the reservoirs are porous media, the absolute stresses are not of great importance and the effective stress is what actually matters.

The effective vertical stress can be estimated by subtracting the pore pressure from the absolute stress as shown in equation (5.17),

$$\sigma_{eff} = \sigma_v - \alpha P_p, \tag{5.17}$$

where, α is the poroelastic constant and P_p is the reservoir pore pressure.

5.3.3 Effective Horizontal Stress

An effective horizontal stress can be estimated by the use of the effective vertical stress as shown in equation (5.18). However, this method should be avoided when possible as it may lead to incorrect breakdown pressure (Craig, 2014).

$$\sigma h_{eff} = \frac{\nu}{(1-\nu)} \sigma_{eff}, \tag{5.18}$$

where ν is Poisson's ratio and σ_{eff} is the effective vertical stress.

5.3.4 Biot's Poroelastic Constant

The poroelastic constant, which is also known as the Biot's poroelastic term, was introduced by Biot in the late 1930s, and since then, multiple procedures have been proposed to measure it (Biot, 1955, 1956, 1962). There are direct and indirect methods to measure the poroelastic constant (Addis et al., 2016). Equation (5.19) shows a direct method to measure the poroelastic constant. In this direct method, the bulk volume change and the pore volume change are measured simultaneously during a triaxial loading test (Addis, et al., 2016).

$$\alpha = \frac{\Delta V_p}{\Delta V}, \tag{5.19}$$

where, ΔV_p is the change of pore volume and ΔV is the change of bulk volume.

Equation (5.20) shows an indirect method to measure the poroelastic constant. In this method, the poroelastic constant can be estimated by measuring the rock bulk modulus and the grain bulk modulus. The grain bulk modulus (K_g) is usually much greater than the rock bulk modulus; hence, the poroelastic constant would be close to unity (Addis et al., 2016). Consequently, with limited information, the poroelastic constant is often taken as one.

Table 5.1 Biot's Poroelastic Constant for North America Unconventional Plays.

Unconventional play	Average Biot's poroelastic constant
Barnett (Yaghoubi & Zoback, 2015)	0.70*
Bakken (He & Ling, 2014)	0.69
Marcellus (Kowan & Ong, 2014)	0.60
Utica (Konstantinovskaya et al., 2012)	0.90
Eagle Ford - US Region (Mokhtari et al., 2014)	0.77
Wolfcamp (Shaochuan & Zoback, 2015)	0.74*
Niobrara (Cui, 2016)	0.80
Woodford (Abousleiman et al., 2008)	0.76
Haynesville (Marsden et al., 2017)	1.00*

* Based on values used in simulation models.

$$\alpha = 1 - \frac{K}{K_g}, \tag{5.20}$$

where K is the rock bulk modulus and K_g is the grain bulk modulus.

There are challenges in measuring the poroelastic constant for unconventional reservoirs. These challenges are due to the extremely low permeability in shale formations as well as the long time that is needed to measure the strain rate during a triaxial loading test as testing one sample could take weeks (Addis et al., 2016). Table 5.1 shows the average Biot's poroelastic constant that is reported in the literature based on data from wells located in certain areas of the unconventional plays. It is important to note that the Biot's poroelastic constant could be different at other locations and at different confining pressure in the same play.

Example 5.3 Estimating the Effective Stresses

Estimate the effective stresses for the formation provided in Example 5.2 assuming an average poroelastic constant of 0.69 and a Poisson's ratio of 0.26 for the Bakken formation.

Solution

From Example 5.2, the absolute vertical stress is $7.45 \times 10^7 Pa$, and the pore pressure is $2.38 \times 10^7 Pa$. Hence, the effective vertical stress can be estimated by the use of equation (5.17) as follows:

$$\sigma_{eff} = \sigma_v - \alpha P_p = 7.45 \times 10^7 - \left(0.69 \times 2.38 \times 10^7\right) = 5.81 \times 10^7\ Pa$$

and

$$\sigma_{eff} = 5.81 \times 10^7\ Pa \equiv 8{,}421.75\ psi.$$

Consequently, the effective horizontal stress can be estimated by the use of equation (5.18) as follows:

$$\sigma_{heff} = \frac{0.26}{1-0.26} \times 5.81 \times 10^7\ Pa = 2.04 \times 10^7\ Pa$$

and

$$\sigma_{heff} = 2.04 \times 10^7\ Pa \equiv 2{,}959\ psi.$$

5.3.5 Horizontal Stresses and Fracturing Pressure

As mentioned earlier, there are three principal stresses. In the previous section, the vertical stress, which is the overburden pressure, was discussed. Since the hydraulic fracturing technique was employed in the petroleum industry in the late 1940s, many researchers investigated how fractures actually occur under the ground (i.e., vertical or horizontal). In the late 1950s, Hubbert and Willis concluded that the horizontal stress is the least principal stress in the relaxed tectonic areas and consequently vertical fractures would occur, and the fractures could be initiated with a pressure that is less than the overburden pressure (Hubbert & Willis, 1957; Smith & Montgomery, 2015). However, in active tectonic areas, the least principal stress could be vertical; thus, the generated fractures would be horizontal, and the fractures would be initiated with a pressure that is equal to or greater than the overburden pressure. Generally, in the development of the unconventional reservoirs, the horizontal stresses are more important than the vertical stress because the least principal stress is horizontal and, consequently, the generated fractures are vertical (Smith & Montgomery, 2015).

Maximum Horizontal Stress

The maximum horizontal stress differs from the minimum horizontal stress (i.e., fracturing pressure) in the tectonically active areas; hence, the horizontal stresses (i.e., $\sigma_{H_{max}}$ and $\sigma_{H_{min}}$), Figure 5.6, would be equal in the inactive tectonic areas (Smith & Montgomery, 2015). The maximum horizontal stress can be estimated by adding the local tectonic stress and the minimum horizontal stress as shown in equation (5.21) (Valkó & Economides, 1995). To accurately estimate the local tectonic stress, which is a complicated task, sophisticated mathematical models have to be used that require the knowledge of the full stress tensor.

$$\sigma_{H_{max}} = \sigma_{H_{min}} + \sigma_{tectonic}. \tag{5.21}$$

The maximum horizontal stress can be related to the minimum horizontal stress as shown in equation (5.22) (Guo et al., 2017),

$$\sigma_{H_{max}} = \delta\sigma_{H_{min}} + \Delta\sigma_h, \tag{5.22}$$

where δ is a factor and $\Delta\sigma_h$is the difference between the two horizontal stresses.

Minimum Horizontal Stress

In order to fracture a formation, the applied stress must be equal to or greater than the least principal stress. Thus, the least principal stress (i.e., $\sigma_{H_{\min}}$) should be estimated before designing a hydraulic fracturing job to select the appropriate equipment with appropriate pump power for the job.

Since the 1950s, several formulas have been proposed to estimate the minimum horizontal stress. Estimating the minimum horizontal stress is important because, in most cases, it is equal to the fracturing pressure (Fakhr, 2016). Hence, the goal of estimating the minimum horizontal stress is actually estimating the fracturing pressure and consequently estimating the fracture gradient. Equation (5.23) shows the main formula to estimate the minimum horizontal stress (i.e., fracturing pressure) (Addis et al., 2016). The other proposed formulas actually differ only on how the effective stress ratio (K_E) is calculated. Also, the earlier proposed formulas did not include the Biot's poroelastic constant (α).

$$\sigma_{H_{\min}} = K_E\left(\sigma_v - \alpha P_p\right) + \alpha P_p, \tag{5.23}$$

where, K_E is the effective stress ratio, σ_v is the vertical stress, α is Biot's poroelastic constant, and P_p is the pore pressure.

In some formulas, the effective stress ratio is a constant value that is usually taken as 1/3; hence, the minimum horizontal stress can be estimated by equation (5.24). This is usually referred to as the Hubbert and Willis formula (Hubbert & Willis, 1957; Fakhr, 2016),

$$\sigma_{H_{\min}} = \frac{\left(\sigma_v - \alpha P_p\right)}{3} + \alpha P_p. \tag{5.24}$$

In other cases, the effective stress ratio is a function of the effective Poisson's ratio as shown in equation (5.25) and, consequently, the fracturing pressure can be estimated by equation (5.26). This is famously known as the Eaton's method, and it is the widely used formula in the industry to estimate the fracturing pressure (Fakhr, 2016).

$$K_{E(E)} = \frac{\nu}{(1-\nu)}, \tag{5.25}$$

where, ν is the effective Poisson's ratio

$$\sigma_{H_{\min}} = \left[\frac{\nu}{(1-\nu)}\right]\left(\sigma_v - \alpha P_p\right) + \alpha P_p. \tag{5.26}$$

Eaton proposed a modified version for offshore wells in the Gulf of Mexico to calculate the effective Poisson's ratio based on empirical correlations and for different ranges of depth (e.g., for wells below 5,000 ft or above that) (Eaton, 1969). For instance, for wells above 5,000 feet in the Gulf of Mexico, equation (5.27) is used to calculate the effective Poisson's ratio, and then it is used in equation (5.26) to estimate the fracturing pressure.

$$\nu = \left(-6.0893\times10^{-9}\times TVD_{Ref}^2\right) + \left(5.7875\times10^{-5}\times TVD_{Ref}\right) + 0.31246 \tag{5.27}$$

and

$$TVD_{Ref} = TVD + WD + AG, \tag{5.28}$$

where, TVD is the true vertical depth, WD is water depth, and AG is the air gap.

Another method to estimate the fracturing pressure is the Matthews and Kelly method (Schlumberger, 2015). In the Matthews and Kelly method, the effective stress ratio K_E, which they also call the matrix-stress coefficient, is calculated based on empirical correlations as shown in equation (5.29).

$$K_{E(MK)} = \left(MK1\times TVD_{Ref}^2\right) + \left(MK2\times TVD_{Ref}\right) + MK3, \tag{5.29}$$

where $MK1$, $MK2$, and $MK3$ are parameters, and their default values are -3×10^{-9}, 8×10^{-5}, and 0.2347, respectively.

The fracturing gradients can be estimated as well based on each method discussed above.

For instance, the fracturing gradient can be estimated based on Hubbert and Willis method and Eaton's method as shown in equations (5.30) and (5.31), respectively (Fakhr, 2016).

$$G_{F(HB)} = \frac{(G_{OB} - \alpha G_p)}{3} + \alpha G_p \quad (5.30)$$

and

$$G_{F(E)} = \left[\frac{\nu}{(1-\nu)}\right](G_{OB} - \alpha G_p) + \alpha G_p. \quad (5.31)$$

Fracture Initiation Pressure

The fracture initiation pressure, breakdown pressure, is slightly different than the fracture propagation pressure because the fracture initiation pressure takes into account the induced stresses near the wellbore. However, the induced stresses become zero after the fracture propagates away from the wellbore (Guo et al., 2017). Basically, to initiate a hydraulic fracture, the pressure should be sufficient enough to overcome the in-situ stresses, the rocks tensile strength as well as the induced stresses. The breakdown pressure can be estimated by equation (5.32),

$$P_{bd} = 3\sigma_{H_{\min}} - \sigma_{H_{\max}} + T_o - P_p, \quad (5.32)$$

where P_{bd} is the breakdown pressure, $\sigma_{H_{\min}}$ is the minimum horizontal stress, $\sigma_{H_{\max}}$ is the maximum horizontal stress, T_o is the rock tensile strength and P_p is the pore pressure.

Example 5.4 Estimate the Minimum Horizontal Stress

Estimate the minimum horizontal stress and the fracture gradient for the formation provided in the previous examples.

a) Hubbert and Willis method
b) Eaton's method
c) Comment on how the fracture is going to be initiated (i.e., vertical or horizontal).

Solution

Data from Examples 5.1, 5.2, and 5.3:

1) Vertical stress, $\sigma_v = 7.45 \times 10^7 Pa$
2) Pore pressure, $P_p = 2.38 \times 10^7\ Pa$
3) Overburden pressure gradient, $G_{OB} = 2.67 \times 10^4\ \frac{Pa}{m}$
4) Pore pressure gradient, $G_p = 8.53 \times 10^3\ \frac{Pa}{m}$
5) Poisson's ratio, $\nu = 0.26$
6) Biot's poroelastic constant, $\alpha = 0.69$

a) The fracturing pressure and fracturing gradient can be estimated by Hubbert and Willis method using equation (5.24) and equation (5.30), respectively, as follows:

$$\sigma_{H_{\min}(HB)} = \frac{(7.45 \times 10^7 - 0.26 \times 2.38 \times 10^7)}{3} + 0.26 \times 2.38 \times 10^7$$
$$= 3.58 \times 10^7\ Pa \equiv 5{,}192\ psi$$

and

$$G_{F(HB)} = \frac{(2.67 \times 10^4 - 0.26 \times 8.53 \times 10^3)}{3} + 0.26 \times 8.53 \times 10^3$$
$$= 1.28 \times 10^4 \frac{Pa}{m} \equiv 0.56 \frac{psi}{ft}.$$

b) The fracturing pressure and fracturing gradient can be estimated by Eaton's method using equation (5.26) and equation (5.31), respectively, as follows:

$$\sigma_{H_{\min}(E)} = \left[\frac{0.26}{(1-0.26)}\right](7.45 \times 10^7 - 0.69 \times 2.38 \times 10^7) + 0.69 \times 2.38 \times 10^7$$
$$= 3.68 \times 10^7\ Pa \equiv 5{,}337\ psi$$

and

$$G_{F(E)} = \left[\frac{0.26}{(1-0.26)}\right](2.67 \times 10^4 - 0.69 \times 8.53 \times 10^3) + 0.69 \times 8.53 \times 10^3$$
$$= 1.32 \times 10^4 \frac{Pa}{m} \equiv 0.58 \frac{psi}{ft}.$$

c) The least principal stress is the horizontal stress as calculated above, and since fractures occur perpendicular to the least principal stress, then the generated fracture would be vertical.

Example 5.5 Estimate the Fracture Initiation Pressure

Estimate the breakdown pressure, which is the fracture initiation pressure, for the formation provided in Example 5.4 knowing that the rocks' tensile strength of the formation is 5 MPa, the offset stress, $\Delta\sigma_h$, is 13.9 MPa, and the factor, δ, is 1.

Solution

Data from Examples 5.4:

1) Pore pressure, $P_p = 2.38 \times 10^7\ Pa$
2) Minimum horizontal stress from Eaton's method $= 3.68 \times 10^7 Pa \equiv 5{,}337\, psi$

First, the maximum horizontal stress should be found by the use of equation (5.22):

$$\sigma_{H_{\max}} = \delta\, \sigma_{H_{\min}} + \Delta\sigma_h,$$
$$\sigma_{H_{\max}} = 3.68 \times 10^7 Pa + 13.9 \times 10^6\ Pa = 5.07 \times 10^7 Pa,$$

and

$$\sigma_{H_{\max}} = 5.07 \times 10^7\ Pa \equiv 7{,}353\ psi.$$

Now, the breakdown pressure can be estimated by the use of equation (5.32) as follows:

$$P_{bd} = 3\sigma_{H_{\min}} - \sigma_{H_{\max}} + T_o - P_p,$$

$$P_{bd} = 3 \times 3.68 \times 10^7\ Pa - 5.07 \times 10^7\ Pa + 5 \times 10^6 - 2.38 \times 10^7\ Pa,$$

and

$$P_{bd} = 4.09 \times 10^7\ Pa \equiv 5{,}932\,psi.$$

Example 5.5 clearly shows that the breakdown pressure is higher than the minimum horizontal stress and lower than the maximum horizontal stress. However, in some formations, the breakdown pressure might even exceed the maximum horizontal stress (Guo et al., 2017).

Measuring the Fracturing Pressure

There are tests that can be conducted to actually measure the fracturing pressure rather than estimate it since the above discussed formulas could significantly underestimate the fracturing gradient if the true Poisson's ratio is used rather than the effective Poisson's ratio, especially in shale formations. Consequently, other methods have been proposed to improve the estimation of the fracturing gradient such as the Eaton-Daines method and Zamora's method (Daines, 1982; Schlumberger, 2015). For instance, in Example 5.4, the fracturing gradient using Eaton's method was found to be 0.58 psi/ft. Several studies have shown that the fracturing gradient for the Bakken formation ranges between 0.5 psi/ft and 0.7 psi/ft (Jarvie et al., 2013; Yang & Zoback, 2014). Hence, the best way to accurately determine the fracturing gradient is by conducting a fracturing-gradient measurement test. The industry relies on two conventional tests to measure the fracturing gradient; the first test is called the Leak-Off Test (LOT), which is also known as the Pressure Integrity Test (PIT), and the second test is the Formation Integrity Test (FIT) (Valkó & Economides, 1995; Addis et al., 2016). Both these tests provide more information than just determining the fracturing pressure. If the purpose of the test is mainly to determine the fracture gradient, then the type of the formation (e.g., soft rock or hard rock) determines which of the two tests (i.e., LOT or FIT) should be conducted (Fakhr, 2016). Both the LOT and FIT are usually performed during drilling, and their procedures are beyond the scope of this chapter.

Fracture Closure Pressure

The fracture closure pressure (i.e., closure stress) is the pressure below which the induced fractures seal. The fracture closure pressure is a very critical parameter in the design of the hydraulic fracturing jobs because it sets the minimum limit of the bottomhole pressure during the injection period; hence, it must be known beforehand. The most conducted test in the unconventional reservoirs (e.g., extremely low permeability formations) to measure the fracture closure pressure is the minifrac test. The minifrac test is usually conducted prior to the main hydraulic fracturing treatment, and it provides several parameters (e.g., fracture gradient, fracture closure pressure, reservoir pressure, and formation permeability) that are essential in the designing process of the main hydraulic fracturing treatment (Fekete, 2017). The minifrac test will be discussed in detail in Chapter 6.

The fracture closure pressure could also be estimated, assuming the tectonic stress is zero, by the use of equation (5.33), hence, the fracture closure pressure would be similar to the minimum horizontal stress. Thus, the minimum horizontal stress could be used as a fracturing criteria, which means above it induced fractures would propagate, and below it induced fracture would close.

$$P_C = \left[\frac{\nu}{(1-\nu)}\right](\sigma_v - \alpha P_p) + \alpha P_p. \tag{5.33}$$

Nowadays, there are advanced downhole tools that are capable of monitoring the pressure and temperature while the hydraulic fracture is being initiated and after the injection is stopped; thus, the fracture propagation pressure and closure stress can be measured accurately. For example, the Modular Formation Dynamics Tester (MDT), which is a mark of Schlumberger, is capable of acquiring real-time accurate measurements of pressure and temperature during the fracture initiation by the use of advanced gauges (Waters et al., 2011).

5.4 Well-Logging Measurements to Determine the Elastic Parameters

The sonic log (i.e., acoustic compressional slowness and acoustic shear slowness) and the bulk density log measurements can be used to determine the dynamic elastic parameters (e.g., dynamic Young's modulus and dynamic Poisson's ratio) of the formation. However, in the hydraulic fracturing design process, the static parameters are used rather than the dynamic parameters (Lacy, 1997). Basically, the moduli are called dynamic when the elastic-wave velocity and bulk density are used to calculate the moduli. On the other hand, the moduli are called static when an actual lab test (e.g., triaxial test) is performed such that a known force is applied on the material (e.g., core samples) and the actual deformation is measured. Several studies have shown that the dynamic moduli for porous materials are higher than the static ones (Ide, 1936; Van Heerden, 1987; Lacy, 1997; Smith & Montgomery, 2015), and they could differ greatly from one another; hence, the dynamic moduli must be corrected before they are used in the hydraulic fracturing design process (Najibi et al., 2015). The ideal way to correlate the dynamic and static moduli is by developing local correlations in the lab for each particular formation by the use of core samples (NSI Technologies, 2017). However, that is an expansive and time consuming procedure, and there are several methods to correlate the dynamic and static moduli. Those methods vary from basic empirical correlations to advanced algorithms that are based on neural networks and the use of artificial intelligence.

5.4.1 Calculating the Dynamic Moduli

The dynamic moduli can be calculated from the dipole sonic logs by use of either the wave velocities or the wave travel time. In fact, the wave velocities can be derived from the wave travel time as shown in equations (5.34) and (5.35):

$$V_p = \frac{1 \times 10^6}{DTC} \tag{5.34}$$

and

$$V_s = \frac{1 \times 10^6}{DTS}, \tag{5.35}$$

where V_P is the compressional wave velocity, V_S is the shear wave velocity, DTC is the compressional wave travel time, and DTS is the shear wave travel time.

The wave travel time (i.e., DTC or DTS) is usually measured in microseconds per meter (μs/m,) and it is commonly provided by the logging service companies as microseconds per foot (μs/ft). Therefore, the factor 1×10^6 in equations (5.34) and (5.35) is basically a unit conversion factor from microseconds to seconds.

Wave Velocity Equations

Equations (5.36) and (5.37) can be used to calculate the dynamic Young's modulus and dynamic Poisson's ratio, respectively, by the use of the compressional wave velocity, shear wave velocity, and bulk density. It is important to note that these equations work for SI units where V_P and V_S are in (m/s) and the bulk density is in (kg/m^3) as well as the field units where V_P and V_S are in (ft/s) and the bulk density is in (g/cm^3). When the SI units are used, the Young's modulus would be in Pa. However, when the filed units are used, it is important to keep track of the units to end up with psi for the dynamic Young's modulus. Thus, when the field units are to be used in calculating the dynamic Young's modulus, it is more convenient to use the wave travel time (i.e., DTC and DTS) equations that are listed in the next section. The dynamic Poisson's ratio is dimensionless; therefore, it does not matter what type of units are used as long as both V_P and V_S have the same units when they are applied in equation (5.37).

$$E_{dynamic} = \frac{\rho_b\left(3\ V_p^2 - 4\ V_s^2\right)}{\left(\frac{V_p}{V_s}\right)^2 - 1} \tag{5.36}$$

and

$$\nu_{dynamic} = \frac{\frac{1}{2}\left(\frac{V_p}{V_s}\right)^2 - 2}{\left(\frac{V_p}{V_s}\right)^2 - 1} = \frac{V_p^2 - 2\,V_s^2}{2\left(V_p^2 - V_s^2\right)}. \tag{5.37}$$

Example 5.6 Calculate the Dynamic Moduli Using Wave Velocity

The compressional wave velocity of a core sample from the Bakken formation was found to be 4,530.67 (m/s) and the wave shear velocity is 2,677.57 (m/s). The bulk density of the core sample is 2,640.3 (kg/m^3). Calculate the following for the core sample:

a) Dynamic Young's modulus ($E_{dynamic}$)
b) Dynamic shear modulus ($G_{dynamic}$)
c) Dynamic bulk modulus ($K_{dynamic}$)
d) Dynamic Poisson's ratio ($\nu_{dynamic}$)

Solution

a) The dynamic Young's modulus is calculated by equation (5.36) as follows:

$$E_{dynamic} = \frac{\rho_b\left(3\ V_p^2 - 4\ V_s^2\right)}{\left(\frac{V_p}{V_s}\right)^2 - 1} = \frac{2{,}640.3 \times (3 \times 4{,}530.67^2 - 4 \times 2{,}677.57^2)}{\left(\frac{4{,}530.67}{2{,}677.57}\right)^2 - 1}$$

and

$$E_{dynamic} = 4.66 \times 10^{10}\ (Pa) \equiv 6.76 \times 10^6 psi \equiv 6.76\ MMpsi.$$

b) The dynamic Poisson's ratio is calculated by equation (5.37) as follows:

$$\nu_{dynamic} = \frac{V_p^2 - 2V_s^2}{2\left(V_p^2 - V_s^2\right)} = \frac{4{,}530.67^2 - (2 \times 2{,}677.57^2)}{2 \times (4{,}530.67^2 - 2{,}677.57^2)}$$

and

$$\nu_{dynamic} = 0.23.$$

c) The dynamic shear modulus can be calculated by the use of equation (5.8) as follows:

$$G_{dynamic} = \frac{E_{dynamic}}{2\left(1 + \nu_{dynamic}\right)} = \frac{4.66 \times 10^{10}}{2 \times (1 + 0.23)}$$

and

$$G_{dynamic} = 1.893 \times 10^{10}\ (Pa) \equiv 2.745\ MMpsi.$$

d) The dynamic bulk modulus can be calculated by the use of equation (5.10) as follows:

$$K_{dynamic} = \frac{E_{dynamic}}{3\left(1 - 2\,\nu_{dynamic}\right)} = \frac{4.66 \times 10^{10}}{3 \times (1 - 2 \times 0.23)}$$

and

$$K_{dynamic} = 2.8958 \times 10^{10}\ (Pa) \equiv 4.2\ MMpsi.$$

Wave Travel-Time Equations

The logging service companies provide the readings of the dipole sonic logs in wave travel time (i.e., DTC and DTS) rather than in wave velocities (i.e., V_P and V_S). In addition, the field units are typically used instead of the SI units. Hence, it is more convenient to calculate the dynamic moduli directly from the DTC and DTS readings without the need to convert the travel time to velocity. Thus, there is a set of equations that are widely used in the industry to calculate the dynamic moduli (equations (5.38) and (5.44) (Schlumberger, 2015). Note that the value 13,474.45 in equations (5.39) and (5.40) is just a conversion factor; therefore, when this set of equations is used with field units, the resulting dynamic moduli are going to be in million psi (MMpsi) and Poisson's ratio is dimensionless.

$$R_o = \frac{DTS}{DTC}, \tag{5.38}$$

$$G_{dynamic} = 13474.45 \frac{\rho_b}{DTS^2}, \tag{5.39}$$

$$K_{dynamic} = 13474.45\, \rho_b \left(\frac{1}{DTC^2}\right) - \frac{4}{3}\, G_{dynamic}, \tag{5.40}$$

$$\nu_{dynamic} = \frac{\left(0.5\, R_o^2 - 1\right)}{\left(R_o^2 - 1\right)}, \tag{5.41}$$

$$\nu_{dynamic} = \frac{3\, K_{dynamic} - 2\, G_{dynamic}}{6\, K_{dynamic} + 2\, G_{dynamic}}, \tag{5.42}$$

$$E_{dynamic} = \frac{9\, G_{dynamic} \times K_{dynamic}}{G_{dynamic} + 3\, K_{dynamic}}, \tag{5.43}$$

and

$$E_{dynamic} = 2\, G_{dynamic}\left(1 + \nu_{dynamic}\right). \tag{5.44}$$

Example 5.7 Calculate the Dynamic Moduli Using Wave Travel Time

For the core sample provided in Example 5.6, calculate the following using the travel-time equations and compare the results:

a) Dynamic shear modulus ($G_{dynamic}$)
b) Dynamic bulk modulus ($K_{dynamic}$)
c) Dynamic Poisson's ratio ($\nu_{dynamic}$)
d) Dynamic Young's modulus ($E_{dynamic}$)

Solution

First, the DTC and DTS values should be calculated from the compressional velocity and shear velocity, respectively, since they are not provided. This can be done by the use of equations (5.34) and (5.35). The bulk density should also be converted to field units as follows:

$$DTC = \frac{1 \times 10^6}{V_p} = \frac{1 \times 10^6}{4{,}530.67} = 220.71\left(\frac{\mu s}{m}\right) = 67.27\left(\frac{\mu s}{ft}\right),$$

$$DTS = \frac{1 \times 10^6}{V_s} = \frac{1 \times 10^6}{2{,}677.57} = 373.47\left(\frac{\mu s}{m}\right) = 113.83\left(\frac{\mu s}{ft}\right),$$

and

$$\rho_b = 2{,}640.3\left(\frac{kg}{m^3}\right) = 2.64\left(\frac{g}{cm^3}\right).$$

The next step in finding the dynamic moduli is to follow the wave travel-time equations as follows:

$$R_o = \frac{DTS}{DTC} = \frac{113.83}{67.27} = 1.69.$$

a) Dynamic shear modulus:

$$G_{dynamic} = 13{,}474.45 \times \frac{2.64}{113.83^2} = 2.745\ (MMpsi).$$

b) Dynamic bulk modulus:

$$K_{dynamic} = 13{,}474.45 \times 2.64 \times \left(\frac{1}{67.27^2}\right) - \frac{4}{3} \times 2.745 = 4.2\ (MMpsi).$$

c) Dynamic Poisson's ratio: There are two ways to calculate Poisson's ratio,

$$\nu_{dynamic} = \frac{(0.5\ R_o^2 - 1)}{(R_o^2 - 1)} = \frac{(0.5 \times 1.692^2 - 1)}{(1.692^2 - 1)} = 0.23$$

and

$$\nu_{dynamic} = \frac{3\ K_{dynamic} - 2\ G_{dynamic}}{6\ K_{dynamic} + 2\ G_{dynamic}} = \frac{3 \times 4.2 - 2 \times 2.74}{6 \times 4.2 + 2 \times 2.74} = 0.23.$$

d) Dynamic Young's modulus using two different methods:

$$E_{dynamic} = 2\ G_{dynamic}\left(1 + \nu_{dynamic}\right) = 2 \times 2.745 \times (1 + 0.2316) = 6.76\ (MMpsi)$$

and

$$E_{dynamic} = \frac{9\ G_{dynamic} \times K_{dynamic}}{G_{dynamic} + 3\ K_{dynamic}} = \frac{9 \times 2.745 \times 4.2}{2.745 + 3 \times 4.2} = 6.76(MMpsi).$$

By comparing Example 5.6 and Example 5.7, it is clear that both the wave velocity method and the wave travel-time method yield the same dynamic moduli; thus, it does not matter which method is used to calculate the dynamic moduli.

5.4.2 Correlations for Static Moduli

As mentioned earlier that the dynamic moduli are not used in the hydraulic fracturing design process, and acquiring the static moduli from lab tests is an expansive procedure (Example 5.1). Hence, in the absence of lab measurements, the industry relies on correlations to obtain the static moduli from the dynamic ones.

Static Young's Modulus

There are several empirical correlations that have been developed to relate the dynamic Young's modulus and static Young's modulus for sedimentary rocks as well as igneous and metamorphic rocks (Najibi et al., 2015; Brotons et al., 2016). Furthermore, some of the correlations were developed for high porosity (e.g., greater than 15%) and high permeability as well. Some of those correlations are also applicable in shales. The majority of the correlations come in similar forms as listed in equations (5.45) to (5.47) (Brotons et al., 2016). Equations (5.48) to (5.50) summarize some of those equations. There are also software packages that support different and more complex correlations (e.g., equation (5.51)).

$$E_{static} = a\,E_{dynamic}^{b}, \tag{5.45}$$

$$E_{static} = a + b\,E_{dynamic}^{c}, \tag{5.46}$$

and

$$E_{static} = a\,E_{dynamic}^{2} + b\,E_{dynamic}, \tag{5.47}$$

where a, b, and c are constant parameters.

The Lacy correlation for sedimentary rocks is reported in equation (5.48) (Lacy, 1997),

$$E_{static} = 0.018\,E_{dynamic}^{2} + 0.422\,E_{dynamic}. \tag{5.48}$$

The Ohen correlation, which is based on the Gulf of Mexico samples with moderate clay content, is applicable in shale formations, and is reported in equation (5.49) (Ohen, 2003),

$$E_{static} = 0.0158\,E_{dynamic}^{2.74}. \tag{5.49}$$

There are correlations that are applicable for a specific range of porosity regardless of the rock type even though they were first derived based on sandstone samples. For instance, if the porosity is less than 25%, equation (5.50) can be used as a general relationship between the static and dynamic Young's modulus (NSI Technologies, 2017). The Morales correlation, equation (5.51), can also be used when the porosity is between 15% and 25% (Morales & Marcinew, 1993).

$$E_{static} = -0.42 + 0.835\,E_{dynamic}^{1} \tag{5.50}$$

and

$$\log E_{static} = 1.829 + 0.692\log\,E_{dynamic}. \tag{5.51}$$

Static Poisson's Ratio

There is no clear relationship between the dynamic Poisson's ratio and the static Poisson's ratio, and a plot between the static and dynamic Poisson's ratios yields scattered points with no clear correlation. However, equation (5.52) is widely used to correlate the dynamic and static Poisson's ratio,

$$\nu_{static} = \nu_{\text{multiplier}}\,\nu_{dynamic}, \tag{5.52}$$

where $\nu_{\text{multiplier}}$ is a Poisson's ratio multiplier, and it is usually taken as 1 when no core data are available to create local correlations.

Static Shear Modulus and Static Bulk Modulus

The static shear modulus and the static bulk modulus can be calculated by substituting the static Young's modulus and static Poisson's ratio in equations (5.8) and (5.10).

Example 5.8 Estimate the Static Young's Modulus

Estimate the static Young's modulus from the calculated dynamic modulus in Example 5.7. Use different correlations to estimate the static Young's modulus, and compare the results.

a) General relationship for porosity less than 25%
b) Lacy's correlation for sedimentary rocks
c) Ohen's correlation for shales

Solution

The dynamic Young's modulus from Example 5.7 is 6.76 (MMpsi).

a) The static Young's modulus by using the correlation for porosity less than 25% can be calculated as follows:

$$E_{static} = -0.42 + 0.835\, E^{1}_{dynamic} = -0.42 + 0.835 \times 6.76^{1} = 5.22(MMpsi).$$

b) Using Lacy's correlation:

$$E_{static} = 0.018\, E^{2}_{dynamic} + 0.422\, E_{dynamic} = 0.018 \times 6.76^{2} + 0.422 \times 6.76$$

and

$$E_{static} = 3.675\ MMpsi.$$

c) Ohen's correlation:

$$E_{static} = 0.0158\, E^{2.74}_{dynamic} = 0.0158 \times 6.76^{2.74} = 2.97(MMpsi).$$

It is very clear that those correlations gave different static Young's modulus with a varying range from 5.22 (MMpsi) to 2.97 (MMpsi) for the same dynamic Young's modulus value. Consequently, it is very critical to choose the correlation that is more representative of the formation. As mentioned earlier, the ideal way to do that is by having core samples from the formation of interest and conducting triaxial tests on them to actually measure the static moduli and then developing a unique local correlation that is representative of different lithofacies present in that formation or use one from the literature that accurately represents different lithofacies of the formation of interest.

5.5 Identifying the Geomechanical Sweet Spots

The term *sweet spot* is commonly used in the unconventional resources to indicate the optimum drilling locations. The optimum drilling locations are identified by studying and characterizing several reservoir quality parameters such as formation porosity, permeability, water saturation, thickness, total organic carbon (TOC), and organic matter maturity indexes. Seismic data are also heavily involved in identifying the sweet spots for drilling. When all the parameters of interest meet a certain criteria at one specific location, then that location is called a sweet spot (i.e., an optimum location for drilling). This section of this chapter focuses on identifying the geomechanical sweet spots, which means locating the optimum locations for hydraulic fracturing treatments by studying the geomechanical properties of the formation.

In the previous chapters, the reservoir quality and geochemical properties of the formation were studied and discussed. Based on those studies, the optimum locations have been identified in terms of their petrophysical and geochemical properties. However, as important as they may be, petrophysical and geochemical properties of the formation are redundant, in the unconventional resources, if hydraulic fracturing is not feasible. The opposite is also true. Some locations might be great candidates for hydraulic fracturing treatments; however, their petrophysical and geochemical properties do not represent significant volumes of oil and mature organic matters.

Identifying the geomechanical sweet spots requires developing sophisticated 3D geomechanical models that are based on seismic data, well logs, and the rock properties of the formation (Addis et al., 2016). The clusters of the natural fractures should be included in the models to accurately locate the geomechanical sweet spots. The 3D models should also be designed to characterize the formation lithology, which is essential in evaluating the formation brittleness index. Visualizing and analyzing the formation geomechanical properties (e.g., stresses and elastic moduli), and thus locating the geomechanical sweet spots, become easier with the use of the 3D geomechanical models (Addis et al., 2016).

In the absences of sophisticated geomechanical models, the geomechanical sweet spots can be identified by studying and characterizing the parameters that have been discussed in this chapter (e.g., elastic moduli, minimum horizontal stress, and closure stress) as well as other indexes such as the brittleness index, which will be discussed next.

5.5.1 Brittleness Index

The term *brittleness index (BI)* is used to indicate if the formation rocks (e.g., shale) are brittle or ductile. It is very important to know if the formation is brittle or ductile because ductile rocks tend to resist and heal the hydraulically induced fractures (Rickman et al., 2008). The brittleness index can be quantified by studying three main indexes (Mathia et al., 2016). The first two indexes are more common because they can be calculated conveniently by the use of well logs. The first index is the elastic brittleness index, which is a combination of Young's modulus and Poisson's ratio. The second index is the mineral brittleness index, which is based on the mineralogy of the formation. The third brittleness index is uncommon because it requires studying the rock static mechanical properties by the use of core samples. However, it considers the confining pressure and the fracture toughness (Hu et al., 2015). Basically, the third BI indicates that at low confining pressure, higher brittleness is achieved.

In general, higher BI values mean the formation is more brittle; thus, a higher chance of a successful hydraulic fracturing treatment. On the other hand, low BI values mean the formation is more ductile; thus, less chance of a successful hydraulic fracturing treatment. However, low BI values mean good trapping locations (i.e., good fracture barrier); hence, less chance of hydrocarbon migration because of the ductile nature of the formation (Rickman et al., 2008).

Elastic Brittleness Index

Since the units of Young's modulus and Poisson's ratio are different, the elastic brittleness index is basically a way to combine both Young's modulus and Poisson's ratio in one parameter in order to have an overall idea if the rock is brittle or ductile. The elastic brittleness index is also called the fracability index.

Different methods have been proposed to calculate the elastic brittleness index. However, there is a common approach that the industry relies on more than any other approach, which is reported by Richman et al. (2008). This approach indicates that higher rock brittleness locations are found in high Young's modulus and low Poisson's ratio locations. Simply, as Young's modulus increases, the more brittle the rock can be. On the other hand, as Poisson's ratio decreases, the more brittle the rock can be. The elastic brittleness index, equation (5.53), is a widely used approach because the needed data to calculate it can be found easily from the dipole sonic logs.

$$BI_{elastic} = \left(\frac{E_n + \nu_n}{2}\right) \times 100, \quad (5.53)$$

where, $BI_{elastic}$ is the elastic brittleness index, E_n is the normalized Young's modulus, and ν_n is the normalized Poisson's ratio.

The normalized Young's modulus and the normalized Poisson's ratio can be calculated by equation (5.54) and equation (5.55), respectively.

$$E_n = \frac{E - E_{min}}{E_{max} - E_{\min}} \quad (5.54)$$

and

$$\nu_n = \frac{\nu - \nu_{min}}{\nu_{max} - \nu_{\min}}, \quad (5.55)$$

where, minimum and maximum values of Young's modulus and Poisson's ratio are found within the zone of interest and vary for different rock types. The first proposed approach by Rickman et al. for normalizing Young's modulus and Poisson's ratio was somewhat based on arbitrary $E_{\min}$, $E_{\max}$, $\nu_{\min}$, and $\nu_{\max}$ values, as shown in equations (5.56) and (5.57).

$$E_n = \left(\frac{E - 1}{(8 - 1)}\right) \times 100 \quad (5.56)$$

and

$$\nu_n = \left(\frac{\nu - 0.4}{(0.15 - 0.4)}\right) \times 100. \quad (5.57)$$

There is no optimum brittleness value that can be used for all formations because the formation mechanical properties (e.g., Young's modulus and Poisson's ratio) differ from one to the other. The same thing goes for the mineral-based brittleness index as the lithology changes, too. In general, formations that are rich in quartz and have high Young's modulus and low Poisson's ratio are considered more brittle than clay-rich formations (Mathia et al., 2016).

As a rule of thumb, if the elastic brittleness is found to be more than 40% in a certain location, then that location is considered a good candidate for hydraulic fracturing (Rickman et al., 2008). However, the mineral brittleness index as well as the other discussed parameters must be considered before that location can be called a sweet spot (Jin et al., 2015; Cudjoe et al., 2016).

Example 5.9 Calculate the Elastic Brittleness Index

Calculate the elastic brittleness index for each zone, and decide which zone is a good candidate for hydraulic fracturing (Tables 5.2, 5.3, 5.4). Use the minimum and maximum approach for normalizing Young's modulus and Poisson's ratio, and pick only one minimum value and one maximum value when evaluating the two zones.

Solution

The first step is to find the minimum and maximum Young's modulus and Poisson's ratio. The maximum Young's modulus and Poisson's ratio values are 10.9952 and 0.3059, respectively. The minimum Young's modulus and Poisson's ratio values are 3.1822 and 0.00546, respectively.

Table 5.2 Data for Example 5.9.

Zone A		Zone B	
Poisson's ratio	Young's modulus (MMpsi)	Poisson's ratio	Young's modulus (MMpsi)
0.040	6.60	0.30	7.79
0.045	6.59	0.27	8.79
0.070	6.70	0.28	8.54
0.048	6.55	0.28	8.13
0.069	6.69	0.28	7.48
0.025	6.39	0.26	6.94
0.076	6.67	0.20	10.99
0.062	6.52	0.22	6.56
0.058	6.50	0.21	6.68
0.055	6.50	0.25	7.14
0.056	6.50	0.26	7.70
0.051	6.36	0.26	8.01
0.005	6.06	0.26	7.97
0.047	6.34	0.27	7.77
0.043	6.16	0.28	7.61
0.064	6.28	0.28	7.45
0.210	3.18	0.27	7.35

The next step is to normalize Young's modulus and Poisson's ratio by the use of equations (5.54) and (5.55), respectively. The first point in Zone A can be normalized as follows:

$$E_n = \frac{6.60 - 3.18}{10.99 - 3.18} = 0.43$$

and

$$\nu_n = \frac{0.040 - 0.005}{0.30 - 0.005} = 0.11.$$

The last step is to calculate the elastic brittleness index by the use of equation (5.53) as follows:

$$BI_{elastic} = \left(\frac{0.43 + 0.11}{2}\right) \times 100 = 27\%.$$

By evaluating both zones, it is clear that Zone B is a good candidate for hydraulic fracturing as the elastic brittleness index for Zone B is greater than 40%.

Table 5.3 Solution to Zone A, Example 5.9.

Zone A				
Poisson's ratio	Young's modulus (MMpsi)	Normalized Young's modulus	Normalized Poisson's ratio	Elastic BI (%)
0.04	6.60	0.43	0.11	27
0.04	6.59	0.43	0.13	28
0.07	6.70	0.45	0.22	33
0.04	6.55	0.43	0.14	28
0.06	6.69	0.44	0.21	33
0.02	6.39	0.41	0.06	23
0.07	6.67	0.44	0.23	34
0.06	6.52	0.42	0.18	30
0.05	6.50	0.42	0.17	30
0.05	6.50	0.42	0.16	29
0.05	6.50	0.42	0.16	29
0.05	6.36	0.40	0.15	28
0.005	6.06	0.36	0.00	18
0.04	6.34	0.40	0.13	27
0.04	6.16	0.38	0.12	25
0.06	6.28	0.39	0.19	29
0.21	3.18	0.00	0.68	34

Table 5.4 Solution to Zone B, Example 5.9

Zone B				
Poisson's ratio	Young's modulus (MMpsi)	Normalized Young's modulus	Normalized Poisson's ratio	Elastic BI %
0.30	7.79	0.58	1.00	79
0.27	8.79	0.71	0.90	81
0.28	8.54	0.68	0.92	80
0.28	8.13	0.63	0.92	78
0.28	7.48	0.55	0.92	73
0.26	6.94	0.48	0.85	66
0.20	10.99	1.00	0.67	83
0.22	6.56	0.43	0.71	57
0.21	6.68	0.44	0.70	57
0.25	7.14	0.50	0.81	66
0.26	7.70	0.57	0.87	72
0.26	8.01	0.61	0.87	74
0.26	7.97	0.61	0.87	74
0.27	7.77	0.58	0.91	74
0.28	7.61	0.56	0.92	74
0.28	7.45	0.54	0.91	73
0.27	7.35	0.53	0.89	71

Mineral-Based Brittleness Index

Along with the elastic brittleness index, which is based on the formation elastic moduli, there is a mineral-based brittleness index. The mineral-based brittleness index is an indicator of the brittleness of the formation based on the mineral composition of the formation. The formation mineralogy can be found by advanced mud logging, mineralogy logging, or core measurement techniques. The mineral-based brittleness index (MBI) was first introduced by Jarvie et al. in 2007 as the quartz weight in percentage divided by the weights of quartz, clay, and carbonate as shown in equation (5.58) (Jarvie et al., 2007). The quartz weight was chosen as the main representative for the rock brittleness in equation (5.58) because that equation was derived based on samples from the Barnett shale in which quartz is the dominant mineral.

$$MBI_J = \frac{\text{Quartz}}{Quartz + \text{Carbonates} + Clays}, \quad (5.58)$$

where all the components are in weight percent.

The mineral-based brittleness index is not a good representative of the rock brittleness because it does not really account for the cementation between the grains, which could fully control how rocks fail (Addis et al., 2016). However, several researchers proposed improved mineral-based brittleness indexes to estimate the rocks brittleness, and they took into account more minerals that contribute to the overall rock brittleness and not just quartz (Wang & Gale, 2009; Jin et al., 2015; Cudjoe et al., 2016; Bratovich & Walles, 2016; and Addis et al., 2016). Several reported equations to calculate the mineral-based brittleness index are shown in equations (5.59) and (5.61).

$$MBI_1 = \frac{Quartz + Dolomite}{Quartz + Dolomite + Limestone + Clays + TOC}, \quad (5.59)$$

where all the components are in weight %.

$$MBI_2 = \frac{Quartz + Dolomite + Calcite}{Quartz + Dolomite + Calcite + Clays + TOC}, \quad (5.60)$$

where, all the components are in weight %, and

$$MBI_3 = \frac{Quartz + Dolomite + Calcite + Feldspars}{Quartz + Dolomite + Calcite + Feldspars + Siderite + TOC}, \quad (5.61)$$

where all the components are in weight %.

The above MBI equations differ based on the mineralogy of the formation. Thus, developing localized MBI formulas for each individual reservoir by including all the components that contribute to the overall brittleness is the ideal way to have a realistic measure of the formation brittleness. Refer to Example 5.9 for further illustration of how the MBI method is not a great indicator of the rock brittleness without the use of a proper MBI formula.

Example 5.10 Calculate the Mineral-Based Brittleness Index

Calculate the mineral-based brittleness index for each zone. Use the first and second MBI methods (i.e., MBI_1 and MBI_2,) and decide which zone is a better candidate for hydraulic fracturing. See Tables 5.5–5.8.

Solution

The mineral-based brittleness index for the first point of Zone A can be found by the use of equations (5.59) and (5.60) as follows:

$$MBI_1 = \frac{0.2325 + 0.0792}{0.2325 + 0.0792 + 0.6384 + 0.0500 + 0.0000} = 0.3116$$

Table 5.5 Rock Composition of Zone A for Example 5.10.

Rock composition Zone A				
Calcite	**Dolomite**	**Kerogen**	**Quartz**	**Clay**
0.6384	0.0792	0.0000	0.2325	0.0500
0.6142	0.1005	0.0000	0.2253	0.0600
0.5842	0.1367	0.0000	0.2490	0.0300
0.5556	0.0885	0.0463	0.2397	0.0700
0.5083	0.0960	0.0511	0.2646	0.0800
0.6320	0.1042	0.0375	0.1943	0.0320
0.6614	0.0625	0.0374	0.1937	0.0450
0.6852	0.0379	0.0367	0.1902	0.0500
0.7023	0.0157	0.0376	0.1944	0.0500
0.6936	0.0083	0.0369	0.1911	0.0700

Table 5.6 Rock Composition of Zone B for Example 5.10.

Rock composition Zone B				
Calcite	**Dolomite**	**Kerogen**	**Quartz**	**Clay**
0.2858	0.1496	0.0776	0.4015	0.0855
0.2821	0.1514	0.0730	0.3781	0.1155
0.2839	0.1350	0.0710	0.3677	0.1423
0.2826	0.1241	0.0699	0.3617	0.1617
0.3412	0.0483	0.0745	0.3857	0.1502
0.0275	0.0000	0.0840	0.4347	0.4538
0.0001	0.0000	0.0875	0.4528	0.4596
0.0000	0.3130	0.0672	0.3476	0.2722
0.0000	0.2924	0.0652	0.3375	0.3049
0.0000	0.2939	0.0642	0.3320	0.3097

and

$$MBI_2 = \frac{0.2325 + 0.0792 + 0.6384}{0.2325 + 0.0792 + 0.6384 + 0.0500 + 0.0000} = 0.95.$$

Hence, the brittleness index for the first point of Zone A using the two methods MBI_1 and MBI_2 is 31.16% and 95%, respectively. Clearly Zone B is a better candidate for hydraulic fracturing since both BMI_1 and BMI_2 yield brittleness values above 40%.

Table 5.7 Solution to Zone A, Example 5.10.

Rock composition Zone A						
Calcite	**Dolomite**	**Kerogen**	**Quartz**	**Clay**	**MBI 1**	**MBI 2**
0.6384	0.0792	0.0000	0.2325	0.0500	0.3116	0.9500
0.6142	0.1005	0.0000	0.2253	0.0600	0.3258	0.9400
0.5842	0.1367	0.0000	0.2490	0.0300	0.3858	0.9700
0.5556	0.0885	0.0463	0.2397	0.0700	0.3281	0.8837
0.5083	0.0960	0.0511	0.2646	0.0800	0.3606	0.8689
0.6320	0.1042	0.0375	0.1943	0.0320	0.2985	0.9305
0.6614	0.0625	0.0374	0.1937	0.0450	0.2562	0.9176
0.6852	0.0379	0.0367	0.1902	0.0500	0.2281	0.9133
0.7023	0.0157	0.0376	0.1944	0.0500	0.2101	0.9124
0.6936	0.0083	0.0369	0.1911	0.0700	0.1994	0.8931

Table 5.8 Solution to Zone B, Example 5.10.

Rock Composition Zone B						
Calcite	**Dolomite**	**Kerogen**	**Quartz**	**Clay**	**MBI 1**	**MBI 2**
0.2858	0.1496	0.0776	0.4015	0.0855	0.5511	0.8369
0.2821	0.1514	0.0730	0.3781	0.1155	0.5294	0.8115
0.2839	0.1350	0.0710	0.3677	0.1423	0.5027	0.7866
0.2826	0.1241	0.0699	0.3617	0.1617	0.4858	0.7684
0.3412	0.0483	0.0745	0.3857	0.1502	0.4341	0.7753
0.0275	0.0000	0.0840	0.4347	0.4538	0.4347	0.4622
0.0001	0.0000	0.0875	0.4528	0.4596	0.4528	0.4529
0.0000	0.3130	0.0672	0.3476	0.2722	0.6606	0.6606
0.0000	0.2924	0.0652	0.3375	0.3049	0.6299	0.6299
0.0000	0.2939	0.0642	0.3320	0.3097	0.6261	0.6261

In conclusion, from a geomechanical point of view, a zone is considered fracable when all the mentioned parameters in this chapter (e.g., fracturing pressure, closure pressure, and the brittleness indexes) show promising results for hydraulic fracturing. Ultimately, a zone is considered a sweet spot in unconventional reservoirs when the geomechanical parameters, petrophysical properties (e.g., organic versus inorganic porosity, permeability, and hydrocarbon saturation), and geochemical parameters (e.g., S_1, S_2, and thermal maturity) show promising results.

After the sweet spot zones are identified, the design of the real hydraulic fracturing treatment begins while taking into account the geomechanical properties of the formation (e.g., fracture gradient and fracture closure stress). In Chapter 6, the process of designing a hydraulic fracturing treatment will be discussed in detail including the selection process of the right equipment and material (e.g., surface pumps, fracturing fluids, and proppant). It will be clear that the geomechanical parameters play a crucial role in the designing process of the hydraulic fracturing treatments (e.g., effect of closure pressure on proppant selection), and that the geomechanical parameters should be studied rigorously before attempting to do a hydraulic fracturing job.

5.6 General Remarks

1) There are several different formulas to estimate the fracturing pressure and fracture closure stress that are not mentioned in this chapter. The calculation of those parameters could significantly vary from one company to the other, and many competitors do not share their in-house developed methods. There are also field tests that are conducted to actually measure the fracturing pressure and fracture closure stress, and they will be discussed in detail in Chapter 6.
2) The static moduli are used in the designing process of the hydraulic fracturing treatments. Some software packages have built-in correlations to convert the dynamic moduli from log measurements to static moduli. However, those correlations might not work perfectly for certain formations. Thus, it is ideal to use appropriate correlations prior to importing the data on the simulation software for designing and simulating hydraulic fracturing job.
3) If dipole sonic log measurements (i.e., DTC and DTS) are not available, which is usually the case in the old wells, techniques exist that can be used to estimate the acoustic shear slowness measurements (i.e., DTS) by the use of the monopole sonic logs. The DTS values are required to calculate the dynamic moduli.
4) There are several methods to estimate the brittleness of the formation. However, the two methods mentioned in this chapter (i.e., elastic brittleness index and mineral-based brittleness index) are used more frequently in the industry because their requirements can be met just by using well-logging measurements. The other methods require collecting core samples and true triaxial tests.
5) The geomechanical properties that are provided in Table 5.9 are based on reported

Table 5.9 Geomechanical Properties.

Unconventional play	Average Young's modulus (MMpsi)	Average Poisson's ratio
Barnett	5.800	0.250
Bakken - Upper Member	1.610	0.200
Bakken - Middle Member	3.020	0.230
Bakken - Lower Member	1.130	0.237
Marcellus	2.089	0.150
Utica	2.580	0.270

Table 5.9 (Continued)

Unconventional play	Average Young's modulus (MMpsi)	Average Poisson's ratio
Eagle Ford - US Region	2.280	0.206
Permian Basin* - Bone Spring	6.000	0.250
Permian Basin* - Clearfork**	2.500	0.200
Permian Basin* - Wolfcamp A	4.500	0.220
Permian Basin* - Wolfcamp B	4.000	0.230
Permian Basin* - Upper Wolfcamp C	4.500	0.280
Permian Basin* - Lower Wolfcamp C	5.400	0.250
Permian Basin* - Cline (Cisco)**	3.000	0.200
Niobrara - Chalk interval	3.980	0.167
Niobrara - Marl interval	2.820	0.249
Woodford*	6.000	0.200
Fayetteville**	1.190	0.161
Haynesville	4.720	0.239
Muskwa-Canada	5.350	0.200
Eagle Ford* - Mexico region	4.100	0.240
Pimienta Formation_Mexico	4.560	0.263
Vaca Muerta*_Argentina	3.800	0.250
Longmaxi shale in Sichuan region _China	4.300	0.120-0.290
Qusaiba Shale_Saudi Arabia	4.640	0.250
Goldwyer*_Australia	6.240	0.220

* Correlations were used to get the static moduli.
** Based on values that were used in fracture design and simulations.

values in the literature. The values of Young's modulus and Poisson's ratio are based on laboratory tests that are conducted on core samples, which are from a specific region and a certain depth of the mentioned play. It is possible that these values vary in different regions or at different depths of the same play because the lithofacies might be different there. For example, the Young's modulus of the Qusaiba shale play in Saudi Arabia ranges from 3.77 MMpsi to 4.64 MMpsi based on the lithofacies of each region.

5.7 Problems

1 Estimate the breakdown pressure of the following formation, which is at a depth of 2,924 m, and the average density of the overburden formation is 2,700 kg/m^3. The formation has a poroelastic constant of 0.63 and a Poisson's ratio of 0.23. The pore pressure gradient of the formation is $8.03 \times 10^3 \frac{Pa}{m}$ and the rock tensile strength is 4 MPa. Assume a tectonic stress of 12.62 MPa. To estimate the minimum horizontal stress, use Eaton's method.

2 Estimate the wave velocities (i.e., V_P and V_S) in (m/s) knowing that the dynamic Young's modulus is 5.7 (MMpsi,) the dynamic Poisson's ratio is 0.22 and the bulk density is 2,670 (kg/m^3).

3 Based on the log measurements provided in Tables 5.10 and 5.11, find the rock composition as well as the porosity, and report the following. (Hint: use matrix inversion.)

(a) What is the average composition of the Kerogen in the zone? (Answer: 0.0283)

(b) What is the average porosity of the zone?

(c) Plot the rock composition and the porosity using a stacked area chart.

Table 5.10 Log Measurements for Problem 5.3.

Depth (m)	GR	RHOB $\left(\frac{g}{cm^3}\right)$	Uphoto	NPHI
3,800	162.20	2.62	7.61	22.83
3,810	158.89	2.63	7.72	23.22
3,820	156.16	2.63	7.79	23.43
3,830	157.42	2.63	7.78	23.36
3,840	158.29	2.62	7.70	23.26
3,850	164.68	2.61	7.67	23.28
3,860	161.62	2.60	7.71	22.91
3,870	161.85	2.60	7.78	22.47
3,880	153.57	2.61	7.75	21.93
3,890	152.91	2.61	7.70	22.01
3,900	148.72	2.62	7.67	22.24
3,910	152.02	2.62	7.69	22.34
3,920	154.82	2.63	7.75	21.52
3,930	156.40	2.64	7.83	21.63
3,940	158.05	2.64	7.91	21.17

Table 5.11 Log Measurements for Each Component for Problem 5.3.

Log reading	Clay	Quartz	Pyrite	Kerogen	PHIeff
GR	216	32	32	1755	32
RHOB	2.8	2.65	5	1.3	1
Uphoto	11.200	4.780	82.200	0.312	0.500
NPHI	43	−2.1	−1.9	60	100
UNITY	1	1	1	1	1

4 Determine whether the zone provided in Table 5.12 is a good candidate for hydraulic fracturing based on the elastic brittleness index. Use the following correlations to find the static Young's modulus.

(a) Formation porosity is less than 25% correlation.

(b) Ohen's correlation

(c) Is the zone a good candidate for hydraulic fracturing if the correlation in part a is used?

(d) Is the zone a good candidate for hydraulic fracturing if the correlation in part b is used?

Table 5.12 Log Measurements for Problem 5.4

Depth (ft)	DTC $\left(\frac{\mu s}{ft}\right)$	DTS $\left(\frac{\mu s}{ft}\right)$	Bulk density $\left(\frac{g}{cm^3}\right)$
9,235	52.21	93.39	2.68
9,235.5	51.97	92.71	2.68
9,236	52.26	93.94	2.69
9,236.5	52.17	93.71	2.70
9,2370	52.58	93.45	2.71
9,237.5	52.42	93.48	2.71
9,238	52.35	94.41	2.70
9,238.5	52.19	93.92	2.70
9,239	52.06	93.37	2.70
9,239.5	52.16	93.61	2.70
9,240	52.52	94.17	2.70
9,240.5	52.69	93.63	2.70
9,241	52.82	93.45	2.69
9,241.5	52.93	94.43	2.69
9,242	53.00	94.17	2.69
9,242.5	53.08	94.03	2.72
9,243	52.92	94.59	2.71
9,243.5	52.70	94.48	2.71
9,244	52.48	94.28	2.69
9,244.5	52.48	94.31	2.69
9,245	52.42	94.27	2.69
9,245.5	52.28	93.64	2.70
9,246	52.42	94.02	2.71
9,246.5	52.52	94.77	2.71
9,247	52.91	94.80	2.70
9,247.5	53.11	94.91	2.68
9,248	53.06	94.61	2.68
9,248.5	53.04	95.08	2.69
9,249.5	53.01	94.94	2.70
9,250	52.86	94.93	2.70

Additional Reading

Jarvie, D., Hill, R., Ruble, T., & Pollastro, R. (2007). Unconventional shale-gas systems: The Mississippian Barnett shale of northcentral Texas as one model for thermogenic shale-gas assessment. *American Association of Petroleum Geologists Bulletin.*

Zoback, M. (2007). *Reservoir geomechanics.* Cambridge University Press.

Mews, K.S.; Alhubail, M.M.; Barati, R.G. (2019). A Review of Brittleness Index Correlations for Unconventional Tight and Ultra-Tight Reservoirs. *Geosciences.*

References

Abousleiman, Y., Tran, M., & Hoang, S. (2008). *Laboratory characterization of anisotropy and fluid effects on shale mechanical properties using inclined direct shear testing device idstd.* The 42nd U.S. Rock Mechanics Symposium (USRMS), San Francisco, American Rock Mechanics Association.

Addis, M., Bordoloi, S., Franquet, J., Hooyman, P., Hurt, R., Kowan, J., et al. (2016). Role of geomechanical engineering in unconventional resources developments. In U. Ahmed & N. Meehan, *Unconventional Oil and Gas Resources: Exploitation and Development.* CRC Press Taylor & Francis Group.

Ahmadov, R., Vanorio, T., & Mavko, G. (2009). Confocal laser scanning and atomic-force microscopy in estimation of elastic properties of the organic-rich Bazhenov Formation. *The Leading Edge.* doi:10.1190/1.3064141

Ahmed, U., & Meehan, N. (2016). *Unconventional oil and gas resources: Exploitation and development.* Boca Raton: CRC Press Taylor & Francis Group. doi:ISBN 13: 978-1-4987-5941-0

Alshakhs, M., & Rezaee, R. (2018). Sweet-spot mapping through formation evaluation and property modelling using data from the Goldwyer Formation of the Barbwire Terrace, Canning basin. *Petroleum.* doi:10.1016/j.petlm.2018.06.003

Aoudia, K., Miskimins, J., Harris, M., & Mnich, C. (2010). *Statistical analysis of the effects of mineralogy on rock mechanical properties of the Woodford shale and the associated impacts for hydraulic fracture treatment design.* 44th U.S. Rock Mechanics Symposium and 5th U.S.-Canada Rock Mechanics Symposium. Salt Lake City: American Rock Mechanics Association.

Avanzini, A., Balossino, P., Brignoli, M., Spelta, E., & Tarchiani, C. (2016). Lithologic and geomechanical facies classification for sweet spot identification in gas shale reservoir. *Interpretation, 4*(3). doi:10.1190/INT-2015-0199.1

Bale, A., Fossen, H., Berg, E., & Kui, T. (2008). *Comprehensive mini-frac testing in the Gullfaks field as a tool for characterisation of reservoir structure and rock mechanics.* International Petroleum Technology Conference, Kuala Lumpur, International Petroleum Technology Conference. doi:10.2523/IPTC-11968-MS

Biot, M. (1955). Theory of elasticity and consolidation for a porous anisotropic solid. *Journal of Applied Physics, 26*(2), 182–185. doi:10.1063/1.1721956

Biot, M. (1956). General solutions of the equations of elasticity and consolidation for a porous material. *Jounral of Applied Mechanics, 23*(1), 91–96.

Biot, M. (1962). Mechanics of deformation and acoustic propagation in porous media. *Journal of Applied Physics, 33*(4). doi:https://doi.org/10.1063/1.1728759

Bishop, K. (2015). *Mechanical stratigraphy of the Vaca Muerta formation, Neuquen basin, Argentina.* The Colorado School of Mines.

Bratovich, M., & Walles, F. (2016). Formation evaluation and reservoir characterization of source rock reservoirs. In U. Ahmed, & N. Meehan, *Unconventional oil and gas resources: Exploitation and development*. CRC Press Taylor & Francis Group.

Bridges, M. (2016). *Mechanical properties of the Niobrara*. The Colorado School of Mines.

Brinkley, S. (2016). *Petroleum geology of the Utica/ Point Pleasant play in Washington County, Ohio*. East Carolina University.

Brotons, V., Tomás, R., Ivorra, S., Grediaga, A., Martínez-Martínez, J., Benavente, D., & Gómez-Heras, M. (2016). Improved correlation between the static and dynamic elastic modulus of different types of rocks. *Materials and Structures*, *49*(8), 3021–3037. doi: 10.1617/ s11527-015-0702-7

Castagna, J., Batzle, M., & Eastwood, R. (1985). Relationships between compressional wave and shear wave velocities in clastic silicate rocks. *Geophysics*, *50*(4), 571–581. doi:10.1190/ 1.1441933

Cheng, C., & Toksoz, M. N. (1983). *Determination of shear wave velocities in "slow" formations*. SPWLA 24th Annual Logging Symposium. Calgary: Society of Petrophysicists and Well-Log Analysts.

Craig, J. (2014). Basic hydraulic fracturing.

Cudjoe, S., Vinassa, M., Henrique Bessa Gomes, J., & Barati, R. (2016). A comprehensive approach to sweet-spot mapping for hydraulic fracturing and CO_2 huff-n-puff injection in Chattanooga shale formation. *Journal of Natural Gas Science and Engineering*, 1201–1218. doi: 10.1016/j.jngse.2016.03.042

Cui, Q. (2016). *Thesis: Stress dependent compaction in tight reservoirs and its impact on long-term production*. Colorado School of Mines.

Daines, S. (1982). Prediction of fracture pressures for wildcat wells. *Journal of Petroleum Technology*, *34*(4). doi:10.2118/9254-PA

Dong, T. (2016). *Geochemical, petrophysical and geomechanical properties of stratigraphic sequences in Horn River shale, Middle and Upper Devonian, northeastern British Columbia, Canada*. University of Alberta.

Eaton, B. (1969). Fracture gradient prediction and its application in oilfield operations. *Journal of Petroleum Technology*, *21*(10). doi:10.2118/ 2163-PA

Economides, M. J., & Nolte, K. G. (2000). *Reservoir stimulation* (3rd ed.). John Wiley & Sons.

Fakhr, S. (2016). *Formation fracture pressure*. Faman Chemie.

Fekete. (2017). *Minifrac tests*. Retrieved from Fekete: www.fekete.com

Fjar, E., Horsrud, P., Raaen, A., Risnes, R., & Holt, R. (2008). *Petroleum related rock mechanics* (2nd ed.). Amsterdam: Elsevier.

Frash, L., Carey, J., Lei , Z., Rougier , E., Ickes , T., & Viswanathan , H. (2016). High-stress triaxial direct-shear fracturing of Utica shale and in situ X-ray microtomography with permeability measurement. *Journal of Geophysical Research: Solid Earth*, *121*(7). doi:10.1002/2016JB012850

Guo, B., Liu, X., & Tan, X. (2017). *Petroleum production engineering* (2nd ed.). Cambridge: Elsevier.

He, J., & Ling, K. (2014). *A new method to determine biot coefficients of Bakken samples*. 48th U.S. Rock Mechanics/Geomechanics Symposium, Minneapolis, American Rock Mechanics Association.

He, J., Rui, Z., & Ling, K. (2016). A new method to determine Biot's coefficients of Bakken samples. *Journal of Natural Gas Science and Engineering*, *35*, 259–264. doi:10.1016/j. jngse.2016.08.061

Hu, Y., Perdomo, M., Wu, K., Chen, Z., Zhang, K., Ji, D., & Zhong, H. (2015). *A novel model of brittleness index for shale gas reservoirs: Confining pressure effect*. SPE Asia Pacific Unconventional Resources Conference and Exhibition, Brisbane, Society of Petroleum Engineers. doi:10.2118/176886-MS

Hubbert, K., & Willis, D. (1957). Mechanics of hydraulic fracturing. Society of Petroleum Engineers.

Hunnert, K., & Rubey, W. (1959). Role of fluid pressure in mechanics of overthrust faulting: I. Mechanics of fluid-filled porous solids and its application to overthrust faulting. *Geological Society of America Bulletin*, *70*(2), 115–166.

doi:10.1130/0016-7606(1959)70[115:ROFPIM]2.0.CO;2

Ide, J. (1936). Comparison of statically and dynamically determined young's modulus of rocks. *National Academy of Sciences*, *22*(2), 81–92. doi:10.1073/pnas.22.2.81

Jaeger, J., Cook, N., & Zimmerman, R. (2007). *Fundamentals of rock mechanics* (4th ed.). Malden: Blackwell Publishing. doi:ISBN-13: 978-0-632-05759-7

Jarvie, D., Coskey, R., Johnson, M., & Leonard, J. (2013). Chapter 9: The geology and geochemistry of the Parshall area,Mountrail County, North Dakota. In J. Robinson, J. LeFever, & S. Gaswirth (Eds.) , *The Bakken-Three Forks petroleum system in the Williston basin* (pp. 229–282). Denver: The Rocky Mountain Association of Geologists.

Jarvie, D., Hill, R., Ruble, T., & Pollastro, R. (2007). Unconventional shale-gas systems: The Mississippian Barnett shale of north-central Texas as one model for thermogenic shale-gas assessment. *American Association of Petroleum Geologists Bulletin*. doi:10.1306/12190606068

Jin, X., Shah, S., Roegiers, J.-C., & Zhang, B. (2015). An integrated petrophysics and geomechanics approach for fracability evaluation in shale reservoirs. *SPE Journal*, *20*(3). doi:10.2118/168589-PA

Jizba, D., & Nur, A. (1990). *Static and dynamic moduli of tight gas sandstones and their relation to formation properties*. SPWLA 31st Annual Logging Symposium, Lafayette, Society of Petrophysicists and Well-Log Analysts.

Kehle, R. (1964). The determination of tectonic stresses through analysis of hydraulic well fracturing. *Journal of Geophysical Research*, *69*(2). doi:10.1029/JZ069i002p00259

Konstantinovskaya, E., Malo, M., Claprood, M., Tran-Ngoc, T., Gloaguen, E., & Lefebvre, R. (2012). *Structural, petrophysical and geomechanical characterization of the Becancour CO_2 storage pilot site (Quebec, Canada)*.

Kowan, J., & Ong, S. (2014). *Wellbore stability: Special considerations for the Marcellus shale*. Baker Hughes Reservoir Development Services.

Lacy, L. (1997). *Dynamic rock mechanics testing for optimized fracture designs*. SPE Annual Technical Conference and Exhibition, San Antonio, Society of Petroleum Engineers. doi:10.2118/38716-MS

Li, H. (2017). *Effects of water content, mineralogy, and anisotropy on the mechanical properties*. Lafayette: University of Louisiana.

Ma, X., & Zoback, M. (2017). Laboratory experiments simulating poroelastic stress changes associated with depletion and injection in low-porosity sedimentary rocks. *Journal of Geophysical Research: Solid Earth*, *122*(4). doi:10.1002/2016JB013668

Marsden, J., Kostyleva, I., Fassihi, M., & Gringarten, A. (2017). *A conceptual shale gas model validated by pressure and rate data from the Haynesville shale*. SPE Annual Technical Conference and Exhibition, San Antonio, Society of Petroleum Engineers.

Mathia, E., Ratcliffe, K., & Wright, M. (2016). *Brittleness index: A parameter to embrace or avoid?* SPE/AAPG/SEG Unconventional Resources Technology Conference, San Antonio, Unconventional Resources Technology Conference. doi:10.15530/URTEC-2016-2448745

Mokhtari, M., Honarpour, M., Tutuncu, A., & Boitnott, G. (2014). *Acoustical and geomechanical characterization of Eagle Ford shale: Anisotropy, heterogeneity and measurement scale*. SPE Annual Technical Conference and Exhibition. Amsterdam: Society of Petroleum Engineers. doi:10.2118/170707-MS

Morales, R., & Marcinew, R. (1993). *Fracturing of high-permeability formations: Mechanical properties correlations*. SPE Annual Technical Conference and Exhibition. Houston: Society of Petroleum Engineers. doi:10.2118/26561-MS

Mustafa, A. (2014). *Integration of lithofacies and geomechanical characteristics of Lower Silurian Qusaiba shale, Saudi Arabia*. Dhahran: King Fahd University of Petroleum & Minerals.

Najibi, R., Ghafoori, M., Lashkaripour, G., & Asef, M. (2015). Empirical relations between strength and static and dynamic elastic properties of Asmari and Sarvak limestones, two main oil reservoirs in Iran. *Journal of Petroleum Science and Engineering*, 78–82. doi:10.1016/j.petrol.2014.12.010

Nolte, K. (1988). Application of fracture design based on pressure analysis. *SPE Production Engineering, 3*(1). doi:10.2118/13393-PA

NSI Technologies (2017). *Stimplan help section.* NSI Technologies.

Ohen, H. (2003). *Calibrated wireline mechanical rock properties model for predicting and preventing wellbore collapse and sanding.* SPE European Formation Damage Conference. The Hague: Society of Petroleum Engineers. doi:10.2118/82236-MS

Purushothama Raj, P., & Ramasamy, V. (2010). *Strength of materials.* Pearson Education India.

Rahman, K., & Gui, F. (2016). *Geomechanical sweet spot identification in unconventional resources development.* SPE Asia Pacific Oil & Gas Conference and Exhibition. Perth: Society of Petroleum Engineers. doi:10.2118/182247-MS

Rahman, K., Gui, F., & He, W. (2014). *Multistage hydraulic fracturing optimization for a shale oil field integrating geomechanics and production modelling.* Abu Dhabi International Petroleum Exhibition and Conference, Abu Dhabi, Society of Petroleum Engineers. doi:10.2118/172097-MS

Reeves, M. (2017). *Faults and stresses.* University of Saskatchewan: www.engr.usask.ca

Rickman, R., Mullen, M., Petre, J., Grieser, W., & Kundert, D. (2008). A *practical use of shale petrophysics for stimulation design optimization: all shale plays are not clones of the Barnett shale.* SPE Annual Technical Conference and Exhibition. Denver, Society of Petroleum Engineers. doi:10.2118/115258-MS

Sattler, F. (2015). *Lithologic properties of the Upper Ordovician Utica formation, Michigan basin, USA: A geological characterization and assessment of carbon dioxide confinement potential.* Western Michigan University.

Schlumberger (2015). *Techlog user guide.* Schlumberger.

Shaochuan, X., & Zoback, M. (2015). *Analysis of stress variations with depth in the Permian basin Spraberry/Dean/Wolfcamp shale.* 49th U.S. Rock Mechanics/Geomechanics Symposium, San Francisco, American Rock Mechanics Association.

Smith, M. (1985). Stimulation design for short, precise hydraulic fractures. *Society of Petroleum Engineers Journal, 25*(3). doi:10.2118/10313-PA

Smith, M., & Montgomery, C. (2015). *Hydraulic fracturing (emerging trends and technologies in petroleum engineering).* Boca Raton: CRC Press Taylor & Francis Group.

Sone, H. (2012). *Mechanical properties of shale gas reservoir rocks and its relation to the in-situ stress cariation observed in shale gas reservoirs.* Standford University.

Valkó, P., & Economides, M. (1995). *Hydraulic fracture mechanics.* Baffins Lane: John Wiley & Sons.

Van Heerden, W. (1987). General relations between static and dynamic moduli of rocks. *International Journal of Rock Mechanics and Mining Sciences and Geomechanics Abstracts, 24*(6), 381–385. doi: 10.1016/0148-9062(87)92262-5

Villamor Lora, R., Ghazanfari, E., & Asanza Izquierdo, E. (2016). Geomechanical characterization of Marcellus shale. *Rock Mechanics and Rock Engineering.* doi:10.1007/s00603-016-0955-7

Wang, F., & Gale, J. (2009). Screening criteria for shale-gas systems. *The Gulf Coast Association of Geological Societies, 59*, 779–793.

Wang, Y., Dong, D., Yang, H., He, L., Wang, S., Huang, J., et al. (2014). Quantitative characterization of reservoir space in the Lower Silurian Longmaxi shale, southern Sichuan, China. *Science China Earth Sciences.* doi:10.1007/s11430-013-4645-y

Waters, G., Lewis, R., & Bentley, D. (2011). *The effect of mechanical properties anisotropy in the generation of hydraulic fractures in organic shales.* SPE Annual Technical Conference and Exhibition, Denver, Society of Petroleum Engineers. doi:10.2118/146776-MS

Yaghoubi, A., & Zoback, M. (2015). *Hydraulic fracturing modeling using a discrete fracture network in the Barnett shale.* doi:10.13140/RG.2.1.3006.9289

Yang, M., Economides, M., Wei, C., & Gao, C. (2013). *Hydraulic fracture design flaws-proppant selection.* SPE Annual Technical Conference and Exhibition. New Orleans: Society of Petroleum Engineers. doi:10.2118/166299-MS

Yang, Y., & Zoback, M. (2014). The role of preexisting fractures and faults during multistage hydraulic fracturing in the Bakken formation. *Interpretation, 2*(3). doi:10.1190/INT-2013-0158.1

Yang, Y., Sone, H., Hows, A., & Zoback, M. (2013). *Comparison of brittleness indices in organic-rich shale formations.* 47th U.S. Rock Mechanics/Geomechanics Symposium. San Francisco: American Rock Mechanics Association.

Zoback, M. (2007). *Reservoir geomechanics.* Cambridge University Press.

6

Hydraulic Fracturing

Mustafa M. Alhubail and Reza Barati Ghahfarokhi

The University of Kansas, Lawrence, Kansas, USA

6.1 Background

Hydraulic fracturing is a stimulation technique that is widely used in the oil and gas industry to improve well productivity and to attain commercial production rates from low permeability reservoirs (e.g., unconventional reservoirs). Hydraulic fracturing increases the production mostly because (1) it creates new conductive channels and bypasses the damaged zones, (2) it extends the conductive channels to new zones, and (3) it changes the flow pattern from radial to linear (Allen & Roberts, 1989; Economides & Nolte, 2000). In unconventional plays, hydraulic fracturing is utilized to actually generate the stimulated reservoir volume (SRV) by creating complex networks of micro-channels and macro-channels as shown in Figure 6.1. The SRV region contains the hydrocarbon in place being targeted using the horizontal well.

The first hydraulic fracturing treatment, which was an experimental treatment, was conducted in the Hugoton gas field in Kansas in 1947. The Hugoton gas field is one of many oil and gas fields in the Hugoton area in Kansas, Oklahoma, and Texas (KGS, 2001). The experimental hydraulic fracturing treatment in the Hugoton gas field did not yield a significant increase in production. In 1949, a patent was granted to Halliburton Oil Well Cementing Company (HOWCO) to the procedure of hydraulic fracturing (King, 2017). The first two commercial hydraulic fracturing treatments were conducted in 1949 by HOWCO in Oklahoma and Texas. After success in the two commercial hydraulic fracturing jobs, hydraulic fracturing became an attractive procedure. More than 300 wells were hydraulically fractured in the first year alone, and the production out of those wells increased significantly (Montgomery & Smith, 2010).

The hydraulic fracturing treatment is an attractive and unique treatment in the oil and gas industry because it not only increases the production of the wells, it also increases the reserves. As discussed in Chapter 1, hydraulic fracturing has unlocked enormous resources (i.e., unconventional resources) that were thought to be impossible to develop or uneconomical to commercially develop. For example, hydraulic fracturing did increase the production rate from the tight gas formations; however, more important, it has added more than 700 Tcf of natural gas to U.S. reserves since 1950. Today, with all the success in hydraulic fracturing in the past decades, almost 60% of all the drilled wells around the world are hydraulically fractured (Smith & Montgomery, 2015).

Although numerous hydraulic fracturing treatments were conducted in the late twentieth century, however, there are two reasons that prevented hydraulic fracturing from really taking off right away to become the phenomenon

Unconventional Hydrocarbon Resources: Techniques for Reservoir Engineering Analysis, Advanced Textbook Series, First Edition. Edited by Reza Barati Ghahfarokhi and Mustafa M. Alhubail.

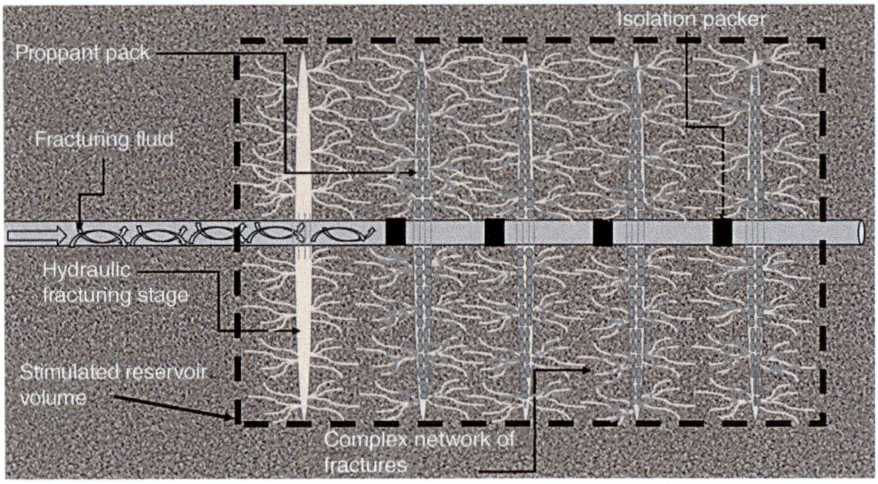

Figure 6.1 Hydraulic fracturing in unconventional reservoirs.

that is known today. The first reason is the abundance of the high-permeability formations back then. Hydraulic fracturing is not practical in high-permeability formation because it does not yield a significant increase in production (Helms, 2008). Hence, it has been mainly applied to bypass the damaged zone and increase the contact surface area between the wellbore and the formation in the conventional reservoirs. As those conventional reservoirs get depleted, the complex and tight formations as well as the source rocks (i.e., unconventional resources) rise to become the next source of oil and gas. Thus, hydraulic fracturing has to be used in a significantly larger size to extract the massive quantity of hydrocarbons out of the unconventional reservoirs. Consequently, hydraulic fracturing has become a first-class stimulation treatment in the oil and gas industry, and it has gained public attention, too. The second and main reason that hydraulic fracturing was not heavily practiced on the large scale that occurs today was the absence of large scale on-shore horizontal drilling techniques. Commercial horizontal drilling was first practiced in the 1980s (Wile, 2012), and it has become a standard practice in the oil and gas industry after years of intensive research, which significantly improved how horizontal drilling is carried out. At first, horizontal drilling was used to target formations that are naturally fractured (e.g., Austin Chalk formation in Texas). In the early 2000s, horizontal drilling and hydraulic fracturing were combined together, and they formed a disruptive innovation (Helms, 2008). This disruptive innovation has led the oil and gas industry to the ultimate stimulation treatment (i.e., multistage fracturing).

6.2 Fundamentals of Hydraulic Fracturing

Hydraulic fracturing is basically the initiation and propagation process of a fracture by a hydraulic load, which is generated by the injected fluid (Alhubail et al., 2017). The hydraulic fracturing treatment consists of two stages. The first stage is called the pad stage, and the second stage is called the slurry stage (Allen & Roberts, 1989). The pad stage is basically the fracture initiation step. The fracture is hydraulically initiated by injecting a suitable fluid into the formation with a pressure that is higher than the least principal stress and also under the right tectonic stress field as discussed in Chapter 5. The injected fluid (i.e., fracturing fluid) does not leak off into the formation fast enough because the injection rate is much higher than the leakoff rate near the wellbore; hence, the pressure builds up in the wellbore (Economides & Nolte, 2000). Consequently, the formation fails and cracks open because the pressure in the wellbore exceeds the formation's least principle stress. As the injection continues at high rates, the initiated fracture propagates and grows in width, height, and length. After the designed pad fluid is injected and a desired fracture geometry is achieved, which is usually known beforehand from mathematical and numerical models as will be discussed later, the second stage begins.

The second stage is the slurry stage. In the slurry stage, sand and some other additives are added to the fracturing fluid. The added sand is called proppant, and it will be discussed later in this chapter. The proppant is blended with the fracturing fluid and injected into the generated fracture. The role of the slurry stage is to fill the fracture with proppant to keep the fracture open after the pumps are shut down. Hence, the proppant must be designed so that it can withstand the formation closure stress. After the pumps are shut down and the injection is stopped, the pressure is released and the generated fracture seals. However, as long as the proppant has an adequate compressive strength so that it does not get embedded or crashed, the proppant will prevent the generated fracture from perfectly sealing because it will act like pillars inside the fracture. Figure 6.1 shows a schematic of the hydraulic fracturing process in the unconventional reservoirs.

6.2.1 Fracture Geometry

It was discussed in Chapter 5 that hydraulic fractures occur perpendicular to the least principal stress, and most of the induced hydraulic fractures are vertical when they are conducted at the depths similar to that of the unconventional reservoirs. Hence, the rock mechanics (i.e., geomechanical properties) of the formation has an important role in how the fracture geometry is going to be. Fracturing fluid and proppant also play crucial roles in how the final fracture geometry look. The fractures' length and height are typically great in value (e.g., 100 ft–1,000+ ft); however, the width of the fractures is very narrow and small in value (e.g., 0.25–2.0 in.). The final fracture geometry cannot be exactly known even with the use of the advanced microseismic imaging technique due to the complexity of the lithology and stress field. Despite some shortcomings, microseismic imaging is quite useful in estimating the stimulated reservoir volume (SRV), and it also provides a reasonably accurate fracture geometry for simple fractures (e.g., geometries of planar hydraulic fractures) (Maxwell, 2011). The stimulated reservoir volume and the microseismic technique will be discussed later in this chapter. Nevertheless, the fracture geometry can be estimated based on several parameters (e.g., geomechanical properties of the formation, injection rate, type of fracturing fluid, and type of proppant) and by the use of simple mathematical semi-analytical 2D models and advanced complex 3D numerical models (Montgomery & Smith, 2010). In order to predict accurate fracture dimensions and fracture conductivity, the advanced software packages require sophisticated geological models with accurate field stress and pumping schedules. Figure 6.2 and Figure 6.3 show an example for a simple hydraulic fracture that is designed using advanced software packages. The final results are provided in Table 6.1.

6.2.2 Fracture Conductivity

The term *fracture conductivity* is used to describe the flow of fluids through fractures, and mathematically it is the product of fracture permeability times the propped fracture width as shown in equation (6.1). The fracture conductivity is usually reported in mD.ft. The higher the fracture conductivity the easier the fluid can flow through a fracture. Thus, the fracture conductivity could be used as a measure of success for the hydraulic fracturing

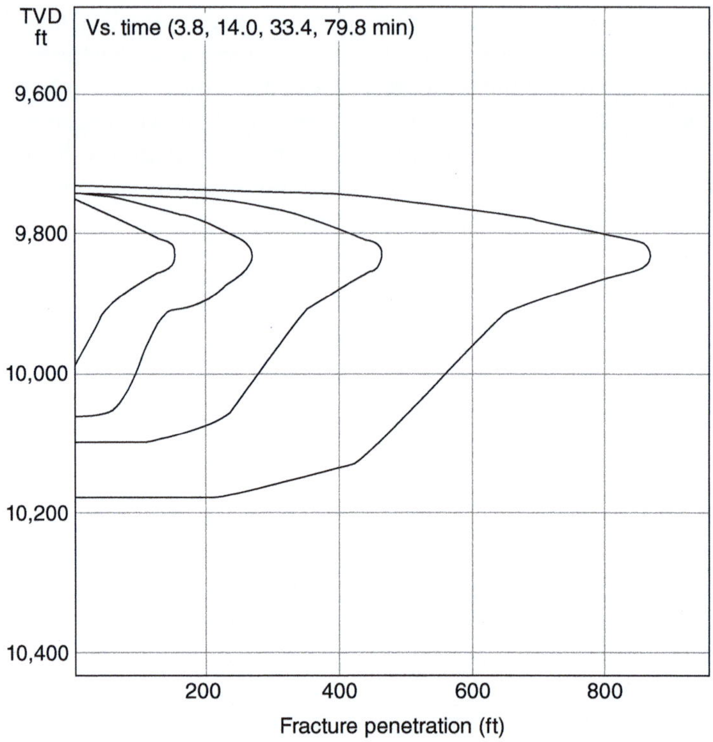

Figure 6.2 Designed fracture length and height using advanced software packages (plot generated using E-Stimplan).

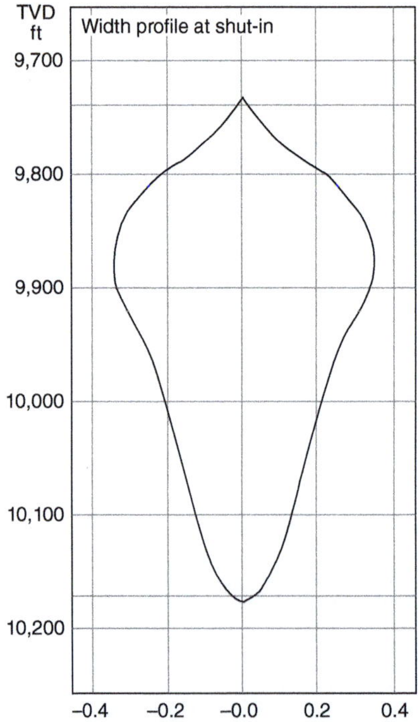

Figure 6.3 Designed fracture width using advanced software packages (plot generated using E-Stimplan).

Table 6.1 Summarized Parameters and Results of a Designed Hydraulic Fracture

Designed parameter	Value
Fracture half length (ft)	882.4
Fracture propped half-length (ft)	521.5
Fracture height (ft)	447.1
Maximum fracture width (inch)	0.69
Average fracture width (inch)	0.32
Average fracture conductivity (mD-ft)	287

stimulation treatment. Extremely high fracture conductivity values lead to linear fracture flow, which is achieved when the permeability of the hydraulic fracture is very high. When the fracture conductivity is not very high, a bilinear fracture flow occurs. That is one linear flow from the matrix to the generated fracture, and the other linear flow is from the generated fracture to the wellbore (Fekete, 2012). Refer to

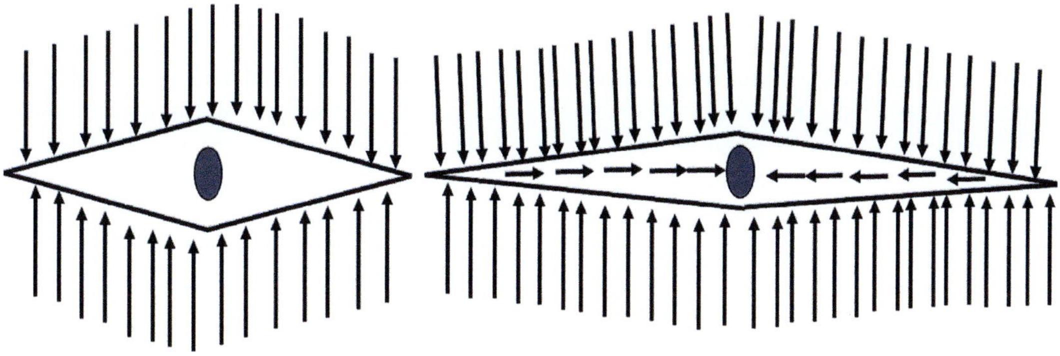

Figure 6.4 Linear (left) and bilinear (right) fracture flow (after Fekete, 2012).

Figure 6.4 for the linear and bilinear fracture flows.

$$F_D = k_f w, \tag{6.1}$$

where F_D is the fracture conductivity, k_f is fracture permeability, and w is the fracture propped width.

The fracture conductivity depends on several factors, and it is heavily influenced by the proppant-pack conductivity, which is also influenced by quite a few factors (e.g., type of proppant and grain size). The pH in the fracture environment also plays a role on the fracture conductivity because the pH affects the proppant carrying capability and thus the proppant-pack permeability as will be discussed later (Mittal et al., 2018). There are also other important factors that affect the fracture conductivity such as fracturing fluid residue, fracture cleanup process, and multiphase flow effects (Davies & Kuiper, 1988; Barati et al., 2009; Tayong et al., 2019).

Example 6.1 Estimate the Fracture Permeability

Based on the data in Table 6.1, estimate the fracture permeability for the designed hydraulic fracture assuming the designed average width is the fracture propped width. The formation permeability is 0.01 mD. Compare the formation permeability and the fracture permeability.

Solution

From Table 6.1, the designed fracture propped width is 0.32 in. and the fracture conductivity is 287 mD.ft. Hence, the permeability of the fracture can be estimated using equation (6.1) but first the fracture propped width should be converted from inch (in.) to foot (ft).

$$w = 0.32\ in. = 0.0267\,ft,$$

$$F_D = k_f w = 287\ mD.ft = k_f \times 0.0267\,ft,$$

and

$$k_f = \frac{287 mD.ft}{0.0267\,ft} = 10,749\ (mD) = 10.749\ Darcy.$$

Clearly, the fracture permeability is significantly greater than the permeability of the formation. With the advancements in the hydraulic fracturing design process, a more important term has emerged to better assess the success of the hydraulic fracturing treatments other than the fracture conductivity. It is the dimensionless fracture conductivity.

Dimensionless Fracture Conductivity

The dimensionless fracture conductivity is basically the fracture conductivity divided by the formation permeability and the fracture half-length, as shown in equation (6.2). The dimensionless fracture conductivity takes into account the formation ability to supply fluids to the generated fracture, and also the fracture capacity to receive the fluid and deliver it to the wellbore (Pearson, 2001).

$$C_{fD} = \frac{k_f w}{k x_f}, \tag{6.2}$$

where C_{fD} is the dimensionless fracture conductivity, k_f is the fracture permeability, w is the fracture propped width, k is the formation permeability, and x_f is the fracture half-length.

Early studies showed that a dimensionless fracture conductivity of 10 is sufficient to maximize production. However, new studies have shown that much higher dimensionless fracture conductivity (e.g., 30 and above) is required for tight formations in order to achieve decent production (Prats, 1961; Cinco-Ley & Samaniego V, 1981; Pearson, 2001). Nevertheless, based on experience, the industry considers a dimensionless fracture conductivity of 10 to be adequate enough, and it is usually the value that is used in the design process.

Example 6.2 Calculate the Dimensionless Fracture Conductivity

Estimate the dimensionless fracture conductivity for the designed hydraulic fracture in Example 6.1. Based on the dimensionless fracture conductivity, is the designed hydraulic fracture considered successful?

Solution

From Example 6.1, the fracture conductivity is 287 mD.ft and the formation permeability is 0.01 mD. From Table 6.1, the fracture propped half-length is 521.5 ft. The dimensionless fracture conductivity can be calculated using equation (6.2) as follows:

$$C_{fD} = \frac{k_f\, w}{k\, x_f} = \frac{287 mD.ft}{0.01 mD \times 521.5 ft} = 55.$$

Thus, the designed hydraulic fracture is considered successful.

Example 6.3 Compare the Dimensionless Fracture Conductivity

Consider the following two hydraulic fracturing designs that are intended to be conducted in the middle Bakken formation. Choose the optimum designed plan (i.e., plan A or plan B) solely based on the dimensionless fracture conductivity. The average permeability of the middle Bakken formation is 0.04 mD. Plan A is designed so that the fracture conductivity is 230 mD.ft and the propped fracture half-length is 625 ft. The fracture conductivity of plan B is the same as plan A, but the propped fracture half-length of plan B is 120 ft.

Solution

First, the dimensionless fracture conductivity should be calculated for each plan by the use of equation (6.2). Then, the dimensionless fracture conductivities can be compared, and the optimum designed plan can be selected, accordingly.

Plan A:

$$C_{fD} = \frac{k_f\, w}{k\, x_f} = \frac{230\ mD.ft}{0.04\ mD \times 625\ ft} = 9.2.$$

Plan B:

$$C_{fD} = \frac{k_f\, w}{k\, x_f} = \frac{230\ mD.ft}{0.04\ mD \times 120\ ft} = 47.91.$$

Clearly, the designed propped fracture half-length of plan A is longer than the one of plan B; however, plan B resulted in a higher dimensionless fracture conductivity. Hence, plan B is the optimum one. It is important to note that a longer propped fracture half-length does not mean a more conductive hydraulic fracture. In terms of production of tight and ultra-tight reservoirs, one must conduct the whole fracture design and production calculation scenarios as there is no height or multiphase flow effect included in the fracture conductivity equation. Furthermore, several studies have proven that a lower conductivity but longer fracture improves production more than a shorter but more conductive fracture due to small reservoir capacity (Voneiff et al., 1996; Barati et al., 2009).

6.2.3 Folds of Increase

The folds of increase (FOI) is a method to assess the successfulness of a hydraulic fracturing treatment. FOI is basically a way to compare the productivity of the well before and after the hydraulic fracturing treatment, and it can be calculated by the use of equation (6.3).

$$FOI = \frac{J_f}{J_o} = \frac{\ln\left(\frac{r_e}{r_w}\right)}{\ln\left(\frac{r_e}{r_w}\right) + S_f} = \frac{\ln\left(\frac{r_e}{r_w}\right)}{\ln\left(\frac{r_e}{r_w'}\right)}, \tag{6.3}$$

where *FOI* is the folds of increase, J_f is the well productivity index after the fracture treatment, J_o is the well productivity index before the fracture treatment, r_e is reservoir radius, r_w is wellbore radius, S_f is the postfracture skin factor, and r_w' is the postfracture equivalent wellbore radius.

The equivalent wellbore radius was first introduced by Prats (1961) as a mean to account for the postfracture skin factor by using an "equivalent wellbore radius" instead of a postfracture skin factor. For fractures with high dimensionless fracture conductivity values (e.g., $C_{FD} > 50$), the postfracture skin factor can be estimated by the use of equation (6.4), and the equivalent wellbore radius can be estimated by the use of equation (6.5). For fractures with low dimensionless fracture conductivity values (e.g., $C_{FD} < 1$), the equivalent wellbore radius can be estimated by the use of equation (6.6) (Prats, 1961; Cinco Ley et al., 1978; Cinco Ley & Samaniego V, 1981; Economides & Nolte, 2000).

$$S_f = 0.7 - \ln\left(\frac{x_f}{r_w}\right), \tag{6.4}$$

where x_f is the fracture half-length

$$r_w' = \frac{x_f}{2} \tag{6.5}$$

and

$$r_w' = 0.28\frac{k_f w}{k}, \tag{6.6}$$

where k_f is the fracture permeability, w is the fracture propped width, and k is the formation permeability.

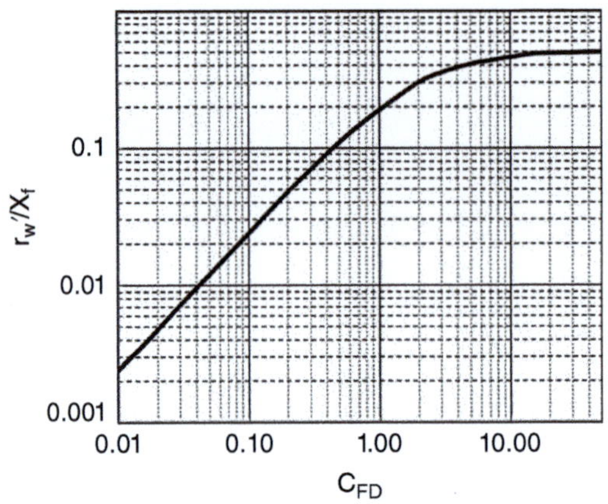

Figure 6.5 The dimensionless equivalent wellbore radius versus dimensionless fracture conductivity (Smith & Montgomery, 2015; after Cinco-Ley & Samaniego-V, 1981).

In general, the postfracture skin factor and the equivalent wellbore radius can be correlated together by equation (6.7) (Economides & Nolte, 2000). Furthermore, based on the work of Prats (1961) and Cinco Ley and Samaniego V (1981), the equivalent wellbore radius can be estimated for different dimensionless fracture conductivity and fracture half-length by using Figure 6.5. It is very clear from Figure 6.5 that at great C_{FD} values, the dimensionless equivalent wellbore radius becomes constant with a value of 0.5, which confirms equation (6.5).

$$S_f = -\ln \frac{r'_w}{r_w}. \tag{6.7}$$

Example 6.4 Find the Folds of Increase

Calculate the FOI for a well in the Bakken formation with a wellbore radius of 0.33 ft. The formation permeability is 0.04 mD, the drainage area is 160 acres, the designed fracture half-length is 580 ft, and the designed fracture conductivity is 70 mD.ft. Based on the calculated FOI, is the designed hydraulic fracture adequate enough to be conducted?

Solution

The FOI can be found by the use of equation (6.3),

$$FOI = \frac{\ln\left(\frac{r_e}{r_w}\right)}{\ln\left(\frac{r_e}{r'_w}\right)}.$$

Both the drainage radius and the equivalent wellbore radius must be found to calculate the FOI. First, the drainage radius (r_e) can be calculated from the drainage area as follows:

$$Drainage\ area = \pi\, r_e^2$$

$$r_e = \sqrt{\frac{160\ acres \times 43,560\ \frac{ft^2}{acres}}{\pi}} = 1,489.46\,ft$$

Next, to find the dimensionless equivalent wellbore radius, the dimensionless fracture conductivity (C_{FD}) has to be calculated using equation (6.2).

$$C_{fD} = \frac{k_f\, w}{k\, x_f} = \frac{70\ mD.ft}{0.04\ mD \times 580\,ft} = 3.02$$

Now, by the use of Figure 6.5, the dimensionless equivalent wellbore radius can be found. At 3.02 dimensionless fracture conductivity, the dimensionless equivalent wellbore radius is 0.37. Hence, the equivalent wellbore radius is

$$r'_w = 0.37 \times 580\,ft = 214.6\,ft.$$

Finally, the FOI is

$$FOI = \frac{\ln\left(\frac{1,489.46\text{ ft}}{0.33\text{ ft}}\right)}{\ln\left(\frac{1,489.46\text{ ft}}{214.6\text{ ft}}\right)} = 4.34$$

Based on the calculated FOI, the postfracturing production would be more than three times the production prior to the fracturing job. Therefore, the designed treatment is adequate, and it is acceptable to be conducted solely based on the calculated FOI.

6.2.4 Multistage Hydraulic Fracturing

It was mentioned earlier in this chapter that horizontal drilling has advanced the hydraulic fracturing technique, and the combination of the two formed a disruptive innovation. Also, it was mentioned in Chapter 1 that one of the reasons that led to the shale phenomenon is multistage hydraulic fracturing. Basically, multistage hydraulic fracturing is a term that indicates multiple hydraulic fracturing treatments are conducted in one well in stages, hence, the name multistage hydraulic fracturing or multistage fracturing. For instance, a drilled lateral is subdivided into sections that are fractured independently in stages. Figure 6.6 illustrates five hydraulic fracturing stages. Fractures have been conducted in stages. In fact, multistage fracturing has been performed in vertical wells. The number of stages that can be conducted has increased over the years. For example, the average hydraulic fracturing stages in a vertical well is 4. In 2011, the average hydraulic fracturing stages in horizontal wells was 16 stages. Nowadays, with the advancements in the completion techniques and the capability of drilling very long laterals, it is common to have more than 40 stages in the horizontal wells.

There are three main completion techniques that the industry relies on, in order to conduct hydraulic fracturing treatments in horizontal wells: open-hole, plug and perf, and sliding sleeve.

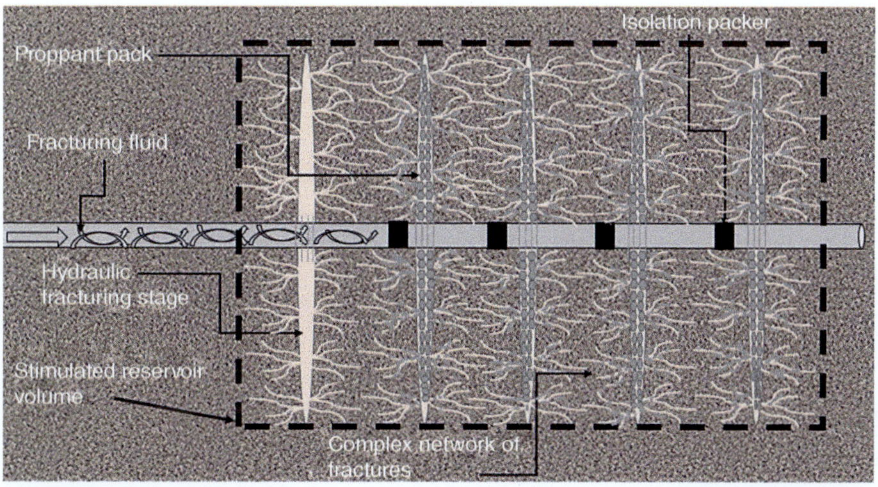

Figure 6.6 Multistage fracturing.

Open-Hole Completion

The open-hole completion, which is also referred to as the barefoot completion, is a single- stage fracturing treatment. In this technique, there are no isolation devices in the wellbore; hence, the wellbore is not divided into multiple segments. The fracturing fluid is injected in the wellbore to actually propagate the existing natural fractures along the entire open-hole segment of the well. This technique is rarely practiced in the unconventional. There are approaches under development to modify this technique so that new fractures are generated in addition to extending the existing ones (Houston et al., 2010). It is important to note that the open wellbore section does not yield any production by itself because the unconventional reservoirs are extremely tight and cannot be produced without fracturing them.

Plug and Perf Completion

The plug and perf (PnP) completion (Fig. 6.7), which is also referred to as the limited entry completion, is the most common and widely practiced fracturing treatment in the unconventional reservoirs. From its name, this completion method involves plugging devices (i.e., isolation packers) and perforations. It is called limited entry because the fracturing fluid comes in contact with the formation through the perforated holes, and the hydraulic fractures would be initiated directly through the perforations. In this technique, the horizontal section of the well is subdivided into several segments where each segment (i.e., fracturing stage) is isolated, perforated, and hydraulically fractured. This makes the plug and perf completion technique a multistage fracturing treatment. Such isolated sections are selected based on the petrophysical, geochemical, and geomechanical parameters of the reservoir. Furthermore, when a segment is perforated, essentially a cluster of perforations is performed with different perf-shot phasing and perf-shot density. Each fracturing stage generates a cluster of fractures and not a single fracture. The plug and perf technique can be used in both cemented and uncemented wellbores and with unlimited fracturing stages. After completing each stage, a plug is placed to isolate the previous stage, and after completing all the stages, the plugs are drilled and the well can be put on production (Maxwell, 2014; Triepke, 2014). The main disadvantage of this technique is the time required to conduct each fracturing stage since each stage requires perforation and plugging in addition to the actual hydraulic fracturing treatment. This is a repetitive and time-consuming process as each stage could take up to 5 hr to complete (Pearson, 2013). So, this technique could be tedious when 40 stages of fracturing are conducted in a horizontal well; yet, it is necessary to do in

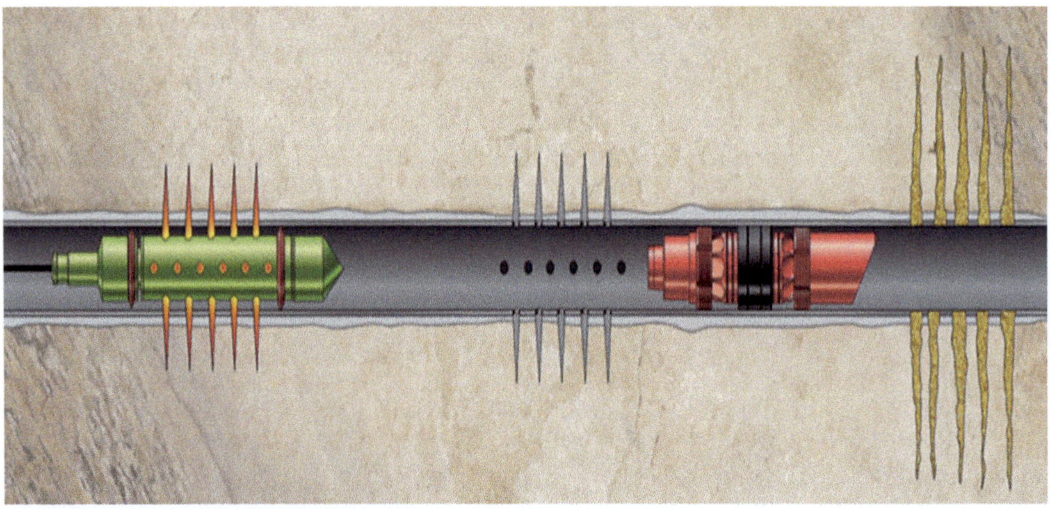

Figure 6.7 Plug and perf completion (courtesy of Halliburton, 2012).

order to have commercial production from the unconventional reservoirs.

Sliding Sleeve Completion

The sliding sleeve technique (Fig. 6.8) is a newer method to conduct multistage hydraulic fracturing treatments, and it is mostly practiced in uncemented, unconventional oil wells. In this method, the casing is deployed in the wellbore with a fixed number of sleeves, which are placed at specified locations and separated by several external packers. The number of sleeves and packers as well as their locations are determined based on the designed hydraulic fracturing stages. After the sleeves are placed in their targeted locations, they need to be activated in order to open the fracture ports. The ball activation method is a common sleeve activation technique. In the ball activation method, a ball with a specific size is dropped, and when it finds its seat in the sleeve, it blocks the casing or tubing causing a back pressure that opens the sleeve. Then, the fracturing fluid is injected, and since it has nowhere else to go but the open port, a hydraulic fracture is initiated. This process is repeated by dropping a bigger sized ball and so on. The sliding sleeve technique is a fast multistage hydraulic fracturing treatment compared to the PnP technique because the sliding sleeve technique does not require perforations. On average, each stage of fracturing using the sliding sleeve completion takes between 1 and 2 hr to complete (Pearson, 2013). The disadvantages of the sliding sleeve technique is the limited number of hydraulic fracturing stages that can be conducted and limited number of fractures that are generated per stage. A typical sliding sleeve treatment can perform 20 stages with a single fracture per stage (Pearson, 2013). However, there are new advanced technologies that are being developed (e.g., multiport opening) that can increase the number of fractures per stage as well as the number of stages (Halliburton, 2012; Triepke, 2014; Maxwell, 2014). A comparison of PnP and sliding sleeve methods is shown in Table 6.2.

Figure 6.8 Sliding sleeves Completion (courtesy of Halliburton, 2012).

Table 6.2 Comparison Between PnP and Sliding Sleeve

	Plug and perf	**Sliding sleeve**
Applicability	Oil and gas wells	Oil wells
Completion type	Cemented and uncemented wells	Mostly uncemented wells
Number of fracturing stages	Unlimited number of stages	20 stages
Number of fractures per stage	A cluster of fractures per stage	A single fracture per stage
Required time per stage	3–5 hours per stage	1–2 hours per stage

Source: Pearson (2013).

6.2.5 Stress Shadow

Stress shadow, or stress shadowing, is basically the change of the stress field due to the hydraulic fracturing treatments. During a hydraulic fracturing job, large volumes of fracturing fluids are injected to initiate and propagate fractures. Then, proppant is also injected to maintain the fractures permanently open. As a result, the stress field around the generated hydraulic fractures changes (Nagel et al.,2013). The effects of the stress shadow could extend to significantly long distances (i.e., more than 3,000 ft) laterally and vertically (Nagel & Sanchez-Nagel, 2011).

The change of the stress field could significantly impact the nearby hydraulic fractures by deviating or limiting their propagation. Hence, the stress shadow could significantly affect the design of the next hydraulic fracturing stages. The main parameters that have an impact on the stress shadow are fracture spacing, fracture cluster, and fracture height. In general, minimizing the fracture spacing in the unconventional plays is desirable because it leads to more efficient development of the reservoirs by creating more fracture surface areas, resulting in a large stimulated reservoir volume. However, reducing the fracture spacing might not be appropriate because of the stress shadow effects, which could lead to frac hits. Frac hits is the interaction between the infill wells and the existing main wells. The stress shadow could also lead to a one-wing symmetric hydraulic fractures. Furthermore, the stress shadow could suppress the fracture height growth of nearby hydraulic fractures, and there is a possibility that it prevents their propagation altogether (Damjanac et al., 2018).

The stress shadow phenomenon has several effects on the hydraulic fracturing treatment, and it impacts the design of the hydraulic fractures (Fisher et al., 2004). The stress shadow increases the least minimum horizontal stress, and it also increases the maximum horizontal stress but with a slower rate (Fisher et al., 2004). In normal fault regime and strike slip fault regime, the minimum horizontal stress dictates the fracture orientation. The stress shadow has an influence on how the hydraulic fractures are oriented. Figure 6.9 shows the change of the minimum horizontal stress due to the stress shadow between two hydraulic fractures. It is clear that a substantial change occurs near the fracture tip.

Another effect of the stress shadow is the influence on the direction of propagation. The stress

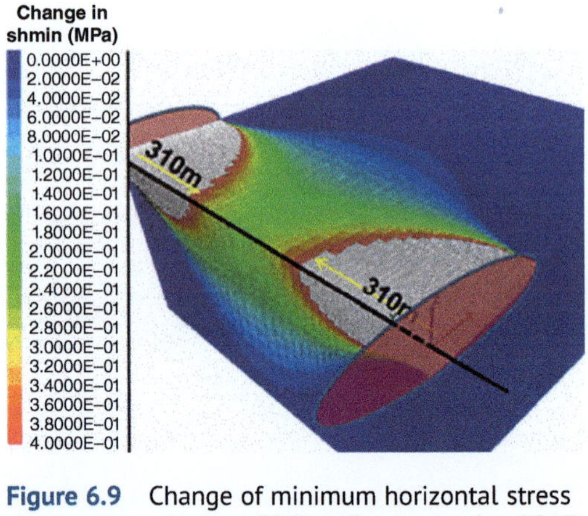

Figure 6.9 Change of minimum horizontal stress due to stress shadow (OilFieldGeomechanics, 2015).

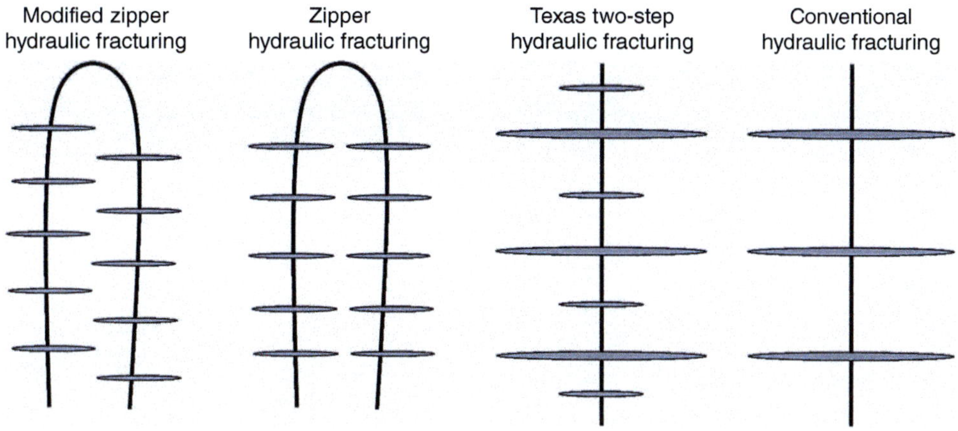

Figure 6.10 Types of hydraulic fracturing (after Jacobs, 2014).

shadow changes the overall stress field around the fractures; hence, the fractures of the following stages might propagate in a different direction than the previous fractures (Skomorowski et al., 2015; Damjanac et al., 2018).

The shadow stress could also be beneficial in encountering the frac hits because it creates a "pressure shield" near the fractures of the parent well by increasing the minimum horizontal stress, especially near the fractures tip. The fractures of the child wells would deviate away from the fractures of the parent well and propagate in the direction of the least principal stress. However, this is a new area of study and a few case studies have concluded that the pressure near the parent well reaches a maximum point where it can no longer be increased (Rainbolt & Esco, 2018). The created "pressure shield" might not be sufficient enough to prevent the occurrence of frac hits.

6.2.6 Zipper Fracturing

Zipper fracturing or modified zipper fracturing (MZF), which is a relatively new technique in the fracturing industry, is the process of fracturing multiple wells simultaneously. For example, two wells are drilled and completed parallel to each other, then the first fracturing stage is conducted on both wells alternatively before the second stage is performed (Gilleland, 2011). This process is called zipper fracturing because the fractures take the shape of a zipper as shown in Figure 6.10. Zipper fracturing can be used to fracture a group of wells at once, not just two wells. Zipper fracturing was practiced in the Barnett shale play, and it doubled the wells production in comparison with the typical hydraulically fractured wells. Zipper fracturing is mostly used to alter the stress distribution for the next fracturing stages and also to create extensively complex fracture networks, hence, a more effective SRV (Rafiee et al., 2012). In addition to that, zipper fracturing actually reduces the risk of fracture hits because the fractures are generated before the pressure is depleted by production (Rassenfoss, 2018). Also, zipper fracturing saves millions of dollars per year because it accelerates the development of the wells as well as minimizes the wireline operation time that is required to plug and perforate the next hydraulic fracturing stage (Jacobs, 2014). That said, there are challenges with the zipper fracturing technique, such as the possibility of proppant exchange between two fractured layers (Kumar et al., 2018). Furthermore, zipper fracturing is not ideal for EOR process (e.g., CO_2 huff-n-puff) in the unconventional reservoirs because large amounts of CO_2 would easily move toward the high-permeability pathways of the adjacent wells (Phan et al., 2018).

The benefits of zipper fracturing (i.e., improvement in production performance and reduction of stand-by time) are significant but those benefits come with challenges. For example, knowing the stress field during zipper fracturing is a serious challenge because the stress shadow affects the stress field after each stage of fracturing. Thus, the conducted hydraulic fractures might not be generated as designed. Another challenge is the existence of natural fractures. During a hydraulic fracturing treatment, the fracturing fluid could leak off to the natural fractures and propagate them. Hence, the natural fractures might overlap with the hydraulic fractures of the other well. That said, the existence of the natural fractures actually help improve the production in zipper fracturing (Jacobs, 2014).

Another challenge that could impact zipper fracturing is the limited number of pumps on the surface. Zipper fracturing demands continuous high-horsepower pumping. Performing maintenance between jobs as in regular fracturing treatments is not possible in zipper fracturing. Thus, more pumps have to be on standby on the surface during the zipper fracturing treatment in case of an emergency. Having sufficient fracturing fluid and proppant to conduct a zipper fracturing job is also a challenge. In zipper fracturing, multiple wells are hydraulically fractured simultaneously. This requires a large amount of fracturing fluid and proppant to be ready on the surface when the treatment is conducted (Jacobs, 2014).

6.2.7 Fracture Hits

Fracture hits, or frac hits, which are also known as interwell fracture interactions (IFI), is the interaction between the infill wells, which are called child wells, and the existing main wells, which are called parent wells (Lindsay et al., 2018). In the fracturing industry, this process is also referred to as the child-parent relationship. The child-parent relationship affects the production of both the child and parent wells; the effects could be either negative or positive (Miller et al., 2016). A positive frac hit is accomplished when the child well creates new conductive paths for the parent well without affecting its existing network of fractures. A negative frac hit, which is common, is identified when the child and parent wells compete for the same amount of resources (Vidma et al., 2018). Figure 6.11

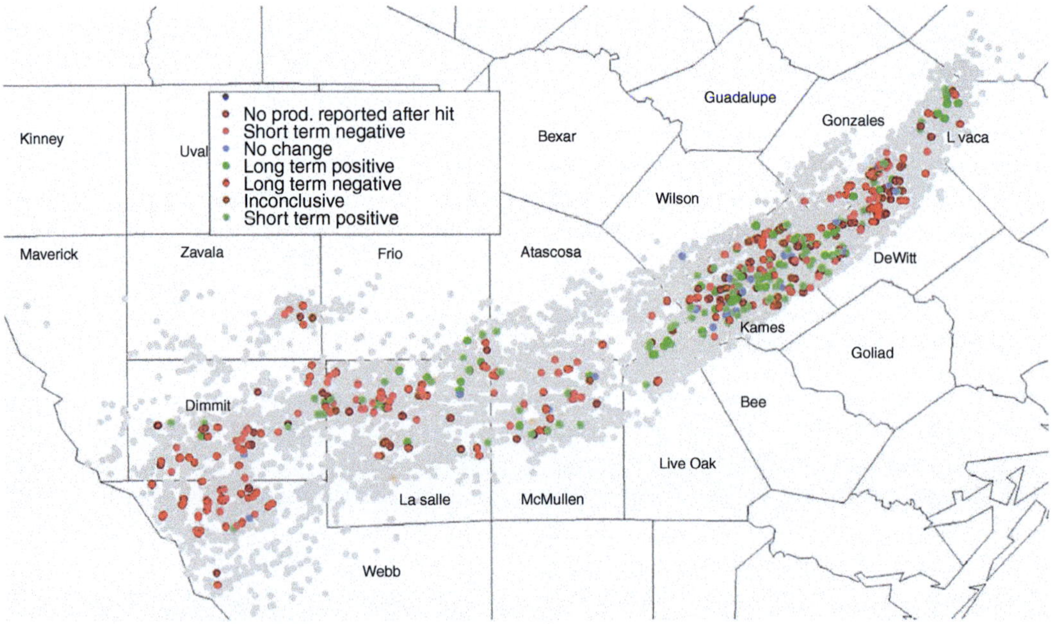

Figure 6.11 Frac hits in the Eagle Ford play (Miller et al., 2016).

shows the identified positive and negative frac hits in the Eagle Ford play. The child-parent relationship has been the focus of many studies lately because it hugely reduces the return of investment of the child well especially if the parent well is depleted (Vidma et al., 2018; Whitfield et al., 2018). The probability of the interwell fracture interaction between multiple wells has increased significantly in the past few years because the unconventional hydraulic fracturing jobs that generate a complex network of fractures has also increased (Sardinha et al., 2014; Kumar et al., 2018). With multiple operators competing in the same unconventional plays and thousands of infill wells being drilled, frac hits are a critical issue because one operator might damage the other operator's already conducted fractures. In fact, several lawsuits have been filed in the United States by different operators over this issue (Jacobs, 2018a,b).

When the fracturing fluids are being injected during a hydraulic fracturing job in a new child well, frac hits occur because there already exist conductive pathways, which were generated during the fracturing job of the parent well. Consequently, the fracturing fluid as well as the proppant, which are being injected in the child well to generate and maintain new pathways, will transfer to the parent well pathways and significantly impact the proppant-pack permeability there. Thus, the fracture hits occur because of the interaction between the pathways of the child and parent wells, and the interaction becomes more severe if the parent well is depleted. Hence, infill wells spacing as well as the spacing between the fracturing stages are very crucial factors that impact the production, and they are the focus of many recent studies (Xu et al., 2018; Daneshy, 2018). Fracture hits can be monitored and tracked by the use of chemical tracers as well as microseismic and pressure data (Kumar et al., 2018). For example, there are operators in the Permian basin that cooperate by sharing information and data such as log data, microseismic data, and production data in order to cope with frac hits. The industry acknowledged it has a serious issue but the question still remains unanswered. Can frac hits be totally prevented?

Frac Hit Classifications

Frac hits can be classified based on the level of severity into three levels. The first level is pressure communication. The pressure communication is identified as detecting a pressure increase in the parent well without any fracturing fluid from the infill well reaching it as shown in Figure 6.12. Hence, there is no fluid transmitting from the child well to the parent well (Pankaj, 2018). The second level of frac hits is when the infill wells' fractures propagate and reach the fracture network of the parent well. This affects the production, and it could also damage the completion equipment of the parent well especially when proppant is produced with the fracturing fluid as shown in Figure 6.13. The third level of frac hits and the most severe ones are experienced when the parent wells are depleted, and the hydraulic fractures that are being conducted in the infill wells are done with high injection rates and high pumping pressure. In this case, the generated hydraulic fractures in the infill well extend and fully communicate with the existing fractures in the parent well, and without a proper pressure control, a blowout could occur in the parent well (EFT, 2015).

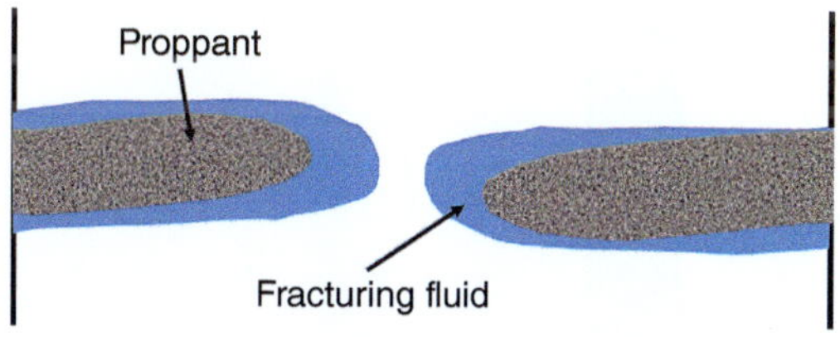

Figure 6.12 Pressure communication.

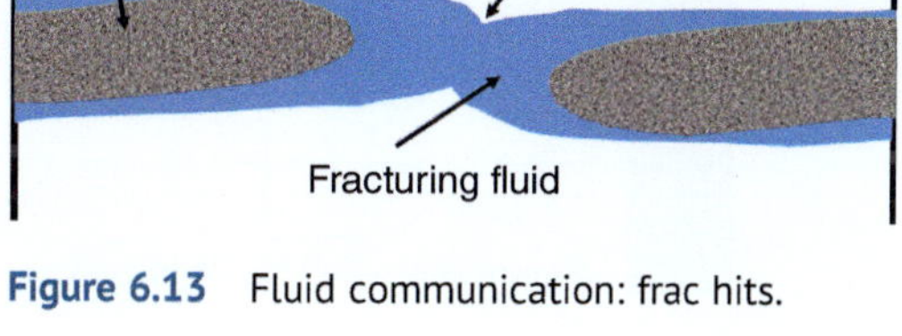

Figure 6.13 Fluid communication: frac hits.

Frac Hit Impacting Factors

There are four major factors that indicate the occurrence of frac hits. The first factor is the well spacing. There is a high risk of having frac hits when the parent well and the child well are not far enough from each other. Figure 6.14 shows the effect of well spacing. It is clear from the left side picture when the well spacing is 600 ft, fracture communication occurs because the two wells are close to each other. On the other hand, when the well spacing is increased to 1320 ft, there is no fracture communication.

The second factor that has an impact on frac hits is the stress field. The stress field dictates how the fractures are oriented; hence, there is a possibility of fracture communication if the fractures of the child well are oriented directly straightaway to the parent well. The third factor is depletion. This factor is somehow connected to the previous one because if the parent well is depleted, then the stress field around it is dropped. Hence, the fractures of the child wells will have an easier path to propagate through the depleted region, and thus the fracture networks communicate. The permeability of the formation, which has a major role on fluid leakoff, also plays a role in frac hits because the pressure waves travel quicker in high permeability reservoirs. The stress field will be affected accordingly (EFT, 2015).

Frac Hit Mitigation

There are a few techniques that the industry relies on in order to mitigate the occurrence of frac hits. The first technique is microseismic. Microseismic is a mapping technique that is utilized to monitor the seismic activity during a hydraulic fracturing treatment, and based on that, visualizing the geometry of the induced hydraulic fractures becomes possible. Hence, by tracking the induced seismic activity during the fracturing treatments, the directions of the future fractures can be detected. Thus, frac hits can be mitigated by modifying the

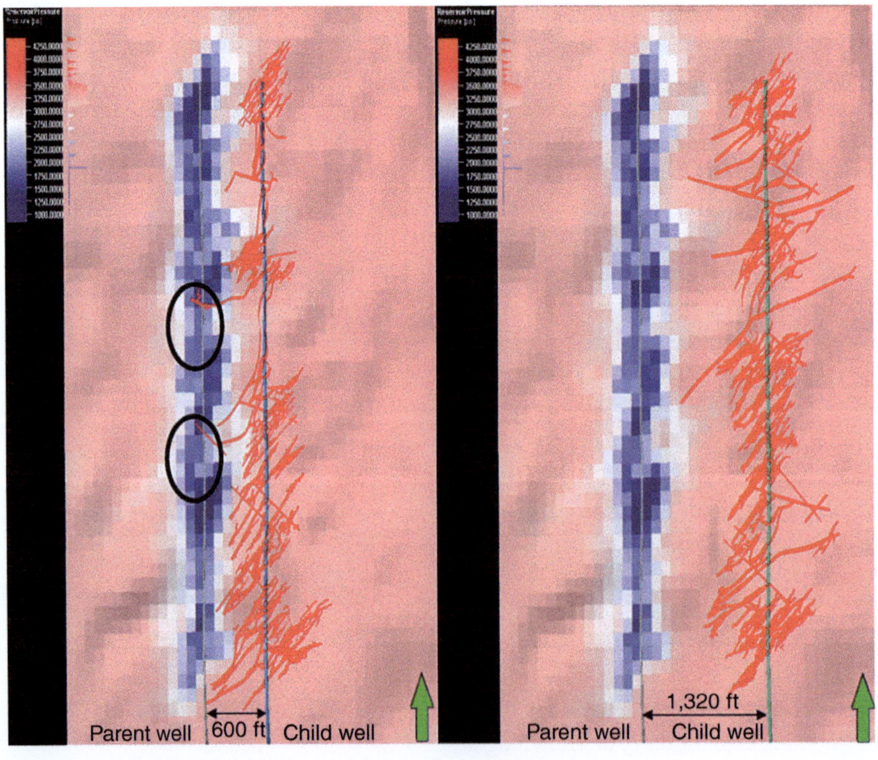

Figure 6.14 Effect of well spacing on frac hits (Ajisafe et al., 2017).

design of the future planned fracturing jobs (e.g., modify pump schedule) (EFT, 2015).

The second technique is well shut-in. Before conducting a hydraulic fracturing job in a child well, the parent well is shut in for a period of time. This practice helps in building up the reservoir pressure around the parent well, mitigating the damage from the frac hits. The well shut-in technique is a preferred technique for mitigating frac hits (EFT, 2015). However, the shut-in period might be extremely long, and another technique has emerged to mitigate the frac hits by repressurizing the parent well.

The repressurizing technique involves injecting a fluid (i.e., water, gas, or foam) into the parent well to increase the pressure and form a protective "pressure shield" around the parent well. The water injection builds up the pressure around the parent well within a few days (Bommer et al., 2017; Bommer & Bayne, 2018; Gala et al., 2018; Jacobs, 2018a,b). In contrast, gas injection takes weeks. However, the use of water might damage the formation and impact the overall production whereas the natural gas, CO_2, or hydrocarbon-based foams actually improve the oil recovery from the parent well. However, these systems are much more expensive than water. The economics play a crucial role in this practice. Another approach to mitigate frac hits and increase the pressure around the parent well is refracturing. Refracturing the parent well involves the injection of large volumes of fracturing fluid; hence, the local stress field around the parent well enhances; thus, moderate frac hits are experienced when the child wells are hydraulically fractured (Jacobs, 2017).

6.2.8 Surface Pumps

After designing the fracturing fluids, it is important to choose the appropriate pumps to successfully conduct the hydraulic fracturing treatment. The hydraulic fracturing pumps are usually selected based on the required pump power, which is the required hydrostatic horsepower (hhp) (Treida, 2012). The required pump power for each job depends on the injection rate and the surface pumping pressure, and it can be estimated by the use of equation (6.8). The surface pumping pressure (SPP) can be estimated by the use of equation (6.9).

$$PP\ (hhp) = \frac{SPP\ (psi) \times pumping\ rate\,(bpm)}{40.8}, \tag{6.8}$$

where *PP* is the required pump power and *SPP* is the surface pumping pressure,

$$SPP = BHFP + FP - HH, \tag{6.9}$$

where *SPP* is the surface pumping pressure; *BHFP* is the bottom hole fracture pressure, which is calculated by equation (6.10); *FP* is the pumping friction pressure, which is calculated by equation (6.11); and *HH* is the hydrostatic head, which is calculated by equation (6.12).

$$BHFP = fracture\ gradient \times TVD, \tag{6.10}$$

$$FP = friction\ gradient \times MD, \tag{6.11}$$

and

$$HH = specific\ gravity \times 0.433\,\frac{psi}{ft} \times TVD\,ft. \tag{6.12}$$

Example 6.5 Find the Required Pump Power

The fracture gradient for a well targeting the Bakken formation is 0.8 psi/ft. The true vertical depth (TVD) and the measured depth (MD) at the perforations are 9,735 ft and 15,962 ft, respectively. Calculate the required pump power knowing that the specific gravity of the fracturing fluid is 1, the friction gradient is 0.2 psi/ft, and the pumping rate is 35 bpm. What would be the required pump power if the pumping rate were increased to 100 bpm? How many 1,000 hhp pumps are required to do the treatment?

Solution

To estimate the required pump power, the surface pumping pressure has to be known. To calculate the surface pumping pressure, the bottom hole fracture pressure, the hydrostatic head and the pumping friction pressure have to be calculated first. The bottom hole fracture pressure (BHFP) is calculated by equation (6.10) as follows:

$$BHFP = 0.8\,\frac{psi}{ft} \times 9{,}735\,ft = 7{,}790\,psi.$$

The pumping friction pressure (FP) is calculated by equation (6.11) as follows:

$$FP = 0.2\frac{psi}{ft} \times 15{,}962\,ft = 3{,}192\,psi.$$

The hydrostatic head (HH) can be estimated by the use of equation (6.12) as follows:

$$HH = 1 \times 0.433\frac{psi}{ft} \times 9{,}735\,ft = 4{,}220\,psi.$$

The surface pumping pressure (SPP) is calculated by equation (6.9) as follows:

$$SPP = 7{,}790\,psi + 3{,}192\,psi - 4{,}220\,psi = 6{,}762\,psi.$$

Hence, the required pump power (PP), in hydrostatic horsepower (hhp), can be estimated by the use of equation (6.8) as follows:

$$PP = \frac{6{,}762\,psi \times 35\,(bpm)}{40.8} = 5{,}801\,(hhp).$$

If the pump rate is increase to 100 (bpm), then the required pump power is

$$PP = \frac{6{,}762\,psi \times 100\,(bpm)}{40.8} = 16{,}573\,(hhp).$$

Thus, if the pumping rate is 35 bpm, then the minimum requirement to do the treatment is six 1,000 HHP pumps. However, if the pumping rate is increased to 100 bpm, then the minimum requirement to do the treatment is seventeen 1,000 hhp pumps or nine 2,000 hhp pumps.

6.2.9 Minifrac and DFIT Tests

The minifrac test is a procedure that is usually performed prior to the main hydraulic fracturing treatment to obtain key parameters that are required in the design process of the actual treatment such as fracturing gradient and fracture closure stress (Smith & Montgomery, 2015).

The minifrac test is conducted in two steps. The first step is the injection period. In the injection period, fracturing fluid is injected to create a fracture (i.e., minifrac). Note that proppant is not used in the minifrac test. To mimic the actual hydraulic fracture, it is ideal to inject at high rates to generate a "big" minifrac. This also helps in obtaining important parameters such as fluid leakoff. However, since this test is mostly performed to measure the fracture closure stress, injecting at high rates is not usually practiced mainly because of the long time that it will take the "big" minifrac to close (Fekete, 2012). Time is of the essence here because the hydraulic fracturing crew is usually on site waiting to perform the actual hydraulic fracturing treatment while the minifrac test is conducted. After the minifrac is created and propagated, the injection is stopped and the falloff step begins.

The second step is the falloff period. The point that the injection is stopped is called the instantaneous shut-in pressure (ISIP). After the injection is stopped, the falloff pressure data are monitored and collected, and consequently, the fracture closure pressure is recorded (see Figure 6.15). By collecting data after the fracture closure pressure is reached, the initial reservoir pressure and the formation permeability can also be estimated. However, in the unconventional reservoirs, it is uncommon to collect data after the fracture closure pressure is recorded because it would require collecting data for weeks before a decent estimation of the formation permeability and the initial reservoir pressure can be made (Fekete, 2012).

The diagnostic fracture injection test (DFIT) is conducted to obtain key parameters (e.g., fracture closure stress) for designing hydraulic fractures. DFIT is very similar to the minifrac test; however, DFIT is practiced more in the unconventional reservoirs (Rizwan & Liu, 2018). One main difference between the two tests is the amount of volume that is injected during the test. In the minifrac tests, usually the amount of fluid that is injected is the same as the amount of fluid that is going to be used in the real treatment. However, because the unconventional reservoirs are extremely tight with low permeability, smaller volumes are injected during DFIT to reduce the time that is needed to reach fracture closure and also to do after-closure analysis (ACA) much quicker (Barree et al., 2015).

After the DFIT is performed, the results are interpreted to get valuable information for the design of the actual hydraulic fracturing treatment (e.g., fracture closure stress, formation permeability, and leakoff rate). There are several pressure-time plots that can be utilized to analyze the DFIT data. The most common method is the G function diagnostic method.

G Function and G plot

The G function, which was developed by Nolte in 1979, is a dimensionless function of time that relates the total time to the total pumping time at a constant injection rate (Nolte, 1988). The start of both times is when the fracture is initiated, not when the pumping is started, and it is calculated using equation (6.13) (Barree et al., 2009),

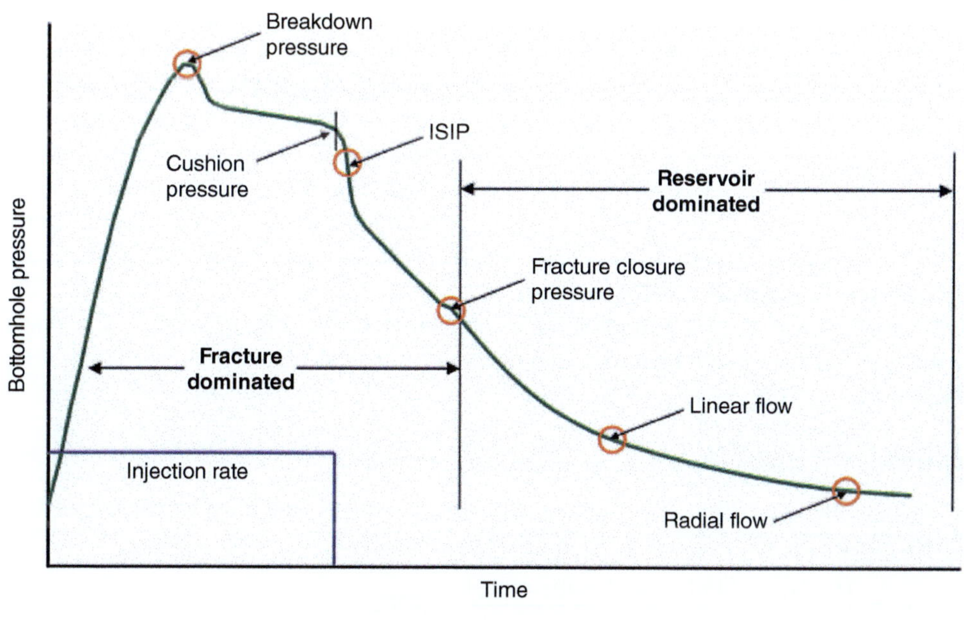

Figure 6.15 Overview of minifrac test courtesy of (Fekete, 2012).

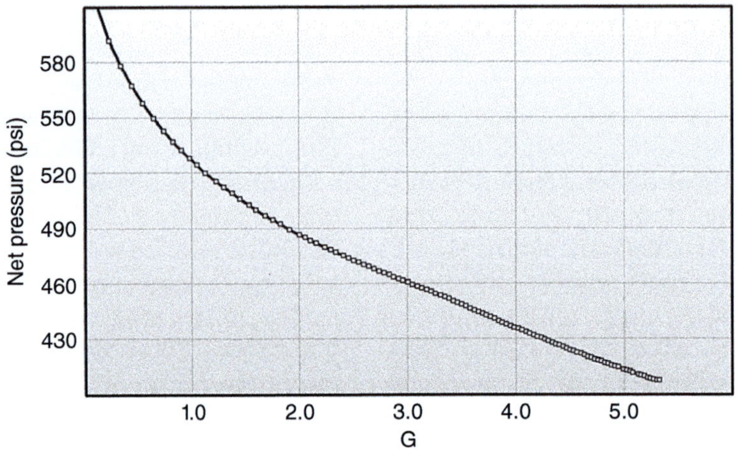

Figure 6.16 G plot for Example 6.1 (plot generated using E-Stimplan).

$$t_D = \frac{(t - t_p)}{t_p}, \tag{6.13}$$

where t is the total time (pumping time and shut-in time) and t_p is the total pumping time.

The G function can be calculated for high efficiency fluids, which gives low fluid leakoff by using equations (6.14)–(6.16):

$$G\,(t_D) = \frac{4}{\pi}\,(\,g(t_D) - g_0), \tag{6.14}$$

$$g(t_D) = \frac{4}{3}\left[(1 + t_D)^{1.5} - t_D^{1.5}\right], \tag{6.15}$$

and

$$g_0 = \frac{4}{3}. \tag{6.16}$$

For low efficiency fluid that gives high fluid leakoff, the G function is calculated using equations (6.17)–(6.19):

$$G\,(t_D) = \frac{4}{\pi}\,(g(t_D) - g_0), \tag{6.17}$$

$$g(t_D) = (1 - t_D)\,\sin^{-1}\left[(1 + t_D)^{-0.5}\right] + t_D^{0.5}, \tag{6.18}$$

and

$$g_0 = \frac{\pi}{2}. \tag{6.19}$$

After the G function is calculated, the recorded net pressure from the DFIT and the calculated G function are plotted. The generated plot is usually called the G function plot or the G plot. The G plot for the designed hydraulic fracture in Example 6.1 is shown in Figure 6.16.

Example 6.6 Calculate the G Function

A DFIT was conducted using a fluid with high efficiency. The total pumping time is 20 min and the total time at the closure point is found to be 44.1 min. Calculate the G function at the closure point.

Solution

To calculate the G function at closure, G_c, the dimensionless time must be calculated first as follows:

$$t_D = \frac{(t - t_p)}{t_p} = \frac{(44.1\,(\min) - 20\,(\min))}{20\,(min)} = 3.205.$$

Since the fluid is with high efficiency and low leakoff, then $g_o = 4/3$. The G function is calculated using equations (6.14) and (6.15) as follows:

$$g(t_D) = \frac{4}{3}\left[(1+t_D)^{1.5} - t_D^{1.5}\right] = \frac{4}{3}\left[(1+3.205)^{1.5} - 3.205^{1.5}\right] = 3.847$$

and

$$G_c\,(t_D) = \frac{4}{\pi}\left(3.847 - \frac{4}{3}\right) = 3.2.$$

Thus, the G function at the fracture closure is 3.2.

Identifying Fracture Closure

After the G function is calculated, more analysis can be done to estimate the fracture closure stress. The G function derivative should be calculated to find the fracture closure stress. The derivatives can be calculated using a finite difference scheme (e.g., central difference). To find the fracture closure stress, the G function derivatives and the G function are plotted next. The point where the straight line section of the G function derivative (i.e., GdP/dG) deviates downward is the fracture closure point. Figure 6.17 shows an example of how the fracture closure point is identified using the G function diagnostic method. A tangent line is drawn at the straight-line portion of the GdP/dG curve. The point at which the GdP/dG curve starts to deviate downward from the drawn tangent line is the fracture closure point. From Figure 6.17, the fracture closure point is located at almost at $G_c = 2.3$ and closure pressure (P_c) is approximately 8,800 psi.

Estimating Formation Permeability

The formation permeability can be estimated using the G function diagnostic method as presented in equation (6.20) (Barree et al., 2009),

$$k = 0.0086\frac{\mu_f\sqrt{0.01P_n}}{\varphi c_t\left(\dfrac{G_c r_p E}{0.038}\right)}, \qquad (6.20)$$

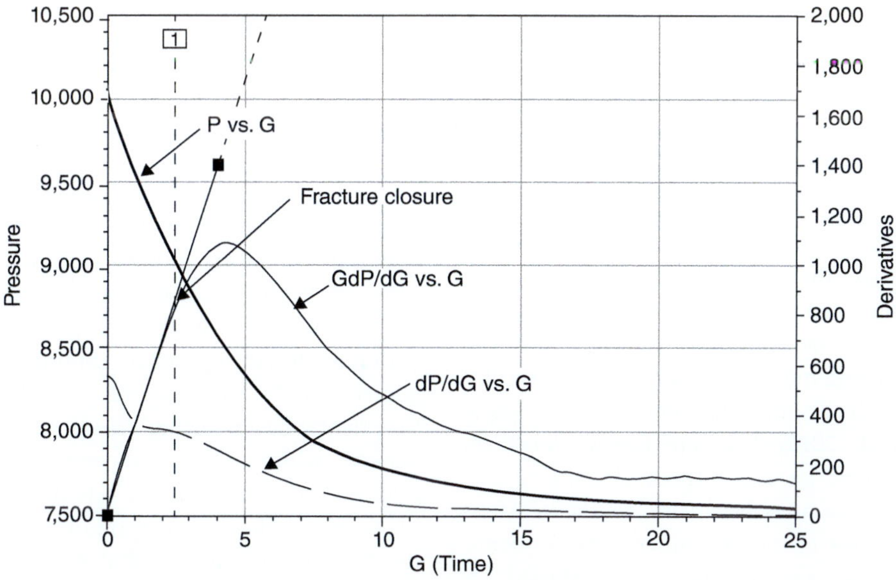

Figure 6.17 Identifying fracture closure from G plot (Barree et al., 2009).

where k is the formation permeability (mD), μ is the fluid viscosity, P_n is the net pressure in the zone (P_n = ISIP – Pc) (psi), φ is the porosity (fraction), c_t is the total compressibility (psi-1), G_c is the G function at closure, r_p is the storage ratio, and E is Young's modulus (Mpsi).

The storage ratio, r_p, is calculated as the permeable area that is opened by the fracture over the total fracture area as shown in equation (6.21) (Smith & Montgomery, 2015),

$$r_p = \frac{Permeable\ fractured\ area}{Total\ fractured\ area}. \tag{6.21}$$

As mentioned before, estimating the formation permeability as well as the initial reservoir pressure for the unconventional reservoirs using the minifrac and DFIT methods is not convenient because a long period of shut-in time is required in order to have a decent estimation. Equation (6.20) is an empirical function that was derived based on numerical simulations. It has been compared with other permeability estimation methods, such as the Horner plot method, and it gave consistent results. However, equation (6.20) should be used with caution as it might not be applicable in extremely tight formations and could lead to misleading results.

Example 6.7 Estimate the Formation Permeability

A DFIT was conducted. The instantaneous shut-in pressure (ISIP) from the test is found to be 7,351 psi and the shut-in pressure (P_c) is 6,567 psi. The formation porosity is 5%, the Young's modulus of the formation is 6.2 Mpsi, and the total compressibility is $4 \times 10^{-5}\ (psi^{-1})$. The filtrate viscosity is 1 cP and the storage ratio is 1. Estimate the permeability of the formation for Example 6.6.

Solution

The permeability of the formation can be estimated using the empirical function. First, the net pressure in the zone should be calculated as follows:

$$P_n = ISIP - P_c = 7,351psi - 6,567psi = 784\ psi.$$

Thus, using the G function at closure from Example 6.6, the formation permeability is calculated using equation (6.20) as follows:

$$k = 0.0086\frac{\mu_f\sqrt{0.01P_n}}{\varphi c_t\left(\frac{G_c r_p E}{0.038}\right)} = 0.0086\frac{1 \times \sqrt{0.01 \times 784}}{0.05 \times 4 \times 10^{-5} \times \left(\frac{3.2 \times 1 \times 6.2}{0.038}\right)} = 0.0567\ (mD).$$

6.2.10 Microseismic Monitoring

The microseismic monitoring is a distinctive technology that has been utilized in the hydraulic fracturing sector to visualize the geometry of the generated hydraulic fractures. Microseismic monitoring can also be used to improve the drilling and completion processes (Wu et al., 2016). The experimental work of microseismic for the purpose of imaging the hydraulic fractures began in the 1980s. After huge success in the experimental trials, commercial microseismic monitoring started in the early 2000s, and shortly after that, the unconventional shale sector adopted the technology (Maxwell, 2014). The controversy that hydraulic fracturing induces earthquakes has seriously begun because microseismic monitoring has become a standard practice in the hydraulic fracturing sector (Smith & Montgomery, 2015). Consequently, people argued that hydraulic fracturing causes major earthquakes without fully understanding why microseismic monitoring is actually used during the hydraulic fracturing treatments. In reality, the seismic intensity that is generated by hydraulic fracturing is very low, and it is not detectable by human beings at the surface. In fact, the seismic intensity levels that are generated by hydraulic fracturing are usually 100,000 times less than the levels that humans can detect (IGU, 2012).

Microseismic monitoring is the process of detecting the seismic energy that is generated from the microseisms (i.e., microearthquakes), which are generated during the hydraulic fracturing treatment. The seismic energy is detected and recorded by the use of receiver arrays, which have geophones, and they are usually placed in a nearby monitoring well. The old wells are typically used as the monitoring wells; however, this is not always ideal because the receiver arrays should be placed in certain positions and within a certain distance to achieve proper results (Warpinski, 2009). Also, the monitoring wells must be completely shut off to eliminate any noise issues. The noise, which could be caused by producing wells or drilling activities near the monitoring wells, can completely ruin the microseismic mapping (Maxwell, 2014). After the receiver arrays are placed, their sensors that acquire the seismic readings should be oriented, and after the seismic data are obtained, the data need to be processed and interpreted, accordingly.

The microseismic hydraulic fracture imaging has been mainly used to determine the dimensions of the generated hydraulic fracture, and it does that to a certain accuracy. Yet, that accuracy becomes less in the presence of a complex network of fractures. Hence, microseismic imaging is quite useful and reliable in determining the dimensions of planar hydraulic fractures but not as much in a complex network of fractures. Figure 6.18 shows the microseismic activity in two different wells. From the microseismic monitoring in Figure 6.18a, it is clear that the generated hydraulic fracture is a planar fracture with two relatively perfect fracture wings. Based on the microseismic map in Figure 6.18a, the dimensions of the generated fracture can be estimated with high accuracy. The microseismic activity in Figure 6.18b is scattered. The scattered microseismic activity is usually referred to as the microseismic "cloud." Consequently, accurate fracture dimensions cannot be determined by interpreting the microseismic cloud. Hence, in Figure 6.18b, a complex network of fractures is generated instead of a simple planar fracture.

Figure 6.19 shows a top view of microseismic mapping of a horizontal well targeting the Bakken play, and Figure 6.20 shows the side view of the same data. The microseismic data are visualized using a software package (e.g., Transform Explorer). The bubbles in the figures illustrate the monitored seismic activity where the size of the bubbles reflects the intensity of the recorded microseisms. The bigger the size, the higher the intensity. The color of the bubbles illustrates the fracturing stages where each color represents a different stage. Thus, Figure 6.19 shows 10 hydraulic fracturing stages. It is clear from Figure 6.19 and Figure 6.20 that the dimensions of the fractures cannot be easily interpreted by looking at the microseismic maps. Therefore, a complex network of fractures is generated rather than planar fractures.

The ultimate goal of the hydraulic fracturing treatments in most unconventional reservoirs is to create a complex network of fractures. However, microseismic imaging is unreliable in determining the dimensions of the complex fractures in the unconventional reservoirs for several reasons (e.g., ubiquitous natural fractures in the unconventional reservoirs). To overcome this inefficiency in microseismic imaging in the complex fractures, the industry has come up with a concept that utilizes and interprets the microseismic imaging where a complex network of fractures exists (Vassilellis et al., 2016). This new concept is called the stimulated reservoir volume (SRV), which is also called stimulated rock volume. The equivalent SRV for the complex network of fractures of the well represented in Figure 6.19 and Figure 6.20 can be seen in Figure 6.22.

6.2.11 Stimulated Reservoir Volume

Fisher et al. (2004) introduced the concept of stimulated reservoir volume (SRV) in 2004

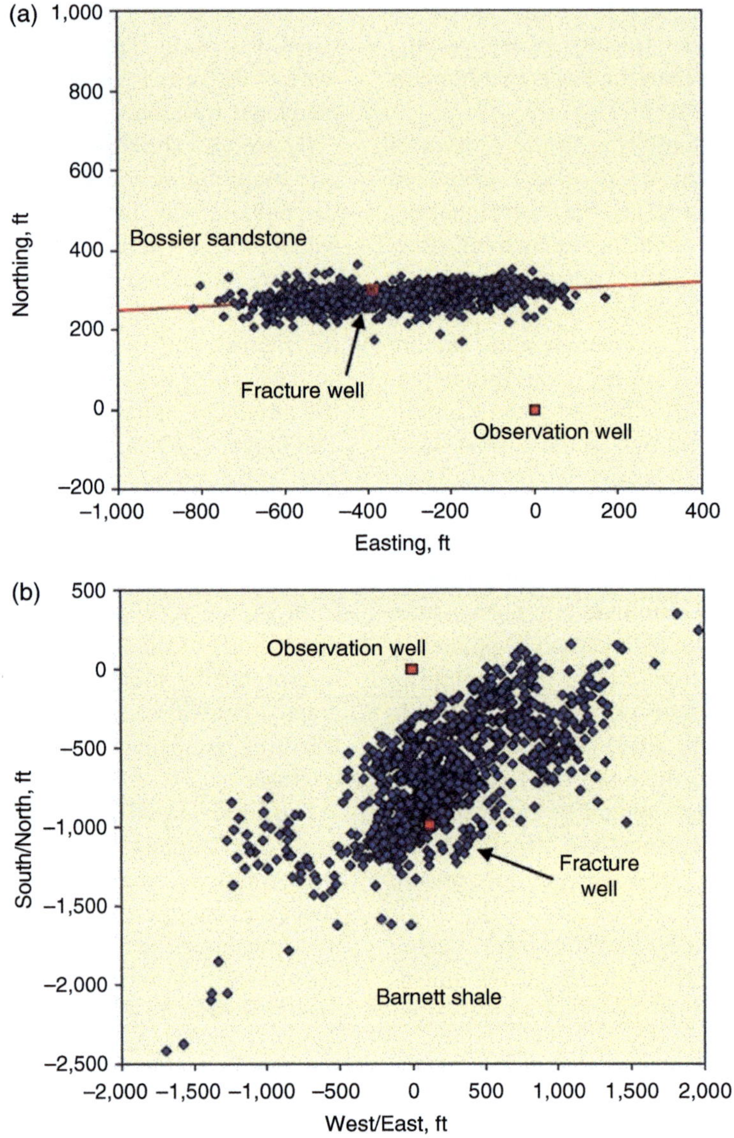

Figure 6.18 Microseismic monitoring of (a) planar fracture and (b) complex network of fractures (Warpinski, 2009, after Sharma et al., 2004; Fisher et al., 2005).

to evaluate the overall well performance after the hydraulic fracturing treatments conducted in the Barnett shale play, and soon after that, the SRV concept took off. As mentioned before, the microseismic imaging technique is not capable of capturing the fracture dimensions when a complex fracture network exists because a microseismic cloud is seen in the microseismic mapping. So, a method was needed to interpret the microseismic cloud in order to measure the effectiveness of the hydraulic fracturing treatment. Therefore, the concept of SRV was introduced to describe the overall stimulated volume after a hydraulic fracturing treatment solely based on the microseismic imaging cloud (Vassilellis et al., 2016). Conceptually, SRV can be defined as the complex fracture network regions that are created by the hydraulic fracturing treatments as shown in Figure 6.21. The SRV based

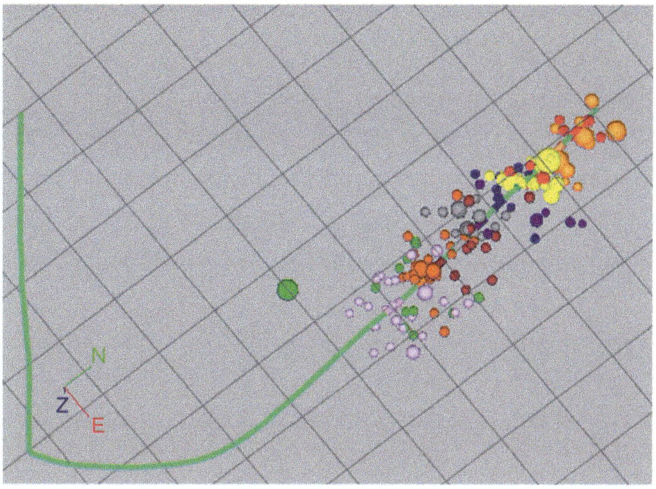

Figure 6.19 Microseismic mapping for a horizontal well (top view, Transform Explorer).

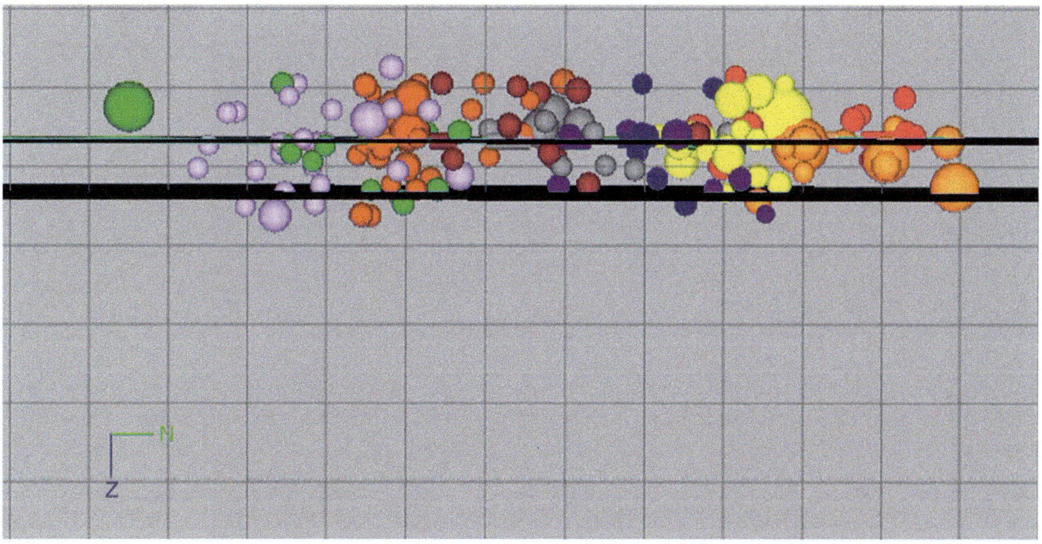

Figure 6.20 Microseismic mapping for a horizontal well (side view, Transform Explorer).

on the microseismic mapping example in Figure 6.19 and Figure 6.20 is represented in Figure 6.22, where each cell block is a 100 x 100 x 100 ft.

6.3 Fracturing Fluids

Fracturing fluid is basically the fluid that is injected in the wellbore to initiate a hydraulic fracture. The fracturing fluid is injected in the wellbore with high pressure and, consequently, the formation cracks because the pressure inside the wellbore exceeds the formation breakdown pressure (Alhubail et al., 2017). Based on the estimations of the formation breakdown pressure, as was discussed in Chapter 5, the injection pressure can be adjusted accordingly.

6.3.1 Purpose

There are two main purposes of using fracturing fluids. The first purpose is to initiate the hydraulic fractures and ideally create a network of fractures. The fracturing fluid is the

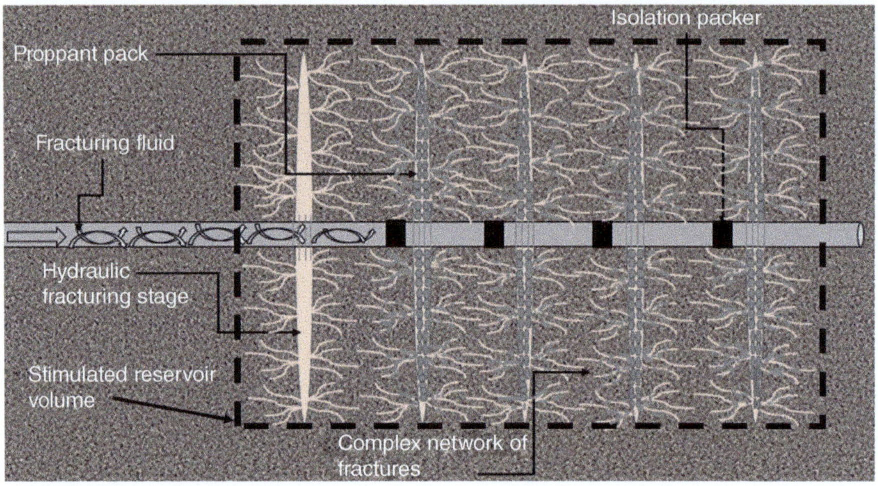

Figure 6.21 Stimulated reservoir volume.

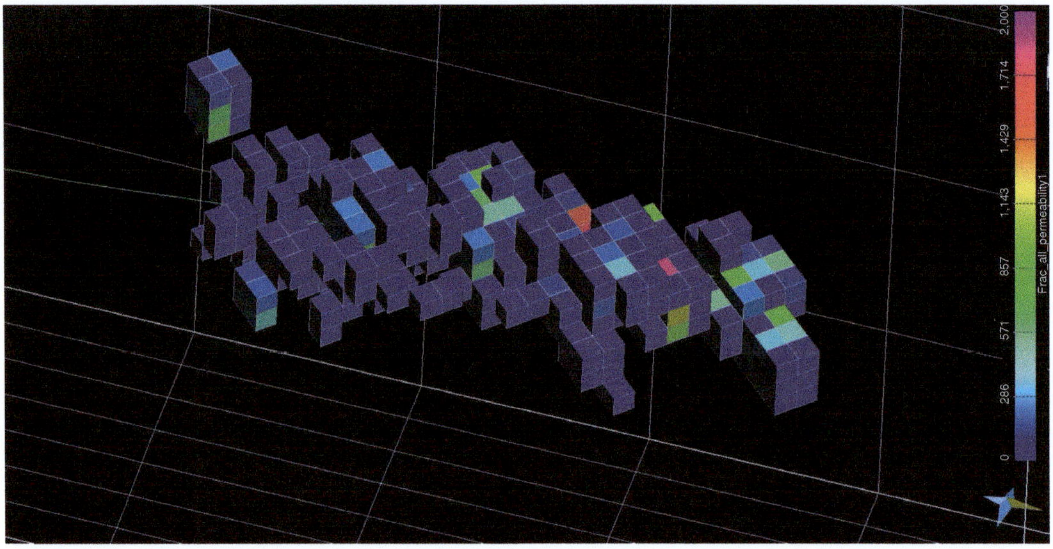

Figure 6.22 Permeability view (in Darcy) of the stimulated reservoir volume based on the microseismic mapping on Figure 6.19.

most significant component of the hydraulic fracturing design process, and the type of the fracturing fluid as well as the additives that will be used in the fracturing treatment should be selected with prior knowledge of the reservoir geomechanical properties and lithology as well as the reservoir fluids (Mata et al., 2016). The second purpose of using fracturing fluids, which is as important as the first one, is transporting the proppant to the generated fractures and placing the proppant inside the fractures in order to keep them open after the injection is stopped (Guo et al., 2017).

6.3.2 Fracturing Fluid Types, Properties and Selection Process

In the very early days of hydraulic fracturing treatments, small amounts of fluid were injected to generate the fractures. Mainly oil-based fluids (e.g., crude oil and diesel) were used back then because it was believed that the oil-based fluids were more compatible with the hydrocarbon-producing formations. Soon after that, the industry realized the necessity of adding other chemical components to the injected fluid to form a proper fracturing fluid and to achieve a more conductive hydraulic fracture. Right after the Second World War, napalm, which is an extremely hazardous material, was very accessible, and it attracted the hydraulic fracturing industry because of its gel mixture behavior. Napalm was used for a short time as a fracturing fluid, and the conducted treatments were actually successful. However, due to napalm's cost and the delicate procedure that is required to perfectly prepare it as well as the hazardous nature of it, the industry moved toward safer and cheaper fluids (i.e., water-based fluids) (Smith & Montgomery, 2015). As more wells were hydraulically fractured, more experience was gained, and in the 1950s, the industry realized that most of the hydrocarbon-baring formations actually contain water and also produce water. More studies were conducted in the 1950s to test water-based fracturing fluids. Shortly after that, with intensive laboratory work, the industry found additives, such as corn starch, that can be used to adjust the viscosity of water-based fracturing fluids. In the late 1950s, the industry found the additive that has made the water-based fracturing fluids more dominant than the oil-based fracturing fluids. That additive is guar gum, which is an excellent gelling agent for the water-based fracturing fluids (Jennings, 1996). The fracturing fluids' additives will be discussed later in this chapter. The water-based fracturing fluids have been improved through the years and more sophisticated water-based fracturing fluids have been used in the hydraulic fracturing treatments. In fact, what has made hydraulic fracturing such a disruptive innovation, in addition to horizontal drilling and multistage fracturing, is the usage of slickwater as a fracturing fluid (Wile, 2012).

Fracturing fluids can be classified into different categories based on the percentage of the fluid type that is used to form the end product of the fracturing fluid. For example, the slickwater fracturing fluid is considered to be a water-based fracturing fluid because more than 98% of it is composed of water. The fracturing fluids are mainly classified as (1) oil-based, (2) water-based, and (3) multiphase-based.

Oil-Based Fracturing Fluids

As mentioned above, the oil-based fracturing fluids were the first fracturing fluids, and actually they are compatible with almost all formations. Today, the oil-based fracturing fluids (e.g., crude oil and gelled oil) are rarely used as fracturing fluids because of their high cost, environmental concerns, and the difficulty of handling them. However, the oil-based fracturing fluids are the best candidates for the water-sensitive formations and also for very cold areas where water could freeze (Guo et al., 2017).

Water-Based Fracturing Fluids

The water-based fracturing fluids, which are currently the most commonly used fracturing fluids, have become popular because of their relatively low cost, and from an operational point of view, they are easy and convenient to work with (Zhao et al., 2018). They also perform quite well. That said, the water-based fracturing fluids should be avoided in water-sensitive formations, especially in low-permeability formations, because they could damage the formation even without the presence of expandable clays (Martin et al., 2016).

Multiphase-Based Fracturing Fluids

The multiphase-based fracturing fluids, which are also referred to as foams and energized fluids, are a mixture of a liquid phase and a gas phase. The stability of the liquid-gas mixture is extremely important for the fracturing fluid system to work well. To ensure the stability of the multiphase fracturing fluids, surfactants become essential to the process of developing the multiphase fracturing fluids. The gas phase of the fracturing fluid mixture is usually carbon dioxide or nitrogen (i.e., CO_2 or N_2) or a combination of both, and the liquid phase can be water, oil, or alcohol (Ward, 1986). To differentiate between foams and energized fluids, the quality of the gas phase has to be known under operating conditions (i.e., pressure and temperature). The quality of the gas phase, which is an indication of the concentration of CO_2 and N_2 in the fracturing fluid, is basically the ratio of the volumes of CO_2 and N_2 to the total fracturing fluid volume under operating conditions. If the multiphase fracturing fluid quality is 52% or higher, then the multiphase fluid is considered as a foam fracturing fluid, and if the quality is less than 52%, then the multiphase fluid is considered as an energized fracturing fluid (Ward, 1986). Multiphase fracturing fluids with quality of 95% or higher are not considered as an appropriate hydraulic fracturing fluid because they become a mist (Ward, 1986). The multiphase fracturing fluids usually do not damage the formation, and they breakdown easily in the fracture providing a rapid fracture cleanup. However, they are costly and require a certain expertise to perfectly prepare them and to maintain their stability (Guo et al., 2017).

Regardless of the fracturing fluid type (i.e., oil-based, water-based, or multiphase), the fracturing fluids must have certain properties before they are selected for the hydraulic fracturing treatment. It is important to note that different types of fracturing fluids might be used in different wells that are targeting the exact same unconventional play because the formation fluid might be different from one area to the other (e.g., Eagle Ford) (Gupta & Valkó, 2007; Donaldson et al., 2013).

Fracturing Fluid Properties

As discussed above, each fluid type has unique properties and certain advantages and disadvantages. However, the fracturing fluids, in general, must have several basic features in order to be considered for a hydraulic fracturing treatment. First and foremost, the fracturing fluids must be compatible with the reservoir fluids and the formation lithology. Second, to achieve an optimum hydraulic fracture, the fracturing fluids must have a reasonable fluid loss control to minimize the fluid leakoff. Fracturing fluids must also have adequate viscosity in order to carry and transport the proppant into the generated fractures. Furthermore, they should exhibit low friction pressure, which also helps reducing the required horsepower. Ideally, the fracturing fluids must be stable and break easily to provide rapid cleanup after the treatment. Additionally, they must be cost effective, environment friendly, and easy to apply. The most preferred properties of the fracturing fluids are shown in Table 6.3.

6.3.3 Rheology of Fracturing Fluids

In order to achieve an optimum fracture geometry and to carry and place the proppant inside the created fractures, the fracturing fluid must have an adequate viscosity. In fact, the viscosity of the fracturing fluid might be the most significant feature of fracturing fluids. Hence, having an idea of the rheology of the fracturing fluid is very important.

Table 6.3 Fracturing Fluids' Most Preferred Properties

Feature	Advantage
Compatibility with reservoir fluid and formation lithology	Minimize damage
Excellent fluid loss control	Optimum fracture geometry
Adequate viscosity	Efficient proppant transport
Stable and easily breakable	Rapid cleanup
Low friction pressure	Reduction of horsepower

In general, the fracturing fluids are either Newtonian fluids or non-Newtonian fluids. The Newtonian fluids (e.g., cross-linked gel, or slickwater) mean that at a constant temperature, the viscosity of the fluid would remain constant. Equation (6.22) shows the flow behavior of Newtonian fluids.

$$\tau = \mu\,\gamma, \tag{6.22}$$

where τ is shear stress, μ is the viscosity, and γ is the shear rate.

The flow behavior of the non-Newtonian fracturing fluids (e.g., cross-linked gel, linear gel, and foams) is mostly described by the power law model, equation (6.23). Most of the non-Newtonian fracturing fluids are shear-thinning fluids. One advantage of the shear-thinning fluids is the reduction of the fluid apparent viscosity as the fluid is being pumped down the wellbore. Reducing the fluid viscosity in the wellbore as the fluid is pumped down the wellbore means reducing the friction pressure, too (Martin & Rylance, 2007). Reducing the friction pressure is beneficial because it helps in sustaining the equipment from wear and tear.

$$\tau = K\,\gamma^n, \tag{6.23}$$

where K is the consistency index and n is the flow behavior index.

The apparent viscosity can be estimated by the use of equations (6.24) and (6.25) for SI units and fluid units, respectively.

$$\mu_a = \frac{K}{\gamma^{(1-n)}} \tag{6.24}$$

and

$$\mu_a = 47,879\,\frac{K}{\gamma^{(1-n)}}, \tag{6.25}$$

where μ_a is the apparent viscosity in cP, K is in lbf/ft^2/s, γ is in 1/s, and n is unitless.

For two parallel planes, the shear rate is defined as the velocity difference between the planes over the distance between the planes as shown in equation (6.26). The shear rate in the annulus can be estimated by the use of equation (6.27). The shear rate at the walls of a fracture can also be estimated by the use of equation (6.28).

$$\gamma = \frac{dv}{dx} \tag{6.26}$$

and

$$\dot{\gamma} = \frac{6\,\overline{v}}{r_2 - r_1}, \tag{6.27}$$

where $\overline{v}$ is the average velocity,

$$\dot{\gamma}_{wf} = \frac{(4 + 2\,n)Q}{h^2 w}, \tag{6.28}$$

where n is the flow behavior index, Q is the pump rate, h is the fracture height, and w is the fracture width.

It is important to note that during a hydraulic fracturing treatment, the shear rate varies inside the fracture with length and time (Figure 6.23). Hence, it is very important to calculate the shear rate during the design process before conducting the actual treatment.

Example 6.8 Calculate the Shear Rate Applied on Fracturing Fluids During Injection

Calculate the shear rate, in the annulus and at the walls of a fracture, for conventional and unconventional wells knowing that the fracturing fluid has a Newtonian fluid flow behavior with a viscosity of 10 cP. The outer radius and inner radius are 6 and 4 in., respectively. The pump rates for the conventional and unconventional cases are 40 BPM and 120 BPM, respectively. The fracture has a half-length of 300 ft. The fracture height is 20 ft and the fracture width is 0.01 ft.

Solution

Since the fracturing fluid is Newtonian, then n is equals to 1. The next step is to find the average velocity $\overline{v}$ for the two cases.

$Q1 = 40\ bpm = 0.106\ \frac{m^3}{s}$,

$Q2 = 120\ bpm = 0.318\ \frac{m^3}{s}$,

$r1 = 6\ in. = 0.1524\ m \quad r2 = 4\ in. = 0.1016\ m$,

and

$w = 0.01\ ft = 0.003048\ m\ H = 20\ ft = 6.096\ m.$

Hence, the average velocities are

$$\overline{v}1 = \frac{Q1}{\pi(r1^2 - r2^2)} = \frac{0.106}{\pi(0.1524^2 - 0.1016^2)} = 2.615\ \frac{m}{s}$$

and

$$\overline{v}2 = \frac{Q2}{\pi(r1^2 - r2^2)} = \frac{0.318}{\pi(0.1524^2 - 0.1016^2)} = 7.845\ \frac{m}{s}.$$

a) For conventional case
Shear rate in the annulus:

$$\dot{\gamma} = \frac{6\,\overline{v}}{r_2 - r_1} = \frac{6 \times 2.615}{(0.1524 - 0.1016)} = 308.85\frac{1}{s}.$$

Shear rate at the walls of the fracture:

$$\dot{\gamma}_{wf} = \frac{(4 + 2\,n)Q}{h^2 w} = \frac{6 \times 0.106}{6.096^2 \times 0.003048} = 5.615\ \frac{1}{s}.$$

b) For unconventional case
Shear rate in the annulus:

$$\dot{\gamma} = \frac{6\,\overline{v}}{r_2 - r_1} = \frac{6 \times 7.845}{(0.1524 - 0.1016)} = 926.547\frac{1}{s}.$$

Shear rate at the walls of the fracture:

$$\dot{\gamma}_{wf} = \frac{(4 + 2\,n)Q}{h^2 w} = \frac{6 \times 0.318}{6.096^2 \times 0.003048} = 16.845\ \frac{1}{s}.$$

It is very clear that the shear rate at the fracture walls is much lower compared with the shear rate in the annulus. It is also obvious that the pump rate for the unconventional well is much higher than the conventional well since slickwater has made higher pumping rates feasible.

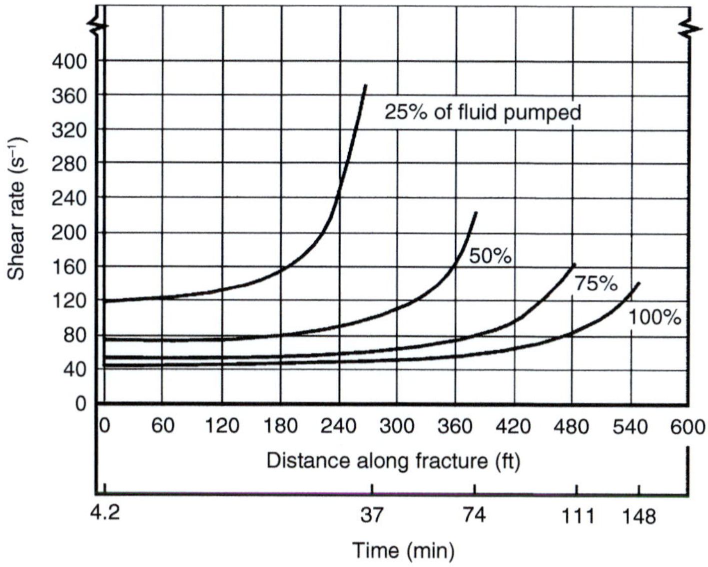

Figure 6.23 Shear rate profile inside a designed hydraulic fracture (Economides & Nolte, 2000, after Worlow, 1987).

6.3.4 Damage of Fracturing Fluid and Fracture Cleanup

One of the fracturing fluid preferred properties is that it is stable and easily breakable. This is essential for the fracture cleanup process. After the hydraulic fracturing treatment is conducted, the generated fracture should be cleaned up properly from all the fracturing fluid residues to minimize the formation damage. During a hydraulic fracturing treatment, there are three different damages that contribute to the overall formation damage: hydraulic damage, chemical damage and mechanical damage. Hydraulic damage is caused by the increase of the water saturation in the invaded zone due to fluid leak-off. Chemical damage is caused by adverse fluid-fluid and adverse rock-fluid interactions, which result in clay swelling, and the chemical damage could also alter the wettability (i.e., from water-wet to oil-wet). The mechanical damage is usually caused when polymers degrade and also from proppant crushing and embedment. The mechanical damage results in plugging of the formation pores. Figure 6.24 shows a schematic of filter cake, proppant distribution, and fluid invasion in a hydraulically induced fracture.

In the unconventional reservoirs where the formation permeability is extremely low, the hydraulic damage and the mechanical damage are the most substantial because they significantly reduce the effective permeability and increase the capillary pressure. Figure 6.25 shows the effect of fracture conductivity on production at several reservoir permeability. Figure 6.26 shows the effect of the fracturing fluid yield stress and viscosity on production. Figure 6.27 shows the effect of the fracture face damage on production at different permeability. It is clear from the figures that any damage in the low-permeability formation is significant. The hydraulic fractures must be cleaned up properly in order to achieve decent production figures from the unconventional reservoirs.

6.3.5 Fracturing Fluids Additives

There are several additives that can modify the fracturing fluid properties, which would help achieve the desired fracturing fluid. The additives might not represent more than 0.5% of the fracturing fluid system; however, it is important to make sure that the used additives do not interfere with each other. The typical fracturing fluid additives are summarized in Table 6.4.

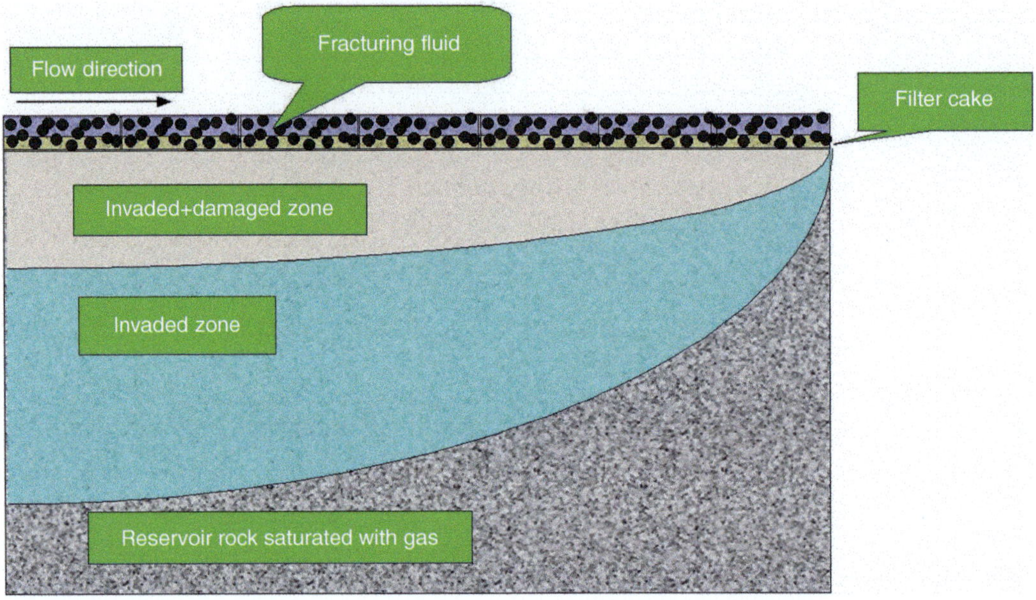

Figure 6.24 Filter cake, proppant distribution and fluid invasion in a hydraulically induced fracture (Barati, 2010).

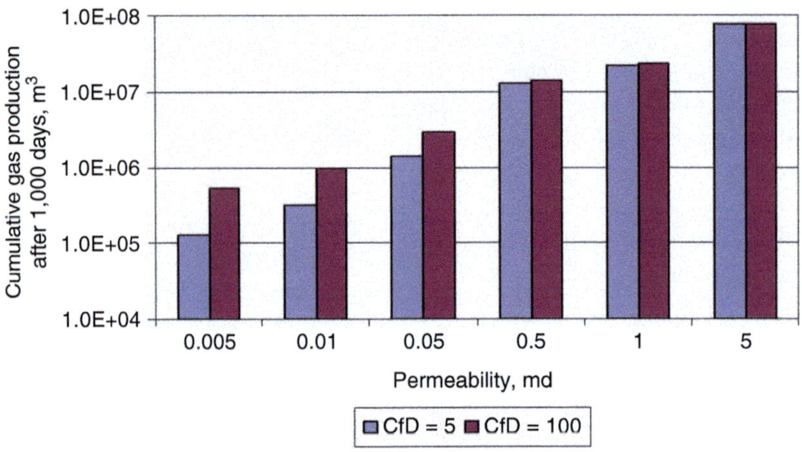

Figure 6.25 Effect of fracture conductivity on gas production using fluid-consistency index K' = 200 mPa·sn', yield stress = 5 Pa, n' = 0.5 (Barati et al., 2009).

Cross-linkers

The cross-linkers are used to enhance the viscosity of the fracturing fluids by forming a 3D network of molecular chains (Economides & Nolte, 2000). Increasing the fracturing fluid viscosity to a specific range is beneficial so that the proppant can be transferred inside the generated hydraulic fracture without any issues. Borate, Ti(IV), Zr(IV), and AI(III) are the metal ions that are typically used in the industry to cross-link the water-soluble polymers. Cross-linking is a complex process because the cross-linking agents work well only within a restricted range of pH and temperature.

The fluid pH and reservoir temperature have major roles in selecting the optimum cross-linker. For example, Borate, which is the most common cross-linker, is active in an 8–12 pH range; hence, it does not work well in acidic environments. Furthermore, the maximum

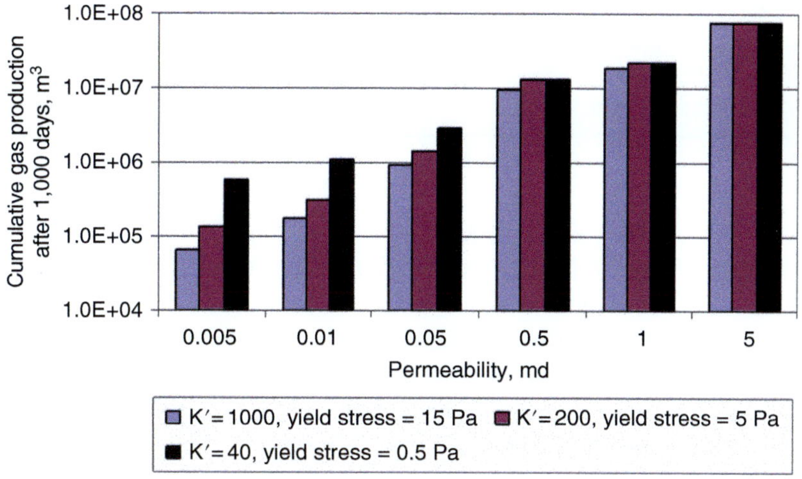

Figure 6.26 Effect of permeability increase on significance of yield stress and viscosity effect when n' = 0.5 and Cfd = 5 (Barati et al., 2009).

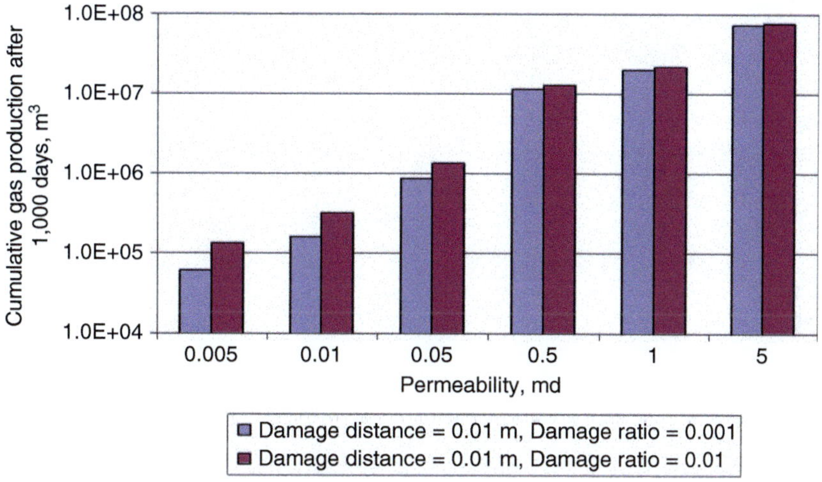

Figure 6.27 Effect of fracture face damage on production when n' = 0.5 and Cfd = 5 (Barati et al., 2009).

temperature for borate cross-linkers is 325°F. When borate cannot be used due to the limited pH and temperature ranges, other cross-linkers (e.g., titanium acetylacetonate and zirconium tetra-acetate) are used to alter the rheology of the fracturing fluid and make it more favorable for the hydraulic fracturing treatments (Barati & Liang, 2014). However, those other cross-linkers have to be carefully selected because they have major shear-sensitivity issues (Economides & Nolte, 2000). In addition, titanium and zirconium exhibit lower fracture conductivity compared to borate cross-linked fluids (Rae & Di Lullo, 1996). The common cross-linkers and their pH and temperature ranges are shown in Table 6.5.

pH Adjusters and Buffers

Altering the pH of the fracturing fluids is usually practiced since some cross-linkers only work at a specific pH range. To facilitate the cross-linking process, pH adjusters and buffers

Table 6.4 Typical Fracturing Fluid Additives

Additives	Features	Example of chemical
Cross-linkers	Increase fluid viscosity	Borate salts
Buffers	Provide pH control	Sodium carbonate
Biocides	Prevent bacteria growth	Glutaraldehyde
Interfacially active agents	Reduce interfacial tension and modify contact angle of leakoff fluid	Methanol/ surfactants
Gel stabilizers	Extend temperature range to stabilize the guar-based fluids	Sodium thiosulfate
Clay stabilizers	Provided control to swelling and fines migration of clays	Potassium chloride
Breakers	Reduce the fluid viscosity	Peroxydisulfates
Fluid loss	Minimize fluid loss	Silica flour and starch
Diverting agents	Temporarily block high-permeability channels	Benzoic acid flakes
Friction reducers	Reduce the friction pressure	Polyacrylamine
Scale inhibitors	Prevent the forming of scales	Phosphonates

Table 6.5 The pH and Temperature Ranges for Common Cross-Linkers

Cross-linking agent	pH range	Temperature range (F)
Zirconates	0–10	100–350
Organic titanates	3–10	100–325
Aluminates	3.5–6	100–200
Conventional borates	8–12	100–225
High temperature borates	8–12	100–350

Source: After Rae and Di Lullo (1996).

are utilized to adjust the pH. The pH adjusters (e.g., caustic soda and hydrochloric acid) are used to change the pH; however, if the pH is required to be maintained at a certain range, buffers must be used. Buffers are used to control the pH of the fracturing fluids and maintain it at a desired range by preventing any rapid change in pH if bases or acids are added to the fracturing fluid.

Biocides

The biocides are added to the fracturing fluids to prevent the growth of bacteria because the bacteria feed on the gelling agents of the fracturing fluids (e.g., guar and polysaccharides) and reduce the viscosity of the fracturing fluid significantly. The biocides kill the bacteria, however, they do not eliminate the already produced enzymes by the bacteria. Hence, biocides are usually added to the fluid tanks when water is added to minimize the enzyme levels, which are produced by the bacteria, in the fracturing fluid. The most common biocide for the fracturing fluid is glutaraldehyde. It is recommended to use different types of biocides each time so the bacteria do not become resistant to a specific type (Gupta & Valkó, 2007).

Surfactants

Surfactants ease the fluid recovery after the hydraulic fracturing treatment by reducing the surface tension and by changing the contact angle of the leakoff fluid. They also reduce the interfacial tension between the fracturing fluid and reservoir fluid.

Gel Stabilizers

Gel stabilizers are added to the fracturing fluids to improve the stability of the guar-based fracturing fluid at high temperatures.

Clay Stabilizers

Clay stabilizers are added to the fracturing fluids to control the clay swelling and fines migration; hence, reducing the formation damage that is caused by the fracturing fluids.

Breakers

Gel breakers are used in the fracturing fluids to enhance the recovery by assuring that the sand pack is cleaned up and fracturing fluids are degraded. The breakers reduce the fluid viscosity by reducing the molecular weight of polymers at the end of the fracturing treatment. The breakers also help with cleaning of the filter cake from the proppant pack; hence, increasing the conductivity. There are two common breakers that the industry relies on: oxidizers and enzymes. The oxidizers are consumed during the reaction, and they react very fast in high temperatures. The oxidizers are not ecofriendly and also not equipment friendly. On the other hand, the enzymes are ecofriendly and easy to operate. In addition, they are not consumed when they break the polymers since they act as catalysts. However, the enzymes are not active in harsh environments (e.g., high salinity and high temperature) (Guo et al., 2017).

Fluid Loss Additives

To conduct a successful hydraulic fracturing treatment, it is important to control and minimize the fluid loss. Fluid loss additives (e.g., silica flour and starch) are added to the fracturing fluid to assist in controlling the fluid loss. In relatively high-permeability formations, the polymers in the fracturing fluid actually help minimize the fluid loss by forming a filter cake.

Diverting Agents

In naturally fractured reservoirs or when there are high-permeability channels in the path of the primary hydraulic fracture, it is important to temporarily block those channels so that the fracturing fluid is not wasted by letting it go to the already high conductive paths, which would consume most of the fracturing fluid. Thus, diverting agents (e.g., rock salts) are added to the fracturing fluid to temporarily block the high-permeability channels; hence, shifting the fracturing fluid to the path of the primary hydraulic fracture for further propagation. During the hydraulic fracturing treatment, the diverting agents should be designed so they are insoluble in the fracturing fluid; however, they should be dissolvable after the treatment is conducted.

Friction Reducers

The friction reducers are mainly used to reduce the friction pressure, which would extend the life of the equipment from wear and tear especially at high pumping rates. The friction reducers also help minimize the required horsepower for pumping the fluids.

Scale Inhibitors

Forming of scale (e.g., calcium carbonate) is a common issue in the industry. Scale is often formed when the formation minerals are dissolved and precipitated in the well and on the equipment. Scales inhibitors (e.g., phosphonates) are usually added to the fracturing fluid, especially the water-based fracturing fluids, to prevent the forming of scales.

6.4 Proppant

Proppant, which is sand or sand-like particles, is one of the major components of the hydraulic fracturing treatments (Roach, 2014). The generated hydraulic fractures would remain open as long as a sufficient pressure is maintained inside the fracture. Once the pumping is stopped, the pressure would drop, the generated fractures would close, and the created permeability would be lost. To prevent the generated fractures from closing and to sustain the created permeability from severely diminishing, proppant must be used. There are recent studies that look into the parameters that may impact the proppant-pack permeability (e.g., proppant embedment and fines) and their effect on the overall fracture conductivity (Karazincir et al., 2018). With the increase of the hydraulic fracturing jobs, the demand for proppant is rapidly increasing (Olmen et al., 2018). The fracturing companies are currently experimenting with different proppant volumes trying to find an optimum proppant quantity to achieve great SRV. Consequently, the demand for proppant is expected to double in the next few years (Srinivasan et al., 2018).

6.4.1 Purpose

The purpose of the proppant is to keep the generated hydraulic fractures open after the pumps are turned off and the formation closure pressure is reached. It was mentioned before that the hydraulic fracturing treatments consist of two stages: the pad stage and the slurry stage. In the first stage, which is the pad stage, the hydraulic fractures are initiated by the fracturing fluid. In the second stage, which is the slurry stage, proppant is blended with the fracturing fluid and pumped down the wellbore. The proppants' main task is to prevent the created fluid flow channels from perfectly sealing, and maintain a conductive path for the fluid to flow through (Donaldson et al., 2013). The proppant properties play a major role in the overall fracture conductivity (Davies & Kuiper, 1988; Martin et al., 2016).

6.4.2 Proppant Characteristics and Selection Process

Before discussing the proppant types, it is important to know the main characteristics required of proppants in order to be considered for a hydraulic fracturing treatment.

Proppant Grain Size

The proppant grain size, or proppant "mesh size," is considered to be the most significant feature of the proppant. The range of the proppant mesh size (e.g., 20/40) is the measure of the proppant grain size. The mesh size range (e.g., 20/40) basically refers to the sieve cut. The lower the numbers of the mesh size range, the larger the proppant grain size. For example, proppant with a 20/40 mesh is larger than proppant with a 40/70 mesh. The proppant is considered to be 20/40 if 90% of the grain particles are contained within that mesh range. Refer to Table 6.6 for different proppant grain sizes.

The grain size and its distribution are important because they directly affect the fracture conductivity as they control the permeability of the proppant pack (Pearson & Brannon, 2007). Larger proppant yields higher proppant pack permeability; however, the larger the grain size, the higher the chance of it to breaking down and crush because the strength of the proppant decreases with increasing grain size. Large proppants should be avoided in high pressure formations. In addition to that, it is more difficult to pump and transport large proppants.

Table 6.6 Proppant Grain Size

Mesh size	Millimeters	Inches
5	4.000	0.1570
10	2.000	0.0787
20	0.841	0.0331
30	0.595	0.0232
40	0.400	0.0165
50	0.297	0.0117
60	0.250	0.0098
70	0.210	0.0083
80	0.177	0.0070
100	0.149	0.0059
120	0.125	0.0049
140	0.105	0.0041
170	0.088	0.0035
200	0.074	0.0029

Source: 911Metallurgist Lab (2017).

Proppant Shape

The proppant shape is basically a measurement or a description of the sphericity and roundness of the proppant particles. The sphericity of a sphere particle is one, and it is less than one if the particle is not a perfect sphere. The roundness describes the corners or curvature sharpness of the proppant particles. The shape of the proppant particles (i.e., sphericity and roundness) is important because it provides valuable knowledge of how the stress load would be distributed on the proppant particles; thus, it provides knowledge of the proppant crush resistance. At high stress levels, higher proppant crush resistance is required. In general, the proppant crush resistance increases as the sphericity and roundness of the proppant particles increase. Higher values of sphericity and roundness would also improve the proppant pack permeability; greater fracture conductivity would be achieved with better sphericity and roundness. The shape of the proppant particles is usually determined by the Krumbein and Sloss chart (Fig. 6.28).

Proppant Bulk Density and Specific Gravity

The proppant density has an important role on proppant transport and placement (Table 6.7). It is difficult to suspend high-density proppant in the fracturing fluid, and it is more difficult to transport it within the generated hydraulic fractures. Hence, the low-density proppant has more benefits than the high-density and intermediate-density proppants in terms of transport and placement. The low-density proppant provides higher conductive fractures because it increases the width of the fractures. The low-density proppant also reduces the overall cost of the hydraulic fracturing treatment because it provides the same reservoir contact with less amount of proppant compared with the intermediate-density proppant (Carbo Ceramics, 2017).

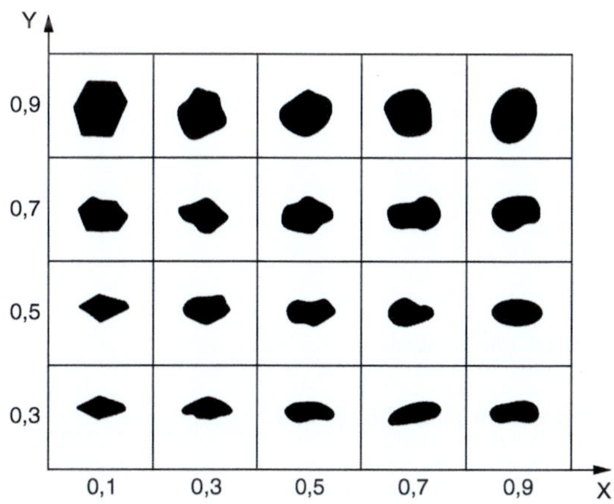

Figure 6.28 The Krumbein and Sloss chart for visual determination of sphericity and roundness (Krumbein & Sloss, 1963).

Table 6.7 Bulk Density of Different Proppant Types

Proppant	Bulk density (g/cm^3)	Specific gravity
Ottawa sand (20/40)	1.6	2.65
Ottawa sand (40/70)	1.58	2.65
Pre-cure RC sand (20/40)	1.59	2.45
Curable RC sand (20/40)	1.53	2.55
LW ceramic (20/40)	1.57	2.71
ISP (20/40)	1.88	3.27
Bauxite (20/40)	2.0	3.55
Ultra LW (14/40)	0.66	1.05

Source: Economides and Martin (2007).

Proppant Quality

The quality of the proppant is identified by measuring the impurities (e.g., clays) in the proppant particles. The proppant quality affects the permeability of the proppant pack, hence the fracture conductivity. The proppant strength also gets affected by the impurities; hence, impurities have a role in the proppant crush resistance. The proppant quality is measured by acid solubility and turbidity tests. The acid solubility test provides knowledge of the impurities in the proppant particles (e.g., carbonates) that would reduce the proppant performance in an acidic environment. Studies have shown that the artificial proppant (e.g., ceramic) loses a great amount of its strength to acid attacks compared with sand proppant, which also suffers loss of strength but not as much as the artificial proppant (Roberts et al., 1990; Assem et al., 2017). The turbidity test measures the amount of impurities (e.g., clays and silt) that are < 200 US mesh in diameter (Assured Testing Services, 2018).

Proppant Strength

The most important feature the proppant should have is an adequate strength. As mentioned before, the main purpose of the proppant is to maintain the generated hydraulic fracture open when the pumps are turned off and the fracture closure pressure is reached. The proppant must have sufficient strength to withstand the fracture closure stress so that it does not crush. The proppant strength is usually measured by crush resistance tests following the ISO 13503-2 standards. Basically, the test is conducted by applying a load, usually 2,000 psi per minute, on the prescreened proppant until the stress of interest is reached. The stress is held at the targeted stress level for 2 min. After that, the proppant is screened again to quantify the generated fines, which are crushed proppant caused by applied stress. The highest stress that the proppant can withstand without producing 10% fines is referred to as the K number. Hence, the higher the K number, the higher the strength of the proppant.

Example 6.9 Find the K Number

Crush resistance tests were conducted on sand proppant. The tests yielded 9.6% and 11.4% fines at 5,000 psi and 6,000 psi, respectively. What is the K number of the proppant?

Solution

The K number of the tested sand proppant is 5K because fewer than 10% fines were produced at 5,000 psi and more than 10% fines were generated at 6,000 psi.

Example 6.10 Find the Minimum Required Compressive Strength for the Proppant

For the example in Chapter 5, Example 5.3, estimate the minimum required compressive strength for the proppant if the production drawdown pressure is 2,000 psi and 3,000 psi. Assume a normal fault regime in the calculation of the stress.

Solution

Data from Example 5.2 and Example 5.3:

Vertical stress (overburden pressure) = 10,805 psi
Pore pressure = 3,455 psi
Poisson's ratio = 0.26
Biot's poroelastic constant = 0.69

First, the initial effective horizontal stress should be calculated:

$$\sigma_{heff} = \frac{\nu}{1-\nu}(\sigma_v - \alpha P_p) = \frac{0.26}{1-0.26} \times (10,805 - 0.69 \times 3,455) = 2,959\ psi$$

Then, the effective horizontal stress at 2,000 psi pressure drawdown:

$$\sigma_{heff} = \frac{\nu}{1-\nu}(\sigma_v - \alpha((P)_p - P_{dd})) = \frac{0.26}{1-0.26} \times (10,805 - 0.69 \times (3,455 - 2,000))$$
$$= 3,444\ psi$$

The effective horizontal stress at 3,000 psi pressure drawdown:

$$\sigma_{heff} = \frac{\nu}{1-\nu}\left(\sigma_v - \alpha\left((P)_p - P_{dd}\right)\right) = \frac{0.26}{1-0.26} \times (10,805 - 0.69 \times (3,455 - 3,000))$$
$$= 3,687\ psi$$

Hence, the minimum required compressive strength for the proppant is 3,687 psi.

Another approach to estimate the required proppant strength was proposed by Sookprasong in 2010.

$$\sigma_{proppant} = \sigma_{heff} + \Delta P_{dd}, \tag{6.29}$$

where ΔP_{dd} is the reservoir drawdown pressure.

So, when the reservoir drawdown pressure is 2,000 psi, the required proppant strength is

$$\sigma_{proppant} = \sigma_{heff} + \Delta P_{dd} = 2,959 + 2,000 = 4,959\ psi.$$

When the reservoir drawdown pressure is 3,000 psi, the required proppant strength is

$$\sigma_{proppant} = \sigma_{heff} + \Delta P_{dd} = 2,959 + 3,000 = 5,959\ psi.$$

The minimum required compressive strength for the proppant is 5,959 psi. Clearly, the difference between the two methods is significant.

6.4.3 Proppant Types

There are two main types of proppant. The first type is naturally occurring sand, and the second type is artificial human-made proppant. The sand proppant is usually used when the fracture closure stress is less than 6,000 psi. On the other hand, the artificial proppant is used for greater fracture closure stresses.

Sand Proppant

The sand proppant, which is also called the natural proppant, is the most common proppant used in the industry, and there are different types or tiers of sand proppant. Most of the sand proppant is located in United States; however, England, Russia, Brazil, and Saudi Arabia also have sand proppant that meets the API specifications for fracturing proppant. There is also sand proppant in the marketplace from Oman, Poland, and China, but these are classified as substandard low-quality sand proppant because they do not meet all the API specifications for fracturing proppant (Pearson & Brannon, 2007). Yet, due to the huge domestic demand recently, China, with government support, is working diligently on improving its proppant quality (Kim, 2012). The sand proppant properties are summarized in Table 6.8.

Ottawa Sand

Ottawa sand, which is the highest quality tier 1 sand proppant, is the predominant proppant used in the industry. Ottawa sand is also called Jordan, Northern, and White sand. The term *Ottawa* is associated with this type of proppant because high concentration of it is located in the district of Ottawa in Illinois. However, the state of Wisconsin holds approximately 75% of the sand proppant market (Guo et al., 2017). Due to its high quality (i.e., roundness,

Table 6.8 Properties of Sand Proppant

Sand proppant	SiO_2 %	Roundness and sphericity	K number
Ottawa	99	≥ 0.7	6K–11K
Brady	97–99	≥ 0.6	4K–8K
Fit-for-purpose	less than 98	< 0.6	3K–5K

Source: Smith and Montgomery (2015).

sphericity, and purity), the Ottawa sand provides decent conductivity and has relatively high K number values.

Brady Sand
Brady sand, which is also referred to as brown sand and Texas brown sand, is a medium quality tier 2 sand proppant, and it is mainly used in the United States. This type of sand proppant is called Brady sand because it is mined near the city of Brady, Texas. This type of sand proppant has relatively low quality compared with Ottawa sand, and it is mostly used when the closure stress is low (Pearson & Brannon, 2007).

Fit-For-Purpose Sand
Due to the high demand of proppant, especially after the boom of the hydrocarbon unconventional resources in the early 2000s, the amounts of traditional sands (i.e., Ottawa sand and Brady sand) were not sufficient to meet the unexpected demand. Fit-for-purpose sand proppant was developed to fill the shortage of proppant. This type of sand is usually a substandard low-quality tier 3 sand proppant and, in most cases, it does not meet all the API specifications for fracturing proppant (Smith & Montgomery, 2015). The fit-for-purpose sand proppant, which is also called nontraditional sand, is used only when it is tested and deemed fit for the fracturing job.

Artificial Proppant
Artificial proppant is human-made, and it is usually manufactured from ceramic materials (e.g., high-quality bauxite and kaolin clay). In general, all ceramic proppant is uniform in size, exhibits good roundness and sphericity, and can handle high closure stresses. Hence, the ceramic proppant yields high fracture conductivity. However, the ceramic proppant is relatively high in cost, and it loses much of its strength in acidic environments compared with the sand proppant. The ceramic proppant is classified into three types based on strength and specific gravity (Economides & Martin, 2007).

Bauxite Proppant
The sintered bauxite proppant, which is also referred to as high-strength proppant (HSP) and high-density ceramic (HDC), was the first ceramic-based proppant, and it was introduced in 1976 in order to achieve decent fracture conductivity in high closure stress locations (Atteberry et al., 1979). The HDC proppant is considered as the strongest proppant available in the market. However, recently, there are proppant manufacturing companies (e.g., CARBO) that provide ultra-high-strength ceramic proppant (Carbo Ceramics, 2017). The HDC proppant provides great fracture conductivity, and it is typically used when the fracture closure stress is greater than 10,000 psi. The HDC proppant has high specific gravity, which is an issue for proppant transport. The fracturing fluid must be designed so that it is able to hold and transfer the high-density proppant.

Intermediate-Strength Ceramic Proppant
The intermediate-strength proppant (ISP), which is also referred to as intermediate-strength ceramic (IDC), has a specific gravity in the range of 2.9 and 3.3, and it was first introduced in 1982. The IDC proppant has lower strength than the HDC proppant, yet, it is strong enough to be used when the fracture closure stress is in the range of 8,000 to 12,000 psi (Pearson & Brannon, 2007).

Lightweight Ceramic Proppant
The lightweight ceramic proppant (LWC) has a specific gravity that is somehow similar to the sand proppant (see Table 6.7). The LWC proppant yields better fracture conductivity compared with the sand proppant. The LWC proppant was first introduced in 1985, and it is typically used when the fracture closure stress is in the range of 6,000 to 10,000 psi. The LWC proppant is attractive because of its low specific gravity and its relatively high strength.

Ultra-Lightweight Proppant
The first generation of the ultra-lightweight proppant (ULW) was first used commercially in 2004. The ULW proppant has a specific

gravity of 1.25, and it can be used for fracture closure stress up to 8,000 psi. The low specific gravity of the ULW proppant has made the slickwater a very popular fracturing fluid (Bestaoui-Spurr & Li, 2018). Due to its low specific gravity, the ULW proppant minimizes the proppant settling. Hence, the ULW proppant can be placed easily in the generated hydraulic fractures even with low viscously fracturing fluids (Pearson & Brannon, 2007).

Resin-Coated Proppant

The resin-coated proppant (RCP) is basically sand or ceramic proppant that is resin coated to improve the proppant strength and minimize the generation of fines (Economides & Martin, 2007). The resin also helps reduce the proppant flowback because it encapsulates the generated fines. There are two types of resin-coated proppant: precured and curable. The proppant is called precured when the resin coating is fully completed during the manufacturing process. The main purpose of precured proppant is to enhance the proppant strength at high fracture closure stress. On the other hand, the curable proppant is partially resin coated by the manufactures, and when the proppant is placed in the fractures, additional curing occurs. The main purpose of the curable proppant is to reduce the proppant flowback. The curable proppant is usually pumped at the end of the fracturing stage, so it bonds all the proppant together to form a strong proppant pack, and the proppant flowback is minimized.

Figure 6.29 shows different proppant types and their recommended closure stress ranges. Figure 6.30 shows different proppant types and their proppant pack permeability at different closure stresses. It is clear from Figure 6.30 that the high-strength proppant (e.g., bauxite) provides excellent proppant pack permeability. It is costly to use, however, and its specific gravity is high. It requires fracturing fluids with adequate properties in order to transport it inside the generated hydraulic fractures.

6.4.4 Proppant Flowback

Proppant flowback is a major concern in the fracturing industry because it could choke the production after the hydraulic fracturing treatments. Proppant flowback is basically the long-term production of proppant from the generated fractures. The purpose of generating fractures is to have a conductive path for

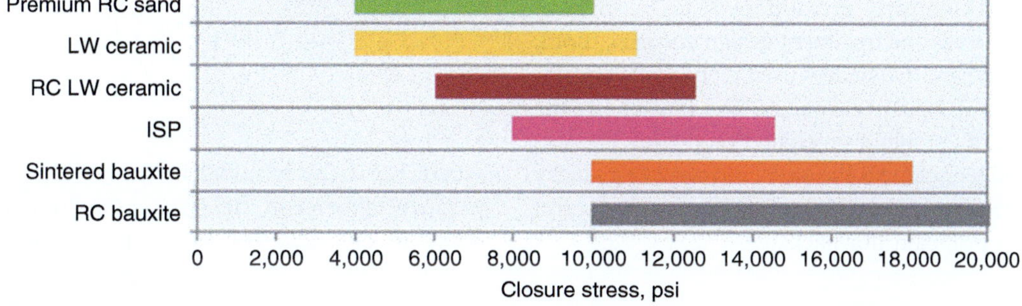

Figure 6.29 Types of proppant and recommended closure stress range (Martin et al., 2016).

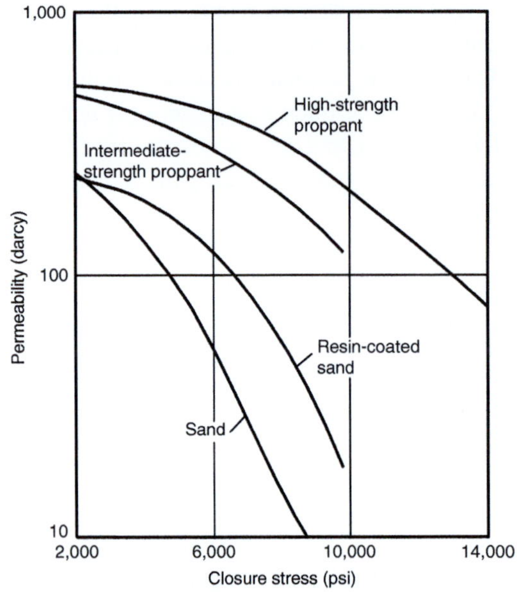

Figure 6.30 Proppant pack permeability for different proppant types at different closure stress (Economides & Nolte, 2000).

the fluid to flow through. The conductive path can be maintained, after fracture closure, only if a decent proppant pack is formed. However, if long-term proppant flowback occurs, the proppant pack would be disturbed, and the production would decline significantly.

During the fracture cleanup process, proppant flowback is substantial because there would be a big difference in pressure between the wellbore and the generated fracture. An insufficient amount of proppant in the generated fracture, especially near the fracture entrance, could lead to complete fracture closure. The proppant flowback can be prevented by taking early precautions (e.g., perforation strategies and treatment design and execution). There are two widely used techniques to minimize proppant flowback. The first one is the use of curable resin-coated proppant. The second technique is the use of additives (e.g., tacky resins and fiber additives) that can help control the proppant flowback.

6.4.5 Proppant Transport

It was mentioned before that the main purpose of the fracturing fluid is to generate fractures and to transport proppant into them. Proppant transport is very important in the hydraulic fracturing treatment because it controls the proppant placement inside the generated fractures. Inefficient proppant transport leads to proppant settling; hence, short fracture length would be achieved. A major factor that impacts the proppant transport is proppant settling velocity. The proppant settling velocity or the terminal velocity can be estimated by the use of Stokes's law. The settling velocity in Newtonian fluid and non-Newtonian fluid for a single particle can be calculated by the use of equations (6.30) and (6.31), respectively.

$$v_\infty = \frac{g\left(\rho_{prop} - \rho_f\right) D_{prop}^2}{18\,\mu}, \tag{6.30}$$

where v_∞ the terminal velocity, g is the gravitational acceleration (9.81 m/s^2), D_{prop} is the proppant diameter, μ is the fluid viscosity, and ρ_{prop} and ρ_f are the density of the proppant and the fluid, respectively.

$$v_\infty = \left[\frac{g\left(\rho_{prop} - \rho_f\right) D_{prop}^{n+1}}{3^{n-1}\,18\,K'}\right]^{\frac{1}{n}}, \tag{6.31}$$

where K' the power law fluid consistency index and n is the power law fluid exponent.

Equation (6.30) in fluid units would become as follows:

$$v_\infty = 1.15 \times 10^3 \left(\frac{D_{prop}^2}{\mu}\right)\left(\gamma_{prop} - \gamma_f\right), \tag{6.32}$$

where v_∞ is the terminal in ft/min, D_{prop} is proppant dimeter in inches, μ is the fluid viscosity in cP, γ_{prop} and γ_f are the proppant specific gravity and the fluid specific gravity, respectively.

Clearly from equation (6.32), the ideal way to improve the proppant transport is by lowering the specific gravity of the proppant. If the proppant specific gravity approaches the fracturing fluid specific gravity, neutral buoyancy would occur, and the proppant will not settle. The proppant transport can also be improved by increasing the fracturing fluid viscosity or decreasing the proppant diameter (Pearson & Brannon, 2007).

6.4.6 Proppant Schedule

It was mentioned before that the proppant is added to the fracturing fluid in the slurry stage of the hydraulic fracturing treatment. The proppant has to be added with a concentration that is compatible with the fluid efficiency. In 1989, Nolte found that the proppant concentration verse time follows the relationship shown in equation (6.33) (Nolte, 1986; Economides et al., 1994).

$$c_p(t) = c_f\left(\frac{t - t_{pad}}{t_t - t_{pad}}\right)^{\epsilon}, \qquad (6.33)$$

where $c_p(t)$ is the slurry concentration in ppg, c_f is the final slurry concentration, t_t is the total time, t_{pad} is the pad time, and ϵ is a variable based on the fluid efficiency.

$$\epsilon = \frac{1 - \eta}{1 + \eta}, \qquad (6.34)$$

where η is the fluid efficiency.

Example 6.11 Find the Proppant Schedule

The fracturing fluid efficiency of a hydraulic fracturing treatment is 0.46, and the final slurry concentration is 2.74 ppg. The total injection time and pad injection time for one fracturing stage are 41 min and 7 min, respectively. What is the proppant concentration at 30 min?

Solution

The first step is to calculate the efficiency variable ϵ:

$$\epsilon = \frac{1 - \eta}{1 + \eta} = \frac{1 - 0.46}{1 + 0.46} = 0.37.$$

The next step is to calculate the concentration by the use of equation (6.33):

$$c_p(30) = c_f\left(\frac{t - t_{pad}}{t_t - t_{pad}}\right)^{\epsilon} = 2.74 \times \left(\frac{30 - 7}{41 - 7}\right)^{0.37} = 2.37\ ppg.$$

The proppant concentration at 30 min is 2.37 ppg.

6.5 Modeling of Hydraulic Fractures

The hydraulic fracturing treatments are very costly jobs: the completion of one well with multistage fracturing could cost more than \$10 million U.S. Refer to Table 1.2 and Table 1.3 in Chapter 1 for more details about the total cost of the wells at different unconventional plays in North America. Before committing to doing a hydraulic fracturing treatment, a comprehensive study must be conducted in order to minimize the risks of failure and to optimize the recovery; one part of that study is modeling. The term *modeling* here means solving a set of mathematical equations (i.e., partial differential equations) to mimic a certain procedure, a process, or a phenomenon. The hydraulic fracturing models are basically a mean to predict the evolution of the fracture dimensions with time (Valkó & Economides, 1995).

6.5.1 Importance of Modeling

Hydraulic fracture modeling is a complex process because developing the models requires strong knowledge in multiple fields of study (e.g., geomechanics, fluid flow, and proppant transport, etc.). Yet, the hydraulic fracturing models are essential to have because they provide substantial information and lead to making remarkable decisions. The hydraulic fracturing models are not used solely to predict the fracture geometry; the advanced models are rather used to determine the optimum fracture designs that yield the highest rate of return on investment (Economides & Nolte, 2000). This is accomplished by varying the primary parameters that would impact the actual treatment (e.g., fluid and proppant types and pump schedule, etc.) and also by conducting a sensitivity analysis on them. Hence, the advanced hydraulic fracturing models are critical, in the overall hydraulic fracturing treatment design process, because they are capable of providing the optimum fracture designs by considering the economics as well as the geometry of the fractures, the number of fracturing stages, and the production performance (Guo et al., 2017). Ultimately, the hydraulic fracturing models help in minimizing risks and cost and lead to making new investments.

6.5.2 Governing Processes of the Models

The modeling of the hydraulic fractures requires the coupling of at least three main processes: (1) rock deformation, (2) fluid flow inside the fractures, and (3) fracture propagation criteria. The advanced models require the coupling of additional processes such as fluid leakoff and proppant transport. The rock deformation process (i.e., a crack opening model) is modeled by the use of the theory of linear elasticity, which shows the nonlocal relationship between the fluid pressure and the fracture width, and it is represented by an integral equation. The second process (i.e., fluid flow) is modeled by relating the velocity of the fluid flow, the pressure gradient, and the fracture width. The combination of those physics is a nonlinear partial differential equation. The third process is the fracture propagation criterion, and it is modeled by the use of the linear elastic fracture mechanics theory (LEFM). LEFM means that the generated fracture will grow as long as stress intensity factor K_I is equal to the rock toughness K_{IC}.

The hydraulic fracturing modeling process is complicated because, in real life, fractures occur on different type of rocks, which have different properties, and on different planes at the same time. Consequently, there are assumptions that are usually made when the hydraulic fracturing models are built. The first assumption that is typically made is that the reservoirs are homogenous, and they are considered to be linear elastic instead of poroelastic (Economides & Nolte, 2000; Adachi et al., 2007). Another assumption is that the reservoirs are considered to be one perfectly bounded layer instead of multilayered reservoirs. The generated fractures are also assumed to be in one single vertical plane. In addition, the models overlook the complex rheologies (e.g., yield stress) considering only simple fluid rheology regimes (e.g., Newtonian and power-law). The effects of proppant transport and proppant settling on the fluid flow are usually simplified. Furthermore, some of the developed models ignore important effects such as the existence of natural fractures and their effects on the generated hydraulic fractures.

6.5.3 Modeling History

The hydraulic fracturing models can be classified, based on their features and complexity, into different categories. In general, the hydraulic fracturing models are classified into simple 1D, 2D, 3D, and most recently the sophisticated unconventional fracture models.

Early Models

The development of the early hydraulic fracturing models, which started in the 1950s, was mostly based on the work of Sneddon and Elliott on the Griffith crack opening solution (Sneddon & Elliott, 1946). The early models were very simplistic, and they required knowing a few parameters in advance. For instance, the Khristianovich and Zheltov model, which is the basis of the famous KGD model, is a simple 1D model that can predict the width of the fracture (Khristianovic & Zheltov, 1955). However, both the length of the fracture and the stress field have to be known beforehand in order to calculate the width. In addition to that, Khristianovich and Zheltov model focused solely on the fracture mechanics overlooking the fluid flow behavior, and they did not consider any material balance. In 1957, Carter accounted for the material balance in the growth of fractures, and he started the pioneering work that led to the famous Carter fluid loss equation and the Carter second equation, which formed the basis of the leakoff equation (Howard & Fast, 1957). The fluid loss equation, which is also called Carter loss, is shown in equation (6.35). The Carter second equation (i.e., Carter equation II) describes the fracture surface area with the consideration of a simple fluid loss model. The Carter equation II, which is also called Carter solution II, is shown in equations (6.36) and (6.37) (Valkó & Economides, 1995). For a simple 1D model, the fracture length can be estimated by the use of Carter equation II as shown in equations (6.38) – (6.42) provided that the width and height of the fracture are known in advance (Smith & Montgomery, 2015).

$$q_{loss} = \frac{2C_t}{\sqrt{t-\tau}}\, dA(\tau), \tag{6.35}$$

where q_{loss} is the rate of fluid loss, C_t is the leakoff coefficient, t is pumping time, dA is the incremental fracture area, and τ is the time at which the fluid first exposed to fracture area.

$$A(t) = \frac{q_i(w + 2S_p)}{4C_t^2\pi}\left[\exp\left(\beta^2\right)erfc(\beta) + \frac{2\beta}{\sqrt{\pi}} - 1\right], \tag{6.36}$$

where $A(t)$ is the fracture surface area at time t, q_i is the injection rate, w is the fracture width, S_p is the spurt-loss coefficient, C_t is leakoff coefficient, and *erfc* is the complementary error function.

$$\beta = \frac{2C_t\sqrt{\pi t}}{w + 2S_p} \tag{6.37}$$

and

$$L = \frac{q_i\, t}{4C_t\, H\sqrt{t} + 2HS_p + wH}, \tag{6.38}$$

where L is the length of the fracture and H is the height of the fracture.

Example 6.12 Find the Fracture Surface Area using Carter Equation II

Calculate the fracture surface area after 41 min of injection knowing that the spurt-loss coefficient is 0, the leakoff coefficient is 0.0023 ft/min$^{0.5}$, the injection rate is 120 bpm, and the average fracture width is 0.23 in.

Solution

Before using Carter equation II, the units should be converted to SI units as follows:

$$w = 0.23\ in. = 0.006\ m$$

and

$$t = 41\ \text{min} = 2,460\ s,\ q_i = 120\ bpm = 0.318\frac{m^3}{s},\ \text{and}\ C_t = 0.0023\frac{ft}{\sqrt{\text{min}}} = 9.05 \times 10^{-5}\frac{m}{\sqrt{s}}.$$

The next step is to calculate β as follows:

$$\beta = \frac{2C_t\sqrt{\pi t}}{w + 2S_p} = \frac{2 \times 9.05 \times 10^{-5} \times \sqrt{\pi \times 2,460}}{0.006} = 2.724.$$

Hence, the exp. function of (β) and the complementary error function of β are as follows:

$$\exp\left(\beta^2\right) = 1.668 \times 10^3$$

and

$$erfc(\beta) = 1.171 \times 10^{-4}.$$

Thus, the fracture surface area after 41 min of injection can be calculated by the use of equation (6.36) as follows:

$$A(t) = \frac{q_i\left(w + 2S_p\right)}{4C_t^2\pi}\left[\exp\left(\beta^2\right)erfc(\beta) + \frac{2\beta}{\sqrt{\pi}} - 1\right]$$

and

$$A(t) = \frac{0.318 \times 0.006}{4 \times \left(9.05 \times 10^{-5}\right)^2 \times \pi}\left[1.668 \times 10^3 \times 1.171 \times 10^{-4} + \frac{2 \times 2.724}{\sqrt{\pi}} - 1\right]$$

$$= 4.095 \times 10^4 \left((m)^2\right).$$

PKN Model

In the 1960s, Perkins and Kern started the development of their 2D model, which is referred to as the PK model. The PK model is a simple analytical model that is capable of predicting the fracture length and width assuming a constant fracture height. However, the PK model does not consider fluid loss (Perkins & Kern, 1961). So, in the 1970s, Nordgren adopted the PK model and considered fluid loss in the model (Nordgren, 1972). Actually, Nordgren derived a new set of equations that has to be solved numerically instead of analytically. The new model is a famous model nowadays, and it is called the PKN model. A general form of the PKN model, which is also called the PKN-C model, where the C stands for Carter's leakoff model, is shown in equations (6.39)–(6.41) (Valkó & Economides, 1995). The PKN model is valid when the fracture length is expected to be greater than the fracture height, and it is the most used 2D model in the industry because it has been found that the PKN model predicts fracture length that is closer to the 3D models (Economides & Nolte, 2000; Rahman & Rahman, 2010).

$$x_f = 2.05\frac{\left(\frac{q_i\,\mu\,x_f}{\grave{E}}\right)^{\frac{1}{4}} q_i}{4C_t^2\pi h_f}\left[\exp\left(\beta^2\right)erfc(\beta) + \frac{2\beta}{\sqrt{\pi}} - 1\right], \tag{6.39}$$

where x_f is the fracture half-length, h_f is the fracture height, C_t is leakoff coefficient, q_i is the injection rate, μ is the fluid viscosity, $\grave{E}$ is the plane strain modulus, and *erfc* is the complementary error function.

$$\beta = \frac{2C_t\sqrt{\pi t}}{2.05\left(\frac{q_i\,\mu\,x_f}{\grave{E}}\right)^{\frac{1}{4}}}, \tag{6.40}$$

where x_f is the fracture half-length, h_f is the fracture height, C_t is leakoff coefficient, q_i is the injection rate, μ is the fluid viscosity, and *erfc* is the complementary error function.

$$\overline{w} = 2.05\left(\frac{q_i\,\mu\,x_f}{\grave{E}}\right)^{\frac{1}{4}}, \tag{6.41}$$

where $\overline{w}$ is the fracture average width.

A simplified analytical version of the PKN model can be achieved for the two extreme cases; without leakoff and with highest leakoff. Also, these analytical equations are considered near the wellbore and that means the fracture width would be the maximum fracture width instead of an average fracture width. In the case of no leakoff and for field units, equations (6.42) and (6.43) can be used to estimate the fracture length and width, respectively. In the case of highest leakoff and for field units, equations (6.44) and (6.45) are used to estimate the fracture length and width, respectively (Guo et al., 2017).

$$x_f = 58.07\left[\frac{Eq_i^3}{(1-\nu^2)\mu h_f^4}\right]^{\frac{1}{5}} t^{\frac{4}{5}}, \quad (6.42)$$

where E is Young's modulus, ν is Poisson's ratio, μ is the fluid viscosity, and t is the pumping time.

$$w_{max} = 0.99\left[\frac{(1-\nu^2)\mu q_i^2}{Eh_f}\right]^{\frac{1}{5}} t^{\frac{1}{5}}, \quad (6.43)$$

where w_{max} is the width near the wellbore and t is the injection time.

$$x_f = 0.89\frac{q_i t^{\frac{1}{2}}}{C_t h_f}, \quad (6.44)$$

where C_t is leakoff coefficient,

$$w_{max} = 0.34\left[\frac{(1-\nu^2)\mu q_i^2}{EC_t h_f}\right]^{\frac{1}{4}} t^{\frac{1}{8}}. \quad (6.45)$$

Example 6.13 Estimate the Fracture Length and Width

Estimate the fracture length and width after 41 min of injection using the PKN model, without leakoff and with highest leakoff, knowing that the fracture height is 300 ft, Young's modulus is 6.26 MMpsi, and Poisson's ratio is 0.26. The viscosity of the fracturing fluid is 420 cP, the leakoff coefficient is 0.0023 ft/min$^{0.5}$, and the injection rate is 120 bpm.

Solution

In the case with no leakoff, equations (6.42) and (6.43) are used to estimate the fracture length and width, respectively.

$$x_f = 58.07\left[\frac{Eq_i^3}{(1-\nu^2)\mu h_f^4}\right]^{\frac{1}{5}} t^{\frac{4}{5}} = 58.07\left[\frac{6.26\times10^6\times120^3}{(1-0.26^2)\times420\times300^4}\right]^{\frac{1}{5}} 41^{\frac{4}{5}} = 1,448\,ft$$

and

$$w_{max} = 0.99\left[\frac{(1-0.26^2)\times420\times120^2}{6.26\times10^6\times300}\right]^{\frac{1}{5}} 41^{\frac{1}{5}} = 0.651\ in..$$

In the case with highest leakoff, equations (6.44) and (6.45) are used to estimate the fracture length and width, respectively.

$$x_f = 0.89\frac{q_i t^{\frac{1}{2}}}{C_t h_f} = 0.89\frac{120\times41^{\frac{1}{2}}}{0.0023\times300} = 991\,ft$$

and

$$w_{max} = 0.34\left[\frac{(1-0.26^2)\mu q_i^2}{6.26\times10^6\times0.0023\times300}\right]^{\frac{1}{4}} 41^{\frac{1}{8}} = 0.578\ in..$$

Hence, when the leakoff is considered, the fracture length decreased more than 31%, and the maximum fracture width decreased more than 11% compared with the case without any leakoff.

KGD Model

In the late 1960s, Geertsma and De Klerk studied the Khristianovich and Zheltov model, and they further developed it to achieve a 2D model (Geertsma & De Klerk, 1969). Their model is usually referred to as the KGD model, also called the GDK model. The KGD model is a well-known 2D model that is applicable only when the height of the fracture is expected to be greater than the length of the fracture (Economides & Nolte, 2000). Geertsma and De Klerk provided analytical solutions to the fracture length and width for the no leakoff case, and they are shown in equations (6.46) and (6.47) (Guo et al., 2017).

$$x_f = 27.81\left[\frac{Eq_i^3}{(1-\nu^2)\mu h_f^3}\right]^{\frac{1}{6}} t^{\frac{2}{3}} \tag{6.46}$$

and

$$w_{max} = 1.54\left[\frac{(1-\nu^2)\mu q_i^3}{Eh_f^3}\right]^{\frac{1}{6}} t^{\frac{1}{3}}. \tag{6.47}$$

Geertsma and De Klerk did not provide a simplified explicit solution to the fracture length in the case of highest leakoff. However, equation (6.44) is applicable for both the PKN and KGD models. The fracture width, for the KGD model with highest leakoff and in flied units, can be estimated by the use of equation (6.48),

$$w_{max} = 0.29\left[\frac{(1-\nu^2)\mu q_i x_f^2}{Eh_f}\right]^{\frac{1}{4}}. \tag{6.48}$$

Example 6.14 Estimate the Fracture Length and Width

Estimate the fracture length and width after 41 min of injection using the KGD model, without leakoff and with highest leakoff, knowing that the fracture height is 465 ft, Young's modulus is 6.26 MMpsi, and Poisson's ratio is 0.26. The viscosity of the fracturing fluid is 420 cP, the leakoff coefficient is 0.0023 ft/min$^{0.5}$, and the injection rate is 45 bpm.

Solution

In the case with no leakoff, equations (6.46) and (6.47) are used to estimate the fracture length and width, respectively.

$$x_f = 27.81\left[\frac{Eq_i^3}{(1-\nu^2)\mu h_f^3}\right]^{\frac{1}{6}} t^{\frac{2}{3}} = 27.81 \times \left[\frac{6.26 \times 10^6 \times 45^3}{(1-0.26^2) \times 420 \times 465^3}\right]^{\frac{1}{6}} 41^{\frac{2}{3}} = 516\,ft$$

and

$$w_{max} = 1.54\left[\frac{(1-\nu^2)\mu q_i^3}{Eh_f^3}\right]^{\frac{1}{6}} t^{\frac{1}{3}} = 1.54\left[\frac{(1-0.26^2) \times 420 \times 45^3}{6.26 \times 10^6 \times 465^3}\right]^{\frac{1}{6}} 41^{\frac{1}{3}} = 0.329\,in..$$

In the case with highest leakoff, equations (6.44), (6.45), and (6.48) are used to estimate the fracture length and width, respectively.

$$x_f = 0.89\frac{q_i t^{\frac{1}{2}}}{C_t h_f} = 0.89\frac{45 \times 41^{\frac{1}{2}}}{0.0023 \times 465} = 240\,ft$$

and

$$w_{max} = 0.29\left[\frac{(1-0.26^2) \times 420 \times 45 \times 240^2}{6.26 \times 10^6 \times 465}\right]^{\frac{1}{4}} = 0.223\,in.$$

Clearly, the KGD model gives shorter fracture length compared with the fracture height when the leakoff is considered.

Both the PKN and KGD models are analytical or semi-analytical limited 2D models because they assume a constant fracture height and a 1D fluid flow equation. They are also derived based on several other assumptions (i.e., linear elastic, homogenous, and isotropic rocks). That said, the PKN and KGD models are still being used today for simple hydraulic fracturing designs (Adachi et al., 2007).

Advanced Models

Due to the limitations of the 2D models, and the need of understanding how the fractures actually propagate, new models have emerged by relaxing the assumption that were made in the PKN and KGD models. These new models are classified as pseudo-3D models (P3D), planar 3D models (PL3D), and fully 3D models.

Pseudo-3D Models

The pseudo three-dimensional models (P3D) are called pseudo 3D because they are not actually 3D models but rather 2D models with modified height instead of a fixed height. In general, the P3D models are classified into two types: cell-based and lumped models. The cell-based P3D models are a series of PKN models with different height along the length of the fracture (Fig. 6.31). The lumped P3D models, which were introduced by Cleary in the 1980, are two half-ellipses that are joined together (Cleary, 1980). The two half-ellipses differ in their vertical extent, and they also change with time (Fig. 6.32). Basically, the lumped P3D models assume a fracture shape and lump several parameters together that would reduce the partial differential equations (PDEs) to ordinary differential equations (ODEs) in time. In order for this model to work well, the lumped parameters have to be selected properly. The cell-based and lumped models also differ in how the fluid flow inside the fracture is handled. For example, the vertical flow is not considered in the cell-based models, whereas a partial vertical flow is considered in the lumped models.

In general, the P3D models are used to capture the fracture geometry with limited computational power. In fact, many software

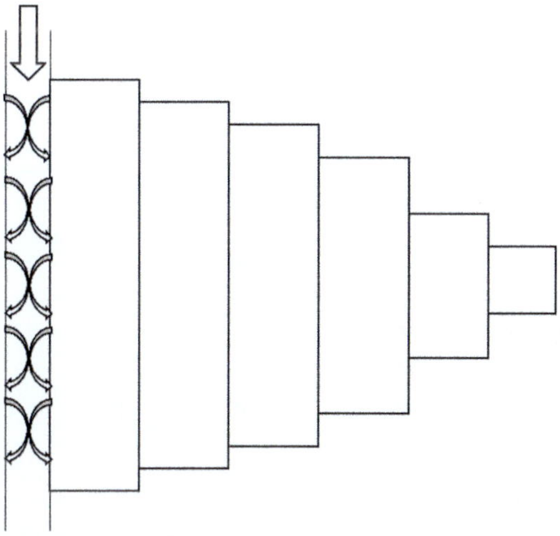

Figure 6.31 Cell-based P3D model.

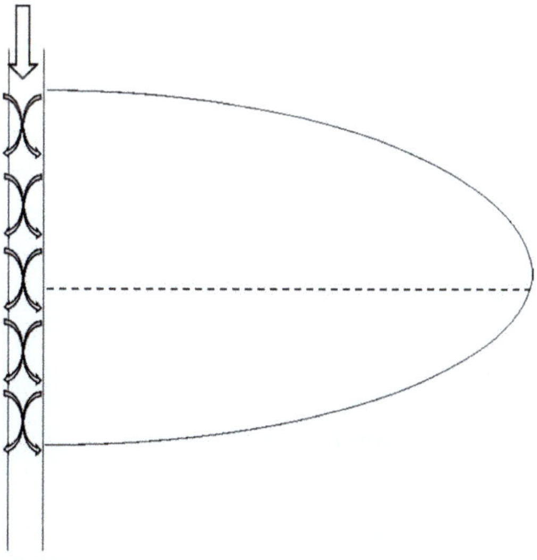

Figure 6.32 Lumped P3D models.

packages (e.g., Stimplan) include sophisticated P3D models, which consider 2D fluid flow as well as proppant transport, and these models are heavily used for preliminary fracture and pump schedule designs. However, it is important to note that the sophisticated P3D models would not yield accurate designs for complex hydraulic fractures.

Planar 3D Models

The planar three-dimensional (PL3D) models, which are used in the industry today, were first developed in the late 1970s (Clifton & Abou-Sayed, 1979; Clifton & Abou-Sayed, 1981). The PL3D models are somehow computationally expansive. However, they have to be used in certain hydraulic fracturing designs where the P3D models fail to work (e.g., when the fracture grows outside the main zone). The PL3D model utilizes the 3D elasticity equations and fully couples and solves the deformation equations and the 2D fluid flow equations, equations (6.49) and (6.50) (Gidley et al., 1990; Economides & Nolte, 2000; Adachi et al., 2007). Hence, the fracture geometry, with time, is determined at each grid point or element based on the pressure distribution inside and around the fracture. Solving the coupled nonlinear PDEs of the PL3D models requires the use of sophisticated numerical methods schemes (e.g., finite element method); thus, the PL3D models are numerically intensive. The PL3D models yield relatively accurate fracture geometry; however, they still assume a planar fracture that is oriented perpendicular to the least principles stress, and the propagation occurs in one plane. There are several software packages in the market that are capable of simulating hydraulic fractures using PL3D models (Gidley et al., 1990).

$$p(x,y,t) - \sigma_c(x,y) = \int_{\Omega(t)} C(x,y,\xi,\eta)\, w(\xi,\eta,t)\, d\xi d\eta \tag{6.49}$$

and

$$\nabla(\rho w \bar{u}) + \frac{\partial}{\partial t}(\rho w) + 2\rho u_L = 0. \tag{6.50}$$

Example 6.15 Design a PL3D Hydraulic Fracture

A geological model for a well that is targeting the Bakken play was built. Based on the designed geological model as well as the well and reservoir data, design a simple PL3D hydraulic fracture using an advanced hydraulic fracturing simulator. The hydraulic fracturing treatment is going to be conducted as a hybrid job, and the pumping schedule is provided in Table 6.9. Summarize the designed parameters.

Table 6.9 Pumping Schedule for Example 6.15

Stage	Proppant type	Proppant concentration (ppg)	Fluid type	Fluid volume (gal)	Pump time (min)	Pumprate (bpm)
1	Pad-stage	0	Slickwater	400	0.40	45
2	Ottawa sand 40-70	3	Slickwater	100	0.13	45
3	Carbo EconoProp 20-40	4	#40 X-Link HPG 140°F	8,500	4.50	45
4	Carbo EconoProp 20-40	5	#40 X-Link HPG 140°F	40,000	21.30	45

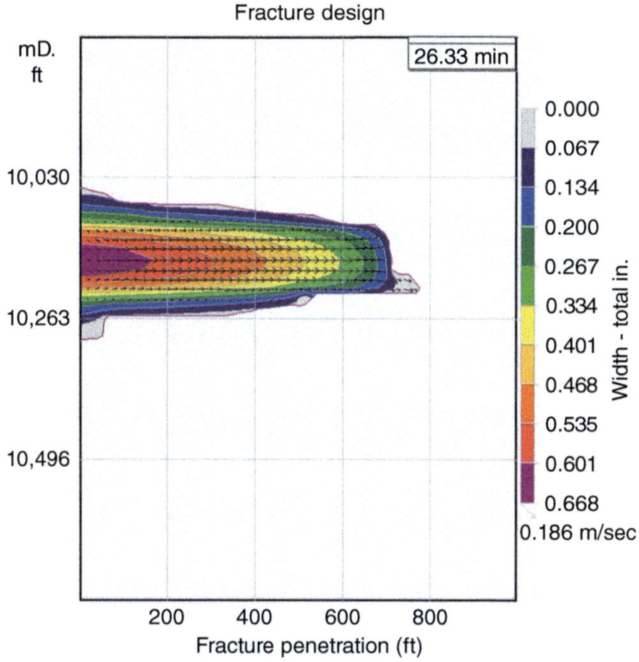

Figure 6.33 Fracture total width distribution (plot generated using E-Stimplan).

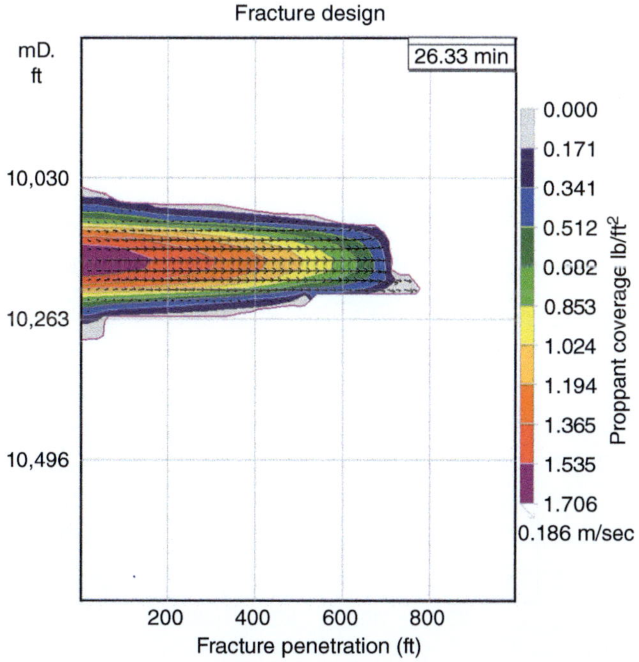

Figure 6.34 Proppant coverage distribution (plot generated using E-Stimplan).

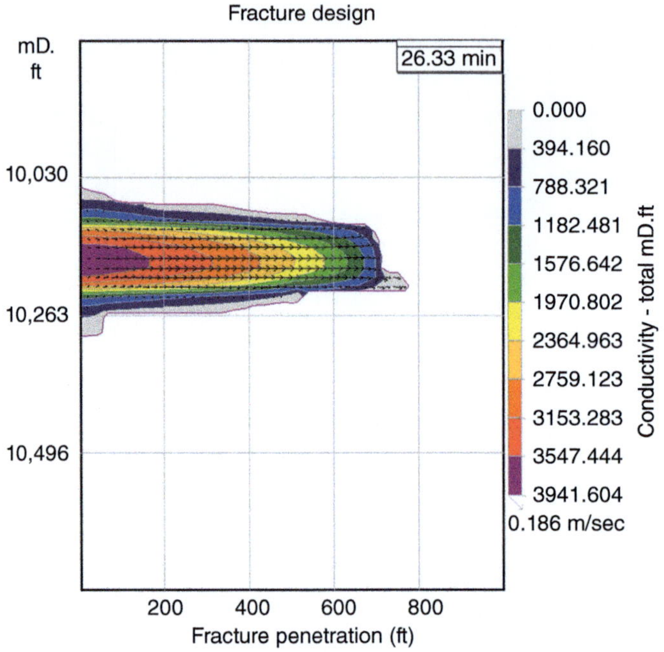

Figure 6.35 Conductivity distribution (plot generated using E-Stimplan).

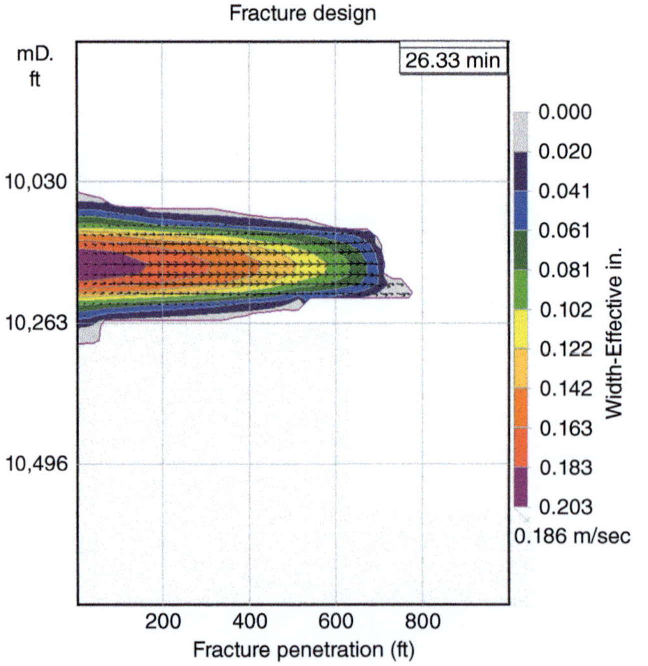

Figure 6.36 Effective width distribution (plot generated using E-Stimplan).

Solution

An advanced software package (E-Stimplan) was used to design a single-stage hydraulic fracture based on the provided pumping schedule. Figures 6.33–6.36 show contour plots for total width, proppant coverage, fracture conductivity, and effective propped width, respectively. The final results are summarized in Table 6.10.

Table 6.10 Summarized Results for Example 6.15

Parameter	Value
Half-length (ft)	776
Propped length (ft)	741
Height (ft)	250
Maximum width (inch)	0.668
Average width (inch)	0.33
Productive surface area (ft^2)	236,719
Average conductivity (mD-ft)	1,834

Fully 3D Models

The fully 3D models, which make no assumption about the fracture orientation, are extremely intensive numerically and computationally, even with the advanced computational power that is available today. For example, simulating the propagation process of a fracture for only 2 in. could take over a month using powerful computer systems (Rahman & Rahman, 2010). These models are very complex because the fractures are allowed to propagate vertically and laterally as well as changing orientation planes based on the local pressure distribution, the rock properties, and the presence of the natural fractures, and so on. In general, the fully 3D models are mostly used in research laboratories and academia

Unconventional Models

Since hydraulic fracturing has become a standard practice in the unconventional reservoirs, the industry has been looking for ways to improve the practice and to maximize the recovery. The complexity of the unconventional reservoirs (e.g., Nano-Darcy permeability and heterogeneity) has shown that the existing hydraulic fracturing models are inadequate and do not yield accurate results when applied in the unconventional reservoirs. New models have been developed that take into account the new discovered effects on the hydraulic fracturing jobs (e.g., natural fractures, fracture interference, etc.). The new models, which are called the unconventional fracture models (UFM), are used to design a network of fractures instead of a single planar fracture. The new models are also referred to as discrete fracture networks (DFNs) models. These new models are capable of modeling the fractures and their propagation, the opening of the natural fractures and their effects on the hydraulic fractures, as well as the effects of the hydraulic fractures themselves on nearby hydraulic fractures (Weng et al., 2011; Profit et al., 2015; McClure et al., 2016). There are also new models that can be used to determine the optimum well spacing (Pankaj et al., 2018). Moreover, there are advanced unconventional models that are currently being developed that design and simulate the SRV instead of the fracture's geometry. The hydraulic fractures are then designed in a way that accounts for the simulated SRV (Alfi et al., 2018).

6.6 Problems

1 Find the required pump power for the hydraulic fracturing treatment knowing that the fracturing gradient and the friction gradient are 0.8 psi/ft and 0.2 psi/ft, respectively. The true vertical depth and the depth to the top of the perforations are 9,735 ft and 15,800 ft, respectively. The specific gravity of the fracturing fluid is 1 and the blender rate is 35 bpm.

2 For the planned hydraulic fracturing treatment, estimate the fracture maximum width, with highest leakoff and no leakoff,

using the PKN and KGD models and assuming the fracture height is 275 ft. Discuss the differences between the two models.

3 Estimate the surface area for the PKN and KGD models with highest leakoff and no leakoff for the designed hydraulic fractures in Problem 2 using the data provided in Table 6.11. Assume the maximum width is the average width and the spurt loss is zero.

Table 6.11 Design Parameters for Problem 2

Design parameters	
Young's modulus (Psi)	4×10^6
Poisson's ratio	0.23
Injection time (minutes)	45
Injection rate (bpm)	80
Fracturing fluid viscosity (cP)	300
Leakoff coefficient $\left(\frac{ft}{\sqrt{min}}\right)$	0.0035

4 A well is drilled in an unconventional reservoir with a permeability of 0.1 mD. The wellbore radius is 0.3 ft and the drainage area is 160 acres. A hydraulic fracture was generated with a permeability of 200 Darcy. The fracture half-length and average width are 1,210 ft and 0.21 in., respectively. Estimate the folds of increase of the well.

5 Estimate the minimum required compressive strength of the proppant. The reservoir data are provided in Table 6.12.

Table 6.12 Reservoir Data for Problem 5

Reservoir data	
Depth (ft)	9,878
Pressure (psi)	6,210
Poisson's ratio	0.24
Biot's poroelastic constant	0.74
Overburden density (lb_m/ft^3)	162
Drawdown production pressure (psi)	3,200

Additional Reading

Valkó, P., & Economides, M. (1995). *Hydraulic fracture mechanics.* Baffins Lane: John Wiley & Sons.

Smith, M., & Montgomery, C. (2015). *Hydraulic fracturing (emerging trends and technologies in petroleum engineering).* Boca Raton: CRC Press Taylor & Francis Group.

Guo, B., Liu, X., & Tan, X. (2017). *Petroleum production engineering* (2nd ed.). Cambridge: Elsevier.

References

Adachi, J., Siebrits, E., Peirce, A., & Desroches, J. (2007). Computer simulation of hydraulic fractures. *International Journal of Rock Mechanics and Mining Sciences.* doi:10.1016/j.ijrmms.2006.11.006

Ajisafe, F., Solovyeva, I., Morales, A., Ejofodomi, E., & Porcu, M. (2017). *Impact of well spacing and interference on production performance in unconventional reservoirs, Permian basin.* SPE/AAPG/SEG Unconventional Resources Technology Conference, Austin, Unconventional Resources Technology Conference. doi:10.15530/URTEC-2017-2690466

Alfi, M., Chai, Z., Pradhan, A., Ramsay, T., Barrufet, M., & Killough, J. (2018). *Time-lapse petro-elastic and seismic modeling to evaluate fracturing efficiency in low-permeability reservoirs.* SPE/AAPG/SEG Unconventional Resources Technology Conference, Houston, Society of Petroleum Engineers. doi:URTEC-2857198-MS

Alhubail, M. (2016). *Acid transport modeling using finite element discretization with weak formulation for simulation of acid fracturing.* Lawrence: The University of Kansas.

Alhubail, M. M., Misra, A., & Barati, R. (2017). *A novel acid transport model with robust finite element discretization.* SPE Kingdom of Saudi Arabia Annual Technical Symposium and Exhibition, Dammam, Society of Petroleum Engineers. doi:10.2118/188030-MS

Allen, T., & Roberts, A. (1989). *Production operations: Well completions, workover and stimulation* (Vol. *2*). Tulsa: Oil & Gas Consultants International.

Assem, A., Nasr-El-Din, H., Fuss, T., Shi, J., & Herskovits, R. (2017). Mud-acid interactions with sand and clay-based ceramic proppants used in gravel-packed and fractured wells. SPE Production & Operations, *32*. doi:10.2118/175833-PA

Assured Testing Services (2018). *Proppant and sand testing*. Retrieved from Assured Testing Services: www.proppantsandtesting.com

Atteberry, R., Tucker, R., & Ritz, J. (1979). *Application of sintered bauxite proppants to stimulation of low permeability south Texas gas reservoirs*. Symposium on Low Permeability Gas Reservoirs, Denver, Society of Petroleum Engineers. doi:10.2118/7924-MS

Badiali, M. (2014). *Two new drilling techniques that will shatter US oil expectations*. Retrieved from *dailyreckoning*: https://dailyreckoning.com/2-new-drilling-techniques-that-will-shatter-us-oil-expectations/

Barati, R. (2010). *Dissertation: Fracturing fluid cleanup by controlled release of enzymes from polyelectrolyte complex nanoparticles*. Lawrence: The University of Kansas.

Barati, R., & Liang, J.-T. (2014). A review of fracturing fluid systems used for hydraulic fracturing of oil and gas wells. *Journal of Applied Polymer Science, 131*, 40735. doi: 10.1002/app.40735

Barati, R., Hutchins, R., Friedel, T., Ayoub, J., Dessinges, M.-N., & England, K. (2009). Fracture impact of yield stress and fracture-face damage on production with a three-phase 2D model. *SPE Production & Operations*. doi:10.2118/111457-PA

Barree, R., Barree, V., & Craig, D. (2009). Holistic fracture diagnostics: Consistent interpretation of prefrac injection tests using multiple analysis methods. *SPE Production & Operations*. doi:10.2118/107877-PA

Barree, R., Miskimins, J., & Gilbert, J. (2015). Diagnostic fracture injection tests: Common mistakes, misfires, and misdiagnoses. *SPE Production & Operations*. doi:10.2118/169539-PA

Bestaoui-Spurr, N., & Li, C. (2018). *Ultra-light weight proppants for better placement*. SPE Annual Technical Conference and Exhibition, Dallas, Society of Petroleum Engineers. doi:10.2118/191585-MS

Bommer, P., & Bayne, M. (2018). *Active well defense in the Bakken: Case study of a ten-well frac defense project, McKenzie County, ND*. SPE Hydraulic Fracturing Technology Conference and Exhibition, The Woodlands, Society of Petroleum Engineers. doi:10.2118/189860-MS

Bommer, P., Bayne, M., Mayerhofer, M., Machovoe, M., & Staron, M. (2017). *Re-designing from scratch and defending offset wells: Case study of a six-well Bakken Zipper Project, McKenzie County, ND*. SPE Hydraulic Fracturing Technology Conference and Exhibition, The Woodlands, Society of Petroleum Engineers. doi:10.2118/184851-MS

Carbo Ceramics. (2017). *Build, enhance and maintain more space to flow*. CARBO.

Cinco L., H., Samaniego V., F., & Dominguez A., N. (1978). Transient pressure behavior for a well with a finite-conductivity vertical fracture. *Society of Petroleum Engineers Journal, 18*(4). doi:10.2118/6014-PA

Cinco-Ley, H., & Samaniego-V, F. (1981). Transient pressure analysis for fractured wells. *Journal of Petroleum Technology, 33*(09). doi:10.2118/7490-PA

Cleary, M. (1980). *Comprehensive design formulae for hydraulic fracturing*. SPE Annual Technical Conference and Exhibition, Dallas, Society of Petroleum Engineers. doi: 10.2118/9259-MS

Clifton, R., & Abou-Sayed, A. (1979). *On the computation of the three-dimensional geometry of hydraulic fractures*. Symposium on Low Permeability Gas Reservoirs, Denver, Society of Petroleum Engineers. doi:10.2118/7943-MS

Clifton, R., & Abou-Sayed, A. (1981). *A variational approach to the prediction of the three-dimensional geometry of hydraulic fractures*. SPE/DOE Low Permeability Gas Reservoirs Symposium, Denver, Society of Petroleum Engineers. doi:10.2118/9879-MS

Damjanac, B., Maxwell, S., Pirayehgar, A., & Torres, M. (2018). *Numerical study of stress shadowing effect on fracture initiation and interaction between perforation clusters*. SPE/AAPG/SEG Unconventional Resources Technology Conference, Houston, Unconventional Resources Technology Conference. doi:10.15530/URTEC-2018-2901800

Daneshy, A. (2018). *Analysis of horizontal well fracture interactions, and completion steps for reducing the resulting production interference*. SPE Annual Technical Conference and Exhibition, Dallas, Society of Petroleum Engineers. doi:10.2118/191671-MS

Davies, D., & Kuiper, T. (1988). Fracture conductivity in hydraulic fracture stimulation. *Journal of Petroleum Technology*. doi:10.2118/17655-PA

Donaldson, E., Alam, W., & Begum, N. (2013). *Hydraulic fracturing explained: evaluation, implementation and challenges*. Houston: Gulf Publishing Company.

Economides, M., & Martin, T. (2007). *Modern fracturing enhancing natural gas production*. Houston: BJ Services Company.

Economides, M., Hill, A., & Ehlig-Economides, C. (1994). *Petroleum production systems*. Prentice Hall.

Economides, M. J., & Nolte, K. G. (2000). *Reservoir stimulation* (3rd ed.). John Wiley & Sons Ltd.

EFT (2015). *What are frac hits?* Retrieved from Eagle Ford Training: https://www.eagleford.training/2015/08/what-are-frac-hits/

Fekete (2012). *Fracture flow capacity*. Retrieved from Fekete: http://www.fekete.com/SAN/TheoryAndEquations/WellTestTheoryEquations/Fracture_Flow_Capacity.htm

Fekete (2012). *Minifrac tests*. Retrieved from Fekete: www.fekete.com

Fisher, M., Heinze, J., Harris, C., Davidson, B., Wright, C., & Dunn, K. (2004). *Optimizing horizontal completion techniques in the barnett shale using microseismic fracture mapping*. SPE Annual Technical Conference and Exhibition, Houston, Society of Petroleum Engineers. doi:10.2118/90051-MS

Fisher, M., Wright, C., Davidson, B., Steinsberger, N., Buckler, W., Goodwin, A., & Fielder, E. (2005). Integrating fracture mapping technologies to improve stimulations in the Barnett shale. *SPE Production & Facilities*. doi:10.2118/77441-PA

Gala, D., Manchanda, R., & Sharma, M. (2018). *Modeling of fluid injection in depleted parent wells to minimize damage due to frac-hits*. SPE/AAPG/SEG Unconventional Resources Technology Conference, Houston, Unconventional Resources Technology Conference. doi:10.15530/URTEC-2018-2881265

Geertsma, J., & De Klerk, F. (1969). A rapid method of predicting width and extent of hydraulic induced fractures. *Journal of Petroleum Technology*. doi:10.2118/2458-PA

Gidley, J., Holditch, S., Nierode, D., & Veatch, R. Jr (1990). *Recent advances in hydraulic fracturing*. Society of Petroleum Engineers.

Gilleland, K. (2011). *Hydraulic fracturing: Game-changing advances in stimulation and production technology are improving well economics*. Hart Energy.

Guo, B., Liu, X., & Tan, X. (2017). *Petroleum production engineering* (2nd ed.). Cambridge: Elsevier.

Gupta, D., & Valkó, P. (2007). Fracturing fluids and formation damage. In M. Economides, & T. Martin, *Modern fracturing enhancing natural gas production*. BJ Services Company.

Halliburton. (2012). *Environmentally friendly diverting agents part of optimized approach to plug and perf fracturing method*. Retrieved from *Drillingcontractor*: http://www.drillingcontractor.org/self-removing-efdas-level-stimulation-access-14457

Helms, L. (2008). Horizontal drilling. *North Dakota Geological Survey Newsletter, 35*(1). Retrieved from www.dmr.nd.gov

Houston, M., McCallister, M., Jany, J., & Audet, J. (2010). *Next generation multistage completion technology and risk sharing accelerates development of the Bakken play*. SPE Annual Technical Conference and Exhibition. Florence: Society of Petroleum Engineers. doi:10.2118/135584-MS

Howard, G., & Fast, C. (1957). Optimum fluid characteristics for fracture extension. *Drilling and Production Practice*. New York: American Petroleum Institute.

IGU (2012). *Shale gas, the facts about the environmental concerns*. Kuala Lumpur: International Gas Union. Retrieved from http://www.igu.org/

Jacobs, T. (2014). The shale evolution: Zipper fracture takes hold. *Journal of Petroleum Technology*. doi:10.2118/1014-0060-JPT

Jacobs, T. (2017). Oil and gas producers find frac hits in shale wells a major challenge. Retrieved from *JPT:* https://spe.org/en/print-article/?art=2819

Jacobs, T. (2018a). Fighting water with water: How engineers are turning the tides on frac hits. Retrieved from *Journal of Petroleum Technology*: https://www.spe.org/en/jpt/jpt-article-detail/?art=4817

Jacobs, T. (2018b). Frac hits to take center stage at SPE Hydraulic Fracturing Conference. Retrieved from *Journal of Petroleum Technology*: https://www.spe.org/en/jpt/jpt-article-detail/?art=4774

Jennings, A. (1996). Fracturing fluids: Then and now. *Journal of Petroleum Technology, 48*(7). doi:10.2118/36166-JPT

Karazincir, O., Li, Y., Zaki, K., Williams, W., Tan, Y., Wu, R., et al. (2018). *Measurement of reduced permeability at fracture face due to proppant embedment and depletion*. SPE Annual Technical Conference and Exhibition, Dallas, Society of Petroleum Engineers. doi:10.2118/191653-MS

KGS (2001). *The Hugoton project*. Retrieved from *Kansas Geological Survey*: www.kgs.ku.edu/Hugoton/background.html

Khristianovic, S., & Zheltov, Y. (1955). *Formation of vertical fractures by means of highly viscous liquid*. 4th World Petroleum Congress, Rome, World Petroleum Congress.

Kim, G. (2012). *Chinese proppant: Supply and demand in China & North America*. AM2F Energy.

King, H. (2017). *Hydraulic fracturing of oil and gas wells drilled in shale*. Retrieved from *Geology*: www.geology.com/articles/hydraulic-fracturing/

Krumbein, W., & Sloss, L. (1963). *Stratigraphy and sedimentation*. W. H. Freeman & Co. doi: ISBN-13: 978-0716702191

Kumar, A., Seth, P., Shrivastava, K., Manchanda, R., & Sharma, M. (2018). *Well interference diagnosis through integrated analysis of tracer and pressure interference tests*. SPE/AAPG/SEG Unconventional Resources Technology Conference, Houston, Unconventional Resources Technology Conference. doi: URTEC-2901827-MS

Lindsay, G., Miller, G., Xu, T., Shan, D., & Baihly, J. (2018). *Production performance of infill horizontal wells vs. pre-existing wells in the major us unconventional basins s*. SPE Hydraulic Fracturing Technology Conference and Exhibition, The Woodlands, Society of Petroleum Engineers. doi:10.2118/189875-MS

Martin, T., & Rylance, M. (2007). Technologies for mature assets. In M. Economides, & T. Martin, *Modern fracturing enhancing natural gas production*. BJ Services Company.

Martin, T., Kotov, S., Nelson, S., & Cramer, D. (2016). Stimulation of unconventional reservoirs. In U. Ahmed, & N. Meehan, *Unconventional oil and gas resources: exploitation and development*. Taylor & Francis Group. doi:ISBN: 13: 978-1-4987-5941-0

Mata, D., Zhou, W., Zee Ma, Y., & Gonzales, V. (2016). Hydraulic fracture treatment, optimization, and production modeling. In Y. Zee Ma, & S. Holditch, *Unconventional oil and gas resources handbook evaluation and development*. Waltham: Elsevier. doi:ISBN: 978-0-12-802238-2

Maxwell, S. (2011). What does microseismic tell us about hydraulic fracture deformation. *Recorder, 36*(8).

Maxwell, S. (2014). *Microseismic imaging of hydraulic fracturing: Improved engineering of unconventional shale reservoirs*. Tulsa: Society of Exploration Geophysicists. doi:ISBN: 978-1-56080-315-7

Mayerhofer, M., Lolon, E., Warpinski, N., Cipolla, C., Walser, D., & Rightmire, C. (2010). What is stimulated reservoir volume? *SPE Production and Operations*. doi:10.2118/119890-PA

McClure, M., Babazadeh, M., Shiozawa, S., & Huang, J. (2016). Fully Coupled hydromechanical simulation of hydraulic fracturing in 3D discrete-fracture networks. *SPE Journal*. doi:10.2118/173354-PA

Miller, G., Lindsay, G., Baihly, J., & Xu, T. (2016). *Parent well refracturing: Economic safety nets in an uneconomic market*. SPE Low Perm Symposium, Denver, Society of Petroleum Engineers. doi:10.2118/180200-MS

Mittal, A., Rai, C., & Sondergeld, C. (2018). *Experimental proppant conductivity observations: evaluating impact of proppant size and fluid chemistry on long-term production in shales*. SPE Annual Technical Conference and Exhibition, Dallas, Society of Petroleum Engineers. doi:10.2118/191741-MS

Montgomery, C., & Smith, M. (2010). Hydraulic fracturing: History of an enduring technology. *Journal of Petroleum Technology, 62*(12). doi:10.2118/1210-0026-JPT

Nagel, N., & Sanchez-Nagel, M. (2011). *Stress shadowing and microseismic events: A numerical evaluation*. SPE Annual Technical Conference and Exhibition. Denver: Society of Petroleum Engineers. doi:10.2118/147363-MS

Nagel, N., Zhang, F., Sanchez-Nagel, M., Lee, B., & Agharazi, A. (2013). *Stress shadow evaluations for completion design in unconventional plays*. SPE Unconventional Resources Conference Canada. Calgary: Society of Petroleum Engineers. doi:10.2118/167128-MS

911 Metallurgist Lab. (2017). *How to do a sieve analysis*. Retrieved from 911metallurgist: https://www.911metallurgist.com/

Nolte, K. (1986). Determination of proppant and fluid schedules from fracturing-pressure decline. *SPE Production Engineering*. doi:10.2118/13278-PA

Nolte, K. (1988). Principles for fracture design based on pressure analysis. *SPE Production Engineering*. doi:10.2118/10911-PA

Nordgren, R. (1972). Propagation of a vertical hydraulic fracture. *Society of Petroleum Engineers Journal*. doi:10.2118/3009-PA

OilFieldGeomechanics (2015). *Stress shadows explained: What it is, what it isn't, and why you should care*. Richmond: OilField Geomechanics LLC.

Olmen, B., Anschutz, D., Brannon, H., & Stribling, K. (2018). *Evolving proppant supply and demand: the implications on the hydraulic fracturing industry*. SPE Annual Technical Conference and Exhibition, Dallas, Society of Petroleum Engineers. doi:10.2118/191591-MS

Palisch, T., Duenckel, R., Chapman, M., Woolfolk, S., & Vincent, M. (2010). How to use and misuse proppant crush tests: Exposing the top 10 myths. *SPE Production & Operations, 25*. doi:10.2118/119242-PA

Pankaj, P. (2018). Decoding *positives or negatives of fracture-hits: A geomechanical investigation of fracture-hits and its implications for well productivity and integrity*. SPE/AAPG/SEG Unconventional Resources Technology Conference, Houston, Unconventional Resources Technology Conference. doi:10.15530/URTEC-2018-2876100

Pankaj, P., Shukla, P., Kavousi, P., & Carr, T. (2018). *Hydraulic fracture and reservoir characterization for determining optimal well spacing in the Marcellus shale*. SPE Liquids-Rich Basins Conference, North America, Midland, Society of Petroleum Engineers. doi:10.2118/191802-MS

Pearson, C. (2001). Dimensionless fracture conductivity: Better input values make better wells. *Journal of Petroleum Technology*. doi:10.2118/60184-JPT

Pearson, C., & Brannon, H. (2007). Proppants and fracture conductivity. In M. Economides & T. Martin, *Modern fracturing enhancing natural gas production*. Houston: BJ Services Company. doi:ISBN: 9781604616880

Pearson, C. M. (2013). *Hydraulic fracturing of horizontal wells: Realizing the paradigm shift that has been 30 years in development*. SPE Distinguished Lecturer Presentation.

Perkins, T., & Kern, L. (1961). Widths of hydraulic fractures. *Journal of Petroleum Technology*. doi:10.2118/89-PA

Phan, T., Cronk, B., Almasoodi, M., & Reza, Z. (2018). Lithologic and geomechanical control on CO_2 huff-n-puff enhanced oil recovery processes using integrated modeling framework in Wolfcamp. SPE/AAPG/SEG Unconventional Resources Technology Conference, Houston, Unconventional Resources Technology Conference.

Prats, M. (1961). Effect of vertical fractures on reservoir behavior-incompressible fluid case. *Society of petroleum Engineers Journal, 1*(02). doi:10.2118/1575-G

Profit, M., Dutko, M., & Yu, J. (2015). *Developing a framework to simulate the hydraulic fracturing of tight gas reservoirs based on integrative adaptive remeshing and combined finite/discrete element approach*. 49th U.S. Rock Mechanics/

Geomechanics Symposium, American Rock Mechanics Association.

Rae, P., & Di Lullo, G. (1996). *Fracturing fluids and breaker systems: A review of the state-of-the-art.* SPE Eastern Regional Meeting, Columbus, Society of Petroleum Engineers. doi:10.2118/37359-MS

Rafiee, M., Soliman, M., & Pirayesh, E. (2012). *Hydraulic fracturing design and optimization: A modification to zipper frac.* SPE Annual Technical Conference and Exhibition, San Antonio, Society of Petroleum Engineers. doi:10.2118/159786-MS

Rahman, M. M., & Rahman, M. K. (2010). A review of hydraulic fracture models and development of an improved pseudo-3d model for stimulating tight oil/gas sand. *Energy sources, part A: Recovery, utilization, and environmental effects.* doi:10.1080/15567030903060523

Rainbolt, M., & Esco, J. (2018). *Frac hit induced production losses: evaluating root causes, damage location, possible prevention methods and success of remediation treatments, Part II.* SPE Hydraulic Fracturing Technology Conference and Exhibition, The Woodlands, Society of Petroleum Engineers. doi:10.2118/189853-MS

Rassenfoss, S. (2018). *Fracturing wells in pairs creates strong links.* Retrieved from *Journal of Petroleum Technology*: https://www.spe.org/en/jpt/jpt-article-detail/?art=4675

Reisz, M. (1992). *Reservoir evaluation of horizontal Bakken well performance on the southwestern flank of the Williston basin.* International Meeting on Petroleum Engineering, Beijing, Society of Petroleum Engineers. doi:10.2118/22389-MS

Rizwan, Y., & Liu, G. (2018). *Fracture closure determination from two successive dfits (diagnostic fracture injection test) in one formation.* SPE/AAPG/SEG Unconventional Resources Technology Conference, Houston, Unconventional Resources Technology Conference.

Roach, E. (2014). *Proppant: The greatest oilfield innovation of the 21st century.* Retrieved from *Drillinginfo*: https://info.drillinginfo.com/proppant-the-greatest-oilfield-innovation/

Roberts, S., Binder, M., & Lane, R. (1990). *Strength, volume and weight loss of gravels and proppants due to HF-based acids.* SPE Formation Damage Control Symposium, Society of Petroleum Engineers. doi:10.2118/20168-MS

Sardinha, C., Petr, C., Lehmann, J., Pyecroft, J., & Merkle, S. (2014). *Determining interwell connectivity and reservoir complexity through frac pressure hits and production interference analysis.* SPE/CSUR Unconventional Resources Conference, Canada, Calgary, Society of Petroleum Engineers. doi:10.2118/171628-MS

Sharma, M., Gadde, P., Sullivan, R., Sigal, R., Fielder, R., Copeland, D., et al.(2004). *Slick Water and hybrid fracs in the bossier: Some lessons learnt.* SPE Annual Technical Conference and Exhibition, Houston, Society of Petroleum Engineers. doi:10.2118/89876-MS

Siriwardane, H., Gondle, R., & Bromhal, G. (2016). *Extent of hydraulic fractures in shales.* National Energy Technology Laboratory, Department of Energy.

Skomorowski, N., Dussealut, M., & Gracie, R. (2015). *The use of multistage hydraulic fracture data to identify stress shadow effects.* 49th U.S. Rock Mechanics/Geomechanics Symposium, American Rock Mechanics Association.

Smith, M., & Montgomery, C. (2015). *Hydraulic fracturing (emerging trends and technologies in petroleum engineering).* Boca Raton: CRC Press Taylor & Francis Group. doi:ISBN-13: 978-1466566859

Sneddon, I., & Elliott, H. (1946). The opening of a Griffith crack under internal pressure. *Quarterly of Applied Mathematics.* Retrieved from www.jstor.org/sTable/43633558

Sookprasong, P. (2010). *Equation for closure stress on proppant in the fracture is incorrect and has been used for decades.* SPE Eastern Regional Meeting, Morgantown, Society of Petroleum Engineers. doi:10.2118/136431-MS

Srinivasan, K., Ajisafe, F., Alimahomed, F., Panjaitan, M., Makarychev-Mikhailov, S., & Mackay, B. (2018). *Is there anything called too much proppant?* SPE Liquids-Rich Basins Conference, North America, Midland, Society of Petroleum Engineers. doi:10.2118/191800-MS

Tayong, R., Alhubail, M., Maulianda, B., & Barati, R. (2019). Simulating yield stress variation along hydraulic fracture face enhances polymer cleanup modeling in tight gas reservoirs.

Journal of Natural Gas Science and Engineering. doi:10.1016/j.jngse.2019.02.006

Treida, M. (2012). *Hydraulic fracturing and its effect on frac pump technology*. Retrieved from *Pumps & Systems*: https://www.pumpsandsystems.com/

Triepke, J. (2014). *Well completion 101 part 3: Well stimulation*. Retrieved from *Drillinginfo*: https://info.drillinginfo.com/well-completion-well-stimulation/

U.S. EPA (2016). *Review of well operator files for hydraulically fractured oil and gas production wells: Hydraulic fracturing operations*. U.S. Environmental Protection Agency Office of Research and Development. doi: EPA/601/R-14/004

Valkó, P., & Economides, M. (1995). *Hydraulic fracture mechanics*. Baffins Lane: John Wiley & Sons Ltd.

Vassilellis, G., Cipolla, C., Erdle, J., & Ahmed, U. (2016). Production Evaluation and Forecasting. In U. Ahmed, & N. Meehan, *Unconventional oil and gas resources: exploitation and development*. CRC Press Taylor & Francis Group.

Vidma, K., Abivin, P., Dunaeva, A., Panga, M., Nikolaev, M., Usoltsev, D., & Mayo, J. (2018). *Far-field diversion technology to prevent fracture hits in tightly spaced horizontal wells*. SPE Annual Technical Conference and Exhibition. Dallas: Society of Petroleum Engineers. doi:10.2118/191722-MS

Voneiff, G., Robinson, B., & Holditch, S. (1996). The effects of unbroken fracture fluid on gas well performance. *SPE Production & Facilities*. doi:10.2118/26664-PA

Walton, I., & McLennan, J. (2013). The role of natural fractures in shale gas production. In R. Jeffrey, *Effective and sustainable hydraulic fracturing*. Intechopen. doi:10.5772/45724

Ward, V. (1986). N_2 and CO_2 in the oil field: Stimulation and completion applications. *SPE Production Engineering*. doi:10.2118/12594-PA

Warpinski, N. (2009). Microseismic monitoring: Inside and out. *Journal of Petroleum Technology, 61*(11). doi:10.2118/118537-JPT

Warpinski, N., Moschovidis, Z., Parker, C., & Abou-Sayed, I. (1994). Comparison study of hydraulic fracturing models, test case: GRI staged field experiment no. 3. *SPE Production & Facilities*. doi:10.2118/25890-PA

Weng, X., Kresse, O., Cohen, C., Wu, R., & Gu, H. (2011). Modeling of hydraulic-fracture-network propagation in a naturally fractured formation. *SPE Production and Operations*. doi:10.2118/140253-PA

Whitfield, T., Watkins, M., & Dickinson, L. (2018). *Pre-loads: Successful mitigation of damaging frac hits in the Eagle Ford*. SPE Annual Technical Conference and Exhibition, Dallas, Society of Petroleum Engineers. doi:10.2118/191712-MS

Wile, R. (2012). *This is the technological breakthrough that's making people wildly bullish on America*. Retrieved from *Business Insider*: http://www.businessinsider.com/this-technological-innovation-is-why-people-are-wildly-bullish-on-america-2012-4

Worlow, D. (1987). *Development and testing of a standard procedure for determining the viscous properties of cross-linked fracture fluids*. College Station: MS Thesis, Texas A&M University.

Wu, Y., Zhao, X., Zinno, R., Wu, H., Vaidya, V., Yang, M., & Qin, J. (2016). The application of microseismic monitoring in unconventional reservoirs. In Y. Ma, & S. Holditch, *Unconventional oil and gas resources handbook evaluation and development*. Waltham: Gulf Professional Publishing (an imprint of Elsevier). doi:ISBN: 978-0-12-802238-2

Xu, T., Lindsay, G., Zheng, W., Yan, Q., Escobar Patron, K., Alimahomed, F., et al. (2018). *Advanced modeling of production induced pressure depletion and well spacing impact on infill wells in Spraberry, Permian basin*. SPE Annual Technical Conference and Exhibition. Dallas: Society of Petroleum Engineers. doi:10.2118/191696-MS

Zhao, H., Danican, S., Torres, H., Christanti, Y., Nikolaev, M., Makarychev-Mikhailov, S., & Bonnell, A. (2018). *Viscous slickwater as enabler for improved hydraulic fracturing design in unconventional reservoirs*. SPE Annual Technical Conference and Exhibition. Dallas: Society of Petroleum Engineers. doi:10.2118/191520-MS

7

Phase Behavior of Shale Oil and Gas

Xiaoli Li, Jyun-Syung Tsau, Qinwen Fu, and Reza Barati Ghahfarokhi

The University of Kansas, Lawrence, Kansas, USA

7.1 Background

Hydrocarbons produced from shale formations have been changing the oil and gas industry markets considerably. The development of such unconventional shale oil and gas is more complex and expensive compared to conventional petroleum resources. In addition to the techniques of hydraulic fracturing and horizontal drilling, which can create the increasing flow paths for oil and gas, the injection of various gases, such as CO_2, methane, ethane, and natural gas, has also been considered as a feasible enhanced oil recovery (EOR) method for the shale resources. However, the mechanisms of gas-assisted EOR processes in shale formations may be different from that in conventional plays. Therefore, it is fundamental and critical to understand and characterize the properties of shale fluids.

Section 7.2 introduces the compositional analyses of petroleum fluids in shale reservoirs. Then, the volumetric and phase behavior together with their experimental measurements are presented in section 7.3, where the quality check of the phase behavior data are presented. Sections 7.4 and 7.5 provide the equation of state (EOS) modeling and the relating pressure, volume, and temperature (PVT), which are widely used to quantify the phase behavior and properties of petroleum fluids. Finally, the methods of determining the minimum miscible pressure (MMP) for shale oil are introduced in section 7.6. Section 7.7 presents a comprehensive example for the readers to practice the teachings of this chapter as a hands-on project based on a Bakken oil and gas data set.

7.2 Compositional Analyses of Shale Fluids

Petroleum fluids are mixtures containing hundreds or thousands of chemical compounds with a wide range of molecular weights. A majority of them are hydrocarbons with elements of carbon and hydrogen. A few nonhydrocarbons can also be found, for example nitrogen (N_2), helium (He), carbon dioxide (CO_2), and hydrogen sulfide (H_2S), which are impurities in petroleum fluids. Shale gas is mainly composed of lighter alkanes while the shale oil consists of heavier and more complex components. It should be noted that, compared with the conventional reservoir fluids, the shale oil is relatively lighter and less complex. The composition of shale gas and oil collected from different development areas can vary significantly.

Kerogen is an insoluble macromolecular solid organic matter in sedimentary rocks and

Unconventional Hydrocarbon Resources: Techniques for Reservoir Engineering Analysis, Advanced Textbook Series, First Edition. Edited by Reza Barati Ghahfarokhi and Mustafa M. Alhubail.

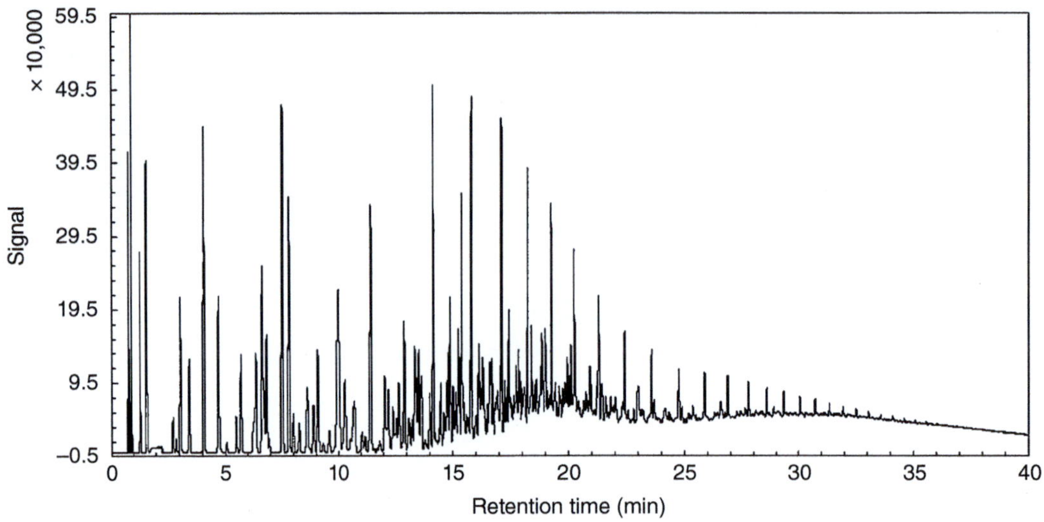

Figure 7.1 GC peaks of a stock tank Bakken oil sample. Oil composition is measured by Core Lab.

is a complex mixture of organic chemical compounds, which plays an important role in shale reservoirs (Hu, 2014). The kerogen can be transformed to hydrocarbon mixtures such as oil and natural gas in a relatively fixed pore space. Due to a large specific surface area, kerogen becomes the major location for the adsorbed and dissolved hydrocarbon fluids. It is reported that the sorbed methane in kerogen makes up nearly 50% of the total storage in Barnett shale samples (Kang et al., 2011). Thermal maturity of source rock measures the degree to which a formation has been exposed to high heat needed to break down the organic matter to generate hydrocarbons (Yenugu & Han, 2013). It is shown that the shales with higher maturity are more brittle and thus are good candidates for hydraulic fracturing (Yenugu & Han, 2013).

The most widely used analytical techniques for compositional analyses are true boiling point (TBP) distillation and gas chromatography (GC). The TBD distillation measures mass quantities, specific gravity, molecular weight, and boiling point of each distillation cut (Whitson & Brule, 2000). It is relatively time-consuming and requires a large amount of fluid sample. The GC technique is a reliable method for reservoir fluids from shale formations, where a small amount of sample can be injected into the GC column. The compounds with various molecular structure and boiling point transport through the column at various rates. Each component can be quantified by analyzing the peak in the GC chromatogram. Figure 7.1 presents the example GC peaks for a Bakken oil collected from a stock tank. The composition of the stock tank Bakken oil is listed in Table 7.1.

7.2.1 Subsurface Sampling

The sampling method impacts the composition of reservoir fluids significantly. The oil and gas can be collected from both subsurface and surface. Subsurface sampling method requires putting a sampling device down the well to collect the fluid at the reservoir pressure and temperature. However, it should be noted that sampling in-situ fluids from liquid-rich shale wells is very difficult. Usually, the producing oil-gas ratio of liquid-rich shale wells drops rapidly as the well is drawn down to the flowing bottom-hole pressure at the heel. Whitson and Sunjerga (2012) recommended that sampling as early as possible with low drawdowns should provide the fluid sample's composition

Table 7.1 Compositional Analysis of a Downhole Bakken Fluid Sample

Component			Atmospheric gas (vapor)	Atmospheric oil (liquid)		Recombined reservoir fluid
Symbol	Name	Molecular weight	Mole%	Mole %	Weight %	Mole%
N_2	Nitrogen	28.01	2.142	0.000	0.000	1.509
CO_2	Carbon dioxide	44.01	0.551	0.000	0.000	0.388
C_1	Methane	16.04	49.964	0.013	0.001	35.198
C_2	Ethane	30.07	20.504	0.143	0.024	14.485
C_3	Propane	44.10	12.960	0.669	0.163	9.327
iC_4	i-Butane	58.12	1.603	0.274	0.088	1.210
nC_4	n-Butane	58.12	5.736	1.676	0.539	4.536
iC_5	i-Pentane	72.15	1.300	1.071	0.427	1.232
nC_5	n-Pentane	72.15	2.122	2.944	1.174	2.365
C_6	Hexanes	86.18	1.543	5.810	2.769	2.804
C_7	Heptanes	95.27	1.057	10.116	5.345	3.735
C_8	Octanes	109.36	0.396	11.756	7.122	3.754
C_9	Nonanes	122.91	0.093	8.696	5.911	2.636
C_{10}	Decanes	134.07	0.029	7.968	5.907	2.376
C_{11}	Undecanes	147.00		6.159	5.006	1.821
C_{12}	Dodecanes	161.00		5.137	4.574	1.519
C_{13}	Tridecanes	175.00		4.811	4.656	1.422
C_{14}	Tetradecanes	190.00		3.979	4.181	1.176
C_{15}	Pentadecanes	206.00		3.365	3.833	0.995
C_{16}	Hexadecanes	222.00		2.926	3.592	0.865
C_{17}	Heptadecanes	237.00		2.317	3.037	0.685
C_{18}	Octadecanes	251.00		2.162	3.001	0.639
C_{19}	Nonadecanes	263.00		1.926	2.801	0.569
C_{20}	Eicosanes	275.00		1.753	2.665	0.518
C_{21}	Heneicosanes	291.00		1.538	2.474	0.455
C_{22}	Docosanes	305.00		1.190	2.007	0.352
C_{23}	Triacosanes	318.00		1.156	2.033	0.342
C_{24}	Tetracosanes	331.00		1.026	1.878	0.303
C_{25}	Pentacosanes	345.00		0.898	1.713	0.265
C_{26}	Hexacosanes	359.00		0.807	1.602	0.239
C_{27}	Heptacosanes	374.00		0.796	1.647	0.235
C_{28}	Octacosanes	388.00		0.660	1.415	0.195
C_{29}	Nonacosanes	402.00		0.592	1.315	0.175
C_{30+}	Triacontanes plus	545.69		5.666	17.100	1.675
Molecu-lar weight			30.38	180.84		74.85

Note: Sampling method: subsurface sampling; GOR: 1,422 scf/STB; Reservoir condition: 6,840 psi and 241°F; Flash from 5,000 psia and 241°F to atmospheric pressure and 80°F.

closer to in-situ reservoir fluid. However, that period may be affected by the three-phase cleanup of hydraulic fractures. The subsurface sampling method allows one to directly collect the sample with the most accurate fluid composition in the sampler. The sampler is then prepared for flash in a laboratory and each split phase (liquid and gas) is injected into GC for compositional analysis. Figure 7.2 is a schematic for analyzing a downhole fluid sample.

A case study is demonstrating the compositions of a shale fluid and introducing how to understand a compositional analysis report. A downhole oil sample collected from a Bakken shale play at 5,000 psia and 241°F has been flashed at the atmospheric pressure and 80°F. The atmospheric gas phase, determined by GPA 2286 method using a multicolumn gas chromatograph and the atmospheric oil phase, is determined by temperature programmed capillary chromatography. The compositions of each flashed phase and the recombined downhole fluid are tabulated in Table 7.1. The GOR is measured to be 1,422 scf/STB and the average molecular weight is measured to be 183.68.

The following four types of information can be obtained from the compositional analysis report: components, molecular weight of components, molar and weight compositions of the flashed gas and liquid, and recombined reservoir fluid.

1) Components
The first column (Table 7.1) lists the symbol and name of components that can be measured in this fluid sample. Two impurities of N_2 and CO_2 are found in the Bakken oil sample. The hydrocarbons with the carbon number from C_1 to C_{29} are present, where the iso and normal C_4 and C_5 can be distinguished. C_{30+} is the plus fraction consisting of the heavy components, which cannot be separated into the individual components. The plus fraction can be characterized up to C_{42+} or even higher depending on the capability of the CG equipment.
2) Molecular weight of components
The next column (Table 7.1) shows the molecular weight of each component. N_2, CO_2, and C_1 through C_6 are well-defined components with molecular weights of 28.01, 44.01, 16.04, 30.07, 44.10, 58.12, 72.15, and 86.18. The carbon number heavier than C_7 usually cannot be identified quantitatively by the GC analysis. Starting from C_7, the carbon number fraction within a given temperature interval is assigned to a certain n-paraffin by the boiling point. The diversity of hydrocarbon components in each fraction should be considered when assigning the molecular weight (Pedersen & Christensen, 2006). As a result, the molecular weights for the carbon numbers higher than C_6 are not the

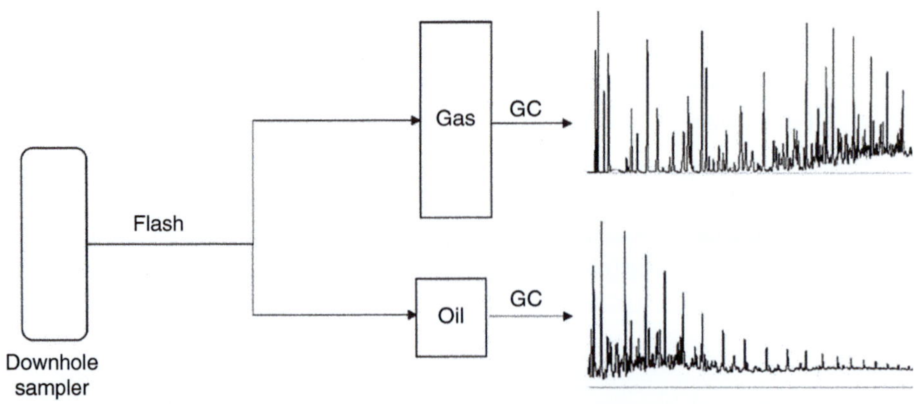

Figure 7.2 Downhole fluid sample analysis.

same as their corresponding normal paraffin. For example, the fraction of C_{13} (tridecanes) has a molecular weight of 175, instead of 184, which is for $C_{13}H_{28}$. Gas Processors Association (GPA) has issued a standard GPA-2145-16-Table of Physical Properties of Hydrocarbons and Other Compounds (G. P. A., 1990). The molecular weights for C_7 through C_{10} in Table 7.1 are from GPA standard. The molecular weights of heavier fractions adopt the data from Katz and Firoozabadi (1978), which are widely accepted in the petroleum industry. The molecular weight of the plus fraction C_{30+} ($M_{C_{30+}}$) is calculated using the formula

$$M_{C_{30+}} = \frac{M_{oil} w_{C_{30+}}}{1 - M_{oil} \sum_{i=1}^{n-1} \frac{w_i}{M_i}}, \tag{7.1}$$

where M_{oil} is the molecular weight of the oil, M_i and w_i are the molecular weight and weight fraction of component i, and n is the number of components.

3) Molar and weight compositions of the flashed gas and liquid
The third column in Table 7.1 presents the composition of the flashed gas phase at atmospheric pressure and 80°F. It can be seen that the gas is mainly composed of C_1, C_2, and C_3 accounting for 49.964, 20.504, and 12.960 mol%, respectively. The heaviest hydrocarbon found in the gas phase is C_{10}. As for the flashed liquid phase, both molar and weight compositions are given. In fact, the compositions measured using GC or TBP techniques are directly given in weight percentage. However, molar compositions are usually required in PVT computations. Thus, the weight percentage (w_i) has to be converted to a molar percentage (Z_i) by using the formula

$$z_i = \frac{w_i / M_i}{\sum_{i=1}^{n} w_i / M_i}. \tag{7.2}$$

The following equation (7.3) is used to convert molar percentage to weight percentage:

$$w_i = \frac{z_i M_i}{\sum_{i=1}^{n} z_i M_i}. \tag{7.3}$$

There are no N_2 and CO_2 in the flashed oil as indicated in Table 7.1. C_7 and C_8 are the main components. The carbon numbers heavier than C_{30} are grouped into C_{30+} accounting for 5.666 mol%.

4) Recombined reservoir fluid
The last column (Table 7.1) shows the molar composition of the recombined reservoir fluid, which is the true composition of the downhole sampled fluid. The reservoir fluid molar composition (z_i) can be calculated from equation (7.4),

$$z_i = \frac{n_{gas} y_i + n_{Liquid} x_i}{n_{total}}, \tag{7.4}$$

where n_{gas}, n_{Liquid}, and n_{total} are the mole numbers of the flashed gas, flashed liquid, and recombined fluid, respectively; y_i and x_i are the molar fractions of gas and liquid phase, respectively. The last row provides the molecular weights of the flashed gas, flashed liquid, and the recombined fluid.

7.2.2 Surface Sampling

Alternatively, surface sampling method can be used to collect oil and gas from the surface separator or stock tank at low pressure and temperature. The entire sampling process can be conducted at the surface, which is convenient and easy. It is necessary to recombine the collected oil and gas in the laboratory in order to reproduce the reservoir fluid composition at the reservoir condition. The gas-oil ratio (GOR) is important in recombination calculation, which is defined as the volume of gas from the separator measured at standard condition divided by volume of oil at separator conditions.

The procedure of analyzing surface samples is illustrated in Figure 7.3. The separator pressure is usually higher than atmospheric pressure at surface temperature. The separator gas is collected and then flashed at the standard

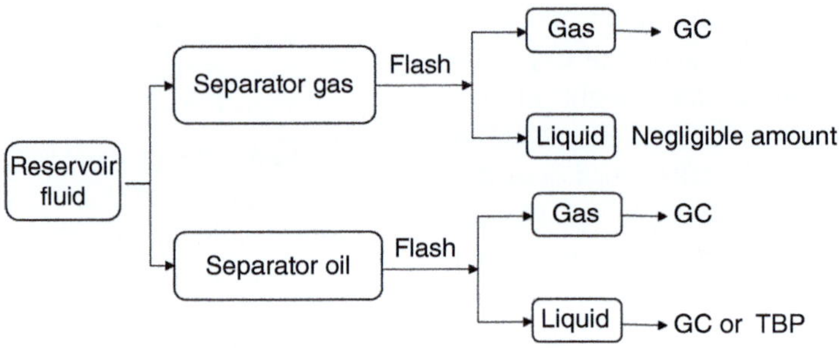

Figure 7.3 Separator fluid sample analysis.

condition in the laboratory. The flashed gas is dominant while the flashed liquid can be negligible. The flashed gas is used in performing the compositional analysis and the volume of the flashed gas should be recorded. The volume of separator oil should be recorded and also flashed at the standard condition in the laboratory, resulting in a flashed gas phase and a flashed oil phase. The volume and the compositional analysis of each phase are measured. The following procedure is used to recombine the reservoir fluid composition.

1) The composition of the separator oil $x_i^{sep,oil}$,

$$x_i^{sep,oil} = \frac{n_{gas}^{flashed} y_i^{flashed} + n_{liquid}^{flashed} x_i^{flashed}}{n_{oil}^{Sep}}, \tag{7.5}$$

$$n_{oil}^{sep} = n_{gas}^{flashed} + n_{liquid}^{flashed}, \tag{7.6}$$

$$n_{gas}^{flashed} = \frac{v_{gas}^{flashed}}{v_M^{flashed}}, \tag{7.7}$$

and

$$n_{liquid}^{flashed} = \frac{v_{oil}^{flashed} \times \rho_{oil}^{flashed}}{M_{oil}^{flashed}}, \tag{7.8}$$

where $n_{gas}^{flashed}$ is mole number of the flashed gas from the separator oil, $n_{liquid}^{flashed}$ is mole number of the flashed liquid from the separator oil, n_{oil}^{Sep} is mole number of the separator oil, $y_i^{flashed}$ is mole fraction of the component i in the flashed gas, $x_i^{flashed}$ is mole fraction of component i in the flashed oil, $v_{gas}^{flashed}$ is volume of the flashed gas from the separator oil, $v_M^{flashed}$ is molar volume of flashed gas from the separator oil, $v_{oil}^{flashed}$ is volume of the flashed oil from the separator oil, $\rho_{oil}^{flashed}$ is density of the flashed oil from the separator oil, and $M_{oil}^{flashed}$ is molecular weight of the flashed oil from the separator oil.

2) Mole number of the separator gas n_{gas}^{Sep},

$$n_{gas}^{Sep} = n_{gas}^{flashed} = \frac{v_{gas}^{flashed}}{v_M^{flashed}} \tag{7.9}$$

and

$$v_{gas}^{flashed} = GOR^{sep} \cdot V_{oil}^{sep}. \tag{7.10}$$

Since the flashed liquid phase from separator gas can be an extremely small amount, the flashed gas composition analyzed by GC can represent the separator gas composition ($y_i^{sep,gas}$). It is noticed that the expression of equation (7.9) is the same as equation (7.7). The difference is that the flashed gas properties are from the separator gas in equation (7.9), that is, $v_{gas}^{flashed}$ is volume of the flashed gas from the separator gas, and $v_M^{flashed}$ is molar volume of flashed gas from the separator gas. GOR^{sep} is the gas-oil ratio where the volume of separator gas is measured at a standard condition and the volume of separator oil is measured at separator condition.

3) Composition of the recombined reservoir fluid, z_i, is calculated using equation (7.11):

$$z_i = \frac{n_{oil}^{Sep} x_i^{sep,oil} + n_{gas}^{Sep} y_i^{sep,gas}}{n_{gas}^{Sep} + n_{oil}^{Sep}}. \quad (7.11)$$

A surface sampling method has also been conducted for the Bakken shale fluid with the reservoir pressure of 6,840 psia and reservoir temperature of 241°F. The separator condition is 143 psia and 43°F. The gas-oil ratio (GOR) is 1,000 SCF/bbl. The separator gas is flashed at the standard condition of 14.7 psi and 60°F and its compositional analysis report can be seen in Table 7.2. The separator liquid is also collected and flashed into an oil phase and a gas phase at atmospheric pressure and 60°F. The gas-oil ratio is measured to be 186 SCF/STB. The compositions of each flashed phase and the recombined separator liquid are listed in Table 7.3. The flashed gas consists mainly of light hydrocarbons of C_1 through nC_4 with an averaged gas molecular weight of 39.78. Heavier hydrocarbons can be found in the flashed oil. The fraction C_{30+} still accounts for 4.746 mol%. The flashed oil has a molecular weight of 162.30. Equation (7.5) is used to combine the flashed gas $\left(y_i^{flashed}\right)$ and liquid $\left(x_i^{flashed}\right)$ into the composition of separator liquid $\left(x_i^{sep,oil}\right)$. Consequently, the compositions of separator gas and separator liquid are combined to reproduce the reservoir fluids by using equation (7.11). The results are presented in Table 7.4. Compared with the subsurface sampling method providing the compositional analysis in Table 7.1, the surface sampling method yields similar compositional results to that presented in Table 7.4. Both tables indicate that methane, ethane, and propane account for the majority of hydrocarbons in Bakken fluid and the impurities of nitrogen and carbon dioxide are present in the reservoir.

Table 7.2 Composition of Separator Gas ($y_i^{sep,gas}$)

Component		Molecular weight	Mole%	Weight%
N_2	Nitrogen	28.013	2.987	3.593
CO_2	Carbon dioxide	44.010	0.682	1.288
H_2S	Hydrogen sulfide	34.082	0.000	0.000
C_1	Methane	16.043	63.481	43.725
C_2	Ethane	30.070	21.647	27.948
C_3	Propane	44.097	8.357	15.821
iC_4	Iso-Butane	58.123	0.598	1.493
nC_4	N-Butane	58.123	1.625	4.056
iC_5	Iso-Pentane	72.150	0.192	0.594
nC_5	N-Pentane	72.150	0.253	0.783
C_6	Hexanes	86.177	0.100	0.372
C_7	Heptanes	92.931	0.055	0.218
C_8	Octanes	105.744	0.017	0.076
C_9	Nonanes	122.170	0.004	0.019
C_{10+}	Decanes plus	159.586	0.002	0.014
Total			100	100
Molecular weight				23.29

Table 7.3 Compositions of the Flashed Gas and Liquid From Separator Liquid

Component			Atmospheric gas, $y_i^{flahsed}$	Atmospheric liquid, $x_i^{flahsed}$		Separator liquid, $x_i^{sep,oil}$	
Symbol	Name	Molecular weight	Mole%	Mole%	Weight%	Mole%	Weight%
N_2	Nitrogen	28.01	0.173	0.000	0.000	0.039	0.008
CO_2	Carbon dioxide	44.01	0.385	0.000	0.000	0.086	0.028
C_1	Methane	16.04	14.364	0.009	0.001	3.205	0.381
C_2	Ethane	30.07	30.487	0.318	0.059	7.036	1.567
C_3	Propane	44.10	33.022	2.532	0.688	9.321	3.044
iC_4	i-Butane	58.12	3.818	1.002	0.359	1.629	0.701
nC_4	n-Butane	58.12	12.440	5.558	1.991	7.090	3.052
iC_5	i-Pentane	72.15	1.679	2.350	1.045	2.201	1.176
nC_5	n-Pentane	72.15	2.232	5.591	2.485	4.843	2.588
C_6	Hexanes	86.18	0.903	7.337	3.896	5.904	3.769
C_7	Heptanes	95.12	0.406	10.211	5.985	8.028	5.656
C_8	Octanes	109.51	0.079	10.442	7.046	8.135	6.598
C_9	Nonanes	122.89	0.007	7.361	5.575	5.724	5.210
C_{10}	Decanes	134.01	0.005	6.642	5.485	5.164	5.126
C_{11}	Undecanes	147.00		5.120	4.637	3.980	4.333
C_{12}	Dodecanes	161.00		4.285	4.250	3.331	3.972
C_{13}	Tridecanes	175.00		4.066	4.384	3.161	4.097
C_{14}	Tetradecanes	190.00		3.260	3.817	2.534	3.566
C_{15}	Pentadecanes	206.00		2.778	3.526	2.159	3.295
C_{16}	Hexadecanes	222.00		2.408	3.293	1.872	3.078
C_{17}	Heptadecanes	237.00		1.950	2.847	1.516	2.661
C_{18}	Octadecanes	251.00		1.794	2.774	1.395	2.593
C_{19}	Nonadecanes	263.00		1.619	2.624	1.259	2.451
C_{20}	Eicosanes	275.00		1.435	2.431	1.115	2.272
C_{21}	Heneicosanes	291.00		1.288	2.310	1.001	2.158
C_{22}	Docosanes	305.00		0.991	1.862	0.770	1.740
C_{23}	Triacosanes	318.00		0.950	1.861	0.738	1.739
C_{24}	Tetracosanes	331.00		0.855	1.744	0.665	1.629
C_{25}	Pentacosanes	345.00		0.767	1.630	0.596	1.524
C_{26}	Hexacosanes	359.00		0.659	1.457	0.512	1.362
C_{27}	Heptacosanes	374.00		0.663	1.528	0.515	1.428
C_{28}	Octacosanes	388.00		0.553	1.323	0.430	1.235
C_{29}	Nonacosanes	402.00		0.460	1.140	0.358	1.065
C_{30+}	Triacontanes plus	545.25		4.746	15.947	3.689	14.899
Total	-	-	100.000	100.000	100.000	100.000	100.000
Molecular weight			39.78	162.30		135.01	

Table 7.4 Recombined Composition of the Reservoir Fluid

Component			Separator gas, $y_i^{sep,gas}$	Separator liquid, $x_i^{sep,oil}$	Reservoir fluid, z_i	
Symbol	Name	Molecular weight	Mole %	Mole%	Mole %	Weight %
N_2	Nitrogen	28.01	2.987	0.039	1.726	0.680
CO_2	Carbon dioxide	44.01	0.682	0.086	0.427	0.264
C_1	Methane	16.04	63.481	3.205	37.694	8.507
C_2	Ethane	30.07	21.647	7.036	15.396	6.512
C_3	Propane	44.10	8.357	9.321	8.769	5.440
iC_4	i-Butane	58.12	0.598	1.629	1.039	0.850
nC_4	n-Butane	58.12	1.625	7.090	3.963	3.240
iC_5	i-Pentane	72.15	0.192	2.201	1.051	1.067
nC_5	n-Pentane	72.15	0.253	4.843	2.217	2.250
C_6	Hexanes	86.18	0.100	5.904	2.583	3.132
C_7	Heptanes	95.10	0.055	8.028	3.466	4.636
C_8	Octanes	109.50	0.017	8.135	3.490	5.375
C_9	Nonanes	122.89	0.004	5.724	2.451	4.237
C_{10}	Decanes	134.02	0.002	5.164	2.210	4.167
C_{11}	Undecanes	147.00		3.980	1.703	3.521
C_{12}	Dodecanes	161.00		3.331	1.425	3.227
C_{13}	Tridecanes	175.00		3.161	1.352	3.329
C_{14}	Tetradecanes	190.00		2.534	1.084	2.898
C_{15}	Pentadecanes	206.00		2.159	0.924	2.677
C_{16}	Hexadecanes	222.00		1.872	0.801	2.501
C_{17}	Heptadecanes	237.00		1.516	0.648	2.162
C_{18}	Octadecanes	251.00		1.395	0.597	2.106
C_{19}	Nonadecanes	263.00		1.259	0.538	1.992
C_{20}	Eicosanes	275.00		1.115	0.477	1.846
C_{21}	Heneicosanes	291.00		1.001	0.428	1.753
C_{22}	Docosanes	305.00		0.770	0.330	1.414
C_{23}	Triacosanes	318.00		0.738	0.316	1.413
C_{24}	Tetracosanes	331.00		0.665	0.284	1.324
C_{25}	Pentacosanes	345.00		0.596	0.255	1.238
C_{26}	Hexacosanes	359.00		0.512	0.219	1.107
C_{27}	Heptacosanes	374.00		0.515	0.220	1.160
C_{28}	Octacosanes	388.00		0.430	0.184	1.004
C_{29}	Nonacosanes	402.00		0.358	0.153	0.865
C_{30+}	Triacontanes plus	545.25		3.689	1.578	12.106
Total	-		100	100	100	100
Molecular weight			23.29	135.01	71.09	

7.3 Phase Behavior and PVT Experiments

Multiple types of fluid can be found in the shale reservoirs, for example, in-situ water, fracturing fluid, and most importantly, oil and gas (hydrocarbons). The hydrocarbons can behave distinctively under various conditions. For instance, a shale oil in the reservoir at high pressure and high temperature is fully liquid. During the transportation from the reservoir to the wellbore, abundant natural gas can be released from the oil forming gas-liquid two-phase coexisting condition due to the pressure reduction. Such a phase change significantly affects the production, reservoir simulation, and field management. Therefore, it is crucial to understand the phase behavior of reservoir fluids.

Gibbs phase rule states that the degrees of freedom, F, must be specified to determine the thermodynamic state of equilibrium for a mixture containing n components distributed in P phases, equation (7.12).

$$F = n - P + 2. \tag{7.12}$$

The degree of freedom is the number of independent intensive variables that need to be specified to fully determine the state of the system. The intensive variables do not depend on the quantity of material in the system, such as pressure, temperature, and concentration. Equilibrium state requires that no net interphase mass transfer can occur. The temperatures and pressures of the coexisting phases must be the same and the chemical potentials of each component in each phase must be equal.

7.3.1 Phase Diagrams

Single-Component Systems

A phase diagram is a plot of pressure (p) versus temperature (T). Figure 7.4 shows a typical phase diagram for a single-component system.

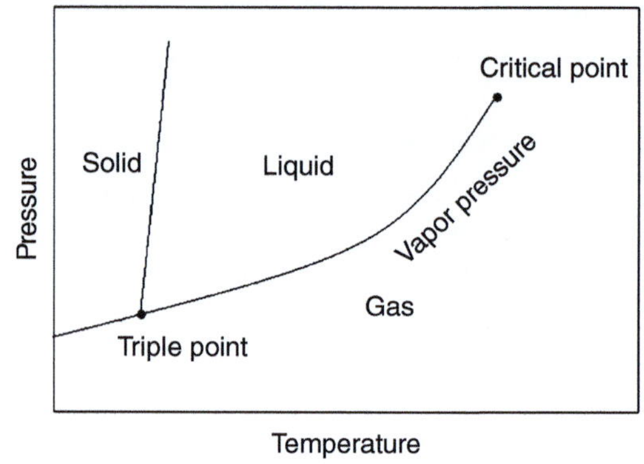

Figure 7.4 A p-T diagram for a single component system.

The vapor pressure line is the boundary between liquid phase and gas phase, which is also called phase boundary. The fluid is behaving liquid-gas two-phase on the vapor pressure line. According to the Gibbs phase rule, for a single-component ($n = 1$) and two-phase ($P = 2$) system, the degree of freedom is equal to 1. It indicates that only one variable, pressure or temperature, needs to be specified to determine the thermodynamic state of the system along the vapor pressure curve. The curve ends at the critical point. Katz and Singleterry (1939) has recommended naming a pure substance single-phase fluid at pressures higher than the critical pressure. The triple point is where liquid, gas, and solid coexist under equilibrium conditions. It is noted that the solid region usually is not the interest of petroleum engineers, especially in the area of shale fluids.

In addition to the p-T diagram, the pressure-volume (p-V) diagram can be plotted at various temperatures. Figure 7.5 shows a p-V diagram for ethane at three temperatures. At 60°F, the pressure curve dramatically decreases with volume in the liquid region due to the extremely small compressibility of the liquefied ethane. Point A is the bubble point where the first gas bubble appears. The horizontal Line AB represents the liquid-gas two-phase coexisting region where the pressure keeps constant with the increasing volume. Point B is the dew point. All ethane becomes vapor beyond point B. Please note that all points on line AB overlap

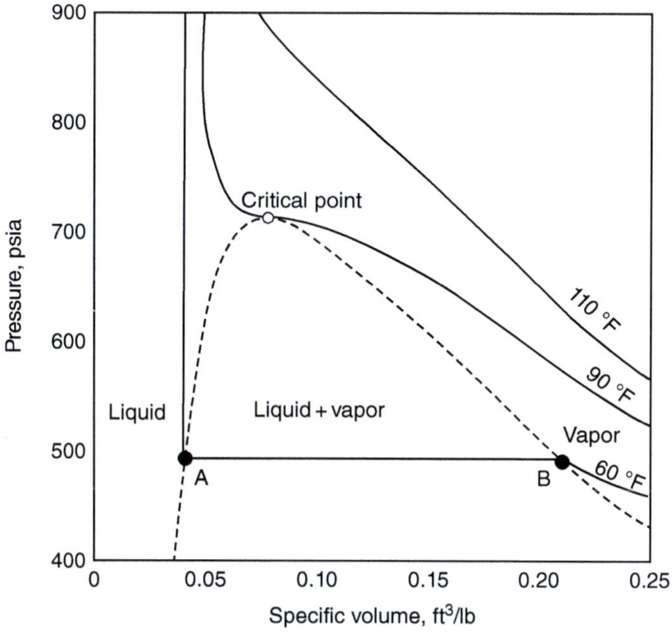

Figure 7.5 A p-V diagram for ethane at 60°F, 90°F, and 110°F (Standing, 1977).

at a specific single point on the vapor pressure line in the corresponding *p-T* diagram. It is due to the fact that the degree of freedom is 1 for a single-component, two-phase system. The horizontal line AB shrinks as temperature increases. At 90°F, points A and B overlaps and the length of the line is reduced to 0. The inflection point is the critical point. The corresponding pressure is called the critical pressure and the corresponding temperature is called the critical temperature. Ethane behaves as a supercritical fluid at 110°F because it is beyond the critical point.

Two-Component Systems

The phase diagram of the multicomponent system is more complex than that of a pure substance. Instead of a single vapor pressure curve (see Fig. 7.4), a wide region can be found for a multicomponent system where liquid and gas coexist at the equilibrium condition in the *p-T* diagram. Figure 7.6 is a phase diagram for the mixture of ethane/n-heptane with various ethane concentrations (Kay, 1938). The curves C and C_4 represent the vapor pressure lines for the pure ethane and n-heptane, respectively. The curves C_1 through C_3 are the two-phase boundaries with 90.22 wt%, 50.25 wt%, 9.78 wt% of ethane in the mixture, respectively. Liquid and vapor coexist within the two-phase boundary. Such a region is also called phase envelope or saturation envelope. The empty circles are the critical points for the systems. Taking the curve C_3 (9.78 wt% ethane + 90.22 wt% n-heptane) as an example, the bubble point line and dew point line join at the critical point. Since the systems contain more n-heptane, the phase envelope is closer to the vapor pressure curve of n-heptane. The dashed curve connecting all the critical points is defined as the locus of critical points of mixtures of ethane and n-heptane. It implies that the critical point of ethane and n-heptane system with any composition should be located on the dashed curve.

The *p-V* diagram of a two-component system is also different from that of a pure substance. Figure 7.7 shows a typical *p-V* diagram of a binary system. Instead of a horizontal line representing the two-phase region as can be

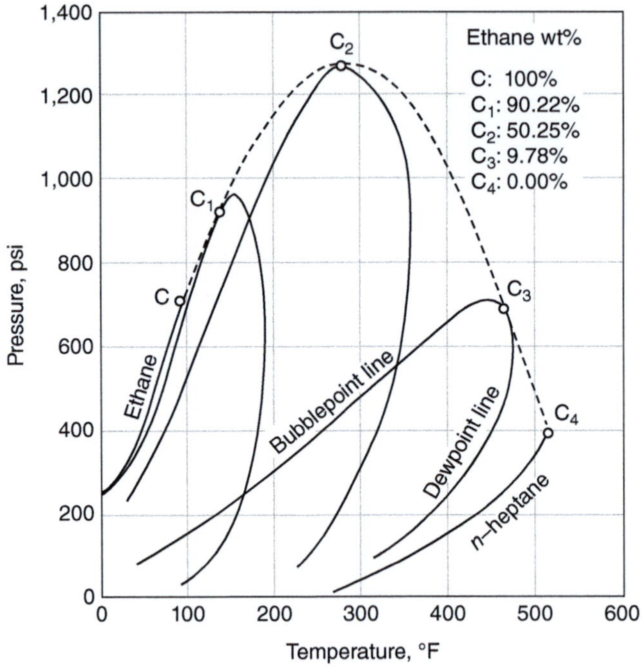

Figure 7.6 A p-T diagram for ethane/n-heptane system (Kay, 1938).

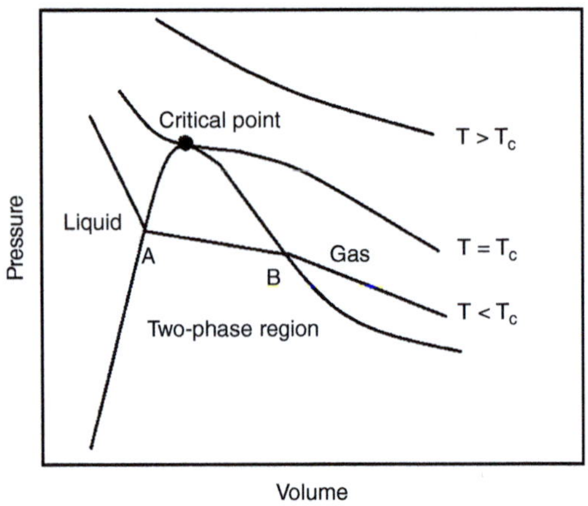

Figure 7.7 A p-V diagram for a binary system.

seen in Figure 7.5, a tilted line AB can be observed in the two-phase region. The length of the line AB shrinks as the temperature increases until the temperature reaches the critical temperature. There is no two-phase region when the temperature is greater than the critical temperature.

Multicomponent Systems, Shale Fluids

The reservoir fluid is a multicomponent mixture. The *p-T* diagram of the Bakken shale fluid is shown in Figure 7.8. The compositional analysis of the Bakken shale fluid can be found in Table 7.1. The *p-T* diagram is calculated by using the Peng-Robinson equation of state (PR EOS), which is introduced in section 7.4.

It can be seen that the cricondenbar (i.e., the greatest pressure on phase envelope) is 3,070 psi and the cricondentherm (i.e., the greatest temperature on phase envelope) is 692°F. The critical pressure is 1,951 psi and the critical temperature is 639°F. The reservoir condition (6,840 psi and 241°F) is located in the liquid region indicating this is a black oil reservoir. The separator condition is within the phase envelope where both oil and gas coexist.

The *p-T* diagram of the recombined Bakken shale fluid sampled from the separator is shown in Figure 7.9. Table 7.4 presents the recombined composition of the Bakken oil sampled from the separator. The phase envelope is calculated by using the Peng-Robinson equation of state (PR EOS), which is introduced in section 7.4. It is noted that the shale fluids used in Figures 7.8 and 7.9 are collected from the Bakken but with different sampling method. The compositions of the two samples are similar with similar molecular weight. As a result, the calculated phase envelope in Figure 7.9 is slightly larger than that in Figure 7.8. The cricondenbar is found to be 3,339 psi. The cricondentherm is 687°F. The critical pressure is 2,139 psi and the critical temperature is 627°F.

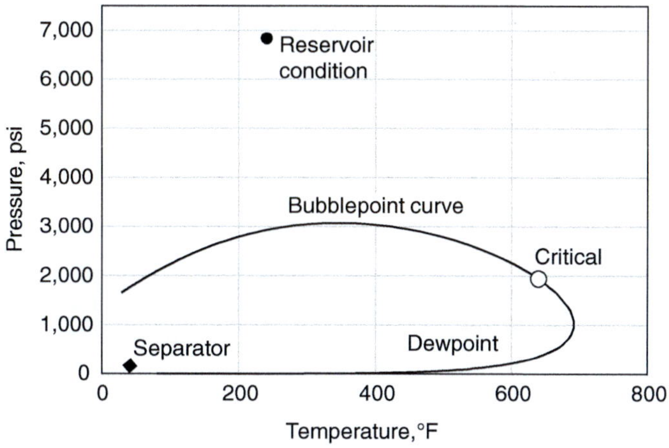

Figure 7.8 A p-T diagram for Bakken shale fluid sampled from downhole. Fluid composition can be seen in Table 7.1.

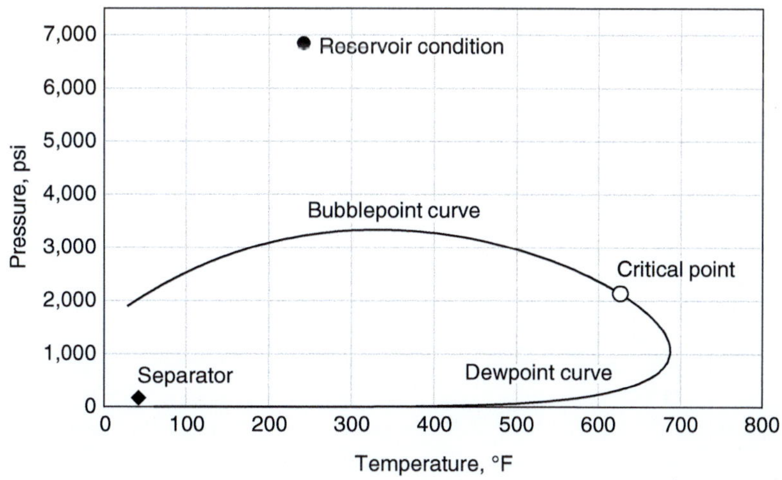

Figure 7.9 A p-T diagram for the recombined Bakken shale fluid sampled from the separator Fluid composition can be seen in Table 7.4.

7.3.2 PVT Experiments and Data Quality Check

Pressure-volume-temperature (PVT) properties are essential knowledge to quantify the phase changes and volumetric behavior of the reservoir fluids. Unlike the well-defined systems, the reservoir fluids produced from different locations have a wide variety of physical and chemical properties and phase behaviors. Such a unique characteristic suggests that the PVT experiments should be conducted for each reservoir fluid sample. The most commonly seen PVT experiments include constant-composition expansion experiment, differential liberation experiment, separator test, and swelling test.

Constant-Composition Expansion (CCE) Experiment

The constant-composition (also called constant-mass) expansion experiment can be applied for oil and gas condensate samples. Saturation pressure at the reservoir temperature and relative volume of gas and oil in the reservoir at various stages can be measured.

Figure 7.10a is the schematic diagram of an oil CCE experiment. A piston is equipped in the PVT cell, which can be moved upward or downward by injecting or withdrawing mercury. A certain amount of oil sample is placed above the piston. The volume of the oil sample changes with the location of the piston. The CCE experiment starts at a pressure higher than the saturation pressure at which the oil is in a single liquid phase, while the volume is recorded. Such condition corresponds to point 1 in Figure 7.10a and b. The piston is then moved downward by withdrawing mercury so that the pressure in oil sample decreases, which is corresponding to point 2. The first gas bubble appears at point 3 indicating the bubble point pressure (or saturation pressure) is reached, which is also the turning point on the *p-V* diagram. More gas can be released as the piston keeps moving downward so that points 4 and

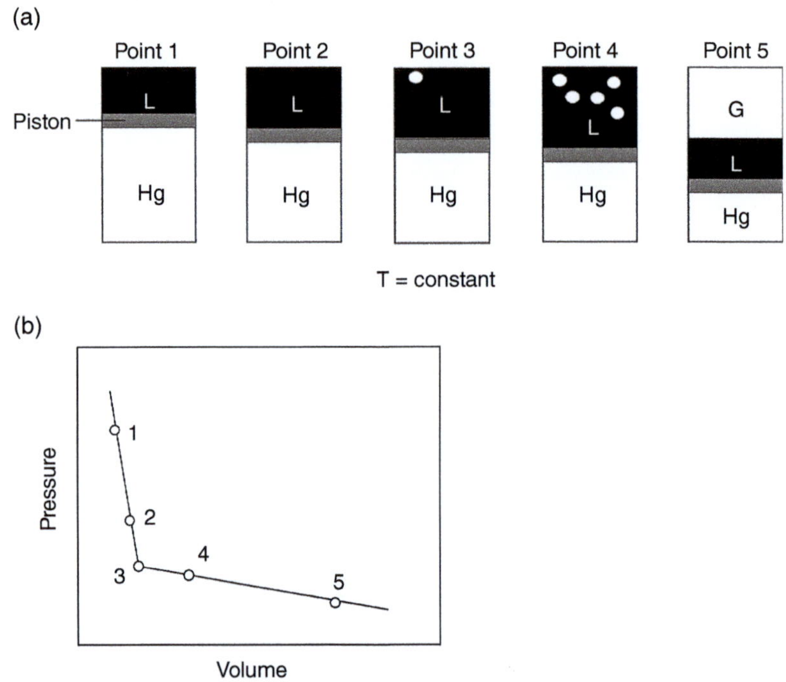

Figure 7.10 Constant composition expansion experiment for oil (a) schematic diagram and (b) p-V diagram.

5 can be measured. Once the saturation pressure is reached, the pressure cannot be reduced significantly with the increasing volume. This is due to the fact that the newly formed gas with high compressibility expands to sustain the cell pressure.

The relative volume (V_r) is defined as the ratio between the total fluid sample volume (V_{total}) and the volume measured at the saturation pressure (V_{sat}),

$$V_r = \frac{V_{total}}{V_{sat}}. \tag{7.13}$$

When the pressure is greater than saturation pressure, the isothermal compressibility of the oil (c_o) can be calculated using

$$c_o = -\frac{1}{V}\left(\frac{\partial V}{\partial p}\right)_T. \tag{7.14}$$

When the pressure is smaller than saturation pressure, the Y-factor can be calculated with equation (7.15),

$$Y = \frac{p_{sat} - p}{p} / \frac{V^{total} - V^{sat}}{V^{sat}}. \tag{7.15}$$

A small value of Y-factor implies that the oil can release a large amount of gas with the decreasing pressure. A large value of Y-factor indicates that the oil can release a small amount of gas with the decreasing pressure.

Table 7.5 is the measured CCE data for the Bakken shale fluid. The fluid composition is listed in Table 7.4. The reservoir pressure is 6,840 psi and the saturation pressure is measured at 3,149 psi, indicating this is an undersaturated oil reservoir. The measured p-V diagram is plotted in Figure 7.11.

Table 7.5 CCE Experiment of the Recombined Bakken Shale Fluid Sampled From the Separator at the Reservoir Temperature of 241°F

Pressure, psi		**Relative volume**	**Density, g/cm^3**	**Oil compressibility, 10^{-6} psi^{-1}**	**Y-function**
7,500		0.914	0.629		
6,840	Reservoir	0.923	0.623	14.55	
6,500		0.928	0.620	15.11	
6,000		0.935	0.615	16.07	
5,500		0.943	0.610	17.37	
5,000		0.952	0.604	18.88	
4,500		0.962	0.598	20.96	
4,000		0.974	0.591	23.56	
3,500		0.988	0.582	29.20	
3,149	Saturation	1.000	0.575	33.41	
3,000		1.020			2.53
2,500		1.116			2.24
2,000		1.290			1.98
1,500		1.619			1.77
1,000		2.376			1.56
500		4.931			1.35

Note: Fluid composition can be seen in Table 7.4.

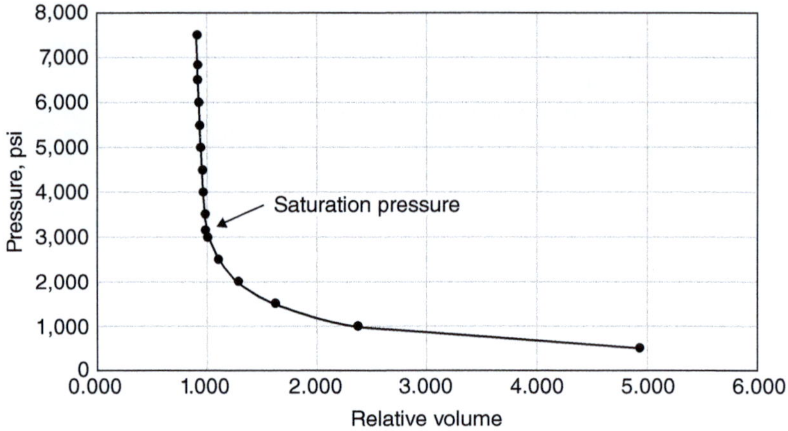

Figure 7.11 Measured p-V diagram of the recombined Bakken shale fluid sampled from the separator at the reservoir temperature of 241°F. Fluid composition can be seen in Table 7.4.

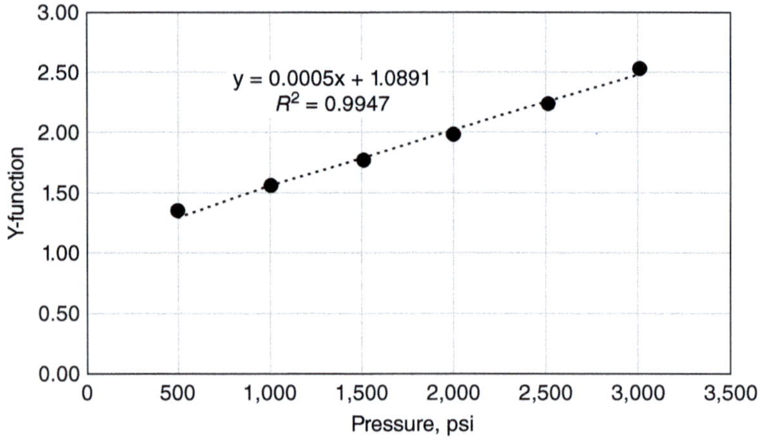

Figure 7.12 Y-function versus pressure of the recombined Bakken tight oil fluid (data from Table 7.5).

Valid CCE data can provide crucial information of the reservoir fluid. The Y-function value can be calculated when pressure is smaller than saturation pressure. The data quality can be assessed by the Y-function linearity test (Ahmed, 2006; Bagoo et al., 2014). If the Y-function follows a linear relationship with pressure, the data are considered to be valid and ready for application. The Y-function data in Table 7.5 plotted against pressure is shown in Figure 7.12. It can be seen there is an excellent linearity with R^2 of 0.9947, indicating reliable CCE data have been measured for this Bakken tight oil reservoir.

Differential Liberation Experiment

Differential liberation (DL) experiment is commonly conducted for oil. The oil formation volume factor (B_o), solution gas-oil ratio (R_s), gas formation volume factor (B_g), and the total formation volume factor (B_t) are measured at different pressure stages.

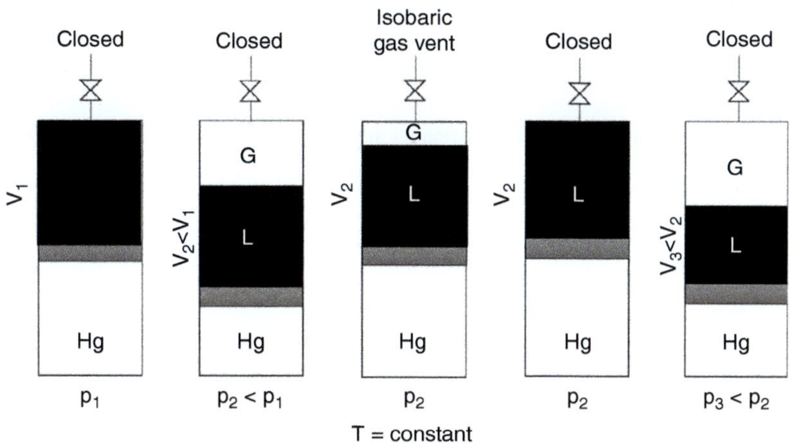

Figure 7.13 Schematic diagram of DL experiment.

The DL experiment is sketched in Figure 7.13. The oil is placed in the PVT cell at the reservoir temperature and the saturation pressure where a single liquid phase is present. The piston is moved downward to reduce the cell pressure so that a certain amount of gas is released. The pressure, p_2, is smaller than p_1 and the oil volume, V_2, is smaller than the original oil volume, V_1. Next step is to isobarically deplete the gas by opening the top valve and moving the piston upward until all gas is removed. Both oil volume and cell pressure are kept constant during the depletion process. The volumes of cell gas, cell oil, and released gas are measured, respectively. Hence, B_o, R_s, B_g, and B_t at the stage p_2 can be calculated. Then the experiment can move to the next pressure stage, p_3. Table 7.6 is the DL experimental data for the recombined Bakken oil. The oil formation volume factor, B_o, and the solution gas-oil ratio, R_s, are plotted against pressure in Figure 7.14.

The inequality test can be used to assess the data quality of DL experiment when pressure is below the saturation pressure. The isothermal oil compressibility at a pressure below the bubble point pressure is defined as

$$C_o = \frac{B_g}{B_o}\frac{dR_S}{dp} - \frac{1}{B_o}\frac{dB_o}{dp}. \tag{7.16}$$

In order to have a positive isothermal oil compressibility, the following condition should be satisfied (Bagoo et al., 2014):

$$B_g \frac{dR_s}{dp} > \frac{dB_o}{dp} \text{ i.e., } C_o > 0. \tag{7.17}$$

An inequality test is performed for the DL experiment of Bakken shale fluid shown in Table 7.6. The test results can be found in Table 7.7. The oil compressibility at the pressure below saturation pressure of 3,149 psi is positive, indicating the DL experiment has been well conducted and the data are valid.

Separator Test

A separator test can be carried out for both oil and gas condensate. The purpose is to measure the volumetric properties at each separator stage.

Figure 7.15 presents a three-stage separator test. The reservoir sample is first placed in a cell at the saturation pressure and reservoir temperature. The fluid is then transferred to the Separator 1 where the fluid is flashed into two phases. Gas is released from the top to the standard condition, at which its volume is measured. The liquid is transferred to the next stage of the separator where the pressure and temperature are lower than the previous stage.

Table 7.6 DL Experiment of the Recombined Bakken Shale Fluid Sampled From the Separator at the Reservoir Temperature of 241 °F

Pressure, psi		Oil density, g/cm^3	Oil formation volume factor, bbl/stb	Solution gas-oil ratio, scf/bbl	Gas Z-factor	Gas formation volume factor, ft^3/scf	Total formation volume factor
7,500		0.629	2.222	1,303			
6,840	Reservoir	0.623	2.243	1,303			
6,500		0.620	2.255	1,303			
6,000		0.615	2.273	1,303			
5,500		0.610	2.293	1,303			
5,000		0.604	2.315	1,303			
4,500		0.598	2.339	1,303			
4,000		0.591	2.367	1,303			
3,500		0.582	2.402	1,303			
3,149	Saturation	0.575	2.431	1,303			
2,500		0.610	2.109	974	0.803	0.006	2.703
2,000		0.634	1.932	778	0.811	0.008	3.130
1,500		0.658	1.792	624	0.828	0.011	3.899
1,000		0.680	1.664	476	0.854	0.017	5.634
500		0.705	1.529	324	0.886	0.035	11.288
15		0.768	1.097	0	1.000	1.348	498.771

Note: Fluid composition is listed in Table 7.4.

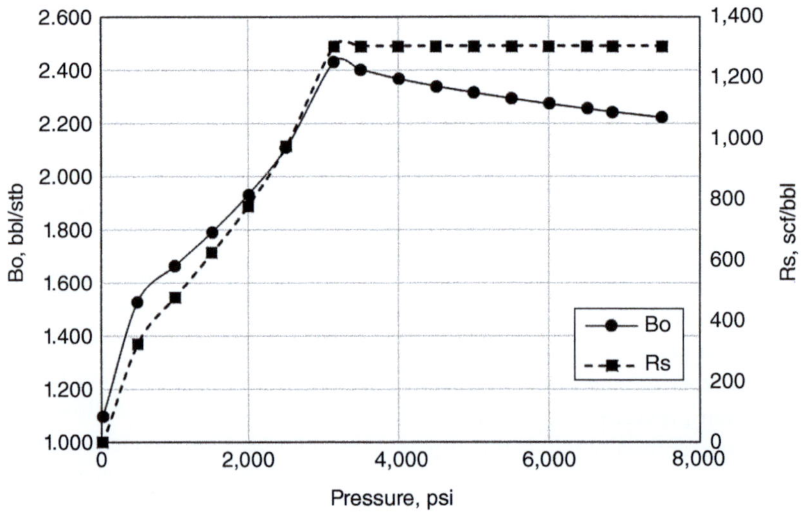

Figure 7.14 Measured oil formation volume factor and solution gas-oil ratio of the recombined Bakken oil. Fluid composition is listed in Table 7.4.

Table 7.7 Inequality Test of DL Experiment for the Recombined Bakken Shale Fluid

Pressure, psi	Oil formation volume factor,bbl/STB	Solution gas-oil ratio, scf/bbl	Gas formation volume factor,ft³/scf	$B_g \frac{dR_s}{dp}$	$\frac{dB_o}{dp}$	C_o
3,149	2.431	1,303				
2,500	2.109	974	0.006	0.000496	0.003042	0.002545
2,000	1.932	778	0.008	0.000354	0.003136	0.002782
1,500	1.792	624	0.011	0.000280	0.003388	0.003108
1,000	1.664	476	0.017	0.000256	0.005032	0.004776
500	1.529	324	0.035	0.000270	0.010640	0.010370
15	1.097	0	1.348	0.000891	0.900520	0.899629

Note: Data from Table 7.6.

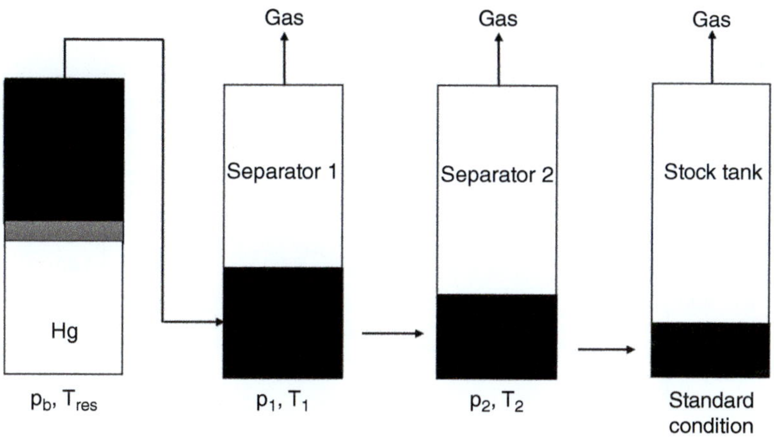

Figure 7.15 Schematic diagram of laboratory separator test.

The conditions for Separators 1 and 2 are set the same as the field condition provided by the operator. Finally, the liquid in Separator 2 is charged into the stock tank where the pressure is 14.7 psi and temperature is 60°F. The stock tank oil cannot liberate gas anymore.

Swelling Test

Gas injection is one of the most important enhanced oil recovery (EOR) methods. The dissolution of gas in oil leads to swelling effect, viscosity reduction, and interfacial tension reduction. Shale oil reservoirs are typically characterized as low-porosity and low-permeability. The most widely used gases include CO_2, N_2, produced natural gas, and hydrocarbons. Gas huff-n-puff has been considered to be an effective approach to improve the contact between the oil and the injected gas. Thus, the solubility of a gas in oil is the key to affect the performance of gas injection, which is highly dependent on pressure. A higher pressure leads to a greater gas solubility and thus an enhanced swollen volume.

Figure 7.16 depicts the procedure for a swelling test. A certain amount of oil sample is placed in the cell. Then a desired amount of gas is injected to mix with the oil. The cell is

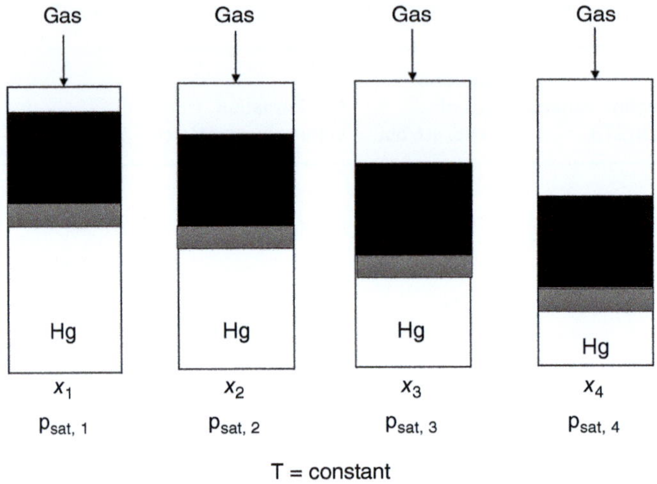

Figure 7.16 Schematic diagram of swelling test.

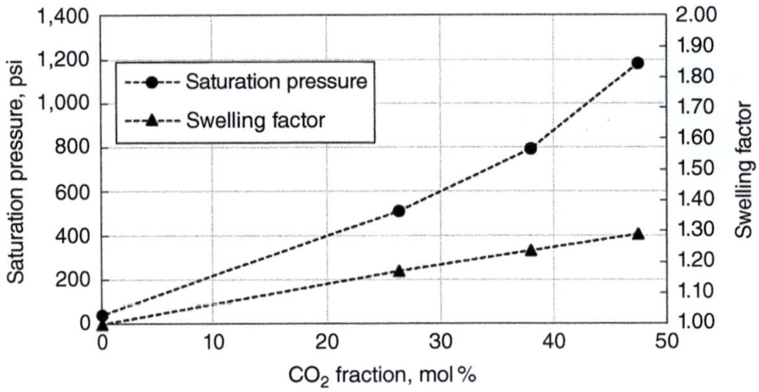

Figure 7.17 Measured swelling test for CO_2-dead Bakken oil system at 256°F.

pressurized by moving the piston upward until all gas dissolved in the oil. The mole fraction of the injected gas in the mixture is denoted as x_1. Accordingly, the CCE is performed to measure the saturation pressure, p_{sat1}, at x_1. Then, more gas is injected forming the mixture with the gas composition of x_2. CCE is performed again to measure $p_{sat,\ 2}$. Likewise, the saturation pressure at various x can be measured. Consequently, the solubility of a gas in oil with pressure can be measured. In addition, swelling factor is an important parameter for designing solvent-based EOR processes. It is defined as the gas saturated oil volume at the saturation pressure and reservoir temperature divided by the dead oil volume at the reservoir temperature.

An example of CO_2 swelling test for a dead oil collected from the Bakken field is plotted in Figure 7.17. It can be seen that 47.4 mol% of CO_2 can be dissolved in the oil at the saturation pressure of 1,180.6 psi. Meanwhile, the oil volume can be swollen with a factor of 1.30.

7.4 Equation of State (EOS)

Reservoir fluids contain a variety of substances of diverse chemical nature that include hydrocarbons and nonhydrocarbons. For conventional reservoirs, in spite of the complexity of hydrocarbon fluids found in the reservoirs, simple cubic equations of state have shown surprising performance in the phase behavior calculation (Firoozabadi, 1999). As for the shale reservoirs, the simple cubic equation of state is not accurate enough because of the effect of confinement. However, different modifications of the cubic equation of state and also some new models are brought out to describe the phase behavior of fluids confined in shale formations.

7.4.1 Cubic Equation of State

Cubic equations of state are simple equations relating pressure, volume, and temperature (PVT). They accurately describe the volumetric and phase behavior of pure compounds and mixtures, requiring only critical properties and an acentric factor of each component. Below are some widely used cubic equations of state.

Van der Waals EOS (1873)

Van der Waals proposed this EOS in 1873. It is the first cubic EOS to give a simple, qualitatively relation between pressure, temperature, and molar volume.

$$p = \frac{RT}{v-b} - \frac{a}{v^2}, \tag{7.18}$$

where a is the attractive parameter, b is the repulsive parameter, and R is the universal gas constant.

Compared with the ideal gas law, the van der Waals equation of state has two main improvements (Whitson & Brule, 2000), one is that the prediction of the liquid phase behavior is more accurate because of the incorporation of parameter b. The van der Waals equation of state also improves the description of non-ideal-gas behavior by considering the term a/v^2, which represents the reduction of system pressure by the attractive component.

Soave-Redlich-Kwong EOS (Soave, 1972)

Redich and Kwong (1949) proposed the equation of state by taking into account the temperature dependencies of the molecular attraction term in 1949.

$$p = \frac{RT}{v-b} - \frac{a}{\sqrt{T}v(v+b)}. \tag{7.19}$$

In 1972, Soave replaced the $1/\sqrt{T}$ term with a function $\alpha(T, \omega)$ involving the temperature and the acentric factor (the resulting equation is known as the Soave-Redlich-Kwong equation of state; SRK EOS). SRK EOS provides an excellent prediction of vapor-liquid equilibrium properties, although it grossly overestimates liquid volumes of petroleum mixtures (Soave, 1972).

$$p = \frac{RT}{v-b} - \frac{a}{v(v+b)}, \tag{7.20}$$

$$a = 0.42747\frac{R^2T_c^2}{p_c}\alpha, \tag{7.21}$$

$$b = 0.08664\frac{RT_c}{p_c}, \tag{7.22}$$

and

$$\alpha = \left[1 + \left(0.480 + 1.574\omega - 0.176\omega^2\right)\left(1 - (T/T_c)^{0.5}\right)\right]^2, \tag{7.23}$$

where T_c and p_c are the critical temperature and pressure of the species, respectively; and ω is the acentric factor for the species.

Peng-Robinson EOS

In 1976, Peng and Robinson proposed a two-constant equation of state to improve the liquid density prediction. Until now, PR EOS (Peng-Robinson Equation of state) has been one of the most widely used thermodynamic models for petroleum fluids in both industrial and academic fields.

$$P = \frac{RT}{v-b} - \frac{a}{v(v+b)+b(v-b)}, \tag{7.24}$$

$$a = 0.45724\frac{R^2T_c^2}{P_c}\alpha, \tag{7.25}$$

$$b = 0.07780\frac{RT_c}{P_c}, \tag{7.26}$$

$$\alpha = \left[1+\kappa\left(1-\sqrt{T_r}\right)\right]^2 \tag{7.27}$$

and

$$\kappa = 0.37464 + 1.54226\omega - 0.26992\omega^2, \tag{7.28}$$

where all the parameters are as introduced before.

Robinson et al. (1979) proposed a modified expression for $\boldsymbol{\kappa}$ in particular consideration of heavier components ($\omega > 0.49$):

$$\kappa = 0.3796 + 1.485\omega - 0.1644\omega^2 + 0.01667\omega^3. \tag{7.29}$$

Recently, a modified alpha function, which can more accurately predict the vapor pressure of both nonhydrocarbon and light-heavy hydrocarbon pure substance, was proposed (Li &Yang, 2011). It can be applied to two-phase and multiphase boundary computations (Li et al., 2013a, 2013b, 2016, 2017a, 2017b).

$$\alpha = \exp\left\{\begin{array}{c}(0.13280 - 0.05052\omega + 0.25948\omega^2)(1-T_r) + \\ 0.81769\ln\left[1 + (0.31355 + 1.86745\omega - 0.52604\omega^2)\left(1-\sqrt{T_r}\right)\right]^2\end{array}\right\}. \tag{7.30}$$

Physically, water inevitably exists in the formation in the form of either initial connate water or injected water, resulting in a range of 5% to 50% water saturation (Li & Nghiem, 1986). However, the original Peng-Robinson EOS (Peng & Robinson, 1976a) is found to produce an erroneous composition prediction for the aqueous phase, which is inconsistent with the experimental data in several orders of magnitude due to the extremely small hydrocarbon solubility in water (Peng & Robinson, 1976b). On the basis of the International Association for the Properties of Water and Steam Formulation 1995 (IAPWS-95), Li and Yang (2013) have developed a new alpha function for water compound to more extensively predict the water vapor pressure in the full temperature range of 273.16 K to 647.10 K,

$$\alpha_{H_2O,\text{modified}} = [c_1 + c_2(1-T_{rw}) - c_3\left(1-T_{rw}{}^{-1}\right) + c_4\left(1-T_{rw}{}^{-2}\right)]^2, \tag{7.31}$$

where, T_{rw} is the reduced temperature of water, $c_1 = 1.00095$, $c_2 = 0.39222$, $c_3 = 0.07294$, and $c_4 = 0.00706$ with the standard errors of 0.00002, 0.00085, 0.00109, and 0.00023, respectively.

As for the mixtures, the following van der Waals mixing rules are recommended for all of the equations of state listed above.

$$a = \sum_i\sum_j x_ix_j\left(a_ia_j\right)^{1/2}\left(1-\delta_{ij}\right) \tag{7.32}$$

and

$$b = \sum_i x_ib_i. \tag{7.33}$$

The term δ_{ij} is the binary interaction coefficient (BIP), whose values can be obtained by fitting the equation of state to vapor-liquid equilibria data for each binary system.

Li et al. (2016) developed a BIP correlation particularly for CO_2-heavy n-alkanes systems (Li et al., 2013),

$$\delta_{ij} = \frac{c_1 + c_2T_{ri} + c_3T_{ri}^2 + d_1\omega_j}{1 + c_4T_{ri} + c_5T_{ri}^2 + d_2\omega_j + d_3\omega_j^2 + d_4\omega_j^3}, \tag{7.34}$$

where δ_{ij} is the BIP between CO_2 (i.e., component i) and n-alkane (i.e., component j), T_{ri} is the reduced temperature of CO_2, and ω_j is the acentric factor of n-alkane. The coefficients c_1 to c_5 and d_1 to d_4 can be found in Table 7.8.

Table 7.8 Coefficient Values in Equation

Coefficient	Value	Coefficient	Value
c_1	0.488890	d_1	−0.023506
c_2	0.470802	d_2	11.982748
c_3	0.120048	d_3	−14.784566
c_4	3.716653	d_4	5.884453
c_5	0.843567		

Two sets of BIPs are required in the PR EOS for modeling the aqueous phase and nonaqueous phase compositions, respectively (Li & Yang, 2013). The following polynomial BIP correlation, which is a function of the reduced temperature, has been proposed to calculate the BIP values for CO_2-water binary (Li & Yang, 2013),

$$\delta_{CO_2}^{AQ} = -1.104324 + 2.040527T_r - 1.417707T_r^2 + 0.379003T_r^3, \quad (7.35)$$

where $\delta_{CO_2}^{AQ}$ is the BIP for CO_2-water pair in the aqueous phase and T_r is the reduced temperature of CO_2.

The following correlation can be used to calculate the BIP for hydrocarbons and water mixture in aqueous phase (Whitson & Brule, 2000):

$$\delta_{ij}^{AQ} = \left(1.112 - 17369\omega_i^{-0.1}\right) + (1.1001 + 0.83\omega_i) T_{ri} - (0.5742 + 1.0988\omega_i)T_{ri}^2, \quad (7.36)$$

where δ_{ij}^{AQ} is the BIP for hydrocarbon-water pair in the aqueous phase, i represents hydrocarbons, and j represents water.

7.4.2 Stability Analysis

Stability analysis is of great significance to decide whether a mixture will split into two (or more) phases under a specific temperature and pressure condition, which is one the most difficult aspects of making the vapor-liquid equilibria calculations.

In 1981, Michelsen (1982) proposed a phase stability test method based on creating a second phase. By locating its compositions with regard to the mixture compositions, whether such a system will split into two phases or stays just one phase is verified. The stability test must be conducted in two parts, considering the second phase to be both vapor-like and liquid-like. A successive-substitution algorithm is summarized in the following procedure (Whitson & Brule, 2000):

1) Calculate the mixture fugacity, f_{zi} with the overall composition.
2) Estimate initial K values by using the Wilson equation,
$$K_i^1 = \frac{\exp\left[5.37(1+\omega_i)\left(1-T_{ri}^{-1}\right)\right]}{P_{ri}}. \quad (7.37)$$
3) Assume the second phase is vapor-like, calculate the second-phase mole numbers, Y_i:
$$Y_i = z_i K_i. \quad (7.38)$$
4) Sum the mole numbers,
$$S_V = \sum_{i=1}^{N} Y_i. \quad (7.39)$$
5) Get the mole fraction by normalizing the second-phase mole numbers,
$$y_i = \frac{Y_i}{S_V}. \quad (7.40)$$
6) Calculate the second-phase fugacity, f_{yi}, with EOS.
7) Calculate the fugacity-ratio corrections for the update of K values,
$$R_i = \frac{f_{zi}}{f_{yi}}\frac{1}{S_V}. \quad (7.41)$$
8) Check whether convergence is achieved,
$$\sum_{i=1}^{N}(R_i - 1)^2 < \varepsilon. \quad (7.42)$$
9) Update the K values if convergence is not obtained,
$$K_i^{(n+1)} = R_i^{(n)}K_i^{(n)}. \quad (7.43)$$
10) Check whether a trivial solution is being approached,

$$\sum_{i=1}^{N} (\ln K_i)^2 < 1 \times 10^{-4}. \quad (7.44)$$

If a trivial solution is approached, stop the procedure; or go back to step 3 for another iteration.

11) Assume the second phase is liquid-like, repeat step 3 to 10 with the equations below:

$$Y_i = z_i / K_i, \quad (7.45)$$

$$S_L = \sum_{i=1}^{N} Y_i, \quad (7.46)$$

$$x_i = \frac{Y_i}{S_L}, \quad (7.47)$$

and

$$R_i = \frac{f_{xi}}{f_{zi}} \frac{1}{S_v}. \quad (7.48)$$

The interpretation of the results of this method follows.

The mixture is stable (single-phase condition prevails) if (1) both tests yield $S < 1$ ($S_L < 1$ and $S_V < 1$), or (2) both tests converge to a trivial solution, or (3) one test converges to the trivial solution and the other gives $S < 1$.

Only one test indicating $S > 1$ is sufficient to determine that the mixture is unstable and that the two-phase condition prevails. The same conclusion is made if both tests give $S > 1$, or if one of the tests converges to the trivial solution and the other gives $S > 1$.

7.4.3 Confinement/Pore Proximity Effect on Phase Behavior

The primary organic pore size in shale formations is in nanoscale. Such meso- and nanopores in shale reservoirs cause the phase behavior of confined fluids to deviate significantly from that of bulk fluids in conventional reservoirs, such as density, saturation pressure, and critical properties (Thommes & Findenegg, 1994). It is well recognized that the phase diagram of confined fluids in shale reservoirs shifts from the bulk fluids in conventional reservoirs due to high capillary pressures. Figure 7.18 presents the two-phase diagram change of an Eagle Ford gas condensate sample by consideration of

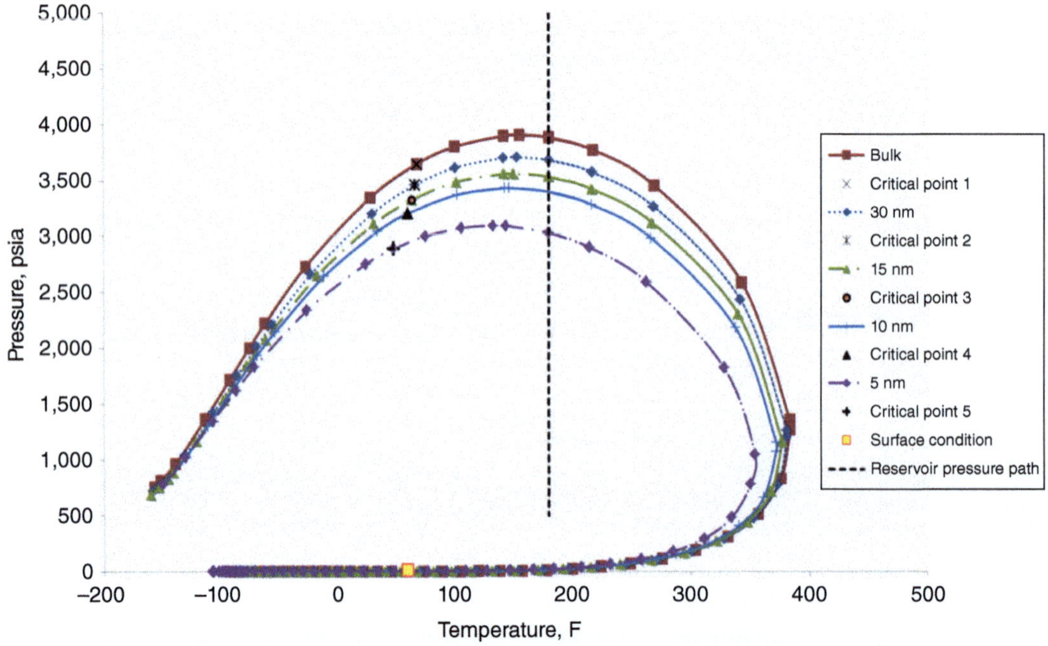

Figure 7.18 Phase diagram change for Eagle Ford gas condensate sample by considering only critical property shift (Sanaei et al., 2014).

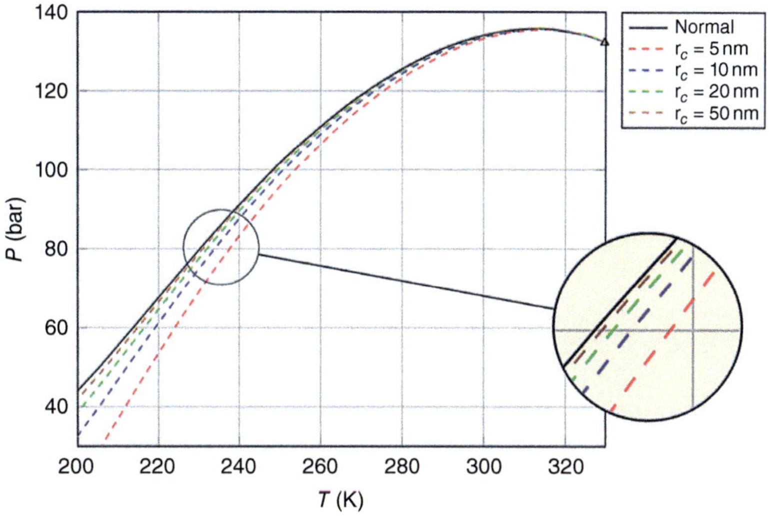

Figure 7.19 Bubblepoint change at different pore sizes for a 70-30 mol% C1-C4 mixture by considering only capillary effect (Sandoval et al., 2016).

critical property shift (Sanaei et al., 2014). Figure 7.19 illustrates the bubble-point pressure change of a hydrocarbon mixture by accounting for the capillarity effect (Sandoval et al., 2016). The confinement effect resulting from increasing molecule-wall interaction and capillary pressure is found to be significant and nonnegligible for unconventional reservoir fluids.

Capillary Pressure in Flash Calculation

It is a fact that the capillary pressure is ignored during the vapor-liquid equilibrium (VLE) calculation for bulk fluids, where the pressures on two sides of the vapor-liquid interface are assumed to be the same. Nevertheless, erroneous results can be made if the capillary pressure is not considered for confined fluids flash calculation due to the nanoscale pores. Therefore, it is necessary to incorporate the pressure in the vapor phase (p_V) and the pressure in the liquid phase (p_L) into the flash calculation. It is assumed that the liquid phase is the wetting phase and the vapor phase is the nonwetting phase. Then, the capillary pressure is

$$p_{cap} = p_V - p_L = \frac{2\sigma \cos\theta}{r_p}, \tag{7.49}$$

where p_{cap} is the capillary pressure, r_p is the pore radius, and θ is the contact angle, which is assumed to be 0° in this work. As for the interfacial tension, it can be calculated with the Danesh model (Danesh et al., 1991):

$$\sigma^{1/E} = \sum_{i=1}^{n} P_i(x_i\rho_L - y_i\rho_V), \tag{7.50}$$

$$E = 3.583 + 0.16(\rho_L - \rho_V), \tag{7.51}$$

and

$$P_i = (8.21307 + 1.97473\omega_i)T_{ci}^{1.30406}p_{ci}^{-0.82636}, \tag{7.52}$$

where x_i and y_i are the compositions of the ith component in liquid and vapor phase, respectively; ρ_L, ρ_V are the densities of liquid and vapor phase, respectively; P_i, ω_i, T_d, and p_d are the parachor, acentric factor, critical temperature, and critical pressure of the ith component.

Figure 7.20 demonstrates a flowchart to introduce the capillary pressure into the flash calculation. It is worth noting that because of the phase behavior shift of confined fluids, the currently used equations of state are not valid for shale reservoir conditions, especially for the microporous media. Modified EOS,

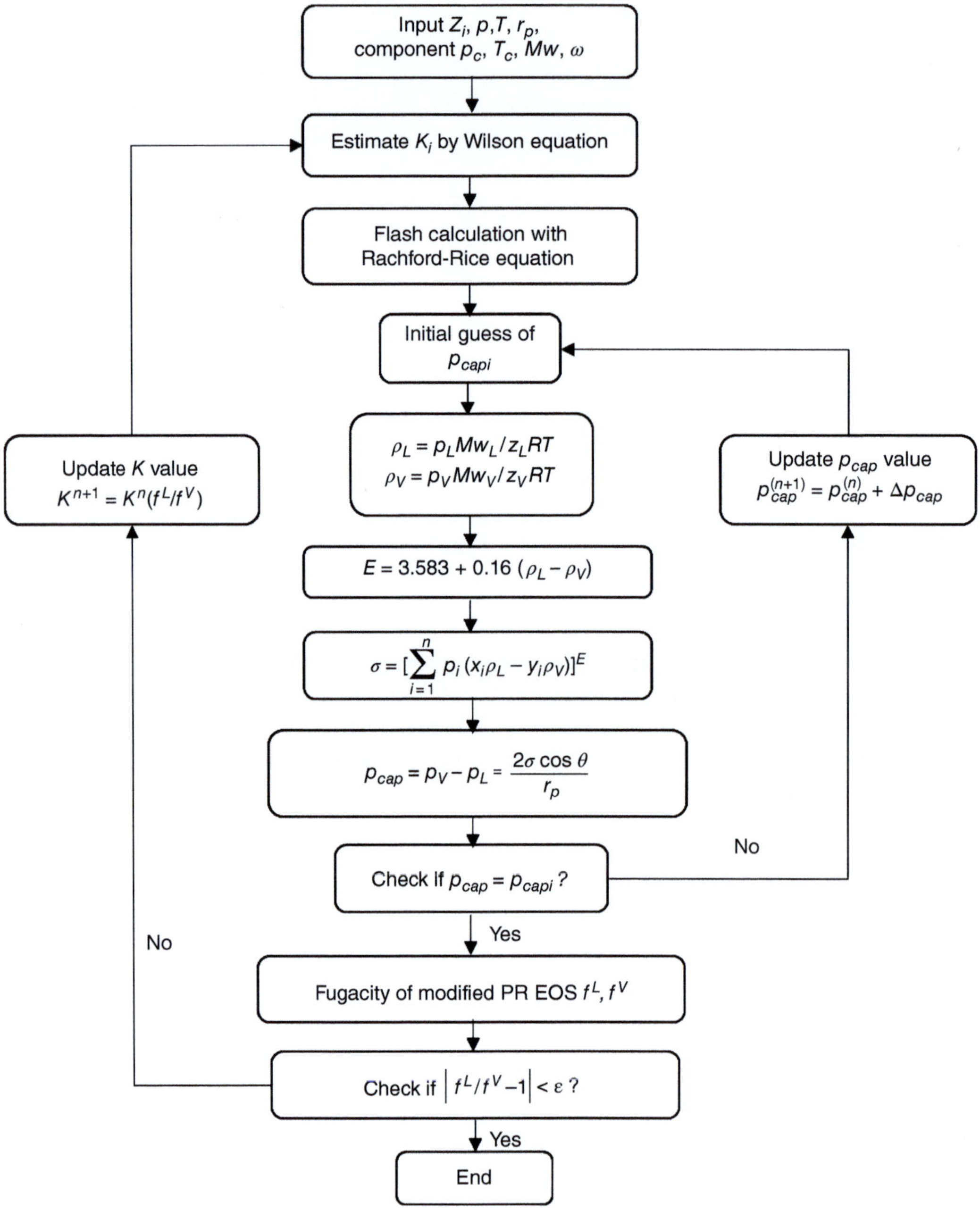

Figure 7.20 Flowchart of the flash calculation with consideration of capillary pressure.

which is applicable for confined fluids, is required for the calculations of fugacity of both vapor and liquid phase.

Taking PR EOS as an example, one way to make the modification is to introduce a new term to represent the effect of strong molecule-wall interaction, which is a significant mechanism underlying the confinement effect. In the original PR EOS, the constant a represents the correction to the internal pressure resulting from intermolecular interactions, which is traditionally interpreted as the attractive component of pressure, while the constant b is referred to as the covolume, physically representing the repulsive component of pressure on a molecular scale (Whitson & Brule, 2000; Peng & Robinson, 1976). As claimed by Derouane (2007), the molecule-wall interaction is a relatively strong effect competing with the intermolecular dispersion force; it plays a role in diminishing the attractive component. Thus, the new term c, representing the correction to the internal pressure resulting from molecule-wall interaction, can be incorporated as follows:

$$p = \frac{RT}{v-b} - \frac{a-c}{v(v+b)+b(v-b)}, \tag{7.53}$$

where c is the term to represent the effect of molecule-wall interaction. The determination of c will be introduced in the next part with a combination of the critical property shift of confined fluids.

Critical Properties Shift in EOS

In nanosized pores, as the molecule size is comparable to the pore space, the molecule-wall interaction, which is usually neglected in bulk fluids, has a significant effect on the phase behavior of confined fluids, especially the critical properties. Several simulated and experimental data have concluded that the critical properties of many compounds change as pore size decreases (Thommes & Findenegg, 1994; Morishige et al., 1997; Singh et al., 2009). Zarragoicoechea and Kuz (2002, 2004) proposed a quadratic correlation between the critical property shift and the dimensionless ratio of σ_{LJ}/r_p, which is shown as follows:

$$\Delta T^* = 0.9409\frac{\sigma_{LJ}}{r_p} - 0.2415\left(\frac{\sigma_{LJ}}{r_p}\right)^2 \tag{7.54}$$

and

$$\Delta P^* = 0.9409\frac{\sigma_{LJ}}{r_p} - 0.2415\left(\frac{\sigma_{LJ}}{r_p}\right)^2, \tag{7.55}$$

where r_p is the pore radius and σ_{LJ} is the Lennard-Jones size parameter (collision diameter).

Jin (2013) suggested that the critical properties should be an exponential function of the ratio of pore size to molecule size in order to properly account for the behavior of confined fluids,

$$\Delta T^* = 0.8493 \times \left(r_p/\sigma_{LJ}\right)^{-1.241} + 0.015 \tag{7.56}$$

and

$$\Delta P^* = 1.8 \times \left(r_p/\sigma_{LJ}\right)^{-0.775}. \tag{7.57}$$

The accuracy of these two models is verified by comparison of the calculated critical temperature shift data with the real data collected in literature.

Table 7.9 lists 15 data points of critical temperature shift (ΔT^*) of three different fluids (CO_2, C_2H_4, and square-well fluid) collected from three references (Burgess et al., 1990; Morishige et al., 1997; Singh & Singh, 2011), including both experimental measurement and molecular simulation results. In this table, a dimensionless pore size is defined as r_p/σ_{LJ}, where r_p is the pore radius and σ_{LJ} is the Lennard-Jones size parameter (collision diameter). As presented in Figure 7.21, the existing models (Zarragoicoechea & Kuz, 2002, 2004; Jin, 2013) cannot match well with the collected data, especially when σ_{LJ}/r_p is greater than 0.2. A more accurate correlation between the critical property shift and the pore size is needed.

Based on the 15 collected data tabulated in Table 7.9, it is found that there is a linear relationship between the natural logarithm of critical temperature shift and the natural logarithm of dimensionless pore size, which is plotted in Figure 7.22. A correlation can be developed with $R^2 = 0.9698$, which is expressed as (Yang et al., 2018)

Table 7.9 Database of the Critical Temperature Shift Data

References	Method	Number of data points	Fluid	r_p/σ_{LJ}	ΔT^*	ln (r_p/σ_{LJ})	ln (ΔT^*)
Burgess et al., 1990	Experiment	1	CO_2	7.8	0.141	2.048	−1.962
Morishige et al., 1997	Experiment	4	C_2H_4	4.1	0.476	1.409	−0.743
				4.8	0.423	1.563	−0.861
			CO_2	3.4	0.431	1.228	−0.841
				4.4	0.359	1.479	−1.024
Singh & Singh, 2011	Molecular simulation	10	Square–well fluid	40	0.009	3.689	−4.709
				30	0.013	3.401	−4.334
				20	0.022	2.996	−3.811
				16	0.030	2.773	−3.496
				12	0.051	2.485	−2.979
				8.0	0.106	2.079	−2.247
				6.0	0.163	1.792	−1.813
				5.0	0.228	1.609	−1.480
				4.0	0.311	1.386	−1.169
				3.0	0.336	1.099	−1.090

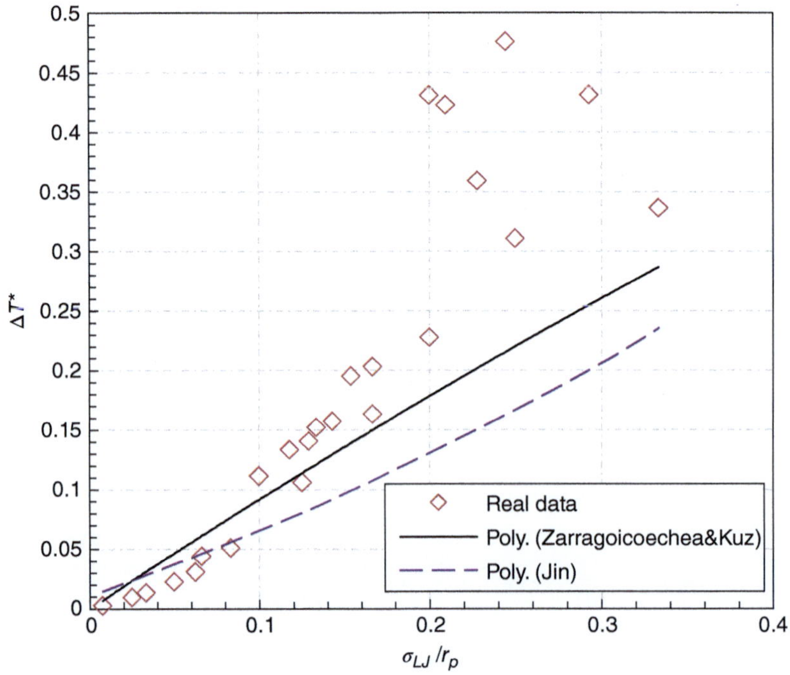

Figure 7.21 Comparison of the real critical temperature shift data with two models.

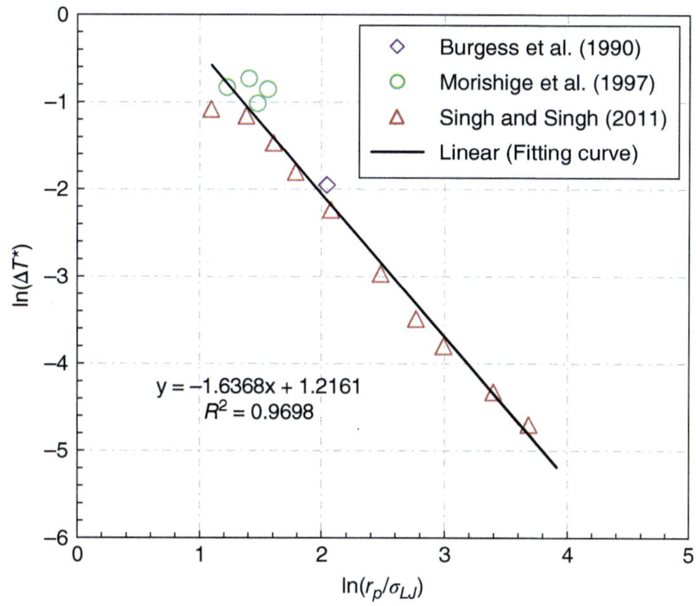

Figure 7.22 Correlation of critical temperature shift with a dimensionless pore size (Yang et al., 2018).

$$\ln(\Delta T^*) = -1.6368 \times \ln(r_p/\sigma_{LJ}) + 1.2161. \quad (7.58)$$

The critical temperature shift can be then derived as

$$\Delta T^* = 3.374 \times (r_p/\sigma_{LJ})^{-1.637}. \quad (7.59)$$

As can be seen from the above correlation and Figure 7.22, the critical temperature shift decreases exponentially with the increase of dimensionless pore size, that is, molecule-wall interaction weakens significantly as the pore size increases. As such, a specific pore size exists where the effect of molecule-wall interaction can be neglected and the original PR EOS can be applied.

Modified PR EOS

The PR EOS can be modified by introduction of the molecule-wall interaction term c, which is shown in equation (7.53). The determination of the c term is introduced below (Yang et al., 2018).

$$p = \frac{RT}{v-b} - \frac{a-c}{v(v+b)+b(v-b)}.$$

For the cubic EOS, an inflection point exists at the critical point. Mathematically, the first and second derivatives of pressure with respect to the volume at a constant temperature are equal to zero at the critical point,

$$\left(\frac{\partial p}{\partial v}\right)_{T=T_c} - \left(\frac{\partial^2 p}{\partial v^2}\right)_{T-T_c} = 0. \quad (7.60)$$

The formulae of critical pressure and critical temperature of the modified PR EOS can be derived as follows:

$$p_{cc} = 0.01324\frac{a-c}{b^2} \quad (7.61)$$

and

$$T_{cc} = 0.17015\frac{a-c}{Rb}, \quad (7.62)$$

where p_{cc} and T_{cc} are the critical pressure and critical temperature for confined fluids. With the critical pressure (p_c) and critical temperature (T_c) of the original PR EOS, the shift of critical properties is defined as follows:

$$\Delta p^* = \frac{p_c - p_{cc}}{p_c} = \frac{c}{a} \tag{7.63}$$

and

$$\Delta T^* = \frac{T_c - T_{cc}}{T_c} = \frac{c}{a}, \tag{7.64}$$

where Δp^* and ΔT^* are the critical pressure shift and the critical temperature shift, respectively.

By comparing equation (7.59) and equation (7.64), it yields,

$$\Delta T^* = \frac{T_c - T_{cc}}{T_c} = \frac{c}{a} = 3.374 \times (r_p/\sigma_{LJ})^{-1.637}. \tag{7.65}$$

Thus, the effect of molecule-wall interaction can be represented by the critical property shift as

$$c = a \times 3.374 \times (r_p/\sigma_{LJ})^{-1.637}. \tag{7.66}$$

Substituting equation (7.66) in equation (7.53), the modified PR EOS with the consideration of critical temperature shift is derived as follows:

$$p = \frac{RT}{v-b} - \frac{a\left[1 - 3.374 \times (r_p/\sigma_{LJ})^{-1.637}\right]}{v(v+b) + b(v-b)}. \tag{7.67}$$

A mixture containing 75 mol% methane, 20 mol% n-butane, and 5 mol% n-octane (C_1+nC_4+nC_8) is applied to the modified EOS model. Figure 7.23 is the phase diagram for the mixture confined in pores with diameters of 5, 10, 20, 50 nm, respectively, by only considering the shift of critical point. It can be seen that both critical pressure and critical temperature decrease as the pore size decreases, resulting in a shrunken phase envelope. Also, when the pore size is greater than 50 nm (a threshold for this case), that is, out of the range of mesopores (2–50 nm), the shift of phase diagram can be neglected, where the modified PR EOS can be degraded into the original PR EOS. One should note that the threshold pore size varies with different fluids. Figure 7.24 plots the effect of capillary pressure on phase envelope without considering the critical point shift. The bubble point

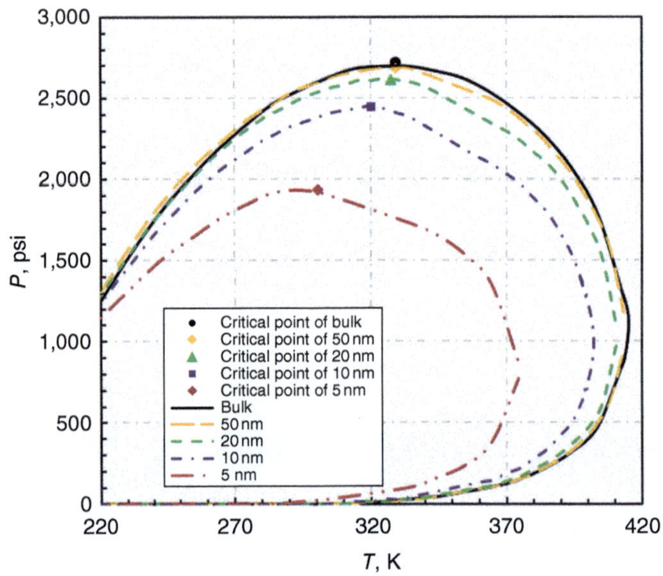

Figure 7.23 Phase diagram of the mixture of C_1 + nC_4 + nC_8 in bulk phase and confined space of 5, 10, 20, 50 nm by considering only the effect of critical point shift.

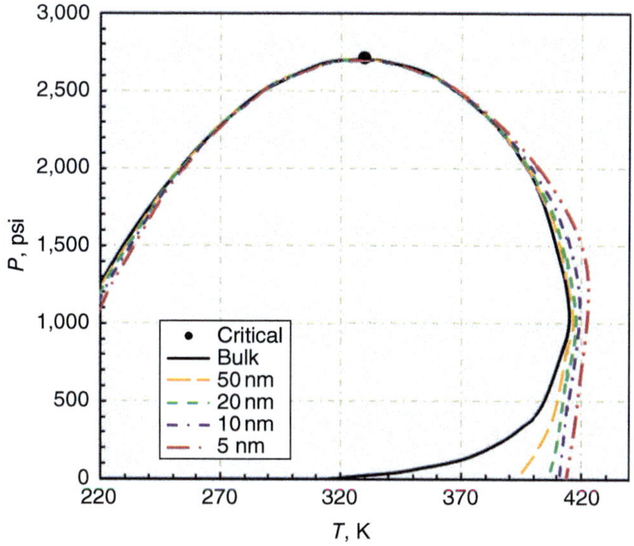

Figure 7.24 Phase diagram for the mixture of C_1 + nC_4 + nC_8 in bulk phase and confined space of 5, 10, 20, 50 nm by considering only the effect of capillary pressure.

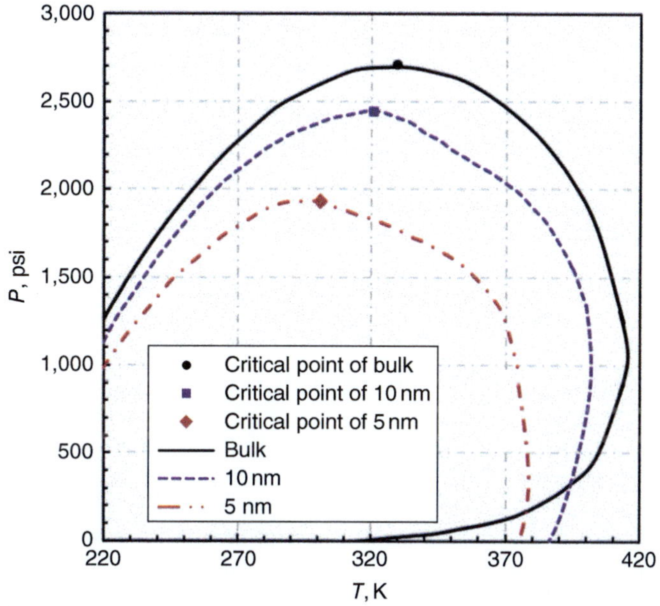

Figure 7.25 Phase diagram of C_1 + nC_4 + nC_8 at various pore sizes by considering both critical point shift and capillary pressure.

curve shifts toward the inside of the bulk phase diagram, whereas the dew point curve shifts toward the outside of the bulk phase diagram. Compared with capillary pressure (Fig. 7.23), the effect of critical point shift (Fig. 7.24) in confined space is more significant. The effect of both capillary pressure and critical property shift is combined and presented in Figure 7.25. It is demonstrated that the bubble point curve, critical point, and the most part of the dew point curve are shrinking in the confined space. The lower part of the dew point curve tends to be wider compared with the bulk fluid phase envelope.

7.4.4 Phase Diagrams of Bakken, Eagle Ford, and Wolfcamp Fluids

This section presents the predicted phase diagrams of three shale reservoir fluids by using the modified EOS with the consideration of confinement effect (see Fig. 7.20 and equation (7.67)), i.e., Bakken shale, Eagle Ford shale, and Wolfcamp shale. The compositions, properties, and binary interaction parameters are tabulated in Table 7.10a–f (Li & Sheng, 2017) (Stimpson, 2017) (Nojabaei et al., 2013). The phase diagrams of Bakken oil, Eagle Ford oil, and Wolfcamp oil at 10 nm, 20 nm, 40 nm, and bulk condition are calculated and plotted in Figure 7.26. The reservoir temperatures are indicated by vertical dashed lines on the graph. It can be seen that the change of phase envelope with the pore size is consistent with the findings for the synthetic mixture of C_1+nC_4+nC_8. Eagle Ford shale oil has the lowest critical point and its shift is the most insignificant. The confinement effect at 40 nm is found to be minor. However, it cannot be neglected when the pore size decreases to 10 nm.

Table 7.10a Compositional Data for Bakken Oil

Component	Molar fraction	P_c, psia	T_c, K	Acentric factor	Molar weight	Parachor coefficient
C_1	0.36736	655.02	186.30	0.0102	16.535	74.8
C_2	0.14885	721.99	305.54	0.1028	30.433	107.7
C_3	0.09334	615.76	369.98	0.1520	44.097	151.9
C_4	0.05751	546.46	421.78	0.1894	58.124	189.6
C_5–C_6	0.06406	461.29	486.38	0.2684	78.295	250.2
C_7–C_{12}	0.15854	363.34	585.14	0.4291	120.562	350.2
C_{13}–C_{21}	0.07330	249.61	740.05	0.7203	220.716	590.0
C_{22}–C_{80}	0.03704	190.12	1,024.72	1.0159	443.518	1,216.8

Source: Nojabaei et al. (2013).

Table 7.10b Binary Interaction Parameters for Bakken Oil

Components	C_1	C_2	C_3	C_4	C_5–C_6	C_7–C_{12}	C_{13}–C_{21}	C_{22}–C_{80}
C_1	0	0.005	0.0035	0.0035	0.0037	0.0033	0.0033	0.0033
C_2	0.005	0	0.0031	0.0031	0.0031	0.0026	0.0026	0.0026
C_3	0.0035	0.0031	0	0	0	0	0	0
C_4	0.0035	0.0031	0	0	0	0	0	0
C_5–C_6	0.0037	0.0031	0	0	0	0	0	0
C_7–C_{12}	0.0033	0.0026	0	0	0	0	0	0
C_{13}–C_{21}	0.0033	0.0026	0	0	0	0	0	0
C_{22}–C_{80}	0.0033	0.0026	0	0	0	0	0	0

Source: Nojabaei et al. (2013).

Table 7.10c Compositional Data for Eagle Ford Oil

Component	Molar fraction	P_c, psia	T_c, K	Acentric factor	Molar weight	Parachor coefficient
C_1	0.5816	667.00	190.56	0.0110	16.040	74.05
C_2	0.0744	706.60	305.33	0.0990	30.070	112.91
C_3	0.0417	616.10	369.83	0.1520	44.100	154.03
C_4	0.0259	542.19	418.71	0.1948	58.120	189.30
CO_2	0.0232	1,069.50	304.11	0.2250	44.010	82.00
C_5–C_6	0.0269	489.75	485.60	0.2398	76.502	247.57
C_7–C_{12}	0.1321	392.81	606.69	0.3548	122.960	402.33
C_{13}–C_{21}	0.0942	226.23	778.31	0.7408	255.280	834.82

Source: Stimpson (2017).

Table 7.10d Binary Interaction Parameters for Eagle Ford Oil

Components	C_1	C_2	C_3	C_4	CO_2	C_5-C_6	C_7-C_{12}	C_{13}-C_{21}
C_1	0	0	0	0	0.12	0	0	0
C_2	0	0	0	0	0.12	0	0	0
C_3	0	0	0	0	0.12	0	0	0
C_4	0	0	0	0	0.12	0	0	0
CO_2	0.12	0.12	0.12	0.12	0	0.12	0.12	0.1
C_5-C_6	0	0	0	0	0.12	0	0	0
C_7-C_{12}	0	0	0	0	0.12	0	0	0
C_{13}-C_{21}	0	0	0	0	0.1	0	0	0

Source: Stimpson (2017).

Table 7.10e Compositional Data for Wolfcamp Oil

Component	Molar fraction	P_c, psia	T_c, K	Acentric factor	Molar weight	Parachor coefficient
C_1	0.3383800	655.02	186.30	0.0102	16.535	74.8
C_2	0.0879662	721.99	305.54	0.1028	30.433	107.7
C_3	0.0969324	615.76	369.98	0.1520	44.097	151.9
C_4	0.0601436	546.46	421.78	0.1894	58.124	189.6
C_5–C_6	0.0881794	461.29	486.38	0.2684	78.295	250.2
C_7–C_{12}	0.1893056	363.34	585.14	0.4291	120.562	350.2
C_{13}–C_{21}	0.0768879	249.61	740.05	0.7203	220.716	590.0
C_{22}–C_{80}	0.0622049	190.12	1,046.94	1.0159	443.513	1,216.8

Source: Li and Sheng (2017).

Table 7.10f Binary Interaction Parameters for Wolfcamp Oil

Components	C_1	C_2	C_3	C_4	C_5–C_6	C_7–C_{12}	C_{13}–C_{21}	C_{22}–C_{80}
C_1	0	0.0500	0.0035	0.0035	0.0037	0.0033	0.0033	0.0033
C_2	0.0500	0	0.0031	0.0031	0.0031	0.0026	0.0026	0.0026
C_3	0.0035	0.0031	0	0	0	0	0	0
C_4	0.0035	0.0031	0	0	0	0	0	0
C_5–C_6	0.0037	0.0031	0	0	0	0	0	0
C_7–C_{12}	0.0033	0.0026	0	0	0	0	0	0
C_{13}–C_{21}	0.0033	0.0026	0	0	0	0	0	0
C_{22}–C_{80}	0.0033	0.0026	0	0	0	0	0	0

Source: Li and Sheng (2017).

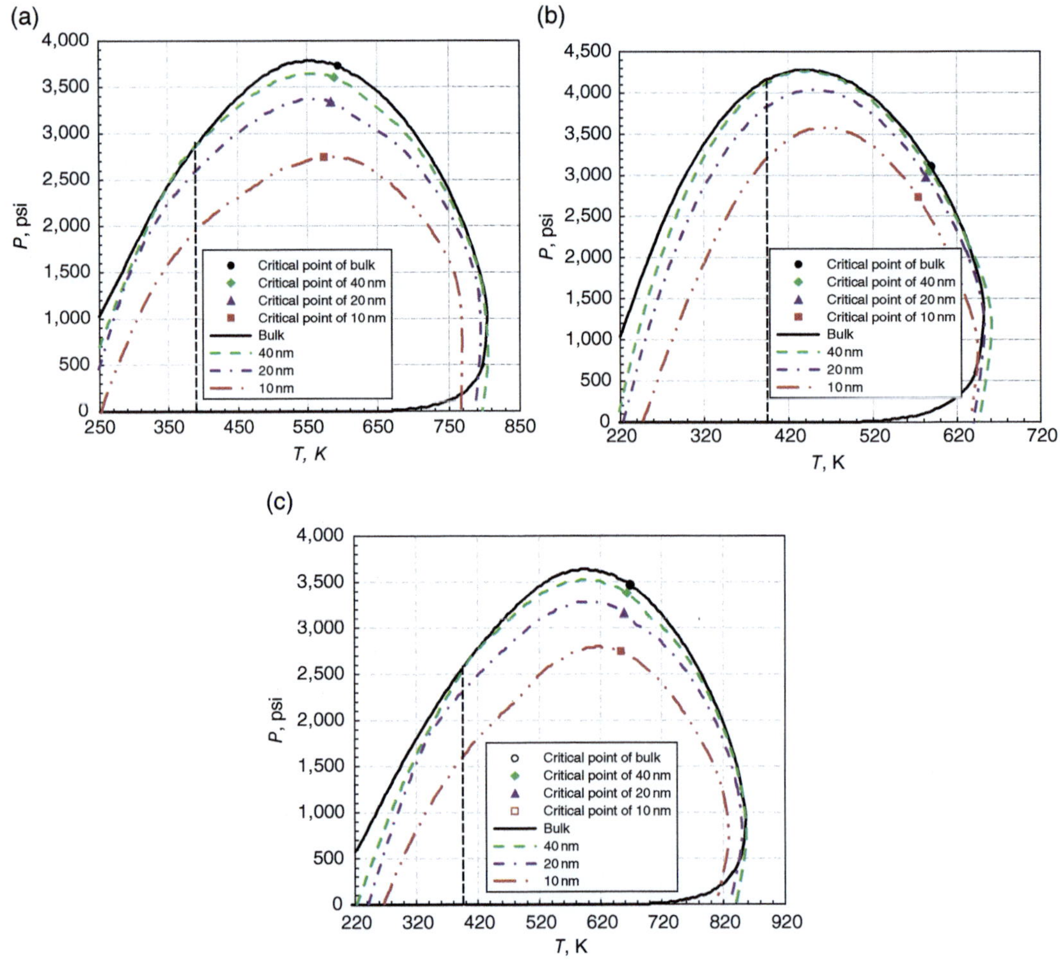

Figure 7.26 Phase diagrams of reservoir fluids at various pore sizes. Vertical dashed lines are reservoir temperatures. (a)Bakken oil, (b) Eagle Ford oil, (c)Wolfcamp oil.

7.4.5 Diffusion Coefficient

Diffusion is the random migration of molecules or particles driven by the tendency of systems to occupy all the possible states. It is a spontaneous tendency of all systems to equalize concentration without any external influence impeding this process. The quantitative study of diffusion dated back to 1850–1855 with the work of Adolf Fick and Thomas Graham. Fick's first equation (Fick, 1855) to describe macroscopically the diffusion process is a linear relationship between the matter flux, J, and the concentration gradient, ∇C:

$$J = -D\nabla C, \tag{7.68}$$

where D is the Fickian diffusion coefficient or transport diffusion coefficient.

Different types of forces under various pore sizes resulted in the difference of the diffusion types in microporous and macroporous materials, such as the gaseous or molecule diffusion, Knudsen diffusion, or liquid diffusion (Roque-Malherbe, 2018). Gaseous flow takes place when the pore diameter is larger than the mean free path of the fluid molecules. The collisions between molecules are more frequent than those between molecules and the pore wall. As the pore size decreases, the fluid molecules tend to collide more with the pore walls instead of among themselves, then the molecules are flowing almost independently from one another, Knudsen diffusion takes over. In the even smaller pore size range, like 2 nm or less, the molecules will constantly experience the interaction with the pore walls, resulting in the diffusion state taking place in the configurational diffusion regime.

Experimental studies of diffusion in microporous media have been carried out by different methods. In microporous adsorbent, there is no clear distinction between molecules on the surface and the molecules in the gas phase because of the adsorption process. The obtained Fickean diffusion coefficient, D, must be corrected in all the cases, in order to get the self-diffusion coefficients, D^*, with the equation below (Barrer & Jost, 1949),

$$D^* = D(1-\theta), \tag{7.69}$$

where $\theta = n_a/N_a$ is the fractional saturation of the adsorbent, in which n_a is the amount adsorbed, and N_a is the maximum amount adsorbed. The calculation of θ can be carried out by the osmotic adsorption isotherm equation (Roque-Malherbe et al., 1995),

$$\theta = \frac{n_a}{N_a} = \frac{Kp^B}{1+Kp^B}, \tag{7.70}$$

where K and B are constants, fitting the experimental isotherms obtained with the equilibrium sorption values.

7.5 EOS Regression to Experimental Data

The EOS is an excellent mathematical method to simulate the PVT experimental data for the well-defined components. The predictive capability of the EOS model is highly dependent on the accuracy of the fluid characterization. However, the petroleum fluids are considered poorly characterized components. The composition, molecular structure, and properties of each fraction can be estimated from various experimental techniques, though there are numerous uncertainties existing in the estimation process, particularly for the plus fraction of a reservoir fluid. It is common that the simulated PVT results from the EOS without tuning deviate from the experimentally measured data leading to an error of ±5%–25%. The objective of regression is mainly to match the phase behavior data (i.e., saturation pressures) and volumetric behavior data (i.e., density, Y-function, swelling factor, formation volume factor).

In addition, it is common that unconventional plays have a range of fluid compositions that can vary spatially, for example, all of fluids types of wet gas, retrograde condensate, volatile oil, and black oil can be found in the Eagle Ford and the Wolfcamp shales. Allouache et al. (2017) have proposed a methodology that a single EOS tuned across the full range of fluid types encountered in a shale play can be used with a reservoir fluid composition correlation to estimate the properties of any fluid within the play. It includes four steps: (1) generate the reservoir fluid composition which is representative to the entire reservoir by measuring of thermal maturity; (2) tune the EOS model to match PVT properties for several individual wells with the same C_{7+} pseudo-component lumping; (3) average the properties for each C_{7+} pseudo-component; and (4) tune C_{7+} pseudo-component EOS parameters to best match condensate/gas ratio (CGR), density, saturation pressure, and viscosity of all individual fluids (Allouache et al., 2017).

In order to improve the predictive capability of the EOS model, it is reasonable to tune the following properties:

1) The physical and critical properties of the plus fraction including p_c, T_c, and ω. Since the values of these properties are commonly calculated from the correlations that were previously developed based on different fluid samples, they may be not the optimum choice for the given fluid sample. The tuning on properties of plus fraction can improve the matching to saturation pressure.
2) Binary interaction parameters (BIP). The BIP for a well-defined binary system can be accurately obtained by using the measured saturation pressure. It is usually a small value for hydrocarbon-hydrocarbon binary due to their similar properties. A large BIP indicates that there is a large property difference between two components. The BIP value for petroleum fluids is most commonly estimated by experience, which can be adjusted to match the measured PVT data. The tuning on BIP can improve the matching of saturation pressure.
3) Volume translation parameter (or volume-shift parameter). Due to the poor liquid density prediction of the EOS, Peneloux et al. (1982) proposed a volume-shift parameter to correct the volume calculated from the EOS. The addition of this parameter has no influence on gas-liquid phase equilibrium calculation results. It means that saturation pressure is not affected by including this parameter in the EOS. The tuning on volume-shift parameter can improve the matching to volume-related data, for example, density, Y-function, and swelling factor.

The absolute average relative deviation (AARD) can be used as the objective function to optimize saturation pressure or volumetric properties,

$$\mathrm{AARD} = \frac{1}{n}\sum_{i=1}^{n}\frac{X_i^{\mathrm{cal}} - X_i^{\mathrm{exp}}}{X_i^{\mathrm{exp}}}, \tag{7.71}$$

where n is the number of data points, X_i^{cal} is the calculated properties, and X_i^{exp} is the measured properties.

Figures 7.27–7.29 present the measured and simulated CCE data (i.e., saturation pressure,

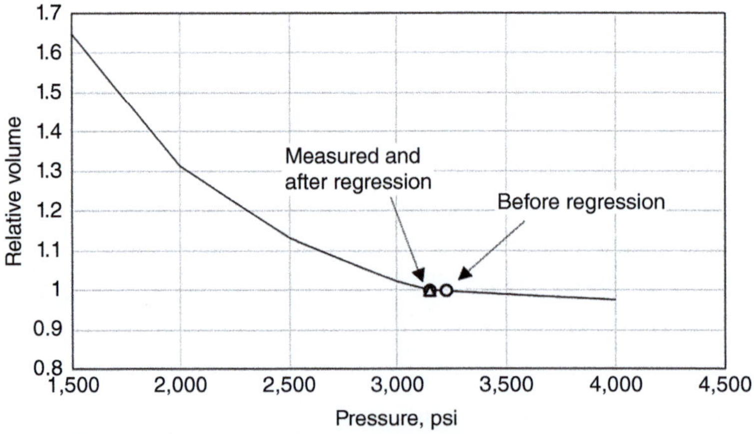

Figure 7.27 Measured and simulated saturation pressure of the Bakken shale oil. Data can be seen in Table 7.5.

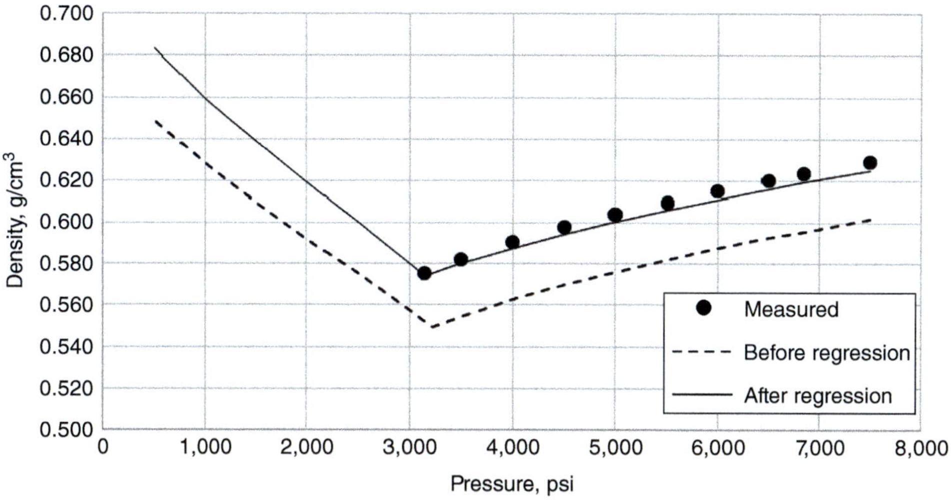

Figure 7.28 Measured and simulated density of the Bakken shale oil. Data can be seen in Table 7.5.

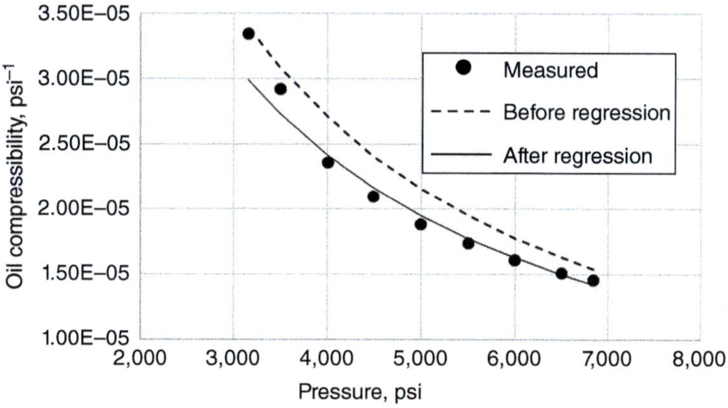

Figure 7.29 Measured and simulated compressibility of the Bakken shale oil. Data can be seen in Table 7.5.

oil density, and oil compressibility) for the Bakken shale oil sample before regression and after regression. The CCE data can be found in Table 7.5. The PR EOS (1978) is selected as the thermodynamic model. The tuning parameters are listed in Table 7.11.

7.6 Minimum Miscibility Pressure

Minimum miscibility pressure is the pressure at which interfacial tension (IFT) between the liquid (oil) and vapor (gas) phases becomes zero and the resulting mixture is a single phase. It is a key parameter governing the displacement efficiency of gas flooding applications. Many gas flooding applications in the oil industry are aimed to operate gas injection at a pressure above MMP for effective enhanced oil recovery. There are many methods to determine the MMP experimentally, analytically, and numerically. The experimental methods for measuring MMP include slim-tube test, rising bubble apparatus (RBA) test, and vanishing interfacial (VIT) technique. The analytical

Table 7.11 Tuning Parameters for Matching Saturation Pressure, Oil Density, and Oil Compressibility

Tuning parameter	Initial value	Final value	Change (%)
p_c of plus fraction	18.14	19.93	9.90
T_c of plus fraction	708.98	724.86	2.24
ω of plus fraction	0.64	0.65	1.70
Volume shift of plus fraction	0.00	0.21	100.00
Volume shift coefficient of plus fraction	0.00	−0.00239	−100.00
Exponent of BIP correlation	1.20	0.03	−97.63

methods include Equation of State Model (EOS) with different algorithms and Method of Characteristic (MOC). The numerical methods include slim-tube compositional simulation and multicontact mixing cell models. In addition, correlation methods have also been used for a quick and screening purpose to estimate MMP based on the composition range in which the correlations were developed.

Gas injection is conducted for improved oil recovery (IOR) applications in shale and tight oil reservoirs. Due to lack of connectivity between different SRV regions, gas injection is conducted in a huff-n-puff manner with an injection time, soaking time, and a production time period. Several cycles are recommended for each IOR project with CO_2 or produced hydrocarbon gases being the injected gas. Produced hydrocarbon gases are a more economically viable option when it comes to gas injection in shale oil reservoirs due to availability and feasibility of the separation process. Even though reaching MMP is an ideal objective for these gas injection projects, it is not practical to achieve full miscibility for most unconventional plays due to the cost and the number of very large and expensive compressors required to inject that volume of gas in such a short period of time.

7.6.1 Experimental Methods

The experimental methods for measuring MMP commonly used in the oil industry are slim tube test, rising bubble apparatus test, and vanishing interfacial tension technique.

Slim-Tube Test

Slim-tube test has been commonly applied and accepted as a standard method to determine MMP by conducting displacement test under reservoir conditions with a slim-tube apparatus. Figure 7.30 is a typical schematic of slim-tube test equipment. The slim tube consists of a coiled, stainless steel tube with 5/16 in. ID, and various length from 20 to 80 ft (Fig. 7.30). The tube is either packed uniformly with fine grade sand or glass beads of a size on the order of 100 mesh. In operation, the MMP of a gas/oil is determined by displacing an oil-saturated slim tube with gas injection for at least four different pressures; then the MMP is defined as the lowest pressure at which near the maximum recoverable of oil was recovered by 1.2 PV of gas injected (Metcalfe et al., 1973). Other definition of MMP has also been used in the slim-tube test. Holm and Josendal defined MMP as the pressure when more than 80% of oil in place was recovered at gas breakthrough, and more than 94% was recovered when the gas oil ratio (GOR) reached 4,000 scf/bbl. In practice, the MMP is defined graphically by the intersection of two lines (containing at least two data points in each line) that define both immiscible and miscible performance regimes on a plot of oil recovery versus pressures. This technique is time-consuming and it may take several weeks to complete one miscibility

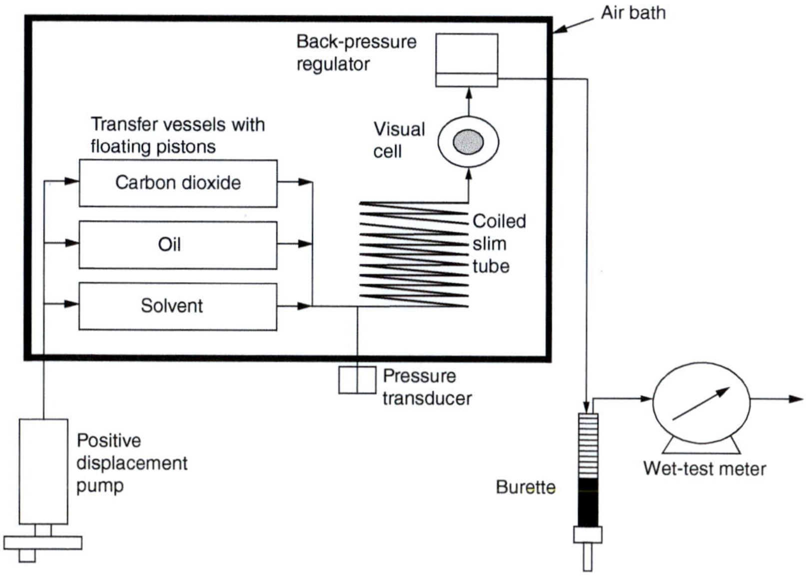

Figure 7.30 Schematic of slim-tube test apparatus (Elsharkawy, 1996).

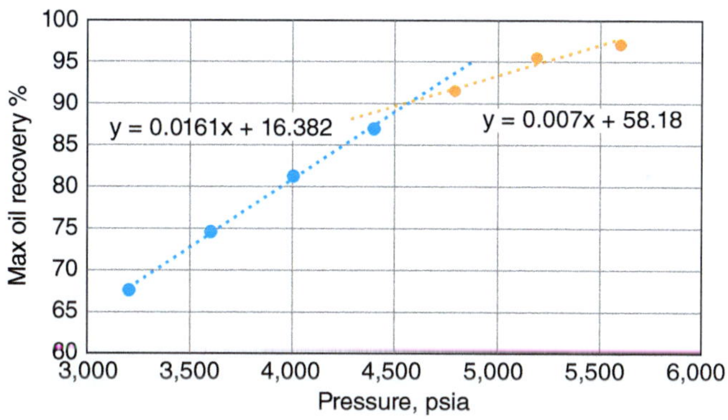

Figure 7.31 Liquid volume matching before and after regression of Eagle Ford sample (Al-Ajmi et al., 2010).

measurement, and therefore is also expensive. Figure 7.31 presents an example of slim-tube test for MMP measurement. The pressure determined at the intersection of two lines shows that 2,665 psia was the MMP of oil and CO_2 at 198°F.

Rising Bubble Apparatus (RBA) Test

Compared with the time-consuming slim-tube test, the RBA method is a relatively faster technique to estimate MMP. Figure 7.32 shows the rising bubble apparatus used to measure MMP. In this test, a gas bubble is injected into the oil column from the bottom of the apparatus at a specified system pressure. The MMP is determined from direct observations of the bubble behavior during the rising process in the test. When the rising gas bubble vanishes in the column of oil, the pressure is defined as the MMP. Figure 7.33 shows an example of bubble behavior for vaporizing gas process. At pressure far below MMP, the gas bubble usually shrinks

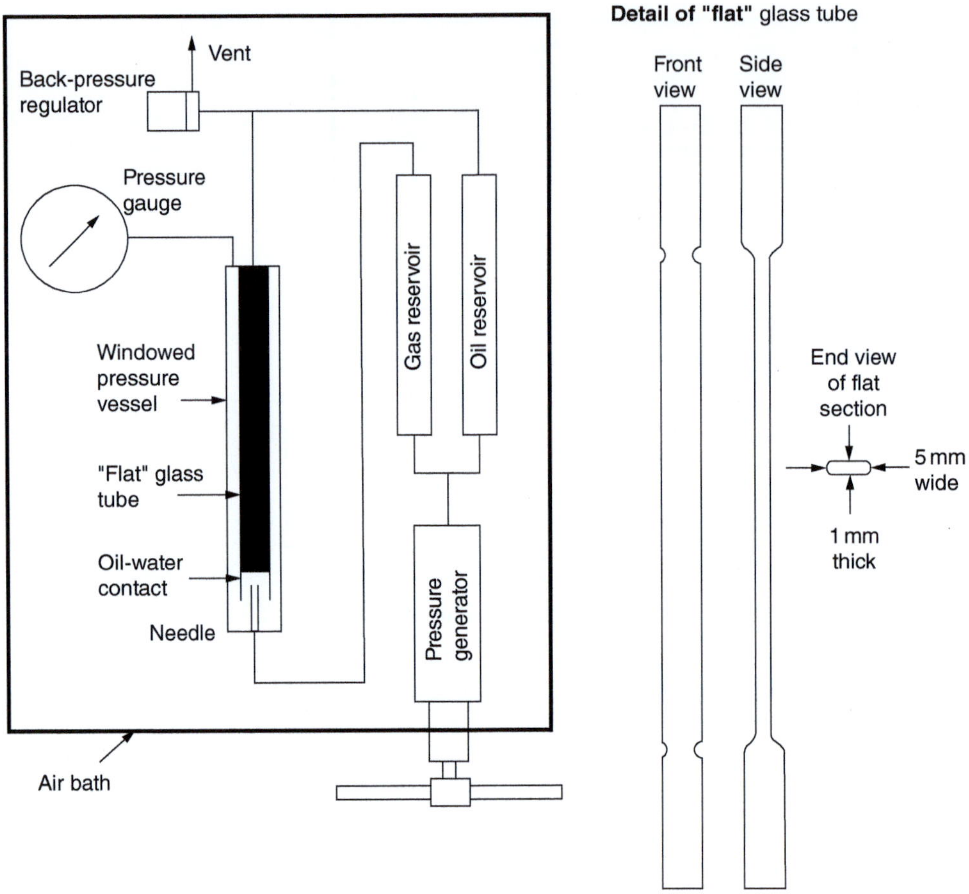

Figure 7.32 Schematic of a rising bubble apparatus (Elsharkawy, 1996).

as it rises to the top of the tube as gas is dissolved in the surrounding oil. At the MMP, the gas bubble degrades as it moves up and quickly disperses into the oil. This method is considerably faster and requires smaller amounts of samples compared with slim-tube testing. However, the determination of MMP with RBA technique is subjective since its interpretation is based on the bubble behavior.

Vanishing Interfacial Tension Technique (VIT)

When miscibility is achieved, the interface between the two fluids disappears and the interfacial tension between the two phases is zero. In VIT method, the gas oil interfacial tension is measured at reservoir temperature and at varying pressures. Figure 7.34 shows the schematic of the high-pressure optical cell and its accessories used in the VIT experiments by Rao and Lee (2002). The pendant drop technique was used to determine the interfacial tension (IFT) between oil and gas by capturing the image of an oil drop hanging from a capillary tube and analyze it for IFT calculation at different pressures. The gas oil miscibility conditions are then determined by extrapolating the plot of interfacial tension against pressure. Figure 7.35 shows an example of MMP estimated by VIT method and slim-tube test where

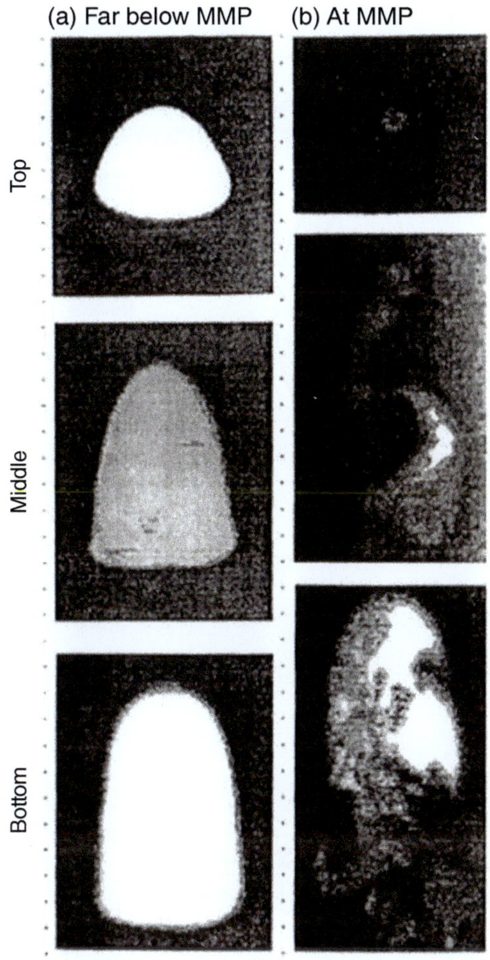

Figure 7.33 Bubble behavior for vaporizing gas process at pressure (a) far below MMP and (b) at MMP (Elsharkawy, 1996).

the MMP for Gilwood stock tank oil with ethane determined by slim-tube test was 7.2–7.4 MPa, while it was 7.53 MPa by VIT method.

Vanishing interfacial tension technique is direct and quick for determination of miscibility, however, it serves only as a single contact technique and may not account for multistages of contact between the fluid phases that are normally observed in the slim tube with conventional multiple contact displacement processes. Recently, Hawthoren et al. (2016) reported a modified VIT method in estimation of MMP for Bakken oil and various injecting gases. They measured the oil heights in three different diameter capillary tubes at various pressures with gas of interest and then extrapolated the oil height versus pressure plot to zero oil height to yield the MMP. Figure 7.36 shows photos with the oil heights in three capillaries (1.12, 0.84, and 0.68 mm inner diameters from right to right in all photos) at low pressure (right photo), the pressure at approximately half of MMP (middle photo), and as the pressure approaches MMP (right photo). Figure 7.37 capillary rise/VIT plot for the MMP of Bakken crude oil with pure CO_2 at 110°C (Hawthoren et al., 2016) shows the estimation of MMP of Bakken crude oil with pure CO_2 at 110°C. The oil heights versus pressure in different capillaries approaches zero at pressure of 17.4 MPa, which is defined as MMP. This method reduces the cost and complexity of the instrumentation to estimate MMP and gave good agreement for a live Bakken oil with the results from slim-tube experiment.

Swelling/Extraction Test

For carbon dioxide and crude oil system, the MMP can be determined through analyzing the experimental data of swelling/extraction tests (Hand & Pinczewski, 1990; Tsau & Willhite, 2010; Abedini et al., 2014). The swelling factor of oil is increased due to the dissolution of carbon dioxide until a pressure at which light hydrocarbon extraction occurs and the swelling factor starts to decrease. Abedini et al. (2014) reported that two distinct regions (upper extraction phase and lower extraction phase) were formed in the extraction process in a swelling factor versus pressure plot. By plotting two lines through these two regions, the pressure at the intersection of these two lines is defined as MMP. The MMP determined by this method is in reasonable agreement with slim-tube and VIT method. Figure 7.38 MMP estimated with swelling test (Abedini et al., 2014) presents an example of MMP estimated with swelling test. The test was conducted at 40°C with CO_2 and oil. The pressure at the intersection of the two lines, 10.22 MPa, is the MMP.

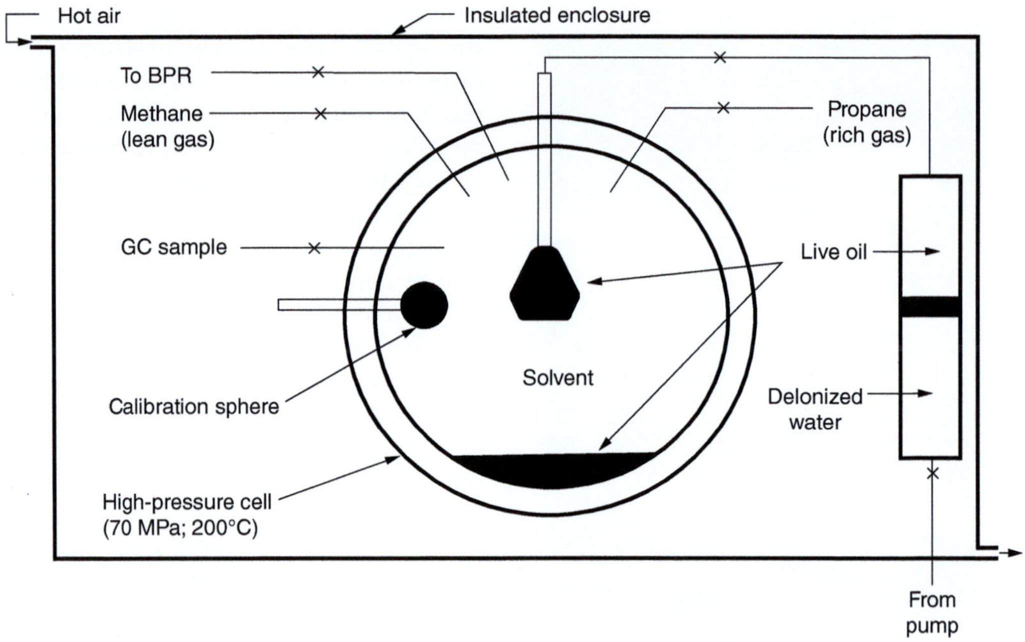

Figure 7.34 High-pressure optical cell used in VIT experiments (Rao & Lee, 2002).

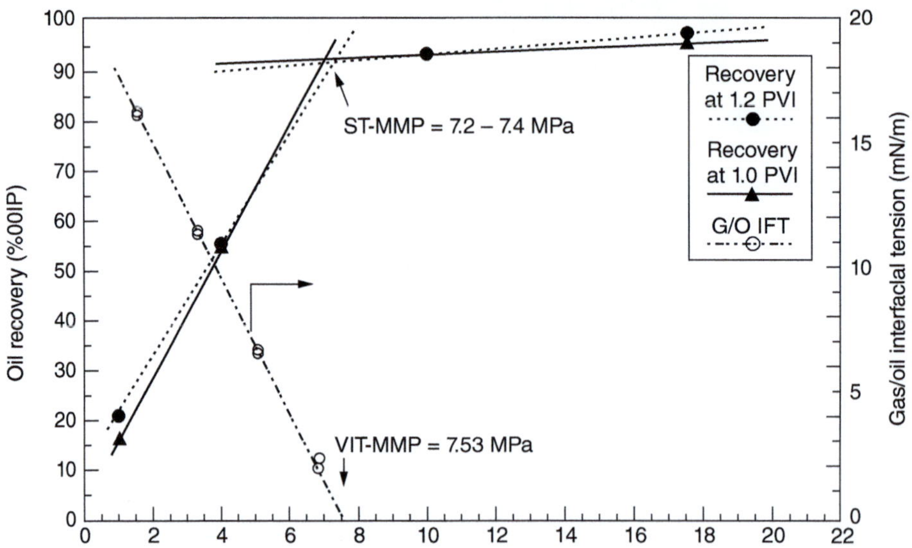

Figure 7.35 MMP estimation from slim-tube and VIT tests for Gilwood stock tank oil with ethane (Rao & Lee, 2002).

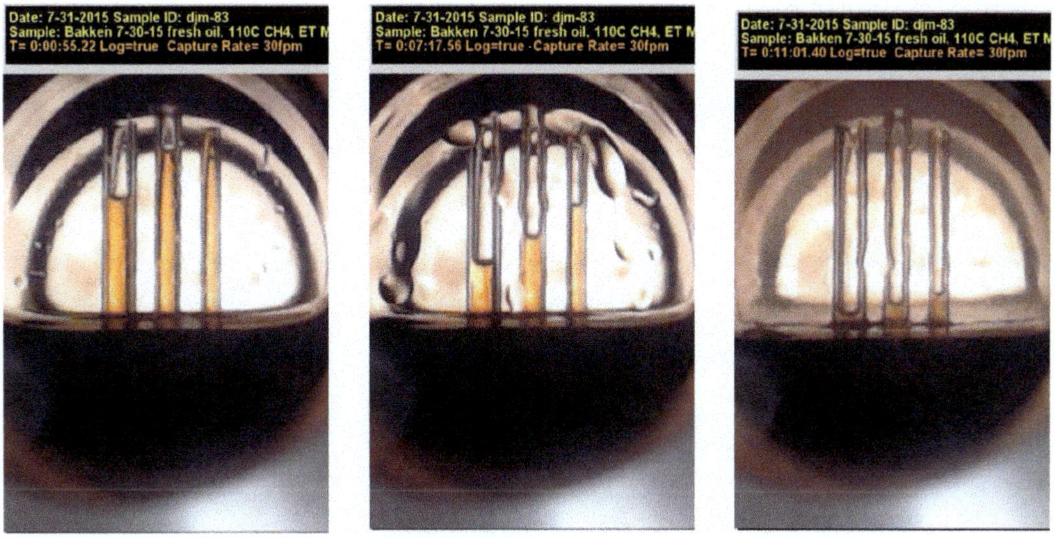

Figure 7.36 Photos of the oil height in three capillaries at (left) low pressure, (middle) one-half of MMP, and (right) the pressure approaches MMP (Hawthoren et al., 2016).

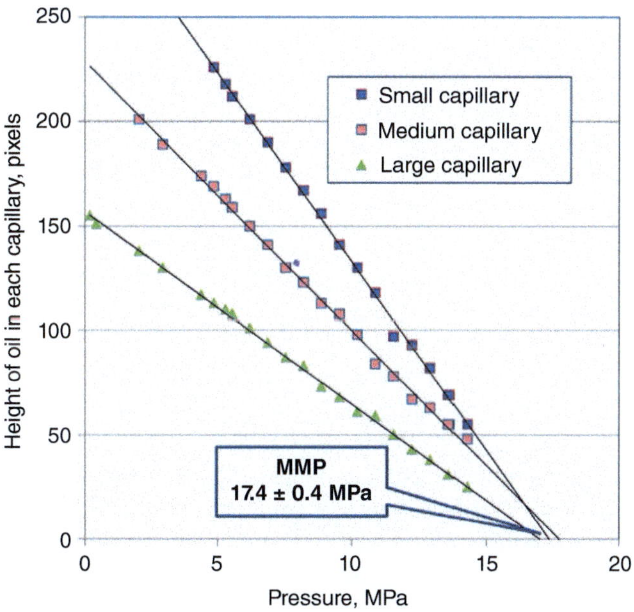

Figure 7.37 Capillary rise/VIT plot for the MMP of Bakken crude oil with pure CO_2 at 110 °C (Hawthoren et al., 2016).

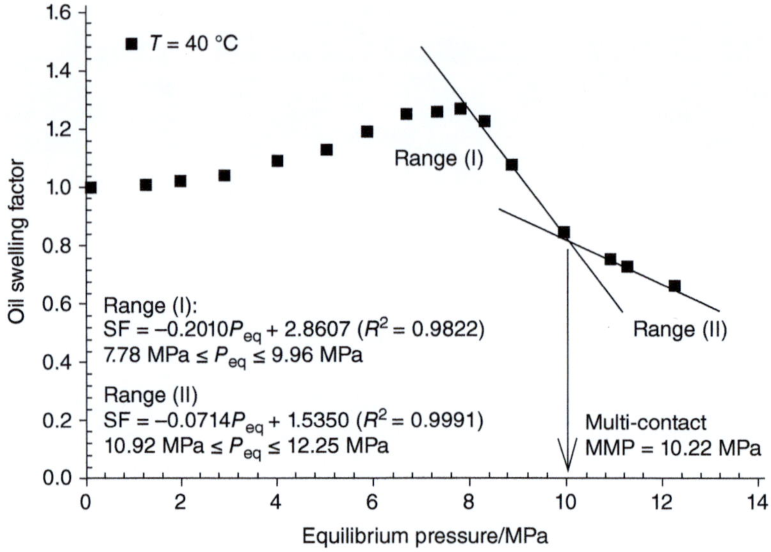

Figure 7.38 MMP estimated with swelling test (Abedini et al., 2014).

7.6.2 Analytical Methods

The analytical MMP calculation method is based on the analytical theory of multicomponent gas-injection processes using appropriate equations of state (EOS) to solve one-dimensional dispersion free flow equation. Several researchers (Dumore et al., 1984; Monroe et al., 1990) presented analytical theory for gas-displacement process and showed that there exists a third key tie line in the displacement route, which is called crossover tie line. The existence of the crossover tie line for condensing/vaporizing process later was confirmed by other researchers (Orr et al., 1993; Johns, 1993). The MMP was demonstrated occurring when one of the key tie lines first intersects a critical point and the tie line length becomes zero. In this approach, method of characteristic (MOC) is used to find the key tie lines that bound the composition paths in a one-dimensional displacement. These key tie lines are found such that each key tie line when extended out of the two-phase region must intersect two neighboring key tie lines. As pressure is increased, the key tie lines are determined until one of them first intersects a critical point. Khorsandi and Johns (2015) present detailed discussion and examples of tie-line solutions for MMP calculation from ternary to five-component displacement processes.

Another method is multiple mixing cell method, which is mixing (analytically) gas and oil in repeated contacts, resulting in new equilibrium compositions. In vaporizing drive (lean-gas injection), the intermediate component in the oil is vaporized into the more mobile gas phase, causing the equilibrium gas composition to move toward the oil tie line and the miscibility is developed at the leading edge of the displacement. On the other hand, when the intermediate component in the gas is condensed in the oil for a condensing gas drive (enriched-gas injection), the miscibility is developed at the trailing edge of the displacement. This method gives reliable MMPs for either condensing or vaporizing drives

(Metcalfe et al., 1973; Zick, 1986). For a combined condensing/vaporizing drive, however, this method fails to give a reliable MMP. The improvement can be made by incorporating MOC method to estimate MMP and, in their report, Ahmadi and Johns (2011) demonstrated a simple and accurate model to estimate the MMP for any drive mechanism.

7.6.3 Numerical Methods

The common numerical methods to estimate MMP are based on simulation of slim-tube model and multiple-mixing-cell model. The assumption of these computational methods is that the phase behavior is accurately described with a tuned EOS. The simulation of slim-tube test is constructed with one-dimensional compositional model and conducted at operation conditions similar to the laboratory experiment. Because of the potential numerical dispersion involved in the simulation, using a sufficient number of grids in the model is always a good practice to mitigate the effect of numerical dispersion on displacement performance. Stalkup (1983) suggested plotting oil recovery versus one over square root of number of grids to minimize the error incurs in determination of MMP due to the numerical dispersion in calculation.

Similarly, the multiple-mixing-cell method also uses the compositional approach to calculate MMP with a reasonable number of mixing cells or steps. In this method, the gas is injected into the first of a sequence of mixing cells. An equilibrium flash calculation is performed and liquid and vapor volumes in excess of the cell volume are moved to the next cell. The sequence of calculation continues until the compositions in the final cell approach a critical composition. The numerical calculations are performed at different pressure and the MMP is defined as the minimum pressure at which the critical composition is reached in the final cell of multiple mixing cells.

7.6.4 Correlation Methods

Correlation methods reported in the literature offer a quick screening means to estimate MMP, given the composition of fluids and test conditions are similar to what correlation was developed. These methods usually provide very limited success but give a fair first guess quickly.

A summary of major saturation pressure, GOR, and MMP for some major international shale oil plays around the world appears in Table 7.12.

7.7 Problems

Given the pure component composition in a Bakken oil sample (Table 7.13), and given the molecular weight and critical properties of each component (Table 7.13):

1. Lump the component into seven pseudo-components and calculate the new critical properties for the new pseudo-components.

2. Using the pseudo-components calculated in the previous section, use CMG's WinProp to tune the equation of state properties. Tune critical temperature for C_{19+} and compare the saturation pressure after regression with the one reported for this oil sample.

3. Continue tuning of the EOS using the Constant Composition Expansion (CCE) lab data reported in Table 7.14. Compare the results before and after tuning

Table 7.12 Summary of major saturation pressure, GOR, and MMP for some of the major international shale oil plays around the world.

Country	Oil	Temper-ature (F)	Method	Gas	MMP (psi)	Saturation P (psi)	GOR (SCF/STB)	Reference
United States	Bakken live oil	237	RBA	CO_2	3,300	1,500 ~2,400	460~870	Kurtoglu, 2013
	Bakken crude oil	220	RBA	CO_2 Methane Ethane 25/75 Ethane/ methane	2,521 4,512 1,358 3,180			Hawthorne et al., 2017
	Bakken live oil	240	Multimixing cell	CO_2	2,200			Zhang et al., 2018
	Bakken live oil	230	Multimixing cell	CO_2	2,540			Sorensen et al., 2018
	Bakken live oil	145.4	WinProp Calculation 1. Cell-to-cell method 2. Multiple-mixing cell	CO_2 CO_2-enriched flue gas Natural gas N_2 CO_2-enriched natural gas	1. / 2. 2,302.5 / 1,819.9 4,768.1 / 5,045.1 3,372.1 / 4,958.6 6,272.9 / 5,039.3 3,843.5 / 3,860.2		563.7	Li & Luo, 2016
	Midland basin	120~211				1,820~3,371	600~1,500	Forrest et al., 2018
	Permian basin live oil	162	RBA	Methane CO_2 Separator gas	3,900 3,600 3,800	3,479	1,230	Liu et al., 2018
	Wolf-camp crude	155	Slim tube exp	CO_2	1,925			Adel et al., 2018
	Eagle Ford live	237	Multimixing cell	CO_2	3,200			Zhang et al., 2018
	Eagle Ford crude	170	Slim tube exp	CO_2	2,132			Adel et al., 2018
Canada	Duvernay shale	230	Slim tube simulation	Rich gas Lean gas	4,700 6,200	3,724		Hamdi et al., 2018
China	Ordos basin	197.6	Slim tube	CO_2	2,872		365.5	Ran et al., 2012

Table 7.13 Pure Component Composition of a Bakken Oil Sample (LHS) in Addition to the Mw and Critical Properties of the Components (RHS) to Be Used for the Problem

Component	Mole %	Z_i	M_i	P_c	T_c	ω
N_2	1.73	1.73	28.01	33.50	126.20	0.04
CO_2	0.43	0.43	44.01	72.80	304.20	0.23
C_1	37.69	37.69	16.04	45.40	190.60	0.01
C_2	15.40	15.40	30.07	48.20	305.40	0.10
C_3	8.77	8.77	44.10	41.90	369.80	0.15
iC_4	1.04	1.04	58.12	36.00	408.10	0.18
nC_4	3.96	3.96	58.12	37.50	425.20	0.19
iC_5	1.05	1.05	72.15	33.40	460.40	0.23
nC_5	2.22	2.22	72.15	33.30	469.60	0.25
C_6	2.58	2.58	86.00	32.46	507.50	0.28
C_7	3.47	3.47	101.60	29.86	563.15	0.29
C_8	3.49	3.49	113.50	27.66	588.38	0.33
C_9	2.45	2.45	121.00	26.94	598.50	0.39
C_{10}	2.21	2.21	134.00	25.01	622.10	0.44
C_{11}	1.70	1.70	147.00	23.17	643.60	0.48
C_{12}	1.42	1.42	161.00	21.63	663.90	0.52
C_{13}	1.35	1.35	175.00	20.43	682.40	0.56
C_{14}	1.08	1.08	190.00	19.33	700.70	0.60
C_{15}	0.92	0.92	206.00	18.25	718.60	0.65
C_{16}	0.80	0.80	222.00	17.15	734.50	0.68
C_{17}	0.65	0.65	237.00	16.35	749.20	0.73
C_{18}	0.60	0.60	251.00	15.65	760.50	0.76
C_{19}	0.54	0.54	263.00	15.06	771.00	0.79
C_{20}	0.48	0.48	275.00	14.36	782.90	0.82
C_{21}	0.43	0.43	291.00	13.83	793.30	0.86
C_{22}	0.33	0.33	300.00	13.26	804.40	0.88
C_{23}	0.32	0.32	312.00	12.83	814.00	0.92
C_{24}	0.28	0.28	324.00	12.38	823.20	0.94
C_{25}	0.26	0.26	337.00	11.84	832.70	0.97
C_{26}	0.22	0.22	349.00	11.48	841.20	0.99
C_{27}	0.22	0.22	360.00	11.13	849.60	1.02
C_{28}	0.18	0.18	372.00	10.76	857.70	1.04
C_{29}	0.15	0.15	382.00	10.49	864.30	1.06
C_{30+}	1.58	1.58	545.69	8.00	900.00	1.20

Table 7.14 Constant Composition Expansion (CCE) Lab Data for the Bakken Sample

Pressure (psia)		Relative volume (V/V_{sat})	Oil density (g/cm^3)	Oil compress (V/V/psi) x 10^6	Y-function (P_{sat}-P)/P(V/V_{sat}-1)
7,500		0.914	0.629		
6,840	Reservoir	0.923	0.623	14.55	
6,500		0.928	0.620	15.11	
6,000		0.935	0.615	16.07	
5,500		0.943	0.610	17.37	
5,000		0.952	0.604	18.88	
4,500		0.962	0.598	20.96	
4,000		0.974	0.591	23.56	
3,500		0.988	0.582	29.20	
3,149	Saturation	1.000	0.575	33.41	
3,000		1.020			2.53
2,500		1.116			2.24
2,000		1.290			1.98
1,500		1.619			1.77
1,000		2.376			1.56
500		4.931			1.35

by adjusting all the volume shift parameters, interaction coefficients for CO_2 and C_{19+}, and hydrocarbon interaction coefficient.

4 Further tune the EOS using the differential liberation data reported in Table 7.15. Compare the simulated parameters with the lab data and report the error.

5 Further tune the model using the viscosity data reported in Table 7.15 and compare the results with the lab measurements.

6 Calculate the swelling factor for different mole fractions of CO_2 as an EOR injectant. Report the simulated saturation pressure as well as the swelling factor.

7 Conduct a MMP, slim-tube simulation study using L = 40 ft, Dy = Dz = 0.0185 ft, phi = 0.3317, and k = 7,100 mD, then report the estimated MMP for the Bakken oil with CO_2 as the injected gas. Run the simulation for at least five different pressures, plot the recovery factor at 1.2 PV CO_2 injection for different pressures, and determine MMP.

Table 7.15 Differential Liberation Data Measured for the Bakken Sample

Pressure		Oil density	Oil compress.	Oil viscosity	Liberated GOR, R_l	Solution GOR, R_{sd}	Oil FVF, B_{od}	Solution GOR, R_s	Sep. Adj. FVF, B_o
(psia)		(g/cm^3)	(V/V/psi) x 10^6	(cP)	(scf/bbl)	(scf/bbl)	(vol/ resid. vol)	(scf/bbl)	(vol/ST vol)
7,500		0.629		0.213	0	2,072	2.222	1,303	1.699
6,840	Reservoir	0.623	14.55	0.205	0	2,072	2.243	1,303	1.715
6,500		0.620	15.11	0.201	0	2,072	2.255	1,303	1.724
6,000		0.615	16.07	0.195	0	2,072	2.273	1,303	1.738
5,500		0.610	17.37	0.188	0	2,072	2.293	1,303	1.753
5,000		0.604	18.88	0.181	0	2,072	2.315	1,303	1.770
4,500		0.598	20.96	0.175	0	2,072	2.339	1,303	1.788
4,000		0.591	23.56	0.167	0	2,072	2.367	1,303	1.810
3,500		0.582	29.20	0.161	0	2,072	2.402	1,303	1.836
3,149	Saturation	0.575	33.41	0.156	0	2,072	2.431	1,303	1.858
2,500		0.610	24.02	0.191	523	1,549	2.109	974	1.674
2,000		0.634	21.83	0.225	835	1,237	1.932	778	1.574
1,500		0.658	20.16	0.269	1,080	992	1.792	624	1.493
1,000		0.680	18.90	0.321	1,315	758	1.664	476	1.421
500		0.705	18.13	0.395	1,557	516	1.529	324	1.343
15		0.768		0.659	2,072	0	1.097	0	1.097
15	at 60°F	0.846	API = 35.61				1.000		

Additional Reading

Firoozabadi, A. (1999). *Thermodynamics of hydrocarbon reservoirs*. New York: McGraw-Hill.

Jin, L., et al. (2013). *Investigating the effect of pore proximity on phase behavior and fluid properties in shale formations*. SPE Annual Technical Conference and Exhibition, New Orleans.

Forrest, J., Zhu, T., Xiong, H., & Pradhan, Y. (2018). *The effect of initial conditions and fluid PVT Properties on unconventional oil and gas recoveries in the Wolfcamp formation in the Midland basin*. Unconventional Resources Technology Conference, Houston, Texas.

References

Abedini, A., Mosavat, N., & Torabi, F. (2014). Determination of minimum miscibility pressure of crude oil-CO_2 system by oil swelling/ extraction test. *Energy Technology*, *2*, 431–439.

Adel, I. A., Tovar, F. D., Zhang, F., & Schechter, D. S. (2018). *The impact of MMP on recovery factor during CO_2-EOR in unconventional liquid reservoirs*. SPE Annual Technical Conference and Exhibition, Dallas, Texas.

Ahmadi, K., & Johns, R. T. (2011). Multiple-mixing-cell method for MMP calculations. *SPE Journal*, 733–742.

Ahmed, T. (2006). *Equations of state and PVT analysis*. Houston: Gulf Publishing Company.

Al-Ajmi, M., Alomair, O., & Elsharkawy, A. (2010). Planning miscibility tests and gas injection projects for four major Kuwaiti reservoirs. *Kuwait Journal of Science & Engineering*.

Allouache, A., Harrison, R., & Turek, E. (2017). *developing equations of state for unconventional source rock fluids*. SPE Annual Technical Conference and Exhibition, San Antonio.

Bagoo, D. E., Ramnarine, M. L., & Hernandez, M. (2014). *Validation and analysis of past PVT studies from a complex and mature offshore asset in Trinidad*. SPE Energy Resources Conference, Port of Spain.

Barrer, R. M., & Jost, W. (1949). A note on interstitial diffusion. *Transactions of the Faraday Society*, *45*, 928–930.

Burgess, C. G., Everett, D. H., & Nuttall, S. (1990). Adsorption of carbon dioxide and xenon by porous glass over a wide range of temperature and pressure-applicability of the Langmuir case VI equation. *Langmuir*, *6*(12), 1734–1738.

Danesh, A. S., Dandekar, A. Y, Todd, A. C., & Sarkar, R. (1991). *A modified scaling law and parachor method approach for improved prediction of interfacial tension of gas-condensate systems*. SPE-22710. SPE Annual Technical Conference and Exhibition, Dallas.

Derouane, E. G. (2007). On the physical state of molecules in microporous solids. *Microporous and Mesoporous Materials*, *104*(1–3), 46–51.

Dumore, J., Hagoort, J., & Risseeuw, A. (1984). An analytical model for one-dimensional, three-component condensing and vaporizing gas drives. *SPE Journal*, 169–179.

Elsharkawy, A. M. (1996). Measuring CO_2 minimum miscibility pressures: Slim-tube or rising-bubble method?, *Energy & Fuels*, *10*, 443–449.

Fick, A. (1999). Ueber diffusion. *Annalen der Physik*, *170*(1), 59–86.

Firoozabadi, A. (1999). *Thermodynamics of hydrocarbon reservoirs*. New York: McGraw-Hill.

Forrest, J., Zhu, T., Xiong, H., & Pradhan, Y. (2018). *The effect of initial conditions and fluid PVT Properties on unconventional oil and gas recoveries in the Wolfcamp formation in the Midland basin*. Unconventional Resources Technology Conference, Houston, Texas.

G. P. A. (GPA) (1990). Table of physical constants of paraffin hydrocarbons and other components of natural gas. Gas Processors Association.

Hamdi, H., Clarkson, C. R., Ghanizadeh, A., Ghaderi, S. M., Vahedian, A., Riazi, N., & Esmail, A. (2018). *Huff-n-puff gas injection performance in shale reservoirs: A case study from Duvernay shale in Alberta, Canada*. Unconventional Resources Technology Conference, Houston, Texas.

Hand, J. L., & Pinczewski, W. V. (1990). Interpretation of swelling/extraction tests. *SPE Researvoir Engineering Journal*, 595–600.

Hawthoren, S. B., Miller, D. J., Jin, L., & Gorecki, C. D. (2016). Rapid and simple capillary-rise/ vanishing interfacial tension method to determine crude oil minimum miscibility pressure: Pure and mixed CO_2, methane, and ethane. *Energy & Fuels*, *30*, 6365.

Hawthorne, S. B., Miller, D. J., Grabanski, C. B., Sorensen, J. A., Pekot, L. J., Kurz, B. A., Gorecki, C. D., et al.(2017). *Measured crude oil MMPs with pure and mixed CO_2, methane, and ethane, and their relevance to enhanced oil recovery from Middle Bakken and Bakken shales*. SPE Unconventional Resources Conference, Calgary, Alberta.

Hu, Y. (2014). *Dependence of organic pore wettability on kerogen maturity: A water droplet microscopic simulation study*. SPE Annual Technical Conference and Exhibition, Amsterdam.

Jin, L., et al. (2013). *Investigating the effect of pore proximity on phase behavior and fluid properties in shale formations*. SPE Annual Technical Conference and Exhibition, New Orleans.

Johns, R., et al. (1993). Analytical theory of combined condensing/vaporizing gas drive. *SPE Advanced Technology Series*, *1*(2).

Kang, S. M., Fathi E., & Ambrose, R. J. (2011). Carbon dioxide storage capacity of organic-rich shales. *SPE Journal*, *16*(4), 842–855.

Katz, D. L., & Firoozabadi, A. (1978). Predicting phase behavior of condensate/crude oil systems using methane interaction coefficients. *Journal of Petroleum Technology*, *30*(11), 1649–1655.

Katz, D. L., & Singleterry, C. C. (1939). Significance of the critical phenomena in oil and gas production. *Transactions of the AIME*, *132*(1), 103–119.

Kay, W. B. (1938). Liquid-vapor phase equilibrium relations in the ethane-n-heptane system. *Industrial and Engineering Chemistry*, *30*(4), 459–465.

Khorsandi, S., & Johns, R. T. (2015). *Tie-line soultions for MMP calculations by equations-of-state*. SPE Annual Technical Conference and Exhibition, Houston, Texas, USA, 28–30 September.

Kurtoglu, B. (2013). Integrated reservior characterization and modeling in support of enhanced oil recovery.

Li, H., & Yang, D. (2011). Modified α function for the Peng-Robinson equation of state to improve the vapor pressure prediction of non-hydrocarbon and hydrocarbon compounds. *Energy & Fuels*, *25*(1), 215–223.

Li, H., Yang, D., & Li, X. (2013). Determination of three-phase boundaries of solvent(s)-CO_2-heavy oil systems under reservoir conditions. *Energy Fuels*, *27*(1), 145–153. doi: 10.1021/ef301549a

Li, L., & Sheng, J. (2017). Nanopore confinement effects on phase behavior and capillary pressure in a Wolfcamp shale reservoir. *Journal of the Taiwan Institute of Chemical Engineers*, *78*, 317–328.

Li, S., & Luo, P. (2016). Experimental and simulation determination of minimum miscibility pressure for a Bakken tight oil and different injection gases. *Petroleum*, 79–86.

Li, X., & Yang, D. (2013). Determination of mutual solubility between CO_2 and water by using Peng-Robinson equation of state with modified alpha function and binary interaction parameter. *Industrial and Engineering Chemistry Research*, *52*(38), 13829–13838.

Li, X, Han, H., Yang, D., Liu, X., & Qin, J. (2017a). Phase behavior of C_3H_8-CO_2-heavy oil systems in the presence of aqueous phase under reservoir conditions. *Fuel*, *209*, 358–370.

Li, X., Li, H., & Yang, D. (2013a). Determination of multiphase boundaries and swelling factors of solvent(s)-CO_2-heavy oil systems at high pressures and elevated temperatures. *Energy Fuels*, *27*(3), 1293–1306.

Li, X., Yang, D., & Fan, Z. (2017b). Vapor-liquid phase boundaries and swelling factors of C_3H_8-

n-C_4H_{10}-CO_2-heavy oil systems under reservoir conditions. *Fluid Phase Equilibria, 434*, 211–221.

Li, X., Yang, D., Zhang, X., Zhang, G., & Gao, J. (2016). Binary interaction parameters of CO_2-heavy-n-alkanes systems by using Peng-Robinson equation of state with modified alpha function. *Fluid Phase Equilibria, 417*, 77–86.

Li, Y. K., & Nghiem, L. X. (1986). Phase equilibria of oil, gas and water/brine mixtures from a cubic equation of state and Henry's law. *The Canadian Journal of Chemical Engineering, 64* (3), 486–496.

Liu, S., Sahni, V., Tan, J., Beckett, D. & Vo, T. (2018). *Laboratory investigation of EOR techniques for organic rich shales in the Permian basin.* Unconventional Resources Technology Conference, Houston, Texas.

Metcalfe, R. S., Fussell, D., & Shelton, J. (1973). Multicell equilibrium separation model for the study of multiple contact miscibility in rich gas drives. *SPE Journal, 3*, 147–155.

Michelsen, M. L. (1982). The isothermal flash problem. Part I. Stability. *Fluid Phase Equilibria, 9*(1), 1–19.

Monroe, W., Silva, M., Larsen, L., & Orr, F. (1990). Composition paths in four component systems: Effect of dissolved methane on 1-D CO_2 flood performance. *SPERE*, 423–432.

Morishige, K., Fujii, H., Uga, M., & Kinukawa, D. (1997). Capillary critical point of argon, nitrogen, oxygen, ethylene, and carbon dioxide in MCM-41. *Langmuir, 13*(13), 3494–3498.

Nojabaei, B., Johns, R., & Chu, L. (2013). Effect of capillary pressure on phase behavior in tight rocks and shales. *SPE Reservoir Evaluation & Engineering, 16*(03), 281–289.

Orr, F., Johns, R., & Dindoruk, B. (1993). Development of miscibility in four-component CO_2 floods. *SPE Reservoir Engineering, 8*(2), 135–142.

Pedersen, K. S., & Christensen, P. L. (2006). *Phase behavior of petroleum reservoir fluids.* London: CRC Press.

Peneloux, A., Rauzy, E., & Freze, R. (1982). A consistent correction for Redlich-Kwong-Soave volumes. *Fluid Phase Equilibria, 8*(1), 7–23. doi:10.1016/0378-3812(82)80002-2

Peng, D. Y., & Robinson, D. B. (1976a). A new two-constant equation of state. *Industrial & Engineering Chemistry Fundamentals, 15*(1), 59–64.

Peng, D. Y., & Robinson, D. B. (1976b).Two and three phase equilibrium calculations for systems containing water. *Canadian Journal of Chemical Engineering, 54*(5), 595–599.

Ran, X., Zhao,Y., & Liao, X. (2012). *An assessment of a CO_2 flood for EOR and sequestration benefits in the Ordos basin. Carbon Management Technology Conference, Orlando, Florida.*

Rao, D. N., & Lee, J. I. (2002). Application of the new vanishing interfacial tension technique to evaluate miscibility conditions for the Terra Nova offshore project. *Journal of Petroleum Science and Engineering, 35*, 247–262.

Redlich, O., & Kwong, J. N. (1949). On the thermodynamics of solutions. V. An equation of state. fugacities of gaseous solutions. *Chemical Reviews, 44*(1), 233–244.

Robinson, D. B., Peng, D. Y., & Ng, H. J. (1979). Capabilities of the Peng-Robinson programs. 2. Three-phase and hydrate calculations. *Hydrocarbon Processing, 58*(9), 269–273.

Roque-Malherbe, R., Wendelbo, R., Mifsud, A., & Corma, A. (1995). Diffusion of aromatic hydrocarbons in H-ZSM-5, H-Beta, and H-MCM-22 zeolites. *The Journal of Physical Chemistry, 99*(38), 14064–14071.

Roque-Malherbe, R. M. (2018). *Adsorption and diffusion in nanoporous materials, 2nd ed.* CRC Press.

Sanaei, A., Jamili, A., Callard, J., & Mathur, A. (2014). *Production modeling in the Eagle Ford gas condensate window: Integrating new relationships between core permeability, pore size, and confined PVT properties.* SPE Western North American and Rocky Mountain Joint Meeting, Denver.

Sandoval, D. R., Yan, W., Michelsen, M. L., & Stenby, E. H. (2016). The phase envelope of multicomponent mixtures in the presence of a capillary pressure difference. *Industrial & Engineering Chemistry Research, 55*(22), 6530–6538.

Singh, S. K., & Singh, J. K. (2011). Effect of pore morphology on vapor-liquid phase transition and crossover behavior of critical properties from 3D To 2D. *Fluid Phase Equilibria*, *300*(1), 182–187.

Singh, S. K., Sinha, A., Deo, G., & Singh, J. K. (2009). Vapor-liquid phase coexistence, critical properties, and surface tension of confined alkanes. *The Journal of Physical Chemistry*, *113* (17), 7170–7180.

Soave, G. (1972). Equilibrium constants from a modified Redlich-Kwong equation of state. *Chemical Engineering Science*, *27*(6), 1197–1203.

Sorensen, J. A., Pekot, L. J., Torres, J. A., Jin, L., Hawthorne, S. B., Smith, S. A., Jacobson, L. L., & Thomas, D. E. (2018). *Field test of CO_2 injection in a vertical Middle Bakken well to evaluate the potential for enhanced oil recovery and CO_2 storage*. Unconventional Resources Technology Conference, Houston, Texas.

Stalkup, F. I. (1983). Miscible displacement monograph series, 8. Dallas: Society of Petroleum Engineers.

Standing, M. B. (1977). Volumetric and phase behavior of oil field hydrocarbon systems. Society of Petroleum Engineers.

Stimpson, B. C. (2017). Impacts of confined space on production from tight reservoirs. Master's thesis.

Thommes, M., & Findenegg, G. H. (1994). Pore condensation and critical-point shift of a fluid in controlled-pore glass. *Langmuir*, *10*(11), 4270–4277.

Tsau, J. S., Bui, L., & Willhite, G. P. (2010). *Swelling/extraction test of a small sample size for phase behavior study*. Improved Oil Recovery Symposium, April, Tulsa, Oklahoma.

Van der Waals, J. D. (1873). *On the continuity of the gaseous and liquid states*. Doctoral Dissertation, Leiden University. Leiden, The Netherlands.

Whitson, C. H., & Brule, M. R. (2000). *Phase behavior*. Richardson: Society of Petroleum Engineers.

Whitson, C. H., & Sunjerga, S. (2012). *PVT in liquid-rich shale reservoirs*. SPE Annual Technical Conference and Exhibition, San Antonio.

Yang, G., Fan, Z., & Li, X. (2018). *Determination of confined fluid phase behavior using modified Peng-Robinson equation of state*. Unconventional Resources Technology Conference, Houston.

Yenugu, M., & Han, D. (2013). *Seismic characterization of kerogen maturity: An example from Bakken shale*. SEG Annual Meeting, Houston.

Zarragoicoechea, G. J., & Kuz, V. A. (2002). Van der Waals equation of state for a fluid in a nanopore. *Physical Review E*, *65*(2), 021110.

Zarragoicoechea, G. J., & Kuz, V. A. (2004). Critical shift of a confined fluid in a nanopore. *Fluid Phase Equilibria*, *220*(1), 7–9.

Zhang, K., Nojabaei, B., Ahmadi, K., & Johns, R. T. (2018). *Minimum miscibility pressure calculation for oil shale and tight reservoirs with large gas-oil capillary pressure*. Unconventional Resources Technology Conference, Houston, Texas.

Zick, A. (1986). *A combined condensing/vaporizing mechanism in the displacement of oil by enriched gases*. SPE Annual Technical Conference and Exhibition. New Orleans, Louisiana.

8

Fluid Flow Through Nanosized Pores

Mohammad Kazemi[1], *Ali Takbiri-Borujeni*[2], *Sherifa E. Cudjoe*[3], *and Reza Barati Ghahfarokhi*[3]

[1] *Slippery Rock University of Pennsylvania, Slippery Rock, Pennsylvania, USA*
[2] *West Virginia University, Morgantown, West Virginia, USA*
[3] *The University of Kansas, Lawrence, Kansas, USA*

8.1 Background

Shales are complex materials consisting of organic matter (OM) pockets, known as kerogen, that are distributed in inorganic matrix (Curtis et al., 2011) with a complex pore structure containing a wide range of pore sizes from ultramicropores ≤0.7 nm to macropores ≥50 nm based on IUPAC pore classification. The amount of total organic content (TOC) of shale matrix is usually reported in weight or volume percent. Typical TOC of North American shale (Kazemi & Takbiri-Borujeni, 2015) plays are less than 10 wt% including Barnett 4%, Marcellus 1%–10%, Haynesville 0%–8%, Horn River 3%, and Woodford 5% (Takbiri-Borujeni et al., 2019). Shale gas is expected to account for 30% of world natural gas production by 2040 (EIA, 2014).

Although kerogen accounts for a small percentage of total shale volume, almost half of the total hydrocarbons reside within its pores. Due to the extremely small pore sizes, the nature of the species transport is different from those predicted by continuum-based theories from which the Navier-Stokes and Darcy equations are derived. Furthermore, depending on the gas type and intermolecular forces between the molecules of gas and solid molecules at the pore walls, the gas molecules cannot escape from the force field at the pore wall, which causes inhomogeneous fluid distribution in the pores. Moreover, due to the differences in the species sorption properties in natural gas, fluids are not uniformly distributed across the pores; heavier components tend to accumulate in the vicinity of the pore walls and may trap within the pore and remain in the reservoir, while the lighter components are extracted. This sieving effect is mainly observed in shale oil reservoirs. Therefore, the composition of the fluids may change as production progresses. This limits the application of conventional petroleum engineering models to predict the volume of hydrocarbon in place and recovery from these resources.

To study fluid transport and storage in these resources, it is important to develop new models that can predict the fluid dynamics under the confinement and also accurately describe the fluid-solid interactions. The former is known as rarefaction effects and the latter as sorption (adsorption and desorption) phenomena. There have been numerous attempts to incorporate these two phenomena in describing the fluid transport in shale. In the following sections, we will first discuss the pore-size distributions in shale resources followed by a brief discussion of these phenomena.

Unconventional Hydrocarbon Resources: Techniques for Reservoir Engineering Analysis, Advanced Textbook Series, First Edition. Edited by Reza Barati Ghahfarokhi and Mustafa M. Alhubail.

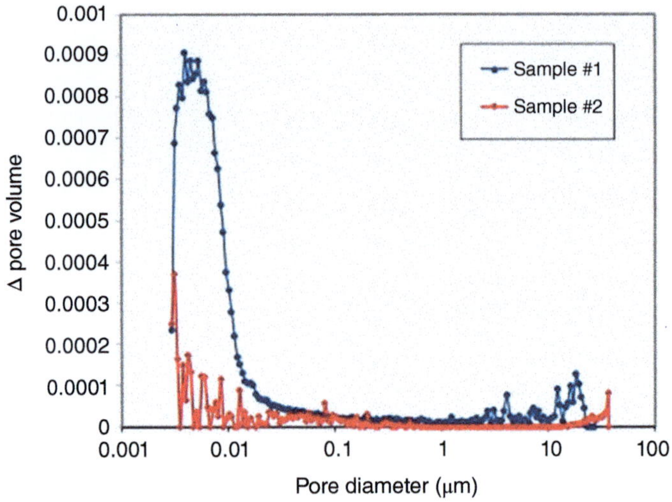

Figure 8.1 Pore size distributions for two Marcellus shale samples (Takbiri-Borujeni et al., 2019).

8.2 Pore Size Distribution

The common methods for determining porosity of shale core samples, such as high-pressure mercury injection, SEM images, and low temperature adsorption measurement, have indicated that pore sizes in shale gas reservoirs have a bimodal distribution ranging from ultra-micropores to macropores corresponding to organic and inorganic materials, respectively. These techniques show that most of the fluids are stored in micropores of organic matter in an adsorbed state despite the fact that organic matter accounts for only a small volume percentage (Collell et al., 2014; Sondergeld et al., 2010; Wang et al., 2009; Adesida et al., 2011). On the other hand, NMR pore size distribution shows a trimodal distribution ranging from micropores (< 2 nm), to mesopores (2 – 50 nm) and macropores including fractures (> 50 nm) in shale oil reservoirs. The micropores are mostly occupied by the bound fluids (clay bound water and bitumen), while the mesopores are predominantly organic pores hosting most of the hydrocarbons. The macropores are the mineral pores in addition to fractures, also occupied by hydrocarbons and wetting fluids. Therefore, measuring the accurate storage and transport parameters in both organic and inorganic materials of shale is important for developing robust and accurate governing equations describing fluid dynamics in these formations. Figure 8.1 depicts pore-size distributions for two Marcellus shale samples obtained using high-pressure mercury injection test that indicates the pores ranging from 5 nm to 20 μm with bimodal distribution (Takbiri-Borujeni et al., 2019). SEM image analysis, a method to differentiate the organic pores from inorganic ones, shows that most of the organic pores are in microrange sizes.

8.3 Adsorption

In organic micropores of shale, molecules of fluid species (i.e., adsorbate) have a tendency to be adsorbed to the surfaces of the pores (i.e., adsorbent) due to the existence of the attraction forces between the fluid and pore wall molecules. The adsorption phenomena in shale result in heterogeneous fluid distributions (i.e., positional-dependent fluid density) within the pores. On the other hand, due to the existence of a wide range of species present in natural gas systems (e.g., methane, ethane, propane, butane, nitrogen, and carbon dioxide), the adsorption selectivity causes the heavier components to accumulate in the vicinity of the pore walls and may trap within the pore and remain in the reservoir, while the lighter components can be recovered. As a result, fluid systems within the organic pores become heavier as gas recovery continues (heavier components accumulate in the pores while lighter components, mostly methane, are recovered). Figures 8.2 and 8.3 depict the species density

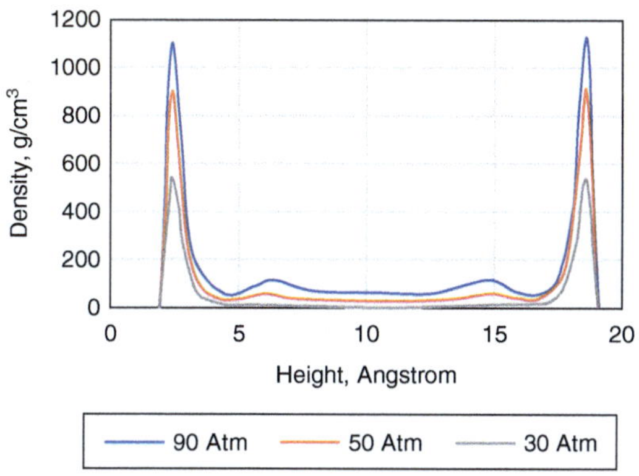

Figure 8.2 Density distribution within carbon nanotubes with 2 nm height at different pressures. The peak densities show the density values at the vicinity of the pore walls (Kazemi & Takbiri-Borujeni, 2016).

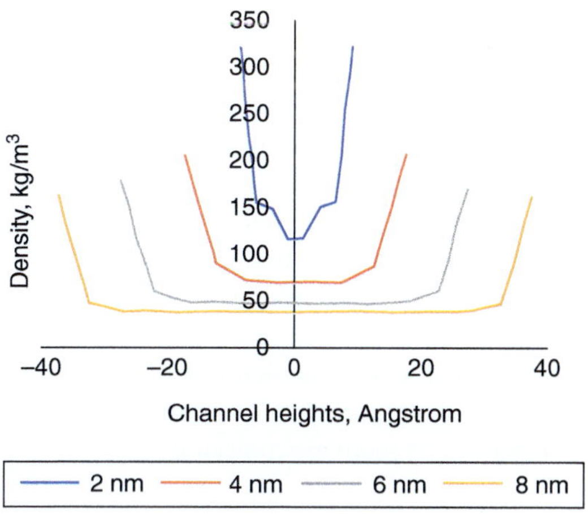

Figure 8.3 Density distributions across carbon nanochannels with 2, 4, 6, and 8 nm.

distribution within carbon nanotubes with 2 nm and 4 nm heights (Kazemi & Takbiri-Borujeni, 2016). The density profiles of methane were determined using Grand Canonical Monte Carlo (GCMC) molecular simulations in a graphite nanochannel. The peak densities show the density values at the vicinity of the pore walls.

Adsorption can be of two types: (1) physisorption, by weak bounds, often by van der Waals and/or dispersion forces due to induced dipole-dipole interactions, and (2) chemisorption, by strong covalent bonding (Keller & Staudt, 2005; Wilcox, 2012). These adsorption types are characterized by the heat of adsorption (Builes, 2013). Generally, physisorption occurs when the heat of adsorption is less than approximately 10–15 kcal/mol, while chemisorption occurs with heat of adsorption greater than 15 kcal/mol (Yang, 2003). Typically, physisorption is a rapid nonactivated and reversible process with a low heat of adsorption and forms monolayer or multilayer coverage of adsorbent by adsorbate. Chemisorption, however, forms monolayer coverages. In shale formations, adsorption is mostly in the form of physisorption.

The equilibrium relationship between adsorbate and adsorbent at a given temperature is represented by an adsorption isotherm. The concentration of adsorbed fluids on adsorbent surfaces is traditionally expressed in terms of the mass adsorbed per unit mass of sorbent material. Shape of the adsorption isotherms is dependent on the interactions of a fluid species with other fluid species and their interactions with the adsorbent.

In a system with adsorption, two quantities are generally defined, *Absolute* and *excess* adsorption, to differentiate between the bulk and adsorbed fluids molecules. Absolute adsorption is the quantity of the adsorbate in the adsorbed state and excess adsorption is defined as the additional adsorbed fluid amount per unit adsorbent volume compared with the adsorbate amount in the same volume in the absence of adsorbent. Figure 8.4 depicts

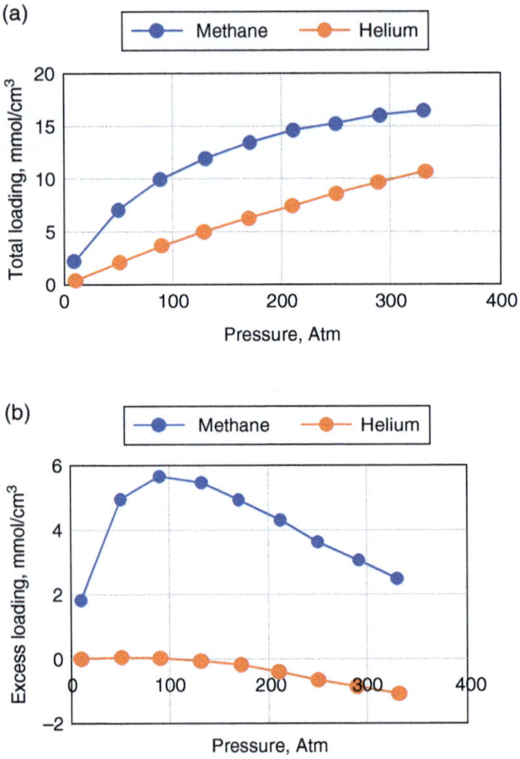

Figure 8.4 Absolute and excess adsorption isotherms of argon, methane, helium, and neon in 2 nm channel height.

the absolute and excess adsorption quantities for two organic (idealized graphite) channels (Kazemi & Takbiri-Borujeni, 2016) using GCMC simulations. As can be seen, the absolute adsorption increases as the pressure increases and tends to saturate at high pressures. The excess adsorption amounts increase to reach a maximum value at an optimum pressure and then decrease. The increase with pressure followed by a subsequent decrease suggests that there exists a maximum adsorbed quantity. Increasing temperature decreases the absolute adsorption amount, however, maximum adsorbed quantity remains constant (Mosher et al., 2013; Michalec & Lísal, 2017).

The most well-known adsorption isotherm is the Langmuir adsorption isotherm (LAI) (Attard & Barnes, 1998; Gregg, 1967; Langmuir, 1918). This isotherm accounts for monolayer adsorption, which takes place at the solid surface consisting of a distribution of noninteracting sites. The Langmuir adsorption isotherm is defined using two parameters, namely, Langmuir volume and Langmuir pressure,

$$V(P) = \frac{V_L P}{P + P_L}, \tag{8.1}$$

where V is the adsorbed gas volume (gas content) in SCF/ton at pore pressure P (psi); V_L is the maximum monolayer adsorption capacity of the sample in SCF/ton; P_L is the Langmuir pressure (psi), which is the pore pressure at which half of the adsorbed sites are taken. To represent experimental data and check graphically their consistency, a linearized relationship between inverse of volume intake and inverse of pressure is often useful,

$$\frac{1}{V} = \frac{1}{V_L} + \left(\frac{P_L}{V_L}\right)\frac{1}{P}. \tag{8.2}$$

Data around this straight line in a 1/p vs. 1/V diagram often indicates experimental consistencies. There are also several generalizations of the LAI that account for real gas adsorption, energetically heterogeneous sorbent materials, and molecules with interactions (Keller & Staudt, 2005).

In order to compare the tendency of different components to be adsorbed to the wall, selectivity of hexane over other gas components is determined (Fig. 8.5), where S is selectivity, xCO_2 is the mole fraction of CO_2 in adsorbed phase, and yCO_2 represents the mole fraction of CO_2 in the bulk gas phase. Selectivity values higher than one show that the CO_2 molecules have more tendency to be adsorbed to walls compared to methane molecules.

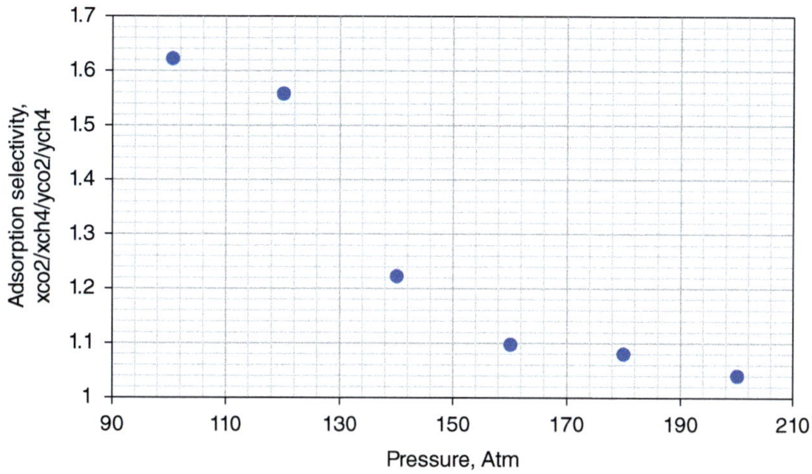

Figure 8.5 Adsorption selectivity: xCO_2 is the mole fraction of CO_2 in adsorbed phase, and yCO_2 represents the mole fraction of CO_2 in the bulk gas phase.

Example 8.1

The following tables (Tables 8.1 and 8.2) show the methane and CO_2 adsorption isotherms experimentally measured using Marcellus shale samples. Check the consistency of the adsorption measurement results, plot the Langmuir isotherm, and determine the values of the constants.

Solution

In order to determine the Langmuir isotherm constants, we need to use equation 8.2. A plot of inverse of pressure as a function of inverse of adsorbed gas volume will result in a straight line. The inverse of the intercept of the linear straight line is the maximum monolayer adsorption capacity of the sample in SCF/ton (V_L) and the slope is the Langmuir pressure (P_L). Tables 8.3 and 8.4 show the inverse of pressure and volume.

Table 8.1 Methane Adsorption Isotherm on Samples From the Marcellus Shale Reservoir

Pressure (psi)	Absolute adsorption (SCF/ton)
45.72	1.27
73.65	3.04
121.88	5.83
243.81	9.38
314.95	9.87
492.79	12.90
703.68	15.42
805.31	16.42
993.33	18.94
1,150.90	18.41
1,232.19	19.93
1,354.17	19.91
1,445.65	20.66

Table 8.2 Carbon Dioxide Adsorption Isotherm on Samples From the Marcellus Shale Reservoir

Pressure (psi)	Adsorption absolute (SCF/ton)
45.60	8.86
91.19	17.28
159.69	24.13
202.83	27.94
339.92	36.06
471.98	41.13
550.73	43.15
687.91	45.93

Table 8.3 Inverse of Pressure and Volume for Measured Valued for Carbon Dioxide Adsorption Isotherm

1/P (1/psi)	1/V (ton/SCF)
0.021931	0.112867
0.010966	0.057883
0.006262	0.041439
0.004930	0.035791
0.002942	0.027733
0.002119	0.024315
0.001816	0.023175
0.001454	0.021773

Table 8.4 Inverse of Pressure and Volume for Measured Valued for Methane Adsorption Isotherm

1/P (1/psi)	1/V (ton/SCF)
0.013578	0.328947
0.008205	0.171527
0.004102	0.10661
0.003175	0.101317
0.002029	0.077519
0.001421	0.064851
0.001242	0.060901
0.001007	0.052798
0.000869	0.054318
0.000812	0.050176
0.000738	0.050226
0.000692	0.048403

The plots of pressure inverse versus inverse of volume (Fig. 8.6) show a linear trend, which shows rhe consistency of the experimental results.

The *y*-intercept demonstrates the values for the inverse of Langmuir volume. For CO_2, Langmuir volume is

$$V_L = \frac{1}{0.0327} = 69.93\frac{SCF}{Ton}. \tag{8.3}$$

For methane, Langmuir volume is

$$V_L = \frac{1}{0.0142} = 30.58\frac{SCF}{Ton}. \tag{8.4}$$

As can be seen, the Langmuir volume for CO_2 is higher than methane in Marcellus shale samples. Slope of the CO_2 trend line is

$$\frac{P_L}{V_L} = 4.39\frac{\text{psi}}{\text{SCF}}. \tag{8.5}$$

Therefore,

$$P_L = 69.93 \times 4.39 = 306.99\ psi. \tag{8.6}$$

And for methane,

$$P_L = 30.58 \times 20.47 = 625.99\ psi. \tag{8.7}$$

8.4 Flow Regimes

Flow of gases in shale nanopores has a complex nature and depends on gas properties, pore geometry, and pore surface properties (e.g., surface roughness and adsorption characteristics). Knudsen number is used to characterize flow regimes in shale gas and oil reservoirs. The Knudsen number is defined using equation 8.8,

$$Kn = \frac{\lambda}{a}, \tag{8.8}$$

where λ is the mean free path of gas molecules and a is the characteristic length of the porous

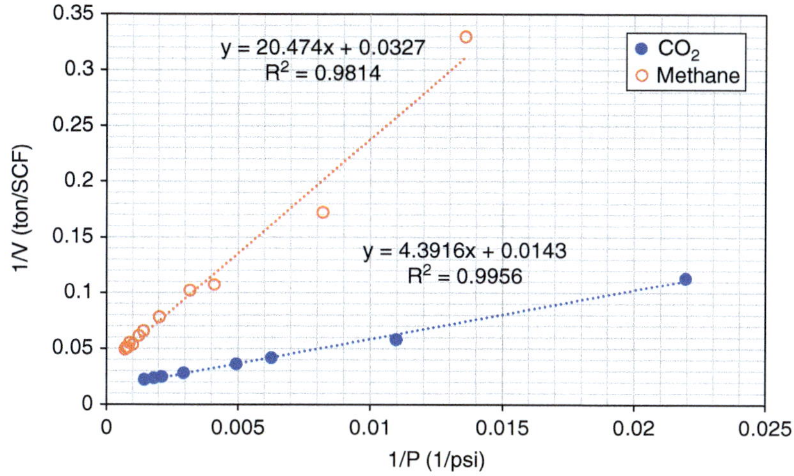

Figure 8.6 The plots of pressure inverse versus inverse of volume show a linear trend, which shows the consistency of the experimental results.

media. The characteristic length depends on shape of the pores. If we assume that pores are cylindrical, the characteristic length of the system is the pore radius. In the case of channel (slit) shaped pores, the characteristic length of the system is channel height. Mean free path is defined as the average distance travelled by molecules before colliding with another molecule. This distance depends on the molecular size and the molecular velocity (Sharipov & Seleznev, 1998). These two parameters cannot be measured. One way to calculate the mean free path is to use viscosity provided by kinetic theory of gases as

$$\lambda = \frac{\mu}{P}\sqrt{\frac{\pi RT}{2M}}, \tag{8.9}$$

where μ is the viscosity in Ps.s, R is the universal gas constant in kg m^2 mol^{-1} K^{-1} s^{-2}, P is the pressure in Pa, M is the gas molecular weight in kg/mol, and T is the temperature in K. Therefore, the Knudsen number can be determined using macroscopic properties of gases, which are measurable.

Usually, three distinct gas flow regimes are identified by Knudsen number. When the mean free path of gas molecules is larger than the characteristic length of the system, gas molecules do not collide with each other frequently and most of collisions occur between molecules and pore wall. Therefore, their movement can be assumed independently and the flow regime is called free molecular regime. In this case, the Knudsen number is larger than one ($Kn > 10$). When the mean free path of gas molecules is much smaller than the characteristic length of the system, the number density (number of molecules per unit volume) in pore is large. Molecular collisions are frequent and the fluid can be considered as a continuous phase. This is why the flow regime is called continuum, or viscous, and Knudsen number is less than 0.01 ($Kn < 0.01$). When we cannot consider gas as either a continuous phase or a free molecular flow regime, the gas flow is considered in transition flow regime ($0.01 < Kn < 10$) (Roy et al., 2003).

In Figure 8.7, a methane molecule is shown moving through a nanoscale channel while colliding with other molecules. The mean free path decreases as the number of gas molecules per unit volume increases.

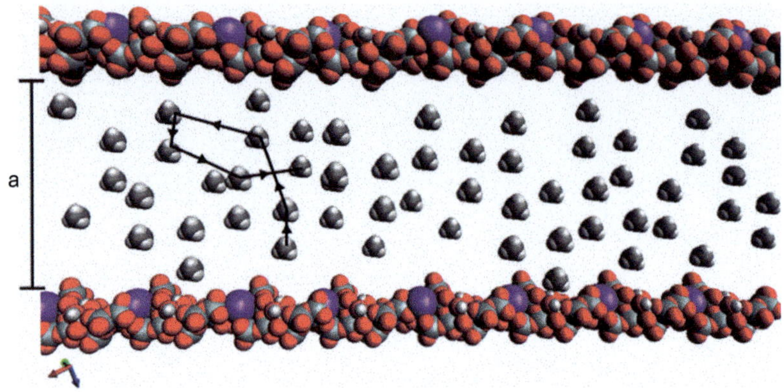

Figure 8.7 Methane molecules inside an idealized nanoscale channel.

Example 8.2
Methane diffuses through the pores of a Marcellus shale sample at a total pressure of 4 MPa and a temperature of 330 K where the average pore diameter is 5 nm. Calculate the gas mean free path and Knudsen number.

Solution

The molecular weight of methane is 16.04 g/mole and its viscosity is 1.2795×10^{-5}. Therefore,

$$\lambda = \frac{\mu}{P}\sqrt{\frac{\pi RT}{2M}} = \frac{1.2795 \times 10^{-5}}{(4 \times 10^6)\left(\frac{\pi \times 8.314 \times 330}{2 \times 16.04}\right)} = 1.66 \times 10^{-9}\ m. \tag{8.10}$$

Therefore, the Knudsen number is casted as

$$Kn = \frac{\lambda}{a} = \frac{1.66 \times 10^{-9}}{5 \times 10^{-9}} = 0.33. \tag{8.11}$$

Mathematical models used for each flow regime are different and, therefore, characterizing the gas flow regime is an important step to determine the gas flow rates through shale nanopores. Significant numbers of pores in shale resources have small pores in nanometer scale. Knudsen

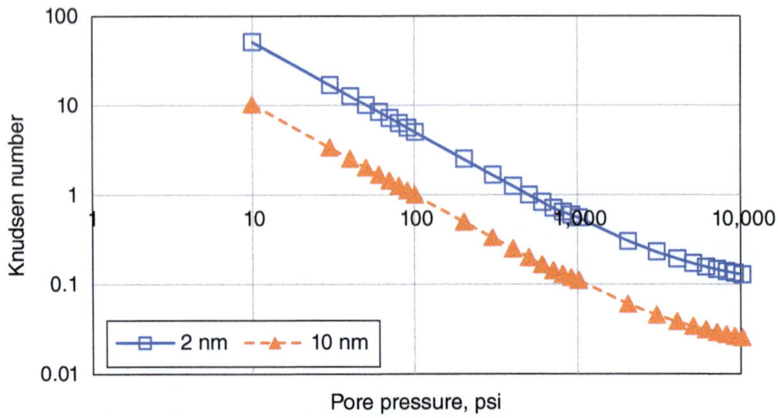

Figure 8.8 Knudsen number as a function of pore pressure for two radii of 2 and 10 nm.

numbers in shale nanopores are determined by assuming that the gas contained on those pores is methane in Figure 8.8. In this figure, it is also assumed that the reservoir temperature is 188°F and the characteristic pore length varies between 2 nm and 10 nm. In a typical shale gas reservoir, the Knudsen number is between 0.01 and 1, which shows that the flow is in transition flow regime.

Example 8.3
Determine the gas flow regime for previous example.

Solution

The Knudsen number calculated from the previous example was 0.33. Therefore, we are in transition flow regime, since $0.01 < Kn < 10$.

8.5 Modeling Techniques

8.5.1 Fluid Transport in Confined Enclosures

Unlike the transport in conventional reservoirs, which can be described using convective transport or Darcy's law, the transport in shale, especially shale gas reservoirs, is highly impacted by sorption characteristics of the organic materials and pore wall confinements of ultramicro and micro pore walls. In the 1940s, Klinkenberg showed that laboratory measurements of porous samples using isothermal gas permeation experiments lead to higher apparent permeability due to the gas slippage phenomena.

$$K_{app} = K_{int}\left(1 + \frac{b}{\overline{P}}\right), \tag{8.12}$$

where K_{app} is the apparent permeability at a Knudsen number, K_{int} is the intrinsic permeability, $\overline{P}$ is the average pressure, and b, in the pressure units, is the slope of the straight line of the plot of K_{app} versus inverse of the average pressure. Later Fathi et al. (2012), showed that apparent permeability of shale matrix could be significantly higher than those predicted by the Klinkenberg slip theory and developed double-slip effect and modified the Klinkenberg equation using lattice Boltzmann simulation (Kazemi& Takbiri-Borujeni, 2015). Analytical models were developed using Regularized-13 moment method to obtain the apparent permeability of nanoscale channels and tubes. More studies have been published on this topic and they all agreed on the fact that the shale permeability measurement is a great function of pressure, temperature, pore size distribution, and gas type that can be investigated with shale permeability as a function of Knudsen number (Cercignani, 1964; Moghadam & Chalaturnyk. 2014; Sakhaee-Pour et al., 2012).

8.5.2 Apparent Permeability of Shale

For nonadsorbing confined enclosures (e.g., inorganic nanopores or microfractures) or when the measurements are performed using nonadsorbing gases (e.g., helium), the effect of adsorption on transport can be less important. In these cases, however, special attention should be devoted to the nanoscale pore sizes and flow under pore confinement. Due to the extremely small sizes of the pores, the nature of the species transport is different from those predicted by continuum-based theories from which the Navier-Stokes and Darcy equations are derived. In 1872, Boltzmann developed an equation describing the state of dilute gases and established a basis for the kinetic theory of gases. The Boltzmann equation has been proved to be valid for all flow regimes (Cercignani, 1964). Solutions to the Boltzmann equation are complex, both numerically and analytically, and thus computational difficulties are the major problems in direct application of the Boltzmann equation. Several reduced-order methods have been proposed to solve Boltzmann equation to predict the fluid transport mechanisms beyond the Navier-Stokes equations. Kazemi and Takbiri-Borujeni (2015) proposed a new model, Regularized-13 apparent permeability (R13 AP), for apparent permeability of nanoscale geometries,

$$K_{app} = K_{int}\left(1 + AKn + BKn^2\right), \quad (8.13)$$

where K_{app} is the apparent permeability at a Knudsen number, K_{int} is the intrinsic permeability, which does not depend on the Knudsen number, A and B are constants of the equation. A values are 6.85 and 4.41, for 2D channels and capillaries, respectively. The B values are 3.39 and 1.35 for 2D channels and capillaries, respectively. At the transition flow regime, the dimensionless permeability substantially increases with the Knudsen number because the rarefaction effects become more dominant.

Example 8.4

Determine the intrinsic permeability of a shale sample with the following gas permeability measurement data (Table 8.5). Temperature is 300 K and the average pore size is 10 nm assuming channel shaped pores.

Solution

We will first try to use the Klinkenberg method to determine the intrinsic permeability of the sample. In order to do that, the permeability of the sample needs to be plotted against the inverse of pore pressure (Table 8.6) according to equation 8.5. The intercept of the linear straight line is the intrinsic permeability.

Table 8.5 Permeability Measurement for a Marcellus Shale Sample at Different Pressures

p (psi)	Permeability (nD)
466.6667	128.3364
352.7132	195.7952
187.2428	434.3693
163.9640	478.7934
143.7599	559.4150
129.6296	603.8391
112.7633	681.1700
103.3504	847.3492

Table 8.6 Permeability as a Function of Inverse of Pressure

Inverse of pressure, 1/psi	Permeability, nD
0.002143	128.3364
0.002835	195.7952
0.005341	434.3693
0.006099	478.7934
0.006956	559.415
0.007714	603.8391
0.008868	681.17
0.009676	847.3492

Note: Data from Table 8.5.

As can be seen from Figure 8.9, the *y*-intercept shows a negative value for the intrinsic permeability. This shows that the traditional Klinkenberg method cannot be applied to shale samples in laboratory conditions. In order to find the intrinsic permeability using R13 AP model, we need to plot the apparent permeability against the Knudsen number and fit equation 8.6 to Table 8.7.

However, using the R13 AP model, the intrinsic permeability is found to be 92 nd by fitting a polynomial equation to the plot of apparent permeability versus Knudsen number.

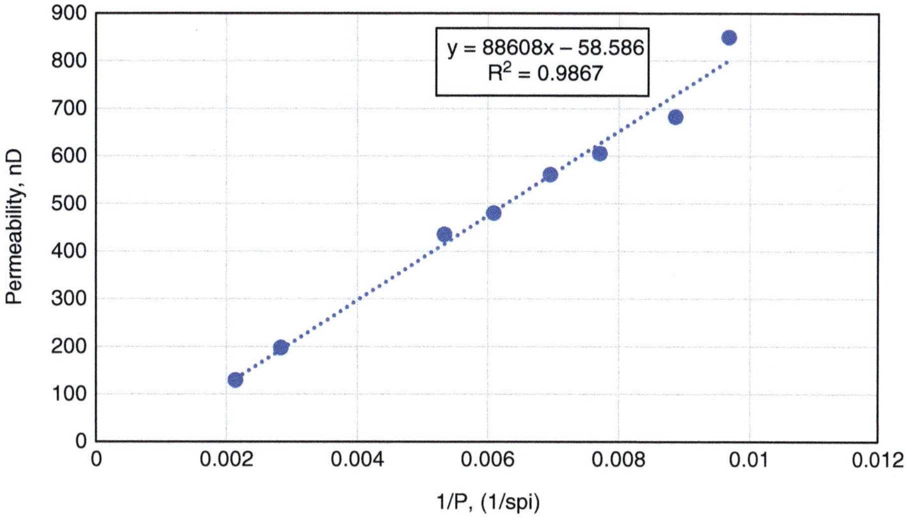

Figure 8.9 Permeability as a function of inverse of pore pressure.

Table 8.7 Viscosity, Mean Free Path, Knudsen Number, and Permeability Versus Pressure

p (psi)	Viscosity (cp)	Mean free path, nm	Knudsen number	Permeability (nD)
466.6667	0.011771	1.80805E-09	0.180805	128.3364
352.7132	0.011612	2.35988E-09	0.235988	195.7952
187.2428	0.011412	4.36879E-09	0.436879	434.3693
163.964	0.011386	4.97768E-09	0.497768	478.7934
143.7599	0.011366	5.66728E-09	0.566728	559.4150
129.6296	0.011351	6.27674E-09	0.627674	603.8391
112.7633	0.011334	7.20477E-09	0.720477	681.1700
103.3504	0.011325	7.85472E-09	0.785472	847.3492

8.5.3 Transport in Organic Nanopores

The mass transfer in organic nanopores of shale is of a diffusional nature (Collell et al., 2014). Therefore, understanding the species diffusion through shale organic pores is of significant importance. Diffusion of pure fluids in organic pores of coal and shale has been extensively studied using molecular simulations (Kazemi et al., 2016; Shahtalebi et al., 2014; Zhang et al., 2014b). However, there are only a few studies on molecular simulation of the diffusion coefficients for fluid mixtures. Collell et al. (2015) studied the transport of hydrocarbon mixtures through kerogen using molecular simulations. They found that the Onsager's autocorrelation coefficients scale linearly with the adsorption loading and inversely with the alkane chain length. They proposed a simple scaling law to predict the transport coefficient of linear alkanes in the mixture. Zhao et al. (2016) computed diffusion coefficients of carbon dioxide-methane mixtures at different temperatures, pressures, and mixture compositions in a bituminous coal model. They showed that the dominating mechanism for the mixture diffusion in coal changes from surface diffusion to configurational diffusion at high pressure.

8.5.4 Molecular Simulations

Molecular simulations are powerful alternatives to physical experiments that are not easy to conduct. They are performed to study the behavior of an ensemble of molecules that is hard to investigate using experiments. The advantage of molecular simulations to experimental measurements is twofold. First, in molecular simulations the system under investigation can be isolated easily, while in experimental measurements it might be a cumbersome task or even impossible. For example, extraction of organic matter from shale samples is a tedious process in studying gas transport and adsorption in organic nanopores of shale. It is an appealing feature of molecular simulations that representative molecular structures of organic matter can be constructed and transport and adsorption properties can be determined. Second, molecular simulations are good substitutes for experiments that require high-temperature and high-pressure environments. Molecular simulations provide the ability to perform numerous sets of tests at different pressures, temperatures, and fluid compositions.

There are two main families of molecular simulations: (1) molecular dynamics and (2) Monte Carlo. In Monte Carlo simulations, the dynamics of system do not mimic the dynamics of real system; however, the equilibrium properties of the system should correspond to the real systems equilibrium properties. The macroscopic and microscopic properties are linked by averaging the property of interest over multiple snapshots in time. The collection of these various states of the system (snapshots) are called statistical ensembles. The probability of the system's being in each state depends on the type of the system under investigation. Depending on the nature of fluid (pure or mixture), the simulations are performed with a constant number of molecules, constant temperature, and constant volume (for pure fluids) or pressure (for mixtures). In molecular dynamics, materials are considered to be a collection of atoms, which exert forces on each other defined by interaction potentials. The atoms in systems move according to Newton's equation of motion. Newton's equation of motion is solved numerically for all atoms in the system, and locations of atoms are determined at different time steps.

$$m_i \frac{\partial^2 r_i}{\partial t^2} = f_i, \quad f_i = \frac{\partial}{\partial^2 r_i} U \tag{8.14}$$

where m_i is the atomic mass, r_i is the position of the atom i, and f_i is the force. The intermolecular forces (f_i) are determined using interaction potential, which is a function of the atomic positions $U(r^N)$ where $r^N = (r_1, r_2, ..., r_n)$. The interatomic interactions can be divided into three main categories: bonded, Van der Waals, and coulombic. The Van der Waals interactions are referred to forces between nonbonded atoms. The most commonly used Van der Waals potential is 12-6 Lennard-Jones potential, which is described as

$$U_{ij}(r_{ij}) = 4\epsilon_{ij}\left[\left(\frac{\sigma_{ij}}{r_{ij}}\right)^{12} - \left(\frac{\sigma_{ij}}{r_{ij}}\right)^{6}\right], \tag{8.15}$$

where r_{ij} is the distance, σ_{ij} is the average diameter, and ϵ_{ij} is the strength of interaction between atoms i and j. Figure 8.10 demonstrates the Lennard-Jones potential (kcal/mol) between four different gases and graphite (C). The higher the value of strength of potential (ϵ_{ij}), the larger the attraction force.

8.5.5 Molecular Structure of Kerogen

Performing molecular simulations requires knowledge of molecular structures of materials under investigation. Furthermore, the interaction potential between atoms of those structures needs to be defined. In transport of gases or oil in shale, organic nanopore in kerogen and nanoscale pores in clay may be of interest. Therefore, the molecular structures and interaction potentials of kerogen and clay depending on their types were investigated by

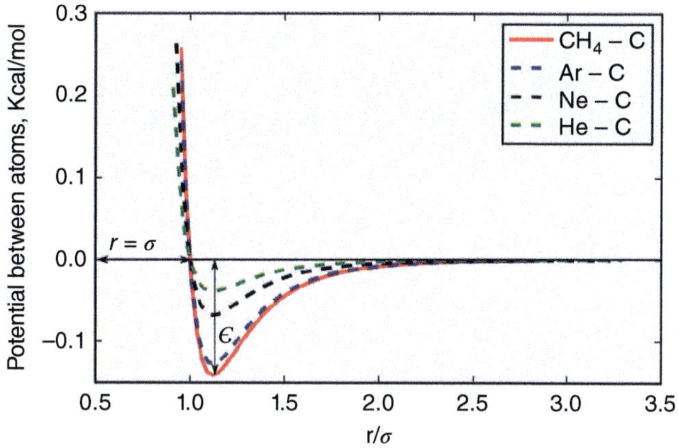

Figure 8.10 Lennard-Jones potential between four different gases and graphite.

many authors (Behar & Vandenbroucke, 1987; Kidena et al., 2008; Siskin et al., 1995; Orendt et al., 2013; Ungerer et al., 2014). The molecular structure and interaction potentials of kerogen (Fig. 8.11) is specifically complex while clays have usually defined structures, and the force fields are known. In order to confirm the molecular structures, one needs to validate them with experimental measurements.

From core plug samples to running MD simulation for determining the transport properties in kerogen, different steps are usually taken. One of the main steps is the sample preparation and separation of the kerogen from the minerals. There are two basic ways to isolate kerogen, physically or chemically. The physical methods of isolation were preferred by some because they did not heavily alter the kerogen chemically. However, physical isolation methods have low recoveries of the organic matter, do not accomplish total isolation of the kerogen, and are not useful when only small amounts of organic matter are available.

Usually pyrolysis is the first step. The 13C NMR setup allows identification of carbon atoms in an organic molecule. Furthermore, it is an important tool in chemical structure elucidation in organic chemistry. Solid-state 13C NMR13-17 and XPS18 are capable of providing detailed information about the amount and form of aromatic carbons present in complex carbonaceous materials. NMR and X-ray methods are used to characterize a suite of kerogens composed of different organic matter types and levels of maturity. Kerogen types in Marcellus shale are II and III.

Adaptation of 13C NMR to solid samples at the beginning of the 80s has shown a great potential for quantitative analysis of kerogen functionalities. Indeed, it is the only method that can give, under controlled experimental conditions, quantitative estimates of the different carbon bond types in kerogens, as opposed to infrared spectroscopy where only qualitative comparisons are possible.

It appears clearly that given the complex macromolecular structure of kerogen, comprising both resistant biomacromolecules and recombined biodegradation products, any detailed chemical model represents only an averaged structure. To fix ideas, a detailed chemical structure, which can be drawn on a single page with a molecular mass of around 25,000, makes an incredibly small amount of substance since the corresponding weight (by division by Avogadro's number) represents 4.10–20 g of kerogen. Any analytical technique used for kerogen characterization works on amounts in the microgram to milligram range, thus, any analytical result is already an average on a large number of kerogen macromolecules. Thus, a real representation of

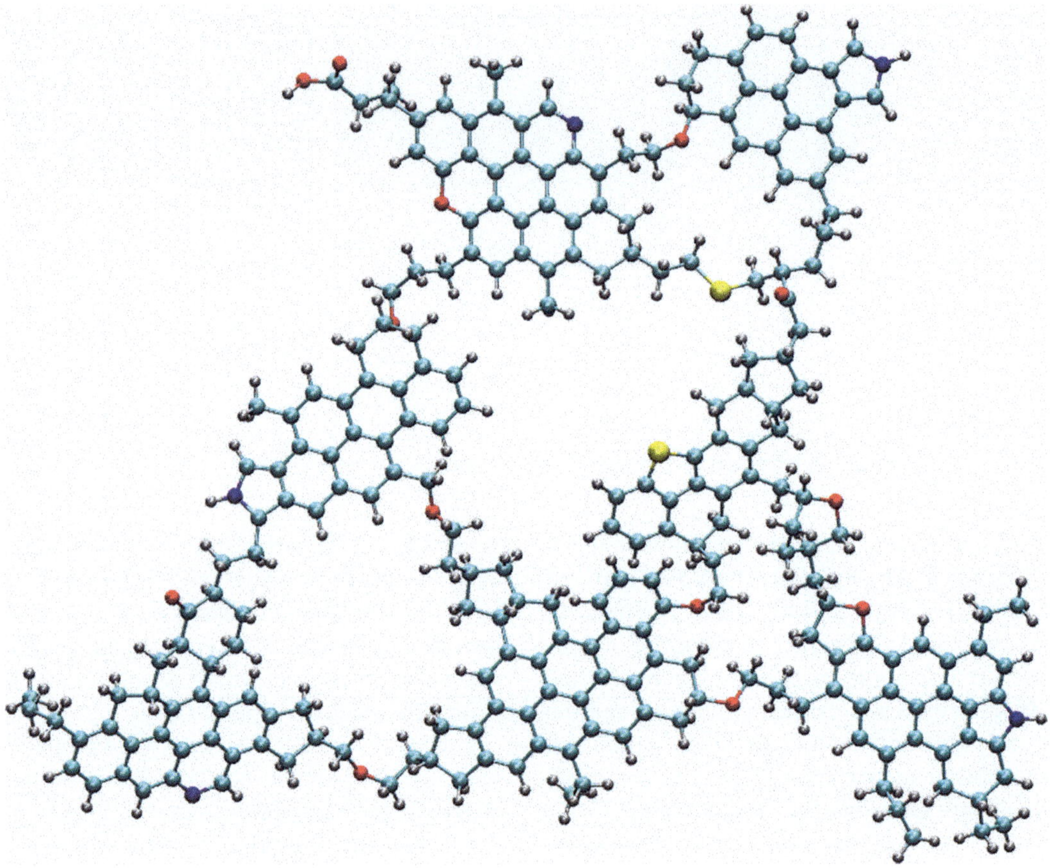

Figure 8.11 Unit cells of kerogen type IIC. Color code: cyan (carbon), silver (hydrogen), red (oxygen), blue (nitrogen), and yellow (sulfur) (Kazemi, 2017).

the chemical structure of kerogen has no meaning and only conceptual models representing at best the largest possible set of physicochemical analytical data, but including also many assumptions, can be obtained.

Although it is very difficult to describe such complex macromolecular mixtures by detailed chemical structures, many structural or chemical models of sedimentary organic matter were published in the past. Their aim was to visualize a number of its properties or analyses in a synthetic form to interpret or understand others. Thus, a model aimed at representing physical or spectral properties can be a structural model, whereas a model for describing chemical interactions or cracking reactions will rather be a molecular model. This rationale is very similar to that of petroleum system modeling, and in the same way the weighting and detail of input and output data will depend on the use for which these models are constructed. The first molecular models of sedimentary organic matter were built by coal scientists as early as 1942 by Fuchs and Sandhoff (1942). Then many other coal models based on an increasing number of analytical results from different techniques were built (Given, 1960; Spiro, 1981; Spiro & Kosky, 1982; Shinn, 1984; Solomon, 1981). Models for other types of sedimentary organic matter were constructed first for oil shales, such as the Green River shale (Burlingame et al., 1969; Yen, 1976; Siskin et al., 1995), then for the three main types of kerogen at various maturity stages (Behar & Vandenbroucke, 1987). Finally, the development of computerized molecular modeling enabled the building of 3D molecular models satisfying not only bond lengths and directions (Faulon et al., 1990) but also minimum energy configurations (Kowalewski et al., 1996).

Building computerized chemical models of kerogen structures may be helpful as a guide to understanding the main chemical characteristics of sedimentary organic matter, and to predicting its evolution during geological burial. Although it will never be possible to represent a true kerogen structure, because it is a mixture of various nonpolymeric macromolecules that cannot be analyzed separately, such models of average kerogen structure, by accounting for a great amount of information from various analyses, can provide a synthetic view of the main resemblances and differences among sedimentary organic matters. They may be used for comparison of organic matter types and evolution during burial, modeling of cracking and compositional yields in source rocks, evaluation of retention or expulsion of petroleum and gas as a function of the thermal maturity, and related thermodynamic properties of the mobile fraction.

The major elements in kerogen concentrates are the following: C, H, N, O, S, and possibly Fe from pyrite, with generally C and N amounting respectively for the highest and the lowest weights. Molecular analyses qualify and quantify small building blocks of a complex structure. They are performed on degradation products of macromolecular kerogen constituents. Either chemical degradation, by oxidation for example, or thermal degradation, both by laboratory pyrolysis (pyrolysates) and by thermal cracking in geological conditions (extracts and oils), provide some of these building blocks. The rationale for analyzing degradation products and, in the ideal case, cracking products, the structure of which is very close to its initial counterpart in kerogen, is that their structure is small enough to be characterized by a number of analytical techniques that cannot be used for solids.

Once quantitative atomic and molecular analyses are obtained on a given kerogen, the quantitative combination of molecular building blocks should ideally fit atomic data. The reality is not so simple because some information is still lacking and many hypotheses have to be made, not only on the types of bonding sites, but also on missing or incomplete data. The reverse situation can also occur, with two different techniques normally giving the same information in fact providing different analytical results. In such situations, one data set is chosen or the data are adapted on the basis of general knowledge of sedimentary organic compounds from the literature.

Randomly selected molecular building blocks from molecular analyses and functional groups from chemical functional analysis were connected by covalent bonds in order to fit elemental analysis and solid state 13C NMR results, removing or adding building blocks and functional groups to improve fitting of calculated and measured atomic parameters at each further step of the construction.

As already noted, the idea was not to represent an exact chemical structure of kerogens since this cannot be analyzed, but rather to show in a synthetic way their main chemical and physical characteristics related to type and maturity, in order to allow visual comparison, even for nonchemists (Fig. 8.11).

8.5.6 Multiscale Modeling Techniques

In shale reservoirs, in addition to matrix organic and inorganic pore structures, a microfracture network also exists that can have a potential to participate in gas transport and production. The FIB/SEM images of different shale samples show that the characteristic length of these complex natural fractures is such that they are only in hydraulic communication with matrix inorganic materials. Convective flow and Darcy's law govern gas transport in these natural fractures.

To account for the transport in both organic and inorganic pores, one technique is to employ discrete multiscale modeling (Xu et al., 2017; Fung et al., 2016), however, these approaches require detailed geometry and special distribution of each continuum. Single-continuum models used in the literature also extremely simplify the problem and neglect the mass exchange and couplings between different continuums. It is also extremely difficult to obtain average storage and transport parameters that can describe the gas dynamic in shale reservoirs as a single continuum. Multi-continuum modeling (Fig. 8.12) can solve the problems associated with discrete and single-continuum models by introducing each continuum (matrix organic, matrix inorganic, and fractures) continuously in space and by providing the opportunity to couple different continuum using series, parallel, or mixed coupling (Akkutlu et al., 2012; Bear & Bachmat, 2012; Swami et al., 2013).

8.6 Lattice Boltzmann Model (LBM)

The lattice Boltzmann model (LBM) considers the trajectory of a group of particles instead of individual particles in molecular dynamics (MD), making it less computationally extensive. Thus, LBM bridges the gap between microscale (MD) and macroscale (Navier-Stokes equation) (Mohamad, 2011). The lattice Boltzmann models originate from the lattice gas cellular automata (LGCA) models developed to replicate the complexities of real fluid flow on triangular or hexagonal lattices. In the LGCA models, fluid is represented by Boolean variables (0 or 1) (Succi, 2001; Wolf-Gladrow, 2000). Based on a set of rules, initial

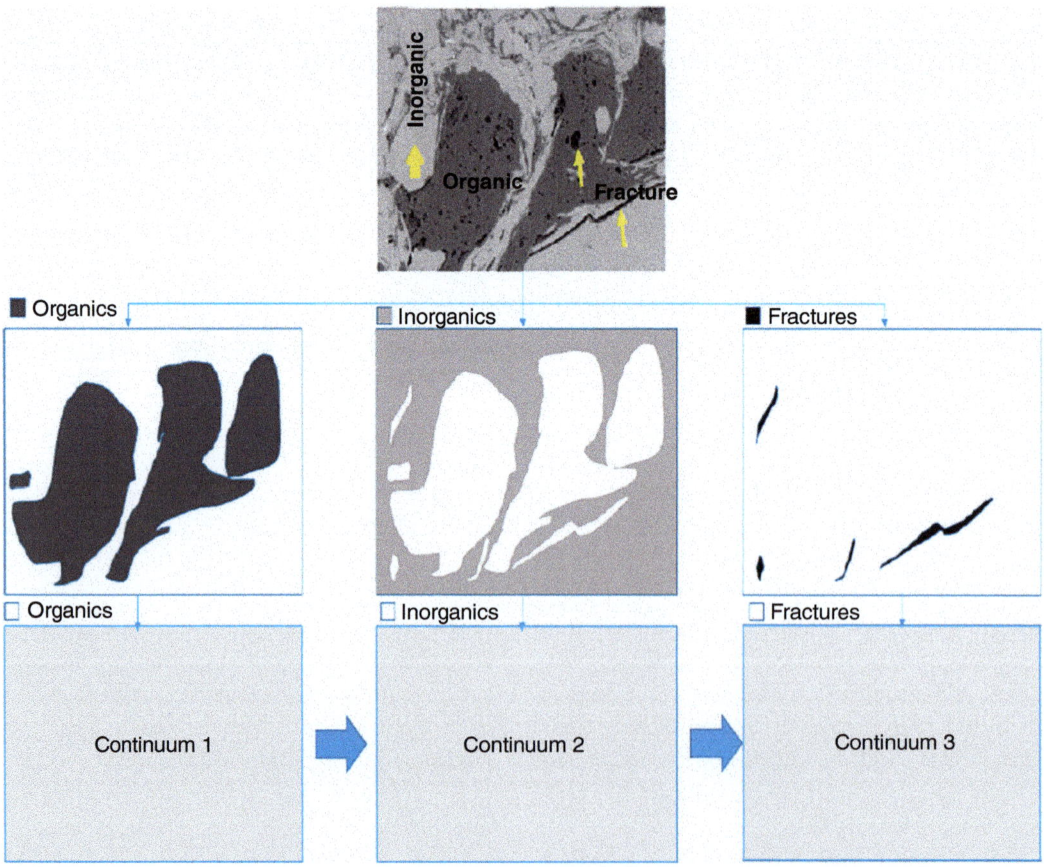

Figure 8.12 Schematic of conceptual model used for multiscale shale matrix modeling (Takbiri-Borujeni, 2019).

and boundary conditions, a cellular automaton (an algorithmic entity), at a given position on a lattice in space interacts with neighboring entities and evolves with time. The interaction at a given time consists of streaming in which the particles propagate to new sites, and collision, where they collide based on the set of collision rules (Sukop & Thorne, 2006; Frisch et al., 1986). However, the LGCA models were riddled with statistical noise due to the discrete Boolean variables (Guo & Shu, 2013).

Subsequently, the LB models were developed by replacing the Boolean variables with distribution functions ($f(x, t)$) assigned to a group of particles confined to a lattice in space and time (McNamara & Zanetti, 1988). The LBM solves the Boltzmann equation (BE) derived from the kinetic theory, where a gas is assumed to be composed of interacting particles, which undergo both streaming and collision to conserve mass and momentum (Wolf-Gladrow, 2000). It simplifies Boltzmann's concept by reducing the number of particle spatial positions and microscopic momenta from a continuum (Guo & Shu, 2013; Sukop & Thorne, 2006). Figure 8.13 shows a two-dimensional Cartesian model with nine velocities including the particle at rest (e_0) in eight directions assuming a single/uniform mass particle with equivalent microscopic velocities and momenta (Sukop & Thorne, 2006; Qian et al., 1992). The velocity

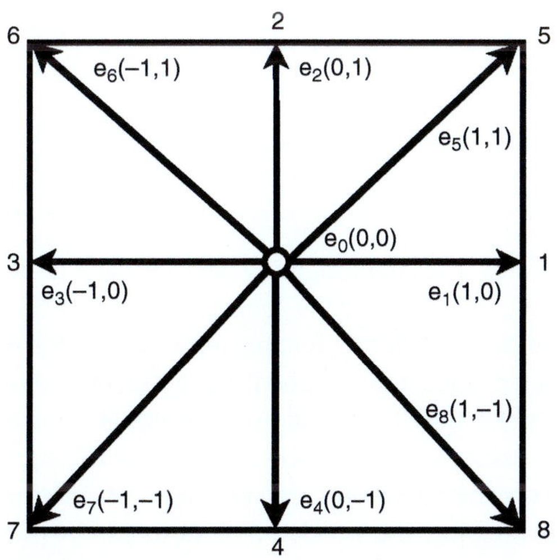

Figure 8.13 D_2Q_9 lattice and velocities.

indexing below is very common, although other forms exist. Lattice unit (lu) is the measure of the length in LB models, while time is measured in time steps (ts). Therefore, the velocity magnitude of e_1 to e_2 (particles along the x- and y-axes) is one lattice unit per time step (lu/ts) and that between e_6 and e_7 (particles along diagonal axes) is $\sqrt{2}$lu/ts. Other commonly used LB models are listed in Table 8.8 below together with their corresponding lattice velocities and associated weighting factors. The speed of sound is dependent on the models. It is also observed that all the x and y components are either 0 or ± 1 (Guo & Shu, 2013; Qian et al., 1992).

The distribution function ($f(x, e, t)$) represents the number of molecules at a time (t) located at positions between x and $x + \Delta x$ within a range of velocities between e and $e + \Delta e$. Particles in a fluid medium reside on lattice nodes and move to their nearest neighbors along the lattice links. Applying an external force will cause a change in the velocity and position of a unit mass gas molecule, and upon collision, a net difference occurs between the number of molecules before and after collision leading to a rate change referred to as the collision operator, Ω (Mohamad, 2011; Sukop & Thorne, 2006). Bhatnagar-Gross-Krook (BGK) (Bhatnagar et al., 1954) introduced a simple linearized collision model to solve the Boltzmann equation. It assumes the distribution is close to local equilibrium state (Perumal & Dass, 2015). The LBM-BGK equation is therefore expressed in equation 8.1 (Wolf-Gladrow, 2000), where the left-hand side (LHS) represents the streaming (advection) term and the right-hand side (RHS) represents the collision process.

$$f_i(x + e_i\Delta t, t + \Delta t) - f_i(x, t) = -\frac{1}{\tau}[f_i(x, t) - f_i^{eq}(x, t)]. \tag{8.16}$$

In this equation, τ is the single relaxation factor, which controls the rate to attain equilibrium, and f^{eq} is the local equilibrium distribution function (EDF) dependent on the LB model in use. The relaxation time (τ) is related to the kinematic viscosity and expressed as

$$\upsilon = c_s^2(\tau - 0.5)\Delta t, \tag{8.17}$$

Table 8.8 Some LB Models and Their Parameters

Model	Lattice velocity (e_i)	Weight (w_i)	Speed of sound (c_s^2)
D_1Q_3	0	2/3	1/3
	±1	1/6	
D_1Q_5	0	6/12	1
	±1	2/12	
	±2	1/12	
D_2Q_7	$(0, 0)$	1/2	1/4
	$\left(\pm 1/2, \pm \sqrt{3}/2\right)$	1/12	
D_2Q_9	$(0, 0)$	4/9	1/3
	$(\pm1, 0), (0, \pm1),$	1/9	
	$(\pm1, \pm1)$	1/36	
D_3Q_{15}	$(0, 0, 0),$	2/9	1/3
	$(\pm1, 0, 0), (0, \pm1, 0), (0, 0, \pm1),$	1/9	
	$(\pm1, \pm1, \pm1)$	1/72	
D_3Q_{19}	$(0, 0, 0)$	1/3	1/3
	$(\pm1, 0, 0), (0, \pm1, 0), (0, 0, \pm1)$	1/18	
	$(\pm1, \pm1, 0), (\pm1, 0, \pm1), (0, \pm1, \pm1)$	1/36	

Source: Modified after Guo (2013).

where, υ is the kinematic viscosity and c_s is the speed of sound. The unit of the kinematic viscosity is lu^2/ts .

The relaxation time (τ) varies for different simulations and some stable values proposed in literature are 0.6 and 1. Although, the single relaxation time (SRT) is commonly used, other relaxation times are reported including the two-relaxation-time (TRT) (Ginzburg et al., 2008) and the multirelaxation time (MRT) (Lallemand & Luo, 2000; d'Humieres, 1992). The MRT is proposed to have higher numerical stability and accuracy than both SRT and TRT.

Subsequently, the macroscopic fluid properties like velocity, density, and pressure can be resolved through the collective behavior (statistical mechanics) of the microscopic particles described by the Navier-Stokes equation (Guo & Shu, 2013).

$$\rho = \sum_{i=0}^{8} f_i \tag{8.18}$$

and

$$u = \frac{1}{\rho}\sum_{i=0}^{8} f_i e_i \tag{8.19}$$

present the estimation of the macroscopic fluid density (ρ) and velocity (u), respectively.

The equilibrium distribution function (EDF) of the D_2Q_9 model is defined as

$$f_i^{eq}(x, t) = w_i \rho \left(1 + \frac{e_i.u}{c_s^2} + \frac{(e_i.u)^2}{2c_s^2} - \frac{u^2}{2c_s^2}\right), \tag{8.20}$$

where, w_i is the weighting factor of each lattice link, c_s is the speed of sound, and u is the fluid velocity. External forcing terms are incorporated into the adjusted fluid velocity term (u^{eq}) in the EDF equation. An example of incorporating gravity starts by considering the equation for Newton's second law of motion, expressing the acceleration term as the rate change of velocity with time ($^{du}/_{dt}$) and with density being proportional to the mass, the resulting expression is given as

$$u^{eq} = u + \Delta u = u + \frac{\tau F}{\rho}, \tag{8.21}$$

where, Δu is the change in velocity, F is the forcing term, and ρ is the fluid density.

The LBM can be applied in complex geometries (Zimny, 2013; Suga et al., 2010; Succi et al., 1989) by implementing different boundary conditions including the bounce-back boundary condition (no-slip) (Chen et al., 1996), hydrodynamic boundary conditions (Zou & He, 1997; Noble et al., 1995), temperature jump and velocity slip (Zheng et al., 2007; Huang et al., 2007) as well as Langmuir slip conditions (Chen & Tian, 2010; Kim et al., 2007; Myong, 2004), where adsorption/desorption, surface diffusion, among other organic nanopore flow behaviors (Ren et al., 2015; Zhang et al., 2014b; Bennett et al. , 2012; Tang et al., 2008) are realized. LBM can also be used to simulate both single and multiphase fluids incorporating particle-particle interactions and particle-surface interactions to understand phase separation and study contact angles (Huang et al., 2015; Huang et al., 2008; Benzi et al., 2006; Sukop & Or, 2004). It is also applicable to thermal and reactive flows and can be used to estimate the apparent permeability (Chen et al., 2014).

8.6.1 LBM Simulation

LBM simulation codes can be generated in any computer language of choice and there are open-source LBM codes online (such as PALABOS), which are easily accessible. This section looks to apply the D_2Q_9 LBM to simulate fluid flow through a pore slit in a shale formation (Fig. 8.14) neglecting adsorption and assuming a no-slip boundary condition (bounce back). The bounce-back boundary condition assumes gas particles hitting the solid surface are reflected back into the fluid medium without slip. As such, the fluid distributions are expressed as below at the wall (Succi, 2001; Wolf-Gladrow, 2000):

$$f_2 = f_4, f_1 = f_3, f_5 = f_7, f_6 = f_8. \tag{8.22}$$

The simulation steps are outlined below to generate a MATLAB code in the Appendix, at the end of this chapter, following guidelines from the works of Guo and Shu (2013), Mohamad (2011), and Sukop and Thorne (2006).

1) Simulation parameters (flow domain, tau, discrete velocities, etc.)
2) Initialization (fluid density, fluid velocity, distribution functions, etc.)
3) Collision
4) Streaming
5) Boundary conditions (bounce back, hydrodynamic, Langmuir slip model, etc.)
6) Fluid variables

The MATLAB code in the Appendix is compared with the analytical Poiseuille velocity profile and shows a good agreement. A maximum velocity of 9.3879e-05 lu/ts is obtained and can be converted to a physical unit of 0.0558 m/s by the scaling factor expressed below.

$$u_{physical}(m/s) = \left(\frac{c_{s-physical}(m/s)}{c_{s-lattice}(lu/ts)}\right) * [u_{lattice}(lu/ts)], \tag{8.23}$$

where, $c_{s-physical}$ is the physical speed of sound given as 343 m/s, $c_{s-lattice}$ is the lattice speed of sound given as $1/\sqrt{3}$, and $u_{lattice}$ is the lattice velocity (x direction) from LBM simulation'

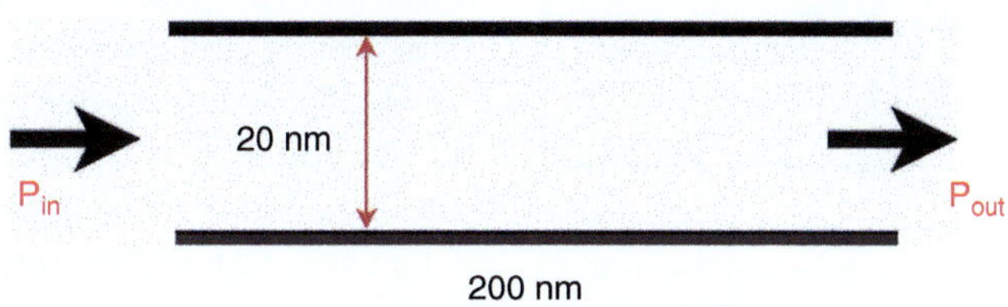

Figure 8.14 Schematic diagram of a nanopore slit of 20 nm by 200 nm.

8.6.2 Implementation of LBM Simulation in Organic Nanopores

Fluid transport in organic-nanopores is determined by permeability, the ability of a medium to transmit fluids. Conventionally, permeability in sedimentary rocks is estimated from Darcy's law. However, due to the transition flow regime associated with organic nanopores, pore wall effects including rarefied gas flow, adsorption, surface and Knudsen diffusions are encountered, which render Darcy's equation inadequate. Figure 8.15 illustrates the pore-wall effects: (a) shows the viscous flow and Knudsen diffusion, and (b) shows surface diffusion. Viscous flow is encountered in macropores (>50 nm), where the collision between the gas molecules is dominant, while Knudsen diffusion occurs as a result of the frequent gas-solid surface interactions when the free mean path is comparable to the thickness/diameter of the domain. On the other hand, surface diffusion occurs in micropores (<2 nm) and serves as the dominant transfer mechanism of strongly adsorbed gas on the pore walls driven by the concentration gradient (Wu et al., 2015).

On the other hand, the LBM provides an effective way of capturing all the pore-wall effects to evaluate permeability (Cudjoe & Barati, 2017; Ren et al., 2015; Fathi & Akkutlu, 2013). The pore-wall effects are incorporated into the LBM simulation by modifying the relaxation time according to the prevailing Knudsen number and this is given as (Suga et al., 2010; Stops, 1970)

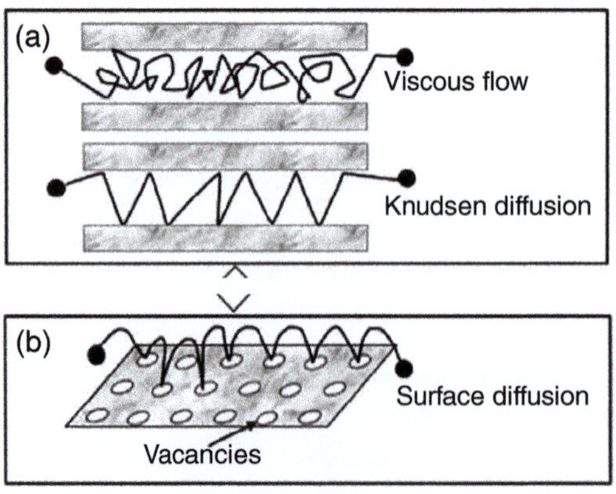

Figure 8.15 Pore-wall effects encountered in organic nanopores: (a) viscous flow and Knudsen diffusion and (b) surface diffusion (modified after Wu, 2015).

$$\psi(Kn) = \frac{2}{\pi}\arctan\left(\sqrt{2}Kn^{-3/4}\right) \qquad (8.24)$$

and

$$\tau_e = \tau\psi(Kn), \qquad (8.25)$$

where, Kn is the Knudsen number in the previous section and τ_e is the effective relaxation time capable of capturing the gas-wall interactions.

In so doing, the relaxation time is observed to decrease with a corresponding increase in the Knudsen number (Zhang et al., 2014b). Thus, with dominant pore-wall interactions, gas molecules undergo collision at short relaxation times.

Boundary Conditions

Incorporating the gas-solid and gas-gas molecular interactions into the LBM requires an effective boundary condition including the Langmuir slip model (Fathi & Akkutlu, 2013; Chen & Tian, 2009) implemented by either the bounce-back/specular-reflection (BSR) (Ren et al., 2015; Succi, 2002) or the nonequilibrium extrapolation scheme (Chen & Tian, 2009).

The Langmuir slip model (LSM) is developed based on the Langmuir theory of adsorption and is proven to be more effective than the use of the accommodation coefficient (σ), which does not capture the real fluid-solid interactions (Fathi & Akkutlu, 2013). Thus, in the LSM, gas molecules are assumed to adsorb to the pore walls for some amount of time before desorbing into the fluid (Myong, 2004). This causes a macroscopic slip velocity to develop, which is captured in the LSM in the equation 8.26 by Fathi et al. (2012) and Chen and Tian (2009), respectively.

$$u_{slip} = (1-\alpha)u_g + \alpha u_w, \qquad (8.26)$$

where, u_{slip} is the slip velocity, u_g is the gas velocity at one mean free path distance away from the wall, u_w is the local wall velocity due the adsorbed-phase transport (surface diffusion), and α is the amount of adsorbed gas. The adsorbed gas is expressed as a function of the Knudsen number (Kn) for monatomic gases as (Kim et al., 2007; Myong, 2004)

$$\alpha = \frac{1}{1 + 4\omega Kn/p}, \tag{8.27}$$

where, ω is determined through experimental data or theoretical consideration of intermolecular force and the gas-surface interaction, Kn is the Knudsen number, and p is the pressure. Thus, the amount of gas adsorbed approaches a value of one, when the Kn approaches zero, and this will result in only the adsorbed-gas velocity (surface diffusion) contributing to the gas slip velocity from equation 8.26.

The adsorbed-gas velocity or surface diffusion (u_w) is also expressed as (Ren et al., 2015)

$$u_w = -D_s \frac{\rho_s M}{\rho_{ads} V_{std}} \frac{V_L p_L}{(p_L + p)^2} \frac{\partial p}{\partial x}, \tag{8.28}$$

where, u_w is the adsorbed-gas velocity (surface diffusion), D_s is the surface diffusion coefficient, ρ_s is the organic solid density, M is the gas molecular weight, V_{std} is the gas volume per mole at standard temperature and pressure, V_L is the Langmuir volume, p_L is the Langmuir pressure, and p is the prevailing gas pressure.

Fluid-Fluid and Fluid-Solid Interaction Forces
Shan and Chen (1993, 1994) introduced the attractive short-range (cohesive) force between neighboring fluid particles in addition to the local collisions and adhesive forces between the fluid-solid phases (Martys & Chen, 1996), respectively, expressed as

$$F(x,t) = -G\psi(\rho(x,t)) \sum_{i=1}^{8} w_i \psi(\rho(x + e_i \Delta t, t)) e_i, \tag{8.29}$$

where, $F(x, t)$ is the cohesive force between fluid-fluid particles, G is the interaction strength, and $\psi(\rho(x, t))$ is the interaction potential expressed as a function of the fluid density.

$$F_{ads} = -G_{ads} \psi(x,t) \sum_i w_i s(x + e_i \Delta t) e_i, \tag{8.30}$$

where, F_{ads} is the adhesive force, G_{ads} is the particle interaction strength between the fluid and solid walls, and $s = 0$ for fluid nodes and $s = 1$ for solid nodes.

Bounce-Back/Specular Reflection Implementation of the Langmuir Slip Model
Bounce-back/specular reflection (BSR) implementation of the Langmuir slip model (LSM) at the lower wall (f_3-f_0-f_1) is expressed in terms of the unknown distribution functions, f_2, f_5, and f_6 as (Ren et al., 2015)

$$f_2 = f_4' + 2r\rho w_2 e_2 . u_w / c_s^2, \tag{8.31}$$

$$f_5 = r f_7' + (1 - r) f_8' + 2r\rho w_5 e_5 . u_w / c_s^2, \tag{8.32}$$

and

$$f_6 = r f_8' + (1 - r) f_7' + 2r\rho w_6 e_6 . u_w / c_s^2, \tag{8.33}$$

where, r is the bounce-back fraction and can assume any number between 0 and 1; a value of 0 assumes a fully specular reflection medium, while a value of 1 results in a fully bounce-back medium. However, a value of 0.6 is commonly used. The post-collision distribution functions (f_4', f_7', f_8') are expressed as

$$f_i' = f_i - (f_i - f_i^{eq})/\tau. \tag{8.34}$$

8.6.3 Apparent Permeability

The gas apparent permeability is calculated by incorporating the pore wall effects as (J. K. Wang et al., 2017)

$$K_{app} = \frac{N_{total} M \mu}{(\rho \Delta p)}, \tag{8.35}$$

where, K_{app} is the gas apparent permeability, N_{total} is the total molar flux of free gas and surface diffusion, ρ is the density of free gas, and Δp is the change in pressure.

8.7 Problems

1 From the MATLAB code in the Appendix, at the end of this chapter, use varying domains and relaxation times and comment on the results after comparing them.

2 Modify the LBM code in the Appendix, at the end of this chapter, and implement the LSM to simulate methane (CH_4) gas flow through a 200 nm pore slit with heights ranging from 5 nm to 50 nm driven by a pressure gradient of 0.1 MPa/m. Additional parameters are outlined in Table 8.9 below.

3 Estimate the apparent permeability with incorporated surface diffusion and compare it with the intrinsic permeability (K_{int}) expressed as $K_{\text{int}} = H/12$ for parallel plates domain. Comment on the permeability effect at 5 nm and 20 nm, respectively.

Table 8.9 Additional Parameters for Methane Flow in a Nano-Pore Slit Considering Pore Wall Effects

Pressure of gas (P) psi	3,500
Temperature of gas (T) °C	25
Dynamic viscosity (μ) Pa.s	21.4e-6
Diffusion coefficient (Ds) m^2/s	1e-5
Molecular weight of methane (M) kg/mol	0.016
Langmuir pressure (pL) MPa	12
Gas volume per mole at STP (V_{std}) m^3/mol	15e-3
Adsorbed-gas density (ρ_{ads}) kg/m^3	320
Organic solid density (ρ_s) kg/m^3	2,500
Langmuir volume (V_L) m^3/kg	1.6e-3
Density (kg/m^3)	173.2

Additional Reading

Bear, J., & Bachmat,Y. (2012). *Introduction to modeling of transport phenomena in porous media*. Vol. *4*. Springer Science & Business Media.

Guo, Z., & Shu, C. (2013). *Lattice Boltzmann method and its applications in engineering*. Singapore: World Scientific Publishing Co. Pte. Ltd.

Huang, H., Sukop, M. C., & Lu, X. (2015). *Multiphase lattice Boltzmann methods: Theory and application*. Chichester, West Sussex: John Wiley & Sons, Ltd.

References

Adesida, A., Akkutlu, I., Resasco, D. E., Shekhar Rai, C., et al. (2011). Characterization of Barnett shale kerogen pore size distribution using DFT analysis and grand canonical monte carlo simulations. SPE Annual Technical Conference and Exhibition.

Akkutlu, I. Y., Fathi, E., et al. (2012). Multiscale gas transport in shales with local kerogen heterogeneities. *SPE Journal*, *17*, 1–2.

Attard, G., & Barnes, C. (1998). Surfaces. *Oxford Chemistry Primers*.

Bear, J., & Bachmat,Y. (2012). *Introduction to modeling of transport phenomena in porous media. Vol. 4*. Springer Science & Business Media.

Behar, F., & Vandenbroucke, M. (1987). Chemical modeling of kerogens. *Organic Geochemistry*, *11*, 15–24.

Bennett, S., Asinari, P., & Dellar, P. J. (2012). A lattice Boltzmann model for diffusion of binary gas mixtures that includes diffusion slip. *International Journal for Numerical Methods in Fluids*, *69*, 171–189.

Benzi, R., Biferale, L., Sbragaglia, M., Succi, S., & Toschi, F. (2006). Mesoscopic modeling of a two-phase flow in the presence of boundaries: The contact angle. *Physical Review E*, *74*, 1–14. doi:10.1103/PhysRevE.74.021509

Bhatnagar, P. L., Gross, E. P., & Krook, M. (1954). A model for collision processes in gases. i. small amplitude processes in charged and neutral one-component systems. *Physical Review*, 511–525.

Builes, S., Sandler, S., & Xiong, R. (2013). Isosteric heats of gas and liquid adsorption. *Langmuir*, *29*, 10416–10422. doi: 10.1021/la401035p

Burlingame, A. L., Haug, P. A., Schnoes, H. K., & Simoneit, B. R. (1969). *Fatty acids derived from the Green River formation oil shale by extractions and oxidations: A review.* Advances in Organic Geochemistry 1968, Proceedings of the 4th International Meeting on Organic Geochemistry, Amsterdam, 16–18 September, 1968, 85–129. doi:10.1016/B978-0-08-006628-8.50008-3

Cercignani, C. (1964). Higher order slip according to the linearized Boltzmann equation. Institute of Engineering Research Report AS-64-19, University of California, Berkeley.

Chen, L., Zhang, L., Kang, Q., Viswanathan, H. S., Yao, J., & Tao, W. (2015). Nanoscale simulation of shale transport properties using the lattice Boltzmann method: Permeability and diffusivity. *Scientific Reports*, *5*, 8089. doi: 10.1038/srep08089

Chen, S., & Tian, Z. (2009). Simulation of microchannel flow using the lattice Boltzmann method. *Physica A*, 4803–4810.

Chen, S., & Tian, Z. (2010). Simulation of thermal micro-flow using lattice Boltzmann method with Langmuir slip model. *International Journal of Heat and Fluid Flow*, *31*, 227–235.

Chen, S., Martinez, D., & Mei, R. (1996). On boundary conditions in lattice Boltzmann methods. *Physics of Fluids*, *8*, 2527. doi: 10.1063/1.869035

Collell, J., Galliero, G., Gouth, F., Montel, F., Pujol, M., Ungerer, P., & Yiannourakou, M. (2014). Molecular simulation and modelisation of methane/ethane mixtures adsorption onto a microporous molecular model of kerogen under typical reservoir conditions. *Microporous and Mesoporous Materials*, *197*, 194–203.

Collell, J., Galliero, G., Vermorel, R., Ungerer, P., Yiannourakou, M., Montel, F., & Pujol, M. (2015). Transport of multicomponent hydrocarbon mixtures in shale organic matter by molecular simulations. *The Journal of Physical Chemistry C*, *119*(39), 22587–22595. doi: 10.1021/acs.jpcc.5b07242

Conti, J. J., Holtberg, P. D., Beamon, J. A., Schaal, A., Ayoub, J. C., & Turnure, J. T. (2014). Annual energy outlook 2014. *US Energy Information Administration*.

Cudjoe, S., & Barati, R. (2017). Lattice Boltzmann simulation of CO_2 transport in kerogen nanopores: An evaluation of CO_2 sequestration in organic-rich shales. *Journal of Earth Science*, 926–932.

Curtis, M., Ambrose, R., Sondergeld, C., et al. (2010). Structural characterization of gas shales on the micro-and nano-scales. Canadian Unconventional Resources and International Petroleum Conference.

Curtis, M., Ambrose, R., Sondergeld, C., Shekhar Rai, C., et al. (2011). Investigation of the relationship between organic porosity and thermal maturity in the Marcellus shale. North American Unconventional Gas Conference and Exhibition.

d'Humieres, D. (1992). Generalized lattice Boltzmann equation in rarefied gas dynamics: theory and simulations. *Progress in Astronautics and Aeronautics*, *159*, 450–458.

EIA (2014). *The annual energy outlook 2014*. U.S. Energy Information Administration.

Facelli, J. C., Pugmire, R. J., Pimienta, I. S., Badu, S., & Orendt, A. (2011). Atomistic modeling of oil shale kerogens and asphaltene along with their interactions with the inorganic mineral matrix. Technical Report.

Fathi, E., & Akkutlu, Y. I. (2013). Lattice Boltzmann method for simulating of shale gas transport in kerogen. *SPE Journal*, 27–37.

Fathi, E., Tinni, A., & Akkutlu, I. Y. (2012). Correction to Klinkenberg slip theory for gas flow in nano-capillaries. *International Journal of Coal Geology*, *103*,: 51–59.

Faulon, J. L., Vandenbroucke, M., Drappier, J. M., Behar, F., & Romero, M. (1990). 3D chemical model for geological macromolecules. *Organic Geochemistry*, 981–993.

Frisch, U., Hasslacher, B., & Pomeau, Y. (1986). Lattice-gas automata for the Navier-Stokes equation. *Physical Review Letters*, *56*, 1505–1508G.

Fuchs, W., & Sandhoff, A. G. (1942). Theory of coal pyrolysis. *Industrial and Engineering Chemistry*, *34*, 567–571.

Fung, L., Du, S., et al. (2016). Parallel-simulator framework for multipermeability modeling with discrete fractures for unconventional and tight gas reservoirs. *SPE Journal*, *21*, 1–370.

Ginzburg, I., Verhaeghe, F., & d'Humieres, D. (2008). Two-relaxation-time lattice Boltzmann scheme: About parametrization, velocity, pressure and mixed boundary conditions. *Communications in Computational Physics*, *3* (2), 427–478.

Given, P. H. (1960). The distribution of hydrogen in coals and its relation to coal structure. *Fuel*, 147–153.

Gregg, S., Sing, K., & Salzberg, H. W. (1967). Adsorption surface area and porosity. *Journal of the Electrochemical Society*, 279C–279C.

Guo, Z., & Shu, C. (2013). *Lattice Boltzmann method and its applications in engineering*. Singapore: World Scientific Publishing Co. Pte. Ltd.

Guo, Z., Shi, B., Zhao, C., et al. (2007). Discrete effects on boundary conditions for the lattice Boltzmann equation in simulating microscale gas flows. *Physical Review E*, 3–7.

Huang, H., Lee, T. S., & Shu, C. (2007). Lattice Boltzmann method simulation gas slip flow in long microtubes. *International Journal of Numerical Methods for Heat & Fluid Flow*, 587–607.

Huang, H., Li, Z., Liu, S., & Lu, X. (2008). Shan-and-Chen-type multiphase lattice Boltzmann study of viscous coupling effects for two-phase flow in porous media. *International Journal of Numerical Methods for Fluids*, 1–14.

Huang, H., Sukop, M. C., & Lu, X. (2015). *Multiphase lattice Boltzmann methods: Theory and application*. Chichester, West Sussex: John Wiley & Sons, Ltd.

Kazemi, M. (2017). *Molecular simulation of transport and storage in shale*. West Virginia University.

Kazemi, M., & Takbiri-Borujeni, A. (2015). An analytical model for shale gas permeability. *International Journal of Coal Geology*, *146*, 188–197.

Kazemi, M., & Takbiri-Borujeni, A. (2016). Flow of gases in organic nanoscale channels: A boundary-driven molecular simulation study. *Energy & Fuels*, *30*, 8156–8163.

Kazemi, M., Maleki, H., Takbiri-Borujeni, A., et al. (2016). Molecular dynamics study of transport and storage of methane in kerogen. SPE Eastern Regional Meeting.

Kazemi, M., Takbiri-Borujeni, A., et al. (2015). Predicting gas apparent permeability of shale samples: A novel analytical approach. SPE Annual Technical Conference and Exhibition.

Keller, J. U., & Staudt, R. (2005). *Gas adsorption equilibria: experimental methods and adsorptive isotherms*. Springer Science & Business Media.

Kidena, K., Murata, S. & Nomura, M. . (2008). A newly proposed view on coal molecular structure integrating two concepts: Two phase and uniphase models. *Fuel Processing Technology*, *89*, 424–433.

Kim, H. M., Kim, D., Kim, W. T., Chung, P. S., & Jhon, M. S. (2007). Langmuir slip model for air bearing simulation using the lattice Boltzmann method. *IEEE Transactions on Magnetics*, *43* (6), 2244–2246.

Kowalewski, I., Vandenbroucke, M., Huc, A. Y., Taylor, M. J. , & Faulon, J. L. (1996). Preliminary results on molecular modeling of asphaltenes using structure elucidation programs in conjunction with molecular simulation programs. *Energy & Fuels*, 97–107.

Lallemand, P., & Luo, L. S. (2000). Theory of the lattice Boltzmann method: dispersion, dissipation, isotropy, Galilean invariance, and stability. *Physical Review E*, *61*, 6546–6562.

Langmuir, I. (1918). The adsorption of gases on plane surfaces of glass, mica and platinum. *Journal of the American Chemical Society*, 1361–1403.

Lille, Ü. (2004). Behavior of Estonian kukersite kerogen in molecular mechanical force field. *Oil Shale*, *21*, 99.

Martys, N. S., & Chen, H. (1996). Simulation of multicomponent fluids in complex three-dimensional geometries by the lattice Boltzmann method. *Physical Review E*, 743–750.

McNamara, G., & Zanetti, G. (1988). Use of a Boltzmann equation to simulate lattice-gas automata. *Physical. Review Letters*, *61* (20), 2332–2335.

Michalec, L., & Lísal, M. (2017). Molecular simulation of shale gas adsorption onto overmature type II model kerogen with control microporosity. *Molecular Physics*, *115*, 1086–1103.

Moghadam, A., & Chalaturnyk, R. . (2014). Expansion of the Klinkenberg's slippage equation to low permeability porous media. *International Journal of Coal Geology*, *123*, 2–9.

Mohamad, A. A. (2011). *Lattice Boltzmann method: fundamentals and engineering applications with computer codes*. London: Dordrecht Heidelberg.

Mosher, K., He, J., Liu, Y., Rupp, E., & Wilcox, J. (2013). Molecular simulation of methane adsorption in micro- and mesoporous carbons with applications to coal and gas shale systems. *International Journal of Coal Geology*, *109*, 36–44.

Myong, R. S. (2004). Gaseous slip models based on the Langmuir adsorption isotherm. *Physics of Fluids*, *16* (1), 104–117.

Ning, Y., He, S., Liu, H., & Wang, H. (2016). *Permeability prediction considering surface diffusion for gas shales by lattice Boltzmann simulations on multiscale reconstructed digital rocks*. International Petroleum Technology Conference (IPTC). Bangkok: IPTC.

Noble, D. R., Chen, S., Georgiadis, J. G., & Buckius, R. O. (1995). A consistent hydro-dynamic boundary condition for the lattice Boltzmann method. *Physics of. Fluids*, *7* (1), 203–209.

Nomura, M., Artok, L., Murata, S., Yamamoto, A., Hama, H., Gao, H., & Kidena, K. (1998). Structural evaluation of Zao Zhuang coal. *Energy & Fuels*, *12*, 512–523.

Orendt, A. M., Pimienta, I., Badu, S., Solum, M., Pugmire, R., Facelli, J., Locke, D., et al. (2013). Three-dimensional structure of the Siskin Green River Oil Shale Kerogen Model: A comparison between calculated and observed properties. *Energy & Fuels*, *27*, 702–710.

Perumal, D. A., & Dass, A. K. (2015). A review on the development of lattice Boltzmann computation of macro fluid flows and heat transfer. *Alexandria Engineering Journal 54*, 955–971.

Qian, Y. H., D'Humieres, D., & Lallemand, P. (1992). Lattice BGK models for Navier-Stokes equation. *Europhysics Letters*, *17* (6), 479–484.

Ren, J., Guo, P., Guo, Z., & Wang, Z. (2015). A lattice Boltzmann model for simulating gas flow in kerogen pores. *Transport in Porous Media*, *106*, 285–301.

Roy, S., Raju, R., Chuang, H., Cruden, B., & Meyyappan, M. (2003). Modeling gas flow through microchannels and nanopores. *Journal of Applied Physics*, *93*, 4870–4879.

Rullkötter, J., & Michaelis, W. (1990). The structure of kerogen and related materials: A review of recent progress and future trends. *Organic Geochemistry*, *16*, 829–852.

Sakhaee-Pour, A., Bryant, S., et al. (2012). Gas permeability of shale. *SPE Reservoir Evaluation \& Engineering*, 401–409.

Schettler, P. D., Parmely, C. R., Lee, W. J., et al. (1989). Gas storage and transport in Devonian shales. *SPE Formation Evaluation*, *4*, 371–376.

Shahtalebi, A., Farmahini, A., Shukla, P., & Bhatia, S. (2014). Slow diffusion of methane in ultra-micropores of silicon carbide-derived carbon. *Carbon*, *77*, 560–576.

Shan, X. , & Chen, H. (1993). Lattice Boltzmann model for simulating flows with multi-phases and components. *Physical Review E*, 1815–1819.

Shan, X., & Chen, H. (1994). Simulation of non-ideal gases and liquid-gas phase transitions by the lattice Boltzmann equation. *Physical Review E*, 2941–2948.

Sharipov, F., & Seleznev, V. (1998). Data on internal rarefied gas flows. *Journal of Physical and Chemical Reference Data*, *27*, 657–706.

Shinn, J. H. (1984). From coal to single-stage and two-stage products: A reactive model of coal structure. *Fuel*, 1187–1196.

Siskin, M., Scouten, C. G., Rose, K. D., Aczel, T., Colgrove, S. G., & Pabst, R. E. (1995). Detailed structural characterization of the organic material in Rundle Ramsay Crossing and Green River oil shales. In *Composition, geochemistry and conversion of oil shales*, 143–158. Springer.

Solomon, P. R., et al.(1981). Coal structure and thermal decomposition. *New Approaches in Coal Chemistry*, 61–71.

Sondergeld, C., Newsham, K., Comisky, J., Rice, M., Shekhar Rai, C., et al. (2010). Petrophysical considerations in evaluating and producing shale gas resources. SPE Unconventional Gas Conference.

Spiro, C. (1981). Space-filling models for coal: a molecular description of coal plasticity. *Fuel*, 1121–1126.

Spiro, C., & Kosky, P.(1982). Space-filling models for coal. 2. Extension to coals of various ranks. *Fuel*, 1080

Stops, D.W. (1970). The mean free path of gas molecules in the transition regime. *Journal of Physics. D: Applied Physics*, 685–696.

Succi, S. (2001). *The lattice Boltzmann equation for fluid dynamics and beyond*. Great Clarendon Street,Oxford: Oxford University Press.

Succi, S. (2002). Mesoscopic modeling of slip motion at fluid-solid interfaces with hetergeneous catalysis. *Physical Review Letters*, 064502.

Succi, S., Foti, E., & Higuera, F. (1989). Three-dimensional flows in complex geometries with the lattice boltzmann method. *EPL, 10*, 433–438.

Suga, K., Takenaka, S., Ito, T., Kaneda, M., Kinjo, T., & Hyodo, S. (2010). Evaluation of a lattice Boltzmann method in a complex nanoflow. *Physical Review E, 82* (016701), 1–10.

Sukop, M. C., & Or, D. (2004). Lattice Boltzmann method for modeling liquid-vapor interface configurations in porous media. *Water Resources Research, 40*, W01509.

Sukop, M. C., & Thorne, D. T. (2006). *Lattice Boltzmann modeling: An introduction for geoscientists and engineers*. Berlin: Heidelberg.

Swami, V., Settari, A., Javadpour, F., et al. (2013). A numerical model for multi-mechanism flow in shale gas reservoirs with application to laboratory scale testing. EAGE Annual Conference & Exhibition Incorporating SPE Europec.

Takbiri-Borujeni, A., Fathi, E., Kazemi, M., & Belyadi, F. (2019a). An integrated multiscale model for gas storage and transport in shale reservoirs. *Fuel, 237*, 1228–1243.

Takbiri-Borujeni, A., Fathi, E., Kazemi, M., & Belyadi, F. (2019b). An integrated multiscale model for gas storage and transport in shale reservoirs. *Fuel, 237*, 1228–1243.

Tang, G. H., Zhang, Y. H., Gu, X. J., & Emerson, D. R. (2008). Lattice Boltzmann modelling Kundsen layer effect in non-equilibrium flows. *EPL, 83* (40008), 1–6.

Ungerer, P., Collell, J., & Yiannourakou, M. (2014). Molecular modeling of the volumetric and thermodynamic properties of kerogen: Influence of organic type and maturity. *Energy & Fuels, 29*, 91–105.

Wang, F. P., Reed, R., et al. (2009). Pore networks and fluid flow in gas shales. SPE Annual Technical Conference And Exhibition.

Wang, J., Kang, Q., Chen, L., & Rahman, S. (2017). Pore-scale lattice Boltzmann simulation of micro-gaseous flow considering surface diffusion effect. *International Journal of Coal Geology*, 62–73.

Wilcox, J. (2012). *Carbon capture*. Springer Science & Business Media.

Wolf-Gladrow, D. A. (2000). *Latttice-gas cellular automata and lattice Boltzmann models: An introduction*. Berlin, Heidelberg: Springer.

Wu, K., Li, X., Guo C., & Chen, Z. (2015). *Adsorbed gas surface diffusion and bulk gas transport in nanopores of shale reservoirs with real gas effect-adsorption -mechanical coupling*. SPE Reservoir Simulation Symposium. Houston, Texas.

Wu, K., Li, X., Wang, C., Yu, W., & Chen, Z. (2015). Model for Surface Diffusion of Adsorbed Gas in Nanopores of Shale Gas Reservoirs. *Ind. Eng. Chem. Res.* 3225–3236.

Xu, Y., Cavalcante Filho, J., Yu, W., Sepehrnoori, K., et al. (2017). Discrete-fracture modeling of complex hydraulic-fracture geometries in reservoir simulators. *SPE Reservoir Evaluation & Engineering, 20*, 403–422.

Yang, R. T. (2003). *Adsorbents: Fundamentals and applications*. John Wiley & Sons.

Yen, T. F. (1976). Structural aspects of organic components in oil shale. In T. F. Yen, *Oil Shale*, 129–148. Amsterdam:.Elsevier.

Zhang, J., Clennell, M. B., Dewhurst, D. N., & Liu, K. (2014). Combined Monte Carlo and molecular dynamics simulation of methane adsorption on dry and moist coal. *Fuel, 122*, 186–197.

Zhang, X., Xiao, L., Shan, X., & Guo, L. (2014). Lattice Boltzmann simulation of shale gas transport in organic nanopores. *Scientific Reports*, 1–6.

Zhao, Y., Feng, Y., & Zhang, X. (2016). Selective adsorption and selective transport diffusion of CO_2–CH_4 binary mixture in coal ultramicropores. *Environmental Science & Technology, 50*, 9380–9389.

Zheng, L., Shi, B., & Chai, Z. (2007). Lattice Boltzmann method for simulating the temperature jump and velocity slip in microchannels. *Communications in Computational Physics, 2* (6), 1125–1138.

Zimny, S., Masilamani, K., Jain, K., & Roller, S. (2013). Lattice Boltzmann simulations on complex geometries. In M. Resch, W. Bez, E. Focht, H. Kobayashi, Y. Kovalenkor (Eds.), *Sustained simulation performance*. Springer, Cham. doi: 10.1007/978-3-319-01439-5_4

Zou, Q., & He, X. (1997). On pressure and velocity boundary conditions for the lattice Boltzmann BGK model. *Physics of Fluids, 9*, 1591.

Appendix

A MATLAB script developed for fluid flow through nanopore slit is presented in this appendix.

% LBM-BGK Simulation of Fluid Flow through a Nanopore slit
% Sherifa Cudjoe

% LBM-BGK Simulation of Fluid Flow through a Nanopore slit
% Sherifa Cudjoe

```
close('all');clear('all')

% Simulation parameters

Ly = 20;     % characteristic length (height) of pore-slit, lu
Lx = 200;    % length of pore-slit

% Assume a Reynolds number (Re)
Re = 100;

% Maximum velocity
Umax = 0.1; % lu/ts

% Average velocity
Uavg = (2/3)*Umax; % lu/ts

% kinematic viscosity
nu = Umax*Ly/Re;  % lu^2/ts

tau = 1./((3*nu)+0.5);     % single relaxation time (SRT)

g = 0.000001; % gravity forcing term

% Setting boundary walls at the top and bottom of pore-slit
s_node=zeros(Ly,Lx);
for i=1:Lx
    s_node(1,i)=1;
    s_node(Ly,i)=1;
end

% Initial fluid density (rho) and distribution functions

rho = ones(Ly,Lx);

f(:,:,1)  =  (4./9. )*rho;
f(:,:,2)  =  (1./9. )*rho;
f(:,:,3)  =  (1./9. )*rho;
f(:,:,4)  =  (1./9. )*rho;
```

```
f(:,:,5) = (1./9. )*rho;
f(:,:,6) = (1./36.)*rho;
f(:,:,7) = (1./36.)*rho;
f(:,:,8) = (1./36.)*rho;
f(:,:,9) = (1./36.)*rho;


% Lattice velocity vectors
ex(1)= 0; ey(1)= 0;
ex(2)= 1; ey(2)= 0;
ex(3)= 0; ey(3)= 1;
ex(4)=-1; ey(4)= 0;
ex(5)= 0; ey(5)=-1;
ex(6)= 1; ey(6)= 1;
ex(7)=-1; ey(7)= 1;
ex(8)=-1; ey(8)=-1;
ex(9)= 1; ey(9)=-1;

% Beginning of main time loop
for ts=1:300

    ts

    % Computing macroscopic density, rho, and velocity, u=(ux,uy).
    for j=1:Ly

        for i=1:Lx

     % Initializing fluid velocity and fluid density
            ux(j,i) = 0.0;
            uy(j,i) = 0.0;
            rho(j,i) = 0.0;

            if ~s_node(j,i)

                for a=1:9

                    rho(j,i)    = rho(j,i)+ f(j,i,a);
                    ux(j,i)     = ux(j,i) + ex(a)*f(j,i,a);
                    uy(j,i)     = uy(j,i) + ey(a)*f(j,i,a);
                end

                ux(j,i) = ux(j,i)/rho(j,i);
                uy(j,i) = uy(j,i)/rho(j,i);
```

```
        end

        %add space matricies for plotting
        x(j,i)=i;
        y(j,i)=j;

    end
end


% Computing the EDF, feq.
c1=3.;
c2=9./2.;
c3=3./2.;

for j=1:Ly

  for i=1:Lx

     if ~s_node(j,i)

          rt0 = (4./9. )*rho(j,i);
          rt1 = (1./9. )*rho(j,i);
          rt2 = (1./36.)*rho(j,i);

          ueqxij  =  ux(j,i)+(tau*g)/rho(j,i);
          ueqyij  =  uy(j,i);

          uxsq    =  ueqxij * ueqxij;
          uysq    =  ueqyij * ueqyij;

          uxuy6   -  ueqxij +  ueqyij;
          uxuy7   = -ueqxij +  ueqyij;
          uxuy8   = -ueqxij + -ueqyij;
          uxuy9   =  ueqxij + -ueqyij;

          usq     =  uxsq + uysq;

        feq(j,i,1) = rt0*( 1. - c3*usq);
        feq(j,i,2) = rt1*( 1. + c1*ueqxij + c2*uxsq - c3*usq);
        feq(j,i,3) = rt1*( 1. + c1*ueqyij + c2*uysq - c3*usq);
        feq(j,i,4) = rt1*( 1. - c1*ueqxij + c2*uxsq - c3*usq);
        feq(j,i,5) = rt1*( 1. - c1*ueqyij + c2*uysq - c3*usq);
        feq(j,i,6) = rt2*( 1. + c1*uxuy6 + c2*uxuy6*uxuy6 - c3*usq);
```

```
            feq(j,i,7) = rt2*( 1. + c1*uxuy7  + c2*uxuy7*uxuy7 -
            c3*usq); feq(j,i,8) = rt2*( 1. + c1*uxuy8 + c2*uxuy8*uxuy8 -
            c3*usq); feq(j,i,9) = rt2*( 1. + c1*uxuy9 + c2*uxuy9*uxuy9 -
            c3*usq);
            end
        end
    end

    % Collision & Bounce-back steps.
    for j=1:Ly
        for i=1:Lx

            if s_node(j,i);  % solid nodes for implementing boundary
            condition

                % Bounce-back boundary condition

              temp   = f(j,i,2); f(j,i,2) = f(j,i,4); f(j,i,4) = temp;
    temp   = f(j,i,3); f(j,i,3) = f(j,i,5); f(j,i,5) = temp;
    temp   = f(j,i,6); f(j,i,6) = f(j,i,8); f(j,i,8) = temp;
    temp   = f(j,i,7); f(j,i,7) = f(j,i,9); f(j,i,9) = temp;

            else

                % Regular collision

                for a=1:9

                    f(j,i,a) = f(j,i,a)-( f(j,i,a) - feq(j,i,a))/tau;

                end
            end
        end
    end

    % Streaming step; Periodicity is implied for infinite length
    for j=1:Ly

        if j>1
            jn = j-1;
        else
            jn = Ly;
        end

        if j<Ly
            jp = j+1;
        else
            jp = 1;
        end
```

```
        for i=1:Lx

            if i>1
                in = i-1;
            else
                in = Lx;
            end
            if i<Lx
                ip = i+1;
            else
                ip = 1;
            end

            ftemp(j,i,1)   = f(j,i,1);
            ftemp(j,ip,2)  = f(j,i,2);
            ftemp(jp,i,3)  = f(j,i,3);
            ftemp(j,in,4)  = f(j,i,4);
            ftemp(jn,i,5)  = f(j,i,5);
            ftemp(jp,ip,6)  = f(j,i,6);
            ftemp(jp,in,7)  = f(j,i,7);
            ftemp(jn,in,8)  = f(j,i,8);
            ftemp(jn,ip,9)  = f(j,i,9);

        end
    end

    f(:,:,:)=ftemp(:,:,:);


    Figure(1)
    % Visualization
    tPlot = 40;
    if (mod(ts,tPlot)==0)
  u = sqrt((ux.^2 + uy.^2));
        imagesc(u); colorbar
    title('Fluid velocity ');
    axis equal off; drawnow
    end


end % end time loop

% Simulation & Analytical Plots
Figure(2)
surf(x,y,ux,uy)

width=Ly-2;
```

```
Figure (3)
% Results of LBM simulation
plot(y(:,1)-1.5-width/2,ux(:,1),'ro')

hold on

%Poiseuille velocity profile as blue line
nu=1/3*(tau-0.5);
u_an = g/(2*nu)*((width/2)^2-(y(:,1)-1.5-width/2).^2);
plot(y(:,1)-1.5-width/2,u_an)
xlabel('characteristic length(lu)')
ylabel('Fluid velocity (lu/ts)')
legend('LBM','Analytical Solution','center',8)
title('Flow through a nanopore slit')
```

9

Decline Curve and Rate Transient Analysis

Mustafa M. Alhubail[1], Mojdeh Rasoulzadeh[2], and Reza Barati Ghahfarokhi[1]

[1] *The University of Kansas, Lawrence, Kansas, USA*
[2] *The University of Alabama, Tuscaloosa, Alabama, USA*

9.1 Background

Completing the horizontal wells in the unconventional plays with multistage hydraulic fracturing is a costly procedure and a risky investment. The main motivation in developing the unconventional plays, which are multi-billion-dollar projects, is profit. In addition to the oil and gas prices, there are several other factors that are examined before an investment decision is made; one important factor that facilitates making a proper investment decision is the expected or estimated ultimate recovery (EUR) of the wells. It might not be possible to accurately estimate the production for the life of the wells (i.e., EUR) because accurate forecasting requires building rigorous reservoir models as well as perfectly processing the observed measured data (i.e., history matching). In general, building a representative reservoir model for the unconventional reservoirs, which requires the capability of handling horizontal wells and multistage hydraulic fracturing, is a laborious task because of the significant data uncertainty. Additionally, the physics and mechanisms, which govern the production in the ultra-tight formations, are not well-defined in the existing commercial reservoir simulation software packages. Thus, in order to estimate the ultimate recovery of the producing wells in the unconventional plays, the industry relies on the decline curves. The decline curve analysis (DCA) technique is the most utilized technique in the industry to do production forecasting because of its simplicity, rapidity, using of commonly available data (flow rates), and reasonable accuracy in assessing the EUR (Zhang et al., 2015).

9.2 Purpose of Decline Curves

Decline curve analysis (DCA) is a rate-time technique that is used to forecast the well performance by graphically projecting the flow rate and cumulative production without the need of pressure data. The DCA is also utilized to detect production issues and to estimate the ultimate recovery for a single well or the entire reservoir by extrapolating the decline curve (Arps, 1956). Figure 9.1 shows the best fit curve for the production data of a gas well. The first blue dot in the decline curve represents the start point (i.e., gas rate 338 MSCFD), the green triangle represents the start of forecasting and the last blue dot represents the point at which the well is abandoned (i.e., gas rate 100 MSCFD).

Unconventional Hydrocarbon Resources: Techniques for Reservoir Engineering Analysis, Advanced Textbook Series, First Edition. Edited by Reza Barati Ghahfarokhi and Mustafa M. Alhubail.

Figure 9.1 Example of decline curve analysis for a single well in a conventional gas reservoir (generated by IHS Harmony).

The decline curves are also utilized to identify the flow regimes (e.g., linear flow and transient flow.) The reservoir characteristics can also be estimated by analyzing the production data and the decline curves (Arps, 1945; Fetkovich, 1980).

9.3 Decline Curve Assumptions and Limitations

The decline curve analysis (DCA) is an empirical technique that is solely based on analyzing the well production data (i.e., flow rate and cumulative production with time). The DCA is not grounded by any fundamental theory; hence, it can be applied whenever it deemed suitable (Satter & Iqbal, 2016). Therefore, the DCA can be used to predict the well future performance by examining the production data of one fluid, multiple fluids, or the ratio of the fluids (e.g., Water-Oil-Ratio and Gas-Oil-Ratio).

There are several assumptions that govern the correctness of the decline curve procedure and the accuracy of the forecast. The first and most important assumption is that the operating conditions (e.g., tubing size and choke settings) and the reservoir conditions (e.g., reservoir characteristics and reservoir drive mechanism) must remain unchanged in order to make a fair prediction of the well EUR (Arps, 1956; Lee & Wattenbarger, 1996). However, if any condition is altered, then the decline curve that is produced prior to the change is not applicable anymore, and a new decline curve should be produced. The second assumption is that the well drainage area is fixed with a boundary dominant flow mechanism and without any influence or interference from the nearby wells. Thus, in the early stages of production, the DCA would not be applicable because of the transient flow regime as the boundaries of the reservoir have not been reached yet (Lee & Wattenbarger, 1996). The third assumption is that the well bottomhole flowing pressure must remain constant (Arps, 1956; Lee & Wattenbarger, 1996). This is the least valid assumption during the production from long horizontal wells with multistages of fractures and complex well flow principles. If any assumption is violated, then the decline curve models would not be reliable in forecasting and predicting the well performance. In addition to the mentioned assumptions, it is also important that the gathered production data are sufficient enough to identify a decline trend that fits the majority of the data. Furthermore, if the well produces at a constant rate for an extended period of time, then the decline curves would not be applicable to analyze the well performance.

Those assumptions are in fact reasonable in the conventional reservoirs as the production from the conventional wells usually abides by the mentioned assumptions. However, those assumptions become invalid in the unconventional reservoirs where the permeability is extremely low, resulting in long transient times; therefore, the traditional decline models are inadequate to analyze the performance of the unconventional wells (Paryani et al., 2016). Nonetheless, it is beneficial to study the traditional decline models before discussing the recent unconventional decline models to have a fundamental understanding of the development process of the decline models.

9.4 Traditional Decline Curve Models

The main goals of developing the decline curve models are to estimate the remaining life of the wells and the remaining reserves based on a specific rate that is economically viable (Paryani et al., 2016). The DCA has been practiced to estimate the hydrocarbon reserves since the 1910s (Lewis & Beal, 1918; Cutler, 1924; Johnson & Bollens, 1927; Allen, 1931). However, in 1945, Arps presented a pioneering work on DCA by classifying the models with detailed derivation and mathematical analysis (Arps, 1945).

9.5 Arps's Models

The Arps's decline models, which are the most common decline models used in the industry, are classified into three types of decline curves, Figure 9.2: (1) exponential, (2) harmonic, and (3) hyperbolic. In fact, the exponential and harmonic decline models are special cases of the hyperbolic decline model. The general form of the Arps's model, in terms of production rate and time, is presented in equation (9.1). It is important to note that the time units must be consistent, and the values of the decline rate (D_i) and the decline exponent (b) depend on the type of decline (Arps, 1945, 1956).

$$q(t) = \frac{q_i}{(1 + b\,D_i\,t)^{\frac{1}{b}}}, \tag{9.1}$$

where q is the flow rate at time (t), q_i is the initial flow rate, t is the time of production, D_i is the decline rate, and b is the decline exponent.

The rate of change or so-called decline rate is defined as shown in equation (9.2):

$$D_i = -\frac{1}{q(t)}\frac{dq(t)}{dt}. \tag{9.2}$$

The decline rate can also be estimated from the decline curve. There are two types of decline rates: the nominal decline rate and the effective decline rate. The nominal decline rate, which is also known as the instantaneous decline, is found from the decline curve by finding the tangent of a single point on the curve. However, the effective decline rate,

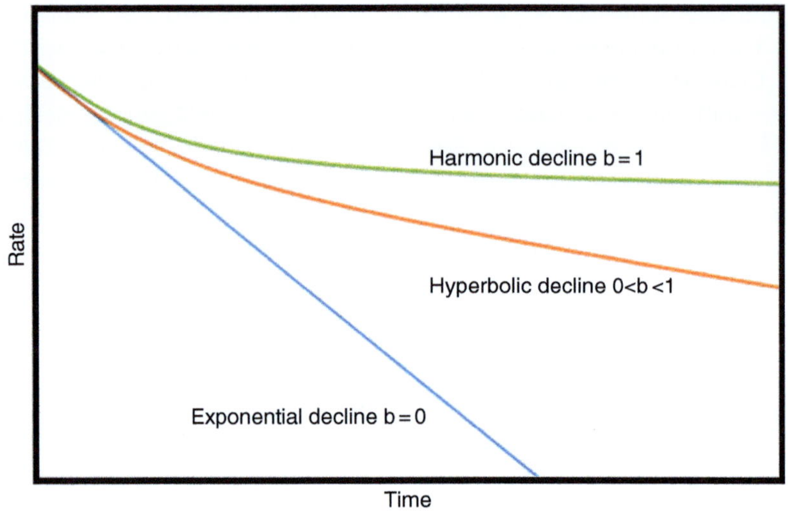

Figure 9.2 Schematic of Arps' decline models.

which is also known as the secant decline rate, is the slope of the line that is drawn between two selected points on the decline curve (SPEE, 2002).

9.5.1 Exponential Decline Model

The exponential decline model, which is also known as the constant-percentage decline and the constant-rate decline, is derived from equation (9.1) when b = 0. Therefore, the exponential decline model represents a special case where the rate of decline is equal to the initial rate of decline, and it stays constant (i.e., the decline rate (D) equals to the initial decline rate (D_i)); hence, the name constant-rate decline. With b = 0, the exponential decline model is represented as shown in equation (9.3):

$$q = q_i\, e^{(-D_i\, t)}. \tag{9.3}$$

By integrating equation (9.3), the cumulative production (Q) can be calculated as shown in equation (9.4):

$$Q = \frac{q_i - q}{D_i}. \tag{9.4}$$

There are two ways to obtain the value of the decline rate (D_i). The first way is graphically by rearranging equation (9.4) and plotting q versus Q; therefore, the slope of the curve is the decline rate, and the intercept is the initial flow rate (q_i). The rearranged equation (9.4) is presented in equation (9.5):

$$q = -(D_i\, Q) + q_i. \tag{9.5}$$

The other way to obtain the decline rate (D_i), is by knowing the values of two flow rates (q_1 and q_2) and their time intervals (t_1 and t_2). Therefore, the decline rate (D_i) can be calculated by the use of equation (9.6):

$$D_i = \left(\frac{1}{t_2 - t_1}\right) \ln\left(\frac{q1}{q2}\right). \tag{9.6}$$

Consequently, the estimated ultimate recovery can be calculated by the use of equation (9.7):

$$EUR = Qi + \left(\frac{q_i - q_f}{D_i}\right), \tag{9.7}$$

where Qi is the cumulative production that occurred before the decline model and q_f is the abandonment rate.

9.5.2 Determination of Exponential Decline Graphically

The exponential decline can be identified graphically by two plots. The first one is plotting flow rate (q) versus time on a semilog graph (i.e., plotting log (q) versus time). The second plot is the flow rate (q) versus the cumulative production (Q). If both plots show a straight-line trend, then the production data are considered to follow Arps's exponential decline model (Arps, 1945; Lee & Wattenbarger, 1996).

Example 9.1 Find the Type of Decline Graphically

Determine the type of decline graphically for the production data that are provided in Table 9.1. Determine the decline rate from the generated plots.

Solution

To identify the type of decline graphically, the two plots of log (q) versus time and (q) versus cumulative production are plotted. If both plots yield a straight-line, then the exponential decline can be adopted for the well.

Both Figure 9.3 and Figure 9.4 show a straight line for the provided production data. Hence, Arps's exponential decline model can be adopted for this well. Figure 9.4 can also be used to find the decline rate as equation (9.5) states that the decline rate (D_i) is the slope of the plot, which is 0.0013 (1/day). The mathematical equation, equation (9.6), gave a decline rate of 0.0015 (1/day). Hence, the two methods are in good agreement.

Table 9.1 Data for Example 9.1

Time (months)	Gas rate (MSCFD)	Cumulative production (MSCF)
0	1,008	0
1	962	30,663
2	934	59,927
3	884	88,340
4	859	115,231
5	831	141,362
6	800	166,641
7	757	190,977
8	720	214,005
9	684	235,907
10	628	256,714

Source: Abu-Khamsin (2012).

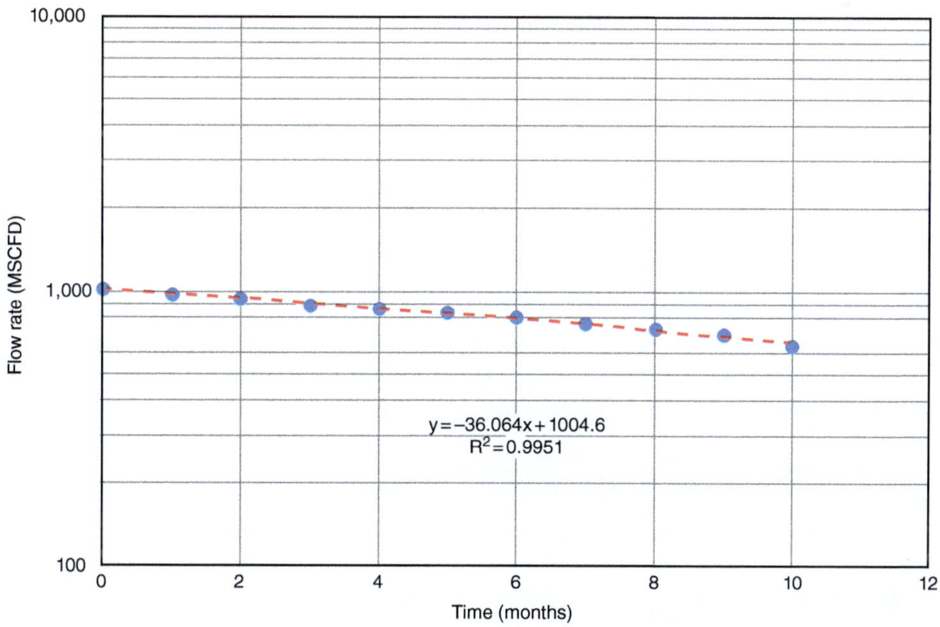

Figure 9.3 Plot of log (q) versus time.

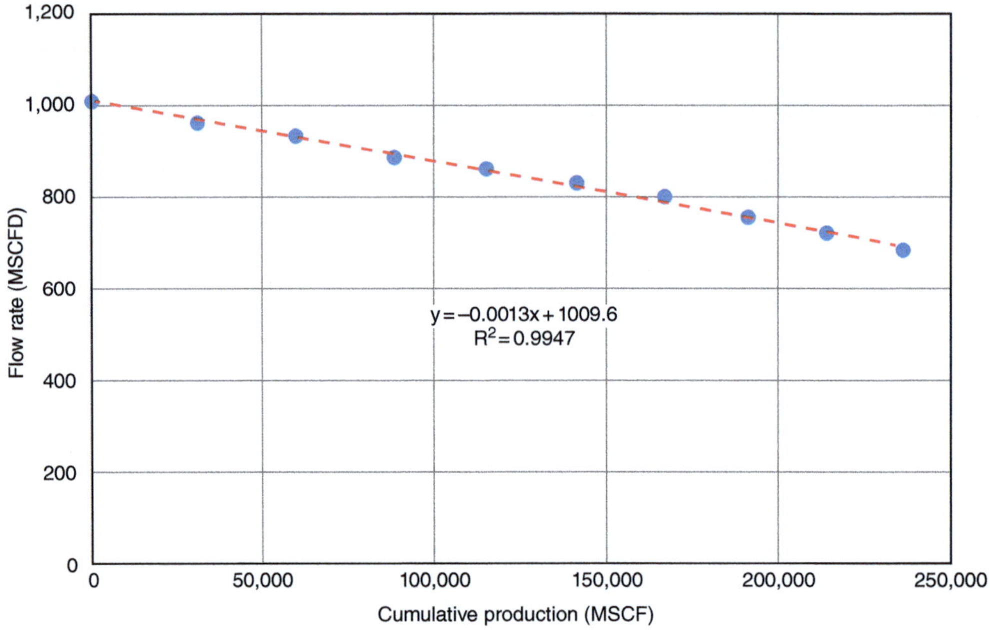

Figure 9.4 Plot of (q) versus cumulative production.

Example 9.2 Predict the Production Flow Rate and EUR of a gas Well

The initial flow rate of a gas well is 1,008 MSCFD as shown in Table 9.1 (Abu-Khamsin, 2012). In the next month, the flow rate of the well dropped to 962 MSCFD. Assume the well follows the exponential decline model. Predict the well flow rate after 1 year and the cumulative production for the first year. What is the EUR of the well if the abandonment rate is 150 MSCFD? Compare the calculated decline rate against the decline rate that was found graphically in Example 9.1.

Solution

The first step is to find the decline rate (D_i) by the use of equation (9.6):

$$D_i = \left(\frac{1}{t_2 - t_1}\right) \ln\left(\frac{q1}{q2}\right) = \left(\frac{1}{1-0}\right) \times \ln\left(\frac{1,008}{962}\right) = 0.0467\left(\frac{1}{month}\right).$$

Hence, the flow rate after 1 year can be found by the use of equation (9.3):

$$q = q_i\, e^{-D_i t} = 1,008 \times e^{-(0.0467 \times 12)} = 575.48\ (MSCFD).$$

The cumulative production after 1 year can be calculated by equation (9.4). However, it is important that the decline rate (D_i) has the same units as the flow rate. So, D_i should be divided by 30.42 to convert its units from months to days as follow:

$$D_i = 0.0467\left(\frac{1}{month}\right) \times \frac{1\ month}{30.42\ day} = 0.0015\left(\frac{1}{day}\right).$$

Thus, the cumulative production after 1 year is calculated as follows:

$$Q = \frac{q_i - q}{D_i} = \frac{1,008 - 575.48}{0.0015} = 281.68\ (MMSCF).$$

Finally, the EUR is calculated by the use of equation (9.7):

$$EUR = \frac{q_i - q_f}{D_i} = \frac{1,008 - 150}{0.0015} = 558.78\ (MMSCF).$$

The decline rate (D_i) from the plot of Example 9.1 is 0.0013 (1/day). The mathematical equations gave a decline rate of 0.0015 (1/day). Hence, the two methods are in good agreement.

9.5.3 Harmonic Decline Model

The harmonic decline, which is the second special case of Arps's equation, is adopted when $b = 1$, and, thus, Arps's equation is simplified to equation (9.8) (Arps, 1956):

$$q = \frac{q_i}{1 + D_i t}. \tag{9.8}$$

By taking the logarithm of both sides of equation (9.8), the equation becomes as shown in equation (9.9):

$$\log(q) = \log(q_i) - \log(1 + D_i t). \tag{9.9}$$

In terms of cumulative production, equation (9.9) can be represented as shown in equation (9.10) (Lee & Wattenbarger, 1996):

$$\log(q(t)) = \log(q_i) - \left(\frac{D_i}{2.303\, q_i}\right) Q(t). \tag{9.10}$$

Equation (9.10) shows that the decline rate can be calculated from the plot of the flow rate and cumulative production (i.e., log (q) versus Q). The decline rate can be calculated by the use of equation (9.11):

$$D_i = -2.303\ slope\ q_i. \tag{9.11}$$

Consequently, the cumulative production and the EUR for the harmonic decline model can be estimated by the use of equation (9.12) and equation (9.13), respectively (Arps, 1945; Arps, 1956):

$$Q = \frac{q_i}{D_i} \ln\left(\frac{q_i}{q}\right) \tag{9.12}$$

and

$$EUR = Q_i + \left[\frac{q_i}{D_i} \ln\left(\frac{q_i}{q_f}\right)\right], \tag{9.13}$$

where Q_i is the cumulative production that occurred before the decline model and q_f is the abandonment rate.

It is important to note that the initial flow rate (q_i) is the intercept from the semilog plot, equation (9.10), which is not necessarily equal to the actual initial flow rate from the production data.

9.5.4 Determination of Harmonic Decline Graphically

The harmonic decline can also be identified graphically by two plots. The first one is plotting flow rate (q) versus time on a log-log graph (i.e., plotting log (q) versus log (time)). The second plot is the flow rate (q) versus the cumulative production (Q) on a semilog graph (i.e., plotting log (q) vs. cumulative production (Q)). If both plots show a straight-line trend, then the production data are considered to follow the Arps's harmonic decline model (Lee & Wattenbarger, 1996).

Example 9.3 Estimate the Production Life and Reserves

The slope and intercept of the production data of a horizontal oil well, which follows the harmonic decline model, are $-2.83 \times 10^{-5} \left(\frac{1}{STB}\right)$ and 105 (STBD), respectively. Estimate the production life of the well and the EUR knowing that the economic viable rate is 25 STBD, and the initial cumulative production before the decline model is 5.3 MSTB.

Solution

To estimate the economically viable production life of the well, first the decline rate must be calculated. The slope and the intercept of the production data are provided. The decline rate can be calculated by the use of equation (9.11) as follows:

$$D_i = -2.303\ slope\ q_i = -2.303 \times -2.83 \times 10^{-5}\left(\frac{1}{STB}\right) \times 105\ (STBD) = 0.00684\left(\frac{1}{day}\right),$$

$$D_i = 0.00684\left(\frac{1}{day}\right) = 2.5\left(\frac{1}{year}\right).$$

Now, the production life of the well can be calculated by rearranging equation (9.8) as follows:

$$t = \frac{1}{D_i}\left(\frac{q_i}{q_f} - 1\right) = \frac{1}{0.00684} \times \left(\frac{105}{25} - 1\right) = 468\ (days).$$

Hence, after 468 days, the well is not producing with an economical rate; so, a workover might be necessary to stimulate the well (e.g., hydraulic fracturing) in order to enhance the production.

Finally, the EUR can be calculated by the use of equation (9.13) as follows:

$$EUR = Q_i + \left[\frac{q_i}{D_i}\ln\left(\frac{q_i}{q_f}\right)\right] = 5.3(MSTB) + \left[\frac{105}{0.00684} \times \ln\left(\frac{105}{25}\right)\right] = 27.3\ (MSTB).$$

It is important to note that the decline rate that is used in the equations is the nominal decline rate, not the effective decline rate. Hence, the value of the decline rate (D_i) which is equivalent to $2.5\left(\frac{1}{year}\right)$ and $250\%\left(\frac{1}{year}\right)$ is reasonable since this is a horizontal well (SPEE, 2002). However, the effective tangent decline rate and the effective secant decline rate for this well are $91.79\%\left(\frac{1}{year}\right)$ and $71.43\%\left(\frac{1}{year}\right)$, respectively.

9.5.5 Hyperbolic Decline Model

The hyperbolic decline model, which is the general case of Arps's equation, is adopted when the decline exponent is between 0 and 1 (i.e., $0 < b < 1$). Hence, the general form of Arps's equation, equation (9.1), can be used to represent the hyperbolic decline model, equation (9.14):

$$q = \frac{q_i}{(1 + bD_i t)^{\frac{1}{b}}}. \quad (9.14)$$

By taking the logarithms of both sides of equation (9.14), the equation becomes as shown in equation (9.15):

$$\log q = \log q_i - \frac{1}{b}\log(1 + bD_i t). \quad (9.15)$$

In terms of cumulative production, the hyperbolic decline model can be represented as shown in equation (9.16):

$$Q = \frac{q_i^b}{D_i(1-b)}\left(q_i^{1-b} - q^{1-b}\right). \quad (9.16)$$

Therefore, the EUR for the hyperbolic decline model can be estimated by the use of equation (9.17) (Arps, 1956):

$$EUR = Q_i + \left[\frac{q_i^b}{D_i(1-b)}\left(q_i^{1-b} - q_f^{1-b}\right)\right], \quad (9.17)$$

where Q_i is the initial cumulative production before the decline model and q_f is the final abandonment rate.

Example 9.4 Estimate the Production Life and Reserves

Figure 9.5 shows the production data of a gas well that follows the hyperbolic decline model. The decline rate and decline exponent are found to be 0.00034 (1/day) and 0.464, respectively. Answer the following questions knowing that the initial flow rate is 338 MSCFD, the abandonment rate is 50 MSCFD and the initial cumulative production is 0.832 BSCF.

1) What would be the flow rate in MSCFD after 3,000 days of production?
2) What would be the cumulative production after 3,000 days of production?
3) What would be the EUR?
4) What is the production life of the well based on the calculated EUR?

Solution

1) Knowing the initial rate (q_i), the decline rate (D_i), and the decline exponent (b), the flow rate after 3,000 days of production can be calculated by the use of equation (9.14) as follows:

$$q = \frac{q_i}{(1 + bD_i t)^{\frac{1}{b}}} = \frac{338}{(1 + 0.464 \times 0.00034 \times 3,000)^{\frac{1}{0.464}}} = 146.6\ (MSCFD).$$

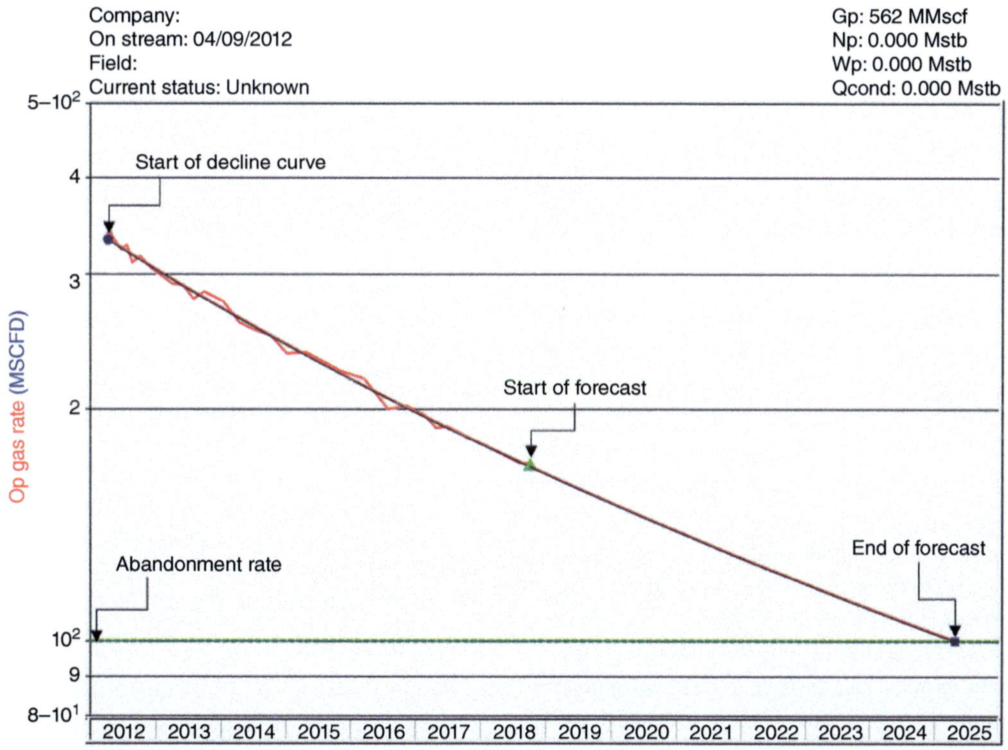

Figure 9.5 Example of decline curve for a gas well. Hyperbolic decline model for example 9.4 (generated by IHS Harmony).

2) The cumulative production after 3,000 days of production can be calculated by the use of equation (9.16) as follows:

$$Q = \frac{q_i^b}{D_i(1-b)}\left(q_i^{1-b} - q^{1-b}\right) = \frac{338^{0.464}}{0.00034 \times (1-0.464)}\left(338^{(1-0.464)} - 146.6^{(1-.464)}\right)$$

$$= 6.69 \times 10^5 (MSCF)$$

and

$$Q = 6.693 \times 10^5 (MSCF) = 0.6693 (BSCF).$$

3) The well EUR can be estimated by the use of equation (9.17) as follows:

$$EUR = Q_i + \left[\frac{q_i^b}{D_i(1-b)}\left(q_i^{1-b} - q_f{}^{1-b}\right)\right]$$

$$= 0.832(BSCF) + \left[\frac{338^{0.464}}{0.00034 \times (1-0.464)}\left(338^{(1-0.464)} - 50^{(1-.464)}\right)\right] = 2.021\ (BSCF)$$

and

$$EUR = 2.021\ (BSCF).$$

4) The production life of the well can be estimated by rearranging equation (9.14) as follows:

$$q = \frac{q_i}{(1 + bD_i t)^{\frac{1}{b}}} = 50 = \frac{338}{(1 + 0.464 \times 0.00034 \times t)^{\frac{1}{0.464}}}$$

and

$$t = 9046\ (days).$$

Thus, the production life of the well is 9,046 days, which is almost 25 years of production.

9.5.6 Determination of Hyperbolic Decline Parameters

There is only one unknown parameter, which is the decline rate (D_i), in the exponential and harmonic decline models. However, the hyperbolic decline model, as can be seen from equations (9.14) to (9.17), is a complicated decline model because there are two unknowns in the model (i.e., decline rate (D_i) and decline exponent (b)). Thus, before the hyperbolic model can be adopted, the unknown parameters must be found.

There are three techniques that can be used to determine the hyperbolic decline parameters. (1) One technique is the simple trial and error. This is conducted by assuming different values for the decline rate (D_i) and the decline exponent (b) until the correct values are found. However, the trial and error method is a tedious and time-consuming technique. (2) Another technique to determine the unknown parameters of the hyperbolic model is by adopting equation (9.16) and plotting the cumulative production versus the term $\left(q_i^{1-b} - q_f{}^{1-b}\right)$ for different values of (b). If a straight-line trend is achieved, then the value of (b) is the correct one, and the decline rate (D_i) can be calculated from the slope of the line, accordingly (Abu-Khamsin, 2012). (3) The third technique that can be utilized to find the two unknowns is the regression routine technique where it is capable of finding the values of the decline rate (D_i) and the decline exponent (b) that best fit the production data. However, the drawback of this technique is that it might yield multiple combinations of the decline rate (D_i) and the decline exponent (b) that fit the production data. Nowadays, with the advanced software packages, the parameters of the hyperbolic decline model can be determined with high accuracy.

Example 9.5 Determine the Decline Model Parameters

The production data of a horizontal well producing from the Mississippian Conglomerate formation in Harvey County, Kansas, are provided in Table 9.2. Determine the type of decline and the model's parameters. Based on the determined parameters and an abandonment rate of 30 STBD, estimate the EUR and the life of the well knowing that the initial oil rate from the decline curve and the initial cumulative production are 2,052 STBD and 285 MSTB, respectively.

Solution

The type of decline can be found by testing the exponential and harmonic decline plots. If the plots do not show a straight line, then the hyperbolic model should be adopted. The exponential decline plot (i.e., rate (q) versus cumulative production) and the harmonic decline plot (i.e., log (q) versus cumulative production) were tested and neither showed a straight-line trend for the data. Hence, the hyperbolic decline model could be adopted. A software package was utilized to determine the model's parameters (i.e., decline rate and decline exponent). The best fit for the data with an R^2 value of 0.951 is shown in Figure 9.6. Based on the best fit shown in Figure 9.6, the decline rate (D_i) and the decline exponent (b) for the hyperbolic decline model are 0.00396 (1/day) and 0.321, respectively. The decline rate (D_i) at the start of the forecasted date is 0.00221 (1/day).

The EUR can be estimated by the use of equation (9.17):

$$EUR = Q_i + \left[\frac{q_i^b}{D_i(1-b)}\left(q_i^{1-b} - q_f^{1-b}\right)\right] = 285\,(MSTB)$$

$$+\left[\frac{2{,}052^{0.321}}{0.00396\times(1-0.321)}\left(2{,}052^{(1-0.321)} - 30^{(1-0.321)}\right)\right]$$

and

$$EUR = 1(MMSTB).$$

Thus, the production life of the well can be estimated by rearranging equation (9.14):

$$q = \frac{q_i}{(1+bD_it)^{\frac{1}{b}}} = 30 = \frac{2{,}052}{(1+0.321\times 0.00396\times t)^{\frac{1}{0.321}}}$$

and

$$t = 2267(days)\tilde{}6.2\,(years).$$

Table 9.2 Production Data for Example 9.5

Time (month-year)	Oil production rate (bbl/d)
Mar 2015	1,943.41
Apr 2015	1,845.19
May 2015	1,789.66
Jun 2015	1,532.88
Jul 2015	1,475.50
Aug 2015	1,324.57
Sep 2015	1,208.37
Oct 2015	821.77
Nov 2015	905.44
Dec 2015	867.41
Jan 2016	710.57
Feb 2016	707.16
Mar 2016	703.04
Apr 2016	520.16
May 2016	472.16
Jun 2016	497.64
Jul 2016	170.27
Aug 2016	498.22
Sep 2016	501.04
Oct 2016	502.33
Nov 2016	271.07
Dec 2016	420.70

Source: KGS (2018).

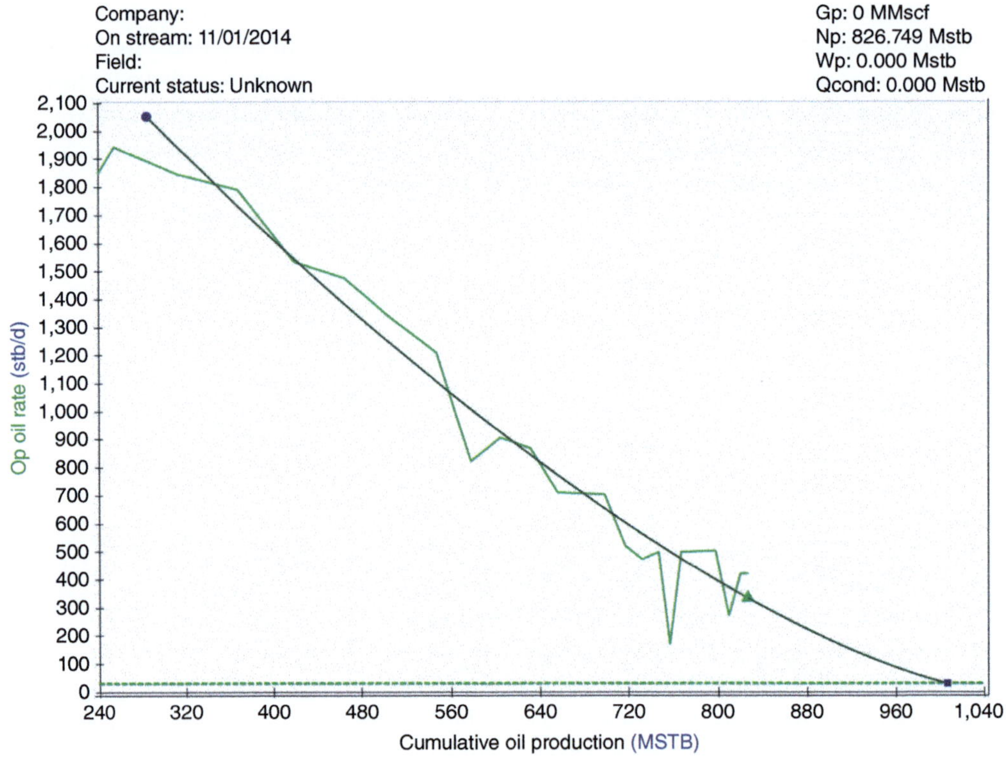

Figure 9.6 Hyperbolic decline model for Example 9.5 (generated by IHS Harmony).

9.6 Modern Decline Curve Models

When it comes to the unconventional reservoirs (i.e., ultra-tight formations), the traditional Arps's decline models become inadequate mostly because of the flow regime in the unconventional reservoirs. There are three fluid-flow regimes that are associated with the horizontal wells and multistage fracturing: linear flow, bilinear flow, and multifracture flow (Okouma et al., 2012). The linear flow occurs when a highly conductive network of fractures is created whereas the bilinear flow occurs when low conductive fractures are created. The linear flow and bilinear flow are discussed in Chapter 6. The multifracture flow is encountered when vertical and horizontal fractures are created. Since the development of the unconventional reservoirs requires the use of the advanced technology (i.e., multistage fracturing) in order to commercially produce from them, then one or multiple flow regimes of the mentioned flow models will be encountered. However, the traditional Arps's decline models are valid only for boundary dominant flow regime as was stated before in the assumptions and limitations section (Lee & Wattenbarger, 1996). Hence, in their current form, they are invalid in the majority of the unconventional reservoirs.

If the general Arps's decline model (i.e., the hyperbolic decline model) is applied in the unconventional reservoirs (e.g., shale gas), it would yield a value of b (i.e., decline exponent) that is greater than 1, which is the only way to fit the production data especially for the early time period (Seshadri & Mattar, 2010; Ilk et al., 2008; Zhang et al., 2015). Multiple researchers have studied different scenarios in which the b value would be greater than 1. One main reason that cause the decline exponent b to be greater than 1 is that the boundary dominant flow has not been reached yet. Another reason is the reservoir characteristics. Layered reservoirs have shown to cause the decline exponent *b* value to be greater than 1 (Gentry & McCray, 1978; Fetkovich et al., 1990). Also by studying the three different flow regimes mentioned above, the decline exponent *b* value could be as high as 4 in order to capture the actual production decline (Okouma et al., 2012).

In the unconventional reservoirs, the issue with the decline exponent *b* being greater than 1 lies in the estimation of the EUR. EUR is one of the main objectives for using the DCA technique (Paryani et al., 2016). However, when the decline exponent *b* is greater than 1, the hyperbolic decline model would almost always result in substantial overestimation of the unconventional reserves (Rushing et al., 2007). It is important to note that this is due to the extrapolation of the decline curve and not the capability of the hyperbolic decline model to capture the production data. To get the EUR, the decline curve can basically be extrapolated to the point at which the abandonment rate is set. However, extrapolating the hyperbolic decline curve, which was generated to capture the production data of the transient flow period, would yield wrong EUR value because the hyperbolic models are limited in the boundary-dominant flow regime (Ilk et al., 2009). In fact, it is believed that the traditional decline models become reliable in the unconventional reservoirs when the boundaries of the reservoir are reached. Hence, the flow regime becomes boundary-dominant flow (Zhang et al., 2015).

Identifying when the transient flow period ends and when the boundary-dominant flow starts is a challenging task. The boundary-dominant flow, for the conventional reservoirs, typically starts after a few days of production. For low-permeability reservoirs with a permeability of 1 to 0.5 (mD), the transient flow period may last several months. On the other hand, the production from the ultra-tight reservoirs, which have extremely low permeability in the range of micro and nano Darcy and require a complex network of fractures to produce from, could last several years before the boundaries of the reservoir are felt (Lee & Wattenbarger, 1996; Satter & Iqbal, 2016). The transient flow period for the unconventional reservoirs takes several years before the boundary-dominant flow regime starts; therefore, the traditional decline curves would not be applicable until then. New decline curve models have been developed to improve the assessment of the EUR in the unconventional reservoirs.

9.6.1 Modified Hyperbolic Model

The modified hyperbolic decline model, which is also referred to as the super-hyperbolic decline and the multisegment decline, is basically a combination of the hyperbolic and exponential decline models. First, the hyperbolic decline model is used to capture the decline of the transient flow regime with an exponent decline *b* value that could be greater than 1. However, if the forecast is continued using just the hyperbolic decline model, then the extrapolation of the curve would overestimate the EUR. To solve this issue, the modified hyperbolic decline model proposes that after a specific period of time, the hyperbolic decline model should be converted to an exponential decline model, and the forecast and the estimation of the EUR value are then continued based on the exponential decline model (Robertson, 1988; Seshadri & Mattar, 2010). By doing so, the estimation of the EUR value would be more reasonable than just using the hyperbolic decline model. However, finding the switch point from the hyperbolic decline to the exponential decline is challenging, and it controls the accuracy of this technique. The switch point is identified by a limiting effective decline rate, which can be converted to a limiting nominal decline rate. The modified hyperbolic decline model is represented by equations (9.18) to (9.21):

$$q = \frac{q_i}{(1 + bD_i t)^{\frac{1}{b}}} \quad D_i > D_{\text{lim}}, \tag{9.18}$$

$$q = q_{\text{lim}}\, e^{\left(-D_{lim}(t - t_{\text{lim}})\right)} \quad D_i \leq D_{\text{lim}}, \tag{9.19}$$

$$q_{\text{lim}} = q_i \left(\frac{D_{lim}}{D_i}\right)^{\left(\frac{1}{b}\right)}, \tag{9.20}$$

and

$$t_{\lim} = \frac{\left(\frac{q}{q_{\lim}}\right)^{b} - 1}{b\, D_i}. \quad (9.21)$$

The cumulative production can also be estimated by the use of equations (9.22) to (9.25):

When b > 1:

$$Q = \frac{q_i^b}{D_i(1-b)}\left(q_i^{1-b} - q^{1-b}\right) D_i > D_{\lim}, \quad (9.22)$$

$$Q = \frac{q_i^b}{D_i(1-b)}\left(q_i^{(1-b)} - q_{lim}{}^{(1-b)}\right) + \frac{q_{lim} - q}{D_{\lim}}\; D_i \le D_{\lim}. \quad (9.23)$$

When b = 1:

$$Q = \frac{q_i}{D_i}\ln\left(\frac{q_i}{q}\right)\; D_i > D_{\lim}, \quad (9.24)$$

$$Q = \frac{q_i}{D_i}\ln\left(\frac{q_i}{q_{\lim}}\right) + \frac{q_{\lim} - q}{D_{\lim}}\; D_i \le D_{\lim}. \quad (9.25)$$

Example 9.6 Comparison of the EUR Between the Traditional Hyperbolic and the Modified Hyperbolic Decline Models

For the data provided in Table 9.2, determine the EUR using the modified hyperbolic decline model and compare the results against the traditional hyperbolic decline model.

Solution

For the modified hyperbolic model, the same values (e.g., initial rate and decline exponent *b* that were used in Example 9.5 are going to be used to fit the first segment of the decline curve, which is based on the hyperbolic decline model. However, after 19.5 months, the hyperbolic decline

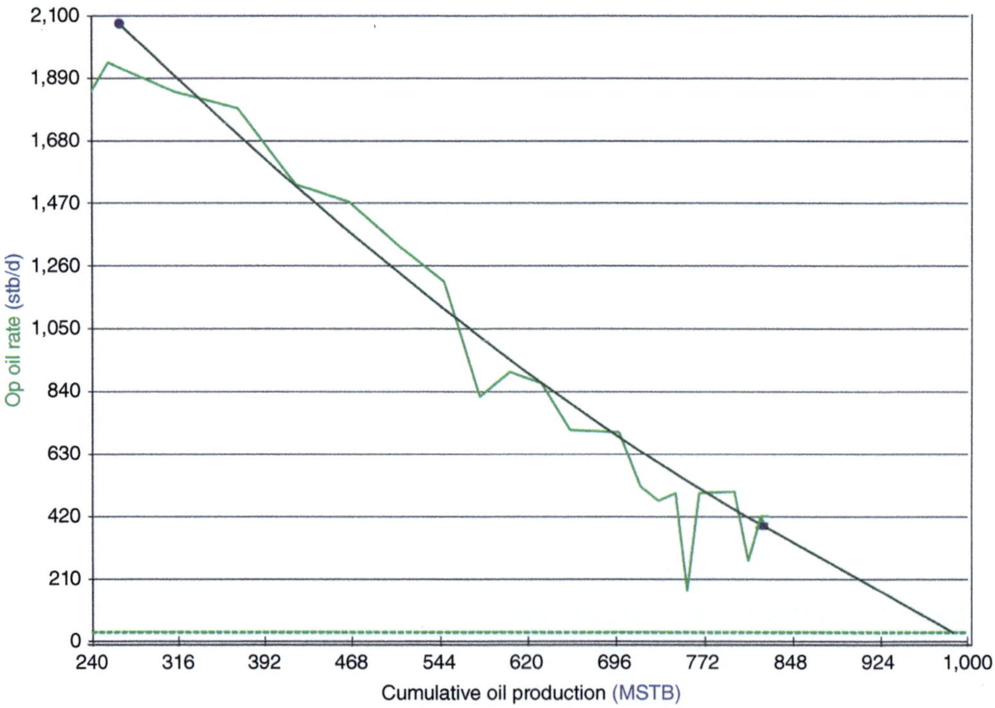

Figure 9.7 Modified hyperbolic decline model for Example 9.6 (generated by IHS Harmony).

model is converted to an exponential decline model. The decline exponent is changed from 0.321 to 0.0, and the exponential decline curve is extrapolated to predict the EUR. Figure 9.7 shows the plot of the modified hyperbolic decline model. Based on the modified hyperbolic model, the EUR value is 987 MSTB. The traditional hyperbolic model gave an EUR value of 1 MMSTB. Thus, the difference is 13 MSTB.

9.6.2 Power-Law Exponential Model

Another decline model was proposed in 2008 by Ilk et al. that addresses the issue of the overestimation of the EUR value in the unconventional reservoirs (Ilk et al., 2008). Instead of modifying the hyperbolic decline model as shown in the previous model, the exponential decline model is rather modified to be a power-law exponential decline. This is accomplished by introducing two more parameters to the exponential decline model. The new parameters are the time exponent n and the decline rate at infinite time D_∞. The power-law exponential decline model is represented by equation (9.26) (Ilk et al., 2008):

$$q = q_i e^{\left(-D_\infty t - \hat{D}_i t^n\right)}, \tag{9.26}$$

where q_i is the rate at $t = 0$, which is found from the intercept of equation (9.26); D_∞ is the decline rate at infinite time; $\hat{D}_i$ is the decline rate and is defined as $\left(\hat{D}_i = \frac{D_1}{n}\right)$; D_1 is the decline rate at $t = 1$; t is the time of production; and n is the time exponent.

The power-law exponential (PLE) decline model includes the decline rate at infinite time D_∞ to prevent the overestimation of the reserves by limiting the decline rate from becoming very low. In fact, the PLE decline model, equation (9.26), can be reduced to the traditional exponential decline model, equation (9.3), by setting the time exponent $n = 1$, and the decline rate at the infinite time $D_\infty = 0$.

Note that equation (9.26) cannot be integrated directly. Hence, there is not an analytical equation that can be used to calculate the cumulative production. In order to find the cumulative production or the EUR for the PLE decline model, a numerical method has to be used (Currie et al., 2010).

Example 9.7 Estimate the Flow Rate Using the PLE Decline Model

An unconventional gas reservoir was hydraulically fractured. Estimate the flow rate of the well after 30 days of production using both the PLE decline model and the modified hyperbolic decline model. The parameters for both models are provided in Table 9.3.

Table 9.3 Data for Example 9.7

Power-law exponential decline model		Modified hyperbolic decline model	
Parameter	**Value**	**Parameter**	**Value**
q_i (MSCFD)	1,000	q_i (MSCFD)	1,000
$\hat{D}_i$ (1/day^n)	1.086	D_i (1/day)	2.738
D_∞ (1/day)	0.00022	b	2
n	0.2	D_{lim} (1/day)	0.0011

Source: Seshadri and Mattar (2010).

Solution

The flow rate for the PLE decline model can be estimated by the use of equation (9.26):

$$q = q_i e^{\left(-D_\infty\, t - \hat{D}_i\, t^n\right)} = 1000 \times e^{\left(-0.00022 \times 30 - 1.086 \times 30^{0.2}\right)} = 116.4\ (MSCFD).$$

The flow rate for the modified hyperbolic decline model for $D_i > D_{lim}$ by the use of equation (9.18):

$$q = \frac{q_i}{(1 + bD_i t)^{\frac{1}{b}}} = \frac{1000}{(1 + 2 \times 2.738 \times 30)^{\frac{1}{2}}} = 77.7\ (MSCFD).$$

The difference between the two models is 38.7 MSCFD. This difference between the two models is not just because of the nature of the two models (i.e., one is modified hyperbolic and the other is modified exponential) but also because the forecast is sensitive to the time exponent n and the decline rate at infinite time "D_∞" as various values of n and "D_∞" would fit the production data for the early transient flow period. However, the same values would not fit the late periods (Seshadri & Mattar, 2010).

In general, the modified hyperbolic decline model is well known in the industry compared with the PLE decline model. Several studies have shown that the PLE decline model yields consistent and realistic EUR results as it is more capable of capturing the transient and boundary-dominant flow periods (Ilk et al., 2008; McNeil et al., 2009).

9.6.3 Stretched Exponential Model

The stretched exponential decline model (SEP) was proposed in its current form by Valko et al. in 2009 (Valko et al., 2009; Valko & Lee, 2010). The basis of this model lies on the stretched exponential decay approach. The SEP decline model basically states that the production decline, for a great number of production data sets, can be represented by the stretched exponential decay providing that a specific distribution of time constants is used (Valko & Lee, 2010).

One of the merits of using the SEP decline model over the other modern decline models for the unconventional reservoirs is that the SEP decline model has just two unknown parameters compared with three in the PLE decline model and the modified hyperbolic model. The SEP decline model also yields bounded EUR values. However, determining the two parameters requires solving nonlinear equations, which could be complicated at times (Tan et al., 2018). Furthermore, in order to get accurate values for the unknown parameters, it is recommended that the production period should be longer than 3 years (Zuo et al., 2016). The flow rate can be estimated based on the SEP decline model by equation (9.27), and the cumulative production and thc rccovery potential can be estimated by equations (9.28) and (9.29), respectively (Valko & Lee, 2010).

$$q = q_i\, e^{-\left(\frac{t}{\tau}\right)^n}, \tag{9.27}$$

where τ is the characteristic time coefficient and n is the time exponent, which is the same for the PLE decline model

$$Q = \frac{q_i^\tau}{n}\left\{\Gamma\left[\frac{1}{n}\right] - \Gamma\left[\frac{1}{n}, \left(\frac{t}{\tau}\right)^n\right]\right\}, \tag{9.28}$$

where Γ(t) is the gamma function and Γ(x, s) is the incomplete gamma function

$$r_p = 1 - \frac{Q}{EUR} = \frac{1}{\Gamma\left[\frac{1}{n}\right]}\Gamma\left[\frac{1}{n}, -\ln\frac{q}{q_i}\right], \tag{9.29}$$

where r_p is the recovery potential

The gamma function, $\Gamma(t)$, is a function that can be represented by different forms. For example, it can be represented by the factorial function as shown in equation (9.30), and also as a definite integral function as shown in equation (9.31). The difference between the incomplete gamma function and the gamma function is the limits of the integral.

$$\Gamma(n) = (n-1)! \tag{9.30}$$

and

$$\Gamma(t) = \int_0^{\infty} x^{t-1} e^{-x}\, dx. \tag{9.31}$$

The model parameters (i.e., τ and n) can be determined by solving the two nonlinear equations, equation (9.32) and equation (9.33). Solving the nonlinear equations, which involve the gamma function and the incomplete gamma function, could be complicated and time-consuming. However, with advanced software packages, this should not be an issue.

$$r_{21} = \frac{\Gamma\left[\frac{1}{n}\right] - \Gamma\left[\frac{1}{n}, \left(\frac{23.5}{\tau}\right)^n\right]}{\Gamma\left[\frac{1}{n}\right] - \Gamma\left[\frac{1}{n}, \left(\frac{11.5}{\tau}\right)^n\right]} \tag{9.32}$$

$$r_{31} = \frac{\Gamma\left[\frac{1}{n}\right] - \Gamma\left[\frac{1}{n}, \left(\frac{35.5}{\tau}\right)^n\right]}{\Gamma\left[\frac{1}{n}\right] - \Gamma\left[\frac{1}{n}, \left(\frac{11.5}{\tau}\right)^n\right]}, \tag{9.33}$$

where r_{21} and r_{31} are the cumulative production ratio for the second and third years, respectively.

The ratios for the second year (r_{21}) and the third year (r_{31}) are considered in finding the parameters of the SEP decline model because they yield relatively stable results (Valko & Lee, 2010).

Example 9.8 Determine the SEP Decline Model Parameters

The weighted mean of the cumulative ratios of a group of wells for the second year (r_{21}) and the third year (r_{31}) are 1.563 and 1.985, respectively (Valko & Lee, 2010). Determine the unknown parameters for the SEP decline model.

Solution

To find the unknown parameters (n and τ), the nonlinear equations, equation (9.32) and equation (9.33), should be solved knowing the values of r_{21} and r_{31} as follow:

$$1.563 = \frac{\Gamma\left[\frac{1}{n}\right] - \Gamma\left[\frac{1}{n}, \left(\frac{23.5}{\tau}\right)^n\right]}{\Gamma\left[\frac{1}{n}\right] - \Gamma\left[\frac{1}{n}, \left(\frac{11.5}{\tau}\right)^n\right]}$$

and

$$1.985 = \frac{\Gamma\left[\frac{1}{n}\right] - \Gamma\left[\frac{1}{n}, \left(\frac{35.5}{\tau}\right)^n\right]}{\Gamma\left[\frac{1}{n}\right] - \Gamma\left[\frac{1}{n}, \left(\frac{11.5}{\tau}\right)^n\right]}.$$

Thus, the value of time exponent (n) and the time coefficient (τ) are 0.229 and 0.544 month, respectively.

Figure 9.8 shows the SEP decline model of a horizontal well with multistage fracturing targeting the Bakken formation. The values of n and τ are 0.234 and 0.32 month, respectively. Based on the SEP decline model, the EUR of this well is approximately 623 MSTB with abandonment rate of 50 STB/day.

The spike of production in Figure 9.8 occurred because the choke size was changed. Adding a separate decline curve segment to model the production data after the change of the choke could lead to a more accurate forecast because the production conditions have changed.

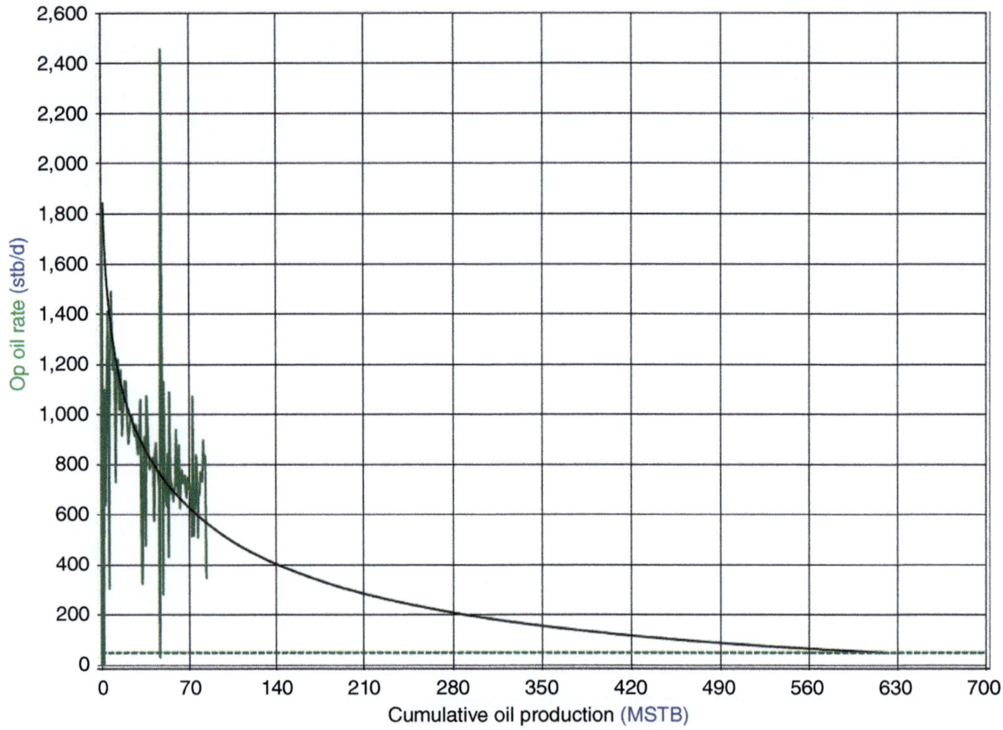

Figure 9.8 SEP decline model (generated by IHS Harmony).

9.6.4 Logistic Growth Model

The logistic growth model (LGM) was adopted by Clark et al. (2011) to model the production decline of extremely low permeability formations . The LGM was first developed to mimic the regrowth of the livers (Spencer & Coulombe, 1966). The idea of the LGM is to cap the cumulative production at a maximum carrying capacity value above which there is no further growth in production. The LGM constrains the EUR value by the carrying capacity parameter (*K*) and the flow rate. The LGM model also assumes that one well, over a sufficient period of time, can actually drain the reservoir (Clark et al., 2011; Paryani et al., 2016). The flow rate and the cumulative production can be estimated for the LGM by equation (9.34) and equation (9.35) (Clark et al., 2011):

$$q = \frac{K\, n_{LGM}\, a_{LGM}\, t^{(n_{LGM}-1)}}{[a_{LGM} + t^{n_{LGM}}]^2}, \tag{9.34}$$

where K is the carrying capacity, n_{LGM} is the time exponent for the LGM, and a_{LGM} is a coefficient for the LGM:

$$Q = \frac{K\, t^{(n_{LGM})}}{[a_{LGM} + t^{n_{LGM}}]}. \tag{9.35}$$

9.6.5 Duong Model

Duong's model is designed specifically for unconventional gas reservoirs to capture the production data when the fracture flow is dominant with negligible contribution from the matrix. Duong's model yields accurate forecast when the flow is linear or bilinear (Duong, 2011; Kanfar & Wattenbarger, 2012). Before using Duong's model, the production data should be screened by removing the noisy data points. The flow rate and the cumulative production for Duong's model can be estimated by equation (9.36) and equation (9.37), respectively (Duong, 2011):

$$q = q_1 t^{-m_{Doung}} e^{\left[\frac{a_{Doung}}{(1-m_{Doung})}\left[t^{(1-m_{Doung})}-1\right]\right]}, \tag{9.36}$$

where q_1 is the flow rate at day 1, m_{Doung} is the time exponent for Duong's model, and a_{Doung} is the model coefficient.

$$G = \frac{q_1}{a_{Doung}} e^{\left[\frac{a_{Doung}}{(1-m_{Doung})}\left[t^{(1-m_{Doung})}-1\right]\right]}, \tag{9.37}$$

where G is the cumulative gas production.

The model's unknown parameters (i.e., m_{Doung} and a_{Doung}) can be found by plotting (q/Q) versus time on a log-log plot where a_{Doung} is the intercept value and m_{Doung} is the slope based on equation (9.38) (Duong, 2011):

$$\frac{q}{G} = a\,t^{-m}. \tag{9.38}$$

The value of q_1 can then be estimated by plotting the flow rate (q) versus t (a,m) on a log-log plot. Where t (a,m) is calculated by the use of equation (9.39) (Satter & Iqbal, 2016):

$$t(a,m) = t^{-m_{Doung}} e^{\left[\frac{a_{Doung}\left(t^{1-m_{Doung}}-1\right)}{(1-m_{Doung})}\right]}. \tag{9.39}$$

A DCA comparison for two oil wells is presented in Table 9.4. One well is targeting the Bakken play, Figure 9.9, and the other well is targeting the Niobrara play, Figure 9.10. The

Table 9.4 DCA Comparison

Formation	Bakken	Niobrara
Decline rate (1/day)	0.00312	0.00506
Decline exponent (b)	1	0.91
Set abandonment rate (STB/day)	15	15
Abandonment date	1/13/2025	1/29/2021
EUR (MSTB)	227.3	127.4

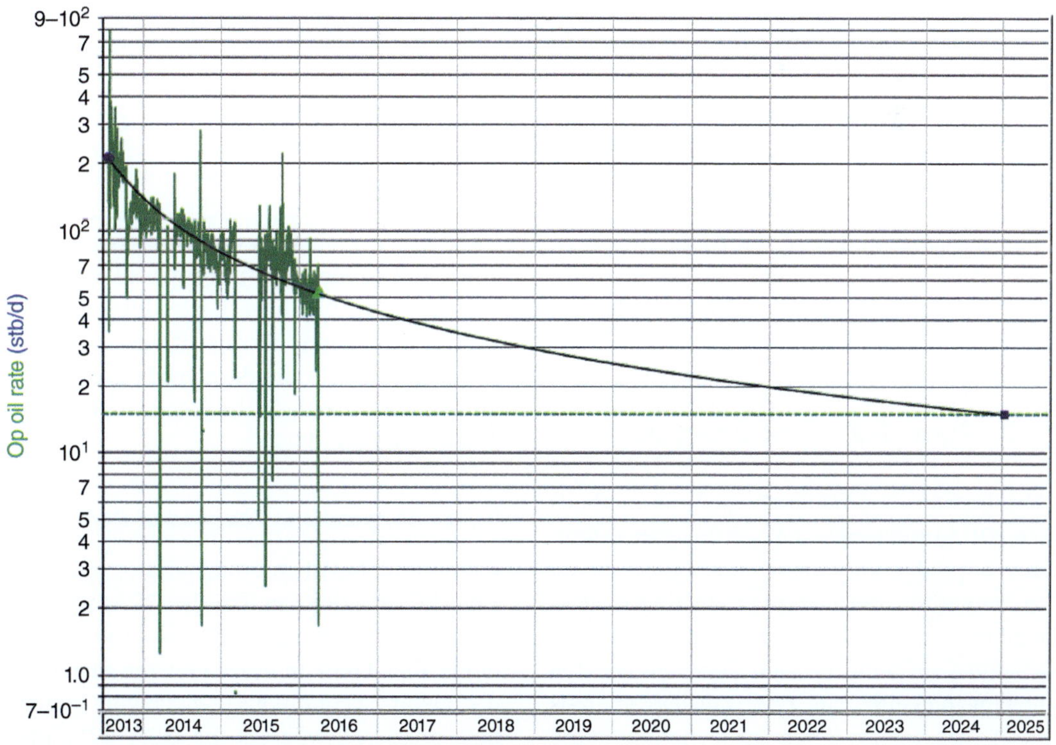

Figure 9.9 Bakken oil well (generated by IHS Harmony).

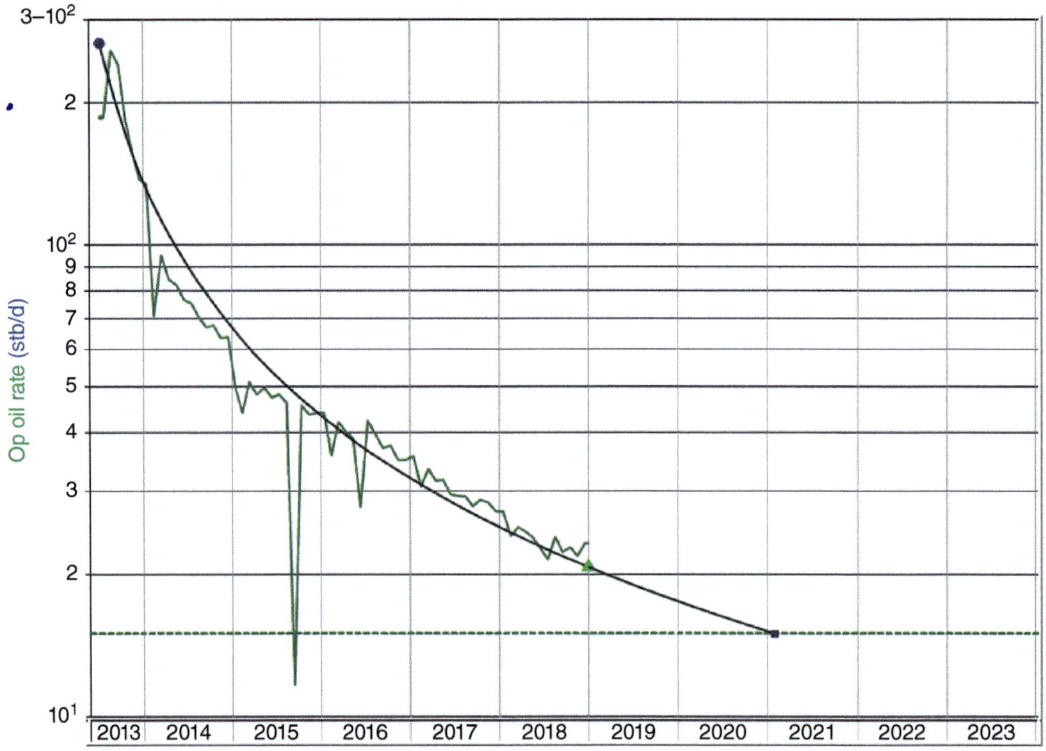

Figure 9.10 Niobrara oil well (generated by IHS Harmony).

decline curve parameters are summarized in Table 9.4. It is clear from the comparison that the well targeting the Niobrara play is declining at a faster pace than the Bakken well. This could be due to that fact that the permeability of the middle Bakken is significantly higher than the permeability of the Niobrara formation. Difference in the SRV of the two wells can be the other cause of this higher estimated ultimate recovery for the Bakken well.

9.7 Rate Transient Analysis

The traditional production analysis methods (i.e., DCA) utilize only the well flow rate to forecast the production and to evaluate the reserves. In comparison, the modern production analysis methods (i.e., rate transient analysis) take advantage of both the well flow rate and the flowing pressure. The rate transient analysis (RTA) has become a popular technique in the industry when both the flow rate and pressure data are available at no extra cost and without the need of shutting in the well to conduct well testing. This became possible due to the development of permanent downhole gauges (Ataei et al., 2018). The RTA should not be mixed up with the pressure transient analysis (PTA). PTA, which involves well testing (e.g., drawdown and buildup tests), is a typical formation evaluation technique that has traditionally been applied for analysis of conventional wells. The RTA and the PTA are developed based on the same theory, however, the data are interpreted differently (Bax, 2017).

9.7.1 Purpose and Features of RTA

RTA is built based on actual physics and mathematical equations unlike the DCA, which is an empirical technique. RTA is a robust technique that takes into account the change of the operating conditions (e.g., flowing bottomhole pressure), whereas the DCA techniques become unreliable when the operating conditions are altered. Additionally, RTA can be used to forecast the production even when a constant rate operating condition is used and predicting the well forecast would otherwise require building an advanced time-consuming reservoir simulator. The purpose of RTA is to forecast the production and to determine the reserves and the reservoir characteristics by making use of the already available production data, which are collected by the permanent downhole gauges.

By utilizing the flowing bottomhole pressure (FBHP) and the flow rate, the RTA techniques are capable of forecasting the production, evaluating the reserves, estimating the original volume in place, and determining the reservoir properties (i.e., permeability and skin). The RTA is also capable of estimating the drainage area and the shape of the reservoir. In addition to that, the RTA can provide insight into the potential of drilling infill wells (Bax, 2017).

In general, PTA yields more accurate estimates of reservoir parameters (e.g., permeability and skin) than the RTA because the PTA tests are typically conducted in a controlled environment (e.g., fixed flow rate and predetermined time of shut-in). However, the RTA provides a long-term understanding of the reservoir. For example, RTA yields better estimates of original volume in place and drainage area and shape because it utilizes the production data over a long production period (Anderson, 2010). RTA is more applicable for tight and ultra-tight unconventional oil and gas reservoirs since the transient time takes a very long time in these wells and it is not practical to keep the well shut for such a long period of time. Shorter versions of such tests are applied for fracture and near wellbore diagnostics as discussed in Chapter 6.

9.7.2 RTA Concept

The fundamental concept of RTA is the integration and interpretation of the transient flow and the boundary-dominant flow. The main boundary-dominant flow equation, which is utilized in RTA to get the volume in place, combines the material balance equation, equation (9.40), and the pseudo-steady-state inflow equation, equation (9.41) (Fetkovich, 1980; Fraim & Wattenbarger, 1987). The combination of the two equations yields the RTA equation, equation (9.42) (Fetkovich, 1980; Doublet et al., 1994). The volume in place (N) can be evaluated from the slope (m_{pss}) of the (P_{i} – P_{wf}) versus (N_p) Plot.

$$\overline{P} = P_i - m_{pss} N_p, \tag{9.40}$$

where $\overline{P}$ is the average reservoir pressure, P_i is the initial reservoir pressure, N_p is the cumulative production, m_{pss} is the slope ($1/c_t\ N$), N is the original volume in place, and c_t is the overall compressibility factor.

$$\overline{P} = P_{wf} + q b_{pss}, \tag{9.41}$$

where P_{wf} is the flowing pressure, q is the flow rate and b_{pss} is the reciprocal of the productivity index.

$$P_i - P_{wf} = m_{pss} N_p + q\ b_{pss}. \tag{9.42}$$

The boundary-dominant flow RTA equation, equation (9.42) can be further derived for the two main operating conditions, constant pressure and constant flow rate, to get the basic analytical RTA models. The derivation of these models is well established in the literature but it is beyond the scope of this chapter (Fetkovich, 1980; Carter, 1985; Fraim & Wattenbarger, 1987; Blasingame et al., 1991; Palacio & Blasingame, 1993; Doublet et al., 1994; Agarwal et al., 1999). Overall, the goal is to get an equation that can be used to understand the flow behavior of the system and forecast the production similar to the DCA equations for each operating condition. The RTA equation for a constant flow rate condition and a constant pressure are represented by equation (9.43) and equation (9.44), respectively (Palacio & Blasingame, 1993; Doublet et al., 1994):

$$\frac{q}{P_i - P_{wf}} = \frac{\frac{1}{b_{pss}}}{\frac{m_{pss} N_p}{b_{pss}\ q} + 1} \tag{9.43}$$

and

$$q(t) = \frac{P_i - P_{wf}}{b_{pss}} e^{-\left(\frac{m_{pss}}{b_{pss}} t\right)}. \tag{9.44}$$

The RTA transient flow equation, which is applied on the transient part of the data to determine the reservoir characteristics (e.g., permeability and skin), is represented by equation (9.45):

$$\frac{q}{P_i - P_{wf}} = \frac{k\,h}{141.2\mu B}\frac{1}{0.5\ln\left(\dfrac{0.0063\,k\,t}{\phi\mu c_t}\right) + 0.4045 + S}, \tag{9.45}$$

where k is the permeability (mD), h is the thickness (ft), B is the volumetric factor (RB/STB), ϕ is the porosity (fraction), μ is the viscosity (cP), c_t is the total compressibility (1/psi), and S is the skin

The transient flow equation is used for the early-time flow or in low permeability reservoirs where it could take years to reach the boundary (e.g., tight formations). The boundary-dominant flow equation is used for the late-time flow. Therefore, by integrating the two RTA equations, the production data can be analyzed for both the transient flow and the boundary-dominant flow. The transient flow and the boundary-dominant flow are combined into one single curve using the concept of type curves developed by pioneers of this field such as (Fetkovich, 1980; Fraim & Wattenbarger, 1987; Palacio & Blasingame, 1993; Doublet et al., 1994; Agarwal et al., 1999).

9.7.3 Type Curves

The type curves are basically an illustration of a dimensionless model of the well, and it is a technique that is used to visualize and identify the flow regimes (i.e., transient flow and the boundary-dominant flow) in one log-log plot (Anderson, 2010). Figure 9.11 shows the type curves for several decline equations with different decline exponent b values where the red line represents Arps's exponential decline model (i.e., $b = 0$), and the dark blue line represents Arps's harmonic decline model (i.e., $b = 1$), and all the potential curves above that belong to the general hyperbolic decline curves with $b > 1$. The type curve technique is unique in a way that if the production data, for example, follows the exponential decline properties, then the data will always match the exponential decline line regardless of the location of the well, the well type (i.e., gas or oil well), or the decline rate of the well. Equation (9.42) can be represented as shown in equation (9.46) in terms of normalized rate and material balance time for oil wells (Palacio & Blasingame, 1993):

$$\frac{P_i - P_{wf}}{q} = m_{pss}\,t_c + b_{pss}, \tag{9.46}$$

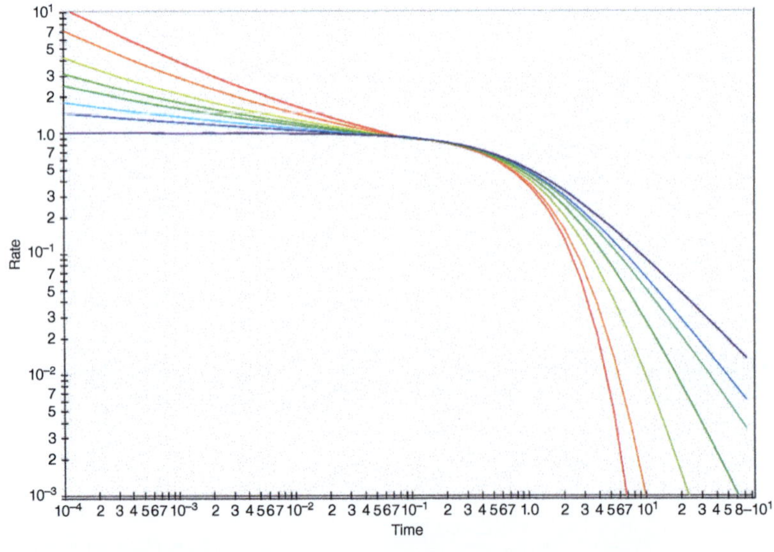

Figure 9.11 Fetkovich type curve for radial system (generated by IHS Harmony).

where m_{pss} is the slope, which is $(1/c_t N)$; N is the original oil in place; t_c is the material balance time, which is also known as the corrected time (days).

$$t_c = \frac{N_p}{q} \tag{9.47}$$

and

$$b_{pss} = \frac{141.2\, B\, \mu}{k\, h}\frac{1}{2}\ln\left(\frac{4}{e^{\gamma}}\frac{A}{C_A r_{wa}^2}\right), \tag{9.48}$$

where A is the drainage area (ft^2), C_A is the reservoir shape factor, r_{wa} is effective wellbore radius (ft), and γ is the Euler's constant = 0.577216.

Note that one would have to use pseudo time and pseudo pressure for gas wells as shown in equation (9.49) (Palacio & Blasingame, 1993):

$$\frac{P_{pi} - P_{pwf}}{q} = m_{a_{pss}} t_{a_c} + b_{a_{pss}}, \tag{9.49}$$

$$m_{a_{pss}} = \frac{1}{G\, c_t}, \tag{9.50}$$

$$t_{a_c} = \frac{\left(\mu_g c_g\right)_i}{q_g}\int_0^t \frac{q_g}{\mu_g c_g}\, dt, \tag{9.51}$$

and

$$b_{a_{pss}} = \frac{141.2 \mu_{gi}\, B_{gi}}{k\, h}\frac{1}{2}\ln\left(\frac{4}{e^{\gamma}}\frac{A}{C_A r_{wa}^2}\right). \tag{9.52}$$

The curves plotted in the early-time period in Figure 9.11 represent the transient flow regime, and the boundary-dominant flow are the curves in the late-time period. Clearly there is a transitioning zone between the transient flow and the boundary-dominant flow where the two curves connect. The point at which the transient flow ends and the boundary-dominant flow starts is an important point because it provides knowledge of when the analytical models stop using the transient flow equations and start using the boundary-dominant flow equations for the production forecast (Fetkovich, 1980). This point in time is usually referred to as the time to pseudo steady state (T_{pss}). The time to pseudo steady state can be calculated by solving the radius of investigation equation for time as shown in equation (9.53) (Anderson, 2010):

$$t_{pss} = \frac{948\, \phi\, \mu\, c_t\, r_i^2}{k}, \tag{9.53}$$

where r_i is the radius of investigation (ft) and t_{pss} is time to pseudo steady state (hours)

By displaying the well production data on the type curves, the reservoir properties (e.g., permeability and skin) can be determined by analyzing the early-time production data, while the reserves and the original volume in place can be determined by analyzing the late-time production data (Doublet et al., 1994).

9.7.4 Type Curve Methods

There are several type curve methods that are used in the industry. There are type curves that were developed for horizontal wells, fractured wells, and water-drive wells and also for different fluids (i.e., gas, oil, and water). The type curves are generally developed by using the dimensionless rate and time. The dimensionless flow rate and the dimensionless time are shown in equation (9.54) and equation (9.55), respectively.

$$q_{Dd} = \frac{141.3\, q(t)\, \mu\, B}{k\, h\, (P_i - P_{wf})} \tag{9.54}$$

and

$$t_{Dd} = \frac{0.0063\, k\, t}{\phi \mu c_t\, r_w^2}, \tag{9.55}$$

where q is the flow rate (STB/day), μ is the viscosity (cP), B is the formation volume factor (RB/STB), k is the permeability (mD), h is the formation thinness (ft), t is time (days), c_t is the total compressibility (1/psi), and r_w is the wellbore radius (ft).

Fetkovich Type Curves

Fetkovich developed his type curves in the 1980s based on Arps's decline models as shown in Figures 9.11 and 9.12 (Fetkovich, 1980). Fetkovich further improved the type curves to account for the change of operating conditions (i.e., flowing pressure) by using the normalized rate (Fetkovich & Vienot, 1984). Since then, several researchers have developed different

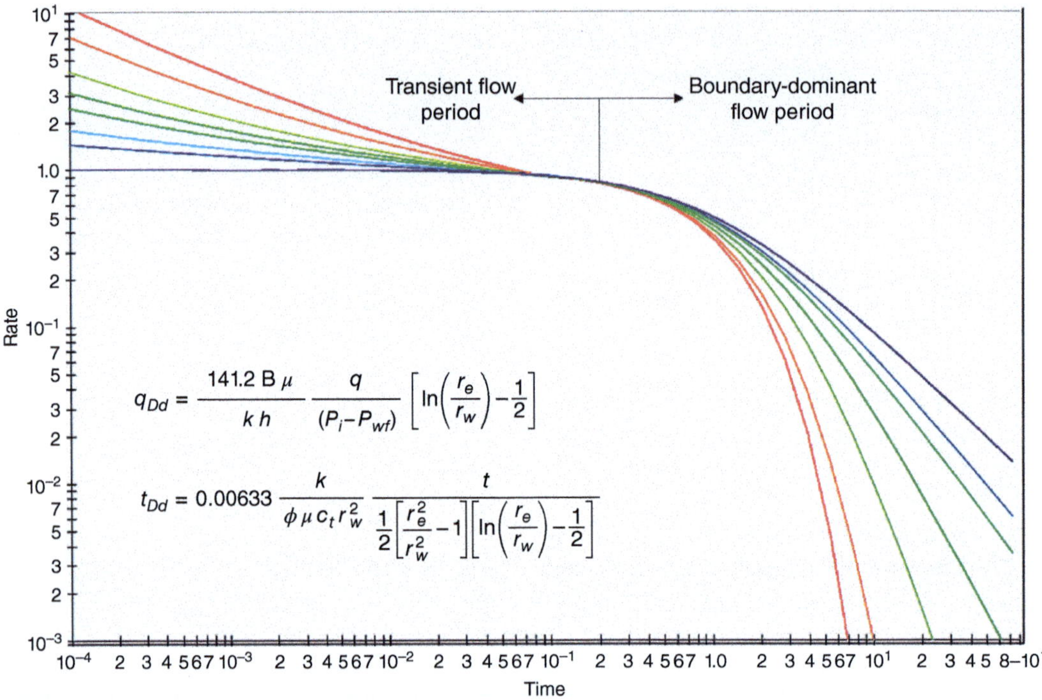

Figure 9.12 Fetkovich type curve for radial system (generated by IHS Harmony) modified after (Fetkovich, 1980).

versions of type curves. That said, the Fetkovich type curves are still being used in the industry, and they are also applied in the unconventional plays. The equations for the Fetkovich type curves are shown in equation (9.56) and equation (9.57):

$$q_{Dd} = \frac{141.2\,B\,\mu}{k\,h}\frac{q}{(P_i - P_{wf})}\left[\ln\left(\frac{r_e}{r_w}\right) - \frac{1}{2}\right] \tag{9.56}$$

and

$$t_{Dd} = 0.00633\frac{k}{\phi\,\mu\,c_t\,r_w^2}\frac{t}{\frac{1}{2}\left[\frac{r_e^2}{r_w^2} - 1\right]\left[\ln\left(\frac{r_e}{r_w} - \frac{1}{2}\right)\right]}. \tag{9.57}$$

Blasingame Type Curves

Blasingame et al. developed type curves that take into account the change of the operating conditions (i.e., well flow rate and flowing pressure). The Blasingame type curves are also unique because there is only one stem that represents the boundary-dominant flow (i.e., depletion stem) as shown in Figure 9.13. The existence of only one depletion stem makes it easier for the users to match the production data on the type curve. This became possible since Blasingame used the normalized rate and the material balance time in generating the type curves plot instead of rate and time as in Fetkovich type curves (Blasingame et al., 1991; Palacio & Blasingame, 1993; Doublet et al., 1994).

Agarwal-Gardner Type Curves

The Agarwal-Gardner type curves are developed for hydraulically fractured wells to estimate the formation properties as well as the original volume in place (Agarwal et al., 1999). The Agarwal-Gardner type curves are also developed by utilizing the normalized rate and the material balance time as in the Blasingame type curves. However, the advantage of the Agarwal-Gardner type curves is that they clearly show the transitioning from the transient flow to the boundary-dominant flow as shown in Figure 9.14.

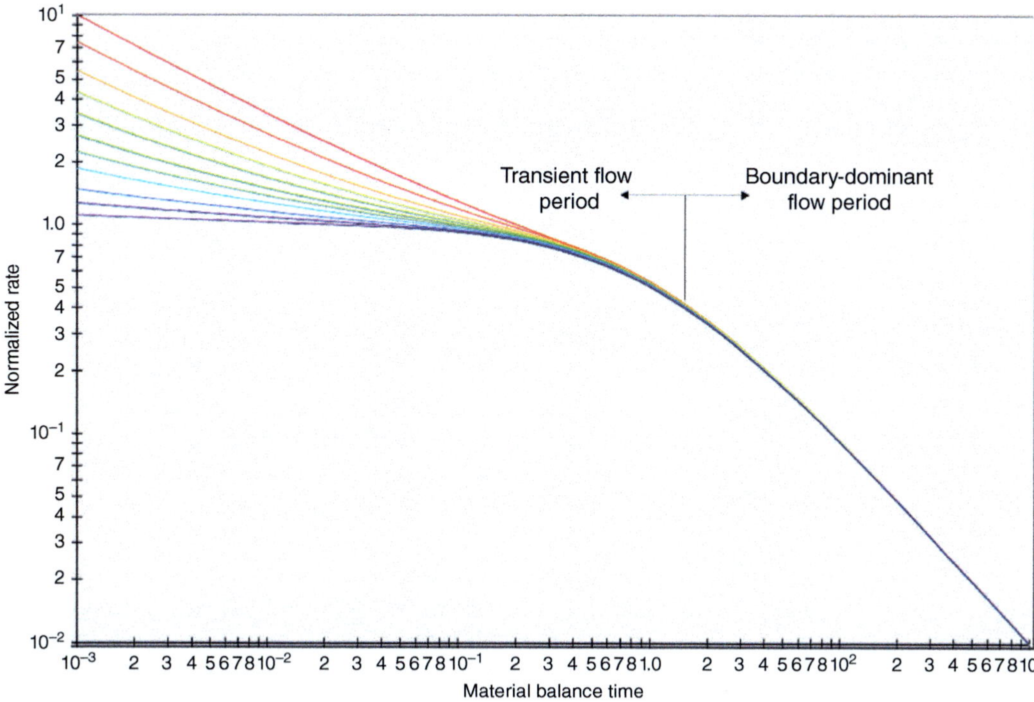

Figure 9.13 Normalized rate versus material balance time (generated by IHS Harmony; after Palacio & Blasingame, 1993).

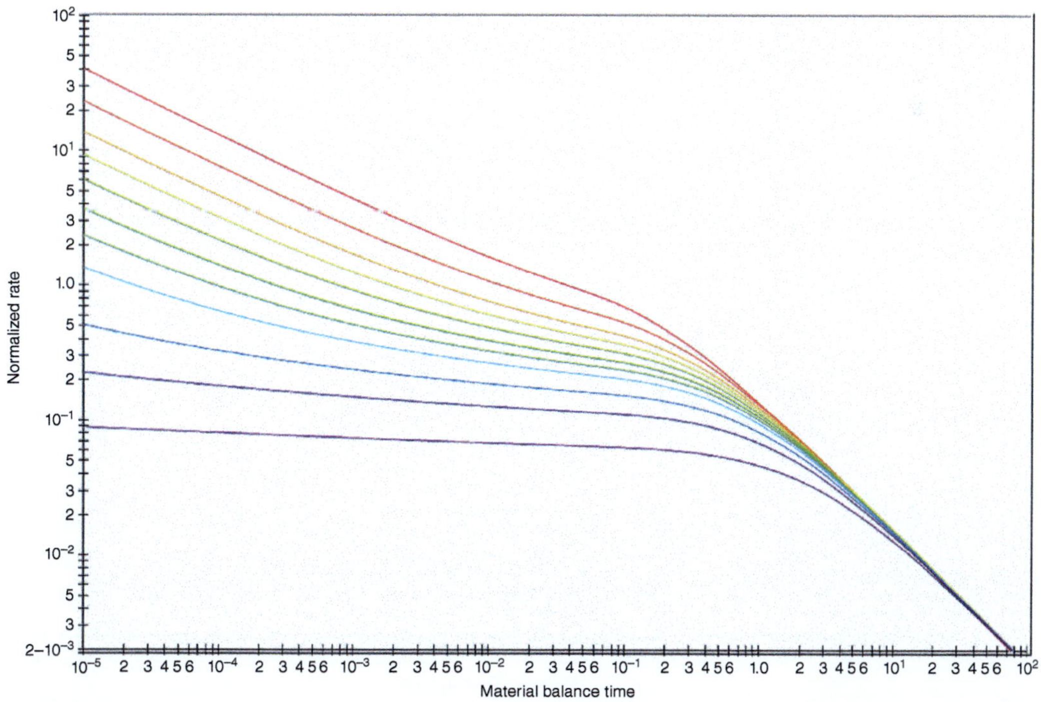

Figure 9.14 Agarwal-Gardner type curve for radial system (generated by IHS Harmony).

The Agarwal-Gardner type curve was developed by the set of equations represented in equations (9.58) to (9.63) (Agarwal et al., 1999).

$$Q_{DA} = \frac{t_{DA}}{P_{wD}} \quad (9.58)$$

and

$$\frac{1}{P_{wD}} = \frac{1422\ T\ q(t)}{k\ h\ (m(P_i) - m(P_{wf}))}, \quad (9.59)$$

where T is reservoir temperature (degree Rankine), and

$$m(P) = 2\int_0^P \frac{P'\ dP}{\mu(P')z(P')}. \quad (9.60)$$

Note the units of $m\ (P)$ is (psi^2/cP),

$$t_{DA} = t_{aD}\left(\frac{r_{wa}^2}{A}\right), \quad (9.61)$$

where A is the drainage area (ft^2),

$$t_{aD} = \frac{0.0063\ k}{\phi\ (\mu\ c_t)_i r_{wa}^2}\ t_a \quad (9.62)$$

and

$$t_a = \frac{1}{q(t)}\ (\mu\ c_g)_i \frac{Z_i G}{2P_i}\ (m(P_i) - m(\overline{P})). \quad (9.63)$$

Normalized Pressure Integral (NPI) Type Curves

The NPI type curves are developed by normalizing the pressure instead of normalizing the rate as in the Blasingame and Agarwal-Gardner type curves. By using the normalized pressure, the inverse of the Agarwal-Gardner type curves is achieved as shown in Figure 9.15. The normalized pressure is typically used in the pressure transient analysis (PTA). Hence, the NPI type curves are the preferred choice of type curves to those with PTA background. The NPI type curves are represented by equation (9.64) and equation (9.65) (Blasingame & Lee, 1988; Blasingame & Lee, 1989):

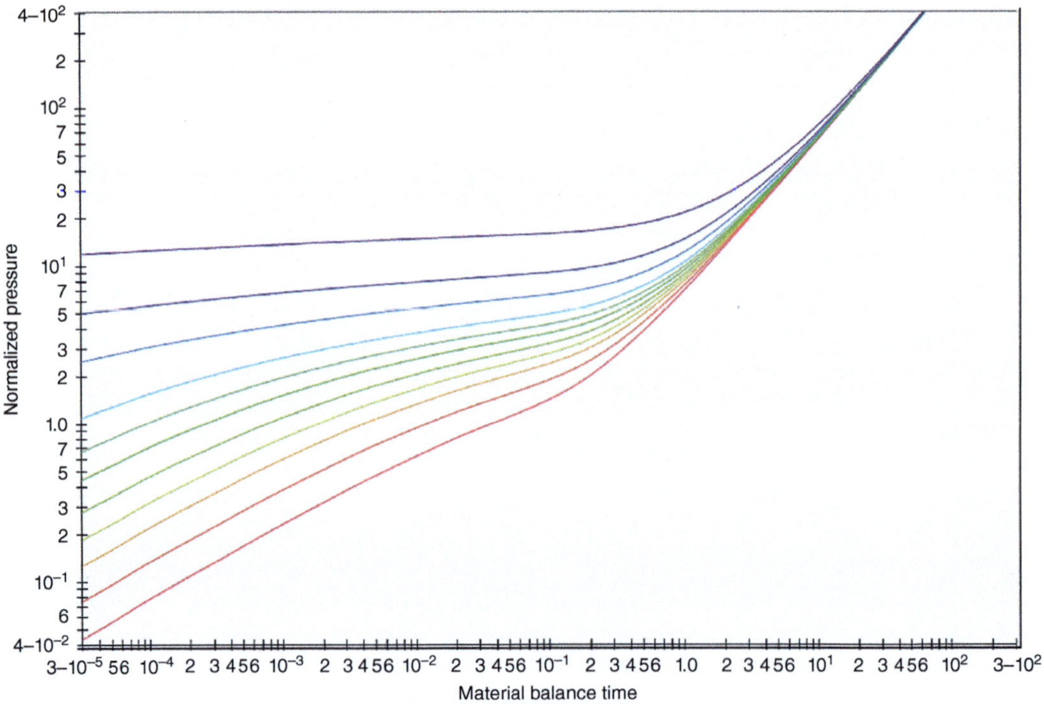

Figure 9.15 Normalized pressure integral (NPI) type curves (generated by IHS Harmony; after Blasingame & Lee, 1988).

$$P_D = \frac{k\,h\,(P_i - P_{wf})}{141.2\,q\,B\,\mu}, \tag{9.64}$$

$$t_D = \frac{0.00633\,k\,t_c}{\phi\,\mu\,c_t\,r_e^2}. \tag{9.65}$$

Wattenbarger Type Curves

The Wattenbarger type curves (Figs. 9.16 and 9.17) are used mainly to analyze the linear flow; they are ideal for tight formations (e.g., shale gas) where the permeability is low and the transient flow period is long and the system is hydraulically fractured. The dimensionless equations for the Wattenbarger type curves are represented by equation (9.66) and equation (9.67) (Wattenbarger et al., 1998):

$$P_{wD} = \frac{1}{q_D} = \frac{k\,h\,(P_i - P_{wf})}{141.2\,q\,B\,\mu}, \tag{9.66}$$

$$t_{Dxf} = \frac{0.00633\,k\,t}{\phi\,\mu\,c_t\,x_f^2}. \tag{9.67}$$

With the use of the advanced RTA software packages (e.g., IHS Harmony), the RTA and type curve analysis become a relatively easy and straightforward procedure, and all it takes is matching the production data to the right type curve stems. Nevertheless, careful engineering judgment must be practiced due to the uncertainty in the data.

9.7.5 Square Root Time and Flowing Material Balance Plots

The square-root time plot is a way to analyze the production data by plotting the normalized pressure versus the square root of time. The square root time plot is used to capture the flow regimes. The linear flow period can be captured from the square root time plot by drawing a

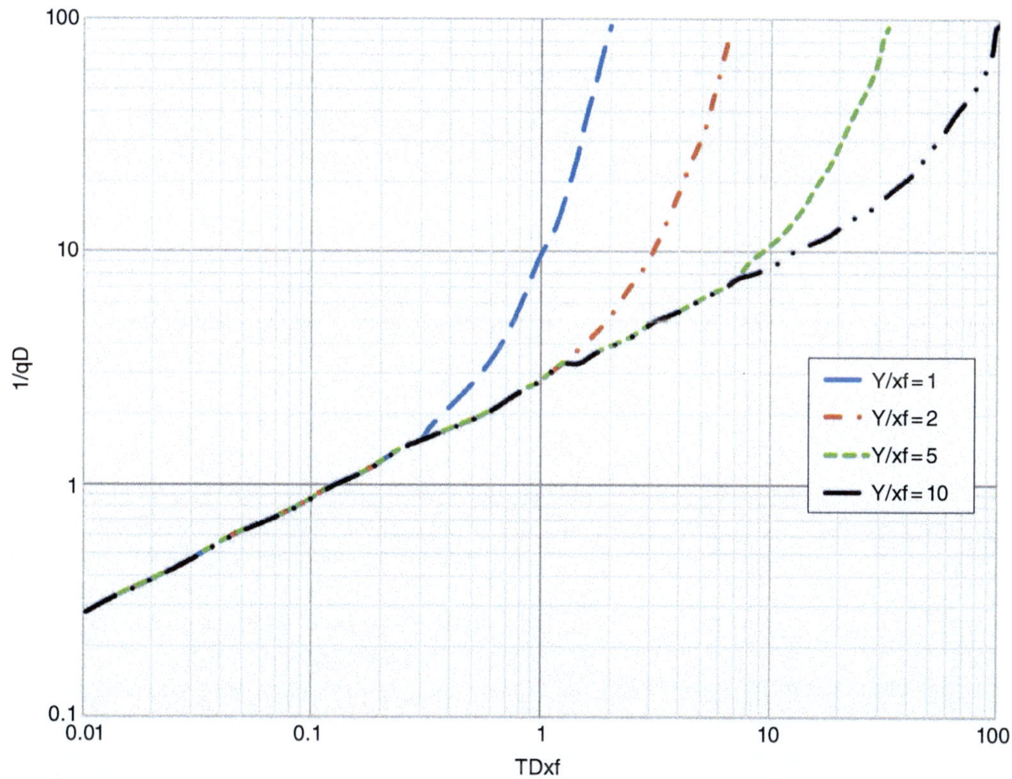

Figure 9.16 Wattenbarger type curve with dimensionless flow rate and time (modified after Wattenbarger et al., 1998).

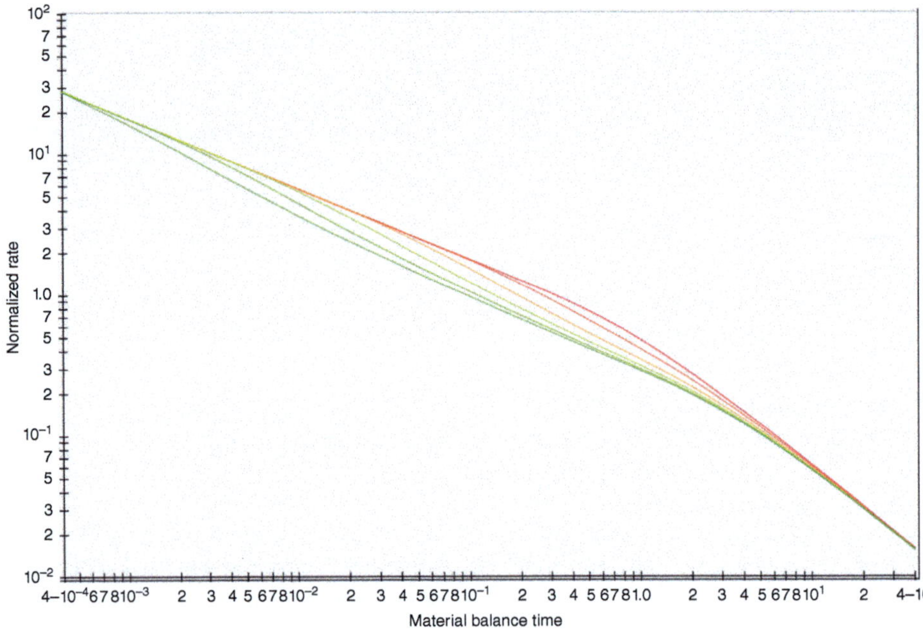

Figure 9.17 Wattenbarger type curve (generated by IHS Harmony).

straight line of the early production data points, and using the slope of that line to determine some of the reservoir's characteristics such as permeability (Anderson et al., 2013). The square root time plot is defined by equations (9.68) to (9.70). An example of the square root time plot for a well targeting the Bakken play is shown in Figure 9.18.

$$\frac{1}{q} = m\sqrt{t} + b, \tag{9.68}$$

$$m = \frac{31.3\,B}{h\,x_f\,\sqrt{k}}\sqrt{\frac{\mu}{\phi\,c_t}}\frac{1}{(P_i - P_{wf})}, \tag{9.69}$$

and

$$b = \frac{141.2\,B\,\mu\,S}{2(P_i - P_{wf})\,k\,x_f}. \tag{9.70}$$

The flowing material balance (FMB) plot is utilized to estimate the original volume in place and the drainage area. The FMB plot is achieved by simply plotting the normalized rate versus the normalized cumulative production as shown in Figure 9.19.

9.7.6 Flow Regimes

Identifying the flow regimes the multifractured horizontal well (MFHW) exhibits is important as the flow regimes play an essential role in estimating the characteristics of the fractures and the reservoirs (Clarkson & Pedersen, 2010). There are several proposed methods to identify the flow regimes in MFHWs (Clarkson & Beierle, 2010; Bello & Wattenbarger, 2010; Blasingame, 2015; Apte & Lee, 2017; Fulford, 2018). The log-log plot of flow rate and time (see Fig. 9.25), the square root time plot (Fig. 9.18), and the Blasingame plot, (see Fig. 9.26), are common plots that can be utilized to identify the flow regimes. In the MFHWs, five flow regimes are typically identified. The first flow regime is the early-time linear flow. The early-time linear flow, which is due to the fractures drainage, occurs when the fracture conductivity is high as shown in Figure 9.20. The early-time linear flow is typically identified by one-half slope in the log-log plot. The second flow regime is the bilinear flow. The bilinear flow occurs when the fracture conductivity is low due to the simultaneous flow in the fractures and

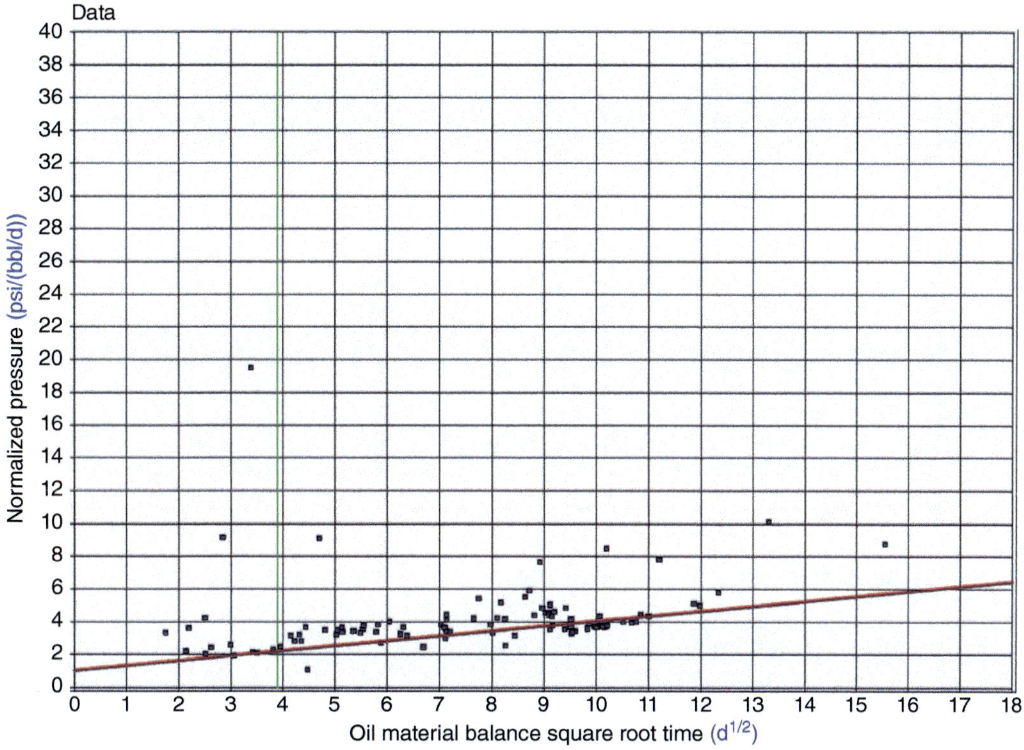

Figure 9.18 Square-root time plot (generated by IHS Harmony).

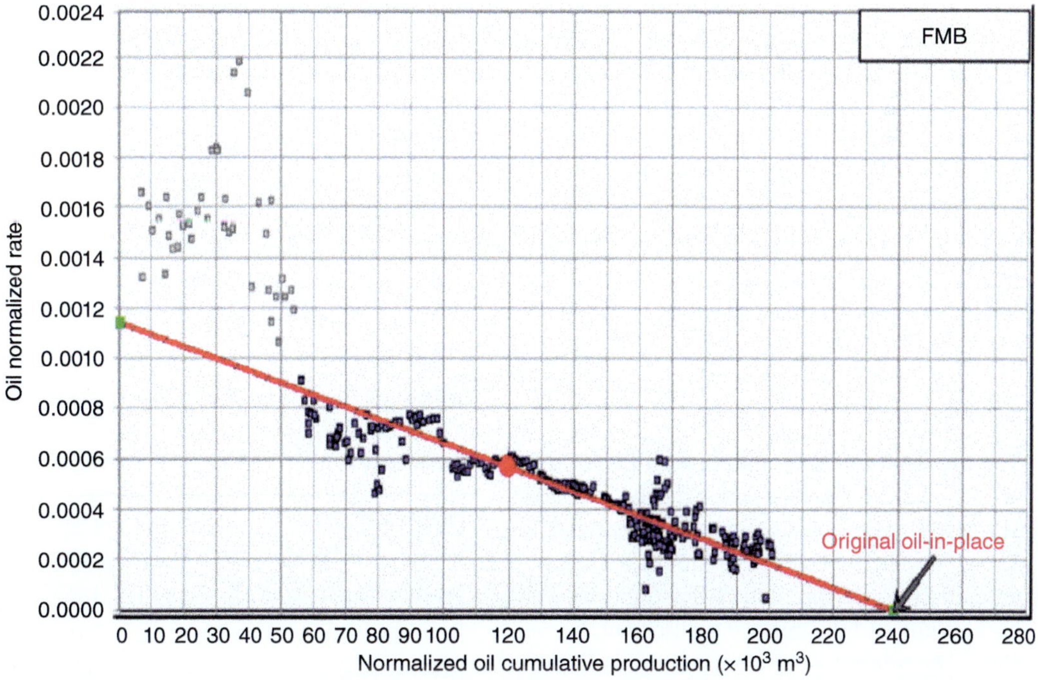

Figure 9.19 Flowing material balance plot (modified after Li et al., 2018).

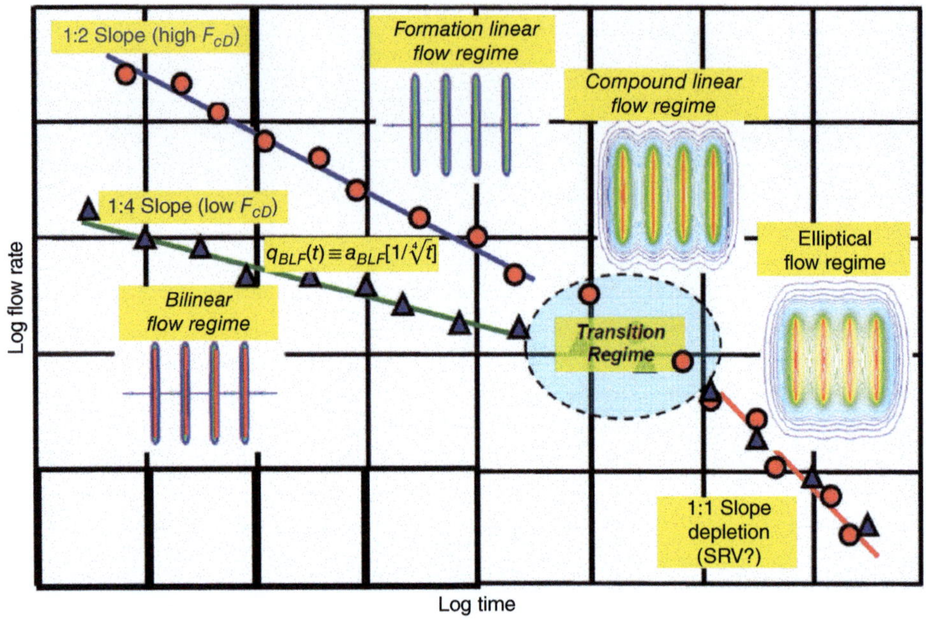

Figure 9.20 Flow regimes for MFHW (modified after Blasingame, 2015).

matrix (Bello & Wattenbarger, 2010). It can be identified by one-fourth slope in the log-log plot. The third flow regime is the compounded linear flow. The compounded linear flow occurs due to the flow from the matrix to the fractures system, and it can be identified by the one-half slope in the log-log plot. Note that both the early-time linear flow and the compounded linear flow are identified by one-half slope in the log-log plot, but they occur at different time periods as shown in Figure 9.20. The fourth flow regime is the SRV depletion, which can be identified by a unit slope in the log-log plot (Clarkson & Beierle, 2010). The fifth flow regime, which is unlikely to occur in the extremely tight unconventional plays, is the infinite acting radial flow (IARF); it shows a slope of zero in the log-log plot.

Example 9.9 Estimate the Matrix Permeability and the Original Oil in Place

The production data of a horizontal well with 45 hydraulic fracturing stages that is targeting the Bakken play are provided. By conducting an RTA study on the data, determine the permeability, the fractures half-length, and the original oil in place.

Solution

The provided production data are utilized using an advanced RTA software package (i.e., IHS Harmony). The results are summarized in Table 9.5. The first step is to input the flow rate data and the pressure data. After that, the wellbore profile and the initial conditions of the well should be entered. A schematic of the wellbore is shown in Figure 9.21, and the schematic of the stimulated reservoir volume is shown in Figure 9.22.

Table 9.5 Summarized Results for Example 9.9

Parameters	Value
OOIP (MMSTB)	4.95
Matrix permeability (mD)	0.079
Fracture half-length (ft)	162
Drainage Area (acres)	253
ASRV (acres)	79
OOIP (analytical model) (MMSTB)	6.26
ASRV (analytical model) (acres)	144
EUR (MSTB)	305.6

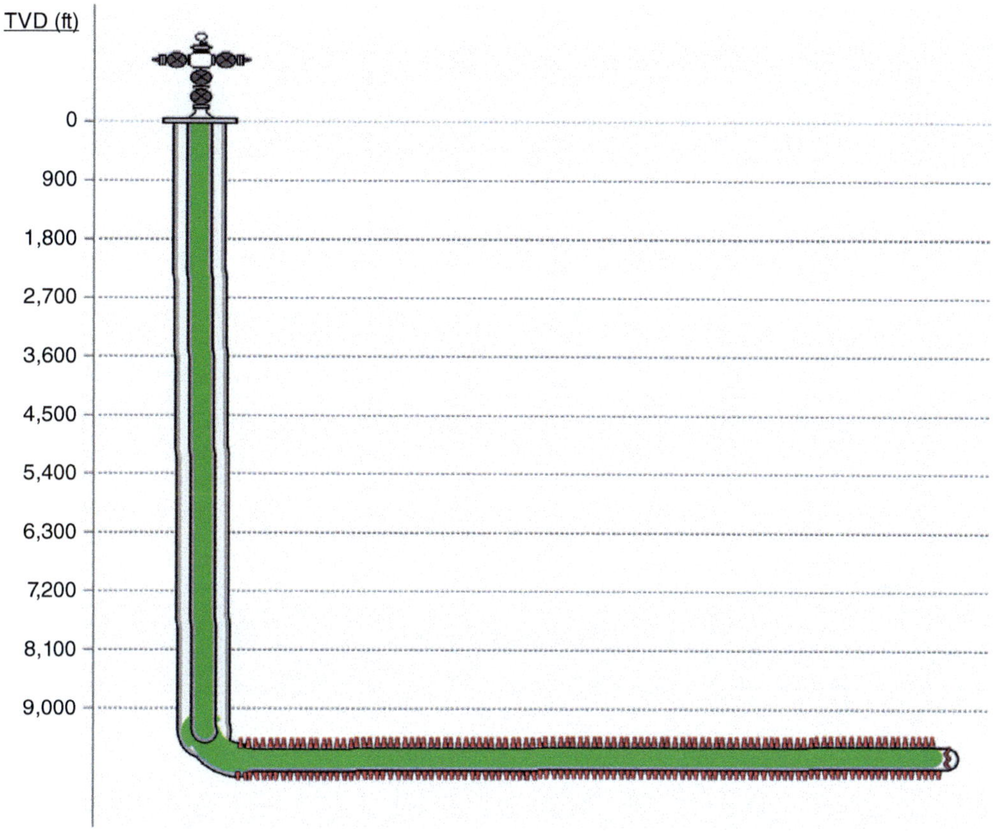

Figure 9.21 Schematic of the wellbore for Example 9.9 (generated by IHS Harmony).

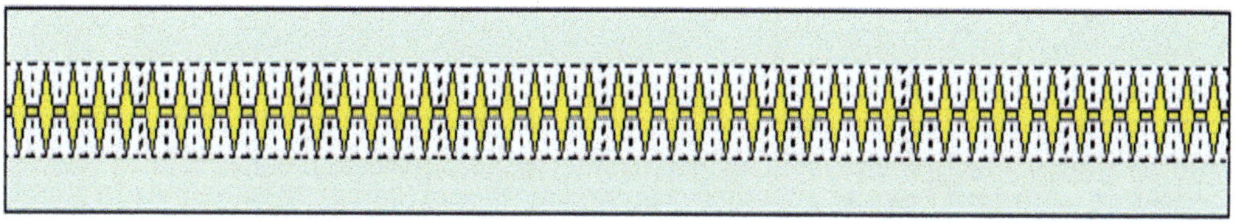

Figure 9.22 Schematic of the stimulated reservoir volume (SRV) for Example 9.9 (generated by IHS Harmony).

The next step is to analyze the production data by plotting the square root time plot, the flowing material balance plot, and the type curve. The square root time plot is generated and shown in Figure 9.23. From the square root time plot, the linear flow can be seen from the early points before the green vertical line. After the green line, the production starts to transition to a different flow regime. Next is to analyze the flowing material balance plot, which is shown in Figure 9.24. From the plot, the original oil in place (OOIP) is estimated to be 4.95 MMSTB. However, the provided production data are insufficient to find a clear trend to accurately estimate the OOIP. Due to insufficient data points as can be seen in Figure 9.24, the OOIP prediction is underestimated. The flowing material balance plot is also used to estimate the drainage area, which is found to be 253 acres. Finally, the type curves are plotted as shown in Figure 9.25 and Figure 9.26. From analyzing

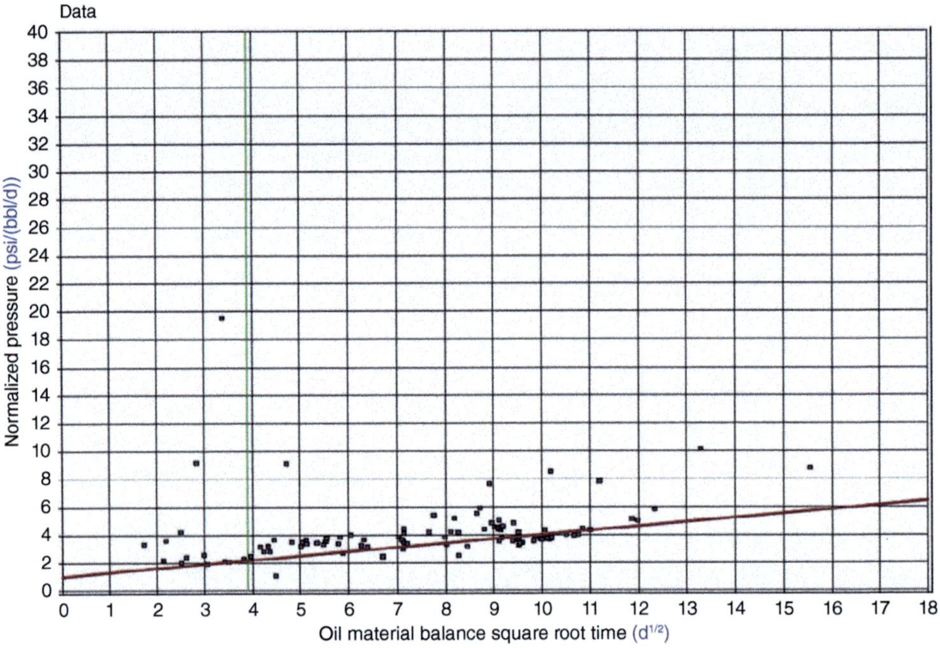

Figure 9.23 Superposition time. Square-root time plot for Example 9.9 (generated by IHS Harmony).

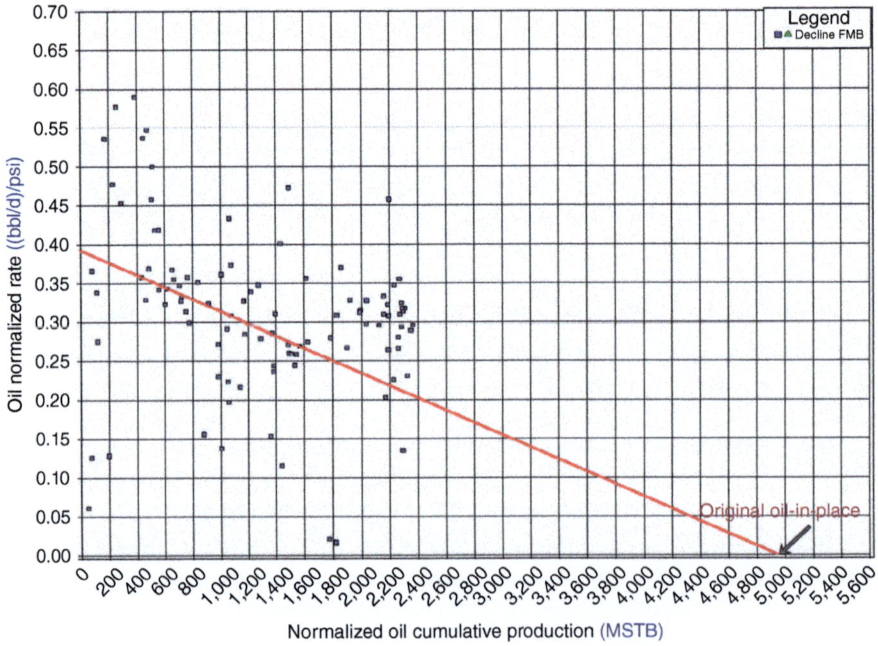

Figure 9.24 Flowing material balance plot for Example 9.9 (generated by IHS Harmony).

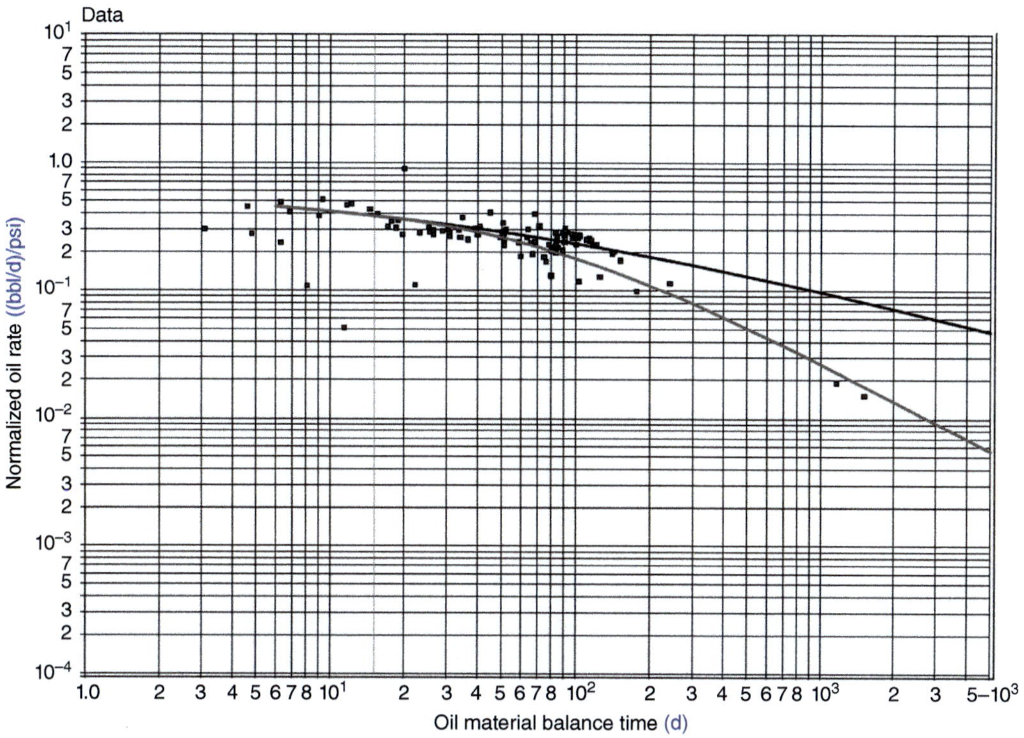

Figure 9.25 Type curve for Example 9.9 (generated by IHS Harmony).

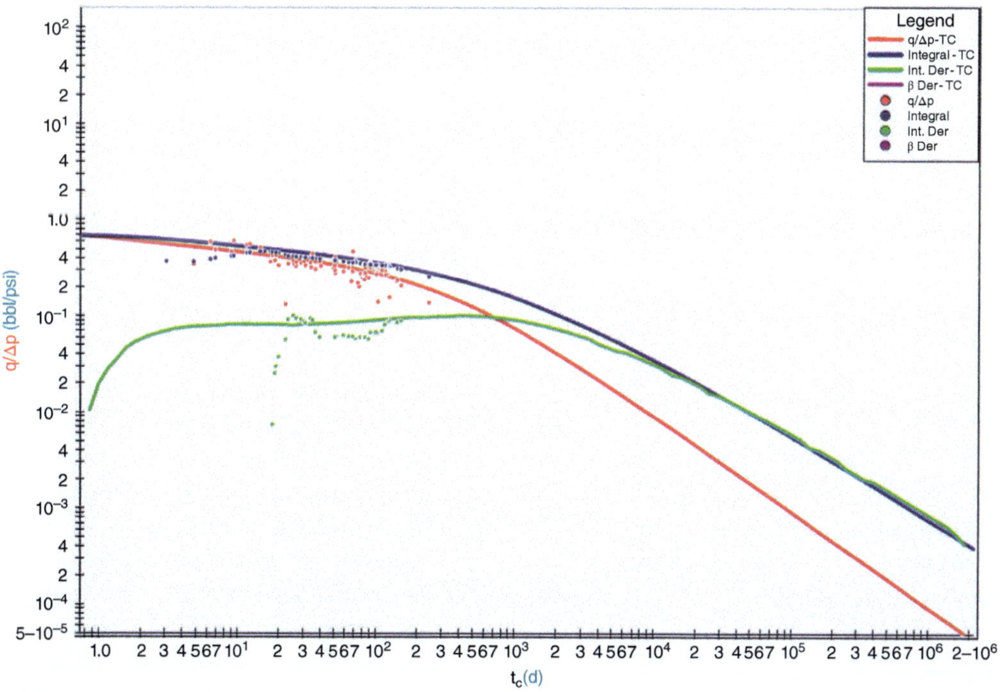

Figure 9.26 Blasingame type curve for Example 9.9 (generated by IHS Harmony).

the plots, the matrix permeability is estimated to be 0.079 mD, the effective fracture half-length is found to be 162 ft, and the area of the stimulated reservoir volume (A_{SRV}) is 79 acres.

After analyzing the data by the above-mentioned plots, an analytical RTA model was built to predict the production and to estimate the EUR and the OOIP. The analytical models can also be utilized to conduct parametric studies. The built analytical model was verified and the history matching error was 9.04%. The analytical RTA model gave an OOIP of 6.26 MMSTB, and the area of the stimulated reservoir volume (A_{SRV}) was found to be 144 acres. Based on the analytical model, the estimated ultimate recovery of the well is 305.6 MSTB.

9.8 Problems

1 From the data provided in Table 9.6, determine the appropriate Arps's traditional decline model that best fits the data, and determine the decline parameters.

Table 9.6 Data for Problem 9.1

Time	Oil rate (bbl/d)
11/09/14	1,491
11/10/14	1,177
11/11/14	1,269
11/12/14	729
11/13/14	1,175
11/14/14	1,220
11/15/14	1,018
11/16/14	1,177
11/17/14	916
11/18/14	1,027
11/19/14	1,134
11/20/14	1,131
11/21/14	964
11/22/14	881
11/23/14	947
11/24/14	998
11/25/14	989
11/26/14	943
11/27/14	914
11/28/14	960

Table 9.6 (Continued)

Time	Oil rate (bbl/d)
11/29/14	879
11/30/14	840
12/01/14	941
12/02/14	1,059
12/03/14	416
12/04/14	322
12/05/14	508
12/06/14	872
12/07/14	910
12/08/14	477
12/09/14	1,075
12/10/14	938
12/11/14	790
12/12/14	781
12/13/14	799
12/14/14	817

2 The production data of a horizontal well with multistage fracturing are provided in Table 9.10. Estimate the OOIP, the permeability, the fracture half-length, the area of the stimulated reservoir volume, and the estimated ultimate recovery for the well. The initial configuration data for the well are provided in Table 9.7 and Table 9.8. Assume that the entire length of the linear is perforated. The well survey data are shown in Table 9.9.

Table 9.7 Casing Data for Problem 9.2

		Data		
Type	**Top depth (ft)**	**Bottom depth (ft)**	**OD (in)**	**Weight (lb/ft)**
Casing	-	1,676	9.5"	36.0
Casing	surface	9,201	7"	26.0
Linear	10,520	14,430	4.5	11.6
Tubing	-	9,681	4.5	13.5

Table 9.8 Well Initialization Data for Problem 9.2

Parameter	Value
Initial pressure (psi)	6,200
Bubble point pressure (psi)	2,000
Formation temperature (°F)	200
Formation thickness (ft)	48
Porosity (%)	8
Oil saturation (%)	75

Table 9.9 Well Survey Data for Problem 9.2

MD depth	TVD depth
9,621	9,619
9,650	9,648
9,674	9,672
9,700	9,698
9,733	9,731
9,764	9,762
9,795	9,792
9,827	9,823
9,858	9,852
9,889	9,880
9,921	9,908
9,951	9,934
9,983	9,961
10,013	9,986

Table 9.9 (Continued)

MD depth	TVD depth
10,045	10,012
10,076	10,036
10,107	10,059
10,139	10,082
10,171	10,104
10,202	10,124
10,233	10,142
10,263	10,158
10,295	10,173
10,325	10,185
10,357	10,197
10,388	10,206
10,419	10,212
10,451	10,217
10,482	10,221
10,520	10,223
10,551	10,224
10,583	10,225
10,614	10,226
10,644	10,226
10,675	10,227
10,706	10,227
10,769	10,227
10,832	10,227

(Continued)

Table 9.9 (Continued)

MD depth	TVD depth
10,895	10,227
10,958	10,226
11,021	10,226
11,084	10,226
11,147	10,226
11,211	10,226
11,273	10,226
11,336	10,225
11,398	10,225
11,461	10,224
11,526	10,223
11,589	10,223
11,652	10,222
11,714	10,221
11,777	10,220
11,840	10,219
11,903	10,219
11,965	10,218
12,028	10,219
12,091	10,219
12,154	10,219
12,218	10,220
12,280	10,221
12,344	10,222
12,407	10,223
12,471	10,223
12,534	10,223
12,597	10,222
12,660	10,222

Table 9.9 (Continued)

MD depth	TVD depth
12,723	10,222
12,786	10,223
12,850	10,223
12,913	10,223
12,975	10,224
13,039	10,224
13,102	10,224
13,166	10,224
13,229	10,224
13,292	10,225
13,354	10,225
13,417	10,225
13,479	10,225
13,542	10,225
13,606	10,224
13,667	10,224
13,730	10,224
13,793	10,224
13,857	10,224
13,918	10,225
13,982	10,225
14,044	10,225
14,107	10,226
14,170	10,227
14,233	10,227
14,296	10,228
14,357	10,228
14,405	10,229
14,460	10,230

Table 9.10 Production Data for Problem 9.2

Time	Oil (bbl/day)	Gas (MSCF/D)	Water (bbl/day)	Pressure of tbg. (Psi)	Pressure of csg. (psi)
01/01/12	43	51	32	90#	40#
01/02/12	42	50	32	90#	40#
01/03/12	45	47	30	90#	40#
01/04/12	38	49	32	90#	40#
01/05/12	42	52	33	90#	40#
01/06/12	48	52	32	90#	40#
01/07/12	40	47	32	90#	40#
01/08/12	45	46	41	90#	40#
01/09/12	40	45	17	80#	45#
01/10/12	38	48	32	80#	45#
01/12/12	42	45	28	80#	45#
01/13/12	42	49	32	80#	45#
01/14/12	40	44	30	80#	45#
01/15/12	45	50	31	80#	45#
01/16/12	40	47	32	80#	45#
01/17/12	40	37	28	80#	45#
01/18/12	38	38	36	80#	45#
01/19/12	45	42	27	80#	45#
01/20/12	56	44	28	80#	45#
01/21/12	23	45	32	80#	45#
01/22/12	40	43	28	80#	45#
01/23/12	45	47	33	80#	45#
01/24/12	43	43	27	80#	45#
01/25/12	43	45	30	80#	45#
01/26/12	38	40	27	80#	45#
01/27/12	45	42	28	80#	45#
01/28/12	38	42	128	80#	45#
01/29/12	40	43	127	80#	45#
02/05/12	40	41	28	90#	45#
02/06/12	38	41	27	90#	45#
02/07/12	0	0	25	90#	45#
02/08/12	77	36	40	90#	45#
02/09/12	42	42	25	90#	45#
02/10/12	42	13	30	90#	45#
02/11/12	43	40	23	90#	45#
02/12/12	37	38	30	90#	45#
02/13/12	43	42	33	90#	45#

(Continued)

Table 9.10 (Continued)

Time	Oil (bbl/day)	Gas (MSCF/D)	Water (bbl/day)	Pressure of tbg. (Psi)	Pressure of csg. (psi)
02/14/12	44	30	30	90#	45#
02/15/12	38	56	28	90#	45#
02/16/12	38	32	22	90#	45#
02/17/12	38	0	27	90#	45#
02/18/12	42	40	28	90#	45#
02/19/12	42	40	115	90#	45#
02/20/12	38	45	28	90#	45#
02/21/12	43	42	3	90#	45#
02/22/12	40	43	27	90#	45#
02/23/12	43	42	32	90#	45#
02/24/12	35	39	10	90#	45#
02/25/12	42	43	38	90#	45#
02/26/12	35	37	30	90#	45#
02/27/12	45	40	25	90#	45#
02/28/12	43	39	27	90#	45#
02/29/12	42	25	10	90#	45#
03/01/12	37	10	28	90#	45#
03/02/12	43	25	28	90#	45#
03/03/12	38	22	32	90#	45#
03/04/12	45	25	29	90#	45#
03/05/12	40	3	22	100#	45#
03/06/12	37	0	28	100#	45#
03/07/12	38	12	15	100#	45#
03/08/12	38	32	38	100#	45#
03/09/12	40	39	32	100#	45#
03/10/12	42	43	33	100#	45#
03/11/12	38	44	27	100#	45#
03/12/12	40	40	32	100#	45#
03/13/12	43	42	27	100#	45#
03/14/12	40	41	30	100#	45#
03/15/12	40	45	28	100#	45#
03/16/12	40	40	30	100#	45#
03/17/12	37	41	25	100#	45#
03/18/12	42	42	28	100#	45#
03/19/12	40	43	30	100#	45#
03/20/12	37	42	28	100#	45#
03/21/12	41	43	25	100#	45#
03/22/12	43	43	33	100#	45#

Table 9.10 (Continued)

Time	Oil (bbl/day)	Gas (MSCF/D)	Water (bbl/day)	Pressure of tbg. (Psi)	Pressure of csg. (psi)
03/23/12	40	45	27	100#	45#
03/24/12	37	39	25	100#	45#
03/25/12	40	42	27	100#	45#
03/26/12	35	36	22	100#	45#
03/27/12	43	39	28	100#	45#
03/28/12	37	39	27	100#	45#
03/29/12	43	43	15	100#	45#
03/30/12	45	43	47	100#	45#
03/31/12	43	36	20	100#	45#
04/01/12	33	40	32	100#	45#
04/02/12	42	38	23	100#	45#
04/03/12	38	40	25	100#	45#
04/04/12	35	39	20	200#	45#
04/05/12	45	42	33	200#	45#
04/06/12	32	35	22	200#	45#
04/07/12	48	45	32	200#	45#
04/08/12	33	37	22	200#	45#
04/09/12	38	41	33	200#	45#
04/10/12	28	28	20	200#	45#
04/11/12	35	40	23	250#	45#
04/12/12	50	37	30	220#	45#
04/13/12	33	35	23	220#	45#
04/14/12	40	40	28	220#	45#
04/15/12	39	37	27	220#	45#
04/16/12	40	40	28	220#	45#
04/17/12	34	36	28	220#	45#
04/18/12	42	38	24	220#	45#
04/19/12	37	33	27	200#	45#
04/20/12	41	47	28	200#	45#
04/21/12	35	33	25	200#	45#
04/22/12	38	38	27	200#	45#
04/23/12	37	36	28	200#	45#
04/24/12	38	41	28	200#	45#
04/25/12	38	38	27	200#	45#
04/26/12	39	37	28	180#	45#
04/27/12	43	44	25	180#	45#
04/28/12	33	35	23	180#	45#

(Continued)

Table 9.10 (Continued)

Time	Oil (bbl/day)	Gas (MSCF/D)	Water (bbl/day)	Pressure of tbg. (Psi)	Pressure of csg. (psi)
04/29/12	37	36	27	180#	45#
04/30/12	40	38	27	180#	45#
05/01/12	37	37	24	180#	45#
05/02/12	47	49	35	180#	45#
05/03/12	33	37	18	180#	45#
05/04/12	30	28	22	180#	45#
05/05/12	41	39	27	150#	45#
05/06/12	47	44	33	150#	45#
05/07/12	28	32	43	170#	45#
05/08/12	42	40	27	170#	45#
05/09/12	33	32	22	170#	45#
05/10/12	40	35	25	170#	45#
05/11/12	34	38	20	170#	45#
05/12/12	43	41	37	170#	45#
05/13/12	38	38	28	190#	45#
05/14/12	33	41	22	160#	45#
05/15/12	45	38	52	160#	45#
05/16/12	30	30	23	160#	45#
05/17/12	40	35	28	160#	45#
05/18/12	35	39	22	160#	45#
05/19/12	33	36	27	160#	45#
05/20/12	42	38	40	160#	45#
05/21/12	40	37	15	160#	45#
05/22/12	37	37	25	160#	45#
05/23/12	40	42	32	160#	45#
05/24/12	32	32	37	160#	45#
05/25/12	40	42	28	160#	45#
05/26/12	40	34	22	160#	45#
05/27/12	30	33	25	160#	45#
05/28/12	38	42	27	160#	45#
05/29/12	32	33	10	160#	45#
05/30/12	37	37	30	160#	45#
05/31/12	37	33	25	160#	45#
06/01/12	40	45	30	160#	45#
06/02/12	35	36	25	160#	45#
06/03/12	37	104	23	160#	45#
06/04/12	42	970	28	160#	45#
06/05/12	37	35	25	160#	45#

Table 9.10 (Continued)

Time	Oil (bbl/day)	Gas (MSCF/D)	Water (bbl/day)	Pressure of tbg. (Psi)	Pressure of csg. (psi)
06/06/12	33	40	35	160#	45#
06/07/12	40	42	23	150#	45#
06/08/12	35	36	23	150#	45#
06/09/12	33	35	25	150#	45#
06/10/12	35	33	23	150#	45#
06/11/12	35	35	25	150#	45#
06/12/12	35	35	23	150#	45#
06/13/12	34	36	18	150#	45#
06/14/12	40	40	33	150#	45#
06/15/12	18	30	20	150#	45#
06/16/12	50	36	25	150#	45#
06/17/12	37	38	26	150#	45#
06/18/12	35	38	27	110#	45#
06/19/12	43	40	28	110#	45#
06/20/12	32	32	22	110#	45#
06/21/12	37	36	20	110#	45#
06/22/12	32	36	28	110#	45#
06/23/12	37	33	27	110#	45#
06/24/12	37	33	22	110#	45#
06/25/12	33	36	23	110#	45#
06/26/12	38	37	25	110#	45#
06/27/12	38	39	27	110#	45#
06/28/12	70	31	23	110#	45#
06/29/12	43	44	32	110#	45#
06/30/12	25	27	22	110#	45#
07/01/12	40	37	25	100#	45#
07/02/12	35	34	30	100#	45#
07/03/12	32	34	17	100#	45#
07/04/12	37	37	25	100#	45#
07/05/12	33	34	27	80#	45#
07/06/12	40	40	25	80#	45#
07/07/12	35	40	27	80#	45#
07/08/12	33	29	30	80#	45#
07/09/12	35	35	35	80#	45#
07/10/12	36	37	13	80#	45#
07/11/12	37	35	33	80#	50#
07/12/12	35	35	15	80#	50#

(Continued)

Table 9.10 (Continued)

Time	Oil (bbl/day)	Gas (MSCF/D)	Water (bbl/day)	Pressure of tbg. (Psi)	Pressure of csg. (psi)
07/13/12	40	41	32	80#	50#
07/14/12	30	28	23	80#	50#
07/15/12	40	36	18	80#	50#
07/16/12	32	35	25	80#	50#
07/17/12	39	35	23	80#	50#
07/18/12	32	36	25	80#	50#
07/19/12	40	37	25	80#	50#
07/20/12	33	31	23	80#	50#
07/21/12	33	35	23	80#	50#
07/22/12	33	34	27	80#	50#
07/23/12	37	36	23	80#	50#
07/24/12	33	36	25	80#	50#
07/25/12	33	35	25	80#	50#
07/26/12	35	34	23	80#	50#
07/27/12	33	35	25	80#	50#
07/28/12	36	35	28	80#	50#
07/29/12	32	31	22	80#	50#
07/30/12	38	35	27	80#	50#
07/31/12	33	33	23	80#	50#
08/01/12	0	824	0	#	#
08/02/12	30	34	27	80#	50#
08/03/12	40	35	23	80#	50#
08/04/12	28	32	23	80#	50#
08/05/12	32	32	23	80#	50#
08/06/12	38	33	23	80#	50#
08/07/12	30	32	23	90#	50#
08/08/12	28	635	18	90#	50#
08/09/12	38	430	22	90#	50#
08/10/12	32	33	33	90#	50#
08/11/12	35	35	25	90#	50#
08/12/12	39	34	23	90#	50#
08/13/12	35	35	25	90#	50#
08/14/12	32	32	20	90#	50#
08/15/12	32	32	18	90#	50#
08/16/12	37	32	5	90#	50#
08/17/12	33	31	13	90#	50#
08/18/12	32	31	20	90#	50#
08/19/12	35	33	18	90#	50#

Table 9.10 (Continued)

Time	Oil (bbl/day)	Gas (MSCF/D)	Water (bbl/day)	Pressure of tbg. (Psi)	Pressure of csg. (psi)
08/20/12	35	32	33	90#	50#
08/21/12	43	34	17	90#	50#
08/22/12	22	32	14	90#	50#
08/23/12	39	35	15	90#	50#
08/24/12	40	40	30	90#	50#
08/25/12	37	42	25	90#	50#
08/26/12	35	43	27	90#	50#
08/27/12	32	40	20	90#	50#
08/28/12	32	38	18	90#	50#
08/29/12	37	45	28	90#	50#
08/30/12	35	43	23	90#	50#
08/31/12	34	43	27	90#	50#
09/01/12	37	47	18	90#	50#
09/02/12	42	47	32	90#	50#
09/03/12	32	41	23	90#	50#
09/04/12	33	45	25	90#	50#
09/05/12	32	45	23	90#	50#
09/06/12	37	40	27	90#	50#
09/07/12	35	43	23	90#	50#
09/08/12	35	40	23	90#	50#
09/09/12	37	41	24	90#	50#
09/10/12	30	40	22	90#	50#
09/11/12	35	43	23	90#	50#
09/12/12	40	42	27	80#	45#
09/13/12	35	42	27	80#	50#
09/14/12	32	44	23	80#	90#
09/15/12	35	39	21	80#	90#
09/16/12	40	40	27	80#	90#
09/17/12	42	42	5	80#	90#
09/18/12	20	39	23	80#	90#
09/19/12	36	22	22	80#	90#
09/20/12	32	40	30	80#	90#
09/21/12	35	43	20	70#	60#
09/22/12	33	42	25	90#	50#
09/23/12	35	38	22	80#	50#
09/24/12	32	39	38	80#	50#
09/25/12	39	42	22	80#	50#

(*Continued*)

Table 9.10 (Continued)

Time	Oil (bbl/day)	Gas (MSCF/D)	Water (bbl/day)	Pressure of tbg. (Psi)	Pressure of csg. (psi)
09/26/12	33	43	26	80#	50#
09/27/12	37	46	23	80#	50#
09/28/12	35	44	38	80#	50#
09/29/12	40	44	27	80#	50#
09/30/12	33	41	30	80#	50#
10/01/12	33	22	13	100#	70#
10/02/12	30	0	20	90#	70#
10/03/12	35	54	25	90#	70#
10/04/12	33	10	25	90#	50#
10/05/12	28	44	27	90#	50#
10/06/12	38	44	18	90#	55#
10/07/12	33	36	23	90#	55#
10/08/12	32	41	28	90#	55#
10/09/12	5	11	2	90#	55#
10/11/12	50	54	39	90#	55#
10/12/12	48	46	25	90#	55#
10/13/12	38	39	25	90#	55#
10/14/12	22	30	13	100#	55#
10/15/12	27	25	12	80#	55#
10/16/12	29	39	33	80#	55#
10/17/12	50	44	25	80#	55#
10/18/12	50	30	25	80#	55#
10/19/12	27	18	30	80#	55#
10/20/12	30	22	22	80#	55#
10/21/12	40	42	27	90#	60#
10/22/12	30	27	22	90#	60#
10/23/12	37	38	27	90#	60#
10/24/12	34	42	29	90#	60#
10/25/12	33	36	20	90#	60#
10/26/12	30	10	22	90#	60#
10/27/12	33	41	23	90#	60#
10/28/12	40	37	23	100#	60#
10/29/12	14	43	25	90#	60#
10/30/12	37	41	25	90#	60#
10/31/12	37	41	22	90#	60#
11/01/12	35	39	32	90#	60#
11/02/12	30	40	23	90#	60#
11/03/12	37	40	20	90#	60#

Table 9.10 (Continued)

Time	Oil (bbl/day)	Gas (MSCF/D)	Water (bbl/day)	Pressure of tbg. (Psi)	Pressure of csg. (psi)
11/04/12	35	36	18	90#	60#
11/05/12	33	40	27	90#	60#
11/06/12	37	40	20	90#	60#
11/07/12	28	33	23	80#	60#
11/08/12	37	40	24	80#	60#
11/09/12	32	41	22	80#	60#
11/10/12	37	39	25	80#	60#
11/11/12	23	37	20	80#	60#
11/12/12	45	39	28	80#	60#
11/13/12	33	35	22	80#	60#
11/14/12	38	39	23	80#	60#
11/15/12	30	39	22	80#	60#
11/16/12	33	41	21	80#	60#
11/17/12	37	40	23	80#	60#
11/18/12	32	39	23	80#	60#
11/19/12	37	42	25	80#	60#
11/20/12	33	37	20	80#	55#
11/21/12	37	47	30	80#	55#
11/22/12	32	36	20	80#	55#
11/23/12	24	38	18	80#	50#
11/24/12	35	41	22	80#	50#
11/25/12	37	41	23	80#	50#
11/26/12	35	41	23	80#	50#
11/27/12	37	40	25	80#	50#
11/28/12	37	42	25	80#	50#
11/29/12	35	40	18	80#	50#
11/30/12	30	36	23	80#	50#
12/01/12	38	41	13	80#	50#
12/02/12	29	37	20	80#	50#
12/03/12	33	37	25	80#	50#
12/04/12	33	38	17	80#	50#
12/05/12	30	32	20	80#	50#
12/06/12	37	44	25	80#	60#
12/07/12	38	37	22	80#	60#
12/08/12	35	40	25	80#	60#
12/09/12	32	34	22	80#	60#
12/10/12	33	36	18	80#	60#

(*Continued*)

Table 9.10 (Continued)

Time	Oil (bbl/day)	Gas (MSCF/D)	Water (bbl/day)	Pressure of tbg. (Psi)	Pressure of csg. (psi)
12/11/12	40	38	23	80#	60#
12/12/12	32	41	14	80#	60#
12/13/12	28	38	33	80#	60#
12/14/12	26	37	23	80#	60#
12/15/12	38	45	22	80#	45#
12/16/12	33	42	23	80#	45#
12/17/12	28	32	17	80#	45#
12/18/12	45	47	28	80#	45#
12/19/12	30	38	22	80#	45#
12/20/12	40	41	22	80#	45#
12/21/12	24	35	18	80#	45#
12/22/12	40	39	13	80#	45#
12/23/12	23	38	30	80#	45#
12/24/12	33	37	17	80#	45#
12/25/12	38	33	22	80#	45#
12/26/12	30	35	23	80#	45#
12/27/12	35	39	23	80#	45#
12/28/12	36	35	20	80#	45#
12/29/12	35	36	23	80#	45#
12/30/12	30	39	22	80#	45#
12/31/12	32	36	22	80#	45#
01/01/13	37	35	23	80#	45#
01/02/13	32	22	17	80#	45#
01/03/13	30	44	15	80#	45#
01/04/13	37	40	25	80#	45#
01/05/13	47	37	28	80#	45#
01/06/13	35	31	22	80#	45#
01/07/13	38	37	18	80#	45#
01/08/13	32	37	20	80#	45#
01/09/13	32	34	18	80#	45#
01/10/13	33	38	18	80#	45#
01/11/13	34	37	33	90#	30#
01/12/13	35	36	23	90#	30#
01/13/13	30	28	20	80#	40#
01/14/13	40	36	23	80#	40#
01/15/13	27	30	18	80#	40#
01/16/13	37	5	18	80#	40#
01/17/13	29	0	18	80#	40#

Table 9.10 (Continued)

Time	Oil (bbl/day)	Gas (MSCF/D)	Water (bbl/day)	Pressure of tbg. (Psi)	Pressure of csg. (psi)
01/18/13	33	94	22	80#	40#
01/19/13	33	916	27	80#	40#
01/20/13	38	30	22	80#	40#
01/21/13	27	28	22	80#	50#
01/22/13	42	34	27	80#	50#
01/23/13	32	34	22	80#	50#
01/24/13	33	35	22	80#	50#
01/25/13	30	34	18	80#	50#
01/26/13	32	32	23	80#	50#
01/27/13	32	37	20	80#	50#
01/28/13	32	33	18	80#	70#
01/29/13	33	37	25	80#	70#
01/30/13	32	34	20	80#	70#
01/31/13	34	39	22	80#	70#
02/01/13	32	47	30	80#	70#
02/02/13	35	34	18	80#	70#
02/03/13	32	36	20	80#	70#
02/04/13	33	37	25	80#	70#
02/05/13	30	35	21	80#	70#
02/06/13	33	34	22	80#	70#
02/07/13	36	38	30	80#	70#
02/08/13	27	32	12	80#	50#
02/09/13	40	44	25	80#	50#
02/10/13	25	29	18	80#	50#
02/11/13	37	39	22	80#	50#
02/12/13	32	36	5	80#	50#
02/13/13	28	33	1	80#	50#
02/14/13	38	38	32	80#	50#
02/15/13	33	36	20	80#	50#
02/16/13	28	35	20	80#	50#
02/17/13	32	37	23	80#	50#
02/18/13	32	41	22	80#	20#
02/19/13	30	7	20	90#	30#
02/20/13	30	19	20	90#	30#
02/21/13	35	71	23	90#	30#
02/22/13	30	25	13	90#	30#
02/23/13	37	16	22	90#	60#

(Continued)

Table 9.10 (Continued)

Time	Oil (bbl/day)	Gas (MSCF/D)	Water (bbl/day)	Pressure of tbg. (Psi)	Pressure of csg. (psi)
02/24/13	34	3	22	90#	60#
02/25/13	28	15	20	90#	60#
02/26/13	33	994	23	90#	60#
02/27/13	28	1	17	90#	40#
02/28/13	37	6	22	80#	40#
03/02/13	33	1	20	80#	40#
03/03/13	30	1	17	80#	60#
03/04/13	35	15	23	80#	60#
03/05/13	33	30	12	80#	60#
03/06/13	30	33	20	80#	60#
03/07/13	30	32	20	80#	60#
03/08/13	33	34	18	80#	80#
03/09/13	32	35	20	80#	70#
03/10/13	30	33	22	80#	80#
03/11/13	32	39	27	80#	80#
03/12/13	38	31	20	80#	80#
03/13/13	23	34	13	80#	80#
03/14/13	32	30	7	80#	80#
03/15/13	35	30	22	80#	80#
03/16/13	25	11	20	80#	80#
03/17/13	38	18	20	80#	80#
03/18/13	32	26	20	80#	80#
03/19/13	32	32	22	90#	40#
03/20/13	51	30	17	90#	40#
03/21/13	22	26	10	90#	40#
03/22/13	27	25	32	90#	40#
03/23/13	33	5	33	90#	40#
03/24/13	30	28	10	90#	40#
03/25/13	30	34	24	90#	40#
03/26/13	33	35	18	90#	40#
03/27/13	33	40	25	90#	110#
03/28/13	32	31	18	80#	60#
03/29/13	32	33	20	80#	60#
03/30/13	33	30	18	80#	60#
03/31/13	25	29	23	80#	60#
04/01/13	33	36	20	80#	60#
04/02/13	0	0	0	80#	60#
04/03/13	0	0	0	80#	60#

Table 9.10 (Continued)

Time	Oil (bbl/day)	Gas (MSCF/D)	Water (bbl/day)	Pressure of tbg. (Psi)	Pressure of csg. (psi)
04/04/13	5	9	2	80#	60#
04/05/13	60	64	45	80#	60#
04/06/13	52	46	23	80#	60#
04/07/13	52	41	25	90#	65#
04/08/13	43	41	25	90#	65#
04/09/13	42	38	20	90#	65#
04/10/13	38	32	20	90#	65#
04/11/13	42	32	20	90#	65#
04/12/13	35	33	13	90#	65#
04/13/13	28	31	30	90#	65#
04/14/13	30	32	23	90#	65#
04/15/13	37	34	20	90#	65#
04/16/13	35	33	23	80#	70#
04/17/13	30	30	20	80#	70#
04/18/13	39	35	25	80#	70#
04/19/13	20	20	13	80#	70#
04/20/13	28	29	15	80#	70#
04/21/13	38	27	27	80#	70#
04/22/13	30	28	20	80#	70#
04/23/13	32	28	20	80#	70#
04/24/13	25	24	20	80#	70#
04/25/13	30	26	21	80#	70#
04/26/13	25	23	20	80#	70#
04/27/13	35	34	15	80#	70#
04/28/13	42	28	13	80#	70#
04/29/13	31	28	13	80#	70#
04/30/13	34	24	13	80#	70#
05/01/13	23	27	20	80#	70#
05/02/13	33	30	25	80#	70#
05/03/13	27	27	17	90#	60#
05/04/13	33	31	22	80#	65#
05/05/13	32	29	22	80#	70#
05/06/13	35	30	18	80#	70#
05/07/13	50	32	25	80#	70#
08/02/13	350	240	207	90#	370#
08/03/13	315	217	198	90#	340#
08/04/13	292	199	185	90#	340#

(Continued)

Table 9.10 (Continued)

Time	Oil (bbl/day)	Gas (MSCF/D)	Water (bbl/day)	Pressure of tbg. (Psi)	Pressure of csg. (psi)
08/05/13	289	197	223	90#	340#
08/06/13	277	192	188	90#	340#
08/07/13	256	169	201	90#	340#
08/08/13	227	153	220	90#	340#
08/09/13	247	161	5	90#	340#

Additional Reading

Lee, J., & Wattenbarger, R. (1996). *Gas reservoir engineering*. Society of Petroleum Engineers.

Blasingame, T. (2015). *Rreservoir engineering aspects of unconventional reservoirs*. Society of Petroleum Evaluation Engineers.

Satter, A., & Iqbal, G. (2016). *Reservoir engineering the fundamentals, simulation, and management of conventional and unconventional recoveries*. Boston: Gulf Professional Publishing.

References

Abu-Khamsin, S. (2012). *Natural gas engineering: Decline curve analysis*. King Fahd University of Petroleum and Minerals.

Agarwal, R., Gardner, D., Kleinsteiber, S., & Fussell, D. (1999). Analyzing well production data using combined-type-curve and decline-curve analysis concepts. *SPE Reservoir Evaluation & Engineering*. doi:10.2118/57916-PA

Allen, R. (1931). Control of California oil curtailment. *Transactions of the AIME*. doi:10.2118/931047-G

Anderson, D. (2010). *RTA theory training course*. IHS Markit.

Anderson, D. (2012). *RTA of unconventional plays*. Fekete.

Anderson, D. (2015). *Practical pointers and rules of thumb for RTA of unconventional plays*. HIS Markit Energy Expert. https://ihsmarkit.com/research-analysis/rpe-practical-pointers-and-rules-of-thumb-for-rta-of-unconventional-plays.html

Anderson, D., & Mattar, L. (2004). *Practical diagnostics using production data and flowing pressures*. SPE Annual Technical Conference and Exhibition, Houston, Society of Petroleum Engineers. doi:10.2118/89939.MS

Anderson, D., Turco, F., Virues, C., & Chin, A. (2013). *Application of rate transient analysis workflow in unconventional reservoirs: Horn River shale gas case study*. SPE Unconventional Resources Conference and Exhibition-Asia Pacific. Brisbane: Society of Petroleum Engineers. doi:10.2118/167042-MS

Apte, S., & Lee, J. (2017). *Elliptical flow regimes in horizontal wells with multiple hydraulic fractures*. SPE Hydraulic Fracturing Technology Conference and Exhibition, The Woodlands, Society of Petroleum Engineers. doi:10.2118/184856-MS

Arps, J. (1945). Analysis of decline curves. *Transactions of the AIME*, *160*(1). doi:10.2118/945228-G

Arps, J. (1956). *Estimation of primary oil reserves*. Petroleum Conference, Economics and Valuation. Dallas, Texas. 29–30 March.

Artus, V., & Houzé, O. (2018). *A physical decline curve for fractured horizontal wells*. SPE/AAPG/SEG Unconventional Resources Technology Conference, Houston, Unconventional

Resources Technology Conference. doi:10.15530/URTEC-2018-2856750

Ataei, A., Motaei, E., Yazdi, M., Masoudi, R., & Bashir, A. (2018). *rate transient analysis RTA and its application for well connectivity analysis: An integrated production driven reservoir characterization and a case study*. SPE Asia Pacific Oil and Gas Conference and Exhibition, Brisbane, Society of Petroleum Engineers. doi:10.2118/192046-MS

Ayala, L., & Morgan, E. (2017). Natural gas production engineering. In M. Riazi, *Exploration and production of petroleum and natural gas*. ASTM International. doi:10.1520/MNL73

Bax, M. (2017). *A study of rate transient analysis: Are you making the most of your data?* Retrieved from IHS Markit: https://ihsmarkit.com/research-analysis/a-study-of-rate-transient-analysis.html

Bello, R., & Wattenbarger, R. (2010). *Multistage hydraulically fractured horizontal shale gas well rate transient analysis*. North Africa Technical Conference and Exhibition. Cairo: Society of Petroleum Engineers. doi:10.2118/126754-MS

Blasingame, T. (2015). R*reservoir engineering aspects of unconventional reservoirs*. Society of Petroleum Evaluation Engineers.

Blasingame, T., & Lee, W. (1986). *Variable-rate reservoir limits testing*. Permian Basin Oil and Gas Recovery Conference, Society of Petroleum Engineers. doi:10.2118/15028-MS

Blasingame, T., & Lee, W. (1988). *The variable-rate reservoir limits testing of gas wells*. SPE Gas Technology Symposium, Society of Petroleum Engineers. doi:10.2118/17708-MS

Blasingame, T., & Lee, W. (1989). Pressure-buildup test analysis-variable-rate case: A new approach. *SPE Formation Evaluation*. doi:10.2118/17052-PA

Blasingame, T., McCray, T., & Lee, J. (1991). *Decline curve analysis for variable pressure drop/variable flowrate systems*. SPE Gas Technology Symposium, Houston, Society of Petroleum Engineers. doi:10.2118/21513-MS

Can, B., & Kabir, S. (2012). Probabilistic production forecasting for unconventional reservoirs with stretched exponential production decline model. SPE Reservoir Evaluation & Engineering. doi:SPE-143666-PA

Carter, R. (1985). Type curves for finite radial and linear gas-flow systems: Constant-terminal-pressure case. *Society of Petroleum Engineers Journal*. doi:10.2118/12917-PA

Clark, A., Lake, L., & Patzek, T. (2011). *Production forecasting with logistic growth models*. SPE Annual Technical Conference and Exhibition, Denver, Society of Petroleum Engineers. doi:10.2118/144790-MS

Clarkson, C. (2013). Production data analysis of unconventional gas wells: Review of theory and best practices. *International Journal of Coal Geology*. doi:10.1016/j.coal.2013.01.002

Clarkson, C., & Beierle, J. (2010). *Integration of microseismic and other post-fracture surveillance with production analysis: A tight gas study*. SPE Unconventional Gas Conference, Pittsburgh, Society of Petroleum Engineers. doi:10.2118/131786-MS

Clarkson, C., & Pedersen, P. (2010). *Tight oil production analysis: Adaptation of existing rate-transient analysis techniques*. Canadian Unconventional Resources and International Petroleum Conference, Calgary, Society of Petroleum Engineers. doi:10.2118/137352-MS

Collins, P. (2016). *Decline curve analysis for unconventional reservoir systems: Variable pressure drop case*. Texas A & M University.

Currie, S., Ilk, D., & Blasingame, T. (2010). *Continuous estimation of ultimate recovery*. SPE Unconventional Gas Conference, Pittsburgh, Society of Petroleum Engineers. doi:10.2118/132352-MS

Cutler, W. (1924). *Estimation of underground oil reserves by oil-well production curves*. U.S. Bureau of Mines.

Doublet, L., Pande, P., McCollum, T., & Blasingame, T. (1994). *Decline curve analysis using type curves–analysis of oil well production data using material balance time: Application to field cases*. International Petroleum Conference and Exhibition of Mexico. Veracruz: Society of Petroleum Engineers. doi:10.2118/28688-MS

Duong, A. (2011). *Rate-decline analysis for fracture-dominated shale reservoirs*. SPE Reservoir Evaluation & Engineering. doi:10.2118/137748-PA

Fekete. (2014). *Traditional decline analysis theory*. Retrieved from Fekete: http://www.fekete.com

Fekete, P., Ekpedekumo, R., Dosunmu, A., Odagme, S., & Bovwe, E. (2016). *Evaluation of the methodologies of analyzing production and pressure data of tight gas reservoir*. SPE Nigeria Annual International Conference and Exhibition, Lagos, Society of Petroleum Engineers. doi:10.2118/184263-MS

Fetkovich, M. (1980). Decline curve analysis using type curves. *Journal of Petroleum Technology*. doi:10.2118/4629.PA

Fetkovich, M., & Vienot, M. (1984). Rate normalization of buildup pressure by using afterflow data. *Journal of Petroleum Technology*. doi:10.2118/12179.PA

Fetkovich, M., Bradley, M., Works, A., & Thrasher, T. (1990). Depletion performance of layered reservoirs without crossflow. *SPE Formation Evaluation*. doi:10.2118/18266-PA

Fetkovich, M., Fetkovich, E., & Fetkovich, M. (1996). Useful concepts for decline curve forecasting, reserve estimation, and analysis. *SPE Reservoir Engineering*. doi:10.2118/28628-PA

Fraim, M., & Wattenbarger, R. (1987). Gas reservoir decline-curve analysis using type curves with real gas pseudopressure and normalized time. *SPE Formation Evaluation*. doi:10.2118/14238-PA

Fulford, D. (2018). *A model-based diagnostic workflow for time-rate performance of unconventional wells*. SPE/AAPG/SEG Unconventional Resources Technology Conference, Houston, Unconventional Resources Technology Conference. doi:10.15530/URTEC-2018-290303

Gentry, R., & McCray, A. (1978). The effect of reservoir and fluid properties on production decline curves. *Journal of Petroleum Technology*. doi:10.2118/6341-PA

Guo, B., Liu, X., & Tan, X. (2017). *Petroleum production engineering (2nd ed.)*. Cambridge: Elsevier. doi:ISBN: 978-0-12-809374-0

Ilk, D., & Blasingame, T. (2013). *Decline curve analysis for unconventional reservoir systems – variable pressure drop case*. SPE Unconventional Resources Conference Canada, Calgary: Society of Petroleum Engineers. doi:10.2118/167253-MS

Ilk, D., Jenkins, C., & Blasingame, T. (2011). *Production analysis in unconventional reservoirs: Diagnostics, challenges, and methodologies*. North American Unconventional Gas Conference and Exhibition, The Woodlands, Society of Petroleum Engineers. doi:10.2118/144376-MS

Ilk, D., Perego, A., Rushing, J., & Blasingame, T. (2008). *Integrating multiple production analysis techniques to assess tight gas sand reserves: Defining a new paradigm for industry best practices*. CIPC/SPE Gas Technology Symposium 2008 Joint Conference, Calgary, Society of Petroleum Engineers. doi:10.2118/114947-MS

Ilk, D., Rushing, J., & Blasingame, T. (2009). *Decline-curve analysis for hp/ht gas wells: theory and applications*. SPE Annual Technical Conference and Exhibition, New Orleans, Society of Petroleum Engineers. doi:10.2118/125031-MS

Ilk, D., Rushing, J., Perego, A., & Blasingame, T. (2008). *Exponential vs. hyperbolic decline in tight gas sands: understanding the origin and implications for reserve estimates using arps' decline curves*. SPE Annual Technical Conference and Exhibition, Denver, Society of Petroleum Engineers. doi:10.2118/116731-MS

Johnson, R., & Bollens, A. (1927). The loss ratio method of extrapolating oil well decline curves. *Transactions of the AIME*. doi:10.2118/927771-G

Kanfar, M., & Wattenbarger, R. (2012). *Comparison of empirical decline curve methods for shale wells*. SPE Canadian Unconventional Resources Conference, Calgary, Society of Petroleum Engineers. doi:10.2118/162648-MS

KGS. (2018). *Master list of oil and gas wells in Kansas*. Retrieved from Kansas Geological Survey: http://www.kgs.ku.edu/Magellan/Qualified/index.html

Kisslinger, C. (1993). The stretched exponential function as an alternative model for aftershock decay rate. *Journal of Geophysical Research: Solid Earth*. doi:10.1029/92JB01852

Lee, J., & Wattenbarger, R. (1996). *Gas reservoir engineering*. Society of Petroleum Engineers. doi: ISBN: 978-1-55563-073-7

Lewis, J., & Beal, C. (1918). Some new methods for estimating the future production of oil wells. *Transactions of the AIME*. doi:10.2118/918492-G

Li, H., Luo, H., Zhang, J., & Wang, J. (2018). Rate-transient analysis in multifractured horizontal wells of tight oil reservoir. *SPE Production & Operations*. doi:10.2118/189992-PA

Mattar, L., & Anderson, D. (2003). *A systematic and comprehensive methodology for advanced analysis of production data*. SPE Annual Technical Conference and Exhibition, Denver, Society of Petroleum Engineers.

McNeil, R., Jeje, O., & Renaud, A. (2009). *Application of the power law loss-ratio method of decline analysis*. Canadian International Petroleum Conference, Calgary, Petroleum Society of Canada. doi:10.2118/2009.159

Okouma, V., Ilk, D., Blasingame, T., Symmons, D., & Hosseinpour-Zonoozi, N. (2012). *Practical considerations for decline curve analysis in unconventional reservoirs - application of recently developed rate-time relations*. SPE Hydrocarbon Economics and Evaluation Symposium, Calgary, Society of Petroleum Engineers. doi: 10.2118/162910-MS

Palacio, J., & Blasingame, T. (1993). *Decline-curve analysis with type curves: Analysis of gas well production data*. Low Permeability Reservoirs Symposium. doi:10.2118/25909.MS

Paryani, M., Ahmadi, M., Awoleke, O., & Hanks, C. (2016). *Using improved decline curve models for production forecasts in unconventional reservoirs*. SPE Eastern Regional Meeting, Canton, Society of Petroleum Engineers. doi:10.2118/184070-MS

Robertson, S. (1988). *Generalized hyperbolic equation*. Society of Petroleum Engineers. doi: SPE-18731-MS

Rushing, J., Perego, A., Sullivan, R., & Blasingame, T. (2007). *Estimating reserves in tight gas sands at HP/HT reservoir conditions: Use and misuse of an Arps decline curve methodology*. SPE Annual Technical Conference and Exhibition, Anaheim, Society of Petroleum Engineers. doi:10.2118/109625-MS

Satter, A., & Iqbal, G. (2016). *Reservoir engineering the fundamentals, simulation, and management of conventional and unconventional recoveries*. Boston: Gulf Professional Publishing. doi:B978-0-12-800219.3.00001

Seshadri, J., & Mattar, L. (2010). *Comparison of power law and modified hyperbolic decline methods*. Canadian Unconventional Resources and International Petroleum Conference, Calgary, Society of Petroleum Engineers. doi:10.2118/137320-MS

SPEE. (2002). *Definition of decline curve parameters*. SPEE Recommended Evaluation Practice.

Spencer, R., & Coulombe, M. (1966). Quantitation of hepatic growth and regeneration. *Growth*.

Tan, L., Zuo, L., & Wang, B. (2018). Methods of decline curve analysis for shale gas reservoirs. *Energies*. doi:10.3390/en11030552

Valko, P. (2009). *Assigning value to stimulation in the Barnett shale: A simultaneous analysis of 7000 plus production hystories and well completion records*. SPE Hydraulic Fracturing Technology Conference, The Woodlands, Society of Petroleum Engineers. doi:10.2118/119369.MS

Valko, P., & Lee, J. (2010). *A better way to forecast production from unconventional gas wells*. SPE Annual Technical Conference and Exhibition, Florence, Society of Petroleum Engineers. doi:10.2118/134231-MS

Van Everdingen, A., & Hurst, W. (1949). The application of the Laplace transformation to flow problems in reservoirs. *Journal of Petroleum Technology*. doi:10.2118/949305-G

Wattenbarger, R., El-Banbi, A., Villegas, M., & Maggard, J. (1998). *Production analysis of linear flow into fractured tight gas wells*. SPE Rocky Mountain Regional/Low-Permeability Reservoirs Symposium. Society of Petroleum Engineers. doi:10.2118/39931-MS

Zhang, H., Cocco, M., Rietz, D., Cagle, A., & Lee, J. (2015). *An empirical extended exponential decline curve for shale reservoirs*. SPE Annual Technical Conference and Exhibition, Houston, Society of Petroleum Engineers. doi:10.2118/175016-MS

Zuo, L., Yu, W., & Wu, K. (2016). A fractional decline curve analysis model for shale gas reservoirs. *International Journal of Coal Geology*. doi:10.1016/j.coal.2016.07.006

10

Petroleum Economics of Unconventional Shale Reservoirs

Mustafa M. Alhubail[1], *Hajar Aghababa*[2], *and Reza Barati Ghahfarokhi*[1]

[1] *The University of Kansas, Lawrence, Kansas, USA*
[2] *PetroEconomic Solutions LLC, Lawrence, Kansas, USA*

10.1 Background

Economic growth is historically correlated to energy prices both nationally and internationally. This is specifically valid for countries that are involved major producers of oil and gas. The world's gross domestic product (GDP), a monetary measure of the final goods and services produced in one year, is typically directly correlated to the oil price in addition to fuel consumption (Fig. 10.1) (Energy Information Administration, 2019). For hydrocarbon-producing countries, oil and gas projects are a source of income in addition to a source of energy.

Depending on their method of managing natural resources, concessionary versus contractual fiscal regimes (Fig. 10.2), governments manage their natural resources differently. The royalty/tax systems are more typical in countries where private ownership of hydrocarbons is legalized. In this system, the operating company owns the hydrocarbon at the wellhead (Karasalihović-Sedlar et al., 2017). Operating companies will have to buy the right to extract and sell resources and pay taxes and royalties despite their profit-making status (Mazeel, 2010). In countries with state-owned oil companies, contractual systems such as production share agreements or service agreements are common (Karasalihović-Sedlar et al., 2017). The hydrocarbon ownership belongs to the government under these sets of contracts. The contractor may own a part of the product and receive it at a location other than the wellhead (Karasalihović-Sedlar et al., 2017). Despite their fiscal regime system, the contractor/operating company works based on an economic model that is based on the calculation of the cash flow and net present value.

The objective of this chapter is to review the commonly known engineering economics concepts based on unconventional oil and gas development examples and present a set of real-world examples based on case studies from major U.S. shale plays.

10.2 Effect of Shale Oil/Gas Developments on Economics and Energy Security

Historical fluctuations in oil and gas prices are correlated with major historical events as hydrocarbon has been a vital source of energy for the whole world and different consuming countries have always been trying to make sure of their secure access to this geopolitical commodity in long term. Figure 10.4 presents a

Unconventional Hydrocarbon Resources: Techniques for Reservoir Engineering Analysis, Advanced Textbook Series, First Edition. Edited by Reza Barati Ghahfarokhi and Mustafa M. Alhubail.

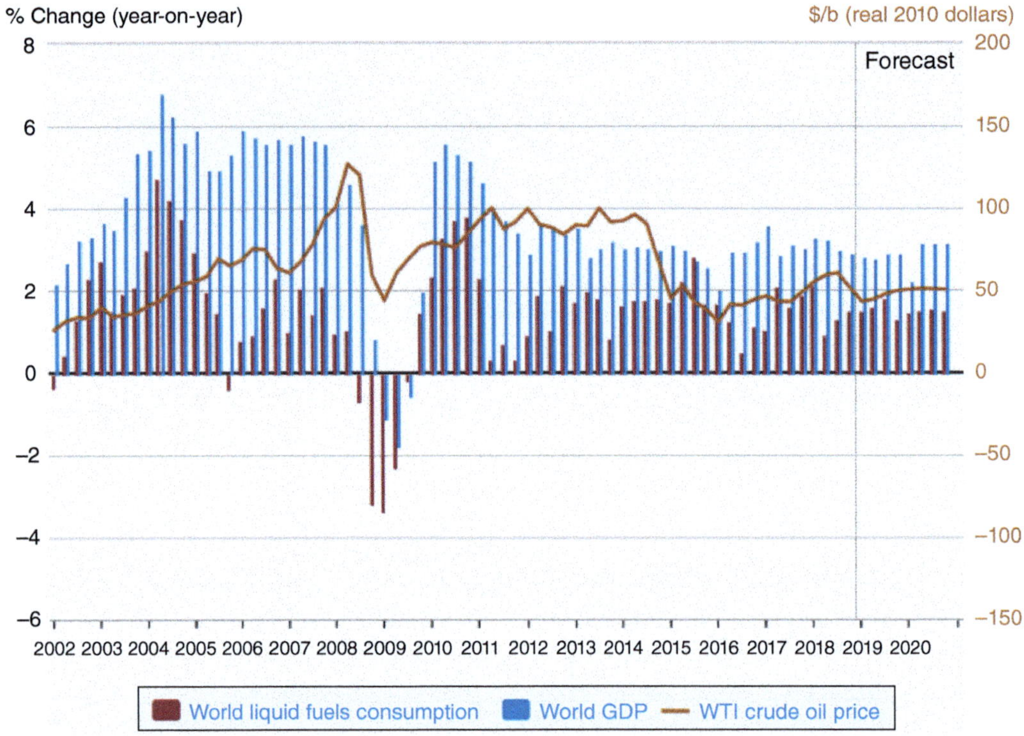

Figure 10.1 World GDP, WTI oil price, and fuel consumption (from EIA, Thomson Reuters, Oxford Economics; Energy Information Administration, 2019).

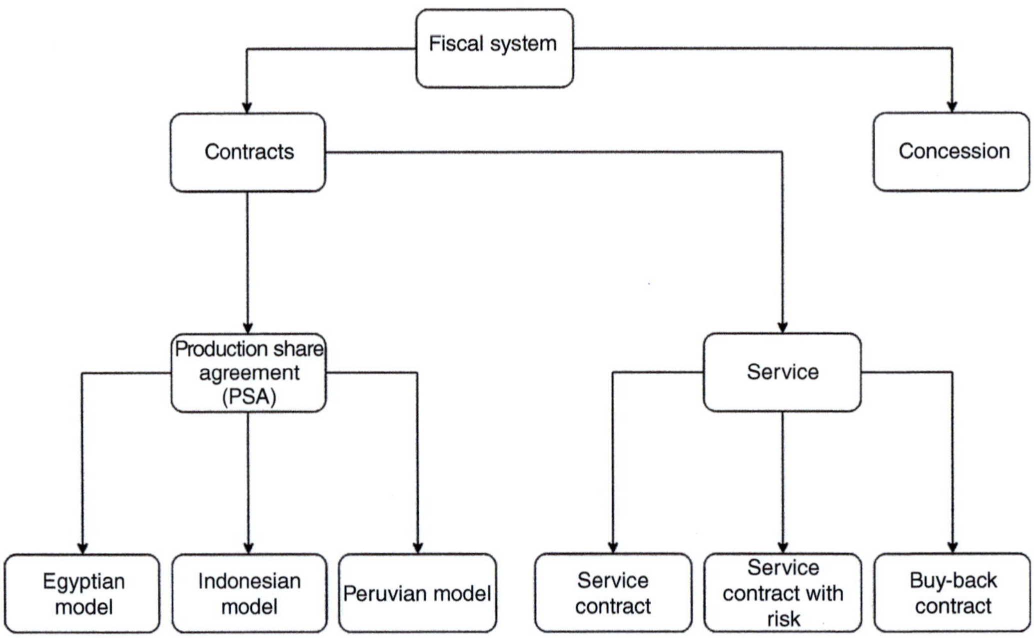

Figure 10.2 Different fiscal regime systems (Karasalihović-Sedlar et al., 2017).

close correlation between the timeline of some major historical events affecting the oil price. Among the causes of major peaks in oil price, the majority of the historical peaks have been caused by an event resulting in uncertainty in continuation of supply, while the recent peaks after mid-2000 have been mainly caused by an increase in demand. This increase in demand is mainly due to the increase in consumption of more fossil fuels in China, India, and other Asian countries. Both supply and demand of crude oil have reached their historically high levels of 100.99 and 100.54 million barrels during the third quarter of 2018 and are forecasted to continue to increase during 2019 (OGJ Editors, 2018).

The conventional oil production reached its peak in the United States in the early 1970s and started to decline sharply in the late 1980s (Figs. 10.3 and 10.4). In the mid 2000s when conventional oil reached its lowest production since the early 1950s, advanced technologies of horizontal drilling and hydraulic fracturing made the United States capable of tapping into previously unproducible resources of shale oil. These technologies had already made the United States independent in terms of natural gas production in the early 2000s and were further modified and used in shale oil development projects. This historically high volume of oil and gas production has made the United States less dependent on the Organization of the Petroleum Exporting Countries (OPEC) countries in terms of energy security and supply. This has strong geopolitical impact in the years to come.

At the state and county level, it has been reported that income, employment, and economic development indexes have shown improvement during the development of unconventional resources. Mineral rights royalty payments, increased job opportunities, increased sales for local businesses, and possible wage growth as a result of increase in the demand for labor are reportedly the various forms of such impacts (Krupnick & Echarte, 2017).

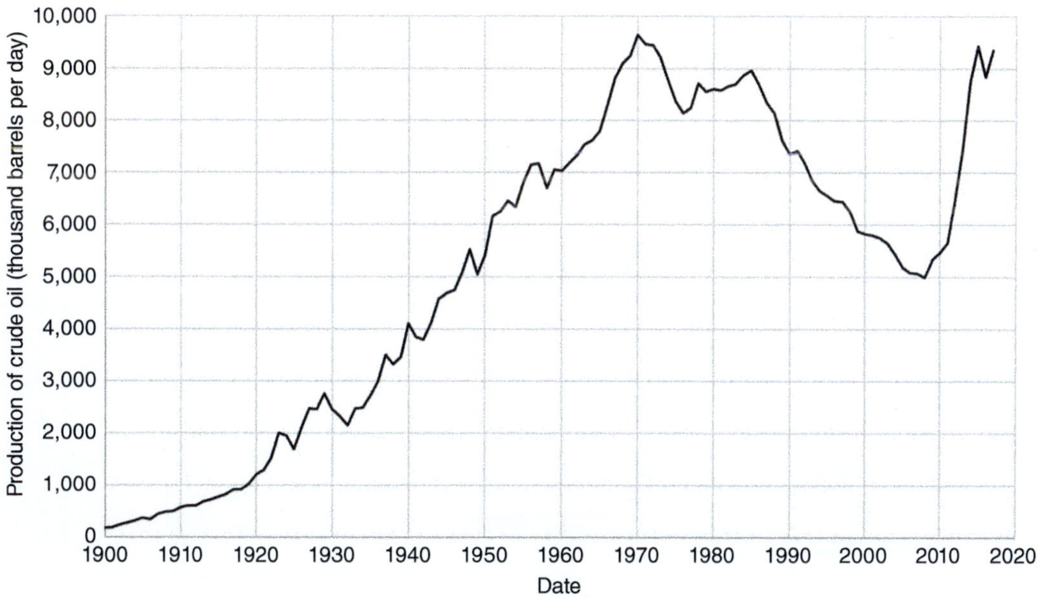

Figure 10.3 Historical U.S. oil production (Energy Information Administration, 2017).

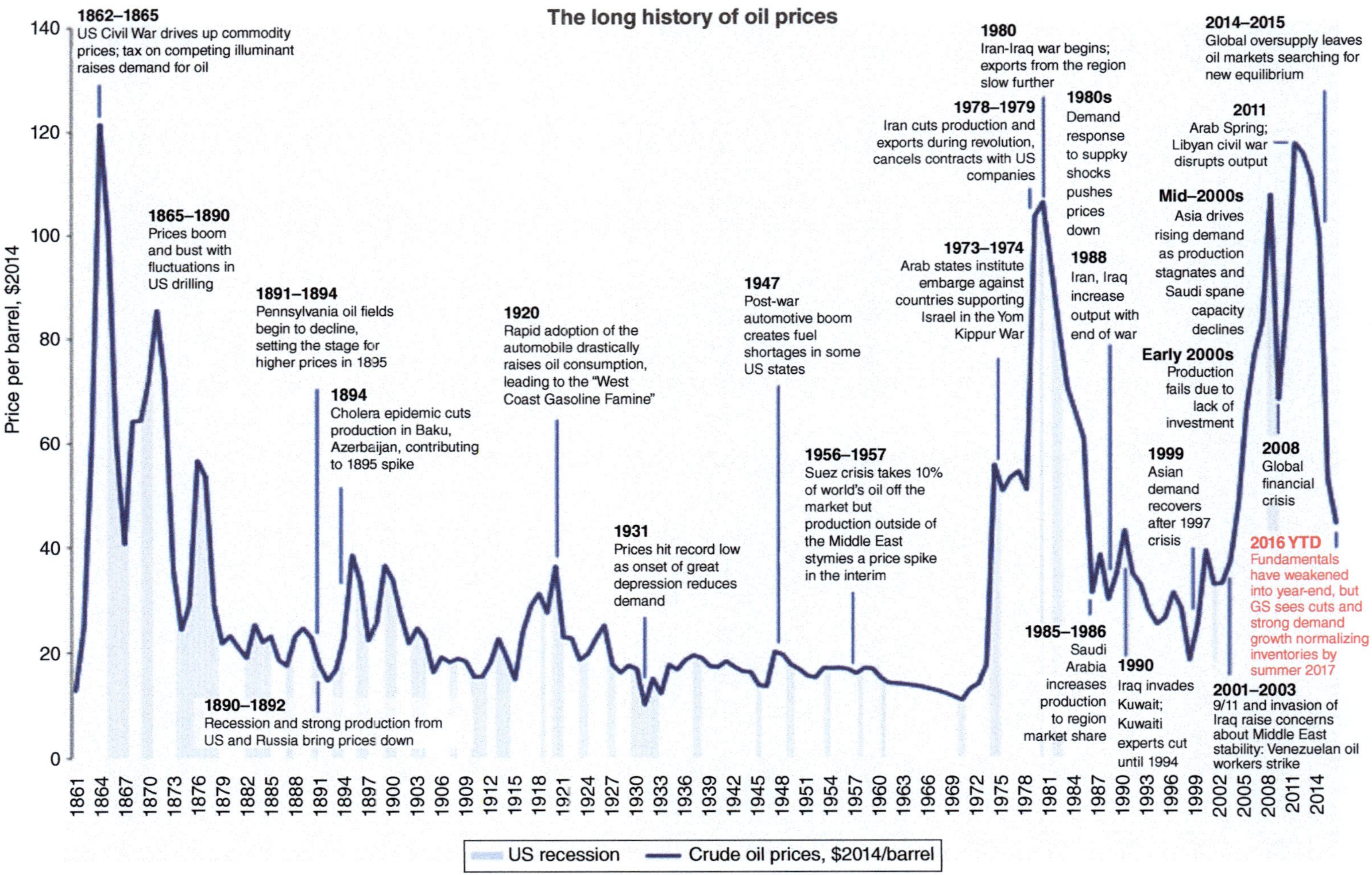

An earlier version of this chart appeared on pg. 16 of top of mind issue #52: OPEC and oil opportunities.
Note: 2016 price shown is YTD average as of Dec. 19. 2016.
Source of data: BP, NBER/Federal reserve bank of St. Louis, haver analytics.
Source for annotations: ©James Hamilton, "Historical oil shocks" University of California, San Diego, February 2011; various news source; Goldman sachs global investment research.

Figure 10.4 A timeline of oil prices and historical events (Holodny, 2016). Note that crude oil prices are reported in 2014 dollar value (with permission from Goldman Sachs).

10.3 Fundamentals of Petroleum Economics

Studying the financial and economic concepts from the petroleum engineering point of view is important as they play a decisive role in the industrial decision-making process. In general, engineers are efficient and knowledgeable when it comes to design and solving issues; thus, they are usually involved in the decision-making process. However, when it comes to the financial aspects, engineers do not contribute much even though they are, most likely, the ones who are familiar with the projects' requirements, and they are also capable of making the best judgments during operation. Yet, they are usually not consulted when it comes down to the financial decisions (Park et al., 2013). This occurs often because the engineers are unfamiliar with the economic concepts. Thus, in this section, the fundamentals of the petroleum economics will be discussed with the purpose of analyzing the economic aspects involved in the development of unconventional reservoirs.

10.3.1 Business Expenditures

Business expenditures can be divided into three parts as shown in Figure 10.5: (1) capital expenditures (CAPEX), (2) operating expenses (OPEX), and (3) government take. CAPEX, which are the front-end costs, are the expenditures that occur at the beginning of a project. In the upstream petroleum industry (i.e., E&P), CAPEX typically includes the cost of drilling wells and sidetracks, installing offshore platforms, buying new assets, and the costs of the field-site facilities (e.g., pipelines, separation units, compressors, etc.). CAPEX must be estimated accurately since underestimating such costs yields to disappointing profit or even loss. However, overestimating the costs could cause rejection of projects and that means missing potential opportunities for investment (Park et al., 2013).

Operating expenses, OPEX, which are the expenditures that are necessary for the day-to-day operations, are divided into two categories: (1) fixed operating costs and (2) variable operating costs. The fixed operating costs are subdivided into direct operating costs and indirect

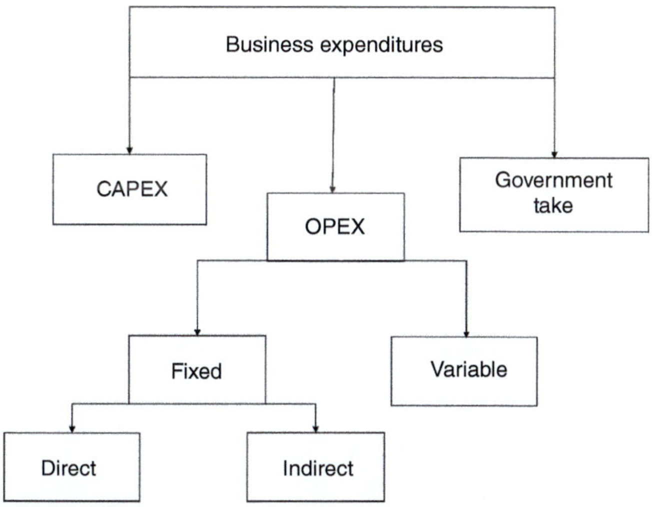

Figure 10.5 Business expenditures.

operating costs. The direct operating costs are the expenses that can be discontinued at any point of time by shutting-in the well. These are considered direct fixed expenses because they directly affect the production, and usually the performed treatments cost a fixed amount of money per well (e.g., well abandonment costs). The fixed indirect expenses, which are referred to as overhead, include employee salaries, office supplies and equipment (e.g., computers and company cars and trucks), and travel costs. The variable operating costs include unscheduled events such as workovers and maintenance services (Husni, 2009).

The third part of business expenditures is government take. *Government take* is a term that is used to refer to royalties, profit sharing, severance, and ad valorem taxes in addition to income taxes.

10.3.2 Basic Cash Flow

A typical cash flow model for an E&P company consists of revenue, investments (e.g., CAPEX), operating expenses (e.g., OPEX), and taxes. Equation (10.1) shows the basic net cash flow (NCF) model:

$$NCF = \text{Revenue} - Investments - Expenses - Taxes. \tag{10.1}$$

The revenue is the total income to the company. For example, it could be the income from selling the hydrocarbons as shown in equation (10.2):

$$\text{Revenue} = Amount\ of\ Hydrocarbon \times Product\ price. \tag{10.2}$$

After deducting the total costs and taxes from the gross revenue, the after-tax cash flow (ATCF) is achieved. The ATCF represents the profit, or the actual value profit. The actual value profit (AVP) is the summation of the NCF over the years as shown in equation (10.3):

$$AVP = \sum NCF. \tag{10.3}$$

The success of any project can be measured by the profit to investment ratio (P/I) and the payout period. The profit to investment ratio can be calculated by equation (10.4):

$$\left(\frac{P}{I}\right) = \frac{AVP}{Investment}. \tag{10.4}$$

Example 10.1 Estimate the Profit

A company decided to invest in drilling and completing a horizontal well with multistage fracturing. The CAPEX is estimated to be $8 million U.S. dollars. The monthly well operating cost is $1,700 U.S. dollars. The royalty, severance tax, state tax, and income tax are 25%, 4.6%, 4.5%, and 12%, respectively. The well is projected to produce oil with an average oil rate of 1,000 (bbl/d) for the first year, 680 (bbl/d) for the second year, 310 (bbl/d) for the third year, and 65 (bbl/d) for the fourth year. The average oil price is expected to be 50 ($/STB) for all four years. Generate the net cash flow (NCF) for this well. When would you recommend abandoning the well and why?

Solution

To generate the NCF, first the revenue must be calculated by the use of equation (10.2):

$$\text{Revenue for first month} = Amount\ of\ Hydrocarbon \times Product\ price$$

$$= 1{,}000\left(\frac{bbl}{day}\right) \times 30.4(day) \times 50\left(\frac{1}{bbl}\right) = 11{,}520{,}000.$$

Then, the NCF can be calculated by the use of equation (10.1):

$$\textit{Total income after taxes and expenses for first month} = \text{Revenue} - Expenses - Taxes$$
$$= \$1,520,000 - \$1,700 - (\$1,520,000 \times 25\%) - (\$1,520,000 \times 4.6\%)$$
$$- (\$1,520,000 \times 4.5\%) - (\$1,520,000 \times 12\%) = \$819,280$$

Finally, the actual value profit (AVP) is calculated by equation (10.3):

$$AVP\ for\ first\ month = -\$8,000,000 - \$819,280 = -\$7,180,720.$$

A sample calculation for the first year is

$$AVP\ for\ year\ 1 = \$2,388,470.40.$$

The results are summarized in Tables 10.1–10.3.

Table 10.1 Calculations for Example 10.1

Date	Oil production (STB)	Gross income ($)	Royalty ($)	Taxes on oil ($)	State tax ($)	Income tax ($)
19 Feb	0	0.00	0.00	0.00	0.00	0.00
19 Mar	30,400	1,520,000.00	380,000.00	69,920.00	68,400.00	182,400.00
19 Apr	30,400	1,520,000.00	380,000.00	69,920.00	68,400.00	182,400.00
19 May	30,400	1,520,000.00	380,000.00	69,920.00	68,400.00	182,400.00
19 Jun	30,400	1,520,000.00	380,000.00	69,920.00	68,400.00	182,400.00
19 Jul	30,400	1,520,000.00	380,000.00	69,920.00	68,400.00	182,400.00
19 Aug	30,400	1,520,000.00	380,000.00	69,920.00	68,400.00	182,400.00
19 Sep	30,400	1,520,000.00	380,000.00	69,920.00	68,400.00	182,400.00
19 Oct	30,400	1,520,000.00	380,000.00	69,920.00	68,400.00	182,400.00
19 Nov	30,400	1,520,000.00	380,000.00	69,920.00	68,400.00	182,400.00
19 Dec	30,400	1,520,000.00	380,000.00	69,920.00	68,400.00	182,400.00
20 Jan	30,400	1,520,000.00	380,000.00	69,920.00	68,400.00	182,400.00
20 Feb	30,400	1,520,000.00	380,000.00	69,920.00	68,400.00	182,400.00
20 Mar	20,672	1,033,600.00	258,400.00	47,545.60	46,512.00	124,032.00
20 Apr	20,672	1,033,600.00	258,400.00	47,545.60	46,512.00	124,032.00
20 May	20,672	1,033,600.00	258,400.00	47,545.60	46,512.00	124,032.00
20 Jun	20,672	1,033,600.00	258,400.00	47,545.60	46,512.00	124,032.00
20 Jul	20,672	1,033,600.00	258,400.00	47,545.60	46,512.00	124,032.00
20 Aug	20,672	1,033,600.00	258,400.00	47,545.60	46,512.00	124,032.00
20 Sep	20,672	1,033,600.00	258,400.00	47,545.60	46,512.00	124,032.00
20 Oct	20,672	1,033,600.00	258,400.00	47,545.60	46,512.00	124,032.00
20 Nov	20,672	1,033,600.00	258,400.00	47,545.60	46,512.00	124,032.00
20 Dec	20,672	1,033,600.00	258,400.00	47,545.60	46,512.00	124,032.00
21 Jan	20,672	1,033,600.00	258,400.00	47,545.60	46,512.00	124,032.00
21 Feb	20,672	1,033,600.00	258,400.00	47,545.60	46,512.00	124,032.00
21 Mar	9,424	471,200.00	117,800.00	21,675.20	21,204.00	56,544.00
21 Apr	9,424	471,200.00	117,800.00	21,675.20	21,204.00	56,544.00
21 May	9,424	471,200.00	117,800.00	21,675.20	21,204.00	56,544.00
21 Jun	9,424	471,200.00	117,800.00	21,675.20	21,204.00	56,544.00
21 Jul	9,424	471,200.00	117,800.00	21,675.20	21,204.00	56,544.00

(Continued)

Table 10.1 (Continued)

Date	Oil production (STB)	Gross income ($)	Royalty ($)	Taxes on oil ($)	State tax ($)	Income tax ($)
21 Aug	9,424	471,200.00	117,800.00	21,675.20	21,204.00	56,544.00
21 Sep	9,424	471,200.00	117,800.00	21,675.20	21,204.00	56,544.00
21 Oct	9,424	471,200.00	117,800.00	21,675.20	21,204.00	56,544.00
21 Nov	9,424	471,200.00	117,800.00	21,675.20	21,204.00	56,544.00
21 Dec	9,424	471,200.00	117,800.00	21,675.20	21,204.00	56,544.00
22 Jan	9,424	471,200.00	117,800.00	21,675.20	21,204.00	56,544.00
22 Feb	9,424	471,200.00	117,800.00	21,675.20	21,204.00	56,544.00
22 Mar	1,976	98,800.00	24,700.00	4,544.80	4,446.00	11,856.00
22 Apr	1,976	98,800.00	24,700.00	4,544.80	4,446.00	11,856.00
22 May	1,976	98,800.00	24,700.00	4,544.80	4,446.00	11,856.00
22 Jun	1,976	98,800.00	24,700.00	4,544.80	4,446.00	11,856.00
22 Jul	1,976	98,800.00	24,700.00	4,544.80	4,446.00	11,856.00
22 Aug	1,976	98,800.00	24,700.00	4,544.80	4,446.00	11,856.00
22 Sep	1,976	98,800.00	24,700.00	4,544.80	4,446.00	11,856.00
22 Oct	1,976	98,800.00	24,700.00	4,544.80	4,446.00	11,856.00
22 Nov	1,976	98,800.00	24,700.00	4,544.80	4,446.00	11,856.00
22 Dec	1,976	98,800.00	24,700.00	4,544.80	4,446.00	11,856.00
23 Jan	1,976	98,800.00	24,700.00	4,544.80	4,446.00	11,856.00
23 Feb	1,976	98,800.00	24,700.00	4,544.80	4,446.00	11,856.00
23 Mar	1,976	98,800.00	24,700.00	4,544.80	4,446.00	11,856.00

Table 10.2 Results for Example 10.1

Date	Income after tax, royalty and cost of operation ($)	Cumulative. income ($)	AVP ($)
19 Feb	0.00	0.00	−8,000,000.00
19 Mar	819,280.00	819,280.00	−7,180,720.00
19 Apr	819,280.00	1,638,560.00	−6,361,440.00
19 May	819,280.00	2,457,840.00	−5,542,160.00
19 Jun	819,280.00	3,277,120.00	−4,722,880.00
19 Jul	819,280.00	4,096,400.00	−3,903,600.00
19 Aug	819,280.00	4,915,680.00	−3,084,320.00
19 Sep	819,280.00	5,734,960.00	−2,265,040.00
19 Oct	819,280.00	6,554,240.00	−1,445,760.00
19 Nov	819,280.00	7,373,520.00	−626,480.00
19 Dec	819,280.00	8,192,800.00	192,800.00
20 Jan	819,280.00	9,012,080.00	1,012,080.00

Table 10.2 (Continued)

Date	Income after tax, royalty and cost of operation ($)	Cumulative. income ($)	AVP ($)
20 Feb	819,280.00	9,831,360.00	1,831,360.00
20 Mar	557,110.40	10,388,470.40	2,388,470.40
20 Apr	557,110.40	10,945,580.80	2,945,580.80
20 May	557,110.40	11,502,691.20	3,502,691.20
20 Jun	557,110.40	12,059,801.60	4,059,801.60
20 Jul	557,110.40	12,616,912.00	4,616,912.00
20 Aug	557,110.40	13,174,022.40	5,174,022.40
20 Sep	557,110.40	13,731,132.80	5,731,132.80
20 Oct	557,110.40	14,288,243.20	6,288,243.20
20 Nov	557,110.40	14,845,353.60	6,845,353.60
20 Dec	557,110.40	15,402,464.00	7,402,464.00
21 Jan	557,110.40	15,959,574.40	7,959,574.40
21 Feb	557,110.40	16,516,684.80	8,516,684.80
21 Mar	253,976.80	16,770,661.60	8,770,661.60
21 Apr	253,976.80	17,024,638.40	9,024,638.40
21 May	253,976.80	17,278,615.20	9,278,615.20
21 Jun	253,976.80	17,532,592.00	9,532,592.00
21 Jul	253,976.80	17,786,568.80	9,786,568.80
21 Aug	253,976.80	18,040,545.60	10,040,545.60
21 Sep	253,976.80	18,294,522.40	10,294,522.40
21 Oct	253,976.80	18,548,499.20	10,548,499.20
21 Nov	253,976.80	18,802,476.00	10,802,476.00
21 Dec	253,976.80	19,056,452.80	11,056,452.80
22 Jan	253,976.80	19,310,429.60	11,310,429.60
22 Feb	253,976.80	19,564,406.40	11,564,406.40
22 Mar	53,253.20	19,617,659.60	11,617,659.60
22 Apr	53,253.20	19,670,912.80	11,670,912.80
22 May	53,253.20	19,724,166.00	11,724,166.00
22 Jun	53,253.20	19,777,419.20	11,777,419.20
22 Jul	53,253.20	19,830,672.40	11,830,672.40
22 Aug	53,253.20	19,883,925.60	11,883,925.60
22 Sep	53,253.20	19,937,178.80	11,937,178.80
22 Oct	53,253.20	19,990,432.00	11,990,432.00
22 Nov	53,253.20	20,043,685.20	12,043,685.20
22 Dec	53,253.20	20,096,938.40	12,096,938.40
23 Jan	53,253.20	20,150,191.60	12,150,191.60
23 Feb	53,253.20	20,203,444.80	12,203,444.80
23 Mar	53,253.20	20,256,698.00	12,256,698.00

Table 10.3 Summarized Results for Example 10.1

Year	AVP ($)
0	−8,000,000.00
1	2,388,470.40
2	8,770,661.60
3	11,617,659.60
4	12,256,698.00

From Table 10.3, the actual value profit for the fourth year is approximately $640,000 more than the third year. Therefore, abandoning the well by the end of the third year or doing an extensive stimulation job (e.g., refracturing) to improve the production are valid options to consider in case the expected profit in the fourth year is not attractive enough for the company.

It is important to note that the time value of money was not considered in Example 10.1. To properly evaluate the future AVP, the discounting NCF (DCF) should be used rather than the NCF. The DCF accounts for the time value of money as a dollar today is worth more than a dollar tomorrow because of the earnings that can be gained due to the interest rate.

10.3.3 Interest Rate

The interest rate can be defined as the percentage that is earned over a specific period of time, which is referred to as the interest period (N). The interest period defines the frequency in which the interest is calculated. There are two types of interest: simple interest and compounded interest (Park et al., 2013). The simple interest is applied only to the initial value (i.e., the principle amount (P)). On the other hand, the compounded interest, which is the most used interest, is applied to the principal amount (P) and to any accumulated amount that has not been withdrawn. The future amount of money (FV) can be calculated for the simple interest and the compounded interest by the use of equations (10.5) and (10.6), respectively (Park et al., 2013). In order to calculate the interest rate accurately, the units of the interest rate (i) should match the units of the interest period (N).

$$FV = P(1 + i)N \tag{10.5}$$

where FV is the future amount of money, P is the principle amount of money, i is the interest rate, and N is the interest period:

$$FV = P\,(1 + i)^N. \tag{10.6}$$

Example 10.2 Determine the Earned Interest

A group of people have decided to invest $1,000,000 U.S. dollars in a bank for one year, and they found two offers. The first offer has a simple yearly interest rate of 2.9%. The second offer has an interest rate of 35% compounded monthly. Determine the earned interest of both offers.

Solution

First, the compounded interest rate should be converted to yearly instead of monthly. This can be done as follows:

$$i = \frac{35\%}{12\%} = 2.917\%.$$

The earned interest can be calculated for the first offer and the second offer by equation (10.5) and equation (10.6), respectively.

$$FV = P + (iP)N = \$1{,}000{,}000 + (0.029 \times \$1{,}000{,}000) \times 1 = \$1,030,000$$

and

$$FV = P\,(1 + i)^N = \$1{,}000{,}000 \times (1 + 0.02917)^1 = \$1,029,170.$$

Hence, the earned interest from the first offer is $30,000 and from the second offer is $29,170.

10.3.4 Future Value

The money has a time value because it has (1) an earning power over time and (2) a purchasing power over time (e.g., inflation). The time value of money is measured by the interest rate or the inflation rate (Park et al., 2013). Equation (10.6) is used to calculate the future value of money with either interest rate or inflation. However, to calculate the future worth of money due to inflation, equation (10.7) should be used:

$$FV = P\,(1 - i)^N. \tag{10.7}$$

Example 10.3 Determine the Future Value of the Capex
In Example 10.1, the cost of completing the horizontal well with multistage fracturing was estimated to be $8 million U.S. dollars. However, the company has decided to postpone the project for one year due to a rig scheduling issue. What would be the future value of the CAPEX if the average year-over-year inflation rate on all services is estimated to be 1.9%? Determine the future worth of the original CAPEX after one year.

Solution

The future value for completing the well after one year with an inflation rate of 1.9% is calculated by the use of equation (10.6):

$$FV = \$8,000,000 \times (1 + 0.019)^1 = \$8,152,000.$$

The future worth of the CAPEX with an inflation rate of 1.9% is calculated by the use of equation (10.7):

$$FV = P\,(1 - i)^N = \$8,000,000 \times (1 - 0.019)^1 = \$7,848,000.$$

Hence, the CAPEX for completing the horizontal well with multistage fracturing would be $8,152,000 U.S. dollars for the next year instead of $8,000,000 U.S. dollars for this year. Thus, the company should invest the money somewhere with an interest rate that is similar to the inflation rate in order to make up for the additional amount of money. Otherwise, the company will have to spend more money because the original CAPEX value (i.e., $8,000,000 U.S. dollars) will be worth less than $7,848,000 U.S. dollars in the next year.

10.3.5 Present Value

The present value, which is also known as the discounted value, is basically the current worth of a future sum of money, and it can be calculated by equation (10.8) (Park et al., 2013):

$$PV = FV\left(\frac{1}{(1 + i)^N}\right) \tag{10.8}$$

where *FV* is the future value of money, *PV* is the present value of money, *i* is the interest rate or the discount rate, and *N* is the number of periods.

Example 10.4 Determine the Required Present Value for the Investment
The CAPEX for completing the horizontal well with multistage fracturing was estimated to be $8,152,000 U.S. dollars for the next year. The company wants to invest an amount of money to cover the CAPEX. The company is looking into another project to potentially invest with a 2.9% interest rate that is compounded annually. Determine the minimum amount of money that the company needs to invest today in order to cover the entire CAPEX of the horizontal well for the next year.

Solution

Since the future value and the interest rate are known, then the required present value for the investment can be calculated by the use of equation (10.8).

$$PV = FV\left(\frac{1}{(1+i)^N}\right) = \$8,152,000 \times \left(\frac{1}{(1+0.029)^1}\right) = \$7,922,255.$$

Hence, the company should invest today at least $7,922,255 U.S. dollars to cover the entire CAPEX of the horizontal well.

10.3.6 Net Present Value (NPV)

The net present value (NPV) measures the present value of the project's cash flow (NCF) to a common reference point in time, which is typically the current year (i.e., year 0) (Park et al., 2013). Hence, NPV is the process of accounting for the time value of money in the calculated NCF. The NPV, which is also referred to as the present value profit (PVP), can be calculated by equation (10.9). The term that is represented in equation (10.10) is referred to as the end of year discount factor.

Note that there is no discount for year (0), which means the discount factor is 1.

$$NPV = C_0 + \sum\left(C_n\left(\frac{1}{(1+i)^N}\right)\right) \tag{10.9}$$

and

$$\text{Discount factor } (\mathbf{DF}) = \left(\frac{1}{(1+i)^N}\right). \tag{10.10}$$

Example 10.5 Calculate the NPV

Based on the generated net cash flow (NCF) for the horizontal well in Example 10.1, calculate the NPV for a discount rate of 5% compounded annually.

Solution

A sample calculation is shown below, and the final results are summarized in Table 10.4. The first step to calculate the NPV is to calculate the monthly discount factor by the use of equation (10.10). Note that the values of the AVP and the NVP for the first month are the same.

$$\text{Monthly Discount factor} = \left(\frac{1}{(1+i)^N}\right) = \left(\frac{1}{\left(1+0.\frac{05}{12}\right)^{\left(\frac{1-month\#}{12}\right)}}\right).$$

Then, the NPV is calculated by equation (10.9):

$$NPV \textit{ for } \boldsymbol{the} \textit{ second } \boldsymbol{month} = C_0 + \sum\left(C_n\left(\frac{1}{(1+i)^N}\right)\right) = -\$8,000,000 + \$1,640,764$$

$$= -\$6,359,236.$$

The actual value profit (AVP) after one year, which was calculated in Example 10.1, is $2,388,470 U.S. dollars. However, when the time value of money is considered, the present value profit (PVP) is calculated, which is equivalent to the NPV. The PVP is $2,378,560 U.S. dollars instead of $2,388,470 U.S. dollars. Therefore, the difference between the AVP and the PVP for the first year is approximately $9,911.

Table 10.4 Summarized Results for Example 10.5

Year	AVP ($)	NVP ($)
0	−8,000,000.00	−8,000,000.00
1	2,388,470.40	2,378,559.73
2	8,770,661.60	8,698,027.03
3	11,617,659.60	11,473,640.68
4	12,256,698.00	12,054,529.99

10.3.7 Rate of Return (ROR)

The rate of return (ROR) is a way to measure the return of the invested value over a period of time. There are different methods to interpret the rate of return.

Return on Investment (ROI)

The return on investment (ROI), which is considered the simple ROR, is a measure of the actual value profit (AVP) relative to the CAPEX over a period of time (Park et al., 2013). The ROI can be calculated by equation (10.11). The main disadvantage of the ROI is that it does not account for the time value of money.

$$ROI = \left(\frac{AVP}{CAPEX}\right) \times 100. \quad (10.11)$$

Example 10.6 Calculate the Return on Investment

Calculate the ROI for the project of completing the horizontal well as stated in Example 10.5. The CAPEX and the AVP are provided in Table 10.4.

Solution

From Table 10.4, the CAPEX of the project and AVP after 4 years are \$8,000,000 U.S. dollars and \$12,256,698 U.S. dollars. Thus, the ROI can be calculated using equation (10.11) as follows:

$$ROI = \left(\frac{net\,profit}{CAPEX}\right) \times 100 = \left(\frac{\$12,256,698}{\$8,000,000}\right) \times 100 = 153\%.$$

The ROI is 15. Since the ROI does not consider the time value of money, the results of high ROI may solely mean long time periods were considered in the calculation.

Discounted Return on Investment (DROI)

The discounted return on investment (DROI) is basically the ROI when the time value of money is considered (Park et al., 2013). Thus, it is the present value profit (PVP) or the NPV relative to the CAPEX of the project, and it can be calculated by equation (10.12):

$$DROI = \left(\frac{PVP}{CAPEX}\right) \times 100. \quad (10.12)$$

Example 10.7 Calculate the Discounted Return on Investment

Calculate the DROI for the project of completing the horizontal well as stated in Example 10.5 considering a discount rate of 5% compounded annually. Compare the DROI against the ROI that was calculated in Example 10.6.

Solution

The DROI can be calculated by knowing the CAPEX and the PVP. The PVP of the project after 4 years with a discount rate of 5%, which was calculated in Example 10.5, is \$12,054,530 U.S. dollars. The CAPEX of the project is \$8,000,000 U.S. dollars. Hence, the DROI can be calculated by the use of equation (10.12) as follows:

$$DROI = \left(\frac{PVP}{CAPEX}\right) \times 100 = \left(\frac{\$12,054,530}{\$8,000,000}\right) \times 100 = 150.68\% \approx 151\%.$$

The DROI, which is 151%, indicates that \$1.51 would be earned for every dollar invested in the project. In comparison, the ROI, which is 153%, indicates that \$1.53 would be earned for every dollar invested. The ROI overestimates the profit.

Internal Rate of Return (IRR)

The internal rate of return (IRR), which is a widely used method to compare the ROR of the projects, is defined as the discount rate value at which the NPV equals zero (Park et al., 2013). It is also defined as the break-even interest rate where the present value of the project's cash outflows equals the present value of its cash inflows. The IRR can be calculated iteratively by equating the NPV (i.e., PVP) to zero and solving for the IRR as shown in equation (10.13) (Park et al., 2013):

$$0 = C_0 + \sum\left(C_n\left(\frac{1}{(1+IRR)^N}\right)\right) \tag{10.13}$$

Example 10.8 Calculate the Internal Rate of Return

Calculate the IRR for the project of completing the horizontal well as stated in Example 10.5.

Solution

To calculate the IRR, the values of NCF for each period of time in addition to the CAPEX should be calculated first. The NCF for the 4 years should be calculated first. Then, equation (10.13) can be used to calculate the IRR as follows:

$$0 = C_0 + \sum\left(C_n\left(\frac{1}{(1+IRR)^N}\right)\right) = -\$8{,}000{,}000 + \left(\$10{,}388{,}470 \times \left(\frac{1}{(1+IRR)^1}\right)\right)$$
$$+ \left(\$6{,}382{,}191 \times \left(\frac{1}{(1+IRR)^2}\right)\right) + \left(\$2{,}846{,}998 \times \left(\frac{1}{(1+IRR)^3}\right)\right)$$
$$+ \left(639{,}038 \times \left(\frac{1}{(1+IRR)^4}\right)\right) = 0.847 \times 100 = 84.7\%.$$

Thus, the IRR is 84.7%.

10.3.8 Payout

The payout, or the payback period, is the length of time in which the CAPEX of a project is expected to be recouped (Kagan, 2018). The payout time period plays a role in accepting or declining a project. If the payout time period of a project is not attractive enough compared with the total lifetime of the project, then the company might just discard the project and invest the money in another project that would yield the same profit over a shorter period of time. Note that the time value of money is not considered in the estimation of the payout time period.

Example 10.9 Estimate the Payout

Estimate the payout time period for the project of completing the horizontal well provided in Example 10.1.

Solution

The NCF for the well was generated based on the amount of the produced oil and the oil price, which were provided in Example 10.1. The payout time can be estimated by calculating the APV for each month as shown in Figure 10.6. The payout time is the point the project starts generating profit. Hence, the payout time is approximately 10 months.

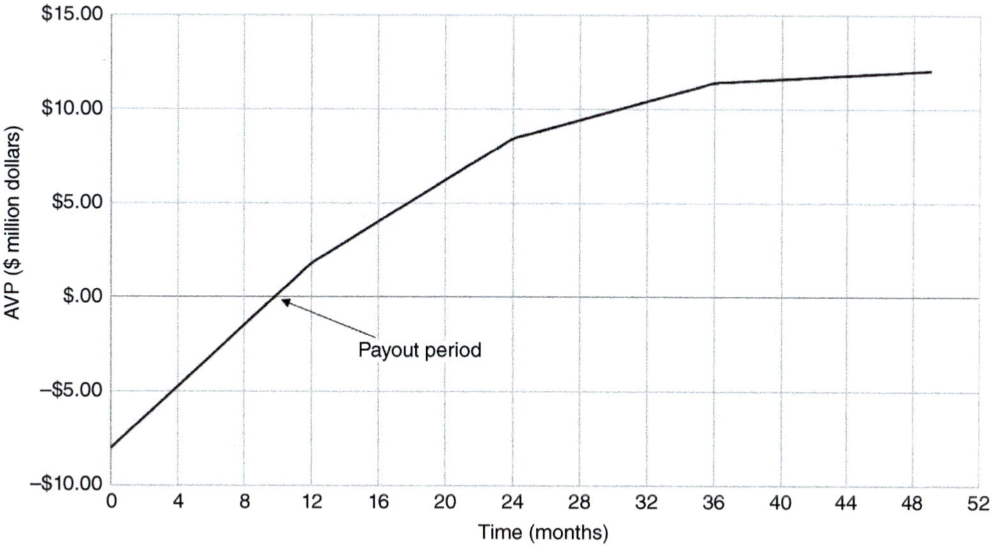

Figure 10.6 Payout time for Example 10.9.

10.4 Fiscal Regimes and Contracts

Fiscal regimes in the mineral industry and hydrocarbon production and exploration are composed of tax systems, contractual agreements, and economic elements. The main fiscal regimes used in the industry worldwide are concessionary and contractual systems (see Fig. 10.9). The contractual system is divided into two main categories: production sharing agreements and service contracts. Different countries in the world use different systems depending on how the ownership rights are defined, how advanced their technologies for production are, and their access to ports of export. Figure 10.7 shows an international distribution of different types of oil and gas agreement contracts (Harraz, 2016).

10.4.1 The Concessionary System (Royalty/Tax System)

In the concessionary system, government, state, or landowner transfers the ownership of the mineral resources to one or multiple commercial entities and, in return, receives royalty and tax. Royalty and taxes are subject to the location of the subsurface resources. All the examples and problems related to the previous sections of this book were valid under a concessionary system, which is the typical system applied in the United States.

10.4.2 The Production Sharing Contracts (PSCs)

In the contractual system, the country owns the hydrocarbon resources and contracts with commercial entities as contractors. The host government allows the contractors to own part of the mineral production bases on the terms and conditions of the contract. Successful production as a result of such contracts will pay the oil company's operational costs while the remaining will be shared, as profit, between the government or one of its agencies such as national oil companies (NOCs) and the foreign oil company (FOC) or international oil company (IOC) (Mian, 2011a; Bindemann, 1999). Figure 10.8 presents a flow diagram of a typical production sharing contract, where taxes and profits are divided between the IOC/FOC and the government entity. The major

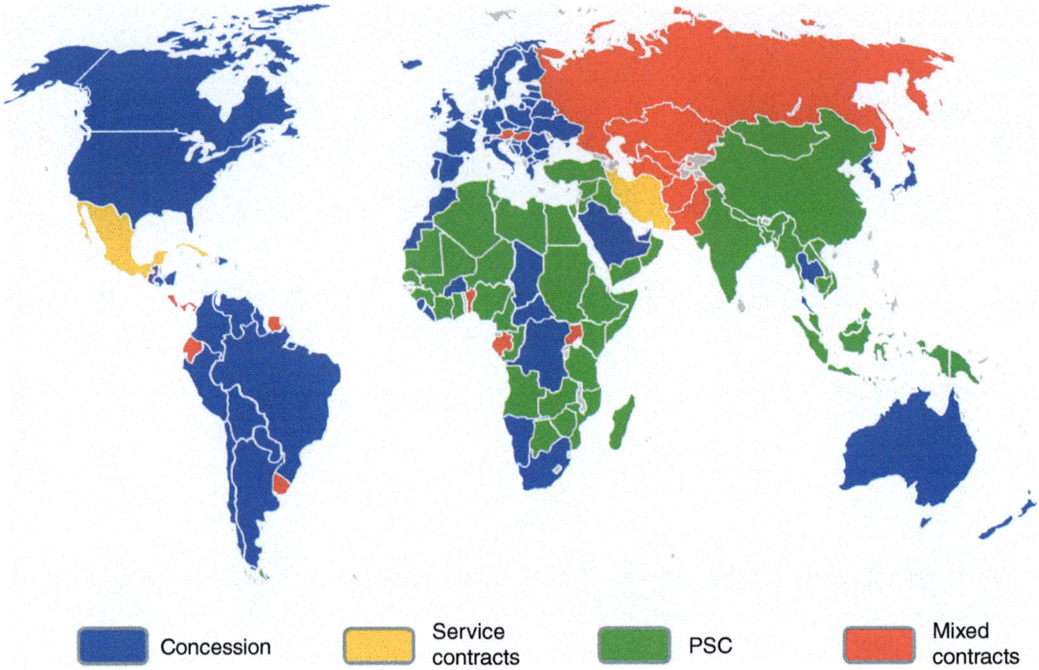

Figure 10.7 International distribution of types of oil and gas agreement contracts (after Harraz, 2016).

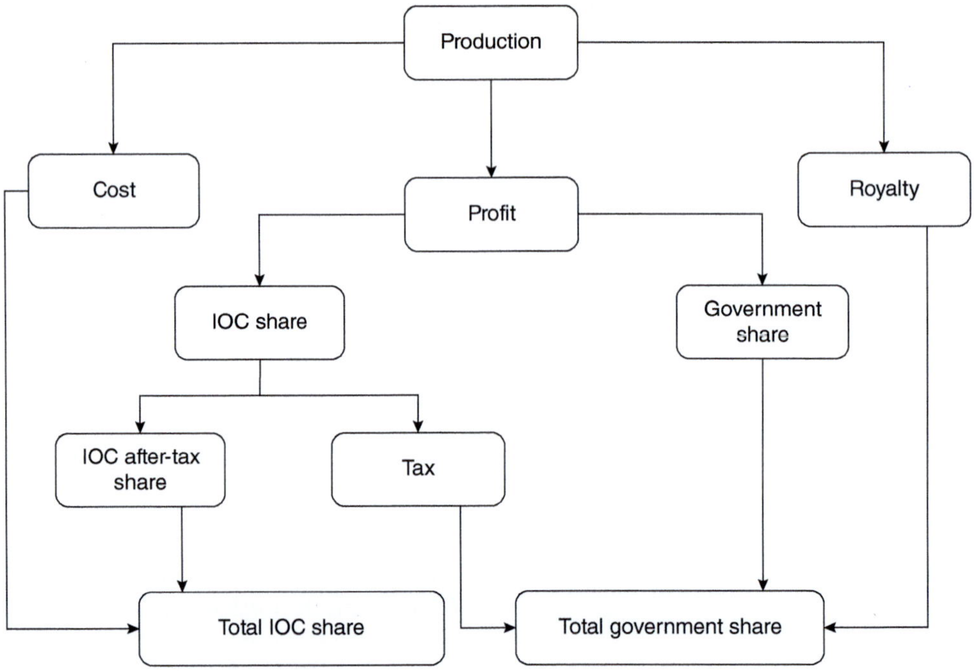

Figure 10.8 A representative diagram of the structure of production sharing agreements (after Colombia, 2007).

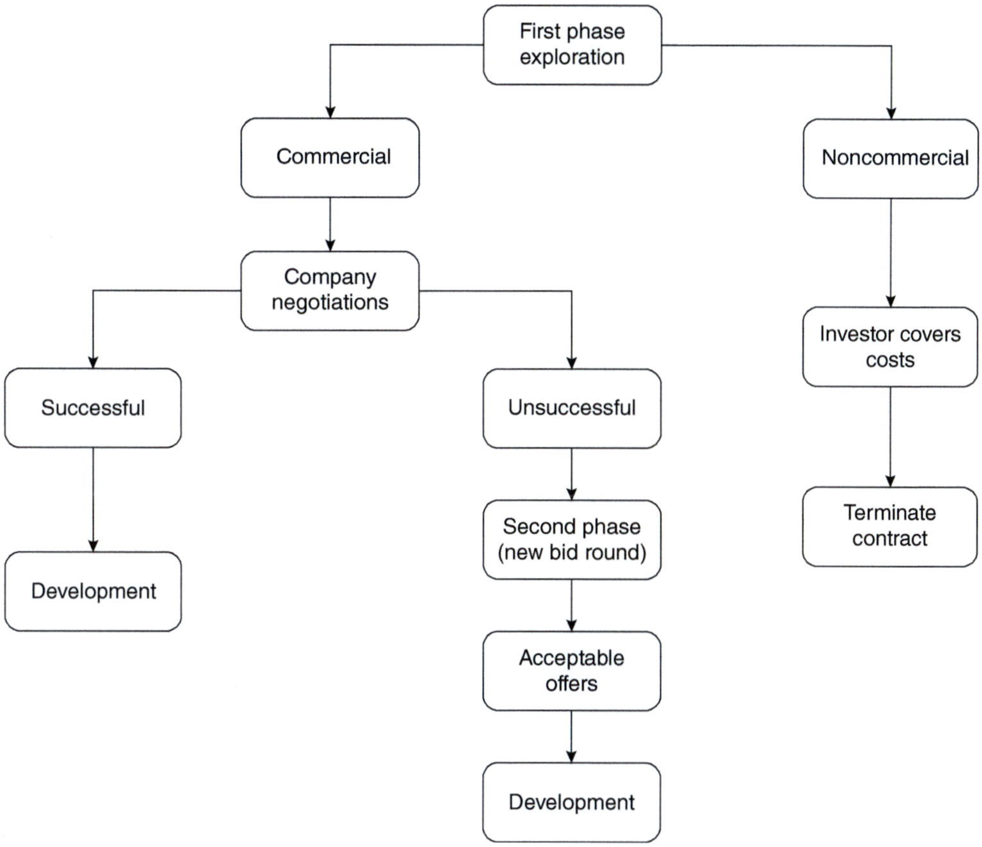

Figure 10.9 Flow diagram for an example buy-back contract (after Bindemann, 1999).

elements included in a PSA calculation sheet are royalty, tax, cost of oil, bonuses, oil profit, excess cost recover, economic rent tax, excess windfall profit tax, and uplifts resulting from increased depreciated capital, which results in tax reduction (Mian, 2011a).

10.4.3 Service Contracts

In service agreement systems, the government or state pays for the service while it has more control over the resources. This requires the host government to be both technologically and financially capable of developing their resources. This type of contract is mainly applied in countries such as Iran, Saudi Arabia, Venezuela, and Kuwait (Karasalihović-Sedlar et al., 2017). Three major types of service contracts reported by Karasalihović-Sedlar et al. (2017) are pure service contracts, service contracts with risk, and buy-back service contracts.

Pure Service Contracts

In pure service contracts, a predetermined fee is considered for the service provided and compensation to the service company can be made in the form of either cash or oil with minimum risk. Service contracts provide the host government with a significant amount of flexibility and are typically looked at less favorably by contractors.

Service Contracts with Risk

With service contracts with risk, discovery and incremental oil production will result in compensation and equity options while failure to have exploitable hydrocarbons will result in lost funds.

Buy-Back Service Contracts

Buy-back service contracts are short term (5–7 yr). Figure 10.9 presents a flow diagram for an example buy-back contract. These contracts begin with exploration and may lead to development contracts with the possibility of more development outcomes or cancellation of project after each phase. Depending on the contract terms, the investor may not be refunded if negotiations with the government do not work out.

Hydrocarbon fiscal systems calculations may vary significantly from country to country even within the same fiscal regime type since different countries have different petroleum tax systems, royalties, cost recovery, and state share of tax (Table 10.5). Hydrocarbon fiscal systems vary even among states within the United States of America due to factors such as availability of resources required for drilling and fracturing, availability of a viable market for the produced product, and state tax and royalties. There is a lack of a comprehensive fiscal system for unconventional reservoirs in the literature.

10.5 Decision, Uncertainty, and Risk Analysis

In the petroleum industry, making project decisions is a difficult task. A decision means distribution of resources (e.g., financial, technical, human, etc.). Decisions are typically made by analyzing the available information, and the quality of the decisions is measured by the outcomes. Nevertheless, if the achieved outcome is negative, it does not necessarily mean that a bad decision has been made, because at that time, the available information might have suggested otherwise. In general, the steps toward making good decisions and achieving favorable outcomes are (1) accurate evaluation of the available information, (2) accurate evaluation of the possible alternatives, and (3) accurate evaluation of the company preferences (Husni, 2009). Abiding by these steps is crucial in multi-billion-dollar projects to make consistent decisions that are based on structured judgments and proper evaluation of the outcomes. The process of decision analysis includes (1) accurate identification of the project, (2) identifying and assessing the most influential variables (e.g., operating cost, oil price, etc.), and (3) risk and uncertainty analysis (Mian, 2011a).

Uncertainty can be defined as the different possible outcomes of a decision where the probability of each outcome cannot be estimated (Mian, 2011a). In the petroleum industry, uncertainty is typically expressed by ranges (e.g., reserves, porosity, permeability, etc.). Uncertainty plays a role in the decision-making process, and it can be evaluated by (1) qualitative and quantitative description by using ranges including optimistic and pessimistic estimates, and (2) understanding how the uncertainty will affect the project's overall outcome. When the probabilities of the decision's outcomes can be estimated, then these scenarios are considered as risky scenarios instead of uncertain scenarios (Mian, 2011a).

The upstream petroleum industry (i.e., E&P) is considered as one of the most uncertain industries due to the high risks in operations, the massive capital investments, and the inconstant political environments as well as the volatile oil and gas prices (Mian, 2011a). Thus, the

Table 10.5 Comparison of Production Sharing Contracts Between the Three Countries of Indonesia, Malaysia, and Nigeria

Type	Nigeria	Malaysia	Indonesia
Fiscal arrangement	PSC	PSC	PSC
Royalty	20% onshore 16.7% deep	10%	(85/15 split) 20% FTP
Cost recovery limit	100%	70%	80% (under review)
State share of profit petroleum	20%–60% (avg. 50%)	Avg. 60% (negotiable)	Avg. 65% (negotiable)
Petroleum tax rate	50%	38%	40% (combined C&D rate)

E&P industry relies on the most advanced techniques to analyze the risks and uncertainty especially when it comes to capital investments. These techniques include (1) sensitivity analysis, (2) scenario analysis, and (3) sophisticated computer simulators (Mian, 2011b). There exist other techniques and approaches that the industry relies on to analyze the risk and uncertainty such as the probability approach.

The sensitivity analysis, which is also known as the "what if" scenario analysis, is a technique that is utilized to evaluate the project's outcomes (e.g., profit) by changing one specific variable while keeping the other variables constant. An example of the sensitivity analysis technique is the estimation of the project's profit when the oil price is at $40/STB, $50/STB, or $60/STB.

When the most influential variables (e.g., cost of operation, production rate, oil price, etc.) are studied together to assess the project's outcome, then this technique is referred to as scenario analysis. In this technique, the variables are examined using high values, low values, and average values. Therefore, by analyzing each outcome, the project's best-case scenario, the worst-case scenario, as well as the most-likely scenario are predicted (Mian, 2011b).

The computer simulation technique, which is considered as the most advanced technique, is used to have a probability distribution of the most influential variables. This is accomplished by simulating hundreds of different combinations. The computer simulation technique relies on sophisticated mathematical and numerical models.

It is important to note that risk and uncertainty exist all over the petroleum industry and not solely during the decision-making process. There is uncertainty in how the reserves are estimated. There is also uncertainty in the decline curve analysis, the rate transient analysis, as well as the advanced simulators and how they are utilized to forecast the production and predict the EUR of the wells.

10.6 Chapter Project

A company drilled and completed a horizontal well with 45 hydraulic fracturing stages, and the well was put onto production in November 2014. The cost of drilling and completing the well and the hydraulic fracture pump units was $3,931,457 U.S. dollars. The company used one pump schedule for all the 45 stages, which is infrequent as shown in Table 10.6. The cost of the fracturing fluids and proppants are provided in Table 10.7. The company decided to go with that specific pump schedule because it showed promising dimensionless fracture

Table 10.6 Pump Schedule

Pump stage	Fluid type	Fluid volume (m-gal)	Proppant type	Proppant volume (m-lb)
1 (pad)	Fluid A	25.3	–	–
2	Fluid A	12.1	Sand (40/70)	5.7
3	Fluid A	34.9	Sand (40/70)	33.4
4	Slickwater	25	Sand (30/50)	84.7
5	Slickwater	20	Sand (30/50)	81.5

Table 10.7 Cost of Fracturing Fluid and Proppant

Sand (30/50)	Sand (40/70)	Fluid A	Slickwater
/lb	/lb	/Gal	/Gal
0.35	0.36	1.17	0.70

Table 10.8 Production Figures for Chapter Project

Month-year	Oil production (STB)	Gas production (MSCF)	Water production (STB)
14 Nov	28,727	10,140	11,243
14 Dec	23,341	10,306	9,645
15 Jan	22,662	4,856	8,599
15 Feb	14,226	5,608	9,460
15 Mar	21,766	4,833	8,811
15 Apr	19,865	4,023	7,985
15 May	19,622	3,707	7,856
15 Jun	18,302	3,269	7,308
15 Jul	18,335	3,125	7,308
15 Aug	17,841	2,919	7,101
15 Sep	16,863	2,663	6,704
15 Oct	17,061	2,611	6,778
15 Nov	16,198	2,410	6,430
15 Dec	16,448	2,384	6,526
16 Jan	16,183	2,290	6,418
16 Feb	14,919	2,066	5,914
16 Mar	15,733	2,136	6,235
16 Apr	15,030	2,002	5,954
16 May	15,343	2,007	6,077
16 Jun	14,679	1,888	5,812
16 Jul	15,005	1,900	5,940
16 Aug	14,849	1,851	5,877
16 Sep	14,230	1,749	5,631
16 Oct	14,568	1,766	5,764
16 Nov	13,974	1,672	5,528
16 Dec	14,317	1,692	5,663
17 Jan	14,198	1,658	5,615
17 Feb	12,727	1,470	5,033
17 Mar	13,988	1,598	5,531
17 Apr	13,439	1,519	5,313
17 May	13,789	1,543	5,452
17 Jun	13,254	1,468	5,239
17 Jul	13,606	1,493	5,378
17 Aug	13,519	1,469	5,343
17 Sep	13,002	1,400	5,139
17 Oct	13,355	1,426	5,278
17 Nov	12,850	1,360	5,078
17 Dec	13,203	1,386	5,217
18 Jan	13,130	1,367	5,188
18 Feb	11,798	1,220	4,662
18 Mar	12,997	1,334	5,135
18 Apr	12,514	1,275	4,944
18 May	12,867	1,301	5,083
18 Jun	12,392	1,244	4,895
18 Jul	12,745	1,271	5,034
18 Aug	12,685	1,256	5,011
18 Sep	12,221	1,203	4,827
18 Oct	12,573	1,229	4,966
18 Nov	12,115	1,177	4,785

conductivity during the modeling phase. The oil, gas, and water production figures of the well for each month are provided in Table 10.8. The company has a contract with a service company to handle the produced water. The water handling cost is \$1.00 per STB. The average oil and gas prices for each month are provided in Table 10.9. Calculate the NCF and AVP for the well. Estimate the payout time period and calculate the return on investment (ROI). Calculate the net present value (NVP) with a discount rate of 8% compounded annually. The well monthly operating cost is \$1,700, and the royalty is 18.75%, the severance tax of oil is 5%, the severance tax of gas is 5%, the state tax is 3%, and the income tax is 10%.

10.6.1 Project Data

Tables 10.6–10.9 summarize project data.

10.6.2 Project Calculations

First, the total cost should be found by adding the cost of the fracturing fluid and proppant of

Table 10.9 Monthly Average Prices of Oil and Gas

Month-year	Average monthly price of oil per STB ($)	Average monthly price of gas per MSCF ($)
14 Nov	76.99	4.27
14 Dec	60.70	3.61
15 Jan	47.11	3.10
15 Feb	54.79	2.98
15 Mar	52.83	2.93
15 Apr	57.54	2.71
15 May	62.51	2.96
15 Jun	61.31	2.88
15 Jul	54.43	2.95
15 Aug	45.69	2.87
15 Sep	46.28	2.76
15 Oct	46.96	2.43
15 Nov	43.11	2.17
15 Dec	36.57	2.00
16 Jan	29.78	2.36
16 Feb	31.03	2.06
16 Mar	37.34	1.79
16 Apr	40.75	1.99
16 May	45.94	1.99
16 Jun	47.69	2.69
16 Jul	44.13	2.92
16 Aug	44.87	2.92
16 Sep	45.04	3.10
16 Oct	49.29	3.09
16 Nov	45.26	2.64
16 Dec	52.62	3.72
17 Jan	53.59	3.42
17 Feb	54.35	2.96
17 Mar	50.90	2.99
17 Apr	52.16	3.21
17 May	49.89	3.27
17 Jun	46.17	3.09
17 Jul	47.66	3.09
17 Aug	49.94	3.01
17 Sep	52.95	3.09
17 Oct	54.92	2.99
17 Nov	59.93	3.12
17 Dec	61.19	2.92
18 Jan	66.23	4.01
18 Feb	63.46	2.77
18 Mar	64.17	2.79
18 Apr	68.79	2.90
18 May	73.43	2.90
18 Jun	71.98	3.08
18 Jul	72.67	2.93
18 Aug	71.08	3.07
18 Sep	75.36	3.11
18 Oct	76.73	3.40
18 Nov	62.32	4.24

Source: Indexmundi (2019).

all 45 fracturing stages to the well completion cost. A sample calculation is shown below and the final results are summarized in Tables 10.10 and 10.11.

$$cost\ of\ fracturing\ fluid\ at\ the\ pad\ stage$$

$$= 25.3\ (MGal.) \times 1000 \times \frac{\$1.17}{Gal}$$

$$= \$29,515.57$$

and

$$cost\ of\ fracturing\ fluid\ and\ proppant\ at$$

$$second\ pumping\ stage = 12.1 \times 1000 \times \frac{\$1.17}{Gal}$$

$$+ 5.7 \times 1000 \times \frac{\$0.36}{lb.} = \$16,168.14.$$

$$The\ total\ cost\ (CAPEX) = \$8,426,731.87$$

$$+ \$3,9314,457 = \$12,358,189.$$

Next, the NCF for the well can be generated by calculating the revenue for each month. A sample calculation is shown here, and the results for both the NCF and the AVP are summarized in Tables 10.12–10.14.

Table 10.10 Cost of a Single Hydraulic Fracturing Stage

Pump stage	Fluid type	Fluid volume (m-gal)	Proppant type	Proppant volume (m-lb)	Costs ($)
1 (pad)	Fluid A	25.3			29,515.57
2	Fluid A	12.1	Sand (40/70)	5.7	16,168.14
3	Fluid A	34.9	Sand (40/70)	33.4	52,739.16
4	Slickwater	25.0	Sand (30/50)	84.7	46,720.85
5	Slickwater	20.0	Sand (30/50)	81.5	42,116.98

Table 10.11 Summation of Total Expenditure

Total cost	Value ($)
1 fracturing stage	187,260.71
45 fracturing stages	8,426,731.87
CAPEX (Fracturing stages and well drilling and completion)	12,358,188.87

Table 10.12 Total Income Calculations for Chapter Project

Date	Gross income from oil ($)	Gross income from gas ($)	Total income before taxes and royalty ($)
14 Nov	0.0000	0.0000	0.0000
14 Nov	2,211,691.7300	43,322.5416	2,255,014.2716
14 Dec	1,416,798.7000	37,191.8806	1,453,990.5806
15 Jan	1,067,606.8200	15,056.6593	1,082,663.4793
15 Feb	779,442.5400	16,690.4735	796,133.0135
15 Mar	1,149,908.5573	14,183.4534	1,164,092.0108
15 Apr	1,143,044.2156	10,888.5311	1,153,932.7467
15 May	1,226,552.8014	10,955.8532	1,237,508.6546
15 Jun	1,122,125.0562	9,424.0693	1,131,549.1255
15 Jul	997,958.1521	9,203.3750	1,007,161.5271
15 Aug	815,173.7899	8,384.7983	823,558.5882
15 Sep	780,399.9067	7,345.6725	787,745.5791
15 Oct	801,163.4308	6,335.8004	807,499.2312
15 Nov	698,283.0229	5,223.2653	703,506.2882
15 Dec	601,518.3665	4,771.3614	606,289.7279

Table 10.12 (Continued)

Date	Gross income from oil ($)	Gross income from gas ($)	Total income before taxes and royalty ($)
16 Jan	481,919.6130	5,414.3844	487,333.9974
16 Feb	462,943.3708	4,263.4596	467,206.8304
16 Mar	587,463.6836	3,832.0054	591,295.6890
16 Apr	612,454.5667	3,986.0621	616,440.6288
16 May	704,857.8762	3,996.0173	708,853.8935
16 Jun	700,038.2361	5,070.8470	705,109.0831
16 Jul	662,162.4140	5,556.2460	667,718.6600
16 Aug	666,272.8487	5,412.9533	671,685.8020
16 Sep	640,931.6441	5,423.0019	646,354.6460
16 Oct	718,073.9607	5,457.3992	723,531.3598
16 Nov	632,440.7580	4,421.3532	636,862.1112
16 Dec	753,351.5341	6,299.0284	759,650.5624
17 Jan	760,896.0716	5,673.8418	766,569.9134
17 Feb	691,733.9210	4,344.5115	696,078.4325
17 Mar	712,009.2521	4,772.5229	716,781.7749
17 Apr	700,967.9587	4,883.1293	705,851.0880
17 May	687,954.6737	5,040.2867	692,994.9603
17 Jun	611,947.5479	4,536.5017	616,484.0496
17 Jul	648,477.4505	4,613.7582	653,091.2086
17 Aug	675,120.6419	4,417.7237	679,538.3656
17 Sep	688,464.0580	4,326.3640	692,790.4220
17 Oct	733,482.4739	4,258.8346	737,741.3085
17 Nov	770,089.8780	4,245.0632	774,334.9412
17 Dec	807,919.3301	4,053.1352	811,972.4653
18 Jan	869,602.4300	5,486.0307	875,088.4607
18 Feb	748,718.3678	3,377.9238	752,096.2916
18 Mar	834,003.4560	3,721.2330	837,724.6890
18 Apr	860,827.1774	3,702.0900	864,529.2674
18 May	944,832.5518	3,777.5836	948,610.1354
18 Jun	891,978.8110	3,831.3832	895,810.1942
18 Jul	926,168.3185	3,730.0164	929,898.3349
18 Aug	901,671.6642	3,855.3171	905,526.9813
18 Sep	920,970.6337	3,742.5330	924,713.1667
18 Oct	964,713.1592	4,180.2714	968,893.4306
18 Nov	755,002.8957	4,992.0454	759,994.9411

Table 10.13 Royalty and Taxes Calculations for Chapter Project

Date	Royalty ($)	Taxes on oil ($)	Taxes on gas ($)	Water handling cost ($)	State tax ($)	Income tax ($)
14 Nov	0.0000	0.00	0.00	0.0000	0.0000	0.0000
14 Nov	422,815.1759	110,584.59	2,166.13	11,243.0000	67,650.4281	225,501.4272
14 Dec	272,623.2339	70,839.94	1,859.59	9,645.0000	43,619.7174	145,399.0581
15 Jan	202,999.4024	53,380.34	752.83	8,599.0000	32,479.9044	108,266.3479
15 Feb	149,274.9400	38,972.13	834.52	9,460.0000	23,883.9904	79,613.3014
15 Mar	218,267.2520	57,495.43	709.17	8,811.4023	34,922.7603	116,409.2011
15 Apr	216,362.3900	57,152.21	544.43	7,985.3094	34,617.9824	115,393.2747
15 May	232,032.8727	61,327.64	547.79	7,855.6610	37,125.2596	123,750.8655
15 Jun	212,165.4610	56,106.25	471.20	7,308.2417	33,946.4738	113,154.9125
15 Jul	188,842.7863	49,897.91	460.17	7,307.6113	30,214.8458	100,716.1527
15 Aug	154,417.2353	40,758.69	419.24	7,101.1197	24,706.7576	82,355.8588
15 Sep	147,702.2961	39,020.00	367.28	6,704.4644	23,632.3674	78,774.5579
15 Oct	151,406.1058	40,058.17	316.79	6,777.5065	24,224.9769	80,749.9231
15 Nov	131,907.4290	34,914.15	261.16	6,430.3424	21,105.1886	70,350.6288
15 Dec	113,679.3240	30,075.92	238.57	6,526.1781	18,188.6918	60,628.9728
16 Jan	91,375.1245	24,095.98	270.72	6,417.6264	14,620.0199	48,733.3997
16 Feb	87,601.2807	23,147.17	213.17	5,914.2046	14,016.2049	46,720.6830
16 Mar	110,867.9417	29,373.18	191.60	6,234.5656	17,738.8707	59,129.5689
16 Apr	115,582.6179	30,622.73	199.30	5,954.0376	18,493.2189	61,644.0629
16 May	132,910.1050	35,242.89	199.80	6,076.5600	21,265.6168	70,885.3893
16 Jun	132,207.9531	35,001.91	253.54	5,812.1497	21,153.2725	70,510.9083
16 Jul	125,197.2488	33,108.12	277.81	5,939.9025	20,031.5598	66,771.8660
16 Aug	125,941.0879	33,313.64	270.65	5,877.0484	20,150.5741	67,168.5802
16 Sep	121,191.4961	32,046.58	271.15	5,631.1950	19,390.6394	64,635.4646
16 Oct	135,662.1300	35,903.70	272.87	5,764.0617	21,705.9408	72,353.1360
16 Nov	119,411.6459	31,622.04	221.07	5,527.9055	19,105.8633	63,686.2111
16 Dec	142,434.4805	37,667.58	314.95	5,662.9733	22,789.5169	75,965.0562
17 Jan	143,731.8588	38,044.80	283.69	5,615.4613	22,997.0974	76,656.9913
17 Feb	130,514.7061	34,586.70	217.23	5,033.1058	20,882.3530	69,607.8432
17 Mar	134,396.5828	35,600.46	238.63	5,531.2153	21,503.4532	71,678.1775
17 Apr	132,347.0790	35,048.40	244.16	5,313.3833	21,175.5326	70,585.1088
17 May	129,936.5551	34,397.73	252.01	5,451.5210	20,789.8488	69,299.4960
17 Jun	115,590.7593	30,597.38	226.83	5,239.4912	18,494.5215	61,648.4050
17 Jul	122,454.6016	32,423.87	230.69	5,378.2558	19,592.7363	65,309.1209
17 Aug	127,413.4435	33,756.03	220.89	5,343.1937	20,386.1510	67,953.8366
17 Sep	129,898.2041	34,423.20	216.32	5,138.7007	20,783.7127	69,279.0422
17 Oct	138,326.4953	36,674.12	212.94	5,277.9949	22,132.2393	73,774.1308

Table 10.13 (Continued)

Date	Royalty ($)	Taxes on oil ($)	Taxes on gas ($)	Water handling cost ($)	State tax ($)	Income tax ($)
17 Nov	145,187.8015	38,504.49	212.25	5,077.8544	23,230.0482	77,433.4941
17 Dec	152,244.8372	40,395.97	202.66	5,217.2964	24,359.1740	81,197.2465
18 Jan	164,079.0864	43,480.12	274.30	5,187.9979	26,252.6538	87,508.8461
18 Feb	141,018.0547	37,435.92	168.90	4,661.5508	22,562.8887	75,209.6292
18 Mar	157,073.3792	41,700.17	186.06	5,134.8425	25,131.7407	83,772.4689
18 Apr	162,099.2376	43,041.36	185.10	4,943.8085	25,935.8780	86,452.9267
18 May	177,864.4004	47,241.63	188.88	5,083.1504	28,458.3041	94,861.0135
18 Jun	167,964.4114	44,598.94	191.57	4,895.2618	26,874.3058	89,581.0194
18 Jul	174,355.9378	46,308.42	186.50	5,034.4316	27,896.9500	92,989.8335
18 Aug	169,786.3090	45,083.58	192.77	5,010.7146	27,165.8094	90,552.6981
18 Sep	173,383.7188	46,048.53	187.13	4,827.1142	27,741.3950	92,471.3167
18 Oct	181,667.5182	48,235.66	209.01	4,965.9278	29,066.8029	96,889.3431
18 Nov	142,499.0514	37,750.14	249.60	4,784.9117	22,799.8482	75,999.4941

Table 10.14 AVP and NVP Calculations for Chapter Project

Date	Income after tax and royalty ($)	Cum. income ($)	AVP ($)	NVP ($)
14 Nov	0.0000	0.0000	−12,358,188.8678	−12,358,188.8678
14 Nov	1,413,353.5268	1,413,353.5268	−10,944,835.3410	−10,944,835.3410
14 Dec	908,304.0422	2,321,657.5690	−10,036,531.2988	−10,030,975.4904
15 Jan	674,485.6506	2,996,143.2196	−9,362,045.6482	−9,351,683.6347
15 Feb	492,394.1311	3,488,537.3507	−8,869,651.5171	−8,854,930.0531
15 Mar	725,776.7945	4,214,314.1452	−8,143,874.7226	−8,125,857.2418
15 Apr	720,177.1529	4,934,491.2981	−7,423,697.5697	−7,403,173.0463
15 May	773,168.5631	5,707,659.8612	−6,650,529.0066	−6,628,470.8065
15 Jun	706,696.5801	6,414,356.4413	−5,943,832.4265	−5,920,838.8217
15 Jul	628,022.0546	7,042,378.4959	−5,315,810.3719	−5,292,315.0295
15 Aug	512,099.6873	7,554,478.1832	−4,803,710.6845	−4,779,831.3884
15 Sep	489,844.6144	8,044,322.7976	−4,313,866.0702	−4,290,045.6897
15 Oct	502,265.7573	8,546,588.5549	−3,811,600.3129	−3,788,455.0587
15 Nov	436,837.3849	8,983,425.9398	−3,374,762.9280	−3,352,413.5047

(Continued)

Table 10.14 (Continued)

Date	Income after tax and royalty ($)	Cum. income ($)	AVP ($)	NVP ($)
15 Dec	375,252.0748	9,358,678.0146	−2,999,510.8532	−2,977,997.1339
16 Jan	300,121.1270	9,658,799.1415	−2,699,389.7263	−2,678,545.0455
16 Feb	287,894.1156	9,946,693.2571	−2,411,495.6106	−2,391,549.4515
16 Mar	366,059.9577	10,312,753.2148	−2,045,435.6530	−2,027,394.3756
16 Apr	382,244.6602	10,694,997.8750	−1,663,190.9928	−1,647,608.6601
16 May	440,573.5276	11,135,571.4026	−1,222,617.4652	−1,210,492.3888
16 Jun	438,469.3454	11,574,040.7480	−784,148.1198	−775,941.7127
16 Jul	414,692.1499	11,988,732.8979	−369,455.9698	−365,387.0980
16 Aug	417,264.2214	12,405,997.1194	47,808.2516	47,255.5592
16 Sep	401,488.1186	12,807,485.2380	449,296.3702	443,856.3950
16 Oct	450,169.5233	13,257,654.7613	899,465.8935	888,083.4928
16 Nov	395,587.3798	13,653,242.1411	1,295,053.2734	1,277,957.0480
16 Dec	473,116.0075	14,126,358.1486	1,768,169.2808	1,743,861.5038
17 Jan	477,540.0090	14,603,898.1576	2,245,709.2898	2,213,610.5248
17 Feb	433,536.5028	15,037,434.6603	2,679,245.7926	2,639,488.4069
17 Mar	446,133.2574	15,483,567.9177	3,125,379.0499	3,077,297.0755
17 Apr	439,437.4299	15,923,005.3476	3,564,816.4798	3,508,031.0629
17 May	431,167.7914	16,354,173.1390	3,995,984.2713	3,930,153.8310
17 Jun	382,986.6701	16,737,159.8092	4,378,970.9414	4,304,447.0382
17 Jul	406,001.9337	17,143,161.7428	4,784,972.8751	4,700,935.7073
17 Aug	422,764.8225	17,565,926.5654	5,207,737.6976	5,113,443.4686
17 Sep	431,351.2412	17,997,277.8066	5,639,088.9388	5,533,919.3760
17 Oct	459,643.3827	18,456,921.1892	6,098,732.3214	5,981,677.3214
17 Nov	482,988.9959	18,939,910.1851	6,581,721.3173	6,451,822.7085
17 Dec	506,655.2879	19,446,565.4730	7,088,376.6052	6,944,632.1222
18 Jan	546,605.4535	19,993,170.9265	7,634,982.0587	7,476,012.3149
18 Feb	469,339.3536	20,462,510.2801	8,104,321.4123	7,931,186.6358
18 Mar	523,026.0233	20,985,536.3034	8,627,347.4356	8,438,365.3978
18 Apr	540,170.9532	21,525,707.2565	9,167,518.3887	8,961,740.3104
18 May	593,212.7602	22,118,920.0168	9,760,731.1490	9,536,355.7050
18 Jun	560,004.6860	22,678,924.7028	10,320,735.8350	10,077,905.4455
18 Jul	581,426.2653	23,260,350.9681	10,902,162.1003	10,639,758.6894
18 Aug	566,035.1011	23,826,386.0691	11,468,197.2013	11,185,974.4072
18 Sep	578,353.9638	24,404,740.0329	12,046,551.1651	11,743,591.1943
18 Oct	606,159.1671	25,010,899.2000	12,652,710.3322	12,327,678.1292
18 Nov	474,211.8885	25,485,111.0885	13,126,922.2207	12,782,628.2579

$$\textit{Oil revenue for Nov.}(2014) = 28727(STB) \times \frac{\$76.99}{STB} = \$2,211,692.$$

$$\textit{Gas revenue for Nov.}(2014) = 10140\ (MSCF) \times \frac{\$4.27}{MSCF} = \$43,322.5.$$

$$\textit{Water handling cost for Nov.}(2014) = 11243 \times \$1 = \$11,243.$$

$$\textit{Total revenue for Nov.}(2014) = \$2,211,692 + \$43,322.5 = \$2,255,014.$$

$$\begin{aligned} &\textit{NCF for Nov.}(2014) \\ &= -\$12,358,189 \\ &+ \$2,243,771 - \$422,815 - \$110,584.6 - \$11,243 - \$67,650 - \$225,501 \\ &= -\$10,944,835. \end{aligned}$$

From the calculations, the payout time or breakeven point can be found. The payout is estimated to be between the months of July and August 2016. Figure 10.10 clearly shows that the payout time period is from 21 to 22 months.

The return on investment (ROI) can be calculated by the use of equation (10.11):

$$ROI = \left(\frac{AVP}{CAPEX}\right) \times 100 = \left(\frac{\$13,126,922.2}{-\$12,358,188.87}\right) \times 100 = 106\%.$$

Hence, when the time value of money is not considered, the ROI indicates that $1.06 was earned for every dollar invested.

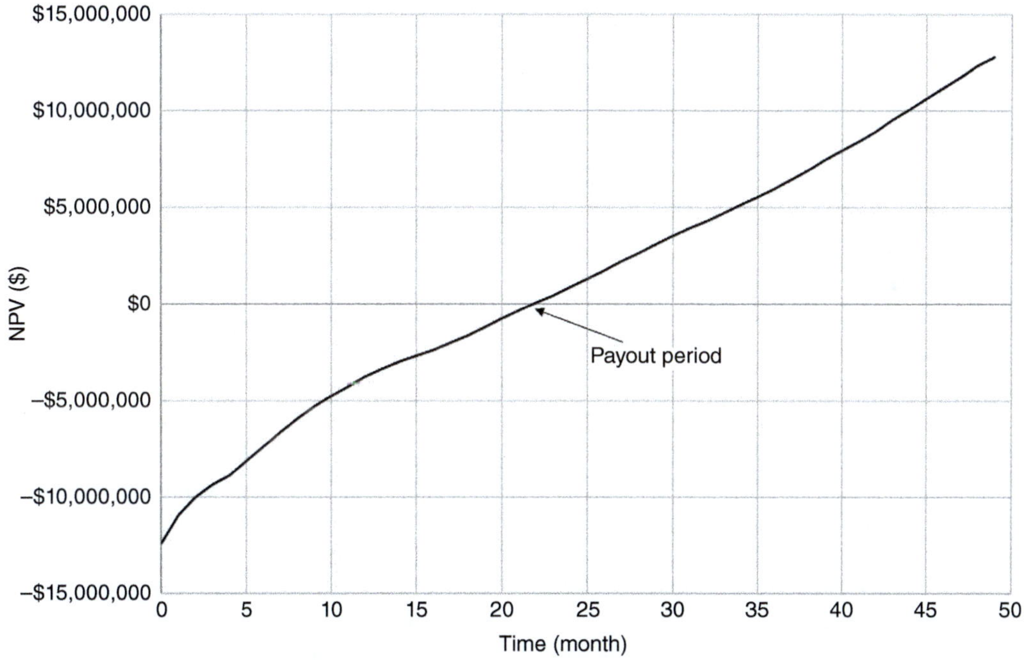

Figure 10.10 Payout time period.

10.7 Problems

1 Generate the NCF for Example 10.1 considering an average oil price of $43 U.S. dollars. Determine the AVP for the well after the 4-year period and estimate the payout period.

2 Determine the IRR for Problem 10.1.

3 Determine the NPV for Problem 10.1 considering the time value of money with a discount rate of 15% compounded annually.

4 Repeat the economic analysis section of the chapter project for a pessimistic scenario of oil price being 25% lower than the historical values reported for that problem, versus an optimistic scenario of oil prices being 25% higher, and compare the major economic factors for the three scenarios.

5 Compare the decline curves reported for an example Niobrara well versus the Bakken well and compare the economics of production assuming everything else being the same.

6 Investigate the history of buy-back contracts applied for different oil and gas projects internationally and report the major characteristics of those contracts.

Additional Reading

Mian, M. (2011). *Project economics and decision analysis. Volume 1: Deterministic models*, 2nd ed. Tulsa: PennWell.

Park, C., Kumar, P., & Kumar, N. (2013). *Fundamentals of engineering economics.* Pearson Education Limited.

Krupnick, A., & Echarte, I. (2017). *Economic impacts of unconventional oil and gas development.* Resources for the Future.

References

Bindemann, K. (1999). *Production sharing agreements: An economic analysis.* Oxford Institute of Energy Studies.

Energy Information Administraiton (EIA) (2017). U.S. Field production of crude oil. Available at https://www.eia.gov/dnav/pet/hist/LeafHandler.ashx?n=pet&s=mcrfpus2&f=a

Energy Information Administration (EIA) (2019). Energy and financial markets. Available at https://www.eia.gov/finance/markets/crudeoil/demand-nonoecd.php

Harraz, H. (2016). *Petroleum laws: Types of international petroleum contracts.* Tanta Universit.

Holodny, E. (2016). TIMELINE: The tumultuous 155-year history of oil prices. *Business Insider.*

Husni, M. (2009). *Petroleum economics.* KFUPM; Dhahran, Saudi Arabia.

OGJ Editors (2018). IEA: Global oil supply will outpace demand throughout 2019. *Oil & Gas Journal*, 14 Nov 2018.

indexmundi (2019). *Crude oil (petroleum) monthly price.* Available at https://www.indexmundi.com/commodities/?commodity=crude-oil&months=60.

Kagan, J. (2018). *Payout.* investopedia.

Karasalihović-Sedlar, D., Barbir, G. & Brkic, V. (2017). Types of fiscal regime in hydrocarbon exploration and production. *The Mining-Geology-Petroleum Engineering Bulletin*, **1**(6), 45–54.

Krupnick, A., & Echarte, I. (2017). Economic impacts of unconventional oil and gas development. *Resources for the Future*, 23 June.

Macrotrends (2018). Crude oil prices: 70 year historical chart. Available at https://www.macrotrends.net/1369/crude-oil-price-history-chart.

Mazeel, M. (2010). *Petroleum fiscla systems and contracts.* Hamburg: Diplomica Verlag.

Mian, M. (2011a). *Project economics and decision analysis. Volume 1: Deterministic models, 2nd ed.* Tulsa: PennWell.

Mian, M. (2011b). *Project economics and decision analysis. Volume II: Probabilistic models, 2nd ed.* Tulsa: PennWell.

Park, C., Kumar, P., & Kumar, N. (2013). *Fundamentals of engineering economics.* Pearson Education Limited.

World Bank. (2007). World Bank Cambodia oil & gas special supplement: Contracts for petroleum development, Part 2 (English). Petroleum Sector Briefing Note; No. 8 (November 2007). Washington, DC: World Bank.

11

Environmental Aspects of Shale Hydrocarbon Reservoir Developments

Reza Barati Ghahfarokhi, Stephen Randtke, and Edward Peltier

The University of Kansas, Lawrence, Kansas, USA

11.1 Background

As a result of exploration and drilling activities that resulted in a substantial increase of shale oil and gas production in the past two decades, use and transport of significantly greater loads of water, chemicals, and equipment (e.g., fracturing pump engines) to production sites has gained much public attention. Pollution caused by fracturing trucks during the injection process, in addition to induced seismicity caused by disposal of large volumes of produced water in a few states, has added even more heat to discussions about the environmental aspects of the recent shale oil and gas development methods (e.g., hydraulic fracturing and horizontal drilling).

Hydraulic fracturing is not a recently invented method and has been applied since the late 1940s. However, as a result of the recent developments in technology and based on the current needs of society to produce sustainable hydrocarbon fuels from resources that have been previously impossible to produce from, a significantly larger volume of water and chemicals is required to complete the current drilling/hydraulic fracturing jobs. This requires access to significantly larger volumes of fresh or low-salinity fresh or produced water, vast amounts of chemicals and sand, and a large number of fracturing engines running for several days to inject the large volumes of fluids required for such fracturing jobs. As an example, the average volume of water used for hydraulic fracturing has increased from 500,000–700,000 gal in conventional reservoirs to 3.2–4.8M gal in unconventional shale oil and gas wells (Freyman, 2014).

The industry has moved toward improving its practices to address these issues by developing new technologies to reduce land impact; implementing energy-efficient equipment and facilities; improving logistics; and reducing emissions, noise, lighting, drilling time, and flaring (Haut et al., 2018).

This chapter discusses the environmental concerns related to water use and reuse, chemicals used in the hydraulic fracturing industry, and the air pollution concerns. Induced seismicity and its relation to the hydraulic fracturing industry will also be addressed in this chapter.

11.2 Water Management and Reuse

11.2.1 Basic Terminology Related to Water Management and Reuse

A portion of the water injected during hydraulic fracturing jobs is returned to the surface during the first days to weeks of production.

Unconventional Hydrocarbon Resources: Techniques for Reservoir Engineering Analysis, Advanced Textbook Series, First Edition. Edited by Reza Barati Ghahfarokhi and Mustafa M. Alhubail.

This fraction, called flowback water, varies among different shale plays depending on their water-sensitive clay content, the complexity of the fracture system, the wettability of the rock, and the role of capillary forces. The flowback water fraction can be very low in dry gas shale plays such as Marcellus, or very high in wetter shale plays such as the Barnett (Table 11.1) (Boschee, 2014). The flowback water production rate typically increases sharply at the beginning of production, reaches a maximum rate, and then goes through a period of decline that can take from a few days to a few weeks (Fig. 11.1) (Li, 2013). Since flowback water is mainly fracturing fluid that has interacted with the reservoir rocks, it may therefore contain minerals, both radioactive and nonradioactive, dissolved from the formation, and it may also undergo mixing with formation water. By the end of the cleanup period, the salinity and composition of the flowback water typically approach those of the in-situ formation water, at which point it contains little or no hydraulic fracturing fluid.

Even though flowback water has been part of the hydraulic fracturing production process

Table 11.1 Percentage of Produced and Flowback Water in Major U.S. Shale Plays

Producing area	Fraction of hydraulic fracturing fluid returned as flowback	% produced water volumes	TDS at the end of flowback time (ppm)
Bakken	15–40	High	150,000–300,000
Eagle Ford	<15–70	Varies	15,000–55,000
Permian basin	20–40	High	20,000–300,000
Marcellus	<10–60	Varies-Low	3,000–260,000
Denver-Julesburg	5–35	Low	20,000–65,000
Barnett	30–60	High	40,000–185,000
Haynesville	10–60	Varies	40,000–250,000

Source: Original table reported by Boschee (2014) including percentage of hydraulic fluid flowback and produced water, modified based on the results reported by Hayes (2011), Blauch (2011), and Kondash et al. (2017).

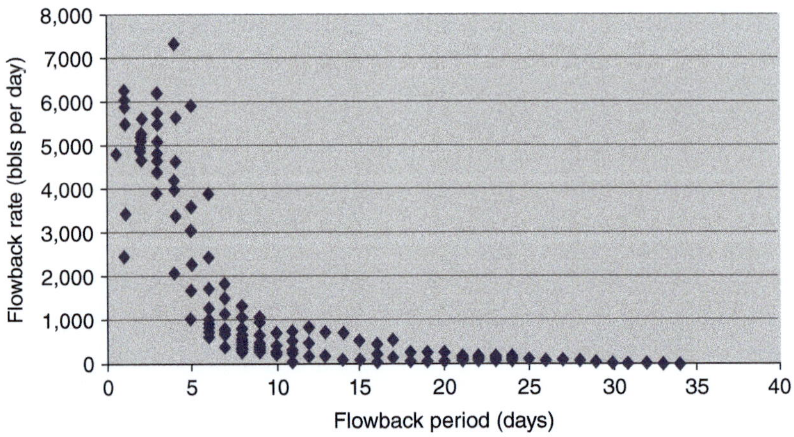

Figure 11.1 A representative flowback rate for the Marcellus shale gas play (Li, 2013).

since this technology was invented, flowback water volumes were historically small for conventional wells. Large volumes of fracturing fluids used in multistage hydraulic fracturing jobs resulted in significantly larger volumes of flowback water that need to be separated and handled during the first weeks of production.

Flowback water is different than the movable water present in-situ within the inorganic matter and microfractures in the formation, or being produced from adjacent water-bearing formations (aquifers) after production starts, and subsequently over much or all of the reservoirs' life. This second type of water is called produced water. Produced water has been generated since the first oil-producing wells were drilled, but greater volumes are now produced in some areas due to the nature of the formations involved and the capacities of the wells made possible using modern drilling technology. The oil and gas industry has typically injected excess produced water that cannot be reused, into water disposal wells in nearby formations that contain no producible hydrocarbons. Produced water and water disposal wells are conventional terms, and the recent unconventional developments have generally not resulted in greater volumes of produced water, per unit of oil or gas produced, compared with conventional wells. In fact, a good number of the current shale oil and gas reservoirs have a very small movable water saturation.

Produced and flowback water are often classified based on their salinity. Classifications vary, but, as an example, Nicot et al. (2012) use the following definitions: (1) fresh water has a total dissolved solids (TDS) concentration of < 1,000 mg/L, (2) brackish water has a TDS value within the 1,000–35,000 mg/L range, (3) saline water has 35,000–50,000 mg/L of TDS, and (4) brine has a TDS of > 50,000 mg/L. (Nicot et al., 2012). Groundwaters having a TDS concentration of <10,000 mg/L of TDS are considered underground sources of drinking water (USDWs) and are protected under several federal statues in the United States (EPA, 2016). However, federal, state, and local regulations, policies, guidelines, and allowable exceptions can differ significantly from one location to another, so the appropriate regulatory authorities should always be consulted regarding water use, reuse, and disposal in a particular setting. A detailed discussion of this topic is beyond the scope of this book, but the reader should note that the 10,000 mg/L guideline is by no means universal and is likely to change in the future. Membrane-based desalination technologies are now being used in many water-short locations around the globe to produce drinking water from water sources containing up to 35,000 mg/L of TDS, and thermal desalination technologies can handle much higher TDS levels.

Water use is defined as the situation where water that is withdrawn for a specific purpose will be returning, partially or completely, to the local hydrologic cycle. Water consumption, on the other hand, is defined as the water that is removed from the local hydrologic cycle following its use due to reasons such as evaporation from storage ponds, retention of water in the subsurface through imbibition, or disposal in Class II wells (Maupin et al., 2014).

11.2.2 Water Cycle in Oil and Gas Production

There are five main steps for the water cycle during the drilling and completion process of a multistage, hydraulically fractured well (Fig. 11.2) (EPA, 2016).

1) Water acquisition for drilling and hydraulic fracturing: Water acquisition from ground or surface resources is the first stage of the hydraulic fracturing water cycle. Water is used extensively during both drilling and hydraulic fracturing processes. Freshwater is normally used, but brackish or produced waters are being increasingly utilized in water-short areas. In some cases, flowback water is filtered and its properties are adjusted for reuse, but the volumes are not sufficient to meet the requirements of another fracturing job of similar size.

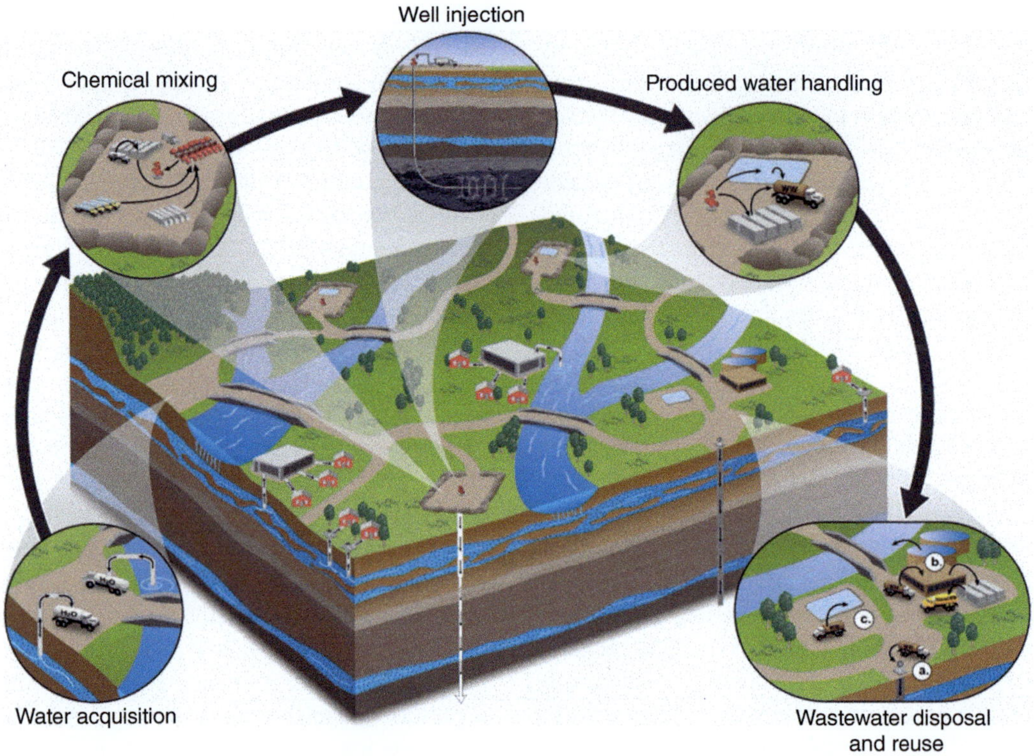

Figure 11.2 Water cycle during drilling and hydraulic fracturing of unconventional wells (EPA, 2016).

2) Chemical mixing for drilling and hydraulic fracturing processes is the next stage in the water cycle. Various chemicals, such as those listed in Chapter 6, are used to prepare hydraulic fracturing fluids. In water-based systems, these chemicals are mixed in water and proppant is added before the fluid is injected.
3) The drilling process requires circulation of the drilling mud back to the surface and reuse of the fluid after drilling cuttings are separated in the shaker, and the mud properties are adjusted by adding different chemicals. Fracturing fluid injection occurs at high pressures to overcome the parting pressure of the rock.
4) Produced and flowback water handling is an important step during the production of hydraulically fractured wells. Produced and flowback water will be separated from the hydrocarbon in multistage separators before being diverted to disposal, reuse, or further treatment.
5) Wastewater disposal, reuse, or treatment are the next stages in the life cycle of water during the postfracturing production phase. Produced and/or flowback water are typically quickly treated before being used for other drilling, fracturing, Frac Hit prevention, or improved oil recovery processes utilizing other conventional and/or unconventional wells in the area. Excess produced water is typically injected into deep wells to be disposed of in formations not containing any usable natural resources such as oil, gas, minerals, or freshwater. In a very rare case where water treatment facilities are available and the practice is economically viable, the excess water may be treated to be used for other industrial or agricultural applications.

11.2.3 Water Acquisition for Hydraulic Fracturing

Historically, the majority of groundwater and surface water resources used for hydraulic fracturing also provided water for public water supplies or private drinking water wells. This has locally impacted some counties, especially in states that heavily rely on agriculture or that have other active industries or population centers requiring access to the same sources of water (Fig. 11.3) (EPA, 2016).

According to EPA (2016), hydraulic fracturing used 36 billion gal of water in 2011 and 52 billion gal in 2012. This results in an annual average of 44 billion gal of water in 2011 and 2012 across all 20 states included in the study. This accounts for only a small percentage of the total volume of water used for all purposes, at both national and state levels (EPA, 2016). USGS has reported the total water usage by mining and oil and gas industry in 2015 to be less than 1% nationally (Fig. 11.4) (Dieter et al., 2018). At the state scale, 94% of the reported water use occurred in 7 of the 20 states studied in the EPA FracFocus 1.0 project database: Texas, Pennsylvania, Arkansas, Colorado, Oklahoma, Louisiana, and North Dakota (Dieter et al., 2018). The distribution is reportedly even more uneven at the county level for

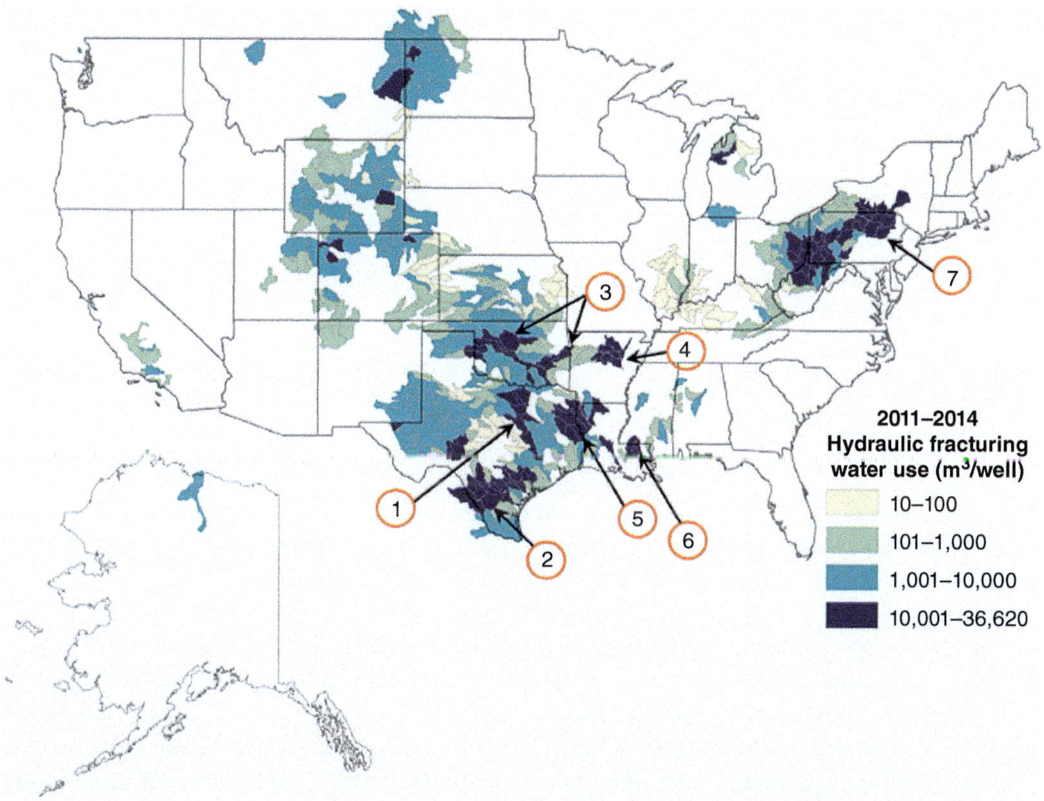

Figure 11.3 Water volumes used per well to hydraulically fracture oil and gas wells (horizontal, directional, and vertical) drilled from January 2011 to August 2014. Shale gas plays are numbered as follows: (1) Barnett (Bend Arch-Ft. Worth Basin), (2) Eagle Ford (Gulf Coast Basins), (3) Woodford (Anadarko and Arkoma Basins), (4) Fayetteville (Arkoma Basin), (5) Haynesville-Bossier (Gulf Coast Basins), (6) Tuscaloosa (Gulf Coast Basins), and (7) Marcellus and Utica (Appalachian Basin) (Gallegos et al., 2015).

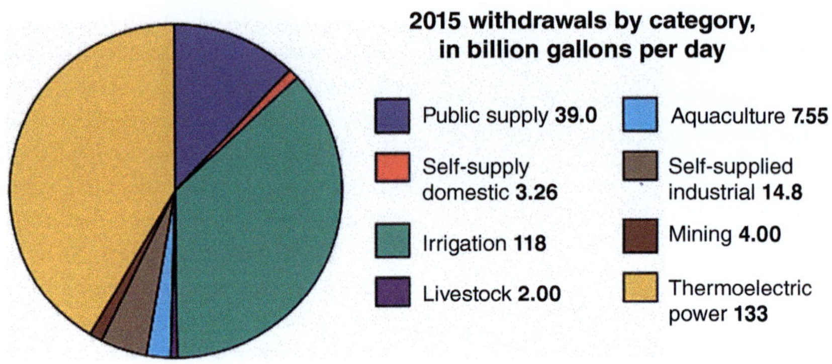

Figure 11.4 Water usage in the United States for year 2015 by category (Dieter et al., 2018).

Table 11.2 Surface Water Versus Groundwater Proportions

Location	Surface water	Groundwater	Year or time period of estimate
Louisiana–Haynesville shale	87%	13%	2009–2012
Oklahoma–Statewide	63%	37%	2011
Pennsylvania–Marcellus shale, Susquehanna River basin	92%	8%	2008–2013
Texas–Barnett shale	50%	50%	2011–2013
Texas–Eagle Ford shale	10%	90%	2011
Texas–TX-LA-MS salt basin	30%	70%	2011
Texas–Permian basin	0%	100%	2011
Texas–Anadarko basin	20%	80%	2011
West Virginia–Statewide, Marcellus shale	91%	9%	2012

Source: Maupin et al. (2014).

some of these states. In Texas, for example, due to the large number of wells drilled, hydraulic fracturing water use was 10% or more compared to total water use in 26 counties, 30% or more in 9 counties, and 50% or more in 4 counties. Note that total consumption estimates follow the same pattern in those counties, with more counties in the higher percentage category. In those counties, total consumption is 10% or more compared to total water consumption in 53 counties; 30% or more in 25 counties; 50% or more in 16 counties; and over 100% in 4 counties (EPA, 2015). This indicates that the reason for the water shortage in counties such as McMullen County was a result of an overall water consumption problem rather than hydraulic fracturing alone.

The water used during hydraulic fracturing may be provided by surface water (rivers, streams, lakes, and reservoirs), groundwater, or reused hydraulic fracturing water. The ratios between ground and surface water vary between different plays (Table 11.2). The source of water used is a function of availability, economics, and local policies. In some states, a percentage of injected water volume is coming from reused hydraulic fracturing wastewater. This decision is driven by economics and the quality of wastewater as well as the availability of groundwater or surface water. Transportation costs (i.e., trucking and piping), availability of Class II wells (wells used to inject salt water, enhanced oil recovery fluids, or hydrocarbon for the purpose of storage), local regulations, and compatibility of produced water with in-situ water are the defining factors (Mantell, 2011; EPA, 2016).

One example of a state that used surface water heavily is Louisiana. According to a report by Louisiana Department of Natural Resources, during the early stages of hydraulic fracturing development in 2008, groundwater from the Carrizo-Wilcox aquifer was used heavily in Louisiana. The state started to encourage surface water use due to concerns related to the sustainability of groundwater resources (Commission, 2012).

Reuse of flowback water is encouraged, but it is limited by the water quality factors affecting the performance of hydraulic fracturing. Factors such as (1) high TDS and ionic strength, (2) high concentrations of cations (calcium, magnesium, iron, barium, strontium) and anions (chloride, bicarbonate, phosphate, and sulfate), and (3) presence of microbial agents can adversely affect the performance of the water-based fracturing fluid systems. Extra treatment or mixing with freshwater may be required to overcome the potential problems caused by these impeding factors (Leiming et al., 2016; Haghshenas & Nasr-El-Din, 2014).

11.2.4 Flowback and Produced Water Quantity and Quality

The injected hydraulic fracturing fluid can be prepared in a brine as simple as a 2%–6% KCl or NaCl solution, a low salinity surface or groundwater, filtered/treated produced/flowback water or a filtered version of produced/flowback water mixed with available lower salinity surface or groundwaters to make the recipe compatible with fracturing fluid formulations. The injected fracturing fluid interacts with the rock and fluids available within the nanosized to microsized pores and fractures during its residence time in the formation. As a result, the flowback water starts with a composition closer to that of the fracturing fluid at the beginning of production, which will gradually change toward the in-situ brine or an equilibrium state version of the fracturing fluid with the reservoir's rock-fluid system (Fig. 11.5). Therefore, it is not practical to use water quality data for a single sample to characterize the composition of the flowback water as the composition changes during the cleanup time. It is common to have significantly higher concentrations of naturally occurring radioactive materials (NORMs) in addition to divalent ions in the flowback water. Table 11.1 presents a range of salinity reported for the flowback/produced water after fracturing for different shale plays in the United States. The range of salinity variation is large and it is partially due to variation of the timing after production when the flowback sample is taken in addition to the fact that in some shale plays a mixture of flowback and produced water is being produced and the contributing ratios of each is nearly impossible to accurately predict, even with the aid of highly capable geochemical models coupled with flow models.

Figure 11.5 summarizes the results of a comprehensive study by Kondash et al. (2017) where they compared the water salinity and water production during the flowback time for different major shale plays across the United States. It is very clear that the TDS values are approaching a plateau toward the end of the fracturing fluid cleanup time. Very long cleanup times are the most highlighted characteristic of all these different plays. Marcellus, being a dry shale gas play, is an exception in terms of having low flowback volumes. The high slope of water production in other plays when TDS is close to reaching a plateau is a good indicator of those wells having a movable water saturation in-situ or having a sort of a connection to a wet layer or an aquifer through natural fractures.

In addition to the inorganic components, small concentrations of semivolatile organic compounds are reported to be present in the flowback water (Table 11.3). These concentrations are

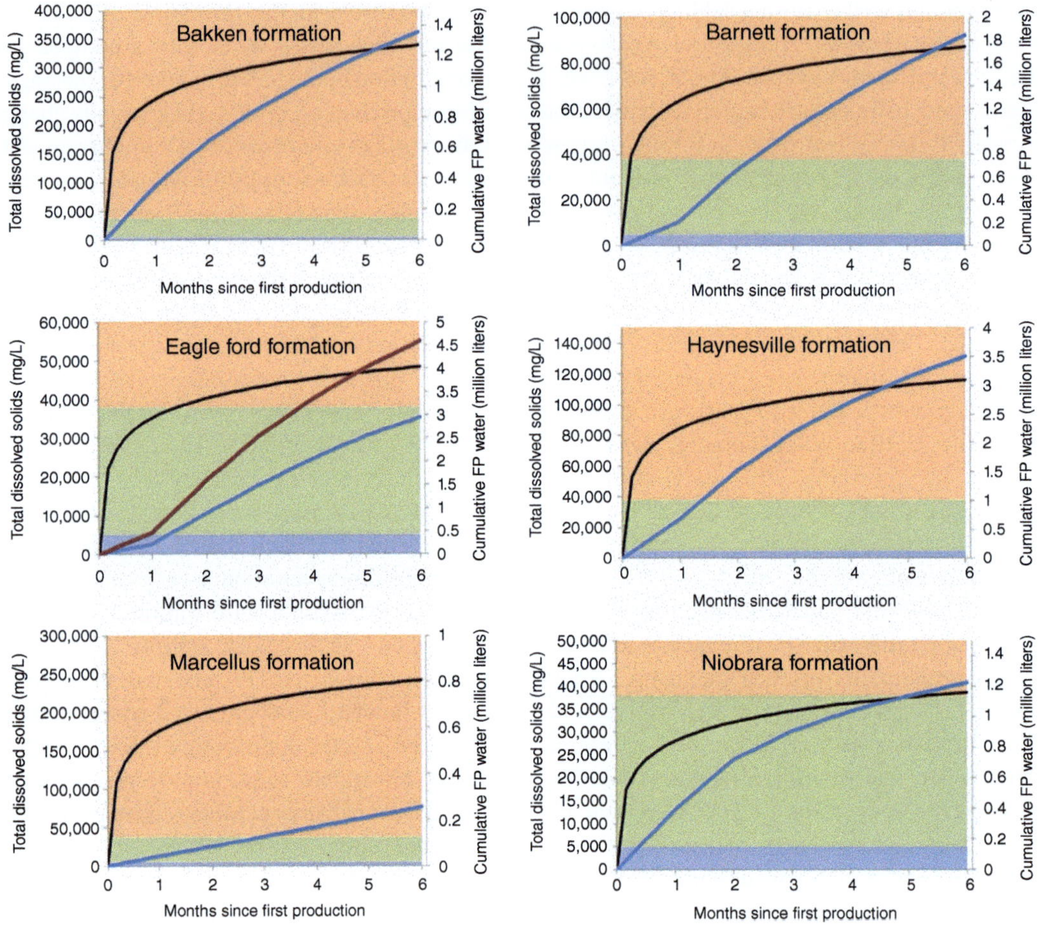

Figure 11.5 A comparative study of water flowback water salinity (black line), and water volume (blue or red) for several major shale oil and shale gas plays in the United States. The blue, green, and orange categories on the chart refer to the brackish, saline, and hypersaline brine categories, respectively (Kondash, 2017).

Table 11.3 Semivolatile Organic Compounds Present in the Flowback Water of Four Major Shale Plays

Parameter	Concentrations in produced formation water by shale play (μg/L)			
	Barnett	**Fayetteville**	**Haynesville**	**Marcellus**
Benzene	< 1–97	<1–140	<1 -3,460	<1–1,950
Ethyl benzene	<1–249	<1–11	<1–1,470	<1–164
Toluene	1–578	<1–117	2.6–11,400	<1–3,190
Xylenes, total	<3–2,480	<3–60.1	12–10,700	<2–1,970
Phenol	ND–1,160	ND	ND–971	ND–459
Pyridine	ND–264	ND–534	ND	ND
No. of wells sampled	3	2	3	4

Source: Coleman (2012).

slightly higher in the produced water samples collected from these formations. Concentrations of these organic compounds are varying significantly from one shale play to another and versus the time of flowback water production. The total organic carbon content in the flowback water decreases versus time to reach a plateau when the majority of the produced sample is actually produced water according to Severin and Hayes (2012).

In reservoirs with movable in-situ water, the composition of the flowback water will eventually transition into the in-situ brine. Table 11.4 summarizes of produced water composition values collected from the literature for six major plays in the United States (Rimassa et al., 2009; Coleman, 2012).

In general, significantly larger volumes of produced water are generated compared with the flowback water volumes. Table 11.5 summarizes the oil and water production for the top 10 states in each category based on the 2012 results published by Veil (2015). Significantly larger volumes of water compared with the produced oil is a very good indication that we are also in the "water business" in addition to being involved with "oil business." It is also worth noting that there are several states such as Kansas and Oklahoma that are higher on the water production ranking compared with their ranking on oil production. These values are of course a function of the total produced volume, which by itself is a function of the oil price. The states of Wyoming, Oklahoma, and Kansas show the highest water-oil-ratios (WOR) compared with the other states. Veil (2015) reported the TDS distribution of their produced-water database for the United States indicating that the majority of those samples had a TDS of below 100,000 mg/L (Fig. 11.6).

11.2.5 Flowback and Produced Water Reuse

Availability of significantly large volumes of produced water and smaller, but still significant, volumes of flowback water has convinced the industry to reuse such waters for hydraulic fracturing and drilling applications in addition to its use for water flooding and EOR processes. While produced water has been historically separated from hydrocarbons and sent to water disposal wells via pipelines or trucks, it will be sent to a water treatment facility for an economically viable treatment before it is stored back at the treated water storage ponds to be used for fracturing purposes in the future. This treated water is typically mixed with other sources of water to make up for the lost volumes during the previous treatments. A schematic of the oil, gas, and water process with water treatment facilities to prepare the produced water for reuse purposes is shown in Figure 11.7.

There have been two major case studies examining replacement of freshwater sources with produced water for different shale plays: the Chesapeake Energy case study (Mantell, 2011) and the State of Oklahoma example (Oklahoma Produced Water Working Group, 2017). In the following paragraphs, a summary of the Chesapeake Energy case study is presented.

1) Barnett shale with produced water flowrates of more than 1,000 gal/MMCF: High volumes of produced water generated in this play demonstrated a significant increase in TDS (50,000–140,000 ppm) and chlorides (25,000–80,000 ppm) over time. With the relatively low total suspended solid (TSS) content of the produced water, filtration of the produced water was not challenging. Furthermore, iron values are relatively low compared with other plays. The reuse rate achieved was 230,000 gal per well, approximately 6% of the total water needed (Mantell, 2011).
2) Fayetteville shale with produced water flowrates of 200–1,000 gal/MMCF: This formation has less water production per MMCF due to its being relatively desiccated. Water quality is very good on both initial and long-term production with very low

Table 11.4 Produced Water Composition Examples from Several Major Shale Plays in the United States

Parameter	Sodium (mg/L)	Magnesium (mg/L)	Calcium (mg/L)	Strontium (mg/L)	Barium (mg/L)	Iron (mg/L)	SO_4^{2-} (mg/L)	HCO_3^- (mg/L)	Cl^- (mg/L)	Potassium (mg/L)	TDS (mg/L)
Fayetteville	5,362.6	77.3	256.3	21	0.8	27.6	149.4	1,281.4	8,042.3	NA	15,219
Marcellus	24,445.0	263.1	2,921.0	347	679.0	25.5	9.1	261.4	43,578.4	NA	72,533
Barnett	12,453.0	253.0	2,242.0	357	42.0	33.0	60.0	289.0	23,797.5	NA	39,570
NorthWattenberg	5,754.8	42.9	380.6	55.4	18.2	80.7	3.5	196.0	10,798.6	NA	18,285
Piceance basin	20,200.0	0	148.0	NA	NA	0	1,600.0	108.0	6,880.0	2,290	33,100
Northeast Texas	49,600.0	1,450.0	14,100.0	NA	NA	62.0	0	NA	80,500.0	1,580	147,000

Source: Data from Coleman (2012) and Rimassa et al. (2009).

Table 11.5 Comparison of the Top 10 States With the Highest Produced Water Volumes With Those With Highest Oil Production

Ranking	State	2012 Water (bbl/yr)	% of Total Water	WOR
1	Texas	7,435,659,000	35	12.23
2	California	3,074,585,000	15	15.55
3	Oklahoma	2,325,153,000	11	25.00
4	Wyoming	2,178,065,000	10	47.99
5	Kansas	1,061,019,000	5	24.26
6	Louisiana	927,635,000	4	11.21
7	New Mexico	769,153,000	4	9.01
8	Alaska	624,762,000	3	3.25
9	Federal offshore	358,389,000	2	0.74
10	Colorado	320,191,000	2	6.49

Ranking	State	2012 Oil (bbl/yr)	% of total oil
1	Texas	608,213,000	26.9
2	Federal offshore	482,774,000	21.3
3	North Dakota	243,272,000	10.7
4	California	197,749,000	8.7
5	Alaska	192,368,000	8.5
6	Oklahoma	92,988,000	4.1
7	New Mexico	85,341,000	3.8
8	Louisiana	82,781,000	3.7
9	Colorado	49,361,000	2.2
10	Wyoming	45,382,000	2.0

Source: Based on Veil (2015).

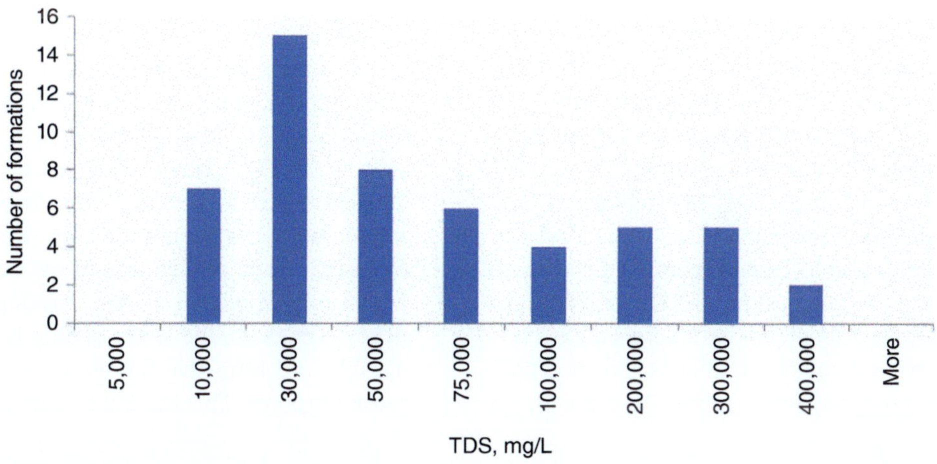

Figure 11.6 Median TDS for produced water in the United States (Veil, 2015).

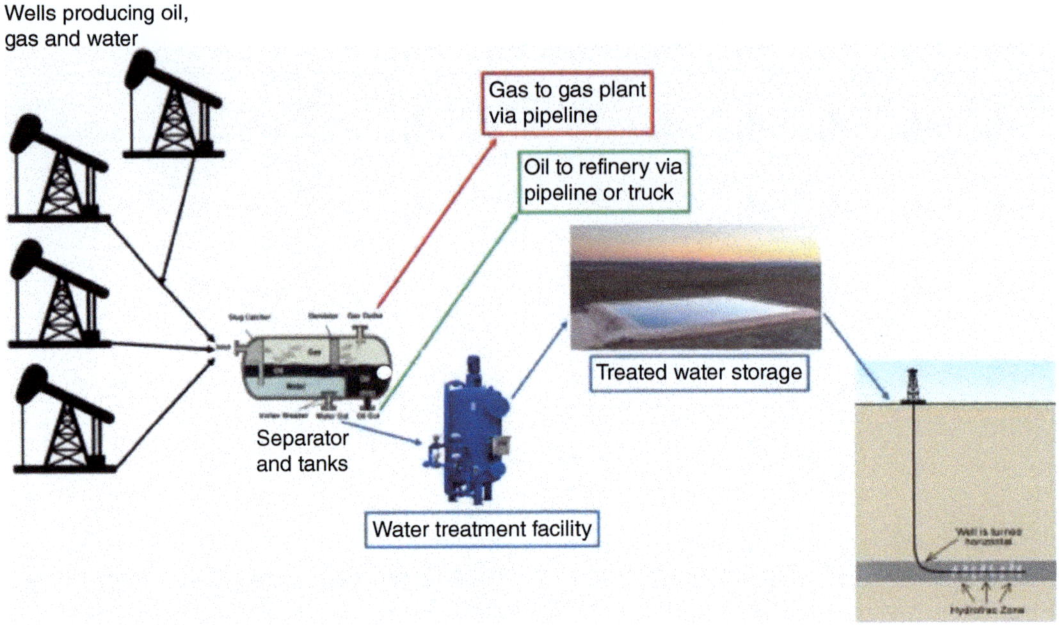

Figure 11.7 Simplified oil, gas, and water process with water treatment facilities for water reuse purposes (Oklahoma Produced Water Working Group, 2017).

chlorides and low TDS of approximately 15,000 ppm. Low calcium and magnesium content of the water resulted in lower scaling tendency. This provides an excellent potential for reuse of both initial and long-term produced water resulting in the reuse of 250,000 gal per well. This is equal to approximately 6% of the total water needed for hydraulic fracturing in that area (Mantell, 2011).

3) Haynesville shale with produced water flowrates of 200–1,000 gal/MMCF: This formation generates relatively little produced water per MMCF because it is relatively desiccated. The water quality is not very good, with high TDS, high chlorides, and high TSS (~350 ppm). The scaling tendency is high due to high calcium (~8,000 ppm) and high magnesium (~500 ppm) concentrations. All of these parameters make Haynesville a relatively unattractive potential for produced water reuse. As a result, no large-scale reuse of produced water was practiced for this shale play (Mantell, 2011).
4) Marcellus shale with produced water flowrates of < 200 gallons/MMCF: Higher water production from Marcellus is experienced in the southern part of the play (West Virginia) and lower water production in the northern part (Pennsylvania). Marcellus shale formation characteristics tend to trap fluids due to its being a highly desiccated formation (very dry). Therefore, water imbibition is dominant during the shut-in time. The produced water exhibits high TDS concentrations of 40,000–90,000 ppm with long-term TDS concentrations of over 120,000 ppm. TSS values are, however, rather low (~160 ppm), making filtration easier. Current concentrations of Ca and Mg cause a moderate to high scaling tendency. That said, the water quality is manageable and attractive for produced water reuse purposes. This has led to a reuse rate

per well of approximately 10% of the total water needed for hydraulic fracturing (Mantell, 2011).

Produced/flowback water was collected and stored in holding tanks onsite before it was pumped through a 100 μm followed by a 20 μm filter to remove suspended solids. It was then stored in a clean-water tank and moved to the next fracturing site. Produced water was mixed with freshwater to reduce TDS as very high TDS affects fracturing fluids and causes extra friction. Water hardness, sulfate, barium, iron content, and bacteria were checked and treated if the values were high, to prevent issues such as scaling, emulsion formation, and plugging of pores (Mantell, 2011).

As a result, Chesapeake Energy reported a reduction in the wastewater volume sent offsite for disposal, reduced freshwater use, and reduced truck traffic, lowering the impact on roads, reducing noise, and improving air quality. Therefore, a significant reduction in the cost of operation by lowering water demand, water disposal, and transportation costs was reported.

Haghshenas and Nasr-El-Din (2014) prepared conventional cross-linked fracturing fluids used by the industry in water samples containing different concentrations of magnesium, calcium, sodium, and potassium demonstrating a maximum acceptable concentration for each ion. Their work demonstrated that guar-based cross-linked fluids can be prepared in a large number of produced and flowback water samples with no significant damaging effect on the performance.

LeBas et al. (2013) reported a successful field application of zirconium-cross-linked CMHPG fluids prepared in produced water with TDS values of 270,000 ppm and significant concentrations of calcium and magnesium in the Delaware basin. They reported the following advantages: (1) Replacement of 8 million barrels of fresh water with produced water; (2) reduction in trucking of 1,400 loads of freshwater; (3) savings of $70,000–$100,000 per well from the water management costs versus freshwater purchasing and transportation.

Based on the latest report by EPA (2015), the average percentage of water reuse from the flowback water across different basins in the United States is only 15%, with a wide distribution depending on the salinity of the produced/flowback water, availability of ground and surface waters for hydraulic fracturing purposes, and the cost of treatment (Table 11.6).

11.3 Chemicals used in Fracturing Fluids

EPA identified 1,173 chemicals associated with hydraulic fracturing operations in the United States, of which 1,076 are (or were) used as additives for fracturing fluids and 134 were reportedly found in flowback or produced water (EPA, 2015). In a different report, EPA summarized the composition of the top 20 most frequently used fracturing fluids used for oil and gas wells based on the data reported by companies to the FracFocus 1.0 database as listed in Tables 11.7, 11.8, and 11.9. In a typical fracturing fluid system, more than 95% (v/v) of the fluid injected is water. The rest of it is mainly sand with 1%–2% (v/v) of chemicals added to the solution to satisfy the necessary conditions required for a fracturing fluid to be successful.

In a comprehensive study, Stringfellow et al. (2014) investigated 81 commonly used hydraulic fracturing fluid additives with reported data. They reported that of the 81 studied chemicals, 54 were organic and 27 were considered biodegradable. They analyzed the contributions of these chemicals to the chemical oxygen demand (COD) of fracturing fluids. COD is a measure of the total amount of chemically oxidizable organic matter in a water sample. They found that only 4 chemicals imparted high

Table 11.6 Reuse of Hydraulic Fracturing Flowback Water in Various States, Basins and Plays as a Percentage of Injected Water Reported by EPA

State, basin, or play	Available estimate	Year of estimate
Texas–Barnett shale	5%	2011
Texas–Eagle Fond shale	0%	2011
Texas–TX-IA-MS salt basin	5%	2011
Texas–Permian basin (far west portion)	0%	2011
Texas–Permian basin (midland portion)	2%	2011
Texas–Anadarko basin	20%	2011
Colorado–Garfield County, Uinta-Piceance basin	100%	NA
Colorado–Wattenberg field, Denver-Julesburg basin	0%	NA
Pennsylvania–Marcellus shale, Susquehanna River basin	18%	2012
West Virginia–Marcellus shale, statewide	15%	2012
California–Monterey shale, statewide	4%	2014
Overall mean	15%	
Overall median	5%	

Source: EPA (2015).
Note: NA = not available.

COD concentrations and 17 imparted moderate COD concentrations. They indicated that higher COD values were mainly caused by corrosion inhibitors and biocides. It was noted in their work that the filtered flowback water may contain high values of these components since the current water treatment membrane technology (i.e., microfiltration) is not capable of removing them.

In terms of toxicity, Stringfellow et al. (2014) reported that most chemicals for which toxicity data are available are of moderate or little concern based on mammalian acute oral toxicity standards. The global harmonized system (GHS) of classification and labeling of chemicals was used to investigate the toxicity of the additives, and it was reported that no hydraulic fracturing chemical was in category 1 (the most severe toxicity), 3 were in category 2, and 10 were in category 3. One should note that no toxicity data were available for one-third of the 81 chemicals.

Steliga et al. (2015) tested samples of hydraulic fracturing fluids and flowback water from shale formations in Poland for their toxicological properties and reported that both cross-linked and slickwater fracturing fluid samples studied by their group were not toxic. In addition, they reported that the flowback water samples after hydraulic fracturing jobs were classified as substances of low toxicity, due to their low-toxicity level.

The EPA reported a very comprehensive study by analyzing 1,173 chemicals and defining a comprehensive score including the toxicity score, an occurrence score, and scores associated with the chemicals' physicochemical properties (mobility, volatility, and persistence) based on the criteria listed in Table 11.10 (EPA, 2015). The results of their study, summarized in Table 11.11, indicate that none of the chemicals used in hydraulic fracturing represents a high overall comprehensively scored risk toward the health of human beings.

Table 11.7 EPA's Analysis of FracFocus 1.0 Database Reported These 20 Most Frequently Reported Water-Based Fracturing Fluid Additives in Oil Wells, Ranked by Their Frequency of Occurrence

		Maximum concentration in hydraulic fracturing fluid (% by mass)				Maximum concentration in additive (% by mass)			
EPA-standardized chemical name	**CASRN**	**Number (%) of disclosures**	**Median**	**5th percentile**	**95th percentile**	**Number (%) of ingredient records**	**Median**	**5th percentile**	**95th percentile**
Methanol	67-56-1	12,484 (72%)	0.0220	0.000640	0.160	26,482 (7.7%)	30.0	0.3900	100
Distillates, petroleum, hydrotreated light*	64742-47-8	10,566 (61%)	0.0870	0.000730	0.390	15,995 (4.6%)	40.0	0.6000	70
Peroxydisulfuric acid, diammonium salt	7727-54-0	10,350 (60%)	0.0076	0.000280	0.067	12,723 (3.7%)	100.0	0.1000	100
Ethylene glycol	107-21-1	10,307 (59%)	0.0230	0.000860	0.098	12,281 (3.5%)	30.0	0.5000	60
Hydrochloric acid	7647-01-0	10,029 (58%)	0.2900	0.013000	1.800	11,817 (3.4%)	15.0	2.9000	50
Guar gum	9000-30-0	9,110 (52%)	0.1700	0.027000	0.430	9,316 (2.7%)	50.0	1.6000	100
Sodium hydroxide	1310-73-2	8,609 (50%)	0.0100	0.000050	0.075	10,300 (3.0%)	10.0	0.0250	45
Quartz*†	14808-60-7	8.577 (49%)	0.0041	0.000040	12.000	12,636 (3.7%)	2.0	0.0200	93
Water*†	7732-18-5	8,538 (49%)	1.0000	0.005000	9.100	23,340 (6.7%)	67.0	15.0000	97
Isopropanol	67-63-0	8,031 (46%)	0.0063	0.000070	0.220	11,975 (3.5%)	15.0	0.1700	100
Potassium hydroxide*	1310-58-3	7,206 (41%)	0.0130	0.000010	0.052	8,050 (2.3%)	15.0	0.1500	50
Glutaraldehyde	111-30-8	5,927 (34%)	0.0065	0.000270	0.020	6,211 (1.8%)	15.0	0.0300	50
Propargyl alcohol	107-19-7	5,599 (32%)	0.00022	0.000030	0.003	6,129 (1.8%)	5.0	0.0029	10
Acetic acid	64-19-7	4,623 (27%)	0.0047	0.000000§	0.047	5,552 (1.6%)	30.0	0.8200	100

(*Continued*)

Table 11.7 (Continued)

		Maximum concentration in hydraulic fracturing fluid (% by mass)				Maximum concentration in additive (% by mass)			
EPA-standardized chemical name	**CASRN**	**Number (%) of disclosures**	**Median**	**5th percentile**	**95th percentile**	**Number (%) of ingredient records**	**Median**	**5th percentile**	**95th percentile**
2-Butoxyethanol*	111-76-2	4,022 (23%)	0.0053	0.000000[§]	0.170	5,096 (1.5%)	10.0	0.2500	100
Solvent naphtha, petroleum, heavy arom.*	64742-94-5	3,821 (22%)	0.0060	0.000000[§]	0.038	4,129 (1.2%)	5.0	0.0000	35
Sodium chloride*	7647-14-5	3,692 (21%)	0.0071	0.000000[§]	0.270	4,445 (1.3%)	25.0	0.0040	100
Ethanol*	64-17-5	3,536 (20%)	0.0260	0.000020	0.160	4,178 (1.2%)	45.0	1.0000	60
Citric acid	77-92-9	3,310(19%)	0.0047	0.000160	0.024	3,491 (1.0%)	60.0	7.0000	100
Phenolic resin	9003-35-4	3,109 (18%)	0.1300	0.019000	2.000	3,238 (0.94%)	5.0	0.9400	20

Source: EPA (2015).
* Chemical has a nonnormal distribution and the median may not represent the central tendency of the data set as well as the median of a normally distributed data set.
† See the text for a discussion of why water and quartz were included in the table.
§ Concentration is less than a millionth of a percentage by mass.

Table 11.8 EPA's Analysis of FracFocus 1.0 Database Reported These 20 Most Frequently Reported Water-Based Fracturing Fluid Additives in Gas Wells, Ranked by Their Frequency of Occurrence

		Maximum concentration in hydraulic fracturing fluid (% by mass)				Maximum concentration in additive (% by mass)			
EPA-standardized chemical name	CASRN	Number (%) of disclosures	Median	5th percentile	95th percentile	Number (%) of ingredient records	Median	5th percentile	95th percentile
Hydrochloric acid	7647-01-0	12,351 (73%)	0.07800	0.006300	0.6700	13,754 (5.3%)	15	2.7000	60
Methanol	67-56-1	12,269 (72%)	0.00200	0.000040	0.0530	19,074 (7.3%)	30	0.5000	90
Distillates, petroleum, hydrotreated light*	64742-47-8	11,897 (70%)	0.01700	0.002100	0.2700	14,289 (5.5%)	30	3.1000	70
Isopropanol	67-63-0	8,008 (47%)	0.00160	0.000010	0.0510	10,326 (3.9%)	30	2.5000	60
Water*†	7732-18-5	7,998 (47%)	0.18000	0.000090	91.0000	17,690 (6.8%)	63	5.0000	100
Ethanol*	64-17-5	6,325 (37%)	0.00230	0.000120	0.0900	7,062 (2.7%)	5	1.0000	60
Propargyl alcohol	107-19-7	5,811 (34%)	0.00007	0.000010	0.0016	5,963 (2.3%)	10	0.0037	40
Glutaraldehyde	111-30-8	5,635 (33%)	0.00840	0.000910	0.0230	5,827 (2.2%)	30	0.1800	60
Ethylene glycol	107-21-1	5,493 (32%)	0.00610	0.000080	0.2400	7,733 (3.0%)	35	1.0000	100
Citric acid	77-92-9	4,832 (28%)	0.00170	0.000050	0.0110	4,885 (1.9%)	60	30.0000	100
Sodium hydroxide	1310-73-2	4,656 (27%)	0.00360	0.000020	0.0880	5,642 (2.2%)	5	1.0000	60
Peroxydisulfuric acid, diammonium salt	7727-54-0	4,618 (27%)	0.00450	0.000050	0.0450	6,402 (2.4%)	100	0.2600	100
Quartz*†	14808-60-7	3,758 (22%)	0.00240	0.000030	11.0000	4,729 (1.8%)	10	0.2000	100

(Continued)

Table 11.8 (Continued)

EPA-standardized chemical name	CASRN	Maximum concentration in hydraulic fracturing fluid (% by mass)				Maximum concentration in additive (% by mass)			
		Number (%) of disclosures	Median	5th percentile	95th percentile	Number (%) of ingredient records	Median	5th percentile	95th percentile
2,2-Dibromo-3-nitrilopropionamide	10222-01-2	3,668 (22%)	0.00180	0.000070	0.0220	3,728 (1.4%)	100	10.0000	100
Sodium chloride*	7647-14-5	3,608 (21%)	0.00910	0.000000§	0.1200	4,176 (1.6%)	30	1.0000	40
Guar gum	9000-30-0	3,586 (21%)	0.10000	0.0005700	0.3800	3,702 (1.4%)	60	1.6000	100
Acetic acid	64-19-7	3,563 (21%)	0.00250	0.000000§	0.0280	3,778 (1.4%)	50	5.0000	90
2-Butoxyethanol*	111-76-2	3,325 (20%)	0.00350	0.0000100	0.0410	4,186 (1.6%)	10	3.0000	40
Naphthalene*	91-20-3	3,294 (19%)	0.00120	0.0000027	0.0050	3,355 (1.3%)	5	0.0071	5
Solvent naphtha, petroleum, heavy arom*	64742-94-5	3,287 (19%)	0.00440	0.0000300	0.0300	3,750 (1.4%)	30	0.0260	30

Source: EPA (2015).

* Chemical has a nonnormal distribution and the median may not represent the central tendency of the data set as well as the median of a normally distributed data set.

† See the text for a discussion of why water and quartz were included in the table.

§ Concentration is less than a millionth of a percentage by mass.

Table 11.9 EPA's Analysis of FracFocus 1.0 Database Reported These 16 Most Frequently Reported Nonaqueous Ingredients in Base Fluids

		Maximum concentration in hydraulic fracturing fluid (% by mass)				Maximum concentration in additive (% by mass)			
EPA-standardized chemical name	**CASRN**	**Number (%) of disclosures**	**Median**	**5th percentile**	**95th percentile**	**Number (%) of ingredient records**	**Median**	**5th percentile**	**95th percentile**
Nitrogen, liquid	7727-37-9	643 (84%)	16.0	3.8	30.0	643 (80%)	100	25	100.0
Carbon dioxide	124-38-9	83 (11%)	32.0	11.0	46.0	83 (10%)	100	100	100.0
Petroleum distillates	8002-05-9	18 (2.4%)	46.0	29.0	67.0	18 (2.2%)	100	100	100.0
Propane	74-98-6	15 (2.0%)	63.0	1.6	79.0	16 (2.0%)	100	2.0	100.0
Isobutane	75-28-5	12 (1.6%)	29.0	8.0	52.0	13 (1.6%)	50	10	100.0
Butane	106-97-8	10 (1.3%)	2.2	1.5	59.0	11 (1.4%)	80	36	100.0
Hexane	110-54-3	4 (0.53%)	14.0	11.0	15.0	4 (0.50%)	20	18	20.0
Pentane	109-66-0	4 (0.53%)	9.8	5.8	14.0	4 (0.50%)	13	10	19.0
Butene	25167-67-3	3 (0.39%)	25.0	8.1	49.0	3 (0.37%)	65	34	65.0
1-Propene	115-07-1	2 (0.26%)	3.0	1.2	4.8	2 (0.25%)	5	5	5.0
2-Methylbutane	78-78-4	2 (0.26%)	16.0	14.0	18.0	2 (0.25%)	25	25	25.0
Benzene	71-43-2	2 (0.26%)	3.3	2.8	3.7	2 (0.25%)	5	5	5.0
Ethane	74-84-0	2 (0.26%)	2.3	1.6	3.1	3 (0.37%)	2	2	9.2
Ethylene	74-85-1	1 (0.13%)	2.1	2.1	2.1	1 (0.12%)	10	10	10.0
Methane	74-82-8	1 (0.13%)	2.1	2.1	2.1	1 (0.12%)	10	10	10.0
White mineral oil, petroleum	8042-47-5	1 (0.13%)	12.0	12.0	12.0	1 (0.12%)	100	100	100.0

Source: EPA (2015).

Table 11.10 EPA Reported Thresholds for Developing a Comprehensive Health Risk, Score Including Toxicity, Occurrence, and Physicochemical Properties

Score	1	2	3	4
Toxicity score:				
Chronic RfV (federal)	>3rd quartile	>2nd quartile to ≤3rd quartile	>1st quartile to ≤2nd quartile	<1st quartile
Occurrence score:				
Percentage of wells nationwide	<1st quartile	≥1st quartile to <2nd quartile	≥2nd quartile to <3rd quartile	≥3rd quartile
Concentration (μg/L)	<1st quartile	≥1st quartile to <2nd quartile	≥2nd quartile to <3rd quartile	≥3rd quartile
Physicochemical properties score:				
Mobility score:				
Log K_{ow}	>5	>3 to 5	>2 to 3	<2
Log K_{oc}	>4.4	>3.4 to 4.4	>2.4 to 3.4	<2.4
Aqueous solubility (mg/L)	<0.1	>0.1 to <100	≥100 to <1000	≥1000
Volatility score:				
Henry's law constant	$>10^{-1}$	$>10^{-3}$ to $\leq 10^{-1}$	$>10^{-5}$ to $\leq 10^{-3}$	$\leq 10^{-5}$
Persistence score:				
Half-life in water (days)	<16	≥16 to <60	≥60 to <180	≥180

Source: EPA (2015).

Note that the scale goes from 1 to 4 with 4 being a high risk.

11.4 Potential Impacts on Drinking Water Resources

One major possibility of chemicals used in hydraulic fracturing having an impact on drinking water resources is a chemical spill. This can happen during the transportation of chemicals to the fracturing site, or at the fracturing site and during the operation. EPA reported three representative examples involving a tank overflow case that impacted surface water, a blender spill due to human error with a soil impact, and a failure in equipment with no impact (Stringfellow et al., 2014). Among the three, the first one could potentially impact the soil and surface water resources. Such accidents can occur once in a while despite all the care being devoted to safety aspects of transportation and operation by service and operating companies. The rate of those incidents is very small compared with the total number of fracturing jobs conducted on a daily basis. The highest number of reported produced and flowback water spills has occurred in the states of Colorado and Pennsylvania. Gross et al. (2015) reported 77 cases of spills in Colorado that represented a very small percentage, less than 0.5%, of nearly 18,000 active wells in the state. The spill volumes ranged from 42 gal to 1,176 gal. Of the 77 spills, 46 contained produced water and oil. Twenty-three of the remaining cases contained only oil, and 8 were only produced water. The oil spill cases

Table 11.11 Comprehensive Health, Risk, Scores of Chemicals Used in Hydraulic Fracturing, Including the Toxicity Score, Occurrence Score, in Addition to Physicochemical Properties Such As Mobility Score, Volatility Score, and Persistence Score Defined Based on the Criteria Listed in Table 11.10

Chemical	CASRN	Physicochemical properties score	Occurrence score	Toxicity score	Total hazard potential score
Propargyl alcohol	107-19-7	1.00	1.00	1.00	3.00
2-Butoxyethanol (EGBE)	111-76-2	1.00	1.00	0.67	2.67
N, *N*-Dimethylformamide	68-12-2	1.00	1.00	0.67	2.67
Acrylamide	79-06-1	1.00	0.67	1.00	2.67
Formaldehyde	50-00-0	1.00	1.00	0.33	2.33
Naphthalene	91-20-3	0.67	1.00	0.67	2.33
Benzyl chloride	100-44-7	0.67	0.67	1.00	2.33
1-Butanol	71-36-3	1.00	0.67	0.67	2.33
Epichlorohydrin	106-89-8	0.67	0.67	1.00	2.33
2-(2-Butoxyethoxy)ethanol	112-34-5	1.00	0.67	0.67	2.33
Methanol	67-56-1	1.00	1.00	0.00	2.00
Ethylene glycol	107-21-1	1.00	1.00	0.00	2.00
Formic acid	64-18-6	1.00	1.00	0.00	2.00
Didecyldimethylammonium chloride	7173-51-5	0.33	1.00	0.67	2.00
1, 4-Dioxane	123-91-1	1.00	0.33	0.67	2.00
(*E*)-Crotonaldehyde	123-73-9	0.67	0.33	1.00	2.00
Aniline	62-53-3	1.00	0.00	1.00	2.00
Furfural	98-01-1	1.00	0.00	1.00	2.00

(*Continued*)

Table 11.11 (Continued)

Chemical	CASRN	Physicochemical properties score	Occurrence score	Toxicity score	Total hazard potential score
1, 2-Propylene glycol	57-55-6	1.00	0.67	0.00	1.67
Hexanedioic acid	124-04-9	1.00	0.67	0.00	1.67
Toluene	108-88-3	0.33	0.67	0.67	1.67
Phenol	108-95-2	1.00	0.33	0.33	1.67
2-Methyl-1-propanol (Isobutanol)	78-83-1	1.00	0.33	0.33	1.67
Dichloromethane	75-09-2	0.67	0.00	1.00	1.67
Ethylenediamine	107-15-3	1.00	0.00	0.67	1.67
Bisphenol *A*	80-05-7	1.00	0.00	0.67	1.67
2-(Thiocyanomethylthio) benzothiazole	21564-17-0	0.67	0.00	1.00	1.67
Benzene	71-43-2	0.67	0.00	1.00	1.67
Dodecylbenzenesulfonic acid	27176-87-0	0.33	1.00	0.00	1.33
Xylenes	1330-20-7	0.33	0.67	0.33	1.33
Ethylbenzene	100-41-4	0.33	0.33	0.67	1.33
Benzoic acid	65-85-0	1.00	0.33	0.00	1.33
Acetophenone	98-86-2	0.67	0.00	0.67	1.33
1, 3-Dichloropropene	542-75-6	0.67	0.00	0.67	1.33
Cumene	98-82-8	0.00	0.33	0.67	1.00
Ethyl acetate	141-78-6	0.67	0.33	0.00	1.00
Acetone	67-64-1	0.67	0.33	0.00	1.00

Source: EPA (2015).

reported organic compound concentrations in the groundwater that were above the acceptable limits reported for drinking water.

Another possibility of water and hydrocarbons communicating with surface and subsurface drinking water resources is when a fracture communicates with shallow groundwater and even surface resources through fault systems in the region. Both of these possibilities are rare due the geological investigations conducted by companies before the development of the area and observation of the fracturing job during the operation. In addition, several states have added regulations regarding the Class II wastewater injection wells located near faults (EPA, 2016). One example of an area at high risk of having such a problem is the San Joaquin basin where there is a high density of faults and already existing wells hydraulically fractured at depths shallower than 1,000 ft. Although no cases of groundwater contamination have been reported, no thorough monitoring has yet been conducted in this region (SAIC Energy, 2011).

The next possibility has to do with the well construction and casing and cement failures. This possibility is most possible one due to the difficulties related to casing and cement placement in deep and challenging environments. A combination of this problem with the presence of natural faults and fractures may result in migration of fracturing fluids and hydrocarbons to the surface and subsurface drinking water resources. The East Mamm Creek drilling case study in Colorado is an example where failure in placement of cement resulted in methane migration through natural faults and fractures (Stringfellow et al., 2014). One example of a flowback water spill due to the wellhead failure is the case reported in Pennsylvania. In this case a wellhead flange failure during operation caused a 10,000 gal spill into a state-designated trout stock fishery and a tributary of the Susquehanna River (Zoback & Walsh, 2015).

Other possible causes of impacts on drinking water resources include not respecting the minimum 2,000 ft separation between the formations targeted using hydraulic fracturing and formations containing drinking water, and applying hydraulic fracturing to formations that contain both hydrocarbon and drinking water resources.

11.5 Induced Seismicity

Reports of induced seismicity caused by an increase in water disposal activities due to the sudden increase in oil and gas development in several states within the United States (Kansas, Oklahoma, Colorado, New Mexico, Arkansas, and Ohio), in addition to a few international reports in this regard, have given the public the mistaken notion that hydraulic fracturing causes induced seismicity (EPA, 2016). As indicated by Zoback and Walsh (2015), some researchers blamed the volumes of flowback water in those states for this increase in seismicity. As presented in the previous sections, the volumes of flowback water are negligible compared with the produced water volumes. Class II water disposal wells are used for the disposal of both flowback and produced water, and the significantly increased volumes of produced water that need to be disposed of when oil production increased resulted in "small- to moderate-sized earthquakes" in some areas. Injection increased volumes of flowback and produced water into disposal wells results in an increase in pore pressure. According to Zoback and Walsh (2015), this increase in pore pressure decreases the effective normal stress, which is the stress preventing fault slips.

Figure 11.8, generated by Zoback and Walsh (2015), illustrates the relationship between the presence of water disposal wells and the significant increase in occurrence of small- to moderate-size earthquakes observed in the state

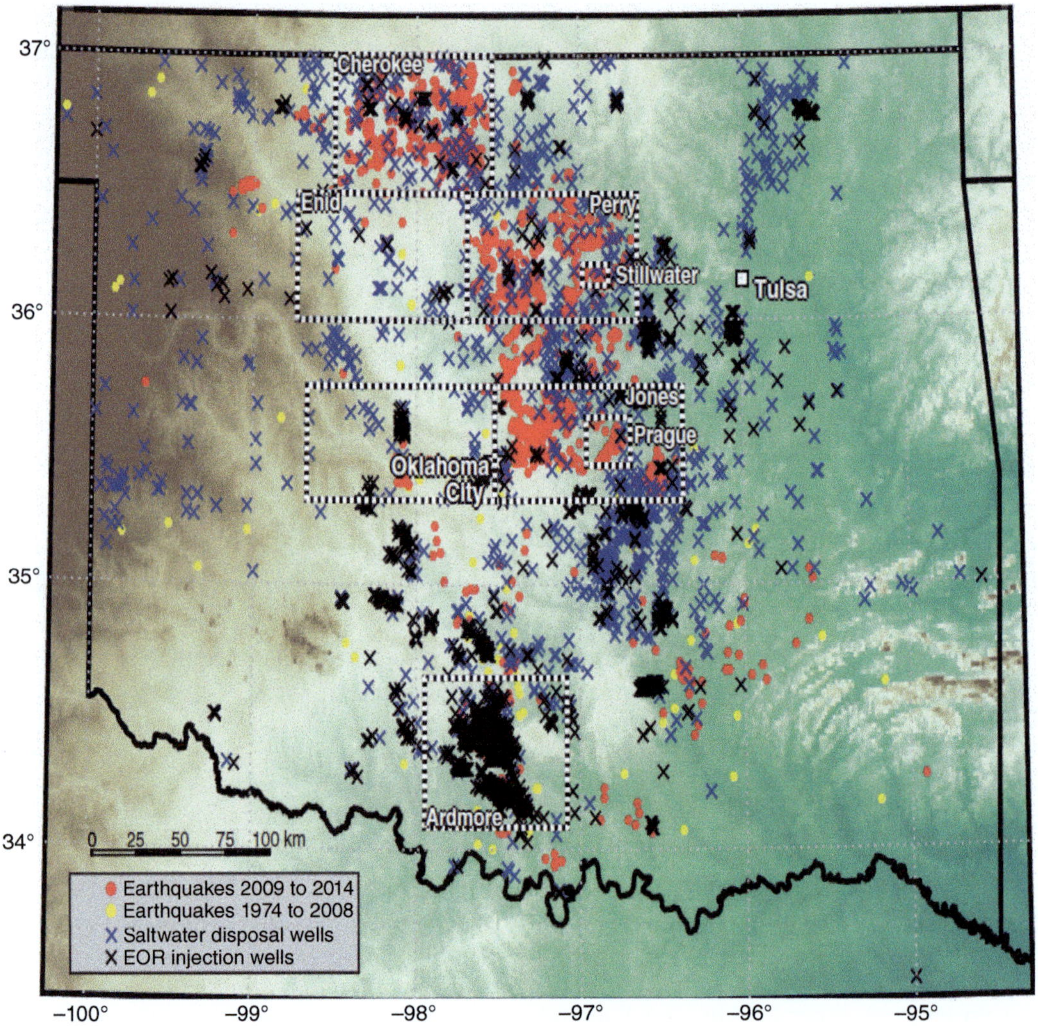

Figure 11.8 This work presented by Zoback and Walsh (2015) shows the earthquake history prior to 2009 and compares it to the one after 2009, demonstrating its relationship with the presence of water disposal wells and lack of relationship to the presence of EOR wells with high volumes of injected water.

of Oklahoma. Their findings were supported by Weingarten et al. (2015). The increase in disposal of produced water due to increased oil and gas production resulted in an increase in pore pressure, which resulted in fault slips. Increases in oil and gas production clearly follow increases in oil price.

One could make a correlation between the number of hydraulic fracturing jobs and the induced seismicity events in a region, but concluding that hydraulic fractures impact faults directly is not a valid conclusion, at least for induced seismicity cases in the United States. In most of the impacted states, incremental drilling activities resulted in more hydraulic fracturing events. Drilled and fractured wells in those regions produced a significant volume of water, which increased with the number of fractures propagated. The incremental volume of water to be disposed on top of what used to be disposed before the sudden increase in production, along with the lack of regulations on disposal volumes and the presence of fault systems, caused a significant increase in minor seismic events.

As a result of tens of earthquakes being felt by residents of the state of Oklahoma, and in response to studies correlating the occurrence of earthquakes to the increase in wastewater disposal volumes, the state of Oklahoma came up with an earthquake response plan for central Oklahoma (Oklahoma Corporation Commission, 2016). Implementation of this plan resulted in a significant (40%) reduction in water disposal volumes for some of the regions (covering 15,000 square miles) marked as being affected by earthquakes due to water disposal. As a result of regulations like this, in addition to the significant lower oil and gas drilling activities in 2017 and 2018, a recent report (Jones, 2018) indicated that the total number of earthquakes declined significantly, from 903 earthquakes of 3.0 and above in 2015 to only 196 in 2018.

One of the most comprehensive studies on the possibility of induced seismicity due to hydraulic fracturing jobs was conducted in Poland as part of a project sponsored by the European Union (López-Comino et al., 2018). In this project, a hybrid seismic network with surface mini-arrays, broadband, and shallow borehole sensors was used to investigate seismic activities during the performance of hydraulic fracturing jobs. The only earthquakes that were observed during the job were two events with M_w of 1.0 and 0.5. This study once more rejected the possibility of any major earthquakes being caused directly by hydraulic fracturing activities.

11.6 Air Pollution

Unconventional development of shale oil and gas plays and the increase in the volume of fluids, sands, and chemicals injected, resulted in a significant increase of the number of trucks injecting fracturing fluids, going to, and returning from fracturing job locations. This increase in activities has contributed to air pollution in the regions that are actively under drilling and fracturing applications. It is the industry's common understanding that hydrocarbon production activities are the main cause of H_2S emissions and contribute to benzene, formaldehyde, xylene, and hexane emissions.

In the absence of a comprehensive and controlled study with baseline measurements over a development area that investigates the effect of hydraulic fracturing activities on the air quality, a study of the oil and gas activities in Kern County in the San Joaquin basin reported the contribution of wells exposed to reservoir stimulation techniques is significantly lower (Jordan et al., 2015).

In another study in western Colorado and near some of the state's oil and gas activities, which lacked an accurate baseline study, Colborn et al. (2014) collected air samples during the drilling, fracturing, and production parts of the wells' history. They reported the highest concentration in the sample to belong to methane, methylene chloride, ethane, methanol, ethanol, acetone, and propane. Formaldehyde and acetaldehydes were also measured in every sample. Polycyclic aromatic hydrocarbons (PAHs) of benzo (a) anthracene, benzo (a) pyrene, benzo (b) fluoranthene, benzo (g,h,i) perylene, benzo (k) fluoranthene, chrysene, dibenzo (a,h) anthracene, indeno (1, 2, 3-cd) pyrene. They reported that the concentrations of the listed chemicals detected in the air were significantly less than OSHA U.S. government safety standards (Occupational Safety and Health Administration, 1993).

11.7 Problems

1 What are the main environmental issues related to hydraulic fracturing?

2 Discuss the impacts of the recent development of the hydraulic fracturing treatments on the water usage.

3 Discuss the impact of regulations as well as the availability of resources on the decision-making process.

4 What is the difference between flowback water and produced water?

5 Explain why the produced water is not directly reused for hydraulic fracturing treatments.

6 Compare the availability of water resources for hydraulic fracturing in the state of Pennsylvania versus those in the state of Louisiana.

7 Discuss the effects of hydraulic fracturing on induced seismicity.

8 Discuss the effects of water disposal wells on induced seismicity.

9 Explain the reason behind the ban of the hydraulic fracturing treatments in England.

Additional Reading

Colborn, T., Schultz, K., Herrick, L., & Kwiatkowski, C. (2014). An exploratory study of air quality near natural gas operations. *Human and Ecological Risk Assessment: An International Journal, 20*(1), 86–105.

EPA (2015). *Assessment of the potential impacts of hydraulic fracturing for oil and gas on drinking water resources*. Washington, DC.

López-Comino, J., Cesca, S., Jarosławski, J., Montcoudiol, N., Heimann, S., Dahm, T., Lasocki, S., et al. (2018). Induced seismicity response of hydraulic fracturing: Results of a multidisciplinary monitoring at the Wysin site, Poland. *Scientific Reports*.

References

Blauch, M. (2011). *Implications, shale frac sequential flowback analyses and reuse*. EPA Technical Workshop, March 30.

Boschee, P. (2014). Produced and flowback water recycling and reuse-economics, limitations, and technology. *Oil and Gas Facilities*, 16–21.

Burden, S., Dean, J., Koplos, J., Meza-Cuadra, C., & Singer, A. T. M. (2015). *Analysis of hydraulic fracturing fluid data from the FracFocus Chemical Disclosure Registry 1*. EPA Report. Washington, DC: EPA.

Colborn, T., Schultz, K., Herrick, L., & Kwiatkowski, C. (2014). An exploratory study of air quality near natural gas operations. *Human and Ecological Risk Assessment: An International Journal, 20*(1), 86–105.

Coleman, N. P. (2012). *Produced formation water sample results from shale plays*. Hydraulic Fracturing Technical Workshop 1-07. EPA's Study of Hydraulic Fracturing and Its Potential Impact on Drinking Water Resources.

Commission, Louisiana Ground Water Resources (2012). Managing Louisiana's groundwater resources: An interim report to the Louisiana legislature, 15 March 2012. Available at http://www.dnr.louisiana.gov/index.cfm?md=pagebuilder&tmp=home&pid=907

Culborn, T., Schultz, K., Herrick, L., & Kwiatkowski, C. (2014). An exploratory study of air quality near natural gas operations. *Human and Ecological Risk Assessment: An International Journal, 20*(1), 86–105.

Dieter, C., Linsey, K. C. R., Harris, M., Ivahnenko, T., Lovelace, J., Maupin, M., & Barber, N. (2018). Estimated use of water in the United States in 2015. *U.S. Geological Survey Circular 1441*. Available at https://doi.org/10.3133/cir1441

EPA (2015). Assessment of the potential impacts of hydraulic fracturing for oil and gas on drinking water resources (external review draft). Washington, DC.

EPA (2016). *Hydraulic fracturing for oil & gas: Impacts from the hydraulic fracturing water cycle on drinking water resources in the United States*. EPA's Study of Hydraulic Fracturing for Oil and Gas and Its Potential Impact on Drinking Water Resources. Washington, DC

Freyman, M. (2014). Hydraulic fracturing & water stress: Water demand by the numbers. *Ceres*, February.

Gallegos, T. J., Varela, B., Haines, S., & Engle, M. (2015). Hydraulic fracturing water use variability in the United States and potential environmental implications. *Water Resources Research*, *51*, 5839–5845. doi: doi.org/10.1002/2015WR017278

Gross, S., Avens, H., Banducci, A., Sahmel, J., Panko, J., & Tvermoes, B. (2013). Analysis of BTEX groundwater concentrations from surface spills associated with hydraulic fracturing operations. *Journal of the Air and Waste Management Association*, *63*(4), 424–432.

Haghshenas, A. (2015). *Reuse of flowback fluids as hydraulic fracturing fluids in tight gas sand reservoirs*.

Haghshenas A., & Nasr-El-Din, H. A. (2014). *Effect of dissolved solids on reuse of produced water and proppant handling in hydraulic fracturing jobs in tight sand gas reservoirs*. SPE Hydraulic Fracturing Technology Conference, 4–6 February, Woodlands, TX. doi: doi.org/10.2118/168614-MS

Haut, R., Wilcox, A., Williams, T., & Burnett, D. (2018). *Overview of industry advancements in environmental awareness*. SPE Annual Technical Conference and Exhibition, Dallas, 24–26 September.

Hayes, T. (2011). *Characterization of Marcellus and Barnett shale flowback waters and technology development for water reuse*. EPA's Study of Hydraulic Fracturing and Its Potential Impact on Drinking Water Resources, Hydraulic Fracturing Technical Workshop 4, Arlington, VA.

Jones, C. (2018). Oklahoma earthquakes decline for third year, but state primed to shake another decade. *Tulsa World*.

Jordan, P., Brandt, A., Ferrar, A., Feinstein, L., & Phillip, S. (2015). A case study of the potential risks associated with hydraulic fracturing in existing oil fields in the San Joaquin basin. Available from osti.gov.

Kondash, A., Albright, E., & Vengosh, A. (2017). Quantity of flowback and produced waters from unconventional oil and gas exploration. *Science of the Total Environment*, *574*, 314–321.

Lebas, R. A., Shahan, T. W., Lord, P., & Luna, D. (2013). *Development and use of high-TDS recycled produced water for cross-linked-gel-based hydraulic fracturing*. SPE Hydraulic Fracturing Technology Conference, The Woodlands, Texas.

Leiming, L., Al-Muntasheri, G., & Liang, F. (2016). A review of cross-linked fracturing fluids prepared with produced water. *Petroleum*, *2*(4), 313–323.

Li, H. (2013). *Produced water quality characterization and prediction for Wattenberg field*. Fort Collins: Colorado State University.

López-Comino, J., Cesca, S., Jarosławski, J., Montcoudiol, N., Heimann, S., Dahm, T., Lasocki, S., et al. (2018). Induced seismicity response of hydraulic fracturing: Results of a multidisciplinary monitoring at the Wysin site, Poland. *Scientific Reports*.

Mantell, M. (2011). *Produced water reuse and recycling challenges and opportunities across major shale plays*. EPA hydraulic fracturing study technical workshop #4.

Maupin, M., Kenny, J., Hutson, S., Lovelace, J., Barber, N., & Linsey, K. (2014). *Estimated use of water in the United States in 2010*. U.S.

Department of Interior & U.S. Geological Survey.

Nicot, J., Reedy, R., Costley, R., & Huang, Y. (2012). Oil and gas water use in Texas: Update to the 2011 mining water use report. Austin, TX.

Occupational Safety and Health Administration (1993). Occupational safety and health standards, toxic and hazardous substances 1910.1000 Table Z-1–Table Z-1 limits for air contaminants. Washington, DC: Occupational Safety and Health Administration.

Oklahoma Corporation Commission (2016). *Media advisory: Regional earthquake response plan for central Oklahoma and expansion of the area of interest.* Oil and Gas Conservation Division.

Oklahoma Produced Water Working Group (2017). Oklahoma water for 2060 produced water reuse and recycling. Oklahoma City.

Rimassa, S. M., Howard, P. R., & Blow, K. A. (2009). *Optimizing fracturing fluids from flowback water.* SPE Tight Gas Completions Conference, San Antonio, Texas.

SAIC Energy (2011). Environment and Infrastructure, LLC, and Groundwater & Environmental Services, Inc., ATGAS investigation initial site characterization and response, 19 April 2011 to 2 May 2011, ATGAS2H Well Pad, permit no. 37-015-21237, Leroy Township, Bradford County, Harrisburg, PA.

Severin, B., & Hayes, T. (2012). Barnett and Appalachian shale water management and reuse technologies. Report No. 08122-05. FINAL.1. Research Partnership to Secure Energy for America.

Steliga, T., Kluk, D., & Jakubowicz, P. (2015). Analysis of chemical and toxicological properties of fluids for shale hydraulic fracturing and flowback water. RPSEA Report No. 08122-05.FINAL.1.

Stringfellow, W., Domen, J., Camarillo, M., Sandelin, W., & Borglin, S. (2014). Physical, chemical, and biological characteristics of compounds used in hydraulic fracturing. *Journal of Hazardous Materials, 32*, 37–54.

U.S. Environmental Protection Agency (2015). Analysis of hydraulic fracturing fluid data from the Frac Focus chemical disclosure registry 1.0: Data management and quality assessment report. Washington, DC.

Veil, J. (2015). U.S. produced water volumes and management practices in 2012. Prepared for Groundwater Protection Council, April.

Weingarten, M., Ge, S., Godt, S., Bekins, B., & Rubinstein, J. (2015). Induced seismicity: High-rate injection is associated with the increase in U.S. mid-continent seismicity. *Science, 348* (624), 1336–1340.

Zoback, M., & Walsh, R. III (2015). Oklahoma's recent earthquakes and saltwater disposal. *Seismology,* June.

Index

a

AARD. *See* Absolute average relative deviation
Abandonment rate
  Arps's decline models 442–43, 445–47, 447f
  modern DCA models 451, 455, 457t
Absolute adsorption 409–10, 410f
Absolute average relative deviation (AARD) 388
Abu Gharadig basin 33
ACA. *See* After-closure analysis
Acetaldehydes 545
Acetone 542t, 545
Acoustic logs 64–65
Actual value profit (AVP)
  in basic cash flow 498–502, 499t–502t
  chapter project of 512–19, 517t–18t
  in NPV 504, 504t
  in ROR 504–5, 504t
AD. *See* Adsorption-desorption
Adamantane units 166–67, 167f
Additives, of fracture fluids 323–27, 325t
Adhesion tests 103
Adsorbed-phase transport 426
Adsorption
  absolute 409–10, 410f
  AD techniques and 94–95
  capacity calculation 110, 110f
  LAI and 410
  nanopores and 408–12, 409f–11f, 411t–12t, 413f
  porosity and 67–69, 68f, 70f
  PSD and 77–79, 79t, 80f
  selectivity 409, 411f
Adsorption-desorption (AD) techniques 94–95
Adsorption isotherm 67
  BET 69, 70f
  carbon dioxide 410–11, 411t–12t
  LAI 68–69, 70f, 109, 410–12, 428t
  methane and 410–12, 410f, 411t–12t, 413f
  osmotic equation 387
Advanced Resources International, Inc. (ARI) 39
Advanced software packages. *See* Image processing
AEO. *See* Annual Energy Outlook
AFM. *See* Atomic force microscopy
Africa
  Frasnian play 31–33, 32f, 32t
  Whitehill play 31, 32t, 112t
After-closure analysis (ACA) 311
After-tax cash flow (ATCF) 498
Agarwal-Gardner type curves 462–64, 463f
AGI. *See* American Geological Institute
Ahnet basin 31, 32f, 33
Air pollution 545
Alabama, U.S. 13
Albian Mowry shale 161
Algeria 31, 32f
Alginite 135t
American Geological Institute (AGI) 132–33

Unconventional Hydrocarbon Resources: Techniques for Reservoir Engineering Analysis, Advanced Textbook Series, First Edition. Edited by Reza Barati Ghahfarokhi and Mustafa M. Alhubail.

Amott-Harvey technique 103, 106–7, 107f–8f
Anadarko basin 16
Analysis software. *See also* Image processing
in X-ray micro-CT 205–7, 206f
Anisotropy 6, 248
Annual Energy Outlook (AEO) 38
Anoxic conditions 129, 130f, 166t
Appalachian basin 11–13, 13f
Apparent permeability 427
of shale 415–16, 416t–17t, 417f
APV. *See* Actual value profit
Archie's model 98–99
Arctic 8
Ardmore basin 16, 22t
Argentina 24
Vaca Muerta in 3, 4f, 20, 27f
Argon ion milling 241, 242f
ARI. *See* Advanced Resources International, Inc.
Arkansas, U.S. 16–17, 525–26, 526t
Arkoma basin 16–17, 17f, 525f
Arps's decline models
abandonment rate in 442–43, 445–47, 447f
exponential decline 442–45, 443t, 444f
harmonic decline 445–46
hyperbolic decline 446–49, 447f, 449t
schematic of 441–42, 442f
Artifacts 192–96, 203, 240–41
Artificial proppant 331
Ash layer detection 49
Asia 495, 496f
petrophysical properties of 112t
unconventional plays 33–36, 34f–35f, 34t
ATCF. *See* After-tax cash flow
Athel formation 30–31, 30t
Atomic force microscopy (AFM) 103
Attenuation coefficient 188, 203
Beer's law and 189, 195
of brine 209
of epoxy 193
linear 189, 195–97, 203, 207
Australia 199, 200f
petrophysical properties of 112t
unconventional plays of 36–37, 36t, 37f
Austral-Magallanes basin 20, 24f
Avizo software 198, 201, 241–43

b

Back-scattered electron (BSE) image
of LEF shale 133, 133f
porosity estimation and 140
Backscattered electrons 216, 217f–18f
Bahrain Petroleum Co. (Bapco) 31
Bahrain Shale 31
Bakken formation 11, 12f, 22t, 274–75
CCE data for 367, 367t
cross-plot of 55, 55f
CT scans 200, 201f
DL experiment and 368–69, 369f–70f, 370t–71t
sample data and 400, 401t
Rock-Eval pyrolysis of 144–49, 144f–45f 146t–49t
Timur-Coates permeability 90, 91t
Bakken shale fluid
binary interaction parameters for 384, 384t
compositional data for 384, 384t
compressibility of 388, 389f
DCA in 457, 457f, 457t
density of 388, 389f
multicomponent systems and 361t, 364–65, 365f
saturation pressure of 388, 388f
Baltic basin 28
Bapco. *See* Bahrain Petroleum Co.
Barber County, Kansas 150–52, 190, 207
SEM image of 217f
Barnett shale 9–10, 17, 22t, 58, 78t
flowback and 529, 534t
fracturing and 281, 315f
groundwater and 526t
imaging of 221, 222f–23f, 228t
Basic cash flow 498–502, 499t–502t
Batra formation 30t, 31
Bauxite proppant 331
Bazhenov formation 25–28, 28t, 29f, 52f

BE. *See* Boltzmann equation
Beer-Lambert law 178
Beer's law 191, 197, 205
Beetaloo basin 36, 36t
Benzene 534t, 545t, 548t, 551
Besa River shale play 18–19
BET. *See* Brunauer-Emmett-Teller
Bhatnagar-Gross-Krook (BGK) 429
BHFP. *See* Bottom hole fracture pressure
BI. *See* Brittleness index
BIB-SEM. *See* Broad ion beam-scanning electron microscopy
Binary image. *See* 2D radiographs
Binary interaction coefficient (BIP) 380, 394, 396t
Binary interaction parameters 390, 390t–92t
Biocides 332t
Biogenic gas 132, 132f
Biological markers 162–63
  hopanes 163t, 165–66, 166t
  steranes 163t, 165–66, 166t
  tetracyclic terpanes 164, 166t
  tricyclic terpanes 164, 166t
Biot's poroelastic constant 268–70, 269t
BIP. *See* Binary interaction coefficient
Bitumen 6
  BRo 137–38
  petroleum geochemistry and 132–34, 133f
  pyrobitumen and 137–40, 138f–39f, 168, 172
  solid reflectance and 137–38
Black Warrior basin 13
Blasingame type curves 462, 463f, 466, 471f, 472
Blue-green algae 129–30
Boltzmann equation (BE) 415, 422–24
Bottom-hole flowing pressure. *See* Flowing bottomhole pressure
Bottom hole fracture pressure (BHFP) 309
Bounce-back/specular reflection (BSR) 427
Boundary-dominant flow 451, 454, 459–62, 462f
Boyle's law 66
Brady sand 330t, 331
Brazil 24, 330
Breakers. *See* Gel breakers
Brine
  imaging techniques and 209, 213–14, 214f
  shale reservoir and 61, 61f
  water quality and 523, 527, 528f, 529
British Columbia 18, 19f
Brittleness index (BI) 278
  elastic brittleness index 279–81, 280t–82t
  MBI and 234–37, 281–85, 283t–84t
BRo. *See* Solid bitumen reflectance
Broad ion beam-scanning electron microscopy (BIB-SEM) 247–49, 248f–49f
Bruker Alpha-R spectrometer 56
Brunauer-Emmett-Teller (BET) isotherms 69, 70f
BSE. *See* Back-scattered electron
BSE1. *See* Singly scattered backscattered electrons
BSE2. *See* Multiple-scattered backscattered electrons
BSR. *See* Bounce-back/specular reflection
Bubblepoint curve 365f, 377f
Buffers 325–26, 325t
Bulk density 149–50, 149t–51t, 156t, 157
Bulk modulus 265, 265f
  dynamic 275–76
Bulk volume water (BVW) 108
Burgos basin 13, 19–20, 21f
Business expenditures 497–98
Buy-back service contracts 509–10, 509f, 510t
BVW. *See* Bulk volume water

C

Cable-tool drilling method 1
Calcite 106–7, 107f, 133, 225
California, U.S. 23t
  San Joaquin basin 17, 18f
  water use and 531t, 534t
Cambrian age 28, 28t, 30, 30t, 35–36, 36t, 129
Canada 2, 9, 19f

Canada (*cont'd*)
  Horn River basin 16, 18, 23t
  Liard basin 18–19
  Quebec province 13, 13f
Canning basin 36, 37f
Capacho shale play 25, 27f
CAPEX. *See* Capital expenditures
Capillary pressure 103
  in flash calculation 377–79, 378f
Capital expenditures (CAPEX) 497–98, 497f
  future value and 503
  payout period and 506, 506f
  ROR and 504–5
Carbon dioxide adsorption isotherms 410–11, 411t–12t
Carboniferous formation 28–29
Carbon nanotubes 409, 409f
Carr-Purcell-Meiboom-Gill (CMPG) pulse sequence 71
Carter equation II 336–37
Carter's leakoff model 337
Catagenesis 130–31, 131f
Catatumbo basin 25, 27f
Cation-exchange capacity (CEC) 100
CBM. *See* Coalbed methane
CCE. *See* Constant-composition expansion
CEC. *See* Cation-exchange capacity
Cell-based P3D model 339, 339f
Center shift 188, 197–98
Centroid path tortuosity (CPT) 246, 246f
Ceramic proppant 331–32, 332f
Chattanooga shale (CS) 13, 150–52
  analysis of 190
  EDS analysis of 230f–31f
  SEM images of 216–19, 217f–20f
  volume fraction estimation of 243, 243t
  whole core of 192, 193f
  X-ray micro-CT 198–201, 198f–201f
Chemical oxygen demand (COD) 533–34
Chemicals, in fracturing fluid 533–40, 534t–42t
Chemisorption 409
Chesapeake Energy 529, 533
Chevron 20, 36
Child-parent relationship 305–6
Child well 304–8, 308f
Chile 20–21, 24f, 27t
China 3, 495
  basins of 33–35, 34t, 35f
  Longmaxi formation 109–10, 112t, 286t
  Ordos basin 398t
  sand proppants 330
Chlorite 48–49
Clavier model 47
Clay. *See also* Illite
  bound water conductivity 99–100
  mineral porosity 58
  stabilizers and 325t, 326
  typing of 48–49
Clean sand 98–100
CMPG. *See* Carr-Purcell-Meiboom-Gill
Coalbed methane (CBM) 7–8, 9f
Coal mining 1
COD. *See* Chemical oxygen demand
Collingham play 31
Collision diameter 379
Colombia 25, 27f
Colorado, U.S., 16, 16f, 19
  water and 525–26, 526t, 540, 543
Combination of logs 53–56, 54t, 55f, 56t
Commercial horizontal drilling 294
Compressive stress 263–64, 264f
Compton scattering 190
Computerized tomography (CT)
  image reconstruction 187
  porosity estimation 207–9, 208f–10f, 208t
Computerized tomography number (CTN) 196–97, 203
  porosity estimation and 207–9, 208f–10f, 208t
  two-phase fluid saturations 211
Concessionary system 507
Cone-beam artifacts 196
Confinement

capillary pressure 377–79, 378f
critical properties shift 379–81, 380f–81f, 380t
EOS in 379–81, 380f–81f, 380t
modified PR EOS 381–83, 382f–83f
Conglomerate formation 448–49
Constant-composition expansion (CCE) 366–68, 366f, 367t, 368f
Bakken sample data of 397–400, 400t
Consulting services 31. *See also* Service contracts
Contact angle 97, 103, 377, 424
of leakoff fluid 325t, 326
measurement 103, 212–13
Contracts, of service 507–10, 509f, 510t
Contrast 188
Conventional resources. *See also* Shale resources
resource triangle and 5–6, 5f, 7f
TRR and 11
Cooper basin 36, 37f
Cordova embayment, British Columbia 18
Core-flooding apparatus 213
Coring 46
Corn starch 318
CPA. *See* Critical path analysis
CPT. *See* Centroid path tortuosity
Cretaceous age 30, 33
Niobrara formation 140
Cricondenbar 364–65
Critical path analysis (CPA) 94
Critical point shift 382–83, 382f–83f
Critical properties shift 379–81, 380f–81f, 380t
Critical temperature shift 379–81, 380f–81f
Cross-linkers 324–25, 325t
Cross-plot techniques 52–53, 53f, 65
CS. *See* Chattanooga shale
CT. *See* Computerized tomography
CTN. *See* Computerized tomography number
Cubic equations of state
PR EOS 373–75, 375t
Soave-Redlich-Kwong EOS 373
Van der Waals EOS 373–74
Cutinite 135t
Cyclohexamantane 166–67

d

Dadas shale play 30t, 31
Darcy equations 407, 414–15, 421, 425
Darcy range 79–80
Darcy's flow 246
DCA. *See* Decline curve analysis
DCF. *See* Discounting net cash flow
Decision, in petroleum economics 510–11
Decline curve analysis (DCA)
Arps's models
abandonment rate in 442–43, 445–47, 447f
exponential decline 442–45, 443t, 444f
harmonic decline 445–46
hyperbolic decline 446–49, 447f, 449t
schematic of 441–42, 442f
assumptions of 440–41
background on 439
Blasingame plot 466, 471f, 472
boundary-dominant flow 451, 454, 459–62, 462f
Duong's model 456–57, 457f–58f, 457t
LGM 455
limitations of 440–41
modern DCA models 449–57, 453f, 453t, 456f–58f, 457t
modified hyperbolic model 451–52, 453f
PLE model 452–54, 453t
purpose of 439–40, 440f
RTA
Agarwal-Gardner type curves 462–64, 463f
Blasingame type curves 462, 463f
features of 457–59
Fetkovich type curves 460f, 461–62, 462f
flow regimes of 466–72, 467f–71f, 469t
matrix permeability in 468–72, 469t
MFHW and 466–68, 468f

Decline curve analysis (DCA) (*cont'd*)
NPI type curves 464–65, 464f
OOIP 469–72, 469t
PTA and 457–59, 464
square-root time plot 465–66, 467f
SRV in 457, 468–69, 468f, 470f, 472
type curves and 460–61, 460f
Wattenbarger type curves 465, 465f–66f
SEP model 454–55, 456f
SRV in 457, 468–69, 468f, 470f, 472
traditional models 441
Deep natural gas 6–7
Deep resistivity 100, 153
DeGolyer and MacNaughton consulting services 31
Degree of freedom 362–63
Delta LogR method 153–57, 153f–56f, 156t–58t
Demerara Rise 221, 224f
Density logs 58–63, 61f–62f, 64f
Density-magnetic resonance (DMR) 60t, 69
Denver-Julesburg basin 16
Department of Energy (DOE) U.S., 8
Depositional kerogen (DK) 169–72, 173t
imaging techniques of 224–26, 225f, 227f
Derivative thermogravimetric (DTG) plot 67
Deutsche Bank 12–13, 17
Devonian age 13, 16, 18, 28, 129, 218–19
Dew point curve 383
DFIT. *See* Diagnostic fracture injection test
DFN. *See* Discrete fracture network
Diagenesis 130, 131f
Diagnostic fracture injection test (DFIT) 310–13, 312f
Diamantanes 166, 167f
Diamondoids 166–68, 167f
Diasterane 163t, 165, 166t
Differential liberation (DL) experiment 368–69, 369f–70f, 370t–71t
Bakken data and 400, 401t
Diffuse reflectance infrared Fourier transform spectroscopy (DRIFTS) 56
of kerogen maturity 174–77, 176f
Diffusion coefficient 88, 387, 416, 428t
Knudsen 245–46
Digital rock analysis (DRA) 93, 108
Digital rock physics (DRP) 93–94
Dimensionless fracture conductivity 297–98
Dipole sonic logging 101
Direct combustion 141
Directional drilling 1–2
Discounted return on investment (DROI) 505
Discounting net cash flow (DCF) 502
Discrete fracture network (DFN) 343
Diverting agents 325t, 326
Diyab formation 30, 30t
DK. *See* Depositional kerogen
DL. *See* Differential liberation
DMR. *See* Density-magnetic resonance
DOE. *See* Department of Energy
Doveton, John 53
Downhole fluid analysis 356, 356f, 365f
DRA. *See* Digital rock analysis
DRIFTS. *See* Diffuse reflectance infrared Fourier transform spectroscopy
Drilling permits 37
for Eagle Ford play 2, 2f
Drinking water 540–43
DROI. *See* Discounted return on investment
DRP. *See* Digital rock physics
DT. *See* Transit time
DTG. *See* Derivative thermogravimetric
Dual-mineral analysis 54–55, 55f
Dual-water (DW) equation 100
Duong's model 456–57, 457f–58f, 457t
DW. *See* Dual-water
Dynamic bulk modulus 275–76
Dynamic moduli 274–77, 285
Dynamic Poisson's ratio 273–77
Dynamic shear modulus 275–76
Dynamic Young's modulus 273–78

e

Eagle Ford oil
binary interaction parameters for 384, 385t
compositional data for 384, 385t
liquid volume matching of 391, 391f

Eagle Ford shale 13–15, 14f–15f. *See also* Lower Eagle Ford
  BSE image of 133, 133f, 236f, 237
  drilling permits for 2, 2f
  frac hits in 305–6, 306f
  horizontal wells in 3, 4f
  of Mexico 19–20, 21f, 23t
  phase diagram for 377, 384
  porosity of 58, 59f
  T_2 spectrum of 72, 72f
  U.S. region 22t
Eagleville field 15, 15f
Early Cambrian age 30
Early Cretaceous age 13, 20
Early Mississippian age 11, 16
Early Neogene age 17
Early Silurian age 28, 30, 33
Earned interest 502
Earthquakes
  induced seismicity and 543–45, 544f
  microseismicity and 38
East Mamm Creek, Colorado 543
Eaton-Daines method 272–73
Eaton's method 270, 271
Echo time (ET) 72
Economics, of petroleum
  chapter project and 511–19, 511t–18t, 519f
  concessionary systems and 507
  decision and 510–11
  fiscal regimes and 507–10, 507f–9f, 510t
  fundamentals of 497f
    business expenditures 497–98
    DROI and 505
    future value 502–3
    interest rate and 502
    IRR and 505–6
    NCF and 498–502, 499t–502t
    NPV 503–4, 504t
    payout and 506, 506f
    present value 503
    return on investment 504–5
    ROR in 504–6
  introduction to 493, 494f
  PSCs and 507–8, 508f
  risk analysis and 510–11
  service contracts and 508–10, 509f, 510t
  shale developments and 493–95, 495f–96f
  uncertainty and 510–11
EDF. *See* Equilibrium distribution function
EDS. *See* Energy-dispersive X-ray spectrometry
Effective horizontal stress 268
Effective medium theory 94
Effective organic porosity 58
Effective stresses 268–69, 269t
Effective vertical stress 268
EIA. *See* Energy Information Administration
Elastic brittleness index 279–81, 280t–82t
Elastic modulus 264–66, 265f
Elastic parameters
  constants 165f, 264–65
  dynamic parameters and 273–74
  Ohen's correlation 277–78
  poroelastic constant and 268–70, 269t
  static bulk modulus 277–78
  static Poisson's ratio 277
  static shear modulus and 277–78
  static Young's modulus 276–77
  wave travel-time equations 275–76
  wave velocity equations 274–75
Electron microscopy (EM)
  EDS and 226–32, 230f–34f
  FIB-SEM 239–41, 240f–44f, 243t–44t
  permeability estimation 247
  PNM 243–47, 245f–46f, 247t
  QEMSCAN 233–39, 238f–39f, 239t
  SEM 215–18, 215f–18f, 218t
  3D segmented volume 241–43, 243f, 243t
  of ultratight formations 218–26, 219f–24f, 227f
EML. *See* Extended Langmuir model
Energy-dispersive X-ray spectrometry (EDS) 226–32, 230f–34f
Energy Information Administration (EIA) U.S., 13, 15, 39
  Monterey play and 17–18
Energy security 493–95, 494f–96f
Energy spectrum 48, 49f

England, sand proppant in 330
Enhanced oil recovery (EOR) methods 5–6
  swelling test 371–72
Environment
  air pollution and 545
  drinking water and 540–43
  fracturing fluid chemicals in 533–40, 534t–42t
  induced seismicity 543–45, 544f
  introduction on 521
  TDS and 523, 527, 529, 530t, 531f, 532–33
  water management
    flowback and 527–29, 528f, 528t
    quality and 530t–31t, 531f
    quality and quantity 527–29, 528f, 528t
    terminology in 521–23, 522f, 522t
    water acquisition 525–27, 525f–26f, 526t
    water cycle and 523–24, 524f
Environmental Protection Agency (EPA)
  FracFocus 1.0 project database 525–26, 526t, 535t–39t
  thresholds for health risk 540t
Enzymes 326
EOR. *See* Enhanced oil recovery
EOS. *See* Equation of state
EPA. *See* Environmental Protection Agency
Equation of state (EOS) 353
  cubic equations of state 373–75, 375t
  diffusion coefficient and 387
  PR EOS 373–75, 375t
  regression to experimental data 387–89, 388f–89f, 390t
  shale reservoir diagrams 384, 384t–86t, 386f
  stability analysis 375–76
  Van der Waals 373–74
Equilibrium distribution function (EDF) 424
Equivalent wellbore radius 299–300, 299f
Errors. *See* Artifacts
Estimated ultimate recovery (EUR) 132, 439
  calculation of 442–45
  decline curve and 451
  harmonic decline and 445–46
  hyperbolic decline and 446–49, 451–52
  modified hyperbolic decline model and 451–52
E-Stimplan software 340–41, 341f–43f, 341t, 343t
Estratos con Favrella shale 20–24, 24f, 27t
ET. *See* Echo time
Ethane 545
  in Bakken fluid 359, 359t–61t
Ethanol 536t–37t, 545
EUR. *See* Estimated ultimate recovery
European plays 25–29, 28t, 29f
  petrophysical properties of 112t
European Union 545
Excess adsorption 409–10, 410f
Exinite 135t
Exponential decline model 442
Extended Langmuir model (ELM) 69
ExxonMobil 26

f

F. *See* Fractional flow
Falcon Oil and Gas Ltd., 36
Fayetteville shale 16–17, 17f, 23t, 529–32
FBHP. *See* Flowing bottomhole pressure
FBP. *See* Filtered back-projection
FEI Helios NanoLab DualBeam 216
Feldspar 47
Fetkovich type curves 460f, 461–62, 462f
FIB. *See* Focused ion beam
FIB-SEM. *See* Focused ion beam-scanning electron microscopy
Fick, Adolf 387
Fickian gas diffusion 246, 387
Field of view (FOV) 187, 189
  of Xradia MicroXCT-400, 194–95, 195t
Filter cake 322, 323f, 326
Filtered back-projection (FBP) 189, 191, 196
  reconstruction of 197f
Filtering. *See also* Imaging techniques
  in X-ray micro-CT 204, 205f
First-order Raman spectra 168–69, 169f
Fischione 1060 SEM Mill 241, 242f
FIT. *See* Formation Integrity Test

Fit-for-purpose sand 330t, 331
Flash calculation 377–79, 378f, 397
Flashed gas phase 357–59, 360t
Flowback
  Marcellus shale 522, 522f, 523–33
  proppants and 333
  in U.S. shale plays 522, 522t
  water quality and 527–29, 528f, 528t, 530t–31t, 531f
  water reuse and 529–33, 532f, 534t
Flowing bottomhole pressure (FBHP) 354, 441, 458
Flowing material balance (FMB) 465–66, 467f
Flow regimes 466–72, 467f–71f, 469t
Fluid communication 307, 307f
Fluid flow. *See* Fracturing fluid; Nanopores
Fluid-fluid interactions 322, 427
Fluid-solid interactions 407, 426–27
FMB. *See* Flowing material balance
FOC. *See* Foreign oil company
Focused ion beam (FIB) 212, 213f, 240, 421
Focused ion beam-scanning electron microscopy (FIB-SEM) 239–41, 240f–44f, 243t–44t
  organic porosity and 138, 138f–39f, 140, 161
Folds of increase (FOI) 299–300, 299f
Foraminifera bodies 132–33
Foraminifera chamber 133, 133f, 169
Forced imbibition 103
Foreign investments 2–3
  in Mexico 20
Foreign oil company (FOC) 508, 508f
Formaldehyde 545
Formation Integrity Test (FIT) 273
Formation permeability 297–99, 313, 322
Fourier transform infrared radiation (FTIR) 174, 177
FOV. *See* Field of view
FracFocus 1.0 database 533, 535t–39t
Frac hit
  child-parent relationship and 305–6
  classification 307, 307f
  in Eagle Ford play 305, 306f
  impacting factors 307, 308f
  mitigation 307–8
  pressure communication 307, 307f
  prevention 524
Fractional conversion parameter 158
Fractional flow (F) 97–98
Fracture cleanup process 297, 322, 333
Fracture closure pressure 273
Fracture conductivity 295–98, 297f
Fracture flow paths 58, 59f
Fracture geometry 295, 296f, 296t
Fracture hits 305–8, 306f–8f
Fracture initiation pressure 271–73
Fracture permeability 295, 297–99
Fracturing fluid. *See also* Nanopores
  additives 323–27
    biocides 325t, 326
    buffers 325–26, 325t
    clay stabilizers 325t, 326
    cross-linkers 324–25, 325t
    diverting agents 325t, 326
    fluid loss 325t, 326
    friction reducers 325t, 327
    gel breakers 325t, 326
    gel stabilizers 325t, 326
    pH adjusters 325–26, 325t
    scale inhibitors 325t, 327
    surfactants 325t, 326
  chemicals 533–40, 534t–42t
  damage and cleanup 322, 323f–24f
  properties of 319–20, 320t
  purpose of 318
  rheology of 320–22, 321f
  slickwater 318, 320, 322, 332, 341t
  types of 318–20, 320t
Fracturing pressure 269–73
France 28–29, 28t
Frasnian shale play 31–33, 32f, 32t
Freznel Zone Plate objective (FZP) 211
Friction reducers 325t, 327
FTIR. *See* Fourier transform infrared radiation
Fugacity 375, 378, 378f

Full width at half maximum (FWHM) 168–70, 169f, 173t
Fusinite 135t
Future value 502–3
FWHM. *See* Full width at half maximum
FZP. *See* Freznel Zone Plate

g

Gamma ray (GR) 149t, 150–52
  logging tool 46
  spectroscopy
    applications of 51, 51t–52t
    ash layer detection and 49
    clay typing 48–49
    energy spectrum and 48, 49f
    mineralogy and 49
    NGT and 48
    uranium-free 48–50, 50f, 51f
Gas chromatography (GC) 46
  compositional analysis and 354–58, 354f, 356f, 358f, 394f
Gas chromatography-mass spectrometry (GC-MS) 163
Gas flow regimes 412–14, 413f–14f
Gas formation volume factor 368–69, 371t
Gas huff-n-puff 371
Gas-oil ratio (GOR) 168
  in compositional analysis 356–59, 390, 398t
Gas physiosorption isotherms 68f
Gas Processors Association (GPA) 356–57
Gas Research Institute (GRI) method 65–66, 80–82, 81f, 108
GC. *See* Gas chromatography
GCMC. *See* Grand Canonical Monte Carlo
GC-MS. *See* Gas chromatography-mass spectrometry
GDP. *See* Gross domestic product
Gel breakers 325t, 326
Gelling agent 318, 326
Gel stabilizers 325t, 326
Geochemistry. *See* Petroleum geochemistry; Petrophysical properties
Geomechanical properties
  background on 263
  BI in 279–81, 280t–82t
  concepts of 263–66, 264f–65f
  elastic BI 279–81, 280t–82t
  elastic constants 165f, 264–65
  elastic parameters and 273–78
  MBI in 234–37, 281–85, 283t–84t, 286t
  stresses in 266–73, 266f, 269t
  sweet spot in 278–85, 280t–84t
Geophones 314
Geopressurized zones 7
Geostratigraphic plays 8–9
G function 311–13
Ghadames basin 31, 32f, 33
GHS. *See* Global harmonized system
Gibbs phase rule 362
Gilsonite 172–73, 173f
Gilwood stock tank oil 393, 394f
Glauconite 48–49
Global harmonized system (GHS) 534
Global oil demand 1–2
Glossary of Geology, AGI 132–33
Goldwyer shale play 36, 36t, 37f
GOR. *See* Gas-oil ratio
Government take 497–98, 497f
GPA. *See* Gas Processors Association
G plot 311–13, 313f
GR. *See* Gamma ray
Graham, Thomas 387
Grain size, of proppant 327–28, 328t
Grand Canonical Monte Carlo (GCMC) 409–10
Graphical determinations
  of exponential decline 442–45, 443t, 444f
  of harmonic decline 445–46
Green River shale 172, 198–99, 199f, 420
  SEM of 204, 206f, 237, 239f
GRI. *See* Gas Research Institute
Gross domestic product (GDP) 493, 493f
Groundwater. *See* Water

GR-TOC method 150–52, 151t–52t
Guar gum 318
Gulf of Mexico 271–72, 277
weather and 7

h

Halliburton Oil Well Cementing Company (HOWCO) 31, 293
Halo plays 8–9
Harmonic Decline Model 445–46
Harvey County, Kansas 448–49
Hassler core holder 106
Haynesville shale play 17, 17f, 23t, 532
HDC. *See* High-density ceramic
Health risk scores 534, 540t–42t
Heliscan micro-CT 196
Helium expansion 65–67, 66f
Hexadecane 105, 105f, 355t, 360t–61t
Hexamantanes 166
Hexane 410
Hexane emissions 545
HH. *See* Hydrostatic head
HI. *See* Hydrogen index
High-density ceramic (HDC) 331
High oil saturation index 143
High-strength proppant (HSP) 331
HI_{pd}. *See* Present-day hydrogen index
Hopanes 163t, 165–66, 166t
Horizontal drilling 1–2, 3–4
workflow development and 37–38
Horizontal stress 268–73
Horizontal wells
in Eagle Ford 3, 4f
in Permian-Midland basin 3, 4f
in Vaca Muerta play 3, 4f
Horn River, Canada 16, 18, 23t
Hounsfield unit (HU) scale 197, 207
HOWCO. *See* Halliburton Oil Well Cementing Company
HSP. *See* High-strength proppant
HT. *See* Hydraulic tortuosity
HU. *See* Hounsfield unit
Hubbert and Willis method 270–71
Hugoton gas field 293
Hydraulic fracturing 29
background on 293–94, 294f
child well in 304–8, 308f
closure pressure 273
DFIT 310–13, 312f
dimensionless conductivity 297–98
equivalent wellbore radius 299–300, 299f
formation permeability 297–99, 313, 322
fracturing fluid
additives 323–27, 325t
damage and cleanup 322, 323f–24f
purpose of 318
rheology of 320–22, 321f
slickwater 318, 320, 322, 332, 341t
types of 318–20, 320t
fundamentals of 294–95
cleanup process 297, 322, 333
DFIT 310–13, 312f
FOI 299–300, 299f
fracture conductivity 295–98, 297f
fracture geometry 295, 296f, 296t
fracture hits and 305–8, 306f–8f
microseismic monitoring 314–16, 315f–16f
minifrac test 310–11, 310f, 313
multistage hydraulic fracturing 300–303, 301f–2f, 303t
MZF and 304–5, 305f
SRV in 316–17, 317f
stress shadow 303–4, 304f
surface pumps 308–9
zipper fracturing 304–5, 305f
G plot and 311–13, 312f
health risk scores of 534, 540t–42t
IFI and 305
ISIP and 310, 310f, 313
LEFM and 335
MFHW and 466–68, 468f
modeling of 334–35

Hydraulic fracturing (*cont'd*)
advanced models 339
cell-based P3D 339, 339f
early models 335–37
E-Stimplan 340–41, 341f–43f, 341t, 343t
fully 3D models 341–42
governing processes of 335
importance of 335
KGD model 338–39
P3D 339–40, 339f–40f
PKN model 337–38
PL3D 340– 341t, 341f–43f, 343t
UFM 343–44
open-hole completion 300–301
origins of 3–4
pad stage of 294, 327, 341f
PnP completion 301–2, 301f, 303t
postfracture skin factor 299
pressure shield and 304, 308
proppants
characteristics of 327–30, 328f, 328t–29t
flowback 333
purpose of 327
schedule 334
transport 333–34
types of 330–33, 330t, 332f–33f
shear rate and 320–22, 321f
sliding sleeve technique 302–3, 302f
slurry stage 294–95, 327, 334
TVD and 296f, 309
UFM and 343–44
well shut-in technique 308
workflow development and 37–38
Hydraulic tortuosity (HT) 246–47, 247t
Hydrocarbon yield. *See also* Total organic carbon
PI and 141, 143, 148–49, 159, 177
Rock-Eval pyrolysis and 159–62, 159t–60t
Hydrogenation 6
Hydrogen index (HI) 37, 71–72, 132
of kerogen 134, 134f
in Rock-Eval pyrolysis 143
Hydrostatic head (HH) 309
Hyperbolic decline model 446–49, 447f, 449t
modified 451–52

i

IAPWS-95. *See* International Association for Properties of Water and Steam Formulation 1995
IARF. *See* Infinite acting radial flow
IDC. *See* Intermediate strength ceramic
IFI. *See* Interwell fracture interactions
IFT. *See* Interfacial tension
IHS Harmony 468
Illite 48, 51f, 54t, 106–7, 107f
in plays and reserves 112t, 228t
Illizi basin 33
ImageJ software 198, 201, 241–42
Image processing 296f, 468
Avizo software 198, 201, 241–43
decline model parameters and 448
E-Stimplan 340–41, 341f–43f, 341t, 343t
ImageJ 198, 201, 241–42
JMicroVision 219
Mango software 201
PerGeos 198, 201, 204, 241–42
XMReconstructor software 189, 191, 196–97
in X-ray micro-CT 196–98, 197f
Imaging techniques
artifacts in 192–96, 203, 240–41
CPT and 246, 246f
electron microscopy
EDS and 226–32, 230f–34f
FIB-SEM 239–41, 240f–44f, 243t–44t
permeability estimation 247
PNM 243–47, 245f–46f, 247t
QEMSCAN 233–39, 238f–39f, 239t
SEM 215–18, 215f–18f, 218t
3D segmented volume 241–43, 243f, 243t
of ultratight formations 218–26, 219f–24f, 227f
Fischione 1060 SEM Mill 241, 242f
HT and 246–47, 247t

introduction to 187–88
ion milling technique 212, 213f, 240–41, 249
linear attenuation coefficient 189, 195–97, 203, 207
MBI and 234–37
reconstruction in 196–98, 197f
REV in 204, 204t
ROI in 196, 203, 209, 212, 240
tortuosity and 245–47, 246f, 247t
watershed algorithm 203f, 204–5, 205f
Xradia MicroXCT-400, 190–91, 190f–91f, 194, 196–97
X-ray micro-CT
analysis of 205–7, 206f
filtering and 204, 205f
image reconstruction in 196–98, 197f
operation of 190–92, 191f–92f
resolution and 188–89, 188f, 190f
sample preparation in 192–94, 193f
scanning procedure in 194–96, 195f, 195t
segmentation in 203–5, 205f–6f
shale sample application 198–201, 199f–201f
visualization and 201–4, 201f–10f, 204t
X-ray nano-CT 211–12, 211f–12f
Improved oil recovery (IOR) 390
India 35f, 495
Indiana limestone 208t
Indirect method
for poroelastic constant 268
for TOC 141
Indonesia
Poupon-Leveaux model 100–102, 101t–2t
PSCs and 509, 510t
Induced seismicity 543–45, 544f
Inequality test 369, 371t
Infinite acting radial flow (IARF) 468
Infra-Cambrian Ara Group 138
Infrared spectroscopy (IR) 177
Initial organic carbon content (TOC_o) 161–62
Inorganic matrix 6
In-situ measurements
delta LogR method 153–57, 153f–56f, 156t–58t
GR-TOC method 150–52, 151t–52t
Schmoker method 149–50, 149t–50t, 151f
In-situ wettability 212–14, 214f
Instantaneous shut-in pressure (ISIP) 310, 310f, 313
Interecho spacing (TE) 71–72, 71f
Interest rate 502
Interfacial tension (IFT) 371, 377, 389–90, 392–93
Intermediate strength ceramic (IDC) 331, 333f
Intermediate strength proppant (ISP) 331, 332f
Internal rate of return (IRR) 505–6
International Association for Properties of Water and Steam Formulation 1995 (IAPWS-95) 374
International oil company (IOC) 508, 508f
Interwell fracture interactions (IFI) 305
Inverse matrix, of log readings 56, 57t
Inverse-square of Pc (MICP) 95, 96f
Inversion recovery free induction decay (IRFID) 71
IOC. *See* International oil company
Ion milling technique 212, 213f, 240–41, 249
IOR. *See* Improved oil recovery
IR. *See* Infrared spectroscopy
Iran 508
IRFID. *See* Inversion recovery free induction decay
IRR. *See* Internal rate of return
ISIP. *See* Instantaneous shut-in pressure
Isothermal compressibility 367, 369
ISP. *See* Intermediate-strength proppant
de Itriago, Araujo 107

j

JMicroVision software 219
Jurassic age 29–30, 33
Haynesville formation 58, 59f
late 17, 20, 25–26

k

Kansas, U.S., 16, 16f, 293
  Barber County 150–52, 190, 207, 217f
  Harvey County 448–49
Kansas Geological Survey 53
Kaolinite 48–49, 133, 225
Karoo basin, South Africa 31
K-C equation. *See* Kozeny-Carman
Kern County, CA 545
Kerogen 6
  classification of, 134–36, 134f–35f, 135t–36t
  compositional analyses and 353–54
  density of 207
  DK and 224–26, 225f, 227f
  DRIFTS and 174–77, 176f
  molecular structure of 418–21, 419f
  permeability estimation of 210, 247
  petroleum geochemistry and 132–34, 133f
  S2-TOC plot and 134–36, 135f
  segmentation and 204
  SEM and 217–18, 221
  in shale formations 228t–29t
  thermal maturity of 168–74, 169f–75f, 173t
  3D segmented volume of 243, 243f
  transformational stages of 130–31, 130f–31f
  X-ray micro-CT of 187, 189, 190f, 192, 192f
Ketuer shale play 34t, 35
K-feldspar 106–7, 107f
Khalij Al-Bahrain basin 31
Khatatba shale play 32t, 33
Khristianovich and Zheltov model 336, 338–39
Kinetic theory of gases 412, 415
Klinkenberg method 415–16
KM. *See* Kubelka-Munk intensity
Knudsen diffusion 245–46, 387
Knudsen number 412–16, 414f, 417t
  in LBM 425–26
Kozeny-Carman (K-C) equation 93, 247
Krumbein and Sloss chart 328, 328f
Kubelka-Munk intensity (KM) 176
Kubelka Munk units 56
Kuwait 508

l

LAI. *See* Langmuir adsorption isotherm
La Luna shale play 25, 27f
Lamalginite 224
Laminated sand/shale model 99
Langmuir adsorption isotherm (LAI) 69, 70f
  pressure 68, 109, 410–11, 428t
  volume 68, 109, 410–12, 428t
Langmuir slip model (LSM) 426–27, 428t
Laplace nonnegative least square fitting 71
Large field of view (LFOV) 211
Larionov equation 47–48
Late Cambrian age 28, 30t, 35
Late Devonian 11
Late Jurassic age 17, 20, 25–26
Late Mississippian 10, 22t
Lateral heterogeneity 6
Late Silurian age 28, 28t, 34t
Lattice Boltzmann model (LBM) 93, 421–24, 423f, 423t
  apparent permeability 427
  Knudsen number and 425–26
  organic nanopores and 425–27, 426f
  permeability estimation and 210, 247
  simulation codes of 425, 425f
Lattice gas cellular automata (LGCA) models 421–22
Lattice unit 422–23
LBM. *See* Lattice Boltzmann model
L. Carboniferous shale play 28
LDT. *See* Litho-Density Tool
Leak-Off Test (LOT) 273
LEFM. *See* Linear elastic fracture mechanics
Lennard-Jones potential 418, 418f
Lennard-Jones size parameter 379
Level of maturity (LOM) 153–57, 153f, 155f
LFOV. *See* Large field of view
LGCA. *See* Lattice gas cellular automata
LGM. *See* Logistic growth model
Liard basin, Canada 18–19
Libya 33
Lightweight ceramic proppant (LWC) 331–32, 332t

Limited entry 301–2
Linear attenuation coefficient 189, 195–97, 203, 207
Linear elastic fracture mechanics (LEFM) 335
Liptinite 135–36, 135t
Liquid diffusion 387
Litho-density tool (LDT) 53, 54t
Lithology evaluation. *See also* Porosity
  combination of logs 53–56, 54t, 55f, 56t
  cross-plot techniques and 52–53, 53f
  DRIFTS 56
  dual-mineral analysis 54–55, 55f
  LDT 53, 54t
  mineral composition and 51–52, 52f
  PSD and
    imaging techniques and 243–47, 245f–46f, 247t
    nitrogen adsorption 77–79, 79t, 80f
    NMR well logs 75–77 76f–77f, 78t
    petrophysical properties and 111t–12t
  radioactive material in 47, 49, 52t
  three-mineral analysis 55–56, 56t–57t, 57f
Liu et al. approach 100–101
Llandovery shale play 28
Logging while drilling (LWD) 46
Logistic growth model (LGM) 455
LOM. *See* Level of maturity
Longmaxi shale play 33–35, 34t
  cores 192, 200f
Longmuir model 69
Los Molles shale play 20, 27t
Los Monos shale play 24, 26f, 27t
LOT. *See* Leak-Off Test
Louisiana, U.S., 17, 525–26, 526t
Low-energy filter 195–96
Lower Cretaceous, porosity and 58, 59f
Lower Eagle Ford (LEF) shale
  BSE images of 133, 133f
  organic porosity and 138–40, 138f–39f, 161
  SEM images 222f, 225–26, 225f
Lower Kyalla shale play 36, 36t, 37f
LSM. *See* Langmuir slip model
Lublin subbasin 28
LWC. *See* Lightweight ceramic proppant
LWD. *See* Logging while drilling

m

Macerals 135–40, 135t, 162, 174
Magnification 187
Malaysia 509, 510t
Mango software 201
Manitoba, Canada 11
Maracaibo basin 25, 27f
Marcellus shale play 11–13, 13f, 17, 22t
  adsorption isotherm 411, 411t
  CT scan of 200, 201f
  flowback of 522, 522f, 523–33
  NMR fluid volume 103f–4f
  organic porosity and 140
  pore size distribution and 408, 408f
Marine phytoplankton 129–30
Martini problem 53
Maryland, U.S., 12, 13f
MATLAB code, in LBM simulation 425
Matrix permeability, estimation of 468–72, 469t
Matrix-stress coefficient 271
Maverick basin, Texas 13
Maximum horizontal stress 270
Maximum temperature (T_{max}) 142–43, 142t
MB. *See* Migrated bitumen
MBI. *See* Mineral-based brittleness index
MD. *See* Molecular dynamics
MDT. *See* Modular Formation Dynamics Tester
Measurement while drilling (MWD) 46
Membrane-based desalination technologies 523
MET. *See* Mineral extraction tax
Metagenesis 130–31, 131f
Methane 545
  adsorption isotherms 410–12, 410f, 411t–12t, 413f
  in Bakken fluid 359, 359t–61t
  biogenic gas and 130–31, 130f–31f
Methane flow 428t
Methane hydrates 8

Methanol 545
Method of characteristic (MOC) 396
Methyldiamantanes 167–68
Methylene chloride 545
Mexico 13
  Eagle Ford Shale in 19–20, 21f, 23t
  foreign investments in 20
  gulf of 270–71, 277
  unconventional plays in 18–20, 21f
MFHW. *See* Multifractured horizontal well
MICP. *See* Inverse-square of Pc
Microscale 95, 96f, 108
  imaging technology and 187, 204t, 216
  LBM and 421–22
Microseismic monitoring 314–16, 315f–16f
Microseismic surveys 38
MID cross-plot 53
Middle Bakken formation 6
  average permeability of 298
Middle Devonian age 11, 129
Middle East
  petrophysical properties of 112t
  unconventional plays 30–31, 30t
Middle Ordovician 13, 22t, 36t
  shale play 35
Migrated bitumen (MB) 169, 172
  SEM and 224–26, 225f, 227f
Mineral-based brittleness index (MBI) 234–37, 281–85, 283t–84t
Mineral extraction tax (MET) 28
Mineralogy 49
Minifrac test 310–11, 310f, 313
Minimum horizontal stress 270–72, 304, 304f
Minimum miscibility pressure (MMP) 353
  analytical methods 396–97
  correlation methods 397, 398t
  experimental methods for 390–93, 391f–96f
  numerical methods 397
Mississippian age 13, 23t, 58, 198, 228t–29t
  Conglomerate formation 448–49
  early 11, 16
  late 10, 22t
MMP. *See* Minimum miscibility pressure
Mobility score 541t–42t
MOC. *See* Method of characteristic
Modeling techniques
  apparent permeability and 415–16, 416t–17t, 417f
  fluid transport and 414–15
  hydraulic fracturing
    dimensionless conductivity 297–98
    fracture conductivity 295–98, 297f
    fracture geometry 295, 296f, 296t
    fracture hits and 305–8, 306f–8f
    postfracture skin factor 299
  molecular simulations 416–18, 418f
  molecular structure of kerogen 418–21, 419f
  multiscale shale matrix 421, 422f
Modern decline curve models 449–50
  boundary-dominant flow 451, 454, 459–62, 462f
  Duong's model 456–57, 457f–58f, 457t
  LGM 455
  modified hyperbolic model 451–52, 453f
  PLE model 452–54, 453t
  SEP model 454–55, 456f
Modified Amott-Harvey technique 103, 106–7, 107f
Modified hyperbolic model 451–54, 453f, 453t
Modified PR EOS 381–83, 382f–83f
Modified pulse decay method 84–85, 84f, 85t
Modified Purcell method 95–97, 96f
Modified Simandoux model 99
Modified van Krevelen plot 218, 218f
  of kerogen 134, 134f
Modified zipper fracturing (MZF) 304–5, 305f
Modular Formation Dynamics Tester (MDT) 273
Modulus of rigidity. *See* Shear modulus
Molecular dynamics (MD) 417–18, 421
Montana, U.S., 9, 11
Monte Carlo simulations 417

Monterey shale play 17–18, 18f, 32t, 63–64
CTN image of 209f–10f
Montmorillonite 99
Montney, Canada 16, 19f
Morales correlation 277
Mouydir basin 33
MRT. *See* Multirelaxation time
Mud logging 46, 281
Multicomponent systems 361t, 364–65, 365f
Multifractured horizontal well (MFHW) 466–68, 468f
Multiphase-based fracturing fluids 319
Multiphase boundary 374
Multiple mixing cell method 396–97
Multiple-scattered backscattered electrons (BSE2) 216, 216f
Multirelaxation time (MRT) 424
Multiscale shale matrix modeling 421, 422f
Multistage fracturing 3–4
open-hole completion 300–301
PnP completion 301–2, 301f, 303t
sliding sleeve technique 302–3, 302f
Murzuq basin 33
Muskwa Besa River 23t
Muskwa shale play 18, 19f
MWD. *See* Measurement while drilling
MZF. *See* Modified zipper fracturing

n

Nanochannels 409, 409f
Nano-computerized tomography (Nano-CT)
in-situ wettability 212–14, 214f
sample preparation for 212, 213f
Nanopores
adsorption and 408–12, 409f–11f, 411t–12t, 413f
flow regimes and 412–14, 413f–14f
introduction to 407
LBM 421–24, 423f, 423t
apparent permeability 427
Knudsen number and 425–26
organic nanopores and 425–27, 426f
permeability estimation and 210, 247
simulation codes of 425, 425f
modeling techniques
apparent permeability and 415–16, 416t–17t, 417f
fluid transport and 414–15
molecular simulations 416–18, 418f
molecular structure of kerogen 418–21, 419f
multiscale shale matrix 421, 422f
pore size distribution 408, 408f
pore wall effects and 425–26, 426f, 428t
transport in organic nanopores 416
Napalm 318
Nappamerri shale play 36, 36t, 37f
National oil companies (NOCs) 508, 508f
Natural Gamma Ray Tool (NGT) 48
Navier-Stokes equation 407, 415, 421, 424
NCF. *See* Net cash flow
Nebraska, U.S., 16, 16f
Net cash flow (NCF) 498–502, 499t–502t
Net present value (NPV) 503–4, 504t
Neutron-density cross-plot 53, 53f
Neutron-density logs 63–65
Neutron-density method 60t
Neutron porosity 149t, 157
Neutron-sonic method 60t
Newtonian fluids 320–21, 333
Newton's equation 418
Newton's second law of motion 424
New York, U.S., 12, 13f
NGT. *See* Natural Gamma Ray Tool
Nigeria, PSCs and 509, 510t
Niobrara shale play 16, 16f, 22t, 140
DCA in 457, 457t, 458f
Nitrogen adsorption
AD and 94–95
PSD and 77–79, 79t, 80f
NMR. *See* Nuclear magnetic resonance
NOCs. *See* National oil companies
Non-newtonian fluids 320, 333
Nonwetting phase 94–95–97–98

Normalized pressure integral (NPI) type curves 464–65, 464f
North American plays 8–20, 21t–22t
petrophysical properties of 111t
TOC of 407
North Dakota, U.S., 9, 11, 525–26, 526t
North West Shelf shale 199, 200f
NPI. *See* Normalized pressure integral
NPV. *See* Net present value
Nuclear magnetic resonance (NMR)
capillary pressure 97
permeability 89–93, 91f–92f, 92t–93t
pore size and 408, 419–21
porosity and 69–73, 71f–73f, 73t–74t
pseudocapillary pressure 97–98
saturation estimation 102
total CMR and 69–70, 76f, 92, 92t
well logs 58, 59f
PSD and 75–77 76f–77f, 78t
wettability and 103–5, 103f–5f

O

Occupational Safety and Health Administration (OSHA) U.S., 545
Occurrence scores 534, 540t–42t
ODE. *See* Ordinary differential equation
OGIP. *See* Original gas in place
Ohen's correlation 277–78
Ohio, U.S., 12, 13f
OI. *See* Oxygen index
Oil-based fracturing fluids 319
Oil formation volume factor 368–69, 370f
Oil lamps 1
Oil prices 3–4
Oil production, in U.S., 495, 495f
Oil saturation index (OSI) 143
Oil shale 6
Oklahoma, U.S., 13, 16, 293, 525–26, 526t, 543–45
OM. *See* Organic matter
Oman 330
OMP. *See* Organic matter pore
Onsager's autocorrelation coefficients 416
OOIP. *See* Original oil in place
OPEC. *See* Organization of Petroleum Exporting Countries
Open-hole completion 300–301
Operating expenses (OPEX) 497–98, 497f
Ordinary differential equation (ODE) 339–40
Ordovician era 78t
middle 13, 22t, 36t
shale play 35
Viola group 140
Organic matter (OM)
catagenesis 130–31, 131f
diagenesis in 130, 131f
evolution of 129–31, 130f–31f
metagenesis in 130–31, 131f
shale pockets of 407–8, 419–20
Organic matter pore (OMP) 219, 221, 221f, 223f
Organic nanopores. *See also* Nanopores
LBM and 425–27, 426f
transport in 416
Organic porosity 138–40, 138f–39f. *See also* Porosity
Rock Eval pyrolysis and 161–62
Organization of Petroleum Exporting Countries (OPEC) 495
Original gas in place (OGIP) 98, 106–7
Original oil in place (OOIP) 98, 106–7, 469–72, 469t
Original total organic content (TOC_o) 159–62, 159t–60t
Oscillating pulse method 85–88, 86f–88f
OSHA. *See* Occupational Safety and Health Administration
OSI. *See* Oil saturation index
Osmotic adsorption isotherm equation 387
Ottawa sand 330t, 331
Overburden Pressure 267–68
Oxidizers 326
Oxygen index (OI) 132
of kerogen 134, 134f
Rock-Eval pyrolysis 143

p

Pad stage 294, 327, 341f
PAH. *See* Polycyclic aromatic hydrocarbon
PAOM. *See* Porosity asserted organic matter
Paraguay 24
Parameter determinations
  for binary interactions 384, 384t–86t
  of decline model 448
  hyperbolic decline model 448–49, 449t
  Lennard-Jones size 379
  volume translation 388
Parana basin 24
Partial differential equation (PDE) 339–40
Patchawarra shale play 36, 37f
Payout 519, 519f
Payout period 506, 506f, 512
PC. *See* Pyrolysis carbon
PDE. *See* Partial differential equation
Pearsall formation 58, 59f
PEMEX 13, 19–20
Peng-Robinson equation of state (PR EOS) 364–65, 373–75, 375t
  modified 381–83, 382f–83f
Pennsylvania, U.S., 12, 13f, 525–26, 526t, 540, 543
Pentamantanes 166
Percolation theory 94
PerGeos software 198, 201, 204, 241–42
Permeability estimation
  apparent permeability 415–16, 416t–17t, 417f, 427
  from CT-scan 210
  Darcy range and 79–80
  of formation 297–99, 313, 322
  fracture 295, 297–99
  LBM and 210, 247
  matrix estimation 468–72, 469t
  NMR capillary pressure 97
  NMR measurements of 89–93, 91f–92f, 92t–93t
  pseudocapillary pressure 97–98
  R13 AP model 415–16
  relative permeability 93–97, 96f
  RTA and 468–72, 469t
  in SEM 247
  single phase 88–89, 89t
  T_2 permeability and 89–92, 91f–92f
  unsteady-state methods
    GRI methods 80–82, 81f
    modified pulse decay 84–85, 84f, 85t
    oscillating pulse 85–88, 86f–88f
    pulse decay 80, 81f, 82–84, 83f, 83t
Permian basin 15–16, 15f, 306
Permian-Carboniferous formation 28–29
Permian formation 33
Permian-Midland basin 3, 4f
Permian Stuart ranges 221, 224f
Persistence score 541t–42t
Petroleum geochemistry. *See also* Total organic carbon
  background on 129
  biological markers in 162–66, 163t, 165f, 166t
  bitumen and 132–34, 133f
  diamondoids and 166–68, 167f
  DRIFTS and 174–77, 176f
  hydrocarbon yield and 159–62, 159t–60t
  kerogen 132–34, 133f
    classification of 134–36, 134f–35f, 135t–36t
  LOM and 153–57, 153f, 155f
  macerals and 135–40, 135t, 162, 174
  organic matter evolution and 129–31, 130f–31f
  organic porosity 138–40, 138f–39f
  pyrobitumen and 132–34, 133f
  solid bitumen reflectance and 137–38
  thermal alteration index and 136, 153
  thermal maturity
    Raman spectroscopy 168–74, 169f–75f, 173t
    source assessment and 162–68, 163t, 165f, 166t, 167f
  TOC guidelines and 131–32, 132t

Petroleum geochemistry (*cont'd*)
TOC$_0$ and 159–62, 159t–60t
vitrinite reflectance and 136–37, 137t
Petroleum operator 5
Petroleum seeps 1–2
Petrophysical properties. *See also* Permeability estimation; Porosity
of Africa 112t
of Asia 112t
of Australian plays 112t
background on 45
BVW and 108
coring 46
of European plays 112t
GR spectroscopy and 48–51, 49f, 50t–52t, 51f
lithology evaluation
combination of logs 53–56, 54t, 55f, 56t
cross-plot techniques and 52–53, 53f
DRIFTS 56
dual-mineral analysis 54–55, 55f
LDT 53, 54t
mineral composition and 51–52, 52f
nitrogen adsorption and 77–79, 79t, 80f
NMR well logs 75–77 76f–77f, 78t
radioactive material in 47, 49, 52t
three-mineral analysis 55–56, 56t–57t, 57f
MICP and 95, 96f
of Middle East 112t
mud logging and 46
nonwetting phase 94–95–97–98
of North American plays 111t
PAOM and 108
pore-throat and 95–96f
pore volume and 107–10, 108f–10f, 110t–12t
reserve estimation and 107–10, 108f–10f, 110t–12t
rock evaluation 46, 47f
shale volume 46–48, 48t
single phase permeability 88–89, 89t
of South American plays 112t
well logging 46, 47f
wettability and 102–7, 103f–8f
wetting phase and 94–95, 97–98
pH adjusters 325–26, 325t
Phase behavior
BIP and 374–75, 388, 390t
CCE and 366–68, 366f, 367t, 368f
Bakken sample data of 397–400, 400t
compositional analysis
of shale fluid 353–54, 354f, 355t
subsurface sampling 354–57, 355t, 356f
surface sampling 357–59, 358f, 359t–61t
confinement on 376–77, 376f–77f
capillary pressure in 377–79, 378f
EOS critical properties shift 379–81, 380f–81f, 380t
modified PR EOS 381–83, 382f–83f
DL experiment data of 400, 401t
downhole fluid analysis 356, 356f, 365f
EOS and
cubic equations of state 373–75, 375t
diffusion coefficient 387
PR EOS 373–75, 375t
regression to experimental data 387–89, 388f–89f, 390t
stability analysis 375–76
Fickian gas diffusion 246, 387
GC and 354–58, 354f, 356f, 358f, 394f
GOR and 356–59, 390, 398t
IFT and 371, 377, 389–90, 392–93
introduction to 353
minimum miscibility pressure
analytical methods 396–97
correlation methods 397, 398t
experimental methods for 390–93, 391f–96f
numerical methods 397
PVT and
CCE and 366–68, 366f, 367t, 368f
data quality check and 360t–71t, 366–72, 367t, 368f–72f

DL experiment 368–69, 369f–70f, 370t–71t
multicomponent systems 361t, 364–65, 365f
phase diagrams 362–65, 362f–65f
p-V diagram 362–63, 363f–64f, 366–67, 366f, 368f
separator test 369–71, 371f
single-component systems 362–63, 362f–63f
swelling test 371–72, 372f
two-component systems 363–64, 364f
Y-function linearity test and 367t, 368, 368f, 388, 400t
shale reservoir diagrams 384, 384t–86t, 386f
Y-function linearity test 367t, 368, 368f, 388, 400t
Phase boundary 362–63, 374
Photoelectric absorption 190
Photosynthetic bacteria 129–30
Phytane 163–64, 166t
PI. *See* Productivity index
Pimienta formation 20, 23t
PIT. *See* Pressure Integrity Test
Pixel 187
PLE. *See* Power-law exponential
Plug and perf (PnP) completion 301–2, 301f, 303t
PNM. *See* Pore network model
PnP. *See* Plug and perf
Podlasie subbasin 28
Poisson's ratio 265–66, 270–71
BI and 279–80
dynamic 273–77
static 277
of unconventional plays 285, 286t
in wave velocity 274–75
Poland 330, 534, 545
Polarization time 69, 71
Polycyclic aromatic hydrocarbons (PAHs) 545
Ponta Grossa shale play 24, 25f, 27t
Pore network model (PNM) 203f, 243–47, 245f–46f, 247t
Pore proximity effect. *See* Confinement
Pore-scale modeling. *See* Digital rock physics
Pore size distribution (PSD) 73–74
imaging techniques and 243–47, 245f–46f, 247t
nitrogen adsorption and 77–79, 79t, 80f
NMR well logs 75–77 76f–77f, 78t
petrophysical properties and 111t–12t
Pore-throat 95–96f
Pore volume 107
volumetric analysis theory 108–10, 109f–10f, 110t–12t
Pore-wall effects 425–26, 426f, 428t
Poroelastic constant 268–70, 269t
Porosity
acoustic logs 64–65
combination of logs 65
components of 57–58, 59f
CT image estimation 207–9, 208f–10f, 208t
density logs 58–63, 61f–62f, 64f
DTG plot and 67
helium expansion 65–67, 66f
neutron-density logs 63–65
nitrogen adsorption and 67–69, 68f, 70f
NMR core and 70–73, 71f–73f, 73t–74t
organic 138–40, 138f–39f, 161–62
TGA plot and 67
total CMR 69–70, 76f, 92, 92t
well logs and 58, 59t
Porosity asserted organic matter (PAOM) 108
Post-collision distribution functions 427
Postfracture skin factor 299
Potassium
gamma ray spectroscopy and 48–51, 50f, 51t
thorium-potassium plot 48–50, 51f, 51t–52t
Poupon-Leveaux model 100–102, 101t–2t
Powder River basin 161
Power-law exponential (PLE) model 452–54, 453t
Precambrian era 36, 36t, 129

Pre-oil-generation bitumen 133
PR EOS. *See* Peng-Robinson equation of state
Present-day hydrogen index (HI_{pd}) 158–61
Present value 503
Present value profit (PVP) 503–5
Pressure communication 307, 307f
Pressure Integrity Test (PIT) 273
Pressure shield 304, 308
Pressure transient analysis (PTA) 458–59, 464
Pressure-volume-temperature (PVT)
  CCE and 366–68, 366f, 367t, 368f
  data quality check and 360t–71t, 366–72, 367t, 368f–72f
  DL experiment 368–69, 369f–70f, 370t–71t
  multicomponent systems 361t, 364–65, 365f
  phase diagrams 362–65, 362f–65f
  p-T diagram 362–64, 362f, 364f–65f
  p-V diagram 362–63, 363f–64f, 366–67, 366f, 368f
  separator test 369–71, 371f
  single-component systems 362–63, 362f–63f
  swelling test 371–72, 372f
  two-component systems 363–64, 364f
  Y-function linearity test and 367t, 368, 368f, 388, 400t
Prince Albert play 31
Pristane 163–64, 166t
Processing, of sample. *See also* Image processing
  in BIB-SEM 248–49, 249f
  software and 241–42
Production life, estimation of 445–49
Production sharing contract (PSC) 507–8, 508f
Productivity index (PI) 141, 148–49, 159, 177
  Rock-Eval pyrolysis 143
Profit to investment ratio 498
Propane 545
  in Bakken fluid 359, 359t–61t
Proppants 295
  Brady sand 330t, 331
  characteristics of
    bulk density 328–29, 329t
    grain size 327–28, 328t
    quality 329
    shape 328, 328f
    strength 329–30
  fit-for-purpose sand 330t, 331
  flowback 333
  Krumbein and Sloss chart 328, 328f
  Ottawa sand 330t, 331
  proppant shape 328, 328f
  purpose of 327
  schedule 334
  specific gravity 328–29, 328f
  transport 333–34
  types of 330, 330t
    artificial proppant 331
    bauxite proppant 331
    ceramic 331–32, 332f
    ISP 331, 332f–33f
    LWC 331–32, 332t
    RCP 332–33, 332f–33f
    ULW 332
Proximity effect. *See* Confinement
PSC. *See* Production sharing contract
PSD. *See* Pore size distribution
Pseudovitrinite 135t
PTA. *See* Pressure transient analysis
Pulse decay method 80, 81f, 82–84, 83f, 83t
Pulse-decay setup 66, 66f
Pure service contracts 508–9
PVP. *See* Present value profit
PVT. *See* Pressure-volume-temperature
Pyrite 133, 225
Pyrobitumen
  organic porosity and 137–40, 138f–39f, 168, 172
  petroleum geochemistry and 132–34, 133f
Pyrolysis 6, 420–21
  Rock-Eval method 37, 63, 110
    of Bakken formation 144–49, 144f–45f, 146t–49t
    hydrocarbon yield and 159–62, 159t–60t
    organic porosity and 161–62

Schmoker method and 149–50, 149t–50t, 151f
TOC and 141, 143
of upper Bakken 148–49, 148t–49t
Pyrolysis carbon (PC) 143

q

QEMSCAN. *See* Quantitative evaluation of minerals by scanning electron microscopy
Qiongzhusi formation 33
Quantitative evaluation of minerals by scanning electron microscopy (QEMSCAN) 192, 233–39, 238f–39f, 239t
Quartz 106–7, 107f, 133, 225
Quebec province, Canada 13, 13f
Qusaiba shale play 30, 30t, 285

r

R13 AP. *See* Regularized-13 apparent permeability
Rachford-Rice equation 378f
Radioactive material 522
in lithology 47, 49, 52t
Raman spectroscopy analysis
first-order 168–69, 169f
thermal maturity 168–74, 169f–75f, 173t
Rate of return (ROR) 504–6
Rate transient analysis (RTA)
Agarwal-Gardner type curves 462–64, 463f
Blasingame type curves 462, 463f
concept of 459–60
features of 457–59
Fetkovich type curves 460f, 461–62, 462f
flow regimes of 466–72, 467f–71f, 469t
FMB plot 466, 467f
matrix permeability in 468–72, 469t
MFHW and 466–68, 468f
NPI type curves 464–65, 464f
PTA and 457–59, 464
purpose of 457–59
square-root time plot 465–66, 467f
SRV in 457, 468–69, 468f, 470f, 472
type curves and 460–61, 460f
Wattenbarger type curves 465, 465f–66f
RBA. *See* Rising bubble apparatus
RC. *See* Residual carbon
RCP. *See* Resin-coated proppant
Reggane basin 31, 32f, 33
Region of interest (ROI) 196, 203, 209, 212, 240
Regression, of EOS 387–89, 388f–89f, 390t
Regularized-13 apparent permeability (R13 AP) 415–16
Relative permeability
adsorption-desorption techniques 94–95
DRP and 93–94
modified Purcell method 95–97, 96f
Renishaw inVia Reflex Spectrometer 169
Representative elementary volume (REV) 204, 204t
Residual carbon (RC) 143
Resin-coated proppant (RCP) 332–33, 332f–33f
Resinite 135t
Resistivity (RT) 100, 149, 153
Resolution 187
Return on investment 335, 504–5, 512, 519
REV. *See* Representative elementary volume
Rising bubble apparatus (RBA) 389
Risk analysis 510–11
Rock Eval pyrolysis 37, 63, 110
of Bakken formation 144–49, 144f–45f, 146t–49t
hydrocarbon yield and 159–62, 159t–60t
organic porosity and 161–62
Schmoker method and 149–50, 149t–50t, 151f
TOC and 141, 143
of upper Bakken 148–49, 148t–49t
ROI. *See* Region of interest
ROR. *See* Rate of return
Rosneft oil company 26
Rotary drilling 1
Royalty 495, 507, 516t–517f

RT. *See* Resistivity
RTA. *See* Rate transient analysis
Rub'Al-Khali basin 30
Russia 16, 496f
 companies 28, 28t
 sand proppant in 330
 west Siberian basin 26, 29f

S

S_1, 132, 141–43, 142f
S_2, 132, 141–43, 142f
 TOC plot 134–36, 135f
S_3, 142–43, 142f
S_4, 142, 142f
Sabinas basin, Mexico 19
Samotlor field 26
Sample preparation
 in BIB-SEM 248–49, 249f
 for nano-CT 212, 213f
 in X-ray micro-CT 192–94, 193f
San Andreas Fault 17, 18f
Sand
 Archie's model and 98–99
 clean sand 98–100
 proppant 330–31, 330t, 332f–33f
Sanish formation 11
San Joaquin basin 17, 18f, 545
Saskatchewan province, Canada 9, 11
Saturation
 Archie's model 98–99
 DW equation 100
 laminated sand/shale model 99
 Liu et al. approach 100–101
 NMR estimation 102
 Poupon-Leveaux model 100–102, 101t–2t
 shaly sand model 99
 Simandoux model 99–102, 101t–2t
 WST equation 100
Saturation pressure
 CCE experiment 366–68, 368f
 EOS regression and 387–88, 388t, 390t
 separator test and 369, 371f
 swelling test and 372, 372f
Saudi Arabia 285, 508
 sand proppant in 330
SCAL. *See* Special core analysis
Scale inhibitors 325t, 327
Scanning electron microscopy (SEM) 103, 215–18, 215f–18f, 218t
 BIB-SEM 247–49, 248f–49f
 DK and 224–26, 225f, 227f
 FIB-SEM 239–41, 240f–44f, 243t–44t
  organic porosity and 138, 138f–39f, 140, 161
 HT and 246–47, 247t
 image analysis 408, 421
 MB and 224–26, 225f, 227f
 permeability estimation in 247
 sample preparation
  in BIB-SEM 248–49, 249f
  for nano-CT 212, 213f
  in X-ray micro-CT 192–94, 193f
 tortuosity and 245–47, 246f, 247t
 of ultratight formations 218–26, 219f–24f, 227f
Schlumberger-Doll Research (SDR) theoretical model 92–93, 93t
Schmoker method 149–50, 149t–50t, 151f
SCI. *See* Spore coloration index
Sclerotonite 135t
SDR. *See* Schlumberger-Doll Research
Second-phase fugacity 375
Second World War 318
Segmentation, in X-ray micro-CT 203–5, 205f–6f
Seismicity 38
 induced 543–45, 544f
 microseismic monitoring 314–16, 315f–16f
SEM. *See* Scanning electron microscopy
Sembar shale play 34t, 35–36
Semifusinite 135t
SEP. *See* Stretched exponential decline model
Separator gas 357–59, 358f–359t, 398t
Separator liquid 359, 360t
Separator test, 369–71, 371f

Service contracts 507–10, 509f, 510t
Shale fluid 7. *See also* Phase behavior
  multicomponent systems and 361t, 364–65, 365f
Shale nanopores 407. *See also* Nanopores
  adsorption and 408–12, 409f–11f, 411t–12t, 413f
  flow regimes and 412–14, 413f–14f
  modeling techniques of 414–21, 416t–17t, 417f–19f
Shale reservoirs 8–9. *See also* Petroleum geochemistry
  apparent permeability of 415–16, 416t–17t, 417f
  characteristics of 228t–29t
  petroleum economics of 493–95, 495f–96f
Shale resources. *See also* Unconventional resources
  in China 3
  economic effects of 493–95, 495f–96f
  global deposits of 6
  phenomenon of 3–4
  price and 3–4
  revolution of 2–5, 2f, 4f
  service companies and 5
  in U.S., 3
Shale volume 46–48, 48t
Shaly sand model 99
Shear modulus 264–65, 265f
  dynamic 275–76
Shear rate 320–22, 321f
Shear stress 263–64, 264f
Shear wave velocity 100–101, 274
Shilaif formation 30, 30t
Sichuan basin 33, 35
Silicon
  carbide paper 248
  oxygen bonded 106, 108f
Silurian age 28, 28t, 30, 30t, 33, 34t, 140
Simandoux model 99–102, 101t–2t
Single-component systems 362–63, 362f
Single phase permeability 88–89, 89t
Single relaxation time (SRT) 424
Singly scattered backscattered electrons (BSE1) 216, 216f
Sirte shale play 32t, 33
Slickwater 318, 320, 322, 332, 341t
Sliding sleeve technique 302–3, 302f
Slim-tube test 389
Sloss chart. *See* Krumbein and Sloss chart
Slurry stage 294–95, 327, 334
Small business 5
Smectite 48–49
Soave-Redlich-Kwong EOS 373
Software packages 468
  Avizo software 198, 201, 241–43
  E-Stimplan 340–41, 341f–43f, 341t, 343t
  ImageJ 198, 241–42
  JMicroVision 219
  PerGeos 198, 204, 241–42
  XMReconstructor 189, 191, 196–97
Solid bitumen reflectance (BRo) 137–38
Sonic logs 273–75, 279, 285
Sonic velocity. *See* Transit time
Source rock. *See also* Petrophysical properties
  biomarkers in 165–66, 166t
  maturity of 163t
South American plays
  petrophysical properties of 112t
  unconventional 20–25, 24f–27f, 27t
South Oman salt basin 138
Sparberry field 15, 15f
Special core analysis (SCAL) 93–94
Specific surface area (SSA) 247
Spin-echo decay 71f
Spontaneous imbibition 103, 212–14, 214f
Spore coloration index (SCI) 136
SPP. *See* Surface pumping pressure
Spriggs well 207
Square-root time plot 465–66, 467f, 470f
SRT. *See* Single relaxation time
SRV. *See* Stimulated reservoir volume
SSA. *See* Specific surface area
Stability analysis 375–76

Standard effective porosity 58
Standard pressure and temperature (STP) 197
Static bulk modulus 277
Static moduli 276–78
Static Poisson's ratio 277
Static shear modulus 277
Static Young's modulus 276–78
Statoil 26
Steiber model 47–48
Steranes 163t, 165–66, 166t
Stigmastanes 167, 167f
Stimulated reservoir volume (SRV) 9
  fracturing and 293, 295, 304, 316–17, 317f
  RTA and 457, 468–69, 468f, 470f, 472
Stock price 4–5
STP. *See* Standard pressure and temperature
Strain 264
Stress 263–64, 264f
Stress shadow 303–4, 304f
Stretched exponential decline model (SEP) 454–55, 456f
Subsurface sampling 354–57, 355t, 356f
Successive-substitution algorithm 375
Surface pumping pressure (SPP) 309
Surface pumps 308–9
Surface sampling 357–59, 358f, 359t–61t
Surface water 525–27, 526t, 533, 540
Surfactants 325t, 326
Susquehanna River, Pennsylvania 543
Sweet spot 129, 278–85, 280t–84t
  brittleness and 52
  MBI and 234
  Niobrara formation and 16
Swelling test 371–72, 372f

t

T_2. *See* Transverse relaxation time
TAI. *See* Thermal alteration index
Tannezuft formation 32t, 33
Tax 507
  ATCF and 498
  calculations 516t–517f
  MET and 28
TBP. *See* True boiling point
TCD. *See* Thermal conductivity detector
TDS. *See* Total dissolved solids
TE. *See* Interecho spacing
Tearing 263–64, 264f
Technically recoverable resource (TRR) 11
Tenappera shale play 36, 37f
Tennessee 13
Tensile stress 263–64, 264f
Terpenoids 164–65
Tetracyclic terpanes (TeT) 164, 166t
Tetramantanes 166
Texas, U.S. *See also* Eagle Ford shale
  Fort Worth basin 9–10
  Maverick basin 13, 17, 19, 293, 525–26, 526t
  Permian basin 15f
Texas two-step hydraulic fracturing 305f
TGA. *See* Thermogravimetric analysis
Theory of linear elasticity 336
Thermal alteration index (TAI) 136, 153
Thermal conductivity detector (TCD) 141
Thermal degradation 420–21
Thermal dissolution 6
Thermal maturity 37
  biomarkers and 165–66, 166t
  diamondoids and 166–68, 167f
  of kerogen 168–74, 169f–75f, 173t
    DRIFTS and 174–77, 176f
  Raman spectroscopy of 168–74, 169f–75f, 173t
  SEM images 222f, 225–26, 225f
  source assessment and 162–68, 163t, 165f, 166t, 167f
Thermogravimetric analysis (TGA) 67
Thorium 48–50, 51f, 51t–52t
Three-dimensional (3D) rock model 241–43, 243f, 243t
  geomechanical models and 278
Three Forks formation 11
Three-mineral analysis 55–56, 56t–57t, 57f
Tight natural gas 6
  in U.S., 7, 8f

Timimoun basin 31, 32f, 33
Timur-Coates equation 90, 91t
Tindouf basin 33
Tithonian La Casita shale play 20, 23t
Tithonian shale play 20, 23t
TK. *See* Tuinstra-Koenig
T_{max}. *See* Maximum temperature
TOC. *See* Total organic carbon
TOC_o. *See* Original total organic content
Tortuosity 245–47, 246f, 247t
Total CMR 69–70, 76f, 92, 92t
Total dissolved solids (TDS)
  environment and 523, 527, 529, 530t, 531f, 532–33
Total formation volume factor 368–69
Total organic carbon (TOC) 11, 37, 407
  delta logR method 153–57, 153f–56f, 156t–58t
  direct combustion and 141
  DRIFTS is 56
  GR-TOC method 150–52, 151t–52t
  guidelines for 131–32, 132t
  indirect method and 141
  in-situ measurements and
    Schmoker method 149–50, 149t–50t, 151f
  LOM and 153–57, 153f, 155f
  porosity and 58
  productivity index and 141, 143, 148–49, 159, 177
  Rock-Eval pyrolysis and 141, 143
  sweet spot and 278
  T_{max} and 142–43, 142t
  TOC_o, 159–62, 159t–60t
Total suspended solid (TSS) 529, 532
Toxicity scores 534, 540t–42t
TPAO. *See* Turkish national energy company
TR. *See* Transformation ratio
Traditional decline curve models 441
Transformation ratio (TR) 131, 143, 158, 161
Transit time (DT) 149, 153
Transverse (T_2) relaxation time
  capillary pressure and 97
  of Eagle Ford shale 72, 72f
  permeability and 89–92, 91f–92f
Triamantanes 166
Triassic age 35
Triaxial loading test 268–69
Tricyclic terpanes (TT) 164, 166t
TRR. *See* Technically recoverable resource
TRT. *See* Two-relaxation-time
True boiling point (TBP) 354, 357, 358f
True resolution 188
True vertical depth (TVD) 296f, 309
TSS. *See* Total suspended solid
TT. *See* Tricyclic terpanes
Tuinstra-Koenig (TK) relationship 169
Tunisia 31, 32f
Turbidity test measures 329
Turkish national energy company (TPAO) 31
TVD. *See* True vertical depth
TW. *See* Waiting time
Two-component system 363–64, 364f
  of Monterey formation 63–64
2D radiographs 191, 197, 202f, 204, 234. *See also* Computerized tomography
Two-dimensional Cartesian model 422–23, 423f
Two-phase boundary 363, 374
Two-phase fluid saturations 210–11
Two-relaxation-time (TRT) 424
Type I kerogen 134f, 136, 136t, 159
  maturity and 174, 174f
Type III kerogen 134f, 136, 136t, 140, 142
  maturity and 174, 175f
Type II kerogen 134f, 136, 136t, 140, 146
  maturity and 174, 175f
Type IV kerogen 134, 134f, 136, 136t
Type curves 460
  Agarwal-Gardner 462–64, 463f
  Blasingame 462, 463f
  Fetkovich 460f, 461–62, 462f
  NPI 464–65, 464f
  Wattenbarger 465, 465f–66f
Type I secondary electrons 215–16, 216f
Type IV isotherms 68

u

UEA. *See* United Arab Emirates
UFM. *See* Unconventional fracture models
Uinta basin 172, 173f
Ukraine 28, 28t
Ultra-lightweight proppant (ULW) 332
Ultratight formations
  characteristics of 228t–29t
  SEM images of 218–26, 219f–24f, 227f
ULW. *See* Ultra-lightweight proppant
Uncertainty 510–11
Unconventional fracture models (UFM) 343–44
Unconventional resources. *See also* Petrophysical properties
  background on 2
  coalbed methane and 7–8
  deep natural gas 6–7
  development stages of 37–38, 38f
  future projection and 38–39
  interpretation workflow of 37–38, 38f
  Methane hydrates 8
  of North America 8–20, 21t–22t
  recent discoveries of 4–5
  resource triangle and 5–6, 5f, 7f
  shale volume 46–48, 48t
  of South America 20–25, 24f–27f, 27t
  tight natural gas 7, 7f
Underground sources of drinking water (USDW) 523
United Arab Emirates (UAE) 30
United States (U.S.)
  CMB in 8, 9f
  median TDS 531f
  oil production in 4–5, 495, 495f
  sand proppant in 330
  shale plays in 3, 9–10, 10f
    flowback in 522, 522f, 523–33
    tight gas in 7, 8f
  water composition examples from 530t
  water usage and 525–27, 525f–26f, 526t, 531t
University of Kansas 169
Unsteady-state methods
  GRI 80–82, 81f
  modified pulse decay 84–85, 84f, 85t
  oscillating pulse 85–88, 86f–88f
  pulse decay 80, 81f, 82–84, 83f, 83t
Upper Bakken 9, 11, 137, 137t
  Rock-Eval pyrolysis and 148–49, 148t–49t
Upper Cambrian age 28t
Upper Cretaceous Austin Chalk 58, 59f
Upper Jurassic age 58, 59f
Uranium 47, 149t, 150–51
  gamma ray spectroscopy and 48
Uranium-free gamma ray 48–50, 50f, 51f
Uruguay 24
U.S. *See* United States
USBM technique 103
USDW. *See* Underground sources of drinking water
U.S. Geological Survey (USGS) 8
  Marcellus shale play and 11–12, 13f
Utah, U.S., 172, 173f
Utica shale play 13, 13f, 22t

v

Vaca Muerta play 20, 27t
  horizontal wells in 3, 4f
Van der Waals interactions 418
  EOS 373–74
Vanishing interfacial technique (VIT) 389
Vapor-liquid equilibrium (VLE) 377
Velkerri shale play 36
Venezuela 25, 27f, 508
Vertical heterogeneity 6
Vertical stress 267–68
Viola group samples 140
Virginia, U.S., 12, 13f
VIT. *See* Vanishing interfacial technique
Vitrinite
  imaging techniques and 198, 218t
  reflectance of 131, 135–38, 137t
    calculation of 148–49
    LOM and 153, 153f, 155

porosity and 139f, 140
T_{max} and 142
VLE. *See* Vapor-liquid equilibrium
Volatility score 541t–42t
Volume fraction estimation 204, 243, 243t–44t
Volume translation parameter 388
Volumetric analysis theory 108–10, 109f–10f, 110t–12t
Voxel 187, 189, 196–97

W

Waiting time (TW) 72
Water
disposal of 524
DW equation 100
hydraulic fracturing usage 5, 525, 525f
management of
flowback and 527–29, 528f, 528t
quality and quantity 527–29, 528f, 528t
terminology in 521–23, 522f, 522t
water acquisition 525–27, 525f–26f, 526t
water cycle and 523–24, 524f
reuse of 529–33, 532f, 534t
saturation
calculating techniques for 98
modified Purcell method and 95, 96f
resistivity logs and 98–102, 108–9
Water-based fracturing fluids 319, 535t–38t
Water-oil-ratio (WOR) 529, 531t
Watershed algorithm 203f, 204–5, 205f
Wattenbarger type curves 465, 465f–66f
Wave travel-time equations 275–76
Wave velocity equations 274–75
Waxman-Smits-Thomas (WST) equation 100
Well logging 46, 47f, 58, 59t
elastic parameters 273–74
dynamic moduli 274–76
static moduli and 276–78
production life 445–49
PSD and 75–77 76f–77f, 78t
Well shut-in technique 308
West Siberian basin 26, 29f
West Virginia, U.S., 12, 13f
Wettability 102
modified Amott-Harvey technique 106–7, 107f–8f
NMR measurements of 103–5, 103f–5f
XPS technique and 106–7, 107f–8f
Wetting phase 94–95, 97–98
Whales 1
Whitehill formation 31, 32t, 112t
Wilson equation 375, 378f
Wireline logs 51–52, 108
Wireline methods 46
Wolfcamp shale play 13, 15–16, 15f, 22t
binary interaction parameters for 384, 386t
compositional data for 384, 385t
phase diagrams of 384, 387
Woodford shale play 13, 16, 16f, 22t
CT scan of 200, 201f
WOR. *See* Water-oil-ratio
WST. *See* Waxman-Smits-Thomas
Wyllie time averaging equation 64
Wyoming, U.S., 16, 16f

X

XMReconstructor software 189, 191, 196–97
XPS. *See* X-ray photoelectron spectroscopy
Xradia MicroXCT-400, 190–91, 190f–91f, 196–97
FOV of 194–95, 195t
X-ray microcomputed tomography (X-ray micro CT)
analysis of 205–7, 206f
EDS and 226–32, 230f–34f
FIB-SEM 239–41, 240f–44f, 243t–44t
filtering and 204, 205f
image visualization and 201–4, 201f–10f, 204t
nano-CT 211–12, 211f–12f, 214t
in-situ wettability 212–14, 214f
sample preparation for 212, 213f
operation of 190–92, 191f–92f
permeability estimation 210
PNM and 243–47, 245f–46f, 247t
porosity estimation 207–9, 208f–10f, 208t

X-ray microcomputed tomography (X-ray micro-CT) (*cont'd*)
PSD in 243–47, 245f–46f, 247t
QEMSCAN 233–39, 238f–39f, 239t
reconstruction in 196–98, 197f
resolution and 188–89, 188f, 190f
sample preparation in 192–94, 193f
scanning procedure in 194–96, 195f, 195t
segmentation in 203–5, 205f–6f
shale sample application 198–201, 199f–201f
3D rock model 241–43, 243f, 243t
two-phase fluid saturations 210–11
X-ray photoelectron spectroscopy (XPS) 106–7, 107f–8f
Xylene 545

y

Y-function linearity test 367t, 368, 368f, 388, 400t
Young's modulus 264–65, 265f
BI and 279–80
dynamic 273–78
static 276–78
of unconventional plays 285, 285t
in wave velocity 274–75

z

Zamora's method 272–73
Zheltov model. *See* Khristianovich and Zheltov model
Zipper fracturing 304–5, 305f